D1674822

Edited by
Wenping Hu

Organic Optoelectronics

Related Titles

Klauk, H. (ed.)

Organic Electronics II

More Materials and Applications

2012

ISBN: 978-3-527-32647-1

Brütting, W. (ed.)

Physics of Organic Semiconductors

2012

ISBN: 978-3-527-41043-8

Wöll, C. (ed.)

Physical and Chemical Aspects of Organic Electronics

From Fundamentals to Functioning Devices

2009

ISBN: 978-3-527-40810-8

Brabec, C., Scherf, U., Dyakonov, V. (eds.)

Organic Photovoltaics

Materials, Device Physics, and Manufacturing Technologies

2008

ISBN: 978-3-527-31675-5

Schwoerer, M., Wolf, H. C.

Organic Molecular Solids

2007

ISBN: 978-3-527-40540-4

Klauk, H. (ed.)

Organic Electronics

Materials, Manufacturing and Applications

2006

ISBN: 978-3-527-31264-1

Edited by Wenping Hu

Organic Optoelectronics

WILEY-VCH Verlag GmbH & Co. KGaA

The Editor

Prof. Wenping Hu
Chinese Academy of Sciences
Laboratory of Organic Solids
Institute of Chemistry
Beijing 100190
China
huwp@iccas.ac.cn

Library of Congress Card No.: applied for

British Library Cataloguing-in-Publication Data
A catalogue record for this book is available from the British Library.

Bibliographic information published by the Deutsche Nationalbibliothek
The Deutsche Nationalbibliothek lists this publication in the Deutsche Nationalbibliografie; detailed bibliographic data are available on the Internet at <http://dnb.d-nb.de>.

Composition Toppan Best-set Premedia Limited, Hong Kong
Printing and Binding Markono Print Media Pte Ltd, Singapore
Cover Design Grafik-Design Schulz, Fußgönheim

Print ISBN: 978-3-527-32968-7
ePDF ISBN: 978-3-527-65348-5
ePub ISBN: 978-3-527-65347-8
mobi ISBN: 978-3-527-65346-1
oBook ISBN: 978-3-527-65345-4

Printed in Singapore
Printed on acid-free paper

Wenping Hu would like to dedicate this book to Prof. Daoben Zhu on the occasion of his 70th birthday, to Prof. Yunqi Liu for his retirement, and to Qiong and Beining, his wife and daughter, for all their patience and encouragement.

Contents

Preface

As a novel emerging science with great applications, organic optoelectronics has attracted the world's attention since the 1990s. Organic optoelectronic materials with special functionalities stem from our increasing ability to manipulate and tune the properties of organic and polymeric materials. This is achieved through a systematic variation of the materials' molecular components, so as to allow for a molecular-level control of the solid-state structure via an arrangement of the functional molecular components into a defined architecture. The optical and electronic processes in organic molecules and polymers govern the behavior of organic semiconductors and their applications in organic optoelectronic devices. Emphasis is placed on the use of organic thin films in active organic devices, including organic light-emitting diodes (OLEDs), organic photovoltaic (OPV) devices, organic field-effect transistors (OFETs), photodetectors, chemical sensors, memory cells and electrochromic devices, as well as xerography and organic non-linear optics. For example, OLEDs have permitted the development of superior flat-panel display technologies that have now been commercialized for cellular telephone applications, and will soon be implemented in large-area, high-definition television screens. Currently, OPV devices have reached a quantum efficiency of over 9%, which makes them attractive for delivering cheap solar power, while the use of OFETs has led to in a revolution in the development of fast and inexpensive integrated circuits on plastic substrates based on organic semiconductor elements. When combined with their advantage of solution processability, organic materials allow for the use of a variety of printing techniques, such as inkjet printing and stamping, to fabricate large-area devices at low cost. Moreover, the mechanical properties of organic semiconductors also allow for flexible electronics. Certainly, the most distinguishing feature of organic semiconductors is their chemical versatility, which permits the incorporation of functionalities by molecular design, for example, to encode factors that help to direct the properties. Clearly, as an exciting research field with many potential practical applications, organic optoelectronics is progressing at an extremely rapid pace.

The intention of this book is to describe the fundamental scientific information and recent breakthroughs relating to both the basic science and real application of organic optoelectronics. Attention will be focused on the optoelectronic behavior of organic semiconductors, and their applications in new optoelectronic devices.

The book covers topics of: (i) organic semiconductors in electronics, such as FETs and circuits; (ii) organic electroluminescent materials and devices (though here only polymer electroluminescent materials and devices are given as examples); (iii) organic photonics, materials, and devices; and (iv) organic semiconductors in photoabsorption and energy conversion, such as organic solar cells and organic thermoelectric power devices. The preparation of functional materials and the fabrication of novel devices–for example, materials synthesis and purification, physical chemical properties, and the basic processes and working principles of the optoelectronic devices–are all emphasized in this book.

We hope that this book will attract the attention of graduate students and young scientists alike, as well as those more senior academic and industrial researchers who are interested in organic optoelectronics. We believe that this book will provide stimulation for the derivation of ideas, methods, and technologies related to chemistry, physics, materials science, semiconductors, electronics, nanotechnology, and biology in this exciting area.

We conclude by thanking all of the authors for their great contributions to the book, notably their hard work, expertise and insightful suggestions. It would have been impossible to complete this volume without their knowledge, dedication, and enthusiasm. Finally, we express our gratitude to Esther Levy and Ulrike Werner at John Wiley & Sons, Ltd for their help and guidance through the editorial process.

Wenping Hu

List of Contributors

Fenglian Bai
Chinese Academy of Sciences
Institute of Chemistry
Zhongguancun North First Street 2
Beijing 100190
China

Huitao Bai
Chinese Academy of Sciences
Institute of Chemistry
Bei Yi Jie No. 2, Zhongguancun
Beijing 100190
China

Thomas Bjørnholm
University of Copenhagen
Nano-Science Center and
Niels Bohr Institute
Universitetsparken 5
2100 Copenhagen ϕ
Denmark

Huanli Dong
Chinese Academy of Sciences
Institute of Chemistry
Zhongguancun North First Street 2
Beijing 100190
China

Karsten Flensberg
University of Copenhagen
Nano-Science Center and
Niels Bohr Institute
Universitetsparken 5
2100 Copenhagen ϕ
Denmark

Hongbing Fu
Chinese Academy of Sciences
Institute of Chemistry
ZhongGuanCun North First St 2
Beijing 100190
China

Xiong Gong
The University of Akron
College of Polymer Science and
Polymer Engineering
250 S Forge Street
Akron, OH 44325
USA

Wenping Hu
Chinese Academy of Sciences
Institute of Chemistry
Zhongguancun North First Street 2
Beijing 100190
China

Lang Jiang
Chinese Academy of Sciences
Institute of Chemistry
Zhongguancun North First Street 2
Beijing 100190
China

Martin Leijnse
University of Copenhagen
Nano-Science Center and
Niels Bohr Institute
Universitetsparken 5
2100 Copenhagen ϕ
Denmark

Hongzhen Lin
Chinese Academy of Sciences
Suzhou Institute of Nano-tech and
Nano-Bionics
398 Ruoshui Road, SEID, SIP
Suzhou 215123
China

Yuze Lin
Chinese Academy of Sciences
Institute of Chemistry
Bei Yi Jie No. 2, Zhongguancun
Beijing 100190
China

Lanchao Ma
Chinese Academy of Sciences
Institute of Chemistry
Bei Yi Jie No. 2, Zhongguancun
Beijing 100190
China

Qing Meng
Chinese Academy of Sciences
Institute of Chemistry
Zhongguancun North First Street 2
Beijing 100190
China

Qinqin Shi
Chinese Academy of Sciences
Institute of Chemistry
Bei Yi Jie No. 2, Zhongguancun
Beijing 100190
China

Qinqxin Tang
North-eastern Normal University
Department of Physics
Changchun 130024
China

Yanhong Tong
North-eastern Normal University
Department of Physics
Changchun 130024
China

Chengliang Wang
Chinese Academy of Sciences
Institute of Chemistry
Zhongguancun North First Street 2
Beijing 100190
China

Xiaowei Zhan
Chinese Academy of Sciences
Institute of Chemistry
Bei Yi Jie No. 2, Zhongguancun
Beijing 100190
China

1
Electronic Process in Organic Solids

Hongzhen Lin, Fenglian Bai

Organic solids, in a broad sense, include all solid-state materials consisting of organic molecules or polymers, namely, compounds with carbon atoms as their essential structural elements [1]. Under this generic term, the category of organic solids covers a wide variety of natural solids such as wood and cotton, and industrial products such as plastics and rubber, many of which are insulators. The scope of this book, however, will be confined to the class of organic solids that can serve as active components in electronic or photonic devices. The functionalities of these materials are mainly based on their capability to carry and transport charges and neutral excitations. For simplicity, hereafter the term "organic solid" will refer specifically to those organic materials showing (semi)conductor properties in the form of crystals, thin films, or glassy state.

Electronic process in organic solids determines the properties of the materials and their potential applications in optoelectronic devices. It is a very complicated process, and has a close relationship with molecular electronic structures, molecular interactions, charge–charge coupling, charge–photon coupling, and exciton–photon coupling, and so on. Consequently, chemists find it difficult to understand due to the complication of electronic process in organic solids. At this point, electronic process will be expounded from the chemist's perspective in order to appreciate the basic concepts and the nature of this complicated process.

1.1
Introduction

Solid-state devices – especially transistors – play a crucial role in modern electronic technology. Whilst the dominant building materials in solid-state electronics are inorganic semiconductors such as silicon or germanium, organic solids – in particular, organic semiconductors – have emerged recently as a new class of electronic materials and have subsequently become a strong competitor of inorganic semiconductors in many aspects of the electronics industries [1]. The key advantages of organic solids over metals and inorganic semiconductors stem from their potential to combine the electrical properties of (semi)conductors with the

Organic Optoelectronics, First Edition. Edited by Wenping Hu.

properties typical of organics – that is, low cost, high processability, mechanical flexibility, and a versatility of chemical synthesis [2]. These materials provide the possibility to realize novel applications such as large areas, flexible displays, low-cost printed integrated circuits, and plastic solar cells [1]. During the past few decades, tremendous progress has been achieved along these lines, such that today a totally new field of research – organic electronics – has emerged and is continuing to develop.

The first realization that organic compounds could carry an electric current was made many years ago. Indeed, the first studies of the conductivity of anthracene crystals, as a prototype of organic semiconductors, date back to the early twentieth century [3, 4]. During the 1950s, various research teams found that polycyclic aromatic compounds could form semi-conducting charge-transfer complex salts with halogens; notably, a high conductivity of $0.12\,S\,cm^{-1}$ was reported in a perylene–iodine complex in 1954 [5]. Subsequently, during the 1960s molecular crystals aroused intense research interest, due to the discovery of electroluminescence [6, 7]. During the 1970s, the successful synthesis of conjugated polymers, and the observation of a controllable conductivity over the range from insulating to metallic [8], led to the establishment of a second important class of organic semiconductors, for which the Nobel Prize in Chemistry was eventually awarded in 2000.

Along with great progress in the academic research of organic solids, tremendous technological developments have also been achieved in the creation of (opto) electronic devices from these materials. Ultimately, the door to "real" applications of organic semiconductors was opened in 1987 by Tang and VanSlyke who, while working at Kodak, successfully fabricated thin-film organic light-emitting diode (OLED) devices from tris(8-hydroxyquinolinato) aluminum (Alq_3), a π-conjugated molecular material [9]. Shortly thereafter, Friend and his group at the Cavendish Laboratory in Cambridge reported a highly efficient polymer-based OLED using a conjugated polymer, poly(*p*-phenylene vinylene) (PPV) [10]. Besides OLEDs, organic semiconductors are also used widely in other devices such as organic solar cells [11, 12], organic field-effect transistors (FETs) [13, 14], chemical sensors [15, 16], and organic lasers [17, 18]. Nowadays, the field of organic electronics has reached a new era and is facing a bright future of industrialization. In particular, OLEDs have made a solid step towards the commercial market, having shown great potential for use in panel displays, digital cameras and mobile phones, and for white light illumination.

Nonetheless, much hard work lies ahead before the large-scale production of organic electronic devices becomes possible, and this will require the extensive collaboration of physicists, chemists, and engineers. Although an incessant effort will clearly be required to develop device fabrication techniques, the most fundamental approach for improving device performance is to create new organic solids with optimal properties as desired. Yet, this raises a major challenge for both organic and materials chemists alike since, in order to guide the design and synthesis of novel materials, it is crucial to acquire a better fundamental understanding of the nature of electronic excitations, charge carriers, and transport phenomena in organic solids [2]. To date, many experimental and theoretical studies have been conducted for this purpose, and comprehensive reviews of the topic are available in books and journals [1, 2, 19–22]. Unfortunately, however, the theoretical interpretations and physical models of organic solids are often "too abstract" for synthetic chemists to understand. Consequently, in this chapter the basic concepts will be introduced from a chemist's perspective, allowing them to capture the nature of the complicated electronic processes in these materials. The chapter is organized as follows: the molecular and supramolecular structural features of some prototype organic solids are described in Section 1.2, after which the fundamental photophysical properties of organic conjugated molecules are introduced in Section 1.3. In Section 1.4, details of the neutral and charged excited states in conjugated polymers are provided, while in Section 1.5 a brief discussion is provided of charge carrier generation and transport in organic solids.

1.2
Structure Characteristics and Properties of Organic Solids

An atom consists of a positively charged nucleus and one or more electrons that are bound by the electric field of the nucleus. In a system having more than one atom, an electron is not necessarily bound to one nucleus but can be shared between different atoms, depending on the interatom interactions and the energy state of the electron. In metals and inorganic semiconductors, the strong interatomic electronic interactions facilitate the delocalization of outer shell electrons over a large number of atoms. But, the situation in organic solids is very different where, depending on the nature of intermolecular interactions, organic solids can be classified into two types [2].:

- In *nonpolar organic solids*, the molecules are held together by van der Waals interactions, which are rather weak compared to covalent bonding. The physical properties of nonpolar organic solids are only slightly changed relative to those of the free molecules, since the intramolecular interactions are dominant.
- *Polar organic solids* are organic solid materials where both ionic bonding and van der Waals forces exist. Examples of this type include organic charge-transfer complexes and radical-ion salts, in which the positive and negative charges are separated and located on different molecules. Nevertheless, the ionic bonding in these ionic molecular crystals is weaker than that in inorganic salts, as molecules are larger than atoms.

Among organic solids can be included photoconductive materials, conductive polymers, electroluminescent materials, and photovoltaic materials, the functionality of which depends essentially on their molecular structures. Although a wide variety of organic semiconductors and conductors have been described, they include in common a conjugated π-electron system in the skeletal structure of the constituent molecules. Typically, a conjugation system is composed of alternating single and multi bonds in which π-electrons are delocalized over the connected p_z-orbitals of contiguous sp^2-hybridized carbon atoms. Other atoms with available p_z-orbitals may also be involved. The characteristic optical and electronic processes that occur in organic (semi)conductors are closely related to which type of conjugation systems are contained, and how those conjugation systems interact one with another. Hence, the feasibility of modifying and altering the conjugation system in a molecule by chemical synthesis offers a wide range of possibilities to tune the optoelectronic properties of these organic solids. Some important classes of organic solid are detailed in following subsections, according to the major conjugation systems that they contain.

1.2.1 Organic Solids

Organic solids are usually classified into two groups according to molecular weight: (i) conjugated small molecules; and (ii) conjugated polymers. Often, the conjugated moiety of a molecule is referred to as a chromophore, one or a few of which are typically contained in a small molecule. In conjugated polymers, the π-electron delocalization is often interrupted by intrinsic or dynamic defects in the polymer chains, and generally persists for only a few tens of repeating units [23]. Consequently, a long conjugated polymer chain can be regarded approximately as an ensemble of weakly coupled chromophores of relatively short conjugation lengths.

The key difference between conjugated polymers and small molecules lies in their crystallinity, and the way in which they are processed to form thin films. For example, small molecules tend to crystallize into ordered arrays when they are deposited from the gas phase by sublimation or evaporation, whereas conjugated polymers can only be processed from solution, such as by spin-coating or printing techniques, so that in general amorphous thin films are formed. The performance of an organic solid in a devices is found to be highly sensitive to the way in which the molecules are arranged in the thin films.

The molecular structures of several prototype organic solids are shown in Figure 1.1. The simplest conjugation system is present in polyenes–compounds which contain one or more sequences of alternating double and single carbon–carbon bonds. A well-known example of this is *trans*-polyacetylene (t-PA) which, despite having the intrinsic properties of an insulator, has demonstrated an enhanced conductivity via chemical reduction/oxidation (redox) [8]. In fact, this finding proved to be a milestone in the development of organic electronic materials, such that the Nobel Prize for Chemistry was awarded to Alan Heeger, Alan MacDiarmid, and Hideki Shirakawa in 2000 for their pioneering research on t-PA and other conductive polymers.

Poly-aromatic hydrocarbons (PAHs) and their substituted derivatives represent another typical class of organic solid, with well-studied examples including anthracene, rubrene, pentacene, fluorene, pyrene, perylene, and coronene. These com-

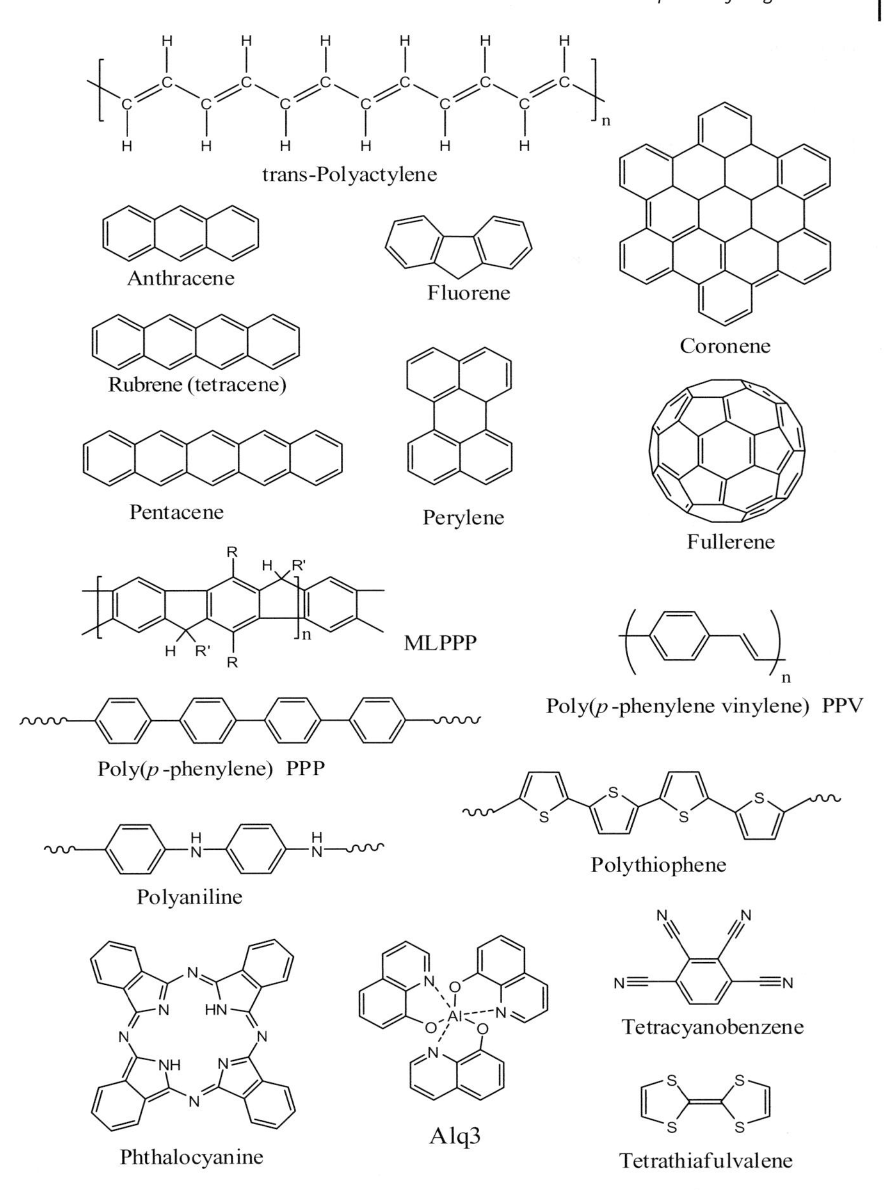

Figure 1.1 Molecular structures of some prototypes of organic solids.

pounds consist of fused aromatic rings, and are recognized as harmful pollutants to the air. Yet, in contrast they have been shown to be good candidates for the construction of electronic devices, with high charge mobilities having been reported in crystallized films of rubrene, pentacene and their derivatives, confirming their potential as organic FETs [24].

Currently, PAHs are also used extensively in OLEDs [25] as emissive dopants or charge-transporting materials. For example, polyacenes such as naphthalene, anthracene, rubrene, and pentacene are frequently employed as model systems in molecular physics and solid-state physics investigations, due to their simple linear structure, their defined conjugation lengths, an ability to form highly ordered crystals, and their well-determined optical and electronic properties. A linear extension of such a fused-ring structure can be demonstrated in ladder-type conjugated polymers, such as methyl-substituted ladder-type poly-*para*-phenylene (MLPPP) [26]. It is also worth noting here that all-carbon materials such as fullerene, carbon nanotubes and graphene, can be visualized as extended PAHs.

Many organic solids are produced from other aromatic hydrocarbons, and in such compounds the conjugation system typically consists of multiple aromatic rings that are connected with each other via a single bond, or through a vinylene or ethynylene group. Conjugated polymers such as poly (*p*-phenylene) (PPP), PPV, and poly (*p*-phenylene ethynylene) (PPE) belong to this group.

Conjugation systems containing heteroatoms such as sulfur and nitrogen are very useful building blocks for organic semiconductors, as the heteroatoms can be incorporated either outside or inside of the aromatic rings. Polyaniline and poly(*p*-phenylene sulfide) are examples of the "outside" case, while the "inside" case includes polythiophene and polypyrrole. As a special type of heteroaromatic compound, tetrapyrrole macrocycles including phthalocyanines, porphyrins and porphyrazines have also been investigated in organic electronics and photonics [27, 28], often appearing as complexes with metals such as copper and zinc.

Other types of organic metal complex have also been identified, a good example being the above-mentioned compound, Alq_3; organic complexes of other metals such as Zn, Pt, Os, Eu, and Ir were also reported [29, 30]. Alq_3 and its derivatives are important electroluminescent materials for OLEDs, and are also used as electron-transporting materials in photovoltaic devices such as solar cells.

In some cases, two different aromatic compounds can form charge-transfer complex crystals with a fixed uniform composition, much like a new compound. When the charge transfer occurs only in an electronically excited state, these are termed "weak" donor–acceptor (D–A) crystals; a good example is anthracene–tetracyanobenzene (TCNB). In the strong D–A complexes, the charge transfer takes place in the electronic ground state; a well-studied example is the compound tetrathiafulvalene–tetracyanoquinodimethane (TTF:TCNQ) [1].

Although, so far, attention has been focused on the skeletal structure of molecules, the conjugated backbones are often decorated with a variety of side groups. These side substituents not only improve the solubility of a compound but also greatly impact on the optical and electronic properties of the material. For instance, the π-π^* energy gap of an aromatic compound can be tuned by introducing electron-donating or withdrawing substituents. Bulky side groups can reduce the formation of non-emitting interchain aggregates in conjugated polymers [31].

Inorganic semiconductors may be either *p*-type or *n*-type, depending on which type of dopant is used. The dominant charge carriers are holes in the *p*-type semiconductors, but electrons in the *n*-type. Likewise, organic semiconductors may also be classified as *p*-type and *n*-type, corresponding to hole-transporting and electron-transporting materials, respectively. It should be noted that the mechanism lying behind this classification of organic semiconductors is actually different from that used for their inorganic counterparts. Whether an inorganic semiconductor is *n*- or *p*-type is determined by the extrinsic dopants, whereas in the case of organic semiconductors this depends much more on the intrinsic chemical structures of the materials. For example, aryl amines are typical hole-transporting materials, whilst Alq_3 is a characteristic electron-transporting material. It should be noted that some organic semiconductors, such as PPVs, possess both hole- and electron-transporting abilities, and consequently their roles in devices will vary from case to case.

1.2.2
Molecular Geometries

The molecules or polymer chains that constitute organic solids may have various geometries. For example, a linear backbone can be found in polyacetylene and PPV, while phthalocyanines adopt a two-dimensional (2-D) planar structure. Organic solids have also been reported with other molecular geometries (Figure 1.2), including star-shaped structures [32], tree-shaped structures such as dendrimers [33] and hyperbranched polymers [34], as well as spiro compounds [35] in which two rings are connected through just one atom. The geometry has a significant effect on the conjugation length, the rigidity, the conformational variety, and the assembling behavior of a molecule, and thus greatly influences the properties of the compound in the solid state.

1.2.3
Aggregations and Assemblies

It should be borne in mind that organic conjugated compounds are used in opto-electronic devices as condensed solid films, but not as independent free molecules.

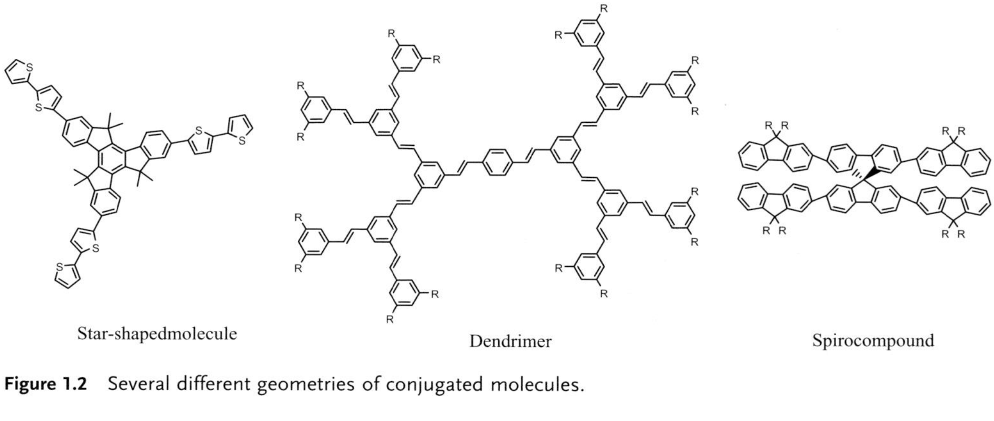

Star-shapedmolecule Dendrimer Spirocompound

Figure 1.2 Several different geometries of conjugated molecules.

The intermolecular forces are crucial for the optical and electronic processes that occur therein, with the effect being either positive or negative depending on the usage. For instance, an efficient π-π stacking is usually good for charge transport, and thus an advantage if the material is to be used in FETs. On the other hand, π-π stacking often results in a quenching of emissive excited states, and therefore should be avoided when the compound is to be used as an electroluminescent material in OLEDs. Intermolecular forces other than π-π stacking include hydrogen bonding, electrostatic forces, and hydrophobic/hydrophilic effect, and these play important roles when the molecules are aggregated or assembled to form a solid thin film. In return, the ultimate morphology of the thin film has a great effect on the way that the molecules interact. In fact, the cooperation mechanism of various intermolecular interactions during molecular assembly, and its effect on device performance, is currently a "hot topic" of research in this field.

1.3 Electronic Processes in Organic Small Molecules

The complex intra- and intermolecular interactions in organic solids create major problems in the mathematical modeling of these materials. Indeed, in order to understand the large-scale properties of organic solids, it is necessary first to have an insight into their molecule-scale properties. Initially, it is both convenient and practical to consider simple model systems, before turning to more complicated cases. In this section, some fundamental optical and electronic processes that take place in an individual molecule, or between two molecules, and which have been investigated for many centuries, will first be revised. In particular, interest will be focused on processes as photogeneration and the relaxation of electronic excited states, and photoinduced charge/energy transfer.

1.3.1 Photophysics of Small Molecules

1.3.1.1 Molecular Orbital Model

A spatially confined particle can only take on certain discrete values of energy, and an electron bound to atoms or molecules falls into this case. These discrete values are termed "energy levels," and in quantum mechanics the states of electrons in atoms or molecules are described by so-called "wavefunctions." Here, the molecular orbital (MO, a model that is much more familiar to chemists) will be used, as MOs represent mathematically the regions in a molecule where an electron is most likely to be found. Any electronic state of a molecule can be described by a certain linear combination of its MOs. According to the Pauli principle, one MO can accommodate at most two electrons; hence, if a MO is fully occupied the two electrons therein must have opposite spin directions. The ground state of the molecule can be reconstructed by filling its electrons to the first few lowest energy MOs, while following the Pauli principle. The π-orbitals of benzene are shown as

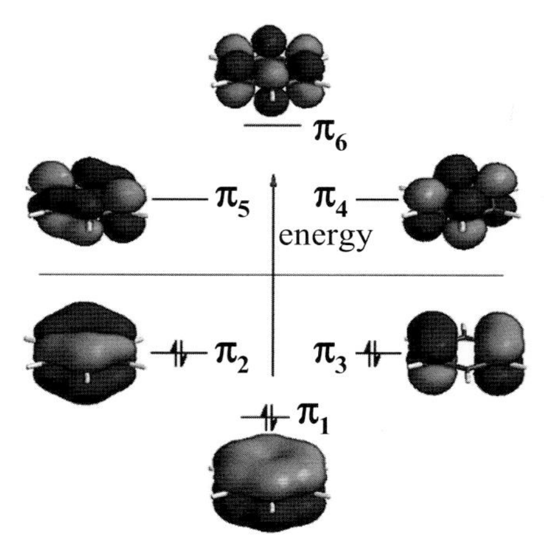

Figure 1.3 π-orbits of benzene. At ground state, the electrons first fill the lower-energy orbits.

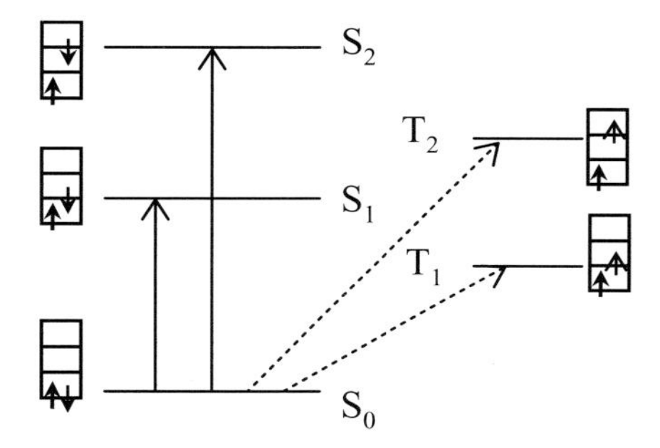

Figure 1.4 A simplified Jablonski diagram of an organic molecule. Solid arrows indicate spin-allowed transitions; dotted arrows indicate spin-forbidden transitions. S, singlet; T, triplet.

an example in Figure 1.3. The electronic transition of a molecule can be expressed as the moving of an electron from one MO to another.

1.3.1.2 **Jablonski Diagram**

Light absorption and emission processes are usually illustrated by the Jablonski diagram, in which the electronic states are vertically arranged according to the relative energy level. A simplified Jablonski diagram for a molecule that is free from intermolecular interactions is shown in Figure 1.4. Such an ideal case can be approximated in gas or in diluted solutions of the molecules in inert solvents.

Not all of the molecular orbitals take part in the light absorption and emission transitions; rather, these transitions are usually dominated by the highest occupied molecular orbital (HOMO) and the lowest unoccupied orbital (LUMO). It should be noted that HOMO and LUMO are defined corresponding to the ground state of a molecule. Commonly, the ground state of an organic molecule is a singlet state in which the HOMO is fully occupied by two electrons of opposite spin. One of the two electrons is then promoted to LUMO or a higher energy level, when the molecule absorbs a photon of proper energy. In most cases, the excited electron will maintain its original spin, resulting in a singlet excited state. In Figure 1.4, the singlet ground, first, and second electronic states are depicted by S_0, S_1, and S_2, respectively. In some rare cases, the excited electron may change its spin direction and a triplet excited state will be then formed (denoted as T_1 and T_2, etc.). Triplet states generally cannot be directly accessed from S_0. An excited state can return to the ground state by emitting a photon of certain energy.

For organic molecules containing a π-conjugated system, the π-bonding is significantly weaker in comparison to the σ-bonds that form the backbone of the molecules. Therefore, the lowest electronic excitations of conjugated molecules are the π-π* transitions, with an energy gap typically between 1.5 and 3 eV, leading to light absorption or emission in the visible spectral range. The energy gap of an organic molecule can be controlled by changing the size of the conjugation system. Typically, the energy gap decreases with increasing conjugation length.

1.3.1.3 Frank–Condon Principle

The energy level scheme in Figure 1.4 does not take into account the motion of the nuclei relative to the molecular coordinate. The displacement of nuclei also corresponds to a series of discrete energy levels, but with a much smaller spacing than the electronic levels; these are termed vibrational energy levels. Each electronic state possesses a series of possible vibrational states; hence, vibrational transition can be thermally activated, and hence a molecule will have the probability to remain at a higher vibrational level, depending on the temperature. To account for this effect, the first few vibration energy levels associated with each electronic state are included into the Jablonski diagram (Figure 1.5, depicted by 0,

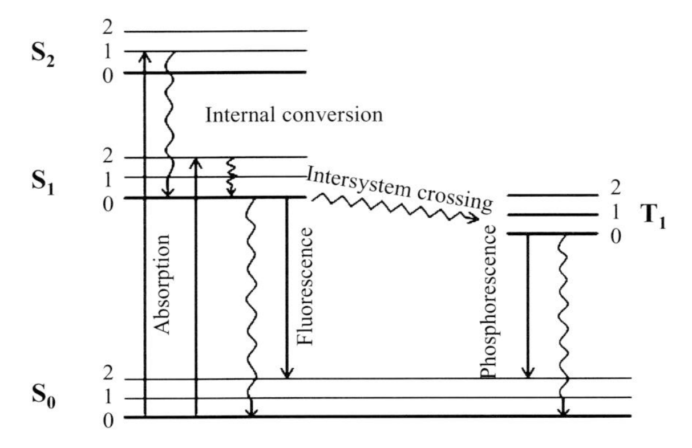

Figure 1.5 Energy level schemes and transitions of an organic molecule. S, singlet; T, triplet.

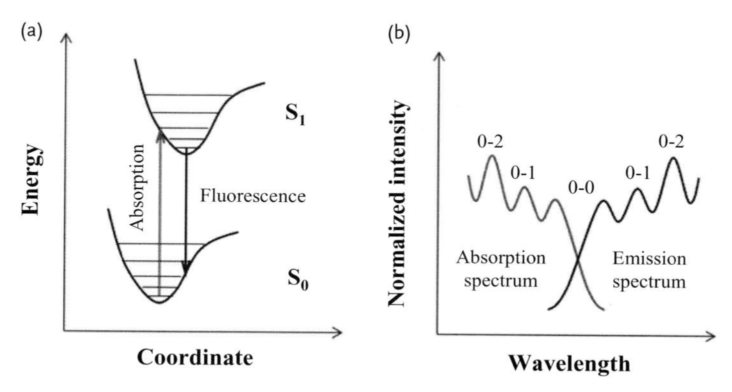

Figure 1.6 (a) Schematic illustration of the Frank–Condon principle. The most probable electronic transition occurs without a change of nuclei coordinates; (b) Schematic illustration of the mirror-image rule.

1, 2, etc.) [36]. At room temperature, the thermal energy is inadequate to significantly populate the excited vibrational states. Absorption and emission occur mostly from molecules with the lowest vibrational energy. The transition between the lowest vibrational levels of the electronic states, for example, between $S_0(0)$ and $S_1(0)$, is termed 0-0 transition.

At this point, it may help to introduce the Frank–Condon principle, which states that an electronic transition is most likely to occur without changes in the positions of the nuclei in the molecule and the surrounding environment. The quantum mechanics description of the principle is that, during an electronic transition, a change from one vibrational energy level to another will be more likely to occur if the two vibrational wavefunctions overlap more significantly. Hence, if the potential energy curves (system energy versus nuclei coordinate) of the initial and final electronic states are plotted, a Frank–Condon transition can be depicted as a vertical line between the two curves, in correspondence to an unchanged coordinate during the transition (Figure 1.6). For this reason, Frank–Condon transition is often termed vertical transition. In the Jablonski diagram, electronic transitions between states are also drawn as vertical lines to illustrate the vertical nature of the transitions. According to the Frank–Condon principle, 0-0 transition is generally not the most probable transition, as the thermally equilibrated molecular geometry (coordinate) of an electronic excited state is often different from that of the ground state.

1.3.1.4 Electronic Absorption

The transition from S_0 to S_1 or higher singlet states is spin-allowed. Such a photon absorption process takes place very rapidly, typically at a time scale of femtoseconds (10^{-15} s). The electronic absorption of a substance at certain light wavelength follows the Lambert–Beer law:

$$I = I_0 e^{-\varepsilon Cl} \tag{1.1}$$

where I_0 and I are the intensity of the incident light and the transmitted light, respectively, ε is the molar extinction coefficient, C is the molar concentration of the molecule, and l is the distance that the light travels through the material.

Equation (1.1) is frequently used for solutions, where l is the thickness of the solution in the light path. For molecular crystals and for thin films consisting of one compound, the concentration C is a constant that depends on the density of the material and the molecular mass. In such cases, the product of ε and C can be replaced by one constant k–that is, $I = I_0e^{-kl}$–where l is the thickness of the crystal or film. More generally, the Lambert–Beer law can be expressed as

$$I = I_0 e^{-\sigma N l} \tag{1.2}$$

where σ is the absorption cross-section of a single molecule and N is the density (number per unit volume) of absorbing molecules. Defining $A = \log(I_0/\mathrm{I})$ leads to

$$A = \log(I_0/I) = \ln 10\varepsilon Cl = \varepsilon' Cl \tag{1.3}$$

where A is termed absorbance, a parameter that is proportional to the concentration of the absorbing molecules. Equation (1.3) is often used to measure the concentration of a substance in solution or in a solid matrix.

The molar extinction coefficient ε (or the molecular absorption cross-section σ) varies with the light wavelength or frequency. The theoretical interpretation of absorption is nontrivial, but some simple facts should be mentioned at this point. First, in order to excite a molecule, the energy of the incident photon (given by $E = \mathrm{h}\upsilon$, where $\hbar$ is Plank's constant and υ is the light frequency) must match the energy gap between the initial state and the final state. This means that only light with certain frequencies (or wavelengths) can be absorbed by the molecule. Second, electronic transition (like absorption) can be seen as a displacement of the negative charge center relative to the positive charge center of the molecule. Hence, a chromophore can be approximately treated as a dipole oscillator with a certain resonance frequency.

The absorption spectrum can be monitored by recording the absorbance as a function of the wavelength (or frequency) of the incident light. The electronic absorption of atoms and molecules generally lies in the ultraviolet or visible region; hence, it is often referred to as "UV-visible absorption." Because of the involvement of vibrational energy levels, different absorption bands may be observed in correspondence to the energy gap between 0-0, 0-1, and 0-2, etc.; this is known as the vibrational structure of the absorption spectrum. The absorption spectrum of an organic compound in a gas state often appears as a series of sharp lines that are greatly broadened in solution because of perturbations from the surrounding solvent molecules; even broader absorption bands may be identified for an amorphous solid. In contrast, molecular crystals often show narrow absorption lines similar to those of a gas, but with a distinctive vibrational structure.

Following light absorption, a molecule is usually excited to a higher vibrational level of either S_1 or S_2. Except for a few rare cases, excited molecules rapidly relax to the lowest vibrational level of S_1; this process is termed internal conversion, and generally occurs within 10^{-12} s or less.

1.3.1.5 Fluorescence and Phosphorescence

Electronic transition from S_1 to S_0 is also spin-allowed; this transition can occur spontaneously by the emission of one photon, when the emitted light is termed fluorescence [36]. A molecule may stay at S_1 for a brief moment before it returns to the ground state; the average time that a molecule spends between its excitation and its return to the ground state is referred to as the fluorescence lifetime, and for a conjugated compound this lies in the range of 10^{-9} to 10^{-8} s. Fluorescence emission generally results from the lowest energy vibrational state of S_1, as the internal conversion (10^{-12} s) is generally complete prior to emission. Following fluorescence emission, the molecule typically returns to a higher vibrational energy level of S_1, which then quickly relaxes to the lowest energy vibrational state through an internal conversion. The return to an excited vibrational state at the level of the S_0 state results in a vibrational structure in the emission spectrum. The fluorescence emission spectrum is a plot of fluorescence intensity as a function of emission wavelength or frequency, in which case the wavelength of the excitation light is fixed. It is also possible to record the fluorescence intensity at a certain emission wavelength while scanning the excitation wavelength, when the resultant spectrum – the fluorescence excitation spectrum – is an analog of the absorption spectrum.

Fluorescence typically occurs at lower energies or longer wavelengths than absorption. This is easy to understand, as some of the excitation energy will be lost during the rapid thermal relaxation of S_1 from higher energy vibrational levels to the lowest, and further lost when S_1 decays to higher vibration levels of S_0. The difference (in wavelength or frequency units) between the positions of the band maxima of the absorption and emission spectra is termed the Stokes shift [36].

With some exceptions, the fluorescence emission spectrum is typically a mirror image of the absorption spectrum of the $S_0 \rightarrow S_1$ transition (see Figure 1.6b). This similarity occurs because electronic excitation does not greatly alter the nuclear geometry; hence, the spacing of the vibrational energy levels of the excited states is similar to that of the ground state. According to the Franck–Condon principle, all electronic transitions occur without any change in the position of the nuclei. Therefore, if a particular transition probability between the 0th and 1st vibrational levels is the largest in absorption, the reciprocal transition is also the most probable in emission. As a result, the vibrational structures seen in the absorption and the emission spectra are similar. For the absorption band of $S_0 \rightarrow S_2$ transition, the corresponding mirror-image emissive band does not exist due to a rapid internal conversion from S_2 to S_1.

An important parameter used to characterize fluorescence is the fluorescence quantum yield (Φ), which is the ratio of the number of photons emitted to the number absorbed. The S_1 state may have decay pathways other than fluorescence, and this leads to Φ values lower than 1. Typically, Φ can be expressed as follows:

$$\Phi = k_f/(k_f + k_{nr}) = \tau/\tau_n \quad (1.4)$$

where k_f is the fluorescence emission rate constant of the S_1 state, and k_{nr} is the non-irradiative decay rate constant of S_1, while $\tau = 1/(k_f + k_{nr})$ is the fluorescence lifetime of the molecule, and $\tau_n = 1/k_f$ is termed the natural lifetime; that is, the fluorescence lifetime in the absence of non-irradiative decay.

If any additional nonirradiative decay pathway is introduced to the molecules, it will result in a decrease of the fluorescence quantum yield; this is termed fluorescence quenching. If the additional nonirradiative decay is caused by a quencher with a quenching rate constant of k_Q, then it is possible to write:

$$\Phi = k_f/(k_f + k_{nr} + k_Q[Q]) = \Phi_0(k_f + k_{nr})/(k_f + k_{nr} + k_Q[Q]) = \Phi_0/(1 + K[Q]) \quad (1.5)$$

where Φ and Φ_0 are the fluorescence quantum yield of the quenched and unquenched sample, respectively, $K = k_Q/(k_f + k_{nr})$ is called the Stern–Volmer quenching constant, and [Q] is the concentration of the quencher. Equation (1.5) is usually expressed in the following form:

$$F_0/F = \Phi_0/\Phi = 1 + K[Q] \quad (1.6)$$

This is the famous Stern–Volmer equation [36], in which F_0 and F are the fluorescence intensity of the unquenched and quenched sample, respectively. This equation is valid for dynamic quenching caused by diffusion-controlled collisions between the fluorophores and quenchers. For static quenching, where a fraction of the molecules are completely quenched as a result of forming nonirradiative stable complexes with quenchers while the others are unaffected, then Eq. (1.6) is still operational but the physical meaning of K is different. For dynamic quenching, the fluorescence lifetime is shortened, whereas for static quenching an unchanged fluorescence lifetime will be observed in correspondence to the fraction of unaffected molecules.

Molecules in the S_1 state can also undergo a spin conversion to the first triplet state T_1. Emission from T_1 is termed phosphorescence, and is generally shifted to longer wavelengths (lower energy) relative to the fluorescence. The conversion of S_1 to T_1 is referred to as intersystem crossing. The transition from T_1 to S_0 is spin-forbidden, and as a result the rate constants for triplet emission are several orders of magnitude smaller than those for fluorescence. The T_1 state typically has a relatively long lifetime, ranging from 10^{-3} to 1 s. Phosphorescence is often too weak to be observed at room temperature; however, the phosphorescence quantum yields can be enhanced by incorporating heavy atoms such as bromine and iodine into the molecules. These heavy atoms facilitate intersystem crossing due to spin–orbit coupling, with high phosphorescence quantum yields being achieved in some metal complexes.

In some cases, an excited molecule and an adjacent ground-state molecule can form an instantaneous complex that emits fluorescence at a longer wavelength (i.e., lower energy) than the excited molecule itself. This complex is called an excimer [37] if the two molecules are of the same type, but an exciplex [38] if the two molecules are different. For example, pyrene forms an excimer in concentrated solution, while anthracene can form an exciplex with aniline. The emission bands of the excimer and exciplex are typically broad and structureless. The

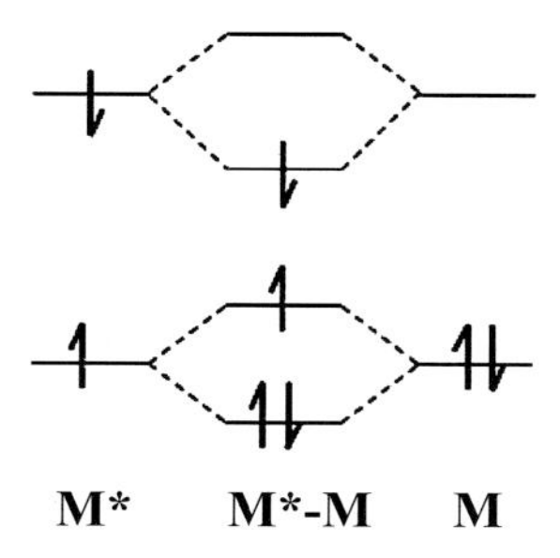

Figure 1.7 Molecular orbital splitting during excimer formation. The excimer emits at a longer wavelength because of a decreased energy gap.

excimer and exciplex are weak charge-transfer complexes having lifetimes of about 10^{-9} s; typically, these complexes dissociate when they return to the ground state. The involved photophysical processes can be expressed as follows.

$D + \hbar\upsilon \rightarrow D^*$ – molecule D absorbs one photon and jumps to excited state
$D^* \rightarrow D + \hbar\upsilon_F$ – the excited state emits one photon (fluorescence) and returns back to ground state
$D^* + A \rightarrow (DA)^*$ – the excited state D* forms a complex with a ground-state molecule A
$(DA)^* \rightarrow D + A + \hbar\upsilon_{FE}$ – the complex emits one photon ($\nu_{FE} < \nu_F$) and returns back to ground state

where D and A can be either the same or different molecules.

The formation of an excimer or exciplex is the result of a charge redistribution between the excited molecule and the ground-state molecule. For such a process to occur, the two molecules must have some overlap between their π-orbitals. The longer wavelength emission of excimer/exciplex can explained as being a result of molecular orbital splitting (Figure 1.7).

1.3.2 Excitation for Charge and Energy Transfer in Small Molecules

1.3.2.1 Photoinduced Electron Transfer

Following excitation, one of the two electrons in the HOMO level jumps to LUMO, or an even higher energy level, so as to provide the molecule with a higher activity in the redox reaction. On the one hand, the excited electron gains more energy and hence becomes easier to be donated; on the other hand, the HOMO becomes only half-occupied and can accept one electron from a reductant. Such a light-driven reduction/oxidation process is termed photoinduced charge transfer (Figure 1.8). The formation of excimer or exciplex is a case of partial charge transfer, where the charge density is slightly redistributed between the two constituent molecules. In general, photoinduced charge transfer refers to the complete transfer of one

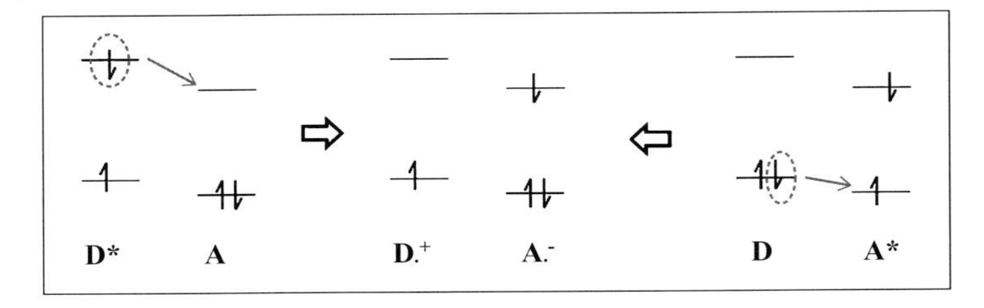

Figure 1.8 Schematic representation of photoinduced charge transfer between the donor (D) and acceptor (A).

charge entity (e.g., an electron or a proton) between two distinct atoms, functional groups, or molecules following photon absorption. At this point, attention will be focused on photoinduced electron transfer (PIET) between two molecules (intermolecular), or between two moieties of the same molecule. In such a process, the molecule or moiety providing an electron is called the donor (D), while the other molecule/moiety receiving the electron is called the acceptor (A). Typically, electron transfer results in the formation of a radical cation of the donor ($D^{\overset{+}{\bullet}}$) and a radical anion of the acceptor ($A^{\overset{-}{\bullet}}$). In general, PIET is a multi-step process, one possible mechanism of which is as follows:

$D + h\nu \rightarrow D^{S_1}$ – Excitation of D to singlet excited state
$D^{S_1} + A \rightarrow [D^{\delta+} \ldots A^{\delta-}]^S$ – Initial charge redistribution between D^{S_1} and A
$[D^{\delta+} \ldots A^{\delta-}]^S \rightarrow [D^{\overset{+}{\bullet}} \ldots A^{\overset{-}{\bullet}}]^S \rightarrow D^{\overset{+}{\bullet}} + A^{\overset{-}{\bullet}}$ – Charge transfer and separation
$D^{S_1} \rightarrow D^{T_1}$ – Conversion to triplet excited state via intersystem crossing
$D^{T_1} + A \rightarrow [D^{\delta+} \ldots A^{\delta-}]^T$ – Initial charge redistribution between D^{T_1} and A
$[D^{\delta+} \ldots A^{\delta-}]^T \rightarrow [D^{\overset{+}{\bullet}} \ldots A^{\overset{-}{\bullet}}]^T \rightarrow D^{\overset{+}{\bullet}} + A^{\overset{-}{\bullet}}$ – Charge transfer and separation

The above steps are not necessarily all involved in each particular case. In principle, electron transfer can occur from either singlet or triplet excited state of the donor. As a rule, the acceptor should have a lower-lying empty energy level in comparison to the donor excited state.

In some cases, electron transfer can take place from a ground-state donor to an excited acceptor:

$$D + A^{S_1 or T_1} \rightarrow [D^{\delta+} \ldots A^{\delta-}]^{S or T} \rightarrow [D^{\overset{+}{\bullet}} \ldots A^{\overset{-}{\bullet}}]^{S or T} \rightarrow D^{\overset{+}{\bullet}} + A^{\overset{-}{\bullet}}$$

This can be understood that, as the excitation of acceptor leaves a "vacancy" in its HOMO level, and the vacancy can accommodate an electron from the donor, this will result in a new vacancy in the donor HOMO. Such a vacancy is termed a "hole". The above process can be viewed as a hole being is transferred from the acceptor to the donor, but in this case the HOMO level of the acceptor should be lower than that of the donor.

It is worth noting that an electronic excited state can be seen as a bound electron-hole pair, and its return to ground state can be seen as recombination of the electron and hole. This concept is important to understand the nature of excitations and charge carriers in the solid state (these will be discussed later in the chapter).

Some of the transient states during PIET, for example $[D^{\dot{+}} \ldots A^{\dot{-}}]^{SorT}$, may lose their energy and return to ground state via nonirradiative relaxation or photon emission. The emission is generally of a low quantum yield because the transition is symmetry-forbidden. An exciplex is actually a stabilized state of $[D^{\delta+} \ldots A^{\delta-}]^{S}$. The complex $[D^{\dot{+}} \ldots A^{\dot{-}}]^{SorT}$ is generally termed a photoinduced charge-transfer state or simply a charge-transfer (CT) state. When the donor and the acceptor are in the same molecule, this is termed an intramolecular charge-transfer (ICT) state. The formation of a CT or ICT state leads to a quenching of the donor emission. In comparison to the locally excited state (D*), the CT state is of a lower energy and emits at a longer wavelength with a broad, structureless emission band. Moreover, the CT state possesses a larger dipole moment as a result of charge separation. Thus, it generally becomes more stable in a polar medium, and its emission becomes remarkably red-shifted with increasing solvent polarity. One special case of ICT is termed twisted intramolecular charge-transfer (TICT), where the donor and acceptor groups are coplanar and electronically coupled at ground state, but twisted relative to each other and hence are decoupled at ICT state. Such a twisted structure will stabilize the separated charges and cause the TICT process to be more favorable from a thermodynamic aspect.

The separated charges, $D^{\dot{+}}$ and $A^{\dot{-}}$, can recombine and eventually return to the neutral ground states D and A; this process is termed charge recombination or electron back transfer.

If the PIET occurs from D^{S_1} to A, it leads to quenching of the fluorescence of the donor. The acceptor molecules act as quenchers. Assuming the rate constant of PIET to be k_{ET}, then:

$$\tau_D = 1/(k_f + k_{nr}) \tag{1.7}$$

$$\tau = 1/(k_f + k_{nr} + k_{ET}) \tag{1.8}$$

and thus

$$k_{ET} = \frac{1}{\tau} - \frac{1}{\tau_D} \tag{1.9}$$

This equation provides a simple means of measuring the rate constant of PIET by recording the fluorescence lifetime of the donor in the case that PIET is the major fluorescence quenching pathway. For a more precise measurement, the formation of radical ions or other transient species should be traced, using for example ultrafast spectroscopy methods.

In organic systems, PIET is generally a short-range interaction. In order for electron transfer to occur, the donor and acceptor should be sufficiently close that their molecular orbitals become overlapped. In quantum mechanics theory, this corresponds to a spatial overlap of the donor and acceptor wavefunctions. Long-range electron transfer may take place when the donor and acceptor are linked through a bridge molecule.

The rate of electron transfer reactions including PIET can be interpreted by the Marcus theory [8–18, 20–40]. This theory was originally developed by Rudolph A. Marcus from classical mechanical considerations, though similar expressions were lately derived from a quantum mechanical viewpoint. The Marcus model takes the donor and acceptor together with the surrounding environment (e.g., solvent molecules) as a whole system when considering the nuclei motions in response to electron transfer. The final equation is expressed as:

$$k_{\mathrm{ET}} = \frac{2\pi}{h}|H|^2 \frac{1}{\sqrt{4\pi k_{\mathrm{B}} T\lambda}} \exp\left[-\frac{(\Delta G^0 + \lambda)^2}{4\lambda k_{\mathrm{B}} T}\right] \quad (1.10)$$

where k_{ET} is the rate constant for electron transfer, h is the reduced Plank constant, $|H|$is the electronic coupling between the initial and final states, λ is the reorganization energy, ΔG^0 is the total Gibbs free energy change for the electron transfer reaction, k_{B} is the Boltzmann constant, and T is the absolute temperature. The reorganization energy is defined as the energy required to "reorganize" the system structure from initial to final coordinates, without making the electron transfer (Figure 1.9). The reorganization is a set of vibrational motions of nuclei in the system.

1.3.2.2 Excitation Energy Transfer

An excited molecule/chromophore (donor) can transfer the excitation energy to another molecule/chromophore (acceptor) under certain circumstances. The donor molecules typically emit at shorter wavelengths that overlap with the absorption spectrum of the acceptor. If the donor and acceptor are denoted as D and A, respectively, then the process can be expressed as:

$$\mathrm{D}^* + \mathrm{A} \rightarrow \mathrm{D} + \mathrm{A}^*$$

where D^* and A^* are electronic excited states of D and A, respectively. Apparently, the energy transfer will compete with the irradiative decay of D^* or, in the other words, it results in a quenching of the fluorescence (or phosphorescence) of the donor. If A^* can decay irradiatively, then fluorescence of the acceptor will be observed following excitation of the donor.

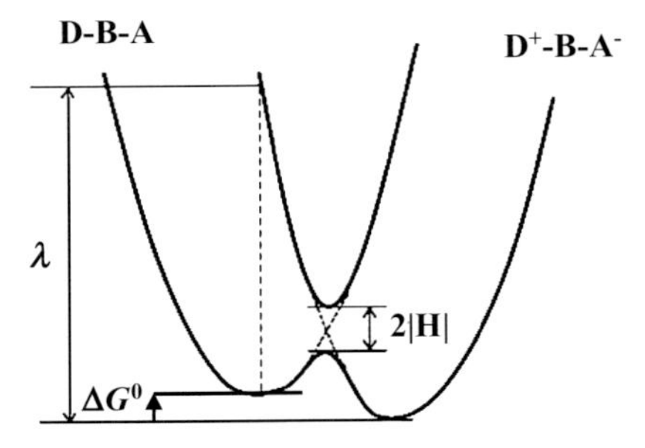

Figure 1.9 Energy potential curve of an electron-transfer reaction.

In the general case, D* and A* are both singlet electronic excited states (S_1), a case that is often termed as singlet–singlet energy transfer. There are other possibilities, however, such as triplet–singlet energy transfer (transfer of excitation from an excited donor in triplet state to produce an excited acceptor in singlet state), and triplet–triplet energy transfer. Energy transfer processes occur typically on time scales that range from picoseconds to nanoseconds for singlet energy transfer up to milliseconds and seconds for triplet energy transfer, because of the much longer lifetimes of triplet states.

Excitation energy transfer between organic molecules is a ubiquitous phenomenon in Nature, a prominent example being the photosynthetic process. In photosynthesis, following photon absorption by light-harvesting complexes, the electronic excitation energy is transferred efficiently towards the photosynthetic reaction center, where the light energy is converted into chemical energy. Energy transfer is also of crucial importance for the application of conjugated materials in organic electronics. For instance, in organic solar cells and photodetectors, the neutral excitations generated by photon absorption must be transferred to particular interfaces in order to dissociate into free charges. In OLEDs, a high-energy gap donor material (host material) is often blended with a low-energy gap acceptor (dopant), so that the singlet and/or triplet excitons created by charge recombination in the host material can be transferred to the dopants, where they are emitted.

Two simple approaches are often used to describe excitation energy transfer in conjugated organic materials, namely the well-known Förster and Dexter models for energy transfer [36].

1.3.2.2.1 The Förster Model

In the framework of Förster theory, energy transfer is mediated via a long-range resonant dipole–dipole interaction between the donor and acceptor molecules. In this case, a chromophore can be roughly seen as a dipole oscillator that is capable of exchanging energy with another dipole having a similar resonance frequency. This is similar to the behavior of coupled oscillators – much like two swings on a common supporting beam [36]. For the above reason, Förster energy transfer is often termed resonance energy transfer or fluorescence resonance energy transfer (FRET). It should be noted that the energy is not actually transferred by fluorescence nor by any other irradiative channel; this case should be distinguished from the situation where the fluorescence of D is reabsorbed by A.

For energy conservation, FRET requires a spectral overlap of the emission spectrum of the donor with the absorption spectrum of the acceptor (Figure 1.10). The extent of energy transfer is then determined by the distance between the donor and acceptor, and the extent of spectral overlap. For convenience, the spectral overlap is described in terms of the Förster radius (R_0). Here, the rate constant of energy transfer k_{DA} is expressed by:

$$k_{DA} = \frac{1}{\tau_D}\left(\frac{R_0}{R}\right)^6 \tag{1.11}$$

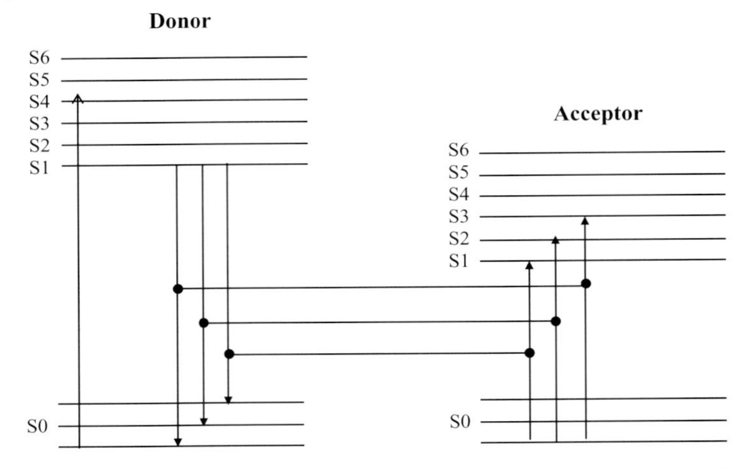

Figure 1.10 Schematic illustration of Förster resonance energy transfer. The donor emission and acceptor absorption should overlap for energy conservation.

where τ_D is the measured fluorescence lifetime of the donor in the absence of the acceptor, and R is the distance between donor and acceptor. If $R = R_0$, the energy transfer rate becomes $k_{DA} = \dfrac{1}{\tau_D} = k_f + k_{nr}$; that is, the energy transfer has the same rate as the excited state decay expressed by the sum of the rates of radiative and nonradiative pathways. Therefore, the Förster radius R_0 is the distance at which the FRET efficiency of a D–A pair is 50%. At this distance, the donor emission would be decreased to half its intensity in the absence of acceptors. Based on dipole approximation, the Förster radius can be derived as:

$$R_0^6 = \frac{9000(\ln 10)k^2\Phi_D}{128\pi^5 N_A n^4 \tau_D} J \tag{1.12}$$

where Φ_D is the fluorescence quantum yield of the donor in the absence of acceptor, n is the refractive index of the medium, and N_A is Avogadro's number. The term k^2 is a factor describing the relative orientation in space of the transition dipoles of the donor and acceptor. J is the spectral overlap, defined as

$$J = \int F_D(\lambda)\varepsilon_A(\lambda)\lambda^4 d\lambda \tag{1.13}$$

where $F_D(\lambda)$ is the emission intensity of the donor at wavelength λ, being normalized so that $\int F_D(\lambda)\, d\lambda = 1$; $\varepsilon_A(\lambda)$ is the molar extinction coefficient of the acceptor at λ.

FRET is a long-range interaction, and does not require a close contact of the donor and acceptor. For some donor–acceptor pairs, the Förster radius may be up to 10 nm, which is much larger than the general molecular radius.

The distance-dependence of FRET allows the measurement of distances between the donors and acceptors. This principle has been proven to be very useful in measuring the distances between two sites on a biologically macromolecule (e.g., a protein) by covalently labeling one of the sites with a donor and the other with an acceptor [36]. Similar approaches have been used to study the conformational dynamics of biomolecules.

It should be noted that the simple dipole approximation may break down in situations where the chromophores are in close proximity, or are linked by bridging moieties. Nevertheless, in many instances the Förster approach can provide a useful, and at least qualitatively correct, physical picture for even complex donor–acceptor systems [41].

1.3.2.2.2 The Dexter Model

In the Dexter model [42], the energy transfer is similar to a bimolecular reaction, and requires an overlap of the involved donor and acceptor molecular orbitals. Electron exchange can only take place in the overlap region. Because the overlap decays exponentially with distance, it is expected that the rate constant k_{DA} decreases even more rapidly with R than was observed in the case of FRET. In comparison to FRET, Dexter energy transfer is a short-range interaction and occurs typically over distances which are similar to the van der Waals distance – that is, $R = 0.5$–$1.0\,\text{nm}$. The rate constant k_{DA} falls exponentially with the distance R between D and A:

$$k_{DA} = KJ\exp(-2R/L) \tag{1.14}$$

where K is a constant in relation to the involved molecular orbitals, J is the spectral overlap between the donor emission and acceptor absorption, and L is the effective average Bohr radius, which is typically on the order of 0.1–0.2 nm.

Dexter energy transfer is a correlated two-electron exchange process. Hence, it allows triplet energy transfer without the additional need for intersystem crossing upon energy transfer of a triplet state. This is in contrast to the Förster energy transfer, which would require a spin-flip for each triplet energy transfer step. For this reason, singlet energy transfer is usually described in the framework of Förster theory, whereas triplet energy transfer is described by the Dexter mechanism [41].

In solution, the donor and acceptor molecules must be close enough for the electron exchange to occur. In this case, the Dexter energy transfer is diffusion-controlled. In contrast, the apparent rate of Förster energy transfer can exceed the diffusion limit. In amorphous films of a donor host doped with a small amount of an acceptor guest, both processes can – in principle – take place, such that the resulting energy transfer mechanism is likely to be a superposition of both modes, depending on the time and distance after the excitation.

In both electron- and energy-transfer cases, the transition mechanism involves vibrational motions driving the reaction coordinates from reactants to products [2]. Therefore, the Marcus model for electron transfer can be implanted into energy transfer cases by considering the energy donor and acceptor and surrounding environment as an entire system.

1.4
Some Basic Concepts of Electronic Process in Conjugated Polymers

In Section 1.3, mention was made of electronic processes in model systems that consisted of only one or two molecules. The situation in the solid state is much more complicated, as the complicity has two aspects: (i) many molecules/atoms interact with each other in a complicated fashion, leading to numerous possible electronic and vibrational states; and (ii) energy and charge transport are typically multistep processes with complex dynamics. In order to describe organic solids, a variety of concepts and terms have been utilized, including energy band, polaron, and exciton, all of which are commonly used in condensed-matter physics. In the following subsections, these basic concepts will be interpreted from a chemist's point of view, with attention focused on conjugated polymers as a representative type of organic semiconductor.

The first step is to provide a brief description of the energy band, which is a very useful concept in solid-state physics. When atoms/molecules are brought together to form a solid, they begin to influence each other. For instance, the outer shell electrons of a molecule will be attracted by the nuclei in other molecules, leading to considerable modifications to their energy levels. This corresponds to a splitting of the atomic/molecular orbitals and a redistribution of the energy levels. For many atoms/molecules, the number of orbitals becomes exceedingly large, and consequently the difference in energy between them becomes very small. Thus, in solids the levels form continuous bands of energy (Figure 1.11) rather than the discrete energy levels of the independent atoms/molecules. Meanwhile, there may be still an energy range left where no electron orbital exists; this is termed the band gap. Typically, an insulator or semiconductor possesses an almost fulfilled band immediately below the band gap, and an almost unoccupied band immediately above the band gap. The former is termed the valence band (VB), and the

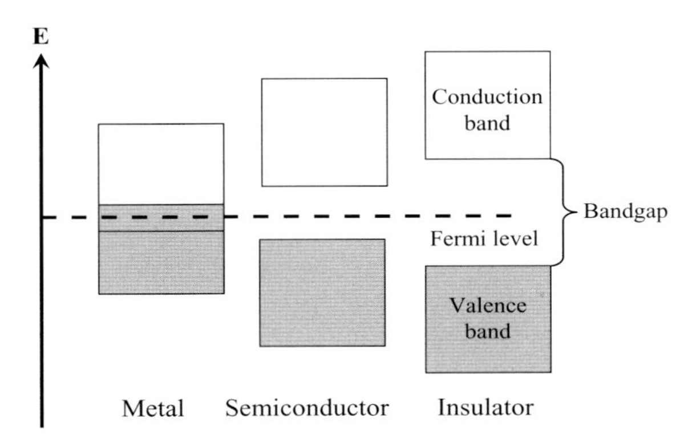

Figure 1.11 Energy band structures of metal, semiconductor, and insulator. The Fermi level is a hypothetical energy level at which an orbital is exactly half-filled.

latter the conduction band (CB). In inorganic semiconductors such as silicon, some of the electrons at the VB may be thermoactivated into the CB, resulting in a small number of mobile "free" electrons in the CB and some mobile holes in the VB. It is for this reason that the materials demonstrate semiconductivity. Whereas, insulators cannot conduct electricity because their band gaps are too large for the thermoexcitation of electrons, metals generally have overlapped VBs and CBs (i.e., no band gap) and thus show good conductivity, even at low temperature.

Traditional polymers such as plastics and rubbers, the backbones of which are mainly composed of saturated carbon atoms linked by single covalent bonds, have good insulating properties. In contrast, conjugated polymers have alternating single and double bonds in their backbones (the molecular structures of some prototype conjugated polymers are shown in Figure 1.1). In the case of t-PA (the simplest conjugated polymer), each carbon atom is connected to one hydrogen atom and to two neighboring carbon atoms through σ bonds. If the carbon–carbon bond lengths were uniform, and there was an unpaired electron on each carbon atom, then the π orbitals would be degenerate and half-filled (Figure 1.12), and the polymer chain would behave like a one-dimensional (1-D) metal. However, this geometry is unstable, and the polymer chain favors a structure with alternating single and double bonds. As a result, the polymer behaves like an insulator rather than a metal. On the other hand, it has been found that charge carriers can be generated in this polymer through chemical reduction/oxidation – that is, by adding electrons into the conjugation system or by taking electrons out [8]. In this way, it has been shown possible to fine-tune the conductivity of t-PA, from insulating to metallic, through redox doping.

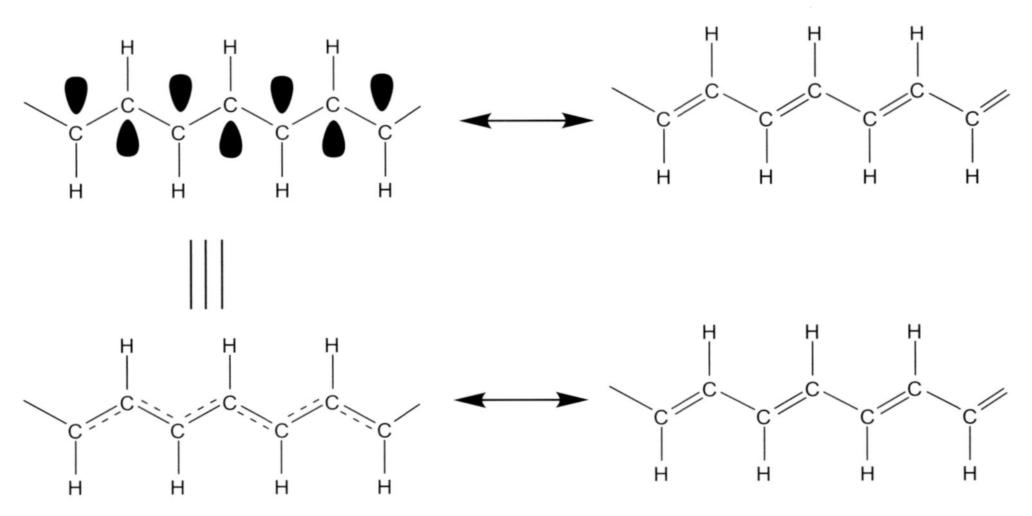

Figure 1.12 Electronic structure of *trans*-polyacetylene: π-orbitals are either half-filled (not stable) or fully filled (stable) forms.

During the 1970s, doping-induced conductivity in conjugated polymers attracted much attention from condensed-matter theorists, and also from chemists and materials scientists. Since the 1980s, however, interest in π-conjugated polymers has been mainly focused on the semiconducting behavior of the pristine systems, rather than on the conducting behavior of the doped materials. While showing much promise in terms of their electronic properties, conjugated polymers can – at least in principle – maintain the good mechanical properties of typical polymers. In fact, conjugated polymer materials have been produced with a conductivity close to that of copper, and with a mechanical performance comparable to that of steel [43]. The development of conjugated polymer-based flexible displays [44] and plastic solar cells [45] has also been demonstrated. Until now, conjugated polymers have shown a wide range of potential, for example in static proofing, radiation protection, corrosion resistance [46], molecular wires and organic circuits [47], lighting and display [48], solar energy conversion and photodetection [49], as well as biological and chemical sensing [16, 50].

1.4.1
Excited States in Conjugated Polymers

There exist various types of charged or neutral excited states (excitations) in conjugated polymers, depending on the molecular structure and the extent of doping. In order to account for their coupling with surrounding environments (lattice distortions), these excited states are generally treated as "qausiparticles" rather than as pure electronic states. This treatment is particularly useful when considering the transport of a charge or an electron-hole pair in a solid. For example, when an electron travels through a solid, its motion is disturbed in complex fashion by its interactions with all other electrons and nuclei; nonetheless, it still behaves (largely) like an electron but with a different mass, traveling unperturbed through free space. This "electron" with a different mass is termed an "electron quasiparticle" (but simply as an electron for convenience). Other terms used include solitons, polarons, bipolarons, and excitons to describe different type of excitations. The chemical nature of these excitations, and their characteristic properties, are described in the following subsections.

1.4.1.1 Soliton

Among conjugated polymers, t-PA is unique as it possesses a degenerate ground state – that is, two geometric structures corresponding exactly to the same total energy [51, 52]. The two structures differ one from another by the exchange of carbon–carbon single and double bonds. It was noted above, that every carbon atom in t-PA has one electron in the p_z orbital, and each two electrons form a π-bond. However, for t-PA containing an odd number of carbon atoms, one unpaired electron will remain unpaired such that a radical will occur in the 1-D chain. By chemical redox, this radical can be turned into a positive or negative charge, so that it can act as a boundary between two segments with opposite ground-state geometries (Figure 1.13). In physics terminology, such a radical or charge associ-

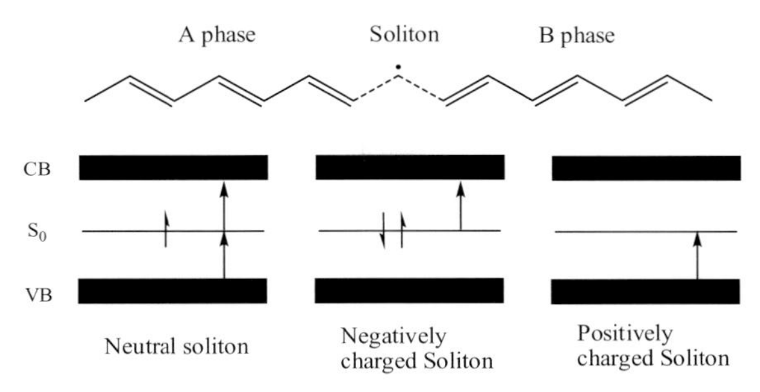

Figure 1.13 Upper: Schematic illustration of a neutral soliton in *trans*-polyacetylene (t-PA); Lower: Band structure for a t-PA chain containing a neutral soliton, a negatively charged soliton, and a positively charged soliton. The allowed transitions are indicated by solid arrows.

ated with a boundary is called a soliton, because it has the properties of a solitary wave that can propagate without deformation and dissipation [51]. The soliton can propagate freely along the polymer chain, as its two sides possess identical energy. In a long chain, the unpaired electron in a neutral soliton (or a charge in a charged soliton) will not be localized on one carbon but rather will be spread over several carbon atoms (up to 14), which causes the soliton to have a width. Although the bond lengths are equal at the middle of the soliton, starting from one side of the soliton the double bonds become gradually longer and the single bonds shorter; consequently, on reaching the other side of the soliton the alteration has completely reversed [51].

The presence of a soliton leads to the appearance of a localized electronic level at mid-gap, which is half-occupied in the case of a neutral soliton and empty (doubly occupied) in the case of positively (negatively) charged soliton (Figure 1.13). Upon increasing the doping level, soliton states at midgap begin to overlap and to form a soliton band. Solitons are one of the primary types of charge carrier in doped t-PA. Solitons only exist in conjugated molecules/polymers with a degenerate ground state; that is, the exchange between the single and double bonds does not alter the total energy of the molecule.

1.4.1.2 **Polaron**

Unlike t-PA, conjugated polymers such as PPP, polypyrrole, and polythiophene possess a nondegenerate ground state, since their ground state corresponds to an aromatic geometry, while the exchange of single bonds with double bonds would result in a quinoid-like structure with a higher total energy [51]. In such polymers, the main charged excitations are radical ions strongly coupled with lattice distortion; these are termed polarons (a positive polaron in the case of a radical cation, and a negative polaron in the case of a radical anion). The net charges in polarons are the result of chemical reduction/oxidation, electrochemical charge injection, or photoinduced charge transfer. The local lattice distortion tends to localize the charge; this corresponds to an upshift of the local HOMO from the valence band,

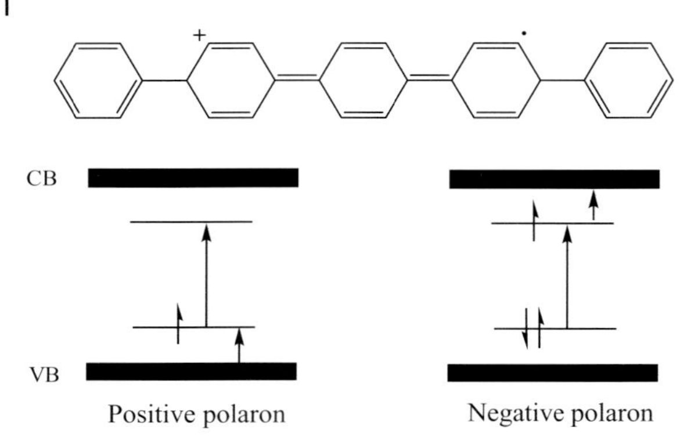

Figure 1.14 Upper: Schematic illustration of a positive polaron in poly(*p*-phenylene) (PPP); Lower: Band structure for a PPP chain containing a positive polaron or a negative polaron. The allowed transitions are indicated by solid arrows.

and a downshift of the LUMO from the conduction band. It should be noted that the valence band remains full and the conduction band remains empty in this case. The geometry of a conjugated polymer PPP in the presence a positive polaron is shown in Figure 1.14. The transition between central anti-symmetric ω_0 and symmetric ω_1 is allowed, but transition from ω_0 to the conduction band is forbidden. Although polarons can propagate along the polymer chain, the propagation length is strongly limited by conjugation interruptions. It is also possible for a polaron to jump from one chain to another if the acceptor chain possesses a similar lattice distortion to the donor (see the Marcus model in Section 1.3). The recombination of a positive polaron with a negative polaron may result in a singlet or triplet excited state. In OLEDs, holes and electrons are injected from opposite sides and recombine in organic semiconductors to form emissive singlet excited states, while triplet states can also be harvested by adding phosphorescent guest molecules.

1.4.1.3 Bipolaron

A bipolaron is a pair of charges of the same sign (dual cations or dual anions) that is coupled to lattice distortion. Bipolarons are similar to polarons, but have larger lattice distortions and begin to dominate at a large doping extent or a high charge injection rate. A positive/negative bipolaron can be seen as taking/adding one electron from/to a positive/negative polaron. Two polarons of the same sign can combine to create a bipolaron, the formation of which implies that the energy gained by the interaction with lattice is larger than the Coulombic repulsion between the two charges of same sign confined in the same location [51].

As the lattice relaxation around two charges is stronger than around only one charge, a bipolaron corresponds to a local shift of HOMO and LUMO even further away from the valence band and conduction band, respectively. Unlike polarons, a bipolaron has no half-filled energy level, and only one transition is allowed for

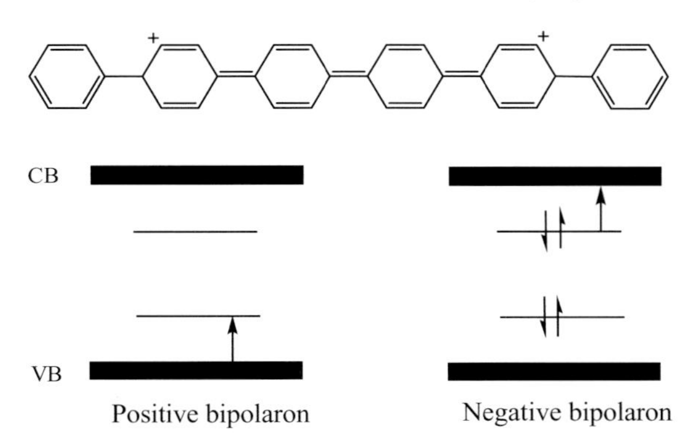

Figure 1.15 Upper: Schematic illustration of a positive bipolaron in PPP; Lower: Band structure for a PPP chain containing a positive bipolaron or a negative bipolaron. The allowed transitions are indicated by solid arrows.

a bipolaron (Figure 1.15). Hence, a bipolaron can easily be distinguished from a polaron by using spectroscopic measurements, such as transient absorption.

1.4.1.4 **Exciton**

The neutral excitations generated following photon absorption or recombination of an electron (negative polaron) and a hole (positive polaron) are termed excitons. These can be regarded as a bound electron–hole pair in a couple with lattice distortion. In general there are three major types of exciton, known as the Frenkel exciton, the Wannier–Mott exciton, and the charge-transfer exciton, respectively. The major differences in these excitons lie in the binding energy between the electron and hole.

- **Frenkel excitons:** These are tightly bound electron–hole pairs, typically with a binding energy of about 1 eV. Due to the strong attraction between the electron and the hole, a Frenkel exciton has a small radius of ~10 . In general, the electron and the hole are located on the same molecule, analogous to molecular excited states (Figure 1.16). Frenkel excitons have well-defined spin states (singlet and triplet), and their transport (diffusion) in solids is generally interpreted by Förster and Dexter energy transfer models.

- **Wannier–Mott excitons:** In comparison to Frenkel excitons, Wannier–Mott excitons possess a much smaller binding energy (typically <0.1 eV), and the electron and hole are only loosely bound in such excitons. In fact, the electron and hole can be treated as two individual quasiparticles (i.e., a negative polaron and a positive polaron) that are bound together by the same lattice distortion. In this sense, Wannier–Mott excitons are also termed neutral bipolarons. A Wannier–Mott exciton has a relatively large radius (~100), while the electron and hole are delocalized (Figure 1.16) and can easily be separated owing to the small Coulombic binding energy. The spin state of Wannier–Mott excitons has

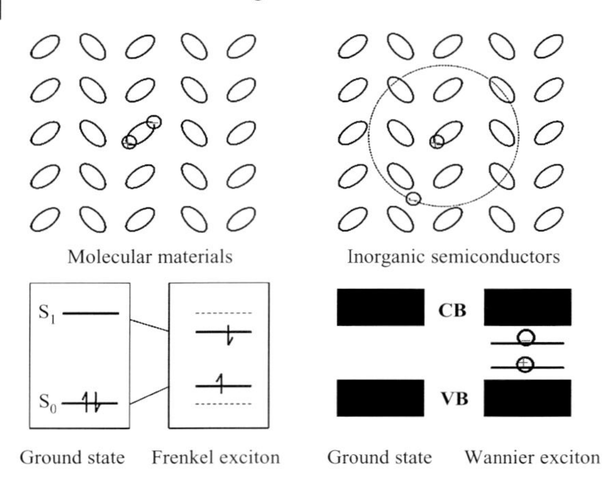

Figure 1.16 Schematic illustration of the difference between a Frenkel exciton and a Wannier exciton.

not been well-defined because of a rapid singlet–triplet exchange. The excitons generated in inorganic semiconductors are usually Wannier–Mott excitons; as these tend to dissociate very rapidly at room temperature they play only a minor role in inorganic semiconductors.

- **Charge transfer excitons:** These are intermediate between Frenkel-type and Wannier–Mott-type excitons. In such excitons, the electron and hole are separated but still tightly bound; the separated charges can be either localized or delocalized in relation to the lattice distortion. Charge transfer excitons are analogous to charge-transfer states observed in D–A molecules. They are often found in charge-transfer crystals, with electrons lying on one of the components and holes on the other component. On the other hand, they may also be formed in solids consisting of only one organic compound.

During recent years there has been much debate concerning the nature of the primary excitations in intrinsic conjugated polymers [53, 54], the main question being whether they are Frenkel excitons, Wannier excitons, charge polarons, or something else? The answer to this point is of crucial importance for the applications of these materials. For example, if the primary excitations are tightly bound excitons, then in electroluminescence devices the injected electron and hole can form a triplet with a spin multiplicity of three, or a singlet with a spin multiplicity of one. Only the latter (corresponding to a formation probability of 25%) will recombine radiatively, unless a triplet-harvesting mechanism is applied. In contrast, with charge polarons as the primary excitations the theoretical maximum efficiency can approach unity. In photovoltaic devices, Frenkel excitons need to diffuse to an interface with built-in electric field (e.g., electron donor–acceptor interface) in order to dissociate into free charges, whereas Wannier–Mott excitons tend to dissociate very rapidly and hence generate free charge carriers locally. In order to determine the nature of the primary excitations, one key parameter is the binding energy between the electron and hole, and that the solution to the problem is clear for several types of conjugated polymer but ambiguous for some others. Polydiacetylenes, for instance, are believed to have an exciton binding energy of about 0.5 eV according to photoconductivity spectrum and electroabsorption measurements [53]. For PPV and its soluble derivatives, however, it has been reported that photoconductivity began to appear at the onset of optical absorption, implying a bind energy <0.1 eV. Yet, a value of ~0.4 eV was derived from other experimental approaches, such as photovoltaic working spectrum, electric field-induced photoluminescence quenching, and magnetic field dependence of photoconductivity. Currently, it is more widely accepted that the lowest energy excitations in conjugated polymers are neutral Frenkel excitons.

The delocalization degree of an exciton in conjugated polymers is prevailingly determined by the conjugation lengths of the segments. As noted above, the effective conjugation length is greatly limited by the chemical and physical defects in the polymer backbone. The most commonly discussed defects include saturated carbon atoms, *cis*-isomerization kinks, and bending and twisting of the backbone. Such defects are unavoidable as they are brought to the polymer chains during material synthesis and processing, and even if an ideal chain without chemical defects could be obtained, physical defects such as twisting and bending would still be present. It has been claimed that the average conjugation length in PPVs is only about 10 repeating units [23], and that a conjugated polymer chain can therefore be viewed as consisting of a series of conjugation segments of different lengths (Figure 1.17). These conjugation segments are generally treated as weakly

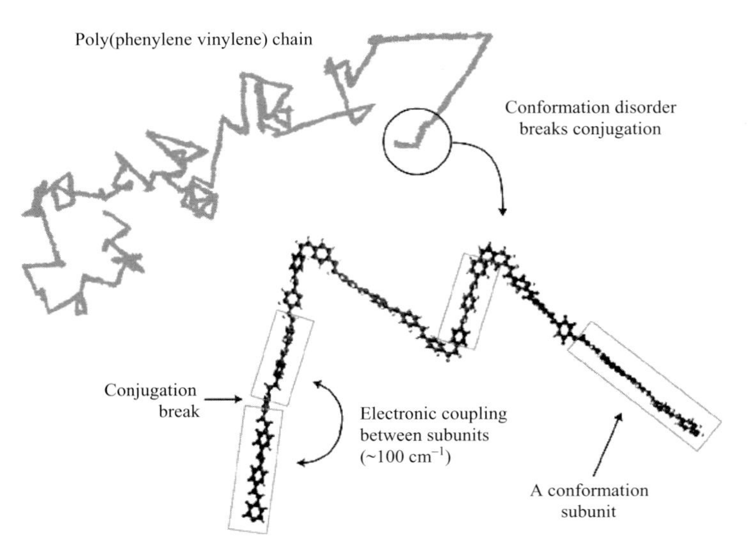

Figure 1.17 Schematic illustration of a PPV chain with defects. Reproduced with permission from Ref. [55]; © 2006, Nature Publishing Group.

coupled chromophores where, typically, shorter segments have absorption and emission in the bluer range, and longer in the redder range. Excitons that are generated on relatively short segments tend to migrate to relatively long segments via for example, resonance dipole coupling or electron exchange. A general belief here is that a migration process consists of a succession of energy-transfer steps, and that the transfer direction of each step obeys the probability law. This is known as the "random-walk model" [56] of energy migration; such a migration process is also referred to as exciton diffusion, as it is similar to the diffusion motion of a molecule/particle in solution.

1.4.2 Interactions between Conjugated Polymer Chains

In the solid state of conjugated polymers, there often exists a strong interchain π–π stacking interaction, especially when the conjugated backbones have a rigid planar structure. Intrachain π–π stacking may also be present in self-folded polymer chains, with two or more adjacent conjugated polymer chains or chain segments forming excimers or interchain/intersegment charge-transfer states as a result of the overlapping of π-orbitals. It should be noted that the degree of chain stacking depends significantly on the way in which the polymer is processed to form the film [57]. Interchain interactions can be evaluated experimentally by comparing the optical spectra of pristine films with those of diluted films (i.e., blends with an inert polymer matrix, such as polystyrene) and solutions. As a replacement for the pristine film, nanosized aggregates of conjugated polymers can be prepared artificially by a controlled precipitation in a poor solvent [58]. The formation of interchain excitations results in a quenching of fluorescence and the appearance of emission bands at longer wavelength regions. Hence, the fluorescence quantum yields of the conjugated polymers in dilute solutions and diluted films are generally larger than those measured in dense films and aggregates. In fact, aggregation-induced fluorescence quenching is a common phenomenon for organic conjugated molecules, although there are some exceptions, with aggregation-induced emission enhancement having been reported in some systems [59–61]. The enhancement mechanisms were generally explained as a suppression of the nonirradiative decay channels of the excited states; for example, decay via a molecular rotational relaxation can be diminished upon aggregation, due to the strong intermolecular steric repulsions.

The interchain excitations reported in conjugated polymers can be categorized as three major types, namely bound polaron pairs, excimers, and ground-state complexes.

1.4.2.1 Bound Polaron Pairs

The first evidence of bound polaron pairs in conjugated polymers was reported by Rothberg *et al.* during the 1990s [62–64]. On investigating the fluorescence of PPV and poly(2-methoxy,5-(2′-ethylhexoxy)-4-phenylenevinylene (MEH-PPV) films at selected excitation wavelengths (300 nm and 500 nm), Rothberg's group showed that the fluorescence quantum yield was lower at a shorter excitation wavelength, but that the emission spectrum and the fluorescence lifetime were maintained in a similar state. These observations indicated that not all of the photons absorbed by PPV would result in singlet excitons, especially at a shorter excitation wavelength (i.e., a higher photon energy). More recently, the same group conducted picosecond transient absorption experiments on PPV, and showed that the generation efficiency of singlet excitons could be much smaller than unity. A transient species that was unrelated to the emission pathway was also observed, but this could not be assigned as either singlet or triplet excitons, nor polarons or bipolarons. Ultimately, the conclusion was reached that bound polaron pair formation was a prominent photophysical pathway in PPV films. Bound polaron pairs, by definition, are Coulombic bound charges on adjacent chains that are formed by the dissociation of hot intrachain singlet excitons (a "hot exciton" is an exciton away from thermodynamic equilibrium state) on a subpicosecond timescale and recombine geminately. It can be seen that bound polaron pairs are essentially charge-transfer excitons. According to Rothberg's interpretations, bound polaron pairs have a lower energy than intrachain singlets at most locations in the polymer, and are basically nonirradiative due to the poor overlapping of wavefunctions of the separated charges. The experimental results of Rothberg showed the generation yield of singlet exciton in PPV to be only about 10%, while 90% of the absorbed photons generated bound polaron pairs [62]. However, this deduction was not supported by follow-up experiments conducted by other groups [65–67], and this in turn led to the conclusion that only a small amount of bound polar pairs was formed in PPV. Moreover, it was also suggested that such a discrepancy had stemmed from the different qualities of PPV samples used, and that the situation was different for the various types of conjugated polymer. For example, according to Sheng *et al.* [67], only about 70% of the photoexcitations in polythiophene are singlet excitons, whereas the other 30% are most likely polarons or polaron pairs; in the case of PPV, the singlet exciton generation yield is close to unity.

1.4.2.2 **Excimers**

Excimer formation in conjugated polymers was studied systematically by Jenekhe *et al.* during the 1990s [68, 69]. Similar to the formation of excimers in small molecules, an excimer in conjugated polymers is generated by an excited chain and a neighboring chain in the ground state. The two chains should have a short interchain distance (0.3–0.6 nm) for efficient π-orbital overlapping. It is possible to have more chains (conjugated segments) involved in excimer formation; that is, the charge density (electron wavefunction) can be distributed over more than two chains. However, in comparison to bound polaron pairs, the extent of charge separation is much smaller in excimers, where the electrons and holes are still tightly bound. Hence, excimers are essentially Frenkel excitons, and are slightly delocalized as a result of a weak intermolecular electronic coupling. When compared to intrachain Frenkel excitons, the emission of the excimer lies in a relatively long wavelength range (see the orbital splitting model in Figure 1.7). It should be borne in mind that an excimer will dissociate when it returns to the ground state, but that any transition between the ground state and excimer state is symmetry-forbidden. Consequently, excimer formation will have no effect on the UV-visible absorption spectrum of a conjugated polymer. For the same reason, excimers possess a lower irradiation rate and a longer lifetime than intrachain excitons.

1.4.2.3 Ground-State Complexes

Similar to the formation of an excimer, molecular orbital splitting will occur if two or more ground-state molecules come sufficiently close together that their inner shell molecular orbitals begin to overlap. The thus-formed ground-state complex has an excited state which may be either an interchain Frenkel exciton or a charge-transfer exciton (bound polaron pair), depending on the degree of charge separation. Due to orbital splitting, the energy gap between the ground state and excited state becomes smaller; consequently, ground-state complexes typically show a red-shifted absorption and red-shifted emission spectra compared to free chromophores. Ground-state complexes are similar to excimers with regards to their emission properties. For example, they both show weak, low-energy, broad and structureless emission bands and a relatively long fluorescence lifetime. The absorption spectral change allows a distinction to be made between ground-state complexes and excimers. It has been proven experimentally that these complexes can be formed in some conjugated polymers [70–74]. The fluorescence spectra of MEH-PPV in films and different solvents are shown in Figure 1.18, where the broad red-shifted bands observed in MEH-PPV films could be assigned to ground-state complexes and excimers. Another good example is a ladder-type conjugated polymer MLPPP (see Figure 1.1), the backbone of which is constituted by fused fluorene rings. MLPPP is planar and highly rigid, so that the polymer chains may be easily arranged in a parallel, face-to-face manner in the solid state. Moreover, the π–π stacking interactions are so strong that ground-state complexes are formed, as proven by the observation of low-energy absorption and emission bands in solid films of MLPPP [70]. As such low-energy bands cannot be observed in dilute solutions, these must have derived from interchain interactions. A similar phenomenon was observed with poly(5-(pyridin-2-yl)vinylene), a heteroatom polymeric derivative of PPV where, in comparison to the phenyl rings in PPV, the pyridine groups have a higher electronegativity that promotes interchain coupling [75].

1.4.3 Photoinduced Charge Transfer between Conjugated Polymers and Electron Acceptors

Photoinduced charge transfer can occur when a conjugated polymer is blended with an electron acceptor. The radical cations (positive polarons) produced are stabilized by the conjugated polymer backbone as the result of a delocalization of charge density over the conjugation units. The energy diagram for photoinduced charge transfer between a conjugated polymer and an electron acceptor is depicted in Figure 1.19.

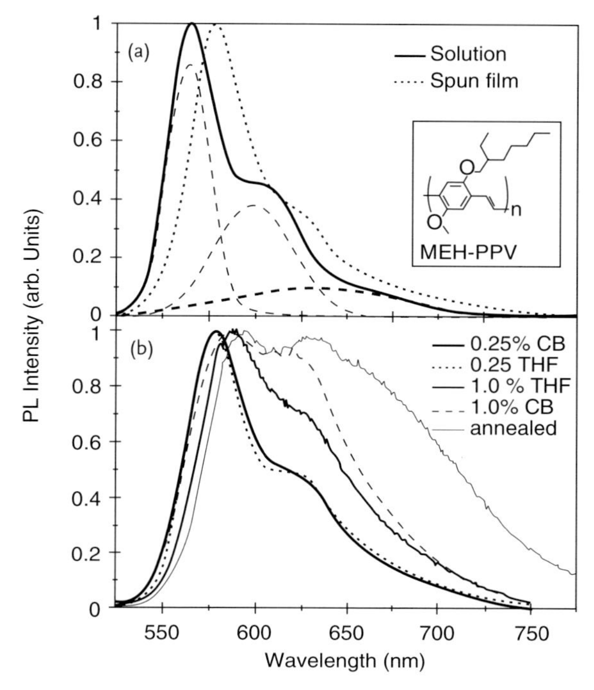

Figure 1.18 Normalized photoluminescence (PL) spectra of MEH-PPV in different environments. (a) PL spectra of a 0.25% (w/v) solution of MEH-PPV in chlorobenzene (CB) (solid curve), and the film resulting from spin-casting the solution (dotted curve). The small dashed curves show Gaussian fits to the three visible peaks of the solution PL; (b) PL spectra of MEH-PPV films cast from a 0.25% (w/v) solution in CB [solid curve, same as dashed curve in panel (a)], a 0.25% (w/v) solution in tetrahydrofuran (THF) (dotted curve), a 1.0% solution in THF (gray solid curve), a 1.0% solution in CB (dashed curve), and the film cast from the 1.0% CB solution after annealing (thin solid curve). Reproduced with permission from Ref. [74]; © 2000, American Chemical Society.

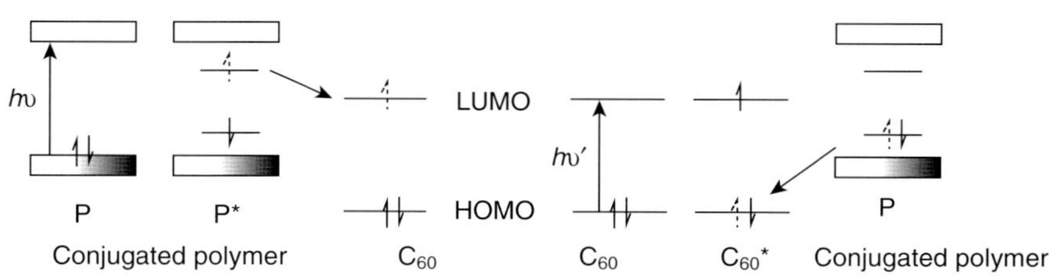

Figure 1.19 Energy diagram illustration of photoinduced charge transfer between conjugated polymer and acceptor (C_{60}). The electron being transferred is depicted as a dashed arrow.

Fullerene (C_{60}) and its derivatives are good electron acceptors, and promising photovoltaic effects in a blend film of MEH-PPV and C_{60} were first described by Sariciftci *et al.* in 1992 [76]. These findings aroused much interest in the development of conjugated polymer-based solar cells that could be used to convert solar energy into electricity. The primary structure of an organic solar cell includes an active layer (the donor–acceptor blend film) that is sandwiched between a transparent conductive anode [e.g., indium tin oxide (ITO) glass] and a metal cathode. Since the early 1990s, major progress has been made in this field, notably in 1995 by Yu *et al.*, who achieved an energy conversion efficiency of 3% by controlling the morphology of the blend film of MEH-PPV and C_{60} to form an interpenetrating network structure [77] (further details are available in Chapter 9). At this point, mention must be made of some of the important steps in energy conversion processes [2].

In organic solar cells, the main steps are as follows.

1) **Light harvesting:** Light is absorbed in the blend film and generates singlet excitons; for solar cells, the absorption should match the solar spectrum as closely as possible.
2) **Exciton diffusion and charge separation:** Excitons must migrate towards the D–A interfaces (heterojunctions); for interpenetrating structures, many local interfaces exist. At the D–A interface, excitons can dissociate into separated charges as a result of electron-transfer processes between the donor and the acceptor; the exciton diffusion length in typical conjugated polymers is about 5 nm, and in the ideal case the heterojunctions should be arranged in such a way that every exciton can locate a heterojunction during its lifetime.
3) **Charge transport:** The separated charges drift in the organic layers, under the driving force of the built-in electric field between the two electrodes.
4) **Charge collection:** The charges must ultimately be collected at the electrodes.

Of the above steps, charge separation is central. As noted in the earliest study on MEH-PPV/C_{60} systems [76], no ground-state complex was formed between the polymer and acceptor, as the electronic absorption spectrum of the blend was simply a superposition of the two components. Moreover, the fluorescence of MEH-PPV was heavily quenched by C_{60}, a very small amount of which quenched such fluorescence by three orders of magnitude. In the meantime, the fluorescence lifetime was decreased from 550 ps to far below 60 ps, indicating an electron transfer rate on the order of $10^{12}\,s^{-1}$ (see Eq. (1.9) for the calculation). This rate was much larger than those of the competing irradiative and nonirradiative processes, and suggested a charge-separation efficiency close to unity. More recently, the formation of charge-separation species was further confirmed by measuring the photoinduced transient absorption spectra and light-induced electron spin resonance (LESR) of the blends [78]. Subsequent ultrafast spectroscopic measurements confirmed that the PIET between MEH-PPV and fullerene occurred at a timescale

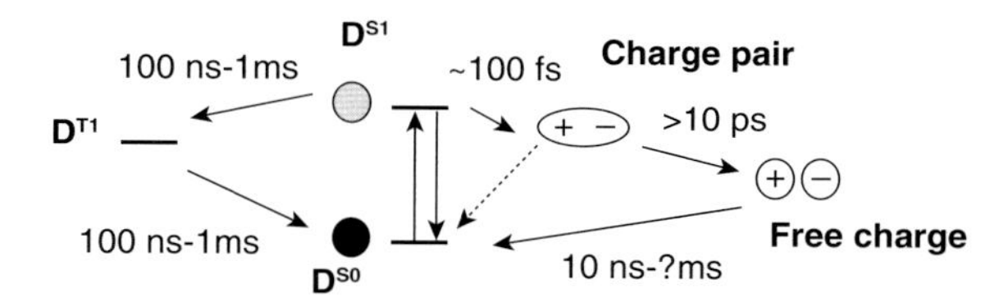

Figure 1.20 Schematic illustration of the photoinduced charge transfer process in conjugated polymer solar cells. Typical time scales for several of the involved steps are also shown.

of ~100 fs [79], which was so rapid that it could even compete with the vibrational relaxation process of the excited states.

Several plausible steps during the photoinduced charge separation process in conjugated polymer/acceptor blends are shown in Figure 1.20; the typical time scales of some steps are also shown. As the efficiency of the charge transfer step is close to unity, the energy conversion efficiency is mainly limited by other processes such as exciton diffusion, charge transport, or charge collection. It should be noted here that charge recombination is an important competing pathway of charge transport.

1.5 Carriers Generation and Transport

The above-described charge carriers (solitons, polarons, bipolarons) exist not only in conjugated polymers but also in other types of organic (semi)conductor. It must be borne in mind that the real charge-carrying particles which can be transported in organic solids are simply electrons. For example, although a polaron is intrinsically a radical ion, its transport is not a translational motion of the entire radical ion but rather is a drift of one electron from the radical ion to another location, where it creates a new radical ion. In this sense, charge transport can be seen as a sequence of redox reactions. From a mathematical aspect, however, it is more convenient and practical to describe the process by using the term "charge carriers."

1.5.1 Charge Carriers

Charge carriers can be generated via optical excitation, electrochemical doping (charge injection from electrodes), or chemical reduction/oxidation reactions. The generation of charge carriers is a precondition for electricity conduction. Taking photoconductors as an example, the main steps include: (i) the generation of conjugated molecules that absorb light and excitons; (ii) the dissociation of excitons into separate charges under the driving force of an external electric field; and

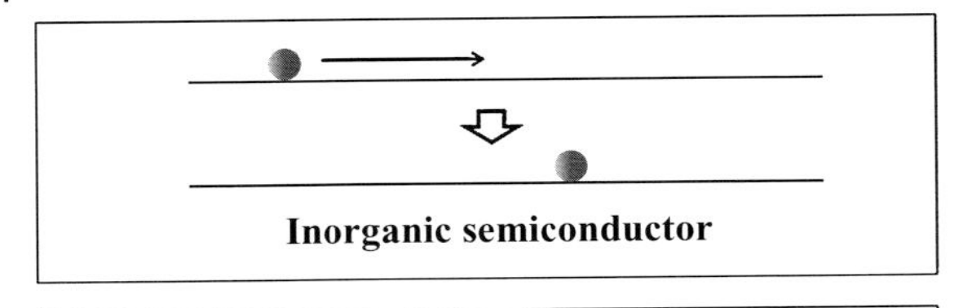

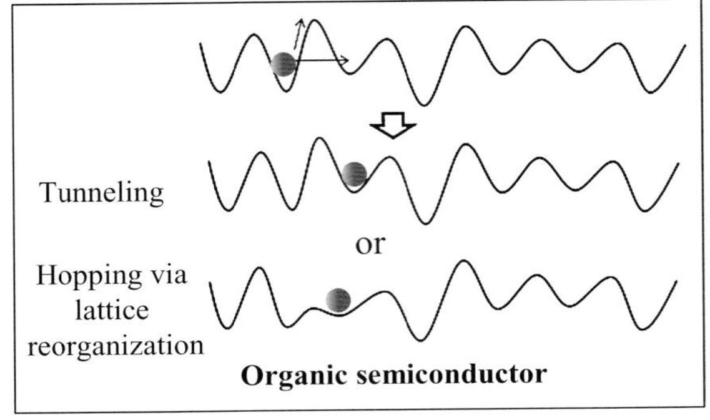

Figure 1.21 Schematic illustration of the transport of a charge carrier (solid ball) in inorganic semiconductors and organic semiconductors. For inorganic semiconductor the energy potential surface is represented as horizontal lines because of negligible energy barriers.

(iii) the generation of a photocurrent as a result of charge transport towards opposite electrodes.

In blending systems consisting of electron donors and acceptors, the charge carrier generation step is often very rapid (100~1000 fs), and its efficiency can approach unity. However, the transport of charge carriers in organic solids is less efficient than in inorganic semiconductors, mainly because charge carriers in organic solids lie within discrete energy levels rather than in continuous energy bands. Likewise, holes and electrons tend to be localized over small regions rather than be delocalized over the entire lattice, as is the case with inorganic semiconductors. Charge transport can only take place through electron hopping or electron tunneling, and it is these processes which essentially determine carrier mobility in organic solids. The energy potential surface for charge transport in inorganic semiconductors and organic semiconductors is illustrated in Figure 1.21.

1.5.2 Carrier Mobility and Its Measurement

The application of an external electric field induces a drift of the charge carriers; the mobility can then be defined as the ratio between the velocity, v, of the charges and the amplitude of the applied electric field, F:

$$\mu = v / F \tag{1.15}$$

The carrier mobility is usually expressed in $cm^2/(V \cdot s)$. *Carrier diffusion* should be seen as a local displacement of the charge around an average position, while drift induces a displacement of the average position. *Drift* is the effect that dominates the migration of the charges across an organic layer in the devices.

Charge mobilities can be determined experimentally by applying various techniques [21], one of the simplest of which is termed time-of-flight (TOF). In this technique, a thin organic layer (of a few microns thickness) is sandwiched between two electrodes, one of which is transparent (e.g., ITO glass). The material to be tested is first irradiated by a laser pulse in the proximity of the transparent electrode, in order to generate charges. Depending on the polarity of the applied bias and the corresponding electric field (in the range of 10^4 to $10^6\,V\,cm^{-1}$), the photogenerated holes or electrons migrate across the material towards the other electrode. The current at that electrode is recorded as a function of time, and the mobility of the holes or electrons is then estimated via:

$$\mu = \frac{v}{F} = \frac{d}{Ft} = \frac{d^2}{vt} \tag{1.16}$$

where d is the distance between the electrodes, F is the electric field, t is the averaged transient time, and V is the applied voltage. For ordered materials a sharp onset will be obtained, whereas for disordered systems such as polymers a broadening of the signal occurs due to a distribution of transient times across the material.

Carrier mobilities can also be extracted from current/current density–voltage curves measured in a FET or a photovoltaic diode. In another technique, termed pulse-radiolysis time-resolved microwave conductivity (PR-TRMC), a pulse of highly energetic electrons is applied to create a low density of free carriers. The carrier mobility can then be derived by recording the pulse-induced change in electrical conductivity as a function of microwave power (further details are available in Ref. [21]).

1.5.3
Mobility-Influencing Factors

An efficient charge transport requires that the charges are able to move from molecule to molecule, and not be trapped or scattered. Therefore, charge carrier mobilities are influenced by many factors that include molecular packing, disorder, and the presence of impurities, temperature, electric field, charge-carrier density, size/molecular weight, and pressure [21]. Moreover, different measuring methods may produce different results, with data derived from methods that measure mobilities over macroscopic distances (~1 mm) often being dependent on the purity and order in the material. Methods that measure mobilities over microscopic distances are less dependent on these characteristics.

Due to the weak intermolecular electronic coupling, the charge carriers must face relatively large energy barriers at the molecular boundaries. Hence, the relative positons (molecular packing) of the interacting molecules are an important influencing factor for carrier mobility. In most instances, unsubstituted conjugated molecules crystallize into a layered herringbone packing, which gives rise to two-dimensional transport within the stacked organic layers, but transport between the layers is less efficient. Static and dynamic disorders in organic solids also greatly impact on the carrier mobilities, because they introduce spatial and temporal variations to intermolecular coupling strength. More recently, attention has also been paid to other influencing factors, such as temperature and pressure (further details of these issues are discussed in Ref. [21]).

References

1 Markus, S. and Hans, C.W. (2007) *Organic Molecular Solids*, Wiley-VCH Verlag GmbH & Co. KGaA, Weinheim.
2 Brédas, J.-L., Beljonne, D., Coropceanu, V., and Cornil, J. (2004) Charge-transfer and energy-transfer processes in pi-conjugated oligomers and polymers: a molecular picture. *Chem. Rev.*, **104** (11), 4971–5003.
3 Koenigsberger, J. and Schilling, K. (1910) On electrical conductivity in fixes elements and compounds I. Minimum resistance, testing on electron behaviour, use of dissociation formulae. *Ann. Phys.*, **32** (6), 179–230.
4 Volmer, M. (1913) The different photoelectrical occurrences on anthracene, their connections to each other, to fluorescence and dianthracene formation. *Ann. Phys.*, **40** (4), 775–796.
5 Naarmann, H. (2000) Polymers, Electrically Conducting, in Ullmann's Encyclopedia of Industrial Chemistry, Wiley-VCH Verlag GmbH & Co. KGaA, Weinheim.
6 Helfrich, W. and Schneider, W.G. (1965) Recombination radiation in anthracene crystals. *Phys. Rev. Lett.*, **14** (7), 229–231.
7 Pope, M., Magnante, P., and Kallmann, H.P. (1963) Electroluminescence in organic crystals. *J. Chem. Phys.*, **38** (8), 2042–2043.
8 Chiang, C.K., Fincher, C.R., Park, Y.W., Jr, Heeger, A.J., Shirakawa, H., Louis, E.J., Gau, S.C., and MacDiarmid, A.G. (1977) Electrical-conductivity in doped polyacetylene. *Phys. Rev. Lett.*, **39** (17), 1098–1101.
9 Tang, C.W. and VanSlyke, S.A. (1987) Organic electroluminescent diodes. *Appl. Phys. Lett.*, **51** (12), 913–915.
10 Burroughes, J.H., Bradley, D.D.C., Brown, A.R., Marks, R.N., Mackay, K., Friend, R.H., Burns, P.L., and Holmes, A.B. (1990) Light-emitting-diodes based on conjugated polymers. *Nature*, **347** (6293), 539–541.
11 Günes, S., Neugebauer, H., and Sariciftci, N.S. (2007) Conjugated polymer-based organic solar cells. *Chem. Rev.*, **107** (4), 1324–1338.
12 Peumans, P., Yakimov, A., and Forrest, S.R. (2003) Small molecular weight organic thin-film photodetectors and solar cells. *J. Appl. Phys.*, **93** (7), 3693–3723.
13 Dodabalapur, A., Torsi, L., and Katz, H.E. (1995) Organic transistors: two-dimensional transport and improved electrical characteristics. *Science*, **268** (5208), 270–271.
14 Horowitz, G. (1998) Organic field-effect transistors. *Adv. Mater.*, **10** (5), 365–377.
15 Bartic, C., Campitelli, A., and Borghs, S. (2003) Field-effect detection of chemical species with hybrid organic/inorganic transistors. *Appl. Phys. Lett.*, **82** (3), 475–477.
16 Thomas, S.W., Joly, G.D., and Swager, T.M. (2007) Chemical sensors based on amplifying fluorescent conjugated polymers. *Chem. Rev.*, **107** (4), 1339–1386.
17 Berggren, M., Dodabalapur, A., Slusher, R.E., and Bao, Z. (1997) Light amplification in organic thin films using

cascade energy transfer. *Nature*, **389** (6650), 466–469.
18 McGehee, M.D. and Heeger, A.J. (2000) Semiconducting (conjugated) polymers as materials for solid-state lasers. *Adv. Mater.*, **12** (22), 1655–1668.
19 Brütting, W. (2005) *Physics of Organic Semiconductors*, Wiley-VCH Verlag GmbH & Co. KGaA, Weinheim.
20 Agranovich, V. (2008) *Excitations in Organic Solids*, Oxford Scientific Publications, Oxford.
21 Coropceanu, V., Cornil, J., da Silva Filho, D.A., Olivier, Y., Silbey, R., and Brédas, J.-L. (2007) Charge transport in organic semiconductors. *Chem. Rev.*, **107** (4), 926–952.
22 Vardeny, Z.V. (2009) *Ultrafast Dynamics and Laser Action of Organic Semiconductors*, CRC Press, Boca Raton.
23 Mukamel, S., Tretiak, S., Wagersreiter, T., and Chernyak, V. (1997) Electronic coherence and collective optical excitations of conjugated molecules. *Science*, **277** (5327), 781–787.
24 Anthony, J.E. (2006) Functionalized acenes and heteroacenes for organic electronics. *Chem. Rev.*, **106** (12), 5028–5048.
25 Kondakov, D.Y., Pawlik, T.D., Hatwar, T.K., and Spindler, J.P. (2009) Triplet annihilation exceeding spin statistical limit in highly efficient fluorescent organic light-emitting diodes. *J. Appl. Phys.*, **106** (12), 124510.
26 Scherf, U. and List, E.J.W. (2002) Semiconducting polyfluorenes – towards reliable structure-property relationships. *Adv. Mater.*, **14** (7), 477–487.
27 van Nostrum, C.F., Picken, S.J., Schouten, A.-J., and Nolte, R.J.M. (1995) Synthesis and supramolecular chemistry of novel liquid crystalline crown ether-substituted phthalocyanines: toward molecular wires and molecular ionoelectronics. *J. Am. Chem. Soc.*, **117** (40), 9957–9965.
28 Holten, D., Bocian, D.F., and Lindsey, J.S. (2002) Probing electronic communication in covalently linked multiporphyrin arrays. A guide to the rational design of molecular photonic devices. *Acc. Chem. Res.*, **35** (1), 57–69.
29 Tsuboyama, A., Iwawaki, H., Furugori, M., Mukaide, T., Kamatani, J., Igawa, S., Moriyama, T., Miura, S., Takiguchi, T., Okada, S., Hoshino, M., and Ueno, K. (2003) Homoleptic cyclometalated iridium complexes with highly efficient red phosphorescence and application to organic light-emitting diode. *J. Am. Chem. Soc.*, **125** (42), 12971–12979.
30 Evans, R.C., Douglas, P., and Winscom, C.J. (2006) Coordination complexes exhibiting room-temperature phosphorescence: evaluation of their suitability as triplet emitters in organic light emitting diodes. *Coord. Chem. Rev.*, **250** (15-16), 2093–2126.
31 Setayesh, S., Grimsdale, A.C., Weil, T., Enkelmann, V., Müllen, K., Meghdadi, F., List, E.J.W., and Leising, G. (2001) Polyfluorenes with polyphenylene dendron side chains: toward non-aggregating, light-emitting polymers. *J. Am. Chem. Soc.*, **123** (5), 946–953.
32 Sun, Y.M., Xiao, K., Liu, Y.Q., Wang, J.L., Pei, J., Yu, G., and Zhu, D.B. (2005) Oligothiophene-functionalized truxene: star-shaped compounds for organic field-effect transistors. *Adv. Funct. Mater.*, **15** (5), 818–822.
33 Burn, P.L., Lo, S.C., and Samuel, I.D.W. (2007) The development of light-emitting dendrimers for displays. *Adv. Mater.*, **19** (13), 1675–1688.
34 Tao, X.-T., Zhang, Y.-D., Wada, T., Sasabe, H., Suzuki, H., Watanabe, T., and Miyata, S. (1998) Hyperbranched polymers for electroluminescence applications. *Adv. Mater.*, **10** (3), 226–230.
35 Katsis, D., Geng, Y.H., Ou, J.J., Culligan, S.W., Trajkovska, A., Chen, S.H., and Rothberg, L.J. (2002) Spiro-linked ter-, penta-, and heptafluorenes as novel amorphous materials for blue light emission. *Chem. Mater.*, **14** (3), 1332–1339.
36 Lakowicz, J.R. (2006) *Principles of Fluorescence Spectroscopy*, Plenum Press, New York, London.
37 Forster, T. (1969) Excimers. *Angew. Chem. Int. Ed. Engl.*, **8** (5), 333–343.
38 Weller, A. (1982) Photoinduced electron transfer in solution: exciplex and radical

ion pair formation free enthalpies and their solvent dependence. *Z. Phys. Chem.*, **133** (1), 93–98.

39 Marcus, R.A. (1956) On the theory of oxidation-reduction reactions involving electron transfer. I. *J. Chem. Phys.*, **24** (5), 966–978.

40 Marcus, R.A. (1993) Electron-transfer reactions in chemistry. Theory and experiment. *Rev. Mod. Phys.*, **65** (3), 599–610.

41 Laquai, F., Park, Y.-S., Kim, J.-J., and Basché, T. (2009) Excitation energy transfer in organic materials: from fundamentals to optoelectronic devices. *Macromol. Rapid Commun.*, **30** (14), 1203–1231.

42 Dexter, D.L. (1953) A theory of sensitized luminescence in solids. *J. Chem. Phys.*, **21** (5), 836–850.

43 Akagi, K., Suezaki, M., Shirakawa, H., Kyotani, H., Shimomura, M., and Tanabe, Y. (1989) Synthesis of polyacetylene films with high-density and high mechanical strength. *Synth. Met.*, **28** (3), D1–D10.

44 He, Y. and Kanicki, J. (2000) High-efficiency organic polymer light-emitting heterostructure devices on flexible plastic substrates. *Appl. Phys. Lett.*, **76** (6), 661–663.

45 Al-Ibrahim, M., Roth, H.K., Zhokhavets, U., Gobsch, G., and Sensfuss, S. (2005) Flexible large area polymer solar cells based on poly(3-hexylthiophene)/fullerene. *Sol. Energy Mater. Sol. Cells*, **85** (1), 13–20.

46 Kinlen, P.J., Silverman, D.C., and Jeffreys, C.R. (1997) Corrosion protection using polyaniline coating formulations. *Synth. Met.*, **85** (1–3), 1327–1332.

47 Sirringhaus, H., Kawase, T., Friend, R.H., Shimoda, T., Inbasekaran, M., Wu, W., and Woo, E.P. (2000) High-resolution inkjet printing of all-polymer transistor circuits. *Science*, **290** (5499), 2123–2126.

48 Sheats, J.R., Antoniadis, H., Hueschen, M., Leonard, W., Miller, J., Moon, R., Roitman, D., and Stocking, A. (1996) Organic electroluminescent devices. *Science*, **273** (5277), 884–888.

49 Brabec, C.J., Sariciftci, N.S., and Hummelen, J.C. (2001) Plastic solar cells. *Adv. Funct. Mater.*, **11** (1), 15–26.

50 McQuade, D.T., Pullen, A.E., and Swager, T.M. (2000) Conjugated polymer-based chemical sensors. *Chem. Rev.*, **100** (7), 2537–2574.

51 Bredas, J.L. and Street, G.B. (1985) Polarons, bipolarons, and solitons in conducting polymers. *Acc. Chem. Res.*, **18** (10), 309–315.

52 Heeger, A.J., Kivelson, S., Schrieffer, J.R., and Su, W.-P. (1988) Solitons in conducting polymers. *Rev. Mod. Phys.*, **60** (3), 781–850.

53 Sariciftci, N.S. (1998) *Primary Photoexcitations in Conjugated Polymers: Molecular Exciton versus Semiconductor Band Model*, World Scientific, Singapore.

54 Rauscher, U., Bässler, H., Bradley, D.D.C., and Hennecke, M. (1990) Exciton versus band description of the absorption and luminescence spectra in poly(para-phenylenevinylene). *Phys. Rev. B*, **42** (16), 9830–9836.

55 Scholes, G.D. and Rumbles, G. (2006) Excitons in nanoscale systems. *Nat. Mater.*, **5** (9), 683–696.

56 Heun, S., Mahrt, R.F., Greiner, A., Lemmer, U., Bassler, H., Halliday, D.A., Bradley, D.D.C., Burn, P.L., and Holmes, A.B. (1993) Conformational effects in poly(*p*-phenylene vinylene)s revealed by low-temperature site-selective fluorescence. *J. Phys. Condens. Matter*, **5** (2), 247–260.

57 Nguyen, T.Q., Doan, V., and Schwartz, B.J. (1999) Conjugated polymer aggregates in solution: control of interchain interactions. *J. Chem. Phys.*, **110** (8), 4068–4078.

58 Collison, C.J., Rothberg, L.J., Treemaneekarn, V., and Li, Y. (2001) Conformational effects on the photophysics of conjugated polymers: a two species model for MEH-PPV spectroscopy and dynamics. *Macromolecules*, **34** (7), 2346–2352.

59 Luo, J., Xie, Z., Lam, J.W.Y., Cheng, L., Chen, H., Qiu, C., Kwok, H.S., Zhan, X., Liu, Y., Zhu, D., and Tang, B.Z. (2001) Aggregation-induced emission of 1-methyl-1,2,3,4,5-pentaphenylsilole. *Chem. Commun.*, (18), 1740–1741.

60 Chen, J., Law, C.C.W., Lam, J.W.Y., Dong, Y., Lo, S.M.F., Williams, I.D., Zhu, D., and Tang, B.Z. (2003) Synthesis, light emission, nanoaggregation, and restricted intramolecular rotation of 1,1-substituted 2,3,4,5-tetraphenylsiloles. *Chem. Mater.*, **15** (7), 1535–1546.

61 Hong, Y.N., Lam, J.W.Y., and Tang, B.Z. (2009) Aggregation-induced emission: phenomenon, mechanism and applications. *Chem. Commun.*, (29), 4332–4353.

62 Yan, M., Rothberg, L.J., Papadimitrakopoulos, F., Galvin, M.E., and Miller, T.M. (1994) Spatially indirect excitons as primary photoexcitations in conjugated polymers. *Phys. Rev. Lett.*, **72** (7), 1104–1107.

63 Yan, M., Rothberg, L.J., Kwock, E.W., and Miller, T.M. (1995) Interchain excitations in conjugated polymers. *Phys. Rev. Lett.*, **75** (10), 1992–1995.

64 Rothberg, L.J., Yan, M., Papadimitrakopoulos, F., Galvin, M.E., Kwock, E.W., and Miller, T.M. (1996) Photophysics of phenylenevinylene polymers. *Synth. Met.*, **80** (1), 41–58.

65 Greenham, N.C., Samuel, I.D.W., Hayes, G.R., Phillips, R.T., Kessener, Y.A.R.R., Moratti, S.C., Holmes, A.B., and Friend, R.H. (1995) Measurement of absolute photoluminescence quantum efficiencies in conjugated polymers. *Chem. Phys. Lett.*, **241** (1–2), 89–96.

66 Halls, J.J.M., Pichler, K., Friend, R.H., Moratti, S.C., and Holmes, A.B. (1996) Exciton diffusion and dissociation in a poly(*p*-phenylenevinylene)/C_{60} heterojunction photovoltaic cell. *Appl. Phys. Lett.*, **68** (22), 3120–3122.

67 Sheng, C.X., Tong, M., Singh, S., and Vardeny, Z.V. (2007) Experimental determination of the charge/neutral branching ratio η in the photoexcitation of π-conjugated polymers by broadband ultrafast spectroscopy. *Phys. Rev. B*, **75** (8), 085206.

68 Jenekhe, S.A. and Osaheni, J.A. (1994) Excimers and exciplexes of conjugated polymers. *Science*, **265** (5173), 765–768.

69 Jenekhe, S.A. (1995) Excited-state complexes of conjugated polymers. *Adv. Mater.*, **7** (3), 309–311.

70 Lemmer, U., Heun, S., Mahrt, R.F., Scherf, U., Hopmeier, M., Siegner, U., Göbel, E.O., Müllen, K., and Bässler, H. (1995) Aggregate fluorescence in conjugated polymers. *Chem. Phys. Lett.*, **240** (4), 373–378.

71 Blatchford, J.W., Gustafson, T.L., Epstein, A.J., Vanden Bout, D.A., Kerimo, J., Higgins, D.A., Barbara, P.F., Fu, D.K., Swager, T.M., and MacDiarmid, A.G. (1996) Spatially and temporally resolved emission from aggregates in conjugated polymers. *Phys. Rev. B*, **54** (6), R3683–R3686.

72 Conwell, E. (1997) Excimer formation and luminescence in conducting polymers. *Trends Polym. Sci.*, **5** (7), 218–222.

73 Grell, M., Bradley, D.D.C., Long, X., Chamberlain, T., Inbasekaran, M., Woo, E.P., and Soliman, M. (1998) Chain geometry, solution aggregation and enhanced dichroism in the liquid-crystalline conjugated polymer poly(9,9-dioctylfluorene). *Acta Polym.*, **49** (8), 439–444.

74 Nguyen, T.-Q., Martini, I.B., Liu, J., and Schwartz, B.J. (2000) Controlling interchain interactions in conjugated polymers: the effects of chain morphology on exciton-exciton annihilation and aggregation in MEH-PPV films. *J. Phys. Chem. B*, **104** (2), 237–255.

75 Nguyen, T.-Q., Kwong, R.C., Thompson, M.E., and Schwartz, B.J. (2000) Improving the performance of conjugated polymer-based devices by control of interchain interactions and polymer film morphology. *Appl. Phys. Lett.*, **76** (17), 2454–2456.

76 Sariciftci, N.S., Smilowitz, L., Heeger, A.J., and Wudl, F. (1992) Photoinduced electron-transfer from a conducting polymer to buckminsterfullerene. *Science*, **258** (5087), 1474–1476.

77 Yu, G., Gao, J., Hummelen, J.C., Wudl, F., and Heeger, A.J. (1995) Polymer photovoltaic cells: enhanced efficiencies via a network of internal donor-acceptor heterojunctions. *Science*, **270** (5243), 1789–1791.

78 Sariciftci, N.S., Heeger, A.J., and Nalwa, H.S. (eds) (1997) *Handbook of Organic*

Conductive Molecules and Polymers, vol. 1, John Wiley & Sons, Inc., New York.

79 Brabec, C.J., Zerza, G., Cerullo, G., De Silvestri, S., Luzzati, S., Hummelen, J.C., and Sariciftci, S. (2001) Tracing photoinduced electron transfer process in conjugated polymer/fullerene bulk heterojunctions in real time. *Chem. Phys. Lett.*, **340** (3–4), 232–236.

2
Organic/Polymeric Semiconductors for Field-Effect Transistors

Qing Meng, Huanli Dong, Wenping Hu

2.1
Introduction

The field-effect transistor (FET) is a three-terminal, solid-state electronic device that can be used to control the electric current or voltage between two of the terminals, by applying an electric current or voltage to the third terminal.

The history of transistors can be traced back to the 1940s when, in 1947, John Bardeen and Walter Brattain, while working at Bell Telephone Laboratories, were investigating the nature of the electrons at the interface between a metal and a semiconductor. Bardeen and Walters realized that, by making two point contacts very close to one another, they could create a three-terminal device – the first "point contact" transistor. This invention was the spark that ignited a huge research effort in solid-state electronics. Subsequently, Bardeen and Brattain received the Nobel Prize in Physics in 1956, together with William Shockley, ". . . for their researches on semiconductors and their discovery of the transistor effect." In 1958 and 1959, Jack Kilby at Texas Instruments and Robert Noyce at Fairchild Camera, devised a solution to the problem of integrating large numbers of transistors for circuits. These investigators showed that, rather than make transistors one-by-one, many could be made at the same time, on the same piece of semiconductor. Moreover, not only transistors but also other electric components such as resistors, capacitors, and diodes could be made using the same process, and the same materials. Since then, over a period of more than 30 years, the number of transistors per unit area has doubled every 1.5 years. This fantastic progression of circuit fabrication is known as Moore's law, after Gordon Moore, one of the early integrated circuit pioneers and founders of the Intel Corporation. In 2000, the Nobel Prize in Physics was awarded to Jack Kilby for the invention of the integrated circuit. It can surely be said that the discovery of the transistor, and the development of the integrated circuit, make the twentieth century the "century of electronics."

The organic field-effect transistor (OFET) is a FET that uses an organic semiconductor instead of an inorganic semiconductor in its channel. The first OFETs were reported by Tsumura *et al.* in 1986, who used polythiophene as the semiconductor layer [1]. Subsequently, OFETs have attracted particular attention and have

Organic Optoelectronics, First Edition. Edited by Wenping Hu.

progressed rapidly for several decades, due mainly to their potential applications in areas such as radiofrequency identification (RFID) tags, flexible displays, electronic papers, and sensors. In this chapter, attention is focused on the development of organic semiconductors, with particular emphasis on the design strategy of novel semiconductors with a high mobility and stability.

2.1.1
Features of Organic/Polymeric Semiconductors

A typical OFET includes an organic semiconductor layer, a gate insulator layer, and three terminals (drain, source, and gate electrodes). Charges are injected from the source electrode into the organic semiconductor, transported in the organic semiconductor, and finally fed out of the organic semiconductor from the drain electrode. The charge transport route in an organic semiconductor is termed the conducting channel, and this locates at the interface between the organic semiconductor and gate insulator with a thickness of one to several molecular layers. Clearly, the organic semiconductor is a key component of an OFET.

Since the first reports of organic thin-film transistors based on polythiophene by Tsumura *et al.* in 1986 [1], a considerable number of organic semiconductors have been designed and synthesized [2]. By comparison to inorganic semiconductors, organic semiconductors possess several merits that inorganic semiconductors cannot approach. For example, due to sufficient resources of raw materials, organic semiconductors can be synthesized at a large scale, and at low cost. Moreover, organic semiconductors can be conveniently purified by applying a variety of techniques that include chromatography, recrystallization, and sublimation, without a need for any harsh conditions. Perhaps the most important aspect, however, is that the physical and chemical properties of organic semiconductors can be directionally tuned by the tailoring of their molecular structures, and this is one of the main benefits of using semiconducting materials for OFETs [3–6]. In terms of device fabrication, organic semiconducting layers (whether thin films or crystals) have been fabricated successfully by employing a variety of vacuum-deposition, physical vapor deposition, and solution-deposition (e.g., spin-coating, drop-casting, LB-film) techniques. Indeed, various investigations have proved that the surface properties of dielectric layers can be modified by the physical or chemical modification of different materials [7]. As a result, the packing mode of organic semiconductors can be controlled by altering the compatibility between a semiconductor and the modified dielectric layers. Likewise, the feasibility of a low-cost, large-scale flexible fabrication of devices has enabled both exploratory studies and practical applications of OFETs in RFID tags, flexible displays, electronic paper, and large-area sensors.

2.1.2
Classification of Semiconductors for Organic Field-Effect Transistors

In terms of molecular weight, organic semiconductors can be subdivided into small molecules and polymers while, on the basis of the main charge carriers transporting in OFET channels, organic semiconductors can be further divided into *p*-type, *n*-type, and bipolar semiconducting materials. While most of these materials exhibit a unipolar field-effect behavior, some demonstrate bipolar field-effect characteristics.

Polymers, which are known to possess a certain degree of solubility, flexibility, and adhesion, have shown great potential in the solution-process of semiconducting layers on both hard and flexible substrates. Unfortunately, one common drawback of these semiconducting layers is the requirement for annealing to improve molecular arrangement and device performance. In addition, most polymers exhibit a poor mobility due to their wide distribution of molecular weight and poor reproducibility. Comparatively speaking, small-molecular weight semiconductors could be synthesized with a high reproducibility, and comprehensively purified by applying various methods such as chromatography, recrystallization, and sublimation. In most applications, the ultra-purified small-molecular weight semiconductors exhibit a better performance when they are used as active layers of FETs.

The energy gaps (E_gs; also known as band gaps) of common organic semiconductors used for OFETs lie in the range of 2 to 3 eV [4]. During the operation of OFETs, the highest occupied molecular orbital (HOMO) and lowest unoccupied molecular orbital (LUMO) energy levels of the semiconductor will shift under a given V_G (positive or negative), and the work function(s) of source/drain electrode(s) will relatively match the shifted HOMO or LUMO energy levels. What accumulates at the semiconductor–insulator interface is the "hole" caused by the vacancy after electron flow from the HOMO energy level into electrodes, or the "electron" caused by the injection from electrodes to the LUMO energy level of semiconductor [3]. In theory, the hole and the electron transport of the same semiconductor could be respectively realized by a reasonable combination of electrode material, insulator modification, and applied test conditions. However, besides compatibility with semiconductors, materials that are suitable for electrodes require an environmental stability and a good transmission performance; hence, the list of choices is finite. When limited by these reasons, most organic semiconductors exhibit a unipolar transport behavior in the applications of OFETs. As most of these semiconductors are not intentionally doped, and the unintentional doping level is always rather low, the device performance may be considered as a reflection of the material's property. Herein, the organic materials which take the hole as a dominant charge carrier and exhibit *p*-channel transistor performance are defined as *p*-type organic semiconductors, whereas the organic materials which take the electron as a dominant charge carrier and exhibit *n*-channel transistor behavior are defined as *n*-type organic semiconductors. Inevitably, however, some materials are both *p*-type and *n*-type semiconductors.

The initial studies of organic semiconductors were focused on only a few compounds, which were found to be stable in air. The subsequent investigations were based on *p*-type small-molecular weight semiconductors, and stimulated by the high performance of pentacene and fused ring compounds. Increasingly, thiophene-fused arenes were shown to be good candidates for high-performance semiconductors, with carrier mobilities greater than that of amorphous silicon.

Hence, with the great demand for *n*-type semiconductors for use in logic circuits, the emphasis of research in this field was not limited to *p*-type semiconductors, and several electron-withdrawing groups were bonded to frameworks that had high electron affinities in order to obtain air-stable *n*-type semiconductors. Many such directionally designed materials exhibited *n*-type field-effect behavior under an inert atmosphere or in a vacuum, among which candidates some possessed typical functional groups. Moreover, the lowest unoccupied molecular orbital (LUMO) of the compounds was very low, such that they exhibited good *n*-type field-effect properties under ambient conditions. As promising frameworks, these functional groups were further developed for use as more air-stable *n*-type semiconductors with a high performance, the details of which have recently been intensively reported.

2.1.3 Main Parameters for the Characterization of Organic/Polymeric Semiconductors

When the terminals (gate, source and drain electrodes) of a transistor are loaded during the operation of OFETs, the process begins with the injection of charge carriers into the conducting channel. The source electrode is often grounded ($V_S = 0\,V$), and the voltage applied between the source and gate electrodes, termed the gate voltage (V_G), could induce an electric field at the semiconductor–insulator interface for charge injection into the conducting channel. The HOMO and LUMO energy levels of organic semiconductors will shift under this electric field. For a given insulator with a stationary capacitance, the number of accumulated charges is proportional to the applied V_G. A conducting channel forms only when V_G is higher than a threshold voltage (V_{Th}), so as to induce sufficient charge after filling the traps in the semiconductor–insulator interface. These additionally induced charges can then flow in the conducting channel under the voltage applied between the source and drain electrodes (V_{DS}).

Mobility (μ), threshold voltage (V_{Th}), and on/off current ratio (I_{on}/I_{off}) are three main parameters to which attention is usually paid during the process of OFETs. To some extent, these are not only the criteria for determining the level of semiconducting materials, but also the reflection of comprehensive quality of devices. Mobility (μ), which is the average drift velocity of charge per unit electric field, indicates the transfer ability of charge in semiconductors. The threshold voltage (V_{Th}) reflects the minimum V_G to form a conducting channel in the specific device, and a low V_{Th} is preferable to reduce the energy consumption of devices. The I_{on}/I_{off} ratio can be defined as the ratio of source-drain current (I_{DS}) between on-state (I_{DS} at maximum V_G) and off-state (I_{DS} at $V_G = 0$), which depends heavily on the parameters chosen to measure the devices. During the past decades, research groups have made great efforts to improve the performance of OFETs with, on the one hand, synthetic chemists synthesizing and characterizing novel semiconductors by the directional tailoring of molecular frameworks and functional groups, thus summarizing a guide ideology for the design of high-performance semiconducting materials [2]. On the other hand, various techniques have been developed to

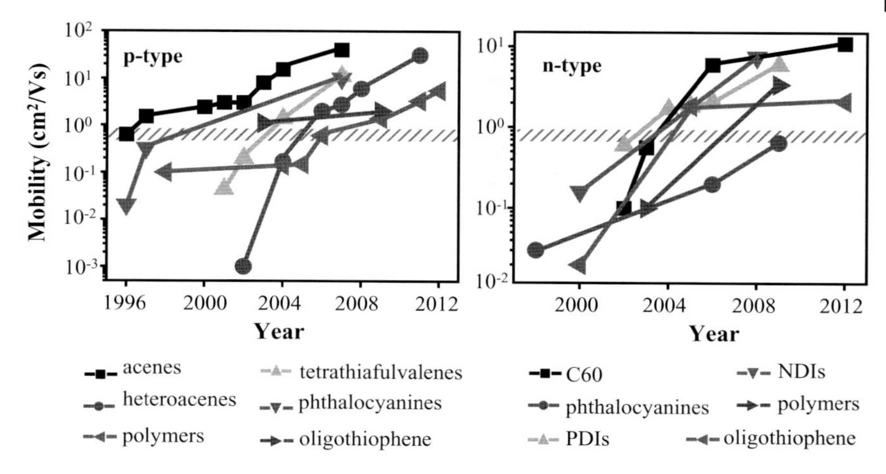

Figure 2.1 Relationship of the mobility of various high-performance organic semiconductors with time (years).

fabricate OFETs based on high-quality thin films or crystals of semiconducting materials. The expectation here would be to improve the mobility of semiconductors by decreasing any external influences such as grain boundaries, trap concentration, and contact resistance. In this way, a "yearbook" of organic semiconductors on mobility was simply depicted, as shown in Figure 2.1. The high-mobility organic semiconductors (i.e., over amphorous silicon) indicate the great prospect of organic electronic devices for potential applications.

2.2 Small-Molecular Semiconductors

2.2.1 P-type Small-Molecular Semiconductors

From a synthetic viewpoint, the frameworks of *p*-type small-molecular semiconductors always possess electron-rich groups, which are reactive to electrophilic substitution reactions. From the perspective of the device, the performance of most *p*-type semiconductor-based devices exhibits a poor correlation with the test circumstances. Consequently, during the past decades an extensive development has been undertaken of *p*-type small-molecular semiconductors.

2.2.1.1 Polycyclic Aromatic Hydrocarbons

Aromatic hydrocarbons (namely arene, or aryl hydrocarbon) consist only of carbon and hydrogen elements, and form compounds with alternating single and double bonds between the carbon atoms. Most polycyclic aromatic hydrocarbons (PAHs) possess planar molecular structures and demonstrate a high degree of intramolecular conjugation; in terms of their molecular structure, they can be subdivided into linear (acenes) and nonlinear forms.

2.2.1.1.1 Acenes and Substituted Acenes

The acenes represent a class of PAHs that are composed of linearly fused benzene rings. Conjugation of these molecules increases with elongation of the molecular structures such that, due to strong intermolecular interactions, this series of molecules usually forms a rigid molecular packing mode in the solid state, which is desirable for improving the charge transport of semiconductors. In general, the poor solubility of the acenes in solvents at room temperature is a common drawback; moreover, the E_g of acenes is decreased as the fused rings are elongated. Inspired by the tight packing modes and high charge transport of acenes, various research groups began to design and synthesize a range of substituted acenes, with the hope of improving the solubility, reactant yields, and chemical stability, while maintaining an efficient charge transport. The chemical structures of some acenes and their derivatives that have been used as semiconductors for OFETs are shown in Figure 2.2.

Anthracene (**1**) consists of three fused benzene rings, and adopts a similar herringbone arrangement as pentacene in solid states. When Aleshin *et al.* examined charge transport in anthracene single crystals, the maximum field-effect mobility was observed as ~0.02 $cm^2/(V \cdot s)$ at ~170–180 K [8]. For enriching the system of acenes, various derivatives based on acenes were designed and synthesized. The long and short axes of the molecular structures were identified as two categories of derivatization; in terms of reactivity, the carbons at the *peri*-positions (short axis) showed a chemical reactive priority over the terminal-positions (long axis).

Later, Pflaum *et al.* grew bulk crystals of 9,10-diphenylanthracene (DPA, **2**) and examined the charge (hole and electron) transport in the molecule by using time-of-flight (TOF) measurements [9]. In this case, the high electron and hole mobilities at room temperature were calculated as 13 and 3.7 $cm^2/(V \cdot s)$, respectively. Hu *et al.* introduced different aryl-acetylene groups (phenylethynyl [**3a**], and naphthalen-2-ylethynyl [**3b**], thiophen-2-ylethynyl [**3c**]) to the short axis of anthracene to study the variety of self-assembly behavior and charge transport property of cruciform molecules [10]. The single crystal structures showed that the structural repulsion between anthracene and adjacent aryl groups could be efficiently deleted by the introduction of acetylene groups, and that these near-planar structures were preferable for enlarging molecular conjugation. Single-crystalline micro/nanometer wires and ribbons of these materials were grown by drop casting and used to fabricate single-crystal field-effect transistors (SCFETs), in which the best performance of the **3a**-, **3b**-, and **3c**-based transistors reached 0.73, 0.52, and 10^{-5} $cm^2/(V \cdot s)$. The poor performance of the **3c**-based transistors was attributed to the thickness and disrupted π stacking of single crystals.

Benefiting from the bridge function of acetylene, the "H"-type anthracene derivative 1,2-di(anthracen-9-yl)ethyne (**4**), which was synthesized by Hu *et al.*, also showed a planar structure in the X-ray diffraction (XRD) analysis of single crystals

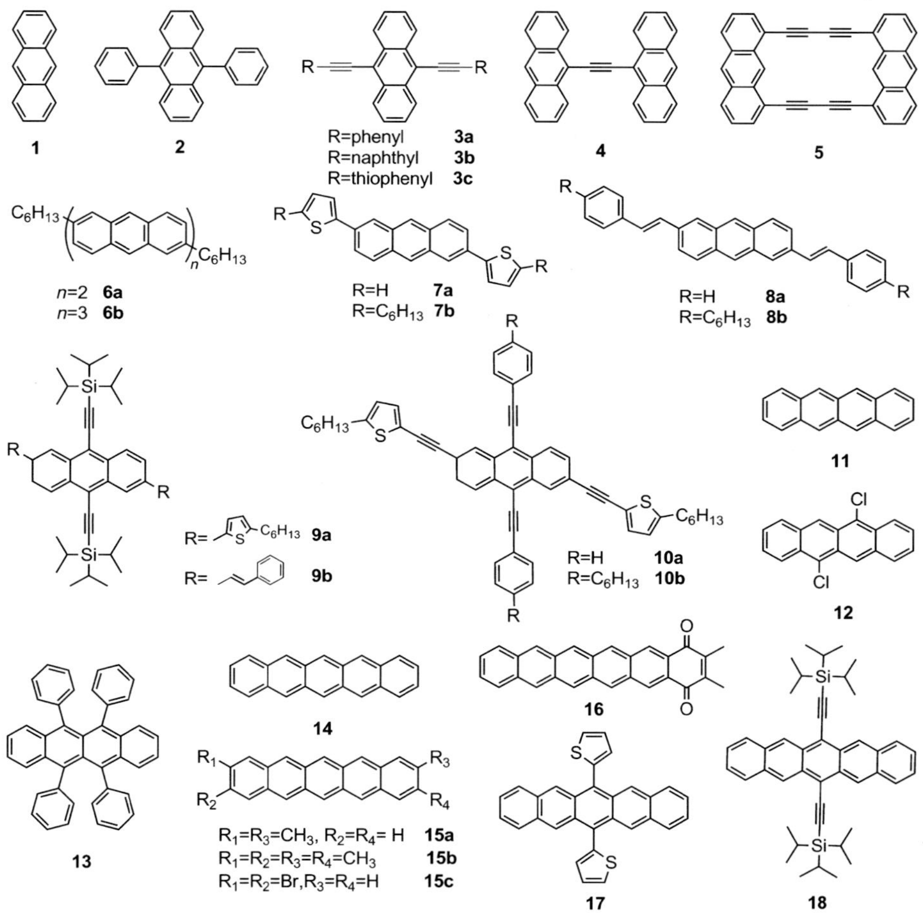

Figure 2.2 Molecular structures of representative acenes and substituted acenes.

[11]. Based on the single-crystalline micro/nanometer ribbons prepared using a physical vapor transport technique, an "organic ribbon mask" method was successfully developed to fabricate transistors. Devices with different channel lengths (micrometer- and even nanometer-size) could easily be fabricated in this way, and the field-effect mobility of **4**-based SCFETs lay in the range of 0.20–0.82 $cm^2/(V \cdot s)$. Similarly, the molecular structure of compound **5** was also composed of two anthracenes as subunits, but with four acetylenes as conjugation paths [12]. By thermally evaporating compound **5** onto octadecyl trichlorosilane (OTS)-modified SiO_2 substrates at 110 °C, the top-contact thin-film transistors (TFTs) fabricated by Miao *et al.* exhibited the best mobility at 0.07 $cm^2/(V \cdot s)$.

Besides *peri*-substituted anthracene derivatives, the derivatives substituted in the molecular long axis were also applied in OFETs. The oligo(2,6-anthrylene)s **6a** and

6b were realized with extended π system by oligomerization of small acenes [13]. Although the alkyl chains at the terminal position of molecular structures did not improve the solubility greatly, they dramatically improved the FET performance at elevated substrate temperatures, with the highest mobilities of **6a**- and **6b**-based TFTs reaching 0.13 and 0.18 $cm^2/(V \cdot s)$, respectively. The thiophene–anthracene oligomers 2,6-di (thiophen-2-yl)anthracene (DTAnt, **7a**) and 2,6-bis(5-hexylthiophen-2-yl)anthracene (DHTAnt, **7b**) were designed as candidates for stable organic thin-film FETs [14]. Both the shelf-life tests (15 months) and continuous operation tests (ca. 2960 double scans) were characterized to confirm the high stability of these devices. By depositing **7b** with a two-stage process, the TFTs exhibited mobilities of up to 0.50 $cm^2/(V \cdot s)$. Di(phenylvinyl)anthracene (DPVAnt, **8a**), which was synthesized via a Suzuki coupling reaction, was designed for the combination of a large carrier mobility with an increased ionization potential and device stability [15]. When compared to pentacene [HOMO: −5.0 eV, E_g: 1.8 eV, μ_{FET}: 1.0 $cm^2/(V \cdot s)$], DPVAnt exhibited a lower HOMO energy level (−5.40 eV), a larger optical energy gap (2.6 eV), and a higher TFT mobility [1.3 $cm^2/(V \cdot s)$], indicating a better environmental stability and superior electrical properties of DPVAnt. A more important point was that the DPVAnt-based TFTs showed an outstanding stability after cycle and shelf-life tests, which pentacene-based TFTs could not match. Hu *et al.* fabricated DPVAnt-SCFETs via an "organic ribbon mask" technique, and the charge transport in these transistors showed an anisotropic behavior [16], with hole mobility up to 4.3 $cm^2/(V \cdot s)$ and an I_{on}/I_{off} ratio of up to 10^7 being observed in the characterization of these SCFETs. The introduction of hexyl groups into DPPVAnt (**8b**) increased the intermolecular interactions and resulted in a more compact packing mode than that of pentacene. In this case, the highest mobility was observed as 1.28 $cm^2/(V \cdot s)$, with an I_{on}/I_{off} ratio of 10^6~10^7 [17].

The X-shaped molecules TIPSAntHT (**9a**) and TIPSAntPV (**9b**) were highly soluble, anthracene-based compounds [18]. When compared to 2-,6-substituted anthracene derivatives, such as DHTAnt and DPPVAnt, the introduction of bulky triisopropylsilylethynyl (TIPS) groups at the 9-,10- position of anthracene deleted edge-to-face contacts between adjacent molecules, promoting the formation of a face-to-face packing mode in the single crystals. Top-contact OFETs of **9a** and **9b** were fabricated by solution processing, and the highest mobilities were calculated as 4×10^{-3} $cm^2/(V \cdot s)$ ($I_{on}/I_{off} = 10^6$) and 3×10^{-4} $cm^2/(V \cdot s)$ ($I_{on}/I_{off} = 10^5$), respectively. Jung *et al.* reported the details of solution-processed OTFTs based on **10a** and **10b**, which formed crystalline thin films without thermal annealing [19]. Hole mobilities of 0.04 $cm^2/(V \cdot s)$ ($I_{on}/I_{off} = 10^5$) and 0.24 $cm^2/(V \cdot s)$ ($I_{on}/I_{off} = 10^6$), respectively, were observed in the OTFTs of as-spun films of **10a** and **10b**.

Tetracene (**11**, also known as naphthacene) possesses a low HOMO energy level (−5.4 eV) and a relatively large E_g (3.0 eV). Both, thin films and single crystals of tetracene were used as active layers in OFETs, and the tetracene-based TFTs [20] and SCFETs [21] exhibited mobilities up to 0.10 and 1.3 $cm^2/(V \cdot s)$, respectively.

Moon *et al.* synthesized a series of halogenated tetracene derivatives, in which the crystal structures were found to be sensitive to the substituents [22]. As XRD analysis showed, 5,11-dichlorotetracene (DCT, **12**) took a slipped π stacking as its packing mode, thus distinguishing itself from other halogenated tetracene derivatives. Those SCFETs based on DCT exhibited the highest mobilities of up to $1.6\,cm^2/(V\cdot s)$. In the case of rubrene (**13**, 5,6,11,12-tetraphenylnaphthacene), which is one of the most popular organic semiconductors used for SCFETs, the crystal structure showed that close interactions between the substituted benzene rings and the neighboring tetracene cores greatly narrowed the distance of adjacent molecules, and this facilitated an overlap of the intermolecular electron cloud. Sundar *et al.* systematically studied the anisotropic charge transport of orthorhombic rubrene by characterizing bottom-contact SCFETs on polydimethylsiloxane (PDMS) substrates [23]. The mobilities along the *a* and *b* crystallographic directions were seen to be as high as 4.4 and $15.4\,cm^2/(V\cdot s)$, respectively. These authors attributed such findings to a faster charge transport along the *b* direction, the closer π–π overlap direction compared to the *a* direction.

Pentacene (**14**), which was first reported as a candidate semiconductor in 1960, has been one of the most widely used semiconductors for OFETs. The HOMO energy level and energy band gap (E_g) of pentacene are −5.14 and 1.77 eV, respectively, and the molecules adopt a typical "herringbone" packing mode in single crystals of pentacene. Due to the compact edge-to-face interactions, the charge transport in highly ordered pentacene-based thin films which are easily obtained by vacuum-deposition are highly efficient. In 2003, Kelley and coworkers fabricated pentacene-based TFTs with poly(α-methylstyrene)-modified insulators [24]. In benefiting from the friendly modified substrates, ordered polycrystalline films of pentacene were formed during the deposition, and the highest mobility of these devices reached $5\,cm^2/(V\cdot s)$. Recently, Hwang and coworkers thermally evaporated pentacene on flexible poly(ethylene terephthalate) substrates which had been covered with solution-processed silk fibroins as gate dielectrics; subsequently, a very high mobility of up to $23.2\,cm^2/(V\cdot s)$ was observed during the operation of these devices [25]. By taking 6,13-pentacenequinone (PQ) films as gate insulators, Palstra *et al.* were able to fabricate SCFETs of pentacene in a top-gate, bottom-contact configuration [26]. The high-quality devices obtained showed typical charge-carrier mobilities in the range of $15\sim40\,cm^2/(V\cdot s)$, which were considered to be an intrinsic property of pentacene crystals.

A substantial effort was spent on the derivatization of pentacene due to the outstanding transfer property of pentacene. Kelley *et al.* adopted alumina as a dielectric, and this was modified by spin-coated alkylphosphonic acid monolayers, to induce the polycrystalline films of 2,9-dimethylpentacene (**15a**) during the vacuum deposition [27]. A high mobility of up to $2.5\,cm^2/(V\cdot s)$ with I_{on}/I_{off} of 10^6 was achieved in these **15a**-based OTFTs. 2,3,9,10-Tetramethyl-pentacene (Me4PENT, **15b**) showed an improved thermal stability compared to pentacene [28]. Meng *et al.* fabricated **15b**-based TFTs on hexamethyldisilazane (HMDS)-modified Si/SiO_2 at different substrate temperatures, and the highest mobility was observed as $0.30\,cm^2/(V\cdot s)$. Okamoto *et al.* introduced electron-withdrawing groups (bromo, cyano, and trifluoromethyl) to the terminal position of pentacene to demonstrate the influence of substituent over the physical and chemical properties of these derivatives, in which the highest field-effect mobility [0.22 $cm^2/(V \cdot s)$] was achieved in the 2,3-dibromopentacene (**15c**)-based TFTs [29]. Asymmetric compound **16** adopted a lamellar packing mode in solid states, and showed a hole mobility of up to 0.052 $cm^2/(V \cdot s)$ when used as an active layer in OTFTs [30].

In order to explore the directing function of substituents in the arrangement of acene cores, Miao *et al.* synthesized and studied the crystal structures of a series of pentacenes derivatives [31]. The electrical properties of these materials were evaluated in thin-film transistors, in which the best mobility was observed for 6,13-di(2′-thienyl)pentacene (**17**)-based TFTs as 0.10 $cm^2/(V \cdot s)$. Among the *peri*-substituted pentacene derivatives, a series of (trialkylsilyl)ethynyl-substituted compounds attracted much attention due to the great regulation of changing the packing mode from herringbone to lamellar, which was thought good for an efficient charge transfer [32, 33]. The environmental stabilities of these compounds were shown to be superior to those of pentacene, as the chemical reactive positions of pentacene were excluded by substituents. In addition, their solubilities in common solvents were also greatly improved. Among these, 6,13-bis((triisopropylsilyl)ethynyl)pentacene (**18**) proved to be the most successful material when used as a semiconducting layer in FETs, exhibiting mobilities of >1 $cm^2/(V \cdot s)$ in OTFTs [34] and 1.42 $cm^2/(V \cdot s)$ in SCFETs [35].

2.2.1.1.2 Nonlinear Polycyclic Aromatic Hydrocarbons (PAHs)

Pyrene (**19**) is a *peri*-fused PAH with a C/H ratio higher than those of the acenes. The hole and electron mobilities in single crystals of pyrene, when characterized by TOF measurements, reached 1.2 and 3.0 $cm^2/(V \cdot s)$ [36]. The thiophene-substituted pyrene **20** adopts a twisted single-crystal structure due to the steric repulsion between the adjacent thienyl and pyrene moieties. The OTFTs based on **20** exhibited a hole mobility as 3.7×10^{-3} $cm^2/(V \cdot s)$ ($I_{on}/I_{off} = 10^4$) [37]. 1-Imino nitroxide pyrene (**21**), a stable organic radical, showed a field-effect mobility of up to 0.1 $cm^2/(V \cdot s)$ with an I_{on}/I_{off} ratio of 10^4 when used as an active layer in OTFTs [38].

As an isomer of pentacene, picene (**22**) possesses a lower HOMO energy level (−5.50 eV) and a larger E_g (3.30 eV) than pentacene, and a better environmental stability when exposed to light and oxygen. The picene-based TFTs fabricated by Okamoto *et al.* exhibited a high *p*-channel mobility of 1.1 $cm^2/(V \cdot s)$ [39]. The further field-effect characterization of devices showed that both mobility and I_{on}/I_{off} were increased after exposure to air (O_2), which indicated that the incorporation of oxygen had promoted the formation of channel transport (Figure 2.3).

When Kotani *et al.* studied the charge transport in thin films and single crystals of perylene (**23**), the performance of single crystals [μ: ~10^{-4} $cm^2/(V \cdot s)$] was much superior to that of thin films [40]. The mobility of single-crystal-based devices was calculated as 0.12 $cm^2/(V \cdot s)$, which was of the same order of magnitude as results obtained with TOF measurements.

Compound **24** is a liquid-crystalline derivative of the discotic molecule hexa-*peri*-hexabenzocoronene (HBC) [41]. Müllen *et al.* invoked a zone-casting tech-

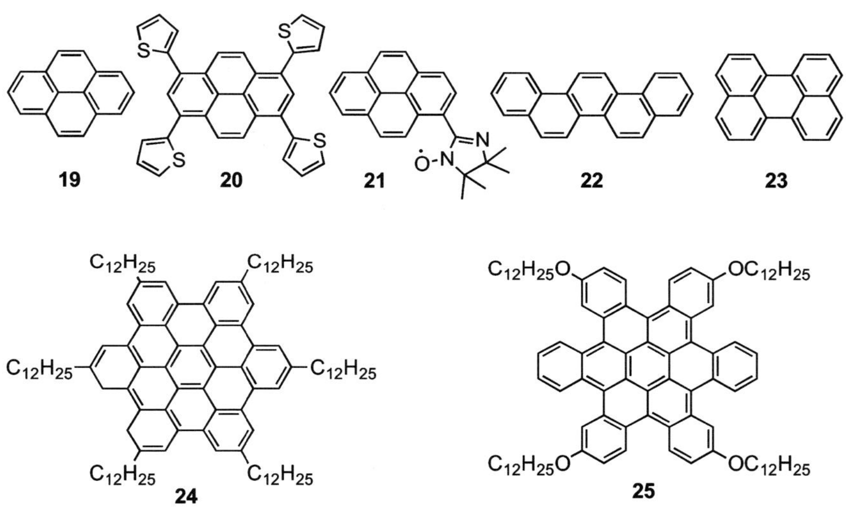

19 20 21 22 23

24 25

Figure 2.3 Molecular structures of some nonlinear polycyclic aromatic hydrocarbons (PAHs).

nique to obtain long-range-oriented films of **24**, on which the OTFTs fabricated and exhibited the highest mobility of $0.01\,cm^2/(V\cdot s)$. Although compound **25** adopts a distorted molecular structure due to the steric congestion in proximal carbon atoms, it possesses a discotic mesophase and a good self-assembled property [42]. XRD studies showed that the molecules of **25** formed highly ordered columns in this intermediate phase, which is important for current flow in TFTs. The solution-processed OTFTs of compound **25** showed field-effect mobilities of up to $0.02\,cm^2/(V\cdot s)$.

2.2.1.2 **Chalcogen-Containing Semiconductors**

2.2.1.2.1 **Oligothiophenes**

Oligothiophene was one of the earliest organic semiconductors to be used in FETs. As the basic unit of molecular frameworks, heterocyclic thiophene consists of a planar, five-membered ring that is considered to be aromatic as the electron pair on the sulfur atom could be efficiently delocalized in the molecular π-electron system. Benefiting from the compact steric structure of thiophene, oligothiophenes adopt planar molecular structures when linked by σ-bonds. With the technical development of device fabrication, the performance of oligothiophene-based FETs has been greatly improved during the past decades; notable, OTFTs based on α-4T (**26a**), α-5T (**26b**), α-6T (**26c**) and α-8T (**26d**) exhibited hole mobilities of up to 0.011 [43], 0.078 [44], 0.03 [45], and $0.33\,cm^2/(V\cdot s)$ [46], respectively. The **26c**-based SCFETs fabricated by Horowitz *et al.* showed the highest mobility, as $0.075\,cm^2/(V\cdot s)$ with an I_{on}/I_{off} ratio $>10^4$ [47–54] (Figure 2.4).

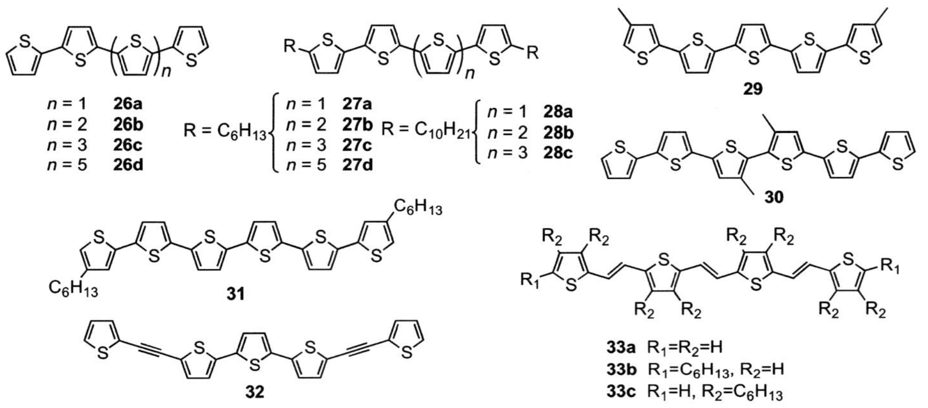

Figure 2.4 Chemical structures of some representative oligothiophenes.

The solubilities of many organic compounds were improved by introducing alkyl chains to the frameworks of molecules. To some extent, the self-assembled properties of these molecules could also be expected to improve in this way [48]. The oligothiophene with alkyl chains substituted in the α-position of thiophene exhibit a planar molecular structure but show no improvement in solubility, whereas the solubility of oligothiophenes with β-substituted alkyl chains could be clearly improved, albeit at the cost of molecular planarity. When a series of oligothiophenes α-terminated by alkyl chains were first used as active layers of OTFTs, the transistors exhibited typical *p*-channel field-effect characteristics (representative hexyl-terminated oligothiophenes are: DHα-4T (**27a**), 0.23 $cm^2/(V \cdot s)$ [49]; DHα-5T (**27b**), 0.10 $cm^2/(V \cdot s)$ [50]; DHα-6T (**27c**), 1.0 $cm^2/(V \cdot s)$ [51]; and DHα-8T (**27d**), 0.02 $cm^2/(V \cdot s)$ [52], while representative decyl-terminated oligothiophenes are: DDα-4T (**28a**), 0.20 $cm^2/(V \cdot s)$; DDα-5T (**28b**), 0.50 $cm^2/(V \cdot s)$; and DDα-6T (**28c**), 0.50 $cm^2/(V \cdot s)$ [53]). Oligothiophenes with β-substituted alkyl chains exhibited relatively lower properties in OTFTs, and the highest mobilities of **29**, **30**, and **31** were calculated as 9.2×10^{-3} [44], 0.02 [54], and 0.06 $cm^2/(V \cdot s)$ [43], respectively. Conjugated groups C=C and C≡C were introduced to the frameworks of oligothiophenes by some research groups for an elongating conjugation of the whole molecules. T5A (**32**), which possessed a similar electronic structure of the frontier orbitals as α-5T, exhibited a much poorer carrier mobility [(8×10^{-4} $cm^2/(V \cdot s)$ with I_{on}/I_{off} of 10], even at elevated substrate temperatures [44]. Videlot *et al.* designed and synthesized C=C-containing oligothiophenes without and with hexyl substitutions (**33a–c**) [55]. Based on glass substrates with aluminum gate electrodes, polymethylmethacrylate (PMMA) was first spun-coated as insulating layers, after which oligothiophenes **33a–c** were respectively sublimed in vacuum at different substrate temperatures. The highest carrier mobilities of **33a–c** were observed as 1.4×10^{-3}, 0.055, and 1.1×10^{-6} $cm^2/(V \cdot s)$, respectively.

2.2.1.2.2 Thiophene-Containing Co-oligomers

Due to the steric repulsion between the hydrogen atoms of adjacent phenylenes, oligo-*p*-phenylenes always adopt nonplanar molecular structures, which are not conducive to charge transfer in solid states. Although oligothiophenes can retain a near-planar structure in single crystals, the HOMO energy levels of electron-rich oligothiophenes are much higher than those of oligo-*p*-phenylenes. Consequently, phenylenes were introduced into the frameworks of oligothiophenes to lower the molecular HOMO energy levels, thus increasing the molecular environmental stabilities. At the same time, the molecular structures of thiophene–phenylene co-oligomers could be kept near-planar, which was beneficial for intermolecular packing and charge transfer in solid states (Figure 2.5).

Hong *et al.* synthesized a series of thiophene-containing oligomers (**34a–c**, **35a–c**, and **39**) and compared their physical and chemical properties with those of oligothiophenes [56]. Electronic structure calculations showed that the HOMO energy levels of these thiophene–phenylene co-oligomers were clearly decreased by the introduction of phenylenes. Ponomarenko *et al.* confirmed the higher oxidation stability of thiophene–phenylene co-oligomers compared to the corresponding oligothiophenes via cyclovoltamperometry measurements [57]. When used as semiconducting materials in OTFTs, the phenylene-centralized oligomers **34a–e** respectively exhibited their highest carrier mobilities as 5×10^{-4}, 0.02, 4×10^{-5} [56], 0.3 [57], and 0.08 $cm^2/(V \cdot s)$ [58], whereas OTFTs based on the thiophene-centralized oligomers **35a–c** [56], **36a–d** [59] and **37** showed relatively closer carrier mobilities of 0.0065 (**35a**), 0.003 (**35b**), 0.033 (**35c**), 0.01 (**36a**), 0.09 (**36b**), 0.054 (**36c**), 0.09 (**36d**), and 0.018 (**37**) $cm^2/(V \cdot s)$. Occasionally, thiophene

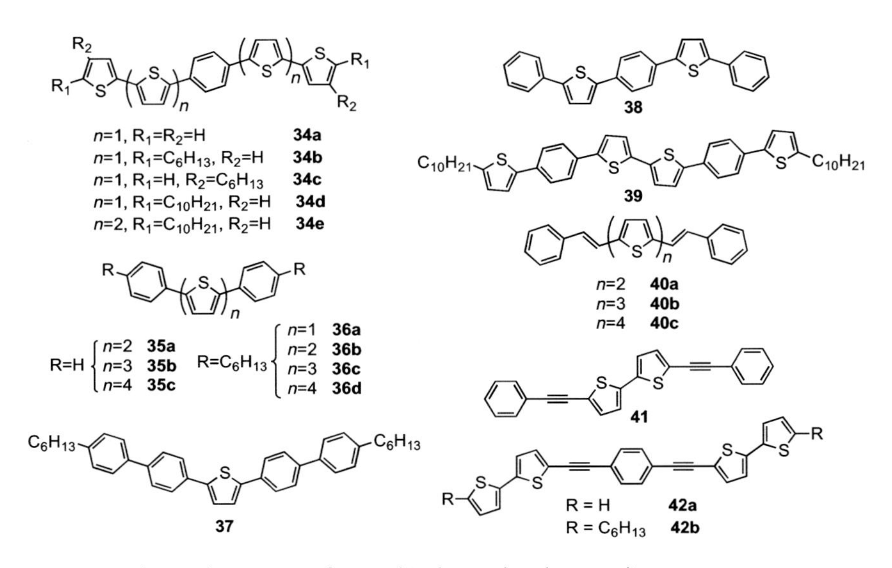

Figure 2.5 Chemical structures of some thiophene–phenylene co-oligomers.

and phenylene monomers were crosslinked in the molecular structures of oligomers, such as compounds **38** and **39**. When vacuum-deposited at 60–80 °C, the **38**-based OTFTs exhibited the best mobility at 0.04 $cm^2/(V \cdot s)$, with an I_{on}/I_{off} ratio of 10^4 [56]. Both, top-contact and bottom-contact OTFTs were fabricated on the basis of **39**, but the latter type of device showed the highest mobility at 0.4 $cm^2/(V \cdot s)$ with an I_{on}/I_{off} ratio of 10^5 [58].

Besides σ-bonds, unsaturated carbon–carbon bonds were also introduced to the backbones of thiophene–phenylene oligomers. For example, Videlot-Ackermann and coworkers reported the syntheses and characterizations of a series of oligothiophenes that had been end-capped with styryl units (**40a–c**) [60]. A terrace morphology of the thin film deposited at elevated substrate temperatures was observed using atomic force microscopy (AFM), and the best mobilities of the **40a–c**-based OTFTs reached 0.02, 0.02, and 0.1 $cm^2/(V \cdot s)$, respectively. As a linear π-conjugated bridge, a carbon–carbon triple bond was introduced to the frameworks of some thiophene–phenylene co-oligomers to elongate the molecular conjugation of the whole molecules; thus, compounds **41**, **42a**, and **42b** exhibited good thermal- and environmental-stabilities [61]. Top-contact OTFTs were fabricated on the basis of these ethynylene-containing semiconductors by thermal deposition, and the highest mobilities of **41**, **42a**, and **42b** reached 0.028, 0.056, and 0.084 $cm^2/(V \cdot s)$, respectively. In addition, an excellent device stability was observed in the characterization of **42b**-based devices after storage for over 12 months (Figure 2.5).

In order to modify the chemical/physical properties of thiophene-containing semiconductors, some other functional groups were chosen as partners of thiophene to form the backbones of novel semiconductors. Tian *et al.* synthesized phenanthrene- and fluorene-containing co-oligomers 2,7-bis(5′-hexyl-2,2′-bithien-5-yl)phenanthrene (DH-TTPhTT, **43**) and 2,7-bis(5′-hexyl-2,2′-bithien-5-yl)fluorene (DH-TTFTT, **45**); subsequently, the **43**-based and **45**-based OTFTs exhibited mobilities of up to 0.067 and 0.01 $cm^2/(V \cdot s)$ when using Ta_2O_5 as dielectric layers [62]. Cho *et al.* synthesized a 9,10-dihydrophenanthrene-centralized thiophene-containing oligomer **44**, and a high mobility of up to 0.42 $cm^2/(V \cdot s)$ was observed in the characterization of **44**-based OTFTs [63]. The fluorene–thiophene co-oligomers CHFTTF (**46a**) [64] and DHFTTF (**46b**) [65] exhibited poor carrier transfer properties when used as the semiconducting materials of FETs [CHFTTF, 0.17 $cm^2/(V \cdot s)$; DHFTTF, 0.12 $cm^2/(V \cdot s)$]. Compound **47**, which included polycyclic 6,12-dihydroindeno[1,2-*b*]fluorene as the substituted parent, exhibited the highest mobility at 0.012 $cm^2/(V \cdot s)$ [66] (Figure 2.6).

2.2.1.2.3 Thienoacenes

Thienoacenes [6], characterized by fused thiophene rings in the backbones of molecules, represent the most popular group of semiconducting materials applied in FETs during recent decades. The various intermolecular interactions (C–H . . . π, π . . . π, S . . . S, S . . . π, etc.) observed in single crystals always make the solid-state packing of thienoacenes compact, which is beneficial for the charge transfer of OFETs. Physical-property comparisons of thienoacenes and acenes with same numbers of fused rings have indicated that the replacement of phenylene with

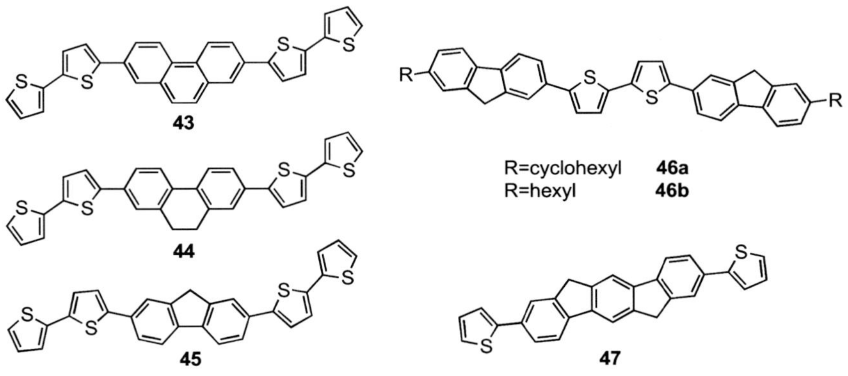

Figure 2.6 Chemical structures of some thiophene-containing co-oligomers.

thiophene causes the HOMO energy levels of the thienoacenes to be much lower than those of acenes [67]. Major improvements were also noted in HOMO–LUMO energy gaps (E_g) of thienoacenes, which led to the thienoacenes possessing higher environmental stabilities than the corresponding acenes. The above-mentioned improvements have led to thienoacenes and their derivatives becoming promising candidates as semiconducting materials in organic FETs.

Pentathienoacene (PTA, **48**), a typical thienoacene that consists of five fused thiophene rings, showed a relatively lower HOMO energy level (–5.33 eV) and a larger E_g (3.29 eV) than pentacene [68]. When Xiao *et al.* fabricated OTFTs on the basis of vacuum-deposited thin films of pentathienoacene, the highest mobility achieved was 0.045 $cm^2/(V \cdot s)$. The heterocyclic circulenes **49a** possessed a HOMO energy level less than –5 eV and a E_g larger than 4 eV [69]. A single-crystal analysis showed that the molecules of **49a** could be packed in column along the *c* axis, where the intermolecular distance was about 3.9 Å, and a mobility of 9×10^{-3} $cm^2/(V \cdot s)$ was observed in the characterization of **49a**-based OTFTs. Compound **49b**, a selenium analog of **49a**, showed a highest mobility of OTFTs at 1×10^{-3} $cm^2/(V \cdot s)$ and an I_{on}/I_{off} ratio of 10^6 [70]. The thienoacenes **50**, **51**, **52a**, **53**, and **54b** were analogs of pentacene, in which one to three phenylene(s) were replaced by thiophene(s) to obtain air-stable, high-performance organic semiconductors. When based on thin films of DBTDT (**50**), which were vacuum-deposited at elevated substrate temperatures, the FETs exhibited charge mobilities of up to 0.51 $cm^2/(V \cdot s)$ [71], whereas FETs based on single crystals of DBTDT exhibited a carrier mobility of up to 1.8 $cm^2/(V \cdot s)$ with an I_{on}/I_{off} ratio of 10^7 [72]. Although the difference in molecular structure between the isomers DBTDT (**50**) and BBTT (**51**) related simply to the direction of the central thiophene ring, the columnar packing mode of BBTT was totally different from the herringbone arrangement of DBTDT in single crystals [73]. A highest mobility of up to 0.6 $cm^2/(V \cdot s)$ was observed during the characterization of BBTT-based SCFETs. The effective ingredient of anthradithiophene (ADT, **52a**) consisted of *syn* and *anti* isomers [74], which could not be separated from each other by using general purification techniques. Consequently, when ADT was thermally evaporated onto substrates at different temperatures, the mobilities of the thin films were of the same order of magnitude [μ_{max}: 0.09 cm^2/(V·s)]. When taking ADT as parent compound, the derivatives with end-substituted dihexyl- (DHADT, **52b**), dodecyl- (DDADT), and octadecyl- (DOADT), were also synthesized and applied as semiconducting materials in OTFTs, while the highest mobility [0.17 cm^2/(V·s)] was achieved by **52b**-based OTFTs. Anthra[2,3-*b*]benzo[*d*]thiophene (ABT, **53**) [75, 76], an asymmetric analogy of pentacene, exhibited a carrier mobility of up to 0.41 cm^2/(V·s) when used as a semiconducting material for OTFTs [77]. Another asymmetric linear analog of pentacene, **54b**, had a carrier mobility of up to 0.47 cm^2/(V·s) [78]. As homologs of **54b**, both **54a** and **54d** showed similar abilities in terms of carrier transport, of >0.1 cm^2/(V·s) [78] and 0.574 cm^2/(V·s) [79]. Subsequently, **54c** was taken as a representative among the derivatives of this series of thienoacenes, exhibiting a carrier transport mobility of 0.85 cm^2/(V·s) when deposited at OTS-modified SiO_2/Si substrate at T_{sub} = 80 °C [80] (Figure 2.7).

Sirringhaus *et al.* utilized acid-induced intramolecular cyclization to synthesize dibenzo[*b*,*b'*]thieno[2,3-*f*:5,4-*f'*]bis-[1]benzothiophene (DBTBT, **55**) [81]. As the coupling reaction towards DBTBT could occur at different positions, two byproducts were inevitably formed during the synthesis. However, purified thin films of DBTBT could be obtained by controlling the sublimation conditions, such that the OTFTs exhibited carrier mobilities of up to 0.15 cm^2/(V·s). Benzoannulated pentathienoacene f-B5TB (**56a**) and its selenium analog f-BT3STB (**56b**), both of

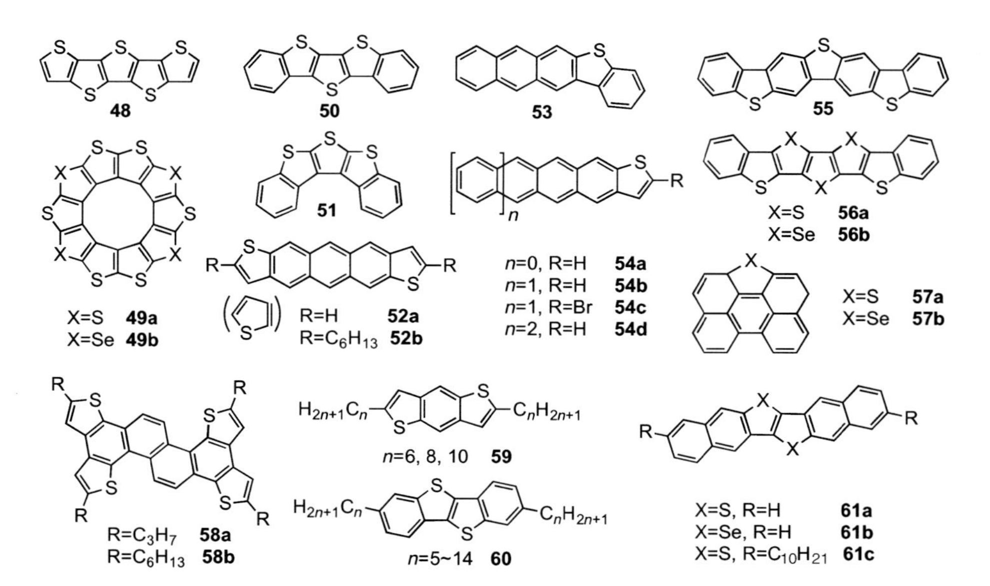

Figure 2.7 Chemical structures of some thienoacenes and their derivatives.

which were synthesized via an intramolecular triple cyclization of bis(*o*-haloaryl) diacetylenes [82], were applied as the semiconducting materials of SCFETs by Yamada and coworkers [83]. Electrical measurements showed that the charge transport mobilities in single-crystals of f-B5TB and f-BT3STB reached 0.5 and 1.1 $cm^2/(V \cdot s)$, respectively. The electrical properties of single crystals of perylo[1,12-*b,c,d*]thiophene (PET, **57a**) [84] and its selenium analog perylo[1,12-*b,c,d*]selenophene (PESE, **57b**) [85], were also tested in SCFETs, when their maximum field-effect mobilities reached 0.8 $cm^2/(V \cdot s)$ [84] and 2.63 $cm^2/(V \cdot s)$ [86], respectively. Nonlinear thienoacenes **58a–b** were derived from chrysene; indeed, the molecular conjugation and intermolecular interactions were greatly enhanced by the introduction of thiophenes to the skeleton of chrysene [87], with layer-by-layer crystal structures and significant intermolecular S . . . S contacts perpendicular to the plane of stacking direction being observed. A hole mobility of up to 0.4 $cm^2/(V \cdot s)$ was observed in both **58a**- and **58b**-based OTFTs.

Takimiya *et al.* designed and synthesized a series of thienoacenes with alkyl chains substituted in the molecular long-axis directions (**59**, **60**, and **61c**), most of which exhibited a prominent performance when used as the active layer of FETs. 2,6-Dialkyl[1,2-*b*:4,5-*b*′]dithiophenes (C_n-BDTs, **59**) were spin-coated to fabricate top-contact OTFTs, and the average field-effect mobility was on the order of 10^{-2} $cm^2/(V \cdot s)$ [88]. In terms of C_n-BTBTs ($n = 5{\sim}14$; **60**)-based OTFTs, the highest carrier transport mobilities of C_{12}-BTBT- and C_{13}-BTBT-based OTFTs reached 3.9 $cm^2/(V \cdot s)$ [89] and 2.75 $cm^2/(V \cdot s)$ [90], respectively. Subsequently, further analogs (**61a–c**) were developed with higher degrees of conjugation, and an improved device performance was confirmed by electrical measurements of the OTFTs and SCFETs (**61a**-OTFT, 2.9 $cm^2/(V \cdot s)$ [91]; **61a**-SCFET, 8.3 $cm^2/(V \cdot s)$ [92]; **61b**-OTFT, 1.9 $cm^2/(V \cdot s)$ [91]; and **61c**-OTFT: 11 $cm^2/(V \cdot s)$ [93]).

As the synthesis of heteroacenes has always been influenced by aspects such as active position, solubility, and yields, many heteroacenes derived from C–C, C=C, and C≡C bonds have been designed and synthesized during recent years. In one such representative semiconductor – the thienoacene-dimers – single-crystal analysis showed that the molecules of the compound α,α′-bis (dithieno[3,2-*b*:2′,3′-*d*] thiophene) (BDT, **62**) would adopt a face-to-face structure as a solid-state packing mode, with the shortest distance between adjacent planes of 3.557 Å. When vacuum-deposited thin films of BDT were used as the active layer of OTFTs, the highest mobility reached was 0.05 $cm^2/(V \cdot s)$ [94, 95]. The dimer of benzo[1,2-*b*:4,5-*b*′]dithiophene (**63**) possessed an exceptional thermal stability, exhibiting a hole mobility of up to 0.04 $cm^2/(V \cdot s)$ when deposited at $T_{sub} = 100\,°C$ [96]. [2,2′]Binaphtha[2,3-*b*]thiophenyl (BNT, **64**), a dimer of naphthothiophene, possessed a low HOMO energy level of −5.73 eV and a large E_g of 2.99 eV [97]. When BNT-films that had been deposited on OTS-modified substrates at 50 °C were used as the active layer of OTFTs, a charge transport mobility of up to 0.67 $cm^2/(V \cdot s)$ was observed. Tan *et al.* synthesized and characterized dimers of dithieno[2,3-*b*:3′,2′-*d*] thiophene with C–C (**65**) and C=C (**66**) bonds [98]. The molecules of **65** adopted a sandwich-herringbone arrangement in single crystals, whereas the single-crystal structure of **66** was a typical herringbone. The **65**-based OTFTs, when fabricated

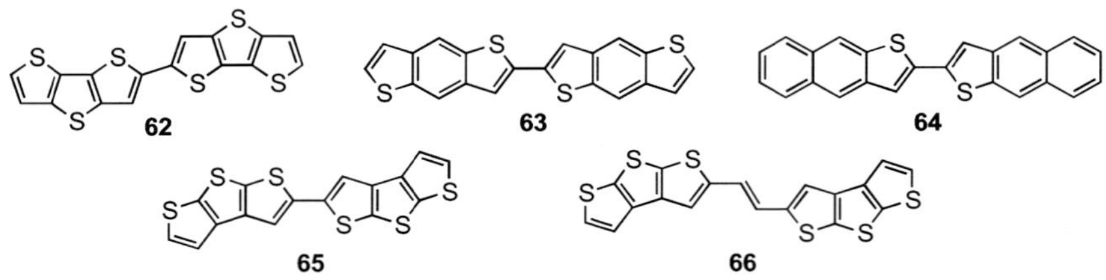

62 63 64 65 66

Figure 2.8 Chemical structures of some thienoacene-dimers.

by vacuum deposition, exhibited a hole mobility of 0.005 $cm^2/(V\cdot s)$, whereas the maximum mobility of the **66**-based OTFTs was 0.89 $cm^2/(V\cdot s)$ (Figure 2.8).

Takimiya *et al.* designed a series of 2,6-diphenylbenzo[1,2-*b*:4,5-*b*′]dichalcogenophenes (**67a–c**) to explore the relationship between device performance and molecular structures with different chalcogen atoms [99]. All three compounds exhibited typical *p*-type field-effect behaviors in the characterization of OTFTs, with the highest mobility being achieved by DPh-BDS (**67b**) at 0.17 $cm^2/(V\cdot s)$. Inspired by these results, the same group further synthesized 2,7-diphenyl-[1]benzothieno[3,2-*b*] benzothiophene (DPh-BTBT, **68a**) and its selenium analog (DPh-BSBS, **68b**) as the semiconducting materials of OTFTs. Subsequently, thin films of DPh-BTBT and DPh-BSBS that had been vacuum-deposited at elevated substrate temperatures exhibited mobilities of up to 2.0 $cm^2/(V\cdot s)$ [100] and 0.31 $cm^2/(V\cdot s)$ [101], respectively. Alternatively, by taking dithieno[3,2-*b*:2′,3′-*d*]thiophene as core, a series of derivatives was synthesized and utilized as semiconductors of OTFTs by Sun *et al.* [102]. Those compounds end-capped by phenyl (**69a**) and biphenyl (**69b**) proved to be outstanding among the derivatives, and exhibited field-effect mobilities of up to 0.42 $cm^2/(V\cdot s)$ and 0.12 $cm^2/(V\cdot s)$ when used as the active materials of OTFTs. Single-crystal analyses showed that while distyryl-bisthienobenzene (**70**) adopted a nonplanar structure in solid states, a vacuum-deposited thin film of this material demonstrated a well-ordered terrace structure [103]. A hole mobility of up to 0.1 $cm^2/(V\cdot s)$ was achieved by **70**-based OTFTs when deposited onto HMDS-modified SiO_2 substrates at $T_{sub} = 80\,°C$. Zhang *et al.* introduced phenyl units (**71a**) and thiophenes (**71b**) to the molecular skeleton of *trans*-1,2-(dithieno[2,3-*b*:3′,2′-*d*]thiophene)ethane, with an expectation of extending the conjugation and providing a tight packing [104]. Vacuum-deposited thin films of **71a** showed a better crystalline form than those of **71b**, and this resulted in a higher charge transport mobility of up to 2.0 $cm^2/(V\cdot s)$. 3,7-Bis(5′-hexyl-thiophen-2′-yl)-dibenzothiophene (3,7-DHTDBTT, **72**), possessing a low HOMO energy level (−5.57 eV) and a large E_g (3.13 eV), was the best semiconductor among the dibenzothiophene derivatives synthesized by Gao and coworkers [105]. Thin films of **72**, when deposited at both room temperature and elevated temperatures, exhibited similar carrier transport mobilities [10^{-2} $cm^2/(V\cdot s)$]. The dithieno[3,2-*b*:2′,3′-*d*] thiophene (DTT)-containing oligomer **73** exhibited a maximum mobility of 0.02 $cm^2/(V\cdot s)$ when used as semiconductor of OTFTs [106] (Figure 2.9).

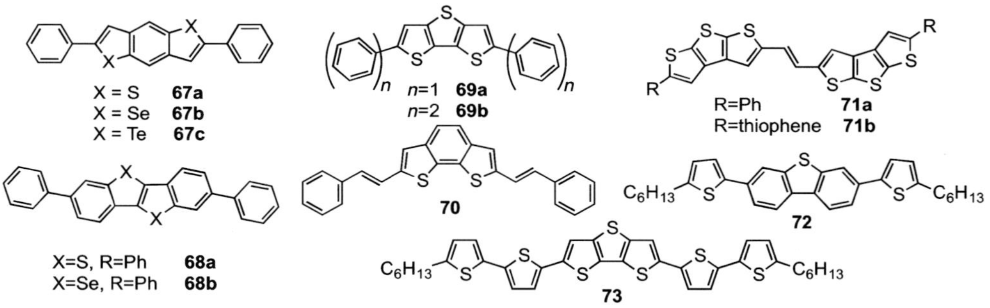

Figure 2.9 Examples of thienoacene-containing oligomers.

Ethynylene, a special π-electron bridge, could introduce aromatic units to the molecular skeleton while retaining molecular planar structures whilst, at the same time, the molecular conjugation could be efficiently enlarged. Meng *et al.* developed benzodithiophene derivatives with aryl-acetylenes, in which the 2,6-bis(phenylethynylene)benzo[1,2-*b*:4,5-*b*′]dithiophene (BPEBDT, **74a**) gave rise to field-effect mobility up to 1.17 $cm^2/(V \cdot s)$ [107]. These authors attributed this property to the well-ordered packing of molecules in solid states, which could be intuitively proved by using AFM and XRD analyses. Besides, the maximum charge transport mobilities of BTEBDT (**74b**)- and BHPEBDT (**74c**)-based OTFTs reached 0.28 and 0.07 $cm^2/(V \cdot s)$, respectively. According to the experience of the derivatization of corresponding acenes, series of (trialkylsilyl)ethynyls were introduced to the short-axis of tetraceno[2,3-*b*]thiophene and anthrax[2,3-*b*]thiophene by Tang and coworkers, who expected to adjust the solid-state packing mode in this way. Of all the tetraceno[2,3-*b*]thiophene derivatives, 5,12-bis(triisopropylsilylethynyl) tetraceno[2,3-*b*]thiophene (**75a**) proved to be the most promising semiconductor during the application test, as it adopted two-dimensional (2-D) π-stacking and exhibited a hole mobility of up to 1.25 $cm^2/(V \cdot s)$ [108]. Further, the same authors functionalized a tetraceno[2,3-*b*]thiophene core with fluorine atoms, to afford 7,8,9,10-tetrafluoro-5,12-bis(TIPSethynyl)tetraceno-tetraceno[2,3-*b*]thiophene (**75b**) with an improved oxidative stability and electron-transport properties [109]. The **75b**-based OTFTs that were deposited on OTS-modified substrates exhibited a hole mobility of up to 0.12 $cm^2/(V \cdot s)$ in air and an electron mobility up to 0.37 $cm^2/(V \cdot s)$ in nitrogen. In investigations conducted by Payne and colleagues, (trialkylsilyl)ethynyls were introduced to the short-axis direction of another thienoacene, anthradithiophenes, which consisted of inseparable *syn*- and *anti*-isomers [110]. Due to the regulation of appropriate sizes of substituents, TIPS-anthradithiophene (**76a**) exhibited the best charge transport in OTFT applications [mobility: 1.0 $cm^2/(V \cdot s)$]. Subramanian further partially fluorizated the core of (trialkyl)ethynyl anthradithiophenes to improve the molecular crystallization and stability [111]. On the basis of improved crystalline films of **76b**, OTFTs were fabricated and demonstrated hole mobilities of up to 1.5 $cm^2/(V \cdot s)$ (Figure 2.10).

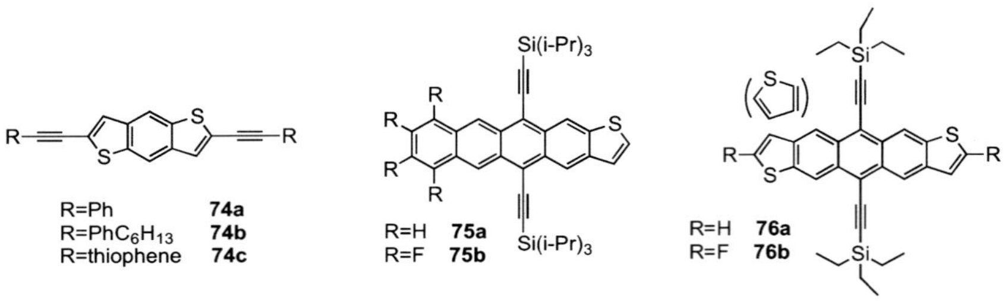

Figure 2.10 Examples of ethynylene-containing thienoacene derivatives.

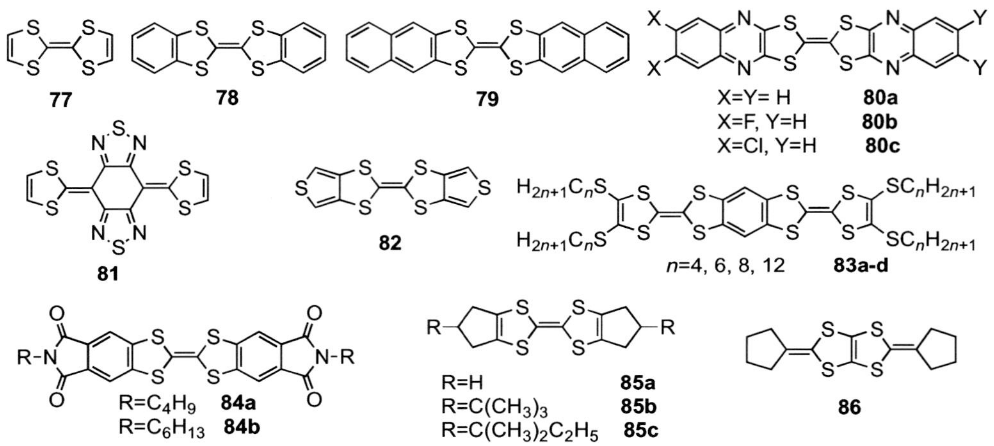

Figure 2.11 Chemical structures of tetrathiafulvalene and derivatives.

2.2.1.2.4 **Tetrathiafulvalene Derivatives**

Tetrathiafulvalene (TTF, 77) was synthesized and reported by Wudl during the 1970s. TTF and its derivatives, which always possess planar and rigid structures, have usually been applied in the field of superconductors. Jiang *et al.* controllably obtained α-phase and β-phase single crystals of TTF and fabricated OTFTs on the basis of these two phases of crystal. Statistical results showed that the α-phase SCFETs had mobilities of up to 1.20 $cm^2/(V \cdot s)$, whereas the β-phase SCFETs exhibited a maximum mobility of 0.23 $cm^2/(V \cdot s)$ [112] (Figure 2.11).

TTF, an electron-rich molecule, is easily oxidized in air. Hence, it is better to functionalize TTF with aromatic and/or electron-withdrawing groups to increase the conjugation and/or to delocalize the electrons of the intact molecules. When Naraso *et al.* introduced aromatic rings (phenylene, and naphthalene) to the long-axis direction of TTF in order to enlarge the molecular conjugation and intermolecular π–π interactions, this resulted in TTF derivatives (**78** and **79**) with lowered HOMO energy levels and improved environmental stabilities [113]. When used as the semiconductors of OTFTs, **78** and **79** demonstrated mobilities of 0.06 and 0.42 $cm^2/(V \cdot s)$, respectively, while **78**-based SCFETs showed field-effect mobilities in the range of 0.1 to 1 $cm^2/(V \cdot s)$ [114]. In addition, when TTF derivatives fused with *N*-heterocyclics (**80a–c**) were synthesized, molecules of **80a** adopted a face-to-face π-stacking in single crystals, and mobilities of up to 0.2 $cm^2/(V \cdot s)$ were observed in the electrical measurement of **80a**-based top-contact OTFTs [113]. A further functionalization of molecular structure with halogen was carried out, in which compounds **80b** and **80c** exhibited film mobilities of up to 0.20 and 0.64 $cm^2/(V \cdot s)$, respectively [115]. When thiadiazole was also introduced to the TTF system as an electron-withdrawing group, OTFTs of DTQBT (**81**) demonstrated mobilities of up to 0.2 $cm^2/(V \cdot s)$ [116]. Single crystals of DT-TTF (**82**), grown by drop-casting, were used as the active layer of SCFETs by Mas-Torrent, and the maximum mobility reached 1.4 $cm^2/(V \cdot s)$ [117]. Benzene-fused bis(tetrathiafulvalene) compounds **83a–d** were synthesized via phosphite-induced crosscoupling reactions in a single step [118, 119]. Subsequent characterization of the crystallinity of spin-coated thin films of **83a–d** was made using XRD analysis, at which point the fabricated OTFTs showed a maximum field-effect mobility of 0.02 $cm^2/(V \cdot s)$. In addition, imide was introduced to the TTF system, affording electron-deficient TTF derivatives **84a** and **84b** with much lower HOMO energy levels [120]. Face-to-face π-stacking with intercolumnar contacts was observed in the single-crystal analysis of **84a**, which was thought to be an effective electronic structure for charge transport. The **84a**- and **84b**-based OTFTs exhibited field-effect mobilities of up to 0.094 and 0.40 $cm^2/(V \cdot s)$, respectively. When Takahashi *et al.* fabricated HMTTF (**85a**)-based SCFETs with TTF-TCNQ films as electrodes, the maximum mobility was seen to be as high as 11.2 $cm^2/(V \cdot s)$ [121]. OFETs fabricated on the basis of thin films of HMTTF exhibited field-effect mobilities up to 3.6 $cm^2/(V \cdot s)$ [122]. A derivatization and device application of HMTTF end-capped with alkyl chains were also carried out, in which the **85b**-based OTFTs and SCFETs exhibited mobilities of up to 0.98 and 2.3 $cm^2/(V \cdot s)$, respectively [122]. In addition, the maximum charge transport mobilities in **85c**-based thin films and single crystals were observed as 0.60 and 1.4 $cm^2/(V \cdot s)$, respectively. Bando *et al.* designed and synthesized TTF derivative **86**, which adopted a herringbone-packing mode in single crystals. Subsequently, a field-effect mobility of up to 0.27 $cm^2/(V \cdot s)$ was observed in the characterization of **86**-based OTFTs [123].

2.2.1.3 Nitrogen-Containing Semiconductors

2.2.1.3.1 *N*-Heterocyclic Acenes

As the planar consistency of molecules can be retained, five- and six-membered nitrogen-containing rings are usually introduced to the molecular skeleton of *p*-type semiconducting materials. Besides an improved environmental stability, the intermolecular interactions of these materials are enriched by the hydrogen bonds.

By taking pentacene and its (trialkyl)ethynyl derivatives as prototypes, Tang *et al.* designed and synthesized 6,13-dihydro-6,13-diazapentacene (DHDAP, **87**) [124], 6,13-bis((triisopropylsilyl)ethynyl)-5,14-dihydro-5,14-diazapentacene (**88**) and 6,13-bis((triisopropylsilyl)ethynyl)-5,14-diazapentacene (**89**) [125]. According to the

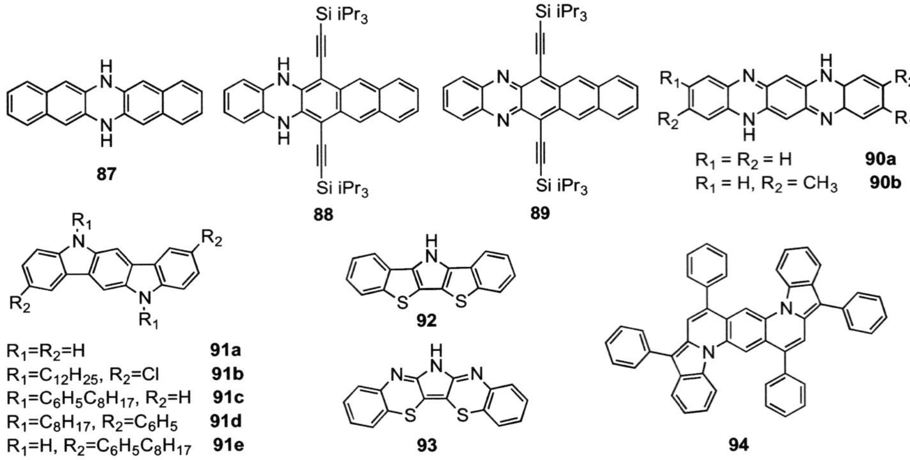

Figure 2.12 Examples of *p*-type *N*-heterocyclic acenes and their derivatives.

(001) spacing (12.9, 13.1, and 13.5 Å, respectively) of the thin films, three crystalline polymorphs of DHDAP could be distinguished. Unlike the latter two phases, the field-effect mobility of OTFTs reached a summit in the 12.9 Å-phase-thin films [0.45 $cm^2/(V \cdot s)$], which was more than 5000-fold higher than the performance of the other two phases. Single-crystal analyses of **88** and **89** showed a similar π-stacking of 2-D brickwork. When thin films of **88** and **89** were vacuum-deposited on octadecyltrimethoxysilane (OTMS)-treated SiO_2/Si substrates at elevated substrate temperatures, the highest mobilities of **88**- and **89**-based OTFTs were observed as 0.07 and 0.05 $cm^2/(V \cdot s)$, respectively. As linear analogs of pentacene, the 5,7,12,14-tetraazapentacenes **90a–b** were synthesized via an improved condensation reaction in high yields [126]. When deposited onto SiO_2 substrates, the thin films of **90a–b** were shown to be amorphous, and to exhibit charge transport mobilities of 0.02 and 0.01 $cm^2/(V \cdot s)$, respectively (Figure 2.12).

Due to its poor solubility and serious decomposition, indolo[3,2-*b*]carbazole (ICZ, **91a**), which was synthesized via a condensation reaction and Fischer indolization, could not be purified by applying common techniques such as chromatography and sublimation. Rather, Zhao and coworkers utilized the diffusion of a poor solvent (ethanol) into a unique good solvent (*N,N*-dimethylformamide) to obtain single crystals of ICZ in multigram quantities [127]. Compact intermolecular N–H...π interactions were observed in the herringbone packing mode of single crystals. In this case, the maximum mobility of ICZ-based OTFTs reached 0.1 $cm^2/(V \cdot s)$. When Ong and colleagues conducted a systematic study of 5,11-didodecyl-indolo[3,2-*b*]carbazoles with halogens substituted at different positions (2,8- or 3,9-), **91b** showed the highest mobility of 0.14 $cm^2/(V \cdot s)$ with an I_{on}/I_{off} ratio of 10^7 [128], but **91c** provided the best transistor performance [μ: 0.12 $cm^2/(V \cdot s)$; I_{on}/I_{off}: $10^6 \sim 10^7$) among the 5,11-alkyl-substituted indolo[3,2-*b*]carbazoles [129]. When

Leclerc and colleagues used indolo[3,2-*b*]carbazoles with 2,8- or 3,9-substituted (hetero)aromatic rings as semiconducting materials to fabricate OTFTs, the highest mobilities were achieved by **91d** [0.20 $cm^2/(V \cdot s)$] [130] and **91e** [0.22 $cm^2/(V \cdot s)$] [131]. Qi *et al.* utilized the Cadogan reaction to synthesize *syn-* and *anti*-dibenzothieno[*b,d*]pyrrole (**92**), but only the *anti*-isomer showed any field-effect characteristics [μ: 0.012 $cm^2/(V \cdot s)$; I_{on}/I_{off}: 10^5) when used as a semiconductor of OTFTs [132]. 6*H*-Pyrrolo[3,2-*b*:4,5-*b*′]bis[1, 4]benzothiazine (PBBTZ, **93**) adopted a sandwich-herringbone packing mode in single crystals, with mobility values of up to 0.34 $cm^2/(V \cdot s)$ being measured in the **93**-based OTFTs [133]. The near-planar heptacyclic bisindoloquinoline derivative **94** was observed to adopt a slipped face-to-face π-stacking in single crystals, on which the SCFETs fabricated showed a high mobility of up to 1.0 $cm^2/(V \cdot s)$ [134].

2.2.1.3.2 Nitrogen-Containing Oligomers

Phthalocyanine and porphyrin are typical nitrogen-containing conjugated macrocyclic molecules, possessing a 2-D molecular plane with a cavity that matches the size of most elements of the Periodic Table. Copper phthalocyanine (CuPc, **95**) is a complex of copper with phthalocyanine, is insoluble in most solvents, and is one of the earliest semiconductors applied to investigations of OFETs. The highest mobilities reported for thin films and single crystals of CuPc were 0.02 $cm^2/(V \cdot s)$ [135] and 1 $cm^2/(V \cdot s)$ [136], respectively. Besides metal elements, phthalocyanine also binds to some metal oxides to form complexes, including titanyl phthalocyanine (TiOPc, **96a**) and vanadyl phthalocyanine (VOPc, **96b**). Notably, the molecular structures of TiOPc and VOPc were no longer coplanar due to the introduction of oxygen atom. However, by controlling the deposition temperatures, Li and coworkers successfully obtained thin films of α-phase TiOPc, in which the molecules of TiOPc adopted a layer-by-layer stacking. In benefitting from the maximum overlap of π molecular orbits, α-phase TiOPc produced a thin-film mobility of up to 10 $cm^2/(V \cdot s)$ [137]. Wang and coworkers employed *para*-sexiphenyl (*p*-6P) to induce the formation of high-quality films of VOPc, on which the OTFTs fabricated showed the highest mobility of >1 $cm^2/(V \cdot s)$ [138]. Spray-coated films of 5,10,15,20-tetra-phenyl porphyrin (**97**) were used as semiconducting layers of OTFTs by Checcoli and coworkers, reaching a maximum hole mobility of 0.012 $cm^2/(V \cdot s)$ [139].

The Langmuir–Blodgett (LB) technique was used to fabricate a semiconducting monolayer of cyclo[8]pyrrole (**98**), a 30-π-electron molecule with very narrow E_g (0.63 eV), on which the OTFTs fabricated exhibited a charge transport mobility of up to 0.68 $cm^2/(V \cdot s)$ [140].

2.2.2 n-Type Small-Molecule Semiconductors

2.2.2.1 Fluorine-Containing Semiconductors

To lower the energy level of the LUMO, which could effectively lower the carrier injection barrier and increase the molecular environmental stability, is a well-accepted strategy in the design of *n*-type semiconductors. Hence, many research groups have introduced electronegative elements (or groups) to molecular skeletons in order to increase electron affinities, with a further expectation of decreasing the mismatch between the LUMO energy level of semiconductors and the Fermi energy level of common noble metals.

Fluorine, the smallest element of the halogen column of the Periodic Table, is a strongly electronegative atom. Indeed, such benefits – including high chemical and thermal stabilities and hydrophobicity – have led to organic fluorine compounds being widely applied to *n*-channel FETs. Perfluoropentacene (**99**) – which is a planar molecule – adopts a herringbone structure in single crystals, and mobility values of up to 0.11 $cm^2/(V \cdot s)$ have been observed in the characterization of perfluoropentacene-based OTFTs [141]. For example, Bao and coworkers have fabricated air-stable OFETs of a series of perfluoro-phthalocyanine derivatives, in which copper hexadecafluorophthalocyanine ($F_{16}CuPc$, **100**)-based OTFTs exhibited the highest electron mobility, up to 0.03 $cm^2/(V \cdot s)$ [142]. Likewise, Tang and coworkers utilized a physical vapor transport technique to fabricate submicrometer- and nanometer-sized single crystals of $F_{16}CuPc$ *in situ*, on which the SCFETs fabricated with an asymmetric drain/source (Au/Ag) showed electron mobilities as high as 0.2 $cm^2/(V \cdot s)$ [143]. In addition, the same group fabricated a range of air/vacuum dielectric OFETs on the basis of single-crystalline submicrometer-ribbons of $F_{16}CuPc$, for which a high mobility of up to 0.35 $cm^2/(V \cdot s)$ and an I_{on}/I_{off} ratio of 10^5 was observed [144]. Various other phenylene–thiophene oligomers and perfluorophenylene–thiophene oligomers were synthesized by Facchetti and colleagues to compare their physical and chemical properties. In this case, 5,5‴-diperfluorophenyl-2,2′:5′,2″:5″,2‴-quaterthiophene (**101**) exhibited an *n*-channel thin-film mobility of up to 0.43 $cm^2/(V \cdot s)$ and an I_{on}/I_{off} ratio of $>10^8$ in an argon-filled environment [145]. The same group also designed and synthesized a series of carbonyl-containing perfluorophenylene–thiophene oligomers [146], a representative of which – 5,5‴-diperfluorophenylcarbonyl-2,2′:5′,2″:5″,2‴-quaterthiophene; DFCO-4T (**102**) – exhibited electron mobilities of up to 0.51 and 0.25 $cm^2/(V \cdot s)$ in thermally vaporized and solution-cast films under vacuum, respectively (Figure 2.13).

By taking dithiazolylbenzothiadiazole as a parent, Akhtaruzzaman and coworkers introduced different electron-withdrawing groups to the long-axis direction of the molecular skeleton to form novel semiconductors, in which the trifluoromethylphenyl derivative (**103**) demonstrated typical *n*-type field-effect characteristics under vacuum. The highest mobility attained by **103**-based OTFTs was 0.068 $cm^2/(V \cdot s)$ [147], and consequently trifluoromethylphenyl groups were retained to accompany different electronegative cores to form high-electron-affinity compounds [148]. Among these derivatives, **104** adopted a π-stacking structure in single crystals, and exhibited a thin-film mobility of 0.30 $cm^2/(V \cdot s)$ in vacuum, whereas the molecules of **105** were arranged in a herringbone structure and exhibited their highest mobility of 0.18 $cm^2/(V \cdot s)$ under vacuum. In subsequent studies, Ando and colleagues synthesized a series of thiazole-containing oligomers that were end-capped with trifluoromethylphenyl groups [149]. In this case, planar

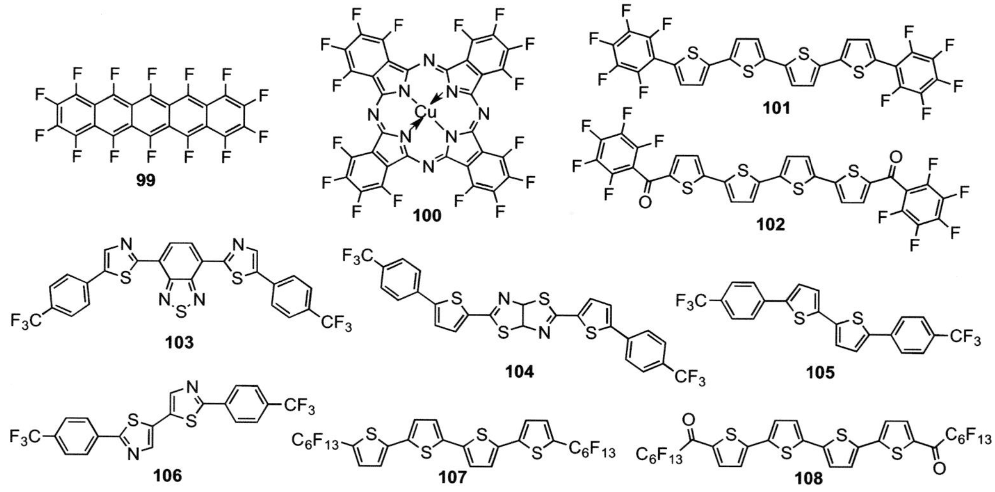

Figure 2.13 Examples of fluorine-containing *n*-type semiconductors.

molecules of **106** formed unique 2-D columnar structure in single crystals, while the shortest intermolecular distance was only 3.37. The OTFTs of **106**, when deposited onto OTS-modified SiO_2 substrates, showed a high electron mobility of up to 1.83 $cm^2/(V \cdot s)$ in vacuum. Perfluoroalkyl chains were also introduced to the skeleton of oligothiophenes to investigate the relationship between chemical structure and properties [43]. The thin-film mobility of DFH-4T (**107**) showed a tendency to increase with an elevation of substrate temperatures, while the maximum electron mobility reached 0.22 $cm^2/(V \cdot s)$ in vacuum when deposited at $T_{sub} = 100\,°C$. Further additions of carbonyl groups into α,ω-diperfluoroalkyl oligothiophenes were carried out as a consequence [150] such that, when deposited onto HMDS-modified SiO_2/Si substrates, the DFHCO-4T (**108**)-based OTFTs exhibited a maximum mobility of 0.6 $cm^2/(V \cdot s)$, under vacuum.

2.2.2.2 **Cyano-Containing Semiconductors**

Cyano is recognized as a strongly electron-withdrawing group, and during a variety of experiments the molecular electron cloud was always stabilized by the introduction of cyano groups, resulting in a much lower LUMO energy level. Tetracyanoquinodimethane (TCNQ, **109**), which had a LUMO energy level as low as −4.8 eV [151], was one of the first compounds to be used as an *n*-type organic semiconductor although, as an electron acceptor it could also form charge-transfer complexes with metal ions and electron-rich molecules. Yamagishi and coworkers fabricated single-crystal transistors based on TCNQ, and observed an *n*-channel field-effect mobility of up to 0.5 $cm^2/(V \cdot s)$ with a very low threshold voltage [152]. 3′,4′-Dibutyl-5,5″-bis(dicyanomethylene)-5,5″-dihydro-2,2′:5′,2″-terthiophene (DCMT, **110**) showed a smaller HOMO–LUMO energy gap (1.85 eV) than terthiophene (3.10 eV),

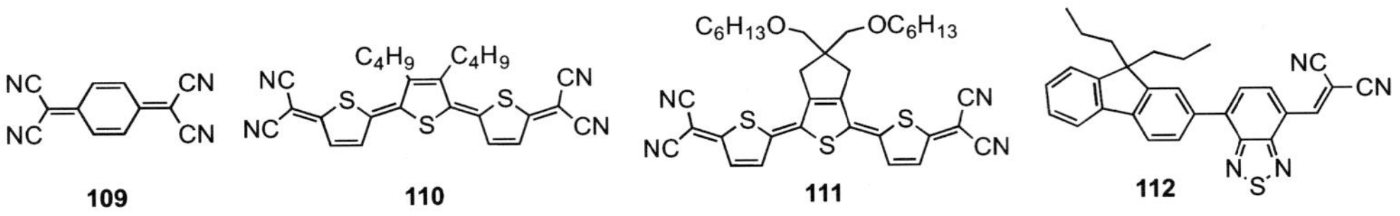

Figure 2.14 Examples of cyano-containing semiconductors.

though the electron affinity of DCMT was much greater than that of terthiophene [153]. An improved mobility of up to 0.2 $cm^2/(V \cdot s)$ was observed in the characterization of DCMT-based OTFTs; notably, compound **111**, as an analog of **110**, exhibited an electron mobility of up to 0.16 $cm^2/(V \cdot s)$ in annealed thin films [154]. Recently, Gui and coworkers synthesized an asymmetric molecule, 2-[(7-(9,9-di-*n*-propyl-9*H*-fluoren-2-yl)benzo[*c*][1,2,5]thiadiazol-4-yl)methylene]malononitrile (K12; **112**), as a candidate for *n*-type semiconductors [155]. Both, spin-coated and thermally evaporated films of K12 were prepared for device fabrication, and an electron mobility of 10^{-3} $cm^2/(V \cdot s)$ was achieved in thin films fabricated using either techniques (Figure 2.14).

2.2.2.3 **Carbonyl and Imide Semiconductors**

1,4,5,8-Naphthalene tetracarboxylic dianhydride (NTCDA, **113**) was utilized as a semiconductor of OFETs by Laquindanum in 1996, with the highest electron mobility reaching 3×10^{-3} $cm^2/(V \cdot s)$ under vacuum [156]. Although the number of electron-withdrawing group of anhydrides is equal to that of imides, the imide-based system shows great potential as the nitrogen positions of imides may be substituted by a variety of groups, without significantly affecting the molecular electronic structure. This would afford novel *n*-type semiconductors with different physical and/or chemical properties.

Naphthalene diimide (NDI) and perylene diimide (PDI) derivatives are two attractive imide-based semiconducting systems that have been studied extensively during recent decades [157, 158]. At this point, some representative derivatives of NDI and PDI are described to highlight the benefits of these materials. For example, Shukla and coworkers synthesized *N,N′*-bis(cyclohexyl)naphthalene-1,4,5,8-bis(dicarboximide) (**114a**), a NDI-derivative that was end-capped by cyclohexyl groups, in which the cyclohexyl groups on the *N*-positions effectively improved the solid-state packing [159]. Thin films of **114a** were vacuum-deposited onto OTS-modified SiO_2/Si substrates, followed by the thermal evaporation of gold as source/drain electrodes; in this case, the maximum electron mobility, when monitored in a constant argon atmosphere, reached 6.2 $cm^2/(V \cdot s)$. For comparison, the *n*-hexyl-substituted NDI derivative (**114b**) was also synthesized and used as semiconducting material of OTFTs; this exhibited a thin-film electron mobility of 0.7 $cm^2/(V \cdot s)$. Subsequently, Katz and coworkers designed and synthesized a series of NDI derivatives that were end-capped with various alkyl and phenyl groups. Among these, **114c** exhibited a thin-film mobility [μ_{max}: 0.16 $cm^2/$

(V·s)] only under vacuum [157], whereas **114d**, which was stabilized by the introduction of perfluorophenyl exhibited mobility up to 0.23 $cm^2/(V \cdot s)$ in air [160], and **114e** exhibited the maximum electron mobility of 0.57 $cm^2/(V \cdot s)$ with an I_{on}/I_{off} ratio as high as 10^8 in air [161].

Similar to NDI-based semiconductors, various alkyl chains and partially fluorinated alkyl chains were introduced to the nitrogen-positions of PDI. The conjugation planar and intermolecular interactions of PDI were greater than those of NDI, and consequently the general performance of the PDI-derivatives was considered promising. When tested in vacuum, the OTFTs based on PTCDI-C_5 (**115a**), PTCDI-C_8 (**115b**), PTCDI-C_{12} (**115c**), and PTCDI-C_{13} (**115d**) showed electron mobilities of up to 0.1 $cm^2/(V \cdot s)$ [162], 1.7 $cm^2/(V \cdot s)$, 0.52 $cm^2/(V \cdot s)$ [163], and 2.1 $cm^2/(V \cdot s)$ [164], respectively. Especially noteworthy was the substitution of partially fluorinated alkyl chains on *N*-positions, which led to **115e** and **115f** exhibiting air-stable electron mobilities of up to 1.24 and 0.37 $cm^2/(V \cdot s)$, respectively [165].

Subsequent investigations confirmed that the introduction of electron-withdrawing groups (e.g., cyano, halogen) on the NDI- or PDI-core was crucial for fluorine-free NDIs and PDIs that exhibited air-stable *n*-channel field-effect characteristics. Among the core-cyanated NDI- and PDI-derivatives, NDI-8CN2 (**116**) exhibited comparable thin-film electron mobilities in air [0.11 $cm^2/(V \cdot s)$] and under vacuum [0.15 $cm^2/(V \cdot s)$] [166], while PDI-8CN2 (**117a**) and PDI-CN2 (**117b**) showed air-stable electron mobilities of up to 0.14 $cm^2/(V \cdot s)$ [167] and 0.10 $cm^2/(V \cdot s)$ [158], respectively. Assisted by the electron-withdrawing of partially fluorinated alkyl chains, PDI-FCN2 (**117c**) showed an even lower LUMO energy level and a high mobility of 0.64 $cm^2/(V \cdot s)$ under ambient atmosphere [158]. The compounds **118a** and **118b** were core-halogenated PDI-derivatives, and exhibited air-stable *n*-channel field-effect mobilities of up to 0.61 $cm^2/(V \cdot s)$ [165] and 0.11 $cm^2/(V \cdot s)$ [168] after comprehensive optimization (e.g., substrate temperature, dielectric layer) of the OTFTs (Figure 2.15).

Besides electron-withdrawing groups, the electron-rich group tetrathiafulvalene (or a moiety of tetrathiafulvalene) was introduced to the imide system by Gao and coworkers. Surprisingly, the LUMO energy levels of core-expanded naphthalene diimide derivatives **119a** and **119b** (both were −4.3 eV) were much lower than that of NDI (−3.4 eV) [169]. The low-lying LUMO energy level was a critical factor for obtaining ambient-stable *n*-channel semiconductors. The OTFTs based on highly ordered solution-processed thin films of **119a** and **119b** exhibited air-stable electron mobilities of up to 0.20 $cm^2/(V \cdot s)$ [169] and 1.20 $cm^2/(V \cdot s)$ [170], respectively.

The semiconductors **120–122** were linear-diimide compounds, in which **120a–c** exhibited respective electron mobilities of up to 0.039, 0.054, and 0.013 $cm^2/(V \cdot s)$, respectively, in air; these mobility values were close to those obtained in vacuum [171]. The anthracenedicarboximides **121–122** demonstrated thin-film *n*-channel field-effect mobilities of 0.02, 0.01, 0.01, and 0.03 $cm^2/(V \cdot s)$ in vacuum, respectively, in which the core-cyanated anthracenedicarboximide **122** exhibited an air-stable electron mobility of up to 0.02 $cm^2/(V \cdot s)$ with an I_{on}/I_{off} ratio of 10^7 [172].

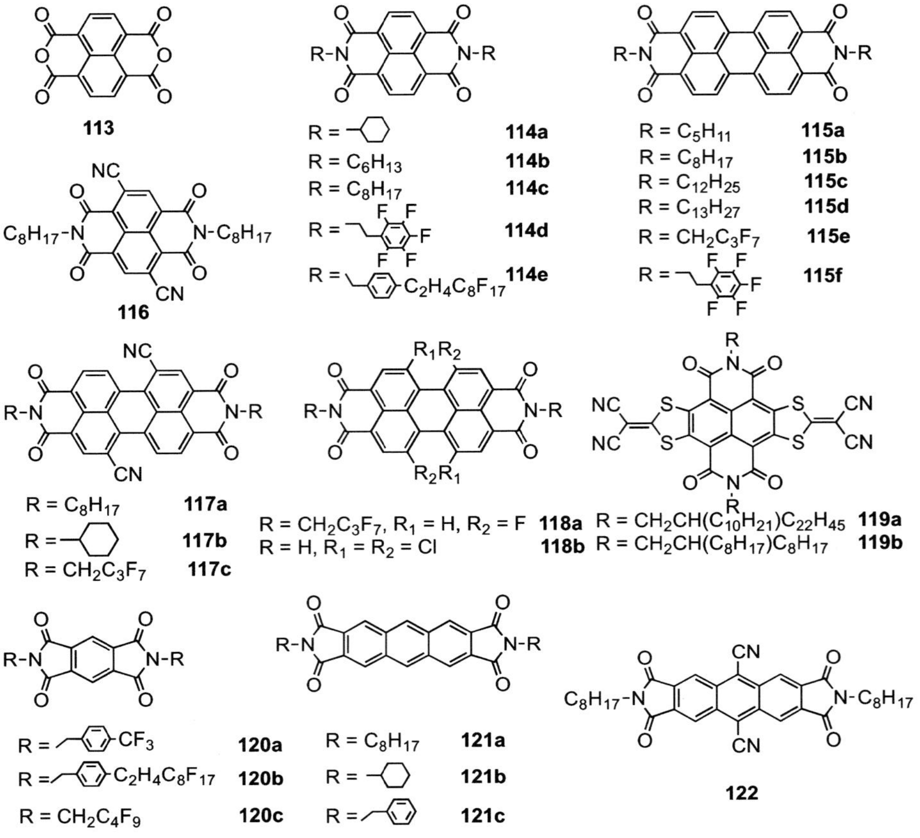

Figure 2.15 Examples of anhydrides and imides (as noted in the text).

2.2.2.4 Fullerenes

Fullerene refers to a type of molecule that is composed entirely of carbon, and which adopts a hollow sphere, ellipsoid, or tube as its existing forms. Buckminsterfullerene (C_{60}, **123**), a representative of spherical fullerene, has been the most popular semiconductor or synthesized parent of fullerene-based semiconducting materials. C_{60} was first used as an *n*-channel semiconductor of OTFTs by Kastner and coworkers in 1993 although, due to the poor air-stability of C_{60} and derivatives, the device performance of C_{60}-based OFETs could be observed only under vacuum. In addition, the solubility of C_{60} in common solvents was too low to process by using solution techniques, while the spherical molecular structure of C_{60} was not good for the formation of continuous ordered thin films. Consequently, various modifications (e.g., dielectric modifications, organic inducing-layers, variety of electrodes) were carried out to obtain high-quality thin films of C_{60}. For example, when Itaka and coworkers utilized an atomically flat monolayer of pentacene-

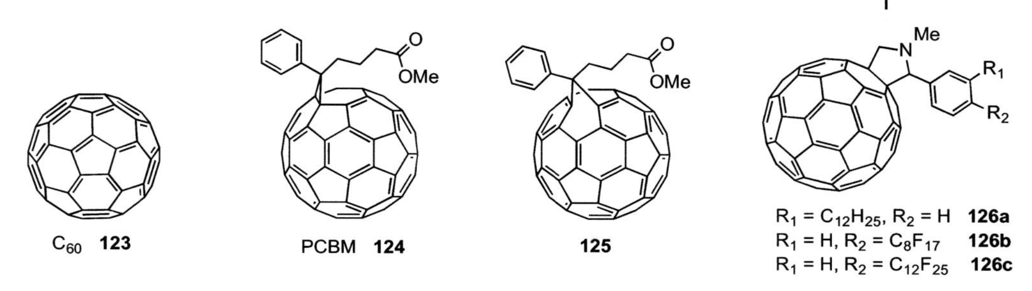

Figure 2.16 Chemical structures of C_{60} and some representatives of C_{60}-derivatives.

induced continuous thin films of C_{60} with hexagonal grains on a large scale, the crystallinity enhancement improved the maximum mobility of C_{60}-based OTFTs to 4.9 $cm^2/(V \cdot s)$ [173]. In later studies conducted by Anthopoulos and colleagues, low work function source/drain electrodes (LiF/Al) were deposited onto the C_{60}-thin films that had been hot-wall epitaxy-grown on divinyl-tetramethyldisiloxane-bis(benzocyclobutene) (BCB)-modified quartz substrates, such that the highest electron mobility reached 6 $cm^2/(V \cdot s)$ [174] (Figure 2.16).

In order to improve the solubility and solution-processability of C_{60} and its derivatives, various alkyl and ester groups were bonded to C_{60}. For example, Singh and coworkers fabricated [6,6]-phenyl-C_{61}-butyric acid methyl ester (PCBM, **124**)-based OTFTs with different dielectrics, and the device performance showed a strong relativity with the category of dielectrics. By using crosslinked BCB and LiF/Al as the dielectric material and source/drain electrodes, respectively, the PCBM-based OTFTs exhibited an optimized electron mobility of up to 0.2 $cm^2/(V \cdot s)$ [175]. Later, Lee and coworkers fabricated **125**-based OTFTs with different types of metal as the source/drain electrodes; devices in which Ca was used in this role showed the highest mobility, at 0.1 $cm^2/(V \cdot s)$ [176]. Chikamatsu and coworkers synthesized a long-chain alkyl-substituted C_{60} ($C_{60}MC_{12}$, **126a**), which subsequently was applied to an OTFT as a highly ordered film. Aided by the self-assembling property of long alkyl chains, the solution-processed $C_{60}MC_{12}$-based OTFTs showed an electron mobility of 0.067 $cm^2/(V \cdot s)$ [177]. Likewise, a series of perfluoroalkyl-substituted C_{60} derivatives was synthesized that were solution-processable and air-stable *n*-type semiconductors. Taking **126b** and **126c** as examples, these solution-processed OTFTs exhibited air-stable electron mobilities of up to 8×10^{-3} and 0.078 $cm^2/(V \cdot s)$, respectively [178].

2.3 Polymer Semiconductors

Polymers became an important category of organic semiconductors following the first reports of macromolecule-based FETs in 1986 [1]. Although the polymers used during the early stages of OFET development focused on polythiophenes that had been polymerized electrochemically, the in-device performance of the poor-quality films obtained in this was far from ideal. Nonetheless, with excellent progress being made in device fabrication, attention is today increasingly being paid to the design of polymer semiconductors, due mainly to their great potential in solution-processed, large-scale, and flexible electronic applications. Unlike small molecules, the molecular weight of a polymer is not a constant value, but is rather a scale. In addition, the chemical structure of a one-pot-synthesized polymer will inevitably be disordered, such that common purification techniques (e.g., sublimation and recrystallization) will not be applicable to these materials. Yet, by ensuring the precise design of the bonding modes of monomers, increasingly regioregular polymers will continue to be produced. The details of some such polymer semiconductors are described in the following subsections.

2.3.1 p-Type Polymer Semiconductors

2.3.1.1 Polythiophenes

Polythiophene (**127**) was the first polymer to be used as a semiconducting material in OFETs. The thin films of polythiophene used in this case were formed by electrochemical polymerization, exhibiting a hole mobility that was on the order of $10^{-5}\,cm^2/(V \cdot s)$ [1]. However, by introducing alkyl chains to the β-positions of thiophenes, the solubility of polythiophene could be greatly improved. Moreover, as there were two β-positions on each thiophene ring available for substitution, three types of regioregularity – head-to-tail (HT), head-to-head (HH), and tail-to-tail (TT) – of the alkylthiophene-dimers in the main chain could be classified according to the relative position of the alkyl side chains on adjacent thiophenes [179]. Thus, side chain-containing polythiophenes could be classified as either regiorandom polythiophenes (which consist of at least two types of regioregularity, as noted above) and regioregular polythiophenes (which consists of only HT, or only HH/TT).

Poly(3-hexylthiophene) (P3HT, **128**), a famously well-known semiconducting material, was originally used extensively in the field of organic solar cells. Assadi and coworkers fabricated OTFTs on the basis of thin films of regiorandom P3HT (**128a**), for which the maximum hole mobility reached only $10^{-4}\,cm^2/(V \cdot s)$ [180]. Subsequently, Sirringhaus and coworkers fabricated thin films of regioregular P3HT (**128b**) with a lamella structure, in which the orientations of self-organized lamella (either parallel or normal to the substrate) could be controlled by the processing conditions employed [181]. Mobility anisotropy was observed in the domains of different packing modes, with the highest mobility of $0.1\,cm^2/(V \cdot s)$ being obtained in the parallel orientation, and indicating that the interchain transport between adjacent molecules was efficient [182]. Further device optimization with a high-k relaxor ferroelectric polymer [P(VDF-TrFE-CFE)] as insulator improved the mobility of P3HT-OFETs to $0.5\,cm^2/(V \cdot s)$ [183] (Figure 2.17).

McCulloch and coworkers designed two polythiophenes (**129a** and **129b**) that were copolymerized by bis(3-octylthiophene) and thiophene, in which the neigh-

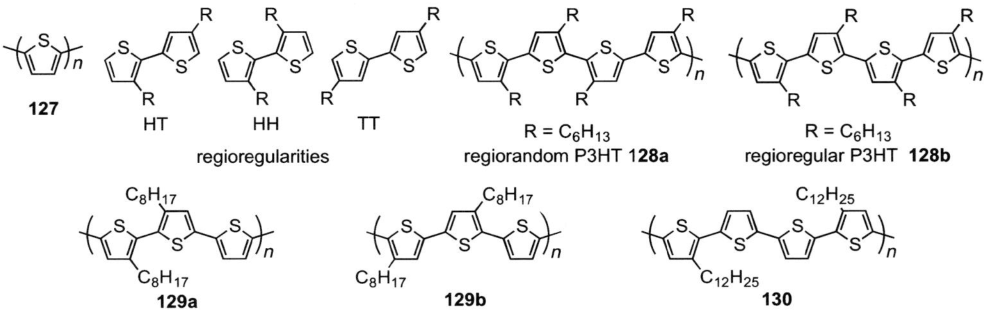

127 HT HH TT regioregularities
R = C_6H_{13} regiorandom P3HT **128a**
R = C_6H_{13} regioregular P3HT **128b**
C_8H_{17} C_8H_{17} **129a**
C_8H_{17} C_8H_{17} **129b**
$C_{12}H_{25}$ $C_{12}H_{25}$ **130**

Figure 2.17 Regioregularities of thiophene-dimers and chemical structures of some polythiophenes.

boring 3-octylthiophenes adopted HH regioregularity in **129a**, but TT regioregularity in **129b**. A great steric repulsion between the HH octyl groups caused the backbone of polymer **129a** to become twisted, thus forcing decreases in intermolecular π–π overlap and ionization potential (5.0 eV) compared to **129b**. This lack of spatial ordering in **129a** led to a reduction in the efficiency of carrier transport, exhibiting a hole mobility as low as $1 \times 10^{-5}\,cm^2/(V \cdot s)$ [184]. OTFTs based on a TT regioregular **129b** (a conjugated polymer with an improved solid-state aggregation and ionization potential of 5.6 eV) showed a carrier transport mobility of up to $0.03\,cm^2/(V \cdot s)$. Later, Ong and coworkers synthesized a regioregular PQT-12 (**130**) with a larger interchain distance, and values of up to $0.14\,cm^2/(V \cdot s)$ were observed in the electrical characterization of PQT-12-based OTFTs [185]. On doping by atmospheric oxygen, the PQT-12-based OTFTs demonstrated device stability after storage in the dark for a one-month period.

2.3.1.2 Thiophene–Heteroacene Copolymers

In order to improve the environmental stability of polythiophenes, notably to increase the molecular ionization potentials, various research groups adopted one of two strategies:

- To increase the rotational freedom of the molecular backbones by decreasing the proportion of substituted alkyl chains in regioregular polymers; this would result in a decreased conjugation of the whole molecules (as in the examples shown in Section 2.3.1.1).
- To introduce heteroacenes to the main chains of regioregular polymers, thus decreasing the intramolecular electron density by localization or electron-withdrawing effects of heteroacenes.

As an example, McCulloch and coworkers designed poly(2,5-bis(3-alkylthiophen-2-yl)thieno[3,2-*b*]thiophenes (**131**) which were copolymerized by 3-alkylthiophenes and thieno[3,2-*b*]thiophene, as semiconducting materials for OFETs. The materials

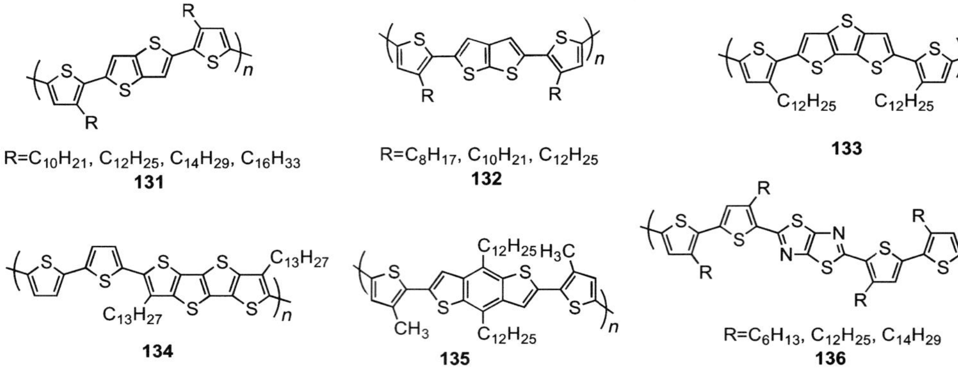

R=$C_{10}H_{21}$, $C_{12}H_{25}$, $C_{14}H_{29}$, $C_{16}H_{33}$
131

R=C_8H_{17}, $C_{10}H_{21}$, $C_{12}H_{25}$
132

133

134

135

R=C_6H_{13}, $C_{12}H_{25}$, $C_{14}H_{29}$
136

Figure 2.18 Chemical structures of some thiophene–heteroacene copolymers.

produced benefited from the localization effect and rotational invariance of thieno[3,2-*b*]thiophene, while highly ordered crystalline domains of thin films were formed from a liquid-crystal phase of **131**, which possessed a lower HOMO energy level than P3HT. A mobility of up to 0.72 $cm^2/(V \cdot s)$ was obtained for the **131**-based OTFTs when tetradecyl groups were used as the side chains of polythiophene [186]. Subsequently, although analogs (**132**) with thieno[2,3-*b*]thiophene as the comonomer were synthesized for comparison, the highest mobility reached only 0.15 $cm^2/(V \cdot s)$, this being achieved by the decyl-substituted polymer [187] (Figure 2.18).

Polythiophene derivatives incorporating dithienothiophene (**133**) [188], tetrathienoacene (**134**) [189], and benzo[1,2-*b*:4,5-*b*′]dithiophene (**135**) [190] were also synthesized and applied as *p*-type semiconductors of OTFTs; these materials exhibited carrier transport mobilities of up to 0.3, 0.33, and 0.15 $cm^2/(V \cdot s)$, respectively. In particular, the thiophene–thiazolothiazole copolymers (**136**), in which the electron-withdrawing group thiazolothiazole was widely used in the backbones of *n*-type small-molecular-weight semiconductors, exhibited respective hole mobilities of 0.05, 0.23, and 0.3 $cm^2/(V \cdot s)$ in the device measurements of hexyl-, dodecyl-, and tetradecyl-substituted copolymers [191, 192].

2.3.1.3 **Other Copolymers**

Fluorene units were introduced into the main chains of some novel copolymers, due to their good stability and high solubility. Poly-9,9′-dioctyl-fluorene-*co*-bithiophene (F_8T_2, **137a**) demonstrated a better oxidative stability than P3HT. With assistance from an alignment layer, high-quality thin films of F_8T_2 with the main chain parallel to the transport direction were solution-processed from a nematic glassy state, on which the OTFTs exhibited the maximum mobility at 0.02 $cm^2/(V \cdot s)$ [193]. Poly(9,9′-*n*-dioctylfluorene-*alt*-biselenophene) (F_8Se_2, **137b**; a selenium analog of F_8T_2) exhibited a hole mobility of 0.012 $cm^2/(V \cdot s)$ with a low threshold voltage of −4 V in solution-processed OTFTs [194]. Later, Chen and coworkers

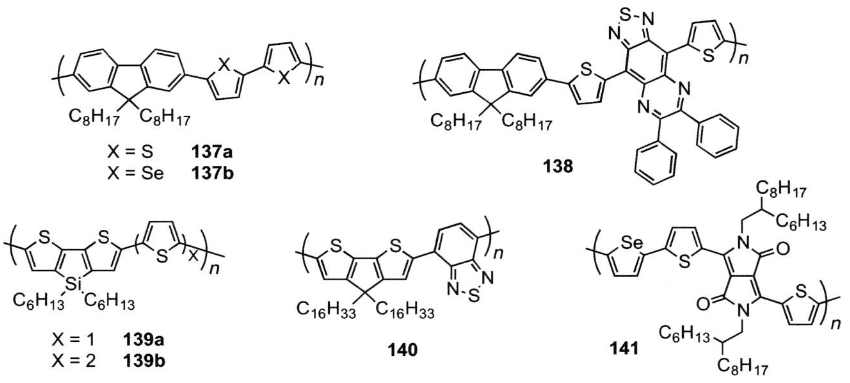

Figure 2.19 Examples of fluorene-containing copolymers and *N*-heterocyclic-thiophene copolymers.

synthesized APFO-Green1 (**138**) that was copolymerized by thiophene, fluorine and thiadiazolo-quinoxaline units as a semiconducting material for OFETs. Solution-processed OTFTs based on air-stable APFO-Green1 showed a field-effect mobility of 0.03 $cm^2/(V \cdot s)$ [195] (Figure 2.19).

Usta and coworkers designed and synthesized the dithienosilole–thiophene copolymers TS6T1 (**139a**) and TS6T2 (**139b**), which exhibited hole mobilities of up to 0.02 and 0.06 $cm^2/(V \cdot s)$, respectively, under ambient conditions [196]. Later, Zhang and coworkers designed and synthesized the benzothiadiazole–cyclopentadithiophene copolymer CDT-BTZ (**140**), which exhibited a hole mobility of 0.17 $cm^2/(V \cdot s)$ in drop-cast thin films [197]. Further device optimization, achieved by improving the macroscopic organization of dip-coated thin films, showed hole mobilities in the range of 1.0 to 1.4 $cm^2/(V \cdot s)$ along the dip-coating direction [198]. Recently, OFETs were fabricated on the basis of highly ordered single fibers of CDT-BTZ, with very high mobilities of up to 5.5 $cm^2/(V \cdot s)$ being observed [199]. When Lin and coworkers synthesized the donor–acceptor-conjugated copolymer PSeDPP (**141**), the optimization of thermal treatment with a high-boiling-point solvent facilitated the formation of ordered thin films with lamellar chain packing, leading to a high hole mobility of up to 1.62 $cm^2/(V \cdot s)$ [200].

2.3.2
n-Type Polymer Semiconductors

Although *n*-type polymer semiconductors have been rarely reported, the electron-withdrawing groups that have been used most widely in *n*-type small-molecular-weight semiconductors have also been incorporated into the main chains of *n*-type polymer semiconductors. As an example, Letizia and coworkers synthesized the phenacyl-quaterthiophene copolymer **142**, with which the OFETs fabricated exhibited an electron mobility of up to 0.01 $cm^2/(V \cdot s)$ [146].

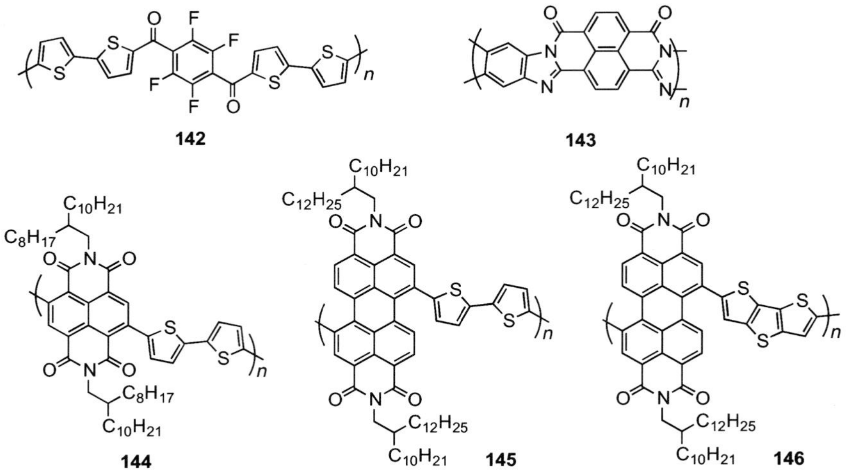

Figure 2.20 Chemical structures of some representative *n*-type polymers.

The ladder polymer poly(benzobisimidazobenzophenanthroline) (BBL, 143) [201], as synthesized by Babel and coworkers, showed a high glass-transition temperature of >500 °C. A high electron mobility of 0.1 $cm^2/(V \cdot s)$ was observed in the electrical measurements of BBL-based OFETs [202]. Chen and coworkers designed and synthesized naphthalenedicarboximide-thiophene (P[NDI2OD-T2], **144**) and perylenedicarboximide-thiophene (P[PDI2OD-T2], **145**) copolymers, with which bottom-gate *n*-channel OFETs were fabricated. Electron mobilities of up to 0.06 and 0.002 $cm^2/(V \cdot s)$, respectively, were observed during the measurement of P(NDI2OD-T2)- and P(PDI2OD-T2)-based OFETs under vacuum [203]. In benefitting from the low LUMO energy levels of P(NDI2OD-T2) (−3.91 eV) and P(PDI2OD-T2) (−3.96 eV), the electron mobilities of P(NDI2OD-T2)- and P(PDI2OD-T2)-based OFETs tested under ambient conditions after storage for 14 weeks and one week were 0.01 and 2×10^{-4} $cm^2/(V \cdot s)$, respectively. Following the optimization of P(NDI2OD-T2)-based OFETs with polymeric dielectrics, the electron mobility was improved to 0.85 $cm^2/(V \cdot s)$ [204]. Zhan and coworkers fabricated OFETs on the basis of a perylenedicarboximide–dithienothiophene copolymer **146**; as a consequence, an electron mobility of 0.013 $cm^2/(V \cdot s)$ was exhibited, with aluminum as the source/drain electrodes under a nitrogen atmosphere [205] (Figure 2.20).

2.4
Normal Synthetic Methods for Organic Semiconductors

As most organic semiconductors possess symmetrical molecular structures, the general synthetic techniques employed in this field will not involve complicated chiral syntheses. Yet, besides the common nucleophilic substitution reactions conducted in the presence of butyl lithium (*n*-BuLi), *tert*-butyl lithium (*t*-BuLi), and/or lithium diisopropylamide (LDA), several named reactions have been used extensively for the synthesis of organic semiconductors, as follows.

2.4.1
Diels–Alder Cycloaddition

The [4π+2π] cyclization between a conjugated diene and an (substituted) alkene to form a cyclohexene derivative is known as the Diels–Alder cycloaddition. Besides alkenes, alkynes, benzynes, and allenes are also good candidates for dienophiles.

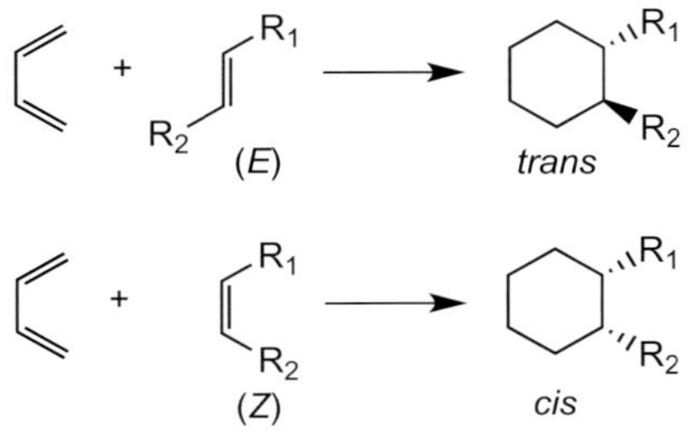

R_1 R_2 (*E*) → *trans*

R_1 R_2 (*Z*) → *cis*

This reaction is widely used in the synthesis of acenes and heteroacenes, especially in the synthetic route of rubrene [206]:

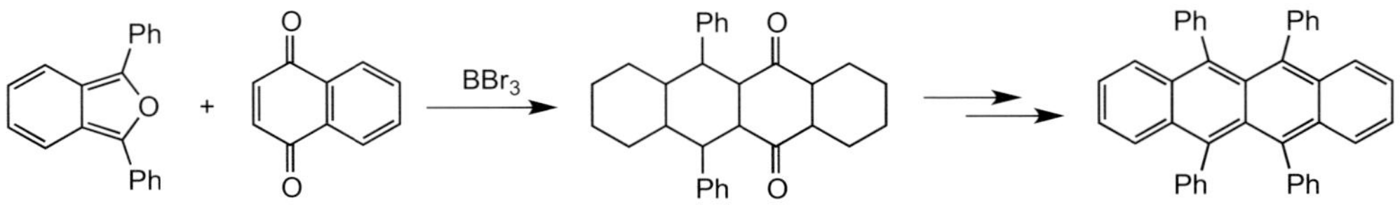

Ph Ph O O BBr_3 Ph O Ph O Ph Ph Ph Ph

2.4.2
Aldol Reaction

The aldol reaction refers to the addition of enol/enolate of carbonyl compounds onto aldehyde or ketone, followed by dehydration to generate the corresponding α,β-unsaturated carbonyl compounds.

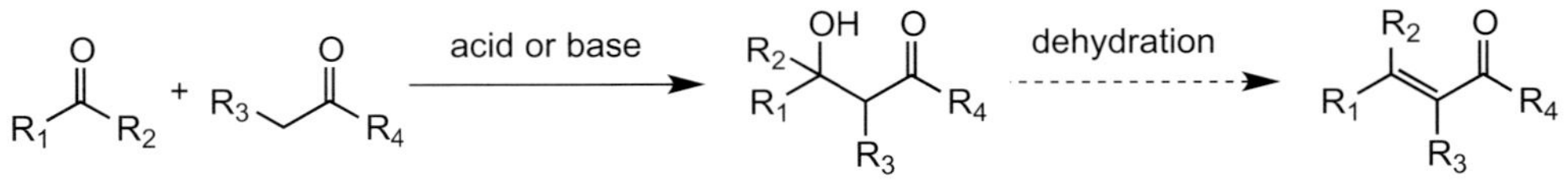

R_1 R_2 O + R_3 R_4 O —acid or base→ R_2 R_1 OH O R_4 R_3 --dehydration→ R_2 O R_1 R_4 R_3

The reaction used in Ref. [207] is an example of the aldol reaction:

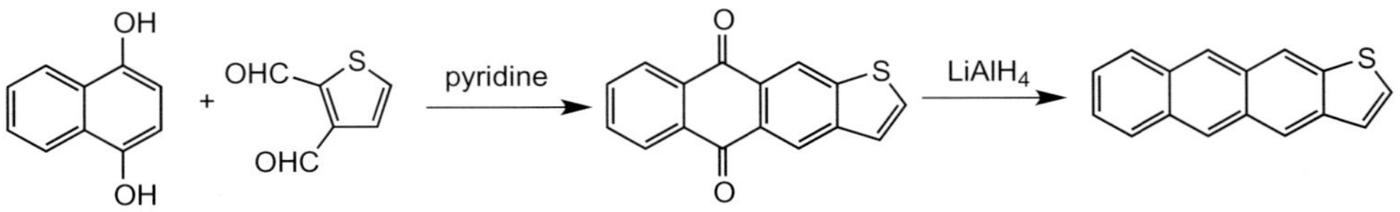

OH OH + OHC OHC S —pyridine→ O O S —$LiAlH_4$→ S

2.4.3 Stille Reaction

The Stille reaction (namely Stille coupling) refers to the Pd-catalyzed coupling reaction of an organic halide with an organic stannyl compound, in which X is Cl, Br, I, or a pseudohalide (triflate, $CF_3SO_3^-$). Although both trimethylstannyl and tributylstannyl compounds are often used, the latter are preferred as the toxicity of about 1000-fold greater than their tributylstannyl counterparts.

$$R{-}X + R'{-}Sn(R)_3 \xrightarrow{\text{Pd catalyst}} R{-}R' + X{-}Sn(R)_3$$

This reaction must be conducted under anhydrous, inert circumstances. Due to the high yield of this reaction, Stille coupling is widely used in polymerization to generate a larger molecular weight. The reaction used in Ref. [203] is an example of the Stille reaction:

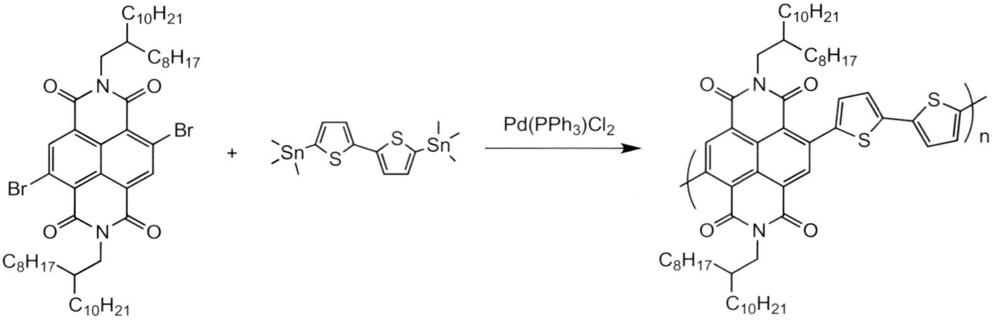

$C_{10}H_{21}$ C_8H_{17} O N O Br Br O N O C_8H_{17} $C_{10}H_{21}$ + Sn S S Sn Pd(PPh$_3$)Cl$_2$ → $C_{10}H_{21}$ C_8H_{17} O N O S S n O N O C_8H_{17} $C_{10}H_{21}$

2.4.4 Suzuki Reaction

The Suzuki reaction, as first described by A. Suzuki and N. Miyaura in 1979, refers to the coupling of an aryl- or vinyl-boronic acid with an organic halide, catalyzed by a palladium(0) complex. Instead of halides, pseudohalides such as triflates (OTf) are also effective in the processing of the Suzuki reaction. The reaction formula of the Suzuki reaction is as follows, where X is Cl, Br, I, or OTf; Y is alkyl, OH, or O-alkyl. The order of reactivity is -I > -OTf > -Br >> -Cl.

$$R{-}X + R'{-}BY_2 \xrightarrow[\text{base}]{\text{Pd catalyst}} R{-}R' + X{-}BY_2$$

The reaction used in Ref. [13] is a representative example of the Suzuki reaction:

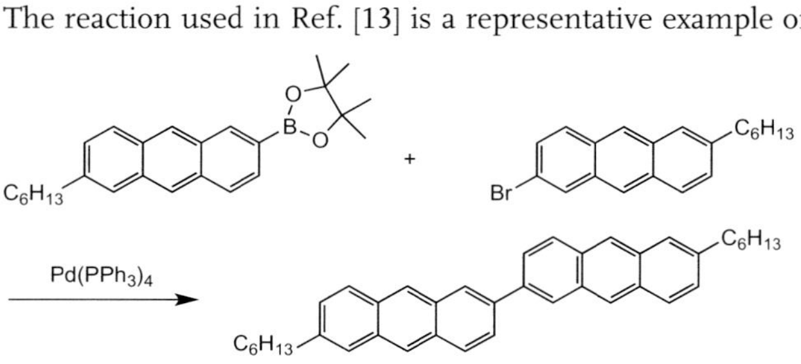

C_6H_{13} O B O + Br C_6H_{13} Pd(PPh$_3$)$_4$ → C_6H_{13} C_6H_{13}

This coupling reaction is generally stereoselective and also regioselective. No rigorous conditions are required in the processing of the Suzuki reaction; moreover, many boronic acids are commercially available and environmentally safe.

2.4.5 Sonogashira Crosscoupling

The Sonogashira reaction refers to the coupling of aryl or vinyl halides with terminal alkynes in the presence of $Pd(PPh_3)_2Cl_2$ and CuI. The reagents and solvents do not need to be rigorously dried, and the reaction can be conducted under mild conditions.

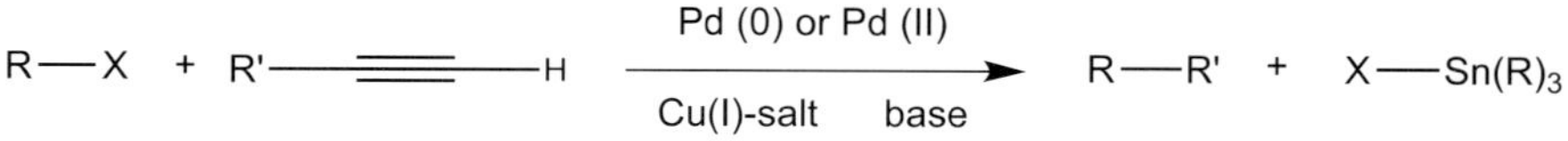

$$\mathrm{R{-}X} + \mathrm{R'{-}C{\equiv}C{-}H} \xrightarrow[\text{Cu(I)-salt} \quad \text{base}]{\text{Pd (0) or Pd (II)}} \mathrm{R{-}R'} + \mathrm{X{-}Sn(R)_3}$$

The reaction used in Ref. [107] is a representative example of the Sonogashira reaction:

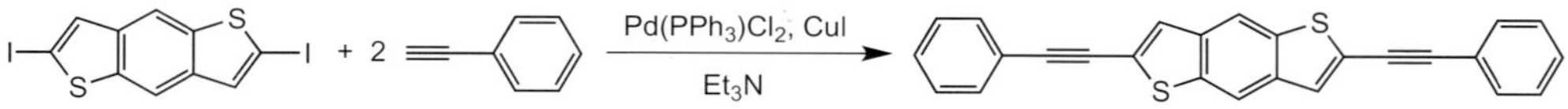

2.4.6 Ullmann Reaction

The copper-catalyzed condensation of two aryl halides is known as the Ullmann reaction (or Ullmann coupling). The reaction can take place either intramolecularly or intermolecularly. The general reactions are as follows, where X is Cl, Br, I, or SCN.

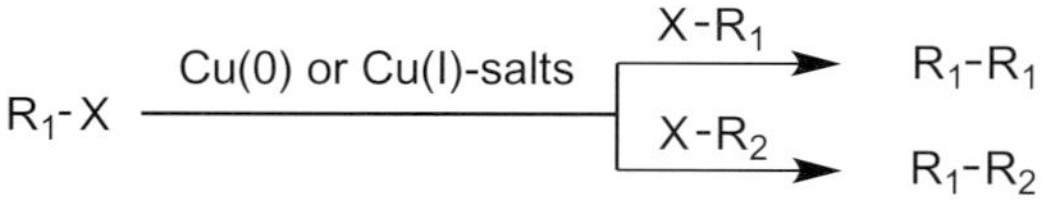

$$\mathrm{R_1\text{-}X} \xrightarrow{\text{Cu(0) or Cu(I)-salts}} \begin{cases} \xrightarrow{\mathrm{X\text{-}R_1}} \mathrm{R_1\text{-}R_1} \\ \xrightarrow{\mathrm{X\text{-}R_2}} \mathrm{R_1\text{-}R_2} \end{cases}$$

The coupling reaction used for the synthesis of 2,2″-bidithieno[2,3-*b*:3′,2′-*d*]thiophene [98] forms part of the Ullmann reaction:

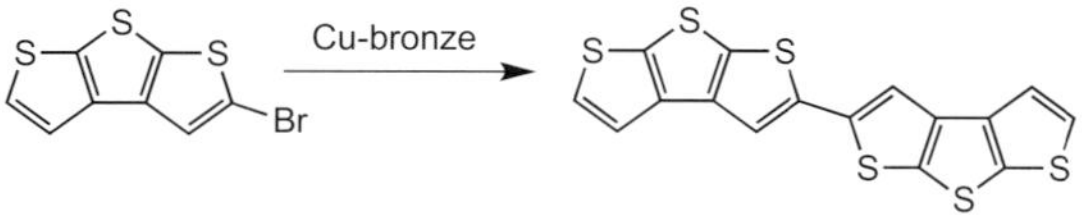

2.4.7 Heck Reaction

The Heck reaction refers to the palladium-catalyzed arylation and alkenylation of olefinic compounds, realized by the coupling of aryl, benzyl or styryl halides with olefinic compounds. The reaction formula is as follows, where X is I, Br, OTf, or Cl.

$$R_1{-}X + (H)(R_2)C{=}C(R_4)(R_3) \xrightarrow{\text{Pd (0) catalyst}} (R_1)(R_2)C{=}C(R_4)(R_3)$$

The migratory insertion of the palladium complex into olefin, as well as the β-hydride elimination, is performed with *syn* stereochemistry; thus, the Heck reaction is stereospecific.

The synthesis towards 3,7-distyryldibenzo[*b,d*]thiophene [208] is an example of the Heck reaction:

Br–(dibenzothiophene)–Br + styrene $\xrightarrow{Pd(OAc)_2}$ 3,7-distyryldibenzo[*b,d*]thiophene

2.5 Purification of Organic Semiconductors

The organic semiconductors obtained from multistep syntheses always require to be further purified, using a variety of techniques:

1) **Recrystallization:** Two types of recrystallization are widely used. In "single-solvent recrystallization," the solubility difference between the target compound and impurities in typical solvents is utilized to gradually eliminate any impurities. This procedure can be carried out using hot filtration or solvent evaporation at room temperature. A sharp decrease in solvent temperature during hot filtration will always result in a rapid precipitation of solute, in which the order of molecular packing is relatively poorer than that after a slow solvent evaporation. In contrast, in "multi-solvent recrystallization," precipitation occurs during the slow diffusion of a poor solvent to a saturated solution of a target compound in a good solvent. Those compounds which showed a minimal association between temperature and solubility are applicable to this technique.

2) **Column chromatography:** This provides an efficient means of purifying an individual compound from a complicated mixture. The stationary phase (adsorbent) and the mobile phase (eluent) are critical components of column chromatography. According to the relative polarities of the target compound and the byproducts, different eluents (single or mixed solvent) and stationary phases (silica gel, alumina, etc.) are chosen to cooperate in the separation of mixtures. With the advancing eluent, individual components that are retained to different degrees by the adsorbent will elute at different rate, and thus be separated from each other.

3) **Physical vapor deposition** (PVD)**:** This technique makes use of the different sublimation points of individual compounds, such that they are separated from one another successively. In most cases, a vacuum or an inert carrier

gas is used to facilitate the procedure, as well as to protect the deposition from atmospheric impurities.

4) **Soxhlet extraction:** This procedure involves the repeated dissolution of a particular compound from a mixture, using a small amount of warmed solvent; the solution produced is then extracted through a siphon. The Soxhlet technique is applicable to compounds with a very low solubility in common solvents, or to compounds that cannot be purified using the above-described methods (especially polymers).

2.6 Outlook

Unlike the situation for fully developed inorganic FETs, many fundamental questions relating to OFETs have yet to be answered. Clearly, a much greater effort will be required regarding the design of novel, high-performance *n*-type and polymeric semiconductors, as well as to understanding the relationship between chemical structure and properties. While many challenges also exist with regards to a precise explanation of carrier transport in the field of OFETs, of particular note is the tremendous development that has occurred recently in this area.

During past decades, the category of semiconductors has increased from *p*-type to ambipolar; the deposition of semiconducting materials has advanced from polymerization *in situ* to thermal evaporation and solution processing; the morphology of active thin films has improved from amorphous to single-crystalline; the conformation of active layers has varied from thin films to single crystals; and the device mobility of organic semiconductors has advanced from $<10^{-5}\,cm^2/(V\cdot s)$ to about $40\,cm^2/(V\cdot s)$. With each of these rapid development having attracted great interest in OFETs and the related synthesis of functional molecules, the widespread use of OFET products can surely be anticipated in the near future.

References

1 Tsumura, A., Koezuka, H., and Ando, T. (1986) Macromolecular electronic device: field-effect transistor with a polythiophene thin film. *Appl. Phys. Lett.*, **49** (18), 1210–1212.

2 Katz, H.E., Bao, Z., and Gilat, S.L. (2001) Synthetic chemistry for ultrapure, processable, and high-mobility organic transistor semiconductors. *Acc. Chem. Res.*, **34** (5), 359–369.

3 Murphy, A.R. and Fréchet, J.M.J. (2007) Organic semiconducting oligomers for use in thin film transistors. *Chem. Rev.*, **107** (4), 1066–1096.

4 Zaumseil, J. and Sirringhaus, H. (2007) Electron and ambipolar transport in organic field-effect transistors. *Chem. Rev.*, **107** (4), 1296–1323.

5 Meng, Q., Dong, H., Hu, W., and Zhu, D. (2011) Recent progress of high performance organic thin film field-effect transistors. *J. Mater. Chem.*, **21** (32), 11708–11721.

6 Takimiya, K., Shinamura, S., Osaka, I., and Miyazaki, E. (2011) Thienoacene-based organic semiconductors. *Adv. Mater.*, **23** (38), 4347–4370.

7 Di, C.A., Liu, Y.Q., Yu, G., and Zhu, D.B. (2009) Interface engineering: an effective approach toward high-performance organic field-effect transistors. *Acc. Chem. Res.*, **42** (10), 1573–1583.

8 Aleshin, A.N., Lee, J.Y., Chu, S.W., Kim, J.S., and Park, Y.W. (2004) Mobility studies of field-effect transistor structures based on anthracene single crystals. *Appl. Phys. Lett.*, **84** (26), 5383–5385.

9 Tripathi, A.K., Heinrich, M., Siegrist, T., and Pflaum, J. (2007) Growth and electronic transport in 9,10-diphenylanthracene single crystals. An organic semiconductor of high electron and hole mobility. *Adv. Mater.*, **19** (16), 2097–2101.

10 Wang, C.L., Liu, Y.L., Ji, Z.Y., Wang, E.J., Li, R.J., Jiang, H., Tang, Q.X., Li, H.X., and Hu, W.P. (2009) Cruciforms: assembling single crystal micro- and nanostructures from one to three dimensions and their applications in organic field-effect transistors. *Chem. Mater.*, **21** (13), 2840–2845.

11 Jiang, L., Gao, J.H., Wang, E.J., Li, H.X., Wang, Z.H., and Hu, W.P. (2008) Organic single-crystalline ribbons of a rigid "H"-type anthracene derivative and high-performance, short-channel field-effect transistors of individual micro/nanometer-sized ribbons fabricated by an "organic ribbon mask" technique. *Adv. Mater.*, **20** (14), 2735–2740.

12 Zhao, W., Tang, Q., Chan, H.S., Xu, J., Lo, K.Y., and Miao, Q. (2008) Transistors from a conjugated macrocycle molecule: field and photo effects. *Chem. Commun.*, (36), 4324–4326.

13 Ito, K., Suzuki, T., Sakamoto, Y., Kubota, D., Inoue, Y., Sato, F., and Tokito, S. (2003) Oligo(2,6-anthrylene)s: acene–oligomer approach for organic field-effect transistors. *Angew. Chem. Int. Ed.*, **42** (10), 1159–1162.

14 Meng, H., Sun, F., Goldfinger, M.B., Jaycox, G.D., Li, Z., Marshall, W.J., and Blackman, G.S. (2005) High-performance, stable organic thin-film field-effect transistors based on bis-5′-alkylthiophen-2′-yl-2,6-anthracene semiconductors. *J. Am. Chem. Soc.*, **127** (8), 2406–2407.

15 Klauk, H., Zschieschang, U., Weitz, R.T., Meng, H., Sun, T., Nunes, G., Keys, D.E., Fincher, C.R., and Xiang, Z. (2007) Organic transistors based on di(phenylvinyl)anthracene: performance and stability. *Adv. Mater.*, **19** (22), 3882–3887.

16 Jiang, L., Hu, W., Wei, Z., Xu, W., and Meng, H. (2009) High-performance organic single-crystal transistors and digital inverters of an anthracene derivative. *Adv. Mater.*, **21** (36), 3649–3653.

17 Meng, H., Sun, F., Goldfinger, M.B., Gao, F., Londono, D.J., Marshal, W.J., Blackman, G.S., Dobbs, K.D., and Keys, D.E. (2006) 2,6-Bis[2-(4-pentylphenyl) vinyl]anthracene: a stable and high charge mobility organic semiconductor with densely packed crystal structure. *J. Am. Chem. Soc.*, **128** (29), 9304–9305.

18 Park, J.-H., Chung, D.S., Park, J.-W., Ahn, T., Kong, H., Jung, Y.K., Lee, J., Yi, M.H., Park, C.E., Kwon, S.-K., and Shim, H.-K. (2007) Soluble and easily crystallized anthracene derivatives: precursors of solution-processable semiconducting molecules. *Org. Lett.*, **9** (13), 2573–2576.

19 Jung, K.H., Bae, S.Y., Kim, K.H., Cho, M.J., Lee, K., Kim, Z.H., Choi, D.H., Lee, D.H., Chung, D.S., and Park, C.E. (2009) High-mobility anthracene-based X-shaped conjugated molecules for thin film transistors. *Chem. Commun.*, (35), 5290–5292.

20 Gundlach, D.J., Nichols, J.A., Zhou, L., and Jackson, T.N. (2002) Thin-film transistors based on well-ordered thermally evaporated naphthacene films. *Appl. Phys. Lett.*, **80** (16), 2925–2927.

21 Goldmann, C., Haas, S., Krellner, C., Pernstich, K.P., Gundlach, D.J., and Batlogg, B. (2004) Hole mobility in organic single crystals measured by a "flip-crystal" field-effect technique. *J. Appl. Phys.*, **96** (4), 2080–2086.

22 Moon, H., Zeis, R., Borkent, E.J., Besnard, C., Lovinger, A.J., Siegrist, T.,

Kloc, C., and Bao, Z.N. (2004) Synthesis, crystal structure, and transistor performance of tetracene derivatives. *J. Am. Chem. Soc.*, **126** (47), 15322–15323.

23 Sundar, V.C., Zaumseil, J., Podzorov, V., Menard, E., Willett, R.L., Someya, T., Gershenson, M.E., and Rogers, J.A. (2004) Elastomeric transistor stamps: reversible probing of charge transport in organic crystals. *Science*, **303** (5664), 1644–1646.

24 Kelley, T.W., Muyres, D.V., Baude, P.F., Smith, T.P., and Jones, T.D. (2003) High performance organic thin film transistors. *Mater. Res. Soc. Symp. Proc.*, **771**, 169–179.

25 Wang, C.-H., Hsieh, C.-Y., and Hwang, J.-C. (2011) Flexible organic thin-film transistors with silk fibroin as the gate dielectric. *Adv. Mater.*, **23** (14), 1630–1634.

26 Jurchescu, O.D., Popinciuc, M., van Wees, B.J., and Palstra, T.T.M. (2007) Interface-controlled, high-mobility organic transistors. *Adv. Mater.*, **19** (5), 688–692.

27 Kelley, T.W., Boardman, L.D., Dunbar, T.D., Muyres, D.V., Pellerite, M.J., and Smith, T.Y.P. (2003) High-performance OTFTs using surface-modified alumina dielectrics. *J. Phys. Chem. B*, **107** (24), 5877–5881.

28 Meng, H., Bendikov, M., Mitchell, G., Helgeson, R., Wudl, F., Bao, Z., Siegrist, T., Kloc, C., and Chen, C.H. (2003) Tetramethylpentacene: remarkable absence of steric effect on field effect mobility. *Adv. Mater.*, **15** (13), 1090–1093.

29 Okamoto, T., Senatore, M.L., Ling, M.M., Mallik, A.B., Tang, M.L., and Bao, Z. (2007) Synthesis, characterization, and field-effect transistor performance of pentacene derivatives. *Adv. Mater.*, **19** (20), 3381–3384.

30 Miao, Q., Lefenfeld, M., Nguyen, T.Q., Siegrist, T., Kloc, C., and Nuckolls, C. (2005) Self-assembly and electronics of dipolar linear acenes. *Adv. Mater.*, **17** (4), 407–412.

31 Miao, Q., Chi, X., Xiao, S., Zeis, R., Lefenfeld, M., Siegrist, T., Steigerwald, M.L., and Nuckolls, C. (2006) Organization of acenes with a cruciform assembly motif. *J. Am. Chem. Soc.*, **128** (4), 1340–1345.

32 Anthony, J.E. (2006) Functionalized acenes and heteroacenes for organic electronics. *Chem. Rev.*, **106** (12), 5028–5048.

33 Anthony, J.E. (2008) The larger acenes: versatile organic semiconductors. *Angew. Chem. Int. Ed.*, **47** (3), 452–483.

34 Park, S.K., Jackson, T.N., Anthony, J.E., and Mourey, D.A. (2007) High mobility solution processed 6,13-bis(triisopropyl-silylethynyl) pentacene organic thin film transistors. *Appl. Phys. Lett.*, **91** (6), 063514.

35 Kim, D.H., Lee, D.Y., Lee, H.S., Lee, W.H., Kim, Y.H., Han, J.I., and Cho, K. (2007) High-mobility organic transistors based on single-crystalline microribbons of triisopropyl silylethynyl pentacene via solution-phase self-assembly. *Adv. Mater.*, **19** (5), 678–682.

36 Ohki, K., Inokuchi, H., and Maruyama, Y. (1963) Charge mobility in pyrene crystals. *Bull. Chem. Soc. Jpn*, **36** (11), 1512–1515.

37 Zhang, H., Wang, Y., Shao, K., Liu, Y., Chen, S., Qiu, W., Sun, X., Qi, T., Ma, Y., Yu, G., Su, Z., and Zhu, D. (2006) Novel butterfly pyrene-based organic semiconductors for field effect transistors. *Chem. Commun.*, (7), 755–757.

38 Wang, Y., Wang, H., Liu, Y., Di, C.-A., Sun, Y., Wu, W., Yu, G., Zhang, D., and Zhu, D. (2006) 1-Imino nitroxide pyrene for high performance organic field-effect transistors with low operating voltage. *J. Am. Chem. Soc.*, **128** (40), 13058–13059.

39 Okamoto, H., Kawasaki, N., Kaji, Y., Kubozono, Y., Fujiwara, A., and Yamaji, M. (2008) Air-assisted high-performance field-effect transistor with thin films of picene. *J. Am. Chem. Soc.*, **130** (32), 10470–10471.

40 Kotani, M., Kakinuma, K., Yoshimura, M., Ishii, K., Yamazaki, S., Kobori, T., Okuyama, H., Kobayashi, H., and Tada, H. (2006) Charge carrier transport in high purity perylene single crystal studied by time-of-flight measurements and through field effect transistor

characteristics. *Chem. Phys.*, **325** (1), 160–169.

41 Pisula, W., Menon, A., Stepputat, M., Lieberwirth, I., Kolb, U., Tracz, A., Sirringhaus, H., Pakula, T., and Müllen, K. (2005) A zone-casting technique for device fabrication of field-effect transistors based on discotic hexa-peri-hexabenzocoronene. *Adv. Mater.*, **17** (6), 684–689.

42 Xiao, S., Myers, M., Miao, Q., Sanaur, S., Pang, K., Steigerwald, M.L., and Nuckolls, C. (2005) Molecular wires from contorted aromatic compounds. *Angew. Chem. Int. Ed.*, **44** (45), 7390–7394.

43 Facchetti, A., Mushrush, M., Yoon, M.-H., Hutchison, G.R., Ratner, M.A., and Marks, T.J. (2004) Building blocks for *n*-type molecular and polymeric electronics. Perfluoroalkyl- versus alkyl-functionalized oligothiophenes (*n*T; $n = 2$–6). Systematics of thin film microstructure, semiconductor performance, and modeling of majority charge injection in field-effect transistors. *J. Am. Chem. Soc.*, **126** (42), 13859–13874.

44 Ostoja, P., Maccagnani, P., Gazzano, M., Cavallini, M., Kengne, J.C., Kshirsagar, R., Biscarini, F., Melucci, M., Zambianchi, M., and Barbarella, G. (2004) FET device performance, morphology and X-ray thin film structure of unsubstituted and modified quinquethiophenes. *Synth. Met.*, **146** (3), 243–250.

45 Dodabalapur, A., Torsi, L., and Katz, H.E. (1995) Organic transistors: two-dimensional transport and improved electrical characteristics. *Science*, **268** (5208), 270–271.

46 Hajlaoui, M.E., Garnier, F., Hassine, L., Kouki, F., and Bouchriha, H. (2002) Growth conditions effects on morphology and transport properties of an oligothiophene semiconductor. *Synth. Met.*, **129** (3), 215–220.

47 Samuel, I.D.W., Ledoux, I., Delporte, C., Pearson, D.L., and Tour, J.M. (1996) Scaling of cubic polarizability with chain length in oligo(3-ethylthiophene ethynylene)s. *Chem. Mater.*, **8** (4), 819–821.

48 Fichou, D. (2000) Structural order in conjugated oligothiophenes and its implications on opto-electronic devices. *J. Mater. Chem.*, **10** (3), 571–588.

49 Katz, H.E., Lovinger, A.J., and Laquindanum, J.G. (1998) α,ω-Dihexylquaterthiophene: a second thin film single-crystal organic semiconductor. *Chem. Mater.*, **10** (2), 457–459.

50 Li, W., Katz, H.E., Lovinger, A.J., and Laquindanum, J.G. (1999) Field-effect transistors based on thiophene hexamer analogues with diminished electron donor strength. *Chem. Mater.*, **11** (2), 458–465.

51 Halik, M., Klauk, H., Zschieschang, U., Schmid, G., Ponomarenko, S., Kirchmeyer, S., and Weber, W. (2003) Relationship between molecular structure and electrical performance of oligothiophene organic thin film transistors. *Adv. Mater.*, **15** (11), 917–922.

52 Hajlaoui, R., Fichou, D., Horowitz, G., Nessakh, B., Constant, M., and Garnier, F. (1997) Organic transistors using α-octithiophene and α,ω-dihexyl-α-octithiophene: influence of oligomer length versus molecular ordering on mobility. *Adv. Mater.*, **9** (7), 557–561.

53 Halik, M., Klauk, H., Zschieschang, U., Schmid, G., Radlik, W., Ponomarenko, S., Kirchmeyer, S., and Weber, W. (2003) High-mobility organic thin-film transistors based on α,α′-didecyloligothiophenes. *J. Appl. Phys.*, **93** (5), 2977–2981.

54 Barbarella, G., Zambianchi, M., Antolini, L., Ostoja, P., Maccagnani, P., Bongini, A., Marseglia, E.A., Tedesco, E., Gigli, G., and Cingolani, R. (1999) Solid-state conformation, molecular packing, and electrical and optical properties of processable β-methylated sexithiophenes. *J. Am. Chem. Soc.*, **121** (38), 8920–8926.

55 Videlot, C., Ackermann, J., Blanchard, P., Raimundo, J.M., Frère, P., Allain, M., de Bettignies, R., Levillain, E., and Roncali, J. (2003) Field-effect transistors based on oligothienylenevinylenes: from

solution π-dimers to high-mobility organic semiconductors. *Adv. Mater.*, **15** (4), 306–310.

56 Hong, X.M., Katz, H.E., Lovinger, A.J., Wang, B.-C., and Raghavachari, K. (2001) Thiophene-phenylene and thiophene-thiazole oligomeric semiconductors with high field-effect transistor on/off ratios. *Chem. Mater.*, **13** (12), 4686–4691.

57 Ponomarenko, S.A., Kirchmeyer, S., Halik, M., Klauk, H., Zschieschang, U., Schmid, G., Karbach, A., Drechsler, D., and Alpatova, N.M. (2005) 1,4-Bis(5-decyl-2,2′-bithien-5-yl)benzene as new stable organic semiconductor for high performance thin film transistors. *Synth. Met.*, **149** (2–3), 231–235.

58 Ponomarenko, S.A., Kirchmeyer, S., Elschner, A., Alpatova, N.M., Halik, M., Klauk, H., Zschieschang, U., and Schmid, G. (2006) Decyl-end-capped thiophene-phenylene oligomers as organic semiconducting materials with improved oxidation stability. *Chem. Mater.*, **18** (2), 579–586.

59 Mushrush, M., Facchetti, A., Lefenfeld, M., Katz, H.E., and Marks, T.J. (2003) Easily processable phenylene-thiophene-based organic field-effect transistors and solution-fabricated nonvolatile transistor memory elements. *J. Am. Chem. Soc.*, **125** (31), 9414–9423.

60 Videlot-Ackermann, C., Ackermann, J., Brisset, H., Kawamura, K., Yoshimoto, N., Raynal, P., El Kassmi, A., and Fages, F. (2005) α,ω-Distyryl oligothiophenes: high mobility semiconductors for environmentally stable organic thin film transistors. *J. Am. Chem. Soc.*, **127** (47), 16346–16347.

61 Meng, Q., Gao, J.H., Li, R.J., Jiang, L., Wang, C.L., Zhao, H.P., Liu, C.M., Li, H.X., and Hu, W.P. (2009) New type of organic semiconductors for field-effect transistors with carbon-carbon triple bonds. *J. Mater. Chem.*, **19** (10), 1477–1482.

62 Tian, H.K., Wang, J., Shi, J.W., Yan, D.H., Wang, L.X., Geng, Y.H., and Wang, F. (2005) Novel thiophene-aryl co-oligomers for organic thin film transistors. *J. Mater. Chem.*, **15** (29), 3026–3033.

63 Cho, N.S., Cho, S., Elbing, M., Lee, J.K., Yang, R., Seo, J.H., Lee, K., Bazan, G.C., and Heeger, A.J. (2008) Organic thin-film transistors based on α,ω-dihexyldithienyl-dihydrophenanthrene. *Chem. Mater.*, **20** (20), 6289–6291.

64 Locklin, J., Li, D., Mannsfeld, S.C.B., Borkent, E.-J., Meng, H., Advincula, R., and Bao, Z. (2005) Organic thin film transistors based on cyclohexyl-substituted organic semiconductors. *Chem. Mater.*, **17** (13), 3366–3374.

65 Meng, H., Zheng, J., Lovinger, A.J., Wang, B.C., Van Patten, P.G., and Bao, Z.N. (2003) Oligofluorene-thiophene derivatives as high-performance semiconductors for organic thin film transistors. *Chem. Mater.*, **15** (9), 1778–1787.

66 Py, C., Gorjanc, T.C., Hadizad, T., Zhang, J., and Wang, Z.Y. (2006) Hole mobility and electroluminescence properties of a dithiophene indenofluorene. *J. Vac. Sci. Technol. A*, **24** (3), 654–656.

67 Dong, H., Wang, C., and Hu, W. (2010) High performance organic semiconductors for field-effect transistors. *Chem. Commun.*, **46** (29), 5211–5222.

68 Xiao, K., Liu, Y., Qi, T., Zhang, W., Wang, F., Gao, J., Qiu, W., Ma, Y., Cui, G., Chen, S., Zhan, X., Yu, G., Qin, J., Hu, W., and Zhu, D. (2005) A highly π-stacked organic semiconductor for field-effect transistors based on linearly condensed pentathienoacene. *J. Am. Chem. Soc.*, **127** (38), 13281–13286.

69 Chernichenko, K.Y., Sumerin, V.V., Shpanchenko, R.V., Balenkova, E.S., and Nenajdenko, V.G. (2006) "Sulflower": a new form of carbon sulfide. *Angew. Chem. Int. Ed.*, **45** (44), 7367–7370.

70 Dadvand, A., Cicoira, F., Chernichenko, K.Y., Balenkova, E.S., Osuna, R.M., Rosei, F., Nenajdenko, V.G., and Perepichka, D.F. (2008) Heterocirculenes as a new class of organic semiconductors. *Chem. Commun.*, (42), 5354–5356.

71 Gao, J.H., Li, R.J., Li, L.Q., Meng, Q., Jiang, H., Li, H.X., and Hu, W.P. (2007) High-performance field-effect transistor based on dibenzo[*d*,*d*′]thieno[3,2-*b*;4,5-*b*′]

dithiophene, an easily synthesized semiconductor with high ionization potential. *Adv. Mater.*, **19** (19), 3008–3011.

72 Li, R., Jiang, L., Meng, Q., Gao, J., Li, H., Tang, Q., He, M., Hu, W., Liu, Y., and Zhu, D. (2009) Micrometer-sized organic single crystals, anisotropic transport, and field-effect transistors of a fused-ring thienoacene. *Adv. Mater.*, **21** (44), 4492–4495.

73 Li, R., Dong, H., Zhan, X., He, Y., Li, H., and Hu, W. (2010) Single crystal ribbons and transistors of a solution processed sickle-like fused-ring thienoacene. *J. Mater. Chem.*, **20** (29), 6014–6018.

74 Laquindanum, J.G., Katz, H.E., and Lovinger, A.J. (1998) Synthesis, morphology, and field-effect mobility of anthradithiophenes. *J. Am. Chem. Soc.*, **120** (4), 664–672.

75 Tedjamulia, M.L., Tominaga, Y., Castle, R.N., and Lee, M.L. (1983) The synthesis of benzo[*b*]phenanthro[*d*]thiophenes and anthra[*b*]benzo[*d*]thiophenes. *Heterocycl. Chem.*, **20** (4), 861–866.

76 Mayer, F., Mombour, A., Lassmann, W., Werner, W., Landmann, P., and Schneider, E. (1931) Farbstoffstudien in der thionaphtenreihe. *Ann. Chem.*, **488** (1), 259–296.

77 Du, C., Guo, Y., Liu, Y., Qiu, W., Zhang, H., Gao, X., Liu, Y., Qi, T., Lu, K., and Yu, G. (2008) Anthra[2,3-*b*]benzo[*d*] thiophene: an air-stable asymmetric organic semiconductor with high mobility at room temperature. *Chem. Mater.*, **20** (13), 4188–4190.

78 Tang, M.L., Okamoto, T., and Bao, Z. (2006) High-performance organic semiconductors: asymmetric linear acenes containing sulphur. *J. Am. Chem. Soc.*, **128** (50), 16002–16003.

79 Tang, M.L., Mannsfeld, S.C.B., Sun, Y.S., Becerril, H.A., and Bao, Z.N. (2009) Pentaceno[2,3-*b*]thiophene, a hexacene analogue for organic thin film transistors. *J. Am. Chem. Soc.*, **131** (3), 882–883.

80 Tang, M.L., Reichardt, A.D., Okamoto, T., Miyaki, N., and Bao, Z. (2008) Functionalized asymmetric linear acenes for high-performance organic semiconductors. *Adv. Funct. Mater.*, **18** (10), 1579–1585.

81 Sirringhaus, H., Friend, R.H., Wang, C., Leuninger, J.R., and Müllen, K. (1999) Dibenzothienobisbenzothiophene – a novel fused-ring oligomer with high field-effect mobility. *J. Mater. Chem.*, **9** (9), 2095–2101.

82 Okamoto, T., Kudoh, K., Wakamiya, A., and Yamaguchi, S. (2005) General synthesis of thiophene and selenophene-based heteroacenes. *Org. Lett.*, **7** (23), 5301–5304.

83 Yamada, K., Okamoto, T., Kudoh, K., Wakamiya, A., Yamaguchi, S., and Takeya, J. (2007) Single-crystal field-effect transistors of benzoannulated fused oligothiophenes and oligoselenophenes. *Appl. Phys. Lett.*, **90** (7), 072102.

84 Sun, Y., Tan, L., Jiang, S., Qian, H., Wang, Z., Yan, D., Di, C., Wang, Y., Wu, W., Yu, G., Yan, S., Wang, C., Hu, W., Liu, Y., and Zhu, D. (2007) High-performance transistor based on individual single-crystalline micrometer wire of perylo[1,12-*b*,*c*,*d*]thiophene. *J. Am. Chem. Soc.*, **129** (7), 1882–1883.

85 Jiang, W., Qian, H., Li, Y., and Wang, Z. (2008) Heteroatom-annulated perylenes: practical synthesis, photophysical properties, and solid- state packing arrangement. *J. Org. Chem.*, **73** (18), 7369–7372.

86 Tan, L., Jiang, W., Jiang, L., Jiang, S., Wang, Z., Yan, S., and Hu, W. (2009) Single crystalline microribbons of perylo[1,12-*b*,*c*,*d*]selenophene for high performance transistors. *Appl. Phys. Lett.*, **94** (15), 153306.

87 Wang, J.Y., Zhou, Y., Yan, J., Ding, L., Ma, Y.G., Cao, Y., Wang, J., and Pei, J. (2009) New fused heteroarenes for high-performance field-effect transistors. *Chem. Mater.*, **21** (13), 2595–2597.

88 Kashiki, T., Miyazaki, E., and Takimiya, K. (2008) 2,6-dialkylbenzo[1,2-*b*:4,5-*b*′] dithiophenes (C_n-BDTs) as soluble organic semiconductors for solution-processed organic field-effect transistors. *Chem. Lett.*, **37** (3), 284–285.

89 Izawa, T., Miyazaki, E., and Takimiya, K. (2008) Molecular ordering of high-performance soluble molecular

semiconductors and re-evaluation of their field-effect transistor characteristics. *Adv. Mater.*, **20** (18), 3388–3392.

90 Ebata, H., Izawa, T., Miyazaki, E., Takimiya, K., Ikeda, M., Kuwabara, H., and Yui, T. (2007) Highly soluble [1] benzothieno[3,2-*b*]benzothiophene (BTBT) derivatives for high-performance, solution-processed organic field-effect transistors. *J. Am. Chem. Soc.*, **129** (51), 15732–15733.

91 Yamamoto, T. and Takimiya, K. (2007) Facile synthesis of highly π-extended heteroarenes, dinaphtho[2,3-*b*:2′,3′-*f*] chalcogenopheno[3,2-*b*] chalcogenophenes, and their application to field-effect transistors. *J. Am. Chem. Soc.*, **129** (8), 2224–2225.

92 Haas, S., Takahashi, Y., Takimiya, K., and Hasegawa, T. (2009) High-performance dinaphtho-thieno-thiophene single crystal field-effect transistors. *Appl. Phys. Lett.*, **95** (2), 022111.

93 Nakayama, K., Hirose, Y., Soeda, J., Yoshizumi, M., Uemura, T., Uno, M., Li, W., Kang, M.J., Yamagishi, M., Okada, Y., Miyazaki, E., Nakazawa, Y., Nakao, A., Takimiya, K., and Takeya, J. (2011) Patternable solution-crystallized organic transistors with high charge carrier mobility. *Adv. Mater.*, **23**, 1626–1629.

94 Li, X.-C., Sirringhaus, H., Garnier, F., Holmes, A.B., Moratti, S.C., Feeder, N., Clegg, W., Teat, S.J., and Friend, R.H. (1998) A highly π-stacked organic semiconductor for thin film transistors based on fused thiophenes. *J. Am. Chem. Soc.*, **120** (9), 2206–2207.

95 Sirringhaus, H., Friend, R.H., Li, X.C., Moratti, S.C., Holmes, A.B., and Feeder, N. (1997) Bis(dithienothiophene) organic field-effect transistors with a high ON/OFF ratio. *Appl. Phys. Lett.*, **71** (26), 3871–3873.

96 Laquindanum, J.G., Katz, H.E., Lovinger, A.J., and Dodabalapur, A. (1997) Benzodithiophene rings as semiconductor building blocks. *Adv. Mater.*, **9** (1), 36–39.

97 Mamada, M., Nishida, J.I., Kumaki, D., Tokito, S., and Yamashita, Y. (2008) High performance organic field-effect transistors based on 2,2′-binaphtho[2,3-*b*]thiophenyl with a simple structure. *J. Mater. Chem.*, **18** (29), 3442–3447.

98 Tan, L., Zhang, L., Jiang, X., Yang, X., Wang, L., Wang, Z., Li, L., Hu, W., Shuai, Z., Li, L., and Zhu, D. (2009) A densely and uniformly packed organic semiconductor based on annelated β-trithiophenes for high-performance thin film transistors. *Adv. Funct. Mater.*, **19** (2), 272–276.

99 Takimiya, K., Kunugi, Y., Konda, Y., Niihara, N., and Otsubo, T. (2004) 2,6-Diphenylbenzo[1,2-*b*:4,5-*b*′] dichalcogenophenes: a new class of high-performance semiconductors for organic field-effect transistors. *J. Am. Chem. Soc.*, **126** (16), 5084–5085.

100 Takimiya, K., Ebata, H., Sakamoto, K., Izawa, T., Otsubo, T., and Kunugi, Y. (2006) 2,7-Diphenyl[1]benzothieno[3,2-*b*] benzothiophene, a new organic semiconductor for air-stable organic field-effect transistors with mobilities up to 2.0 $cm^2 V^{-1} s^{-1}$. *J. Am. Chem. Soc.*, **128** (39), 12604–12605.

101 Takimiya, K., Kunugi, Y., Konda, Y., Ebata, H., Toyoshima, Y., and Otsubo, T. (2006) 2,7-Diphenyl[1] benzoselenopheno[3,2-*b*][1] benzoselenophene as a stable organic semiconductor for a high-performance field-effect transistor. *J. Am. Chem. Soc.*, **128** (9), 3044–3050.

102 Sun, Y.M., Ma, Y.Q., Liu, Y.Q., Lin, Y.Y., Wang, Z.Y., Wang, Y., Di, C.A., Xiao, K., Chen, X.M., Qiu, W.F., Zhang, B., Yu, G., Hu, W.P., and Zhu, D.B. (2006) High-performance and stable organic thin-film transistors based on fused thiophenes. *Adv. Funct. Mater.*, **16** (3), 426–432.

103 Didane, Y., Mehl, G.H., Kumagai, A., Yoshimoto, N., Videlot-Ackermann, C., and Brisset, H. (2008) A “kite” shaped styryl end-capped benzo[2,1-*b*:3,4-*b*′] dithiophene with high electrical performances in organic thin film transistors. *J. Am. Chem. Soc.*, **130** (52), 17681–17683.

104 Zhang, L., Tan, L., Wang, Z., Hu, W., and Zhu, D. (2009) High-performance, stable organic field-effect transistors based on trans-1,2-(dithieno[2,3-*b*:3′,2′-*d*] thiophene)ethene. *Chem. Mater.*, **21** (9), 1993–1999.

105 Gao, J.H., Li, L.Q., Meng, Q., Li, R.J., Jiang, H., Li, H.X., and Hu, W.P. (2007) Dibenzothiophene derivatives as new prototype semiconductors for organic field-effect transistors. *J. Mater. Chem.*, **17** (14), 1421–1426.

106 Iosip, M.D., Destri, S., Pasini, M., Porzio, W., Pernstich, K.P., and Batlogg, B. (2004) New dithieno[3,2-*b*:2′,3′-*d*] thiophene oligomers as promising materials for organic field-effect transistor applications. *Synth. Met.*, **146** (3), 251–257.

107 Meng, Q., Jiang, L., Wei, Z., Wang, C., Zhao, H., Li, H., Xu, W., and Hu, W. (2010) Development of organic field-effect properties by introducing aryl-acetylene into benzodithiophene. *J. Mater. Chem.*, **20**, 10931–10935.

108 Tang, M.L., Reichardt, A.D., Siegrist, T., Mannsfeld, S.C.B., and Bao, Z.N. (2008) Trialkylsilylethynyl-functionalized tetraceno[2,3-*b*]thiophene and anthra[2,3-*b*]thiophene organic transistors. *Chem. Mater.*, **20** (14), 4669–4676.

109 Tang, M.L., Reichardt, A.D., Miyaki, N., Stoltenberg, R.M., and Bao, Z. (2008) Ambipolar, high performance, acene-based organic thin film transistors. *J. Am. Chem. Soc.*, **130** (19), 6064–6065.

110 Payne, M.M., Parkin, S.R., Anthony, J.E., Kuo, C.C., and Jackson, T.N. (2005) Organic field-effect transistors from solution-deposited functionalized acenes with mobilities as high as 1 cm^2/Vs. *J. Am. Chem. Soc.*, **127** (14), 4986–4987.

111 Subramanian, S., Park, S.K., Parkin, S.R., Podzorov, V., Jackson, T.N., and Anthony, J.E. (2008) Chromophore fluorination enhances crystallization and stability of soluble anthradithiophene semiconductors. *J. Am. Chem. Soc.*, **130** (9), 2706–2707.

112 Jiang, H., Yang, X., Cui, Z., Liu, Y., Li, H., Hu, W., Liu, Y., and Zhu, D. (2007) Phase dependence of single crystalline transistors of tetrathiafulvalene. *Appl. Phys. Lett.*, **91** (12), 123505.

113 Naraso, Nishida, J.-I., Ando, S., Yamaguchi, J., Itaka, K., Koinuma, H., Tada, H., Tokito, S., and Yamashita, Y. (2005) High-performance organic field-effect transistors based on π-extended tetrathiafulvalene derivatives. *J. Am. Chem. Soc.*, **127** (29), 10142–10143.

114 Mas-Torrent, M., Hadley, P., Bromley, S.T., Crivillers, N., Veciana, J., and Rovira, C. (2005) Single-crystal organic field-effect transistors based on dibenzo-tetrathiafulvalene. *Appl. Phys. Lett.*, **86** (1), 012110.

115 Naraso, Nishida, J.-I., Kumaki, D., Tokito, S., and Yamashita, Y. (2006) High performance *n*- and *p*-type field-effect transistors based on tetrathiafulvalene derivatives. *J. Am. Chem. Soc.*, **128** (30), 9598–9599.

116 Takada, M., Graaf, H., Yamashita, Y., and Tada, H. (2002) BTQBT (bis-(1,2,5-thiadiazolo)-*p*-quinobis(1,3-dithiole)) thin films; A promising candidate for high mobility organic transistors. *Jpn. J. Appl. Phys. Part 2*, **41**, L4–L6.

117 Mas-Torrent, M., Durkut, M., Hadley, P., Ribas, X., and Rovira, C. (2004) High mobility of dithiophene-tetrathiafulvalene single-crystal organic field effect transistors. *J. Am. Chem. Soc.*, **126** (4), 984–985.

118 Gao, X., Wu, W., Liu, Y., Qiu, W., Sun, X., Yu, G., and Zhu, D. (2006) A facile synthesis of linear benzene-fused bis(tetrathiafulvalene) compounds and their application for organic field-effect transistors. *Chem. Commun.*, (26), 2750–2752.

119 Gao, X., Wu, W., Liu, Y., Jiao, S., Qiu, W., Yu, G., Wang, L., and Zhu, D. (2007) Linear benzene-fused bis(tetrathiafulvalene) compounds for solution processed organic field-effect transistors. *J. Mater. Chem.*, **17** (8), 736–743.

120 Gao, X.K., Wang, Y., Yang, X.D., Liu, Y.Q., Qiu, W.F., Wu, W.P., Zhang, H.J., Qi, T., Liu, Y., Lu, K., Du, C.Y., Shuai, Z.G., Yu, G., and Zhu, D.B. (2007) Dibenzotetrathiafulvalene bisimides: new building blocks for organic electronic materials. *Adv. Mater.*, **19** (19), 3037–3042.

121 Takahashi, Y., Hasegawa, T., Horiuchi, S., Kumai, R., Tokura, Y., and Saito, G. (2007) High mobility organic field-effect transistor based on hexamethylenetetrathiafulvalene with organic metal electrodes. *Chem. Mater.*, **19** (26), 6382–6384.

122 Kanno, M., Bando, Y., Shirahata, T., Inoue, J.I., Wada, H., and Mori, T. (2009) Stabilization of organic field-effect transistors in hexamethylenetetrathiafulvalene derivatives substituted by bulky alkyl groups. *J. Mater. Chem.*, **19** (36), 6548–6555.

123 Bando, Y., Shirahata, T., Shibata, K., Wada, H., Mori, T., and Imakubo, T. (2008) Organic field-effect transistors based on alkyl-terminated tetrathiapentalene (TTP) derivatives. *Chem. Mater.*, **20** (16), 5119–5121.

124 Tang, Q., Zhang, D., Wang, S., Ke, N., Xu, J., Yu, J.C., and Miao, Q. (2009) A meaningful analogue of pentacene: charge transport, polymorphs, and electronic structures of dihydrodiazapentacene. *Chem. Mater.*, **21** (7), 1400–1405.

125 Liang, Z., Tang, Q., Xu, J., and Miao, Q. (2011) Soluble and stable N-heteropentacenes with high field-effect mobility. *Adv. Mater.*, **23** (13), 1535–1539.

126 Ma, Y., Sun, Y., Liu, Y., Gao, J., Chen, S., Sun, X., Qiu, W., Yu, G., Cui, G., Hu, W., and Zhu, D. (2005) Organic thin film transistors based on stable amorphous ladder tetraazapentacenes semiconductors. *J. Mater. Chem.*, **15** (46), 4894–4898.

127 Zhao, H.P., Jiang, L., Dong, H.L., Li, H.X., Hu, W.P., and Ong, B.S. (2009) Influence of intermolecular N-H...π interactions on molecular packing and field-effect performance of organic semiconductors. *ChemPhysChem.*, **10** (13), 2345–2348.

128 Li, Y., Wu, Y., Gardner, S., and Ong, B.S. (2005) Novel peripherally substituted indolo [3,2-*b*]carbazoles for high-mobility organic thin-film transistors. *Adv. Mater.*, **17** (7), 849–853.

129 Wu, Y., Li, Y., Gardner, S., and Ong, B.S. (2004) Indolo[3,2-*b*]carbazole-based thin-film transistors with high mobility and stability. *J. Am. Chem. Soc.*, **127** (2), 614–618.

130 Boudreault, P.-L.T., Wakim, S., Blouin, N., Simard, M., Tessier, C., Tao, Y., and Leclerc, M. (2007) Synthesis, characterization, and application of indolo[3,2-*b*]carbazole semiconductors. *J. Am. Chem. Soc.*, **129** (29), 9125–9136.

131 Boudreault, P.-L.T., Wakim, S., Tang, M.L., Tao, Y., Bao, Z., and Leclerc, M. (2009) New indolo[3,2-*b*]carbazole derivatives for field-effect transistor applications. *J. Mater. Chem.*, **19** (19), 2921–2928.

132 Qi, T., Guo, Y., Liu, Y., Xi, H., Zhang, H., Gao, X., Liu, Y., Lu, K., Du, C., Yu, G., and Zhu, D. (2008) Synthesis and properties of the anti and syn isomers of dibenzothieno[*b,d*]pyrrole. *Chem. Commun.*, (46), 6227–6229.

133 Hong, W., Wei, Z., Xi, H., Xu, W., Hu, W., Wang, Q., and Zhu, D. (2008) 6H-Pyrrolo[3,2-*b*:4,5-*b*′]bis[1,4] benzothiazines: facilely synthesized semiconductors for organic field-effect transistors. *J. Mater. Chem.*, **18** (40), 4814–4820.

134 Ahmed, E., Briseno, A.L., Xia, Y., and Jenekhe, S.A. (2008) High mobility single-crystal field-effect transistors from bisindoloquinoline semiconductors. *J. Am. Chem. Soc.*, **130** (4), 1118–1119.

135 Bao, Z., Lovinger, A.J., and Dodabalapur, A. (1996) Organic field-effect transistors with high mobility based on copper phthalocyanine. *Appl. Phys. Lett.*, **69** (20), 3066–3068.

136 Zeis, R., Siegrist, T., and Kloc, C. (2005) Single-crystal field-effect transistors based on copper phthalocyanine. *Appl. Phys. Lett.*, **86** (2), 022103.

137 Li, L.Q., Tang, Q.X., Li, H.X., Yang, X.D., Hu, W.P., Song, Y.B., Shuai, Z.G., Xu, W., Liu, Y.Q., and Zhu, D.B. (2007) An ultra closely π-stacked organic semiconductor for high performance field-effect transistors. *Adv. Mater.*, **19** (18), 2613–2617.

138 Wang, H.B., Song, D., Yang, J.L., Yu, B., Geng, Y.H., and Yan, D.H. (2007) High mobility vanadyl-phthalocyanine polycrystalline films for organic

field-effect transistors. *Appl. Phys. Lett.*, **90** (25), 253510.

139 Checcoli, P., Conte, G., Salvatori, S., Paolesse, R., Bolognesi, A., Berliocchi, M., Brunetti, F., D'Amico, A., Di Carlo, A., and Lugli, P. (2003) Tetra-phenyl porphyrin based thin film transistors. *Synth. Met.*, **138** (1–2), 261–266.

140 Xu, H., Yu, G., Xu, W., Xu, Y., Cui, G., Zhang, D., Liu, Y., and Zhu, D. (2005) High-performance field-effect transistors based on Langmuir–Blodgett films of cyclo[8]pyrrole. *Langmuir*, **21** (12), 5391–5395.

141 Sakamoto, Y., Suzuki, T., Kobayashi, M., Gao, Y., Fukai, Y., Inoue, Y., Sato, F., and Tokito, S. (2004) Perfluoropentacene: high-performance p–n junctions and complementary circuits with pentacene. *J. Am. Chem. Soc.*, **126** (26), 8138–8140.

142 Bao, Z., Lovinger, A.J., and Brown, J. (1998) New air-stable *n*-channel organic thin film transistors. *J. Am. Chem. Soc.*, **120** (1), 207–208.

143 Tang, Q.X., Li, H.X., Liu, Y.L., and Hu, W.P. (2006) High-performance air-stable *n*-type transistors with an asymmetrical device configuration based on organic single-crystalline submicrometer/nanometer ribbons. *J. Am. Chem. Soc.*, **128** (45), 14634–14639.

144 Tang, Q., Tong, Y., Li, H., and Hu, W. (2008) Air/vacuum dielectric organic single crystalline transistors of copper-hexadecafluorophthalocyanine ribbons. *Appl. Phys. Lett.*, **92** (8), 083309.

145 Yoon, M.-H., Facchetti, A., Stern, C.E., and Marks, T.J. (2006) Fluorocarbon-modified organic semiconductors: molecular architecture, electronic, and crystal structure tuning of arene- versus fluoroarene-thiophene oligomer thin-film properties. *J. Am. Chem. Soc.*, **128** (17), 5792–5801.

146 Letizia, J.A., Facchetti, A., Stern, C.L., Ratner, M.A., and Marks, T.J. (2005) High electron mobility in solution-cast and vapor-deposited phenacyl–quaterthiophene-based field-effect transistors: toward *n*-type polythiophenes. *J. Am. Chem. Soc.*, **127** (39), 13476–13477.

147 Akhtaruzzaman, M., Kamata, N., Nishida, J.-I., Ando, S., Tada, H., Tomura, M., and Yamashita, Y. (2005) Synthesis, characterization and FET properties of novel dithiazolylbenzothiadiazole derivatives. *Chem. Commun.*, (25), 3183–3185.

148 Ando, S., Nishida, J., Tada, H., Inoue, Y., Tokito, S., and Yamashita, Y. (2005) High performance *n*-type organic field-effect transistors based on π-electronic systems with trifluoromethylphenyl groups. *J. Am. Chem. Soc.*, **127** (15), 5336–5337.

149 Ando, S., Murakami, R., Nishida, J., Tada, H., Inoue, Y., Tokito, S., and Yamashita, Y. (2005) *n*-Type organic field-effect transistors with very high electron mobility based on thiazole oligomers with trifluoromethylphenyl groups. *J. Am. Chem. Soc.*, **127** (43), 14996–14997.

150 Yoon, M.H., DiBenedetto, S.A., Facchetti, A., and Marks, T.J. (2005) Organic thin-film transistors based on carbonyl-functionalized quaterthiophenes: high mobility *n*-channel semiconductors and ambipolar transport. *J. Am. Chem. Soc.*, **127** (5), 1348–1349.

151 Hasegawa, T. and Takeya, J. (2009) Organic field-effect transistors using single crystals. *Sci. Technol. Adv. Mater.*, **10** (2), 024314.

152 Yamagishi, M., Tominari, Y., Uemura, T., and Takeya, J. (2009) Air-stable *n*-channel single-crystal transistors with negligible threshold gate voltage. *Appl. Phys. Lett.*, **94** (5), 053305.

153 Chesterfield, R.J., Newman, C.R., Pappenfus, T.M., Ewbank, P.C., Haukaas, M.H., Mann, K.R., Miller, L.L., and Frisbie, C.D. (2003) High electron mobility and ambipolar transport in organic thin-film transistors based on a π-stacking quinoidal terthiophene. *Adv. Mater.*, **15** (15), 1278–1282.

154 Handa, S., Miyazaki, E., Takimiya, K., and Kunugi, Y. (2007) Solution-processible *n*-channel organic field-effect transistors based on dicyanomethylene-substituted terthienoquinoid derivative. *J. Am. Chem. Soc.*, **129** (38), 11684–11685.

155 Gui, K., Mutkins, K., Schwenn, P.E., Krueger, K.B., Pivrikas, A., Wolfer, P., Stingelin Stutzmann, N., Burn, P.L., and Meredith, P. (2012) A flexible *n*-type organic semiconductor for optoelectronics. *J. Mater. Chem.*, **22** (5), 1800–1806.

156 Laquindanum, J.G., Katz, H.E., Dodabalapur, A., and Lovinger, A.J. (1996) *n*-Channel organic transistor materials based on naphthalene frameworks. *J. Am. Chem. Soc.*, **118** (45), 11331–11332.

157 Katz, H.E., Lovinger, A.J., Johnson, J., Kloc, C., Siegrist, T., Li, W., Lin, Y.Y., and Dodabalapur, A. (2000) A soluble and air-stable organic semiconductor with high electron mobility. *Nature*, **404** (6777), 478–481.

158 Jones, B.A., Ahrens, M.J., Yoon, M.-H., Facchetti, A., Marks, T.J., and Wasielewski, M.R. (2004) High-mobility air-stable *n*-type semiconductors with processing versatility: dicyanoperylene-3,4:9,10-bis(dicarboximides). *Angew. Chem. Int. Ed.*, **43** (46), 6363–6366.

159 Shukla, D., Nelson, S.F., Freeman, D.C., Rajeswaran, M., Ahearn, W.G., Meyer, D.M., and Carey, J.T. (2008) Thin-film morphology control in naphthalene-diimide-based semiconductors: high mobility *n*-type semiconductor for organic thin-film transistors. *Chem. Mater.*, **20** (24), 7486–7491.

160 Jung, B.J., Sun, J., Lee, T., Sarjeant, A., and Katz, H.E. (2009) Low-temperature-processible, transparent, and air-operable *n*-channel fluorinated phenylethylated naphthalenetetracarboxylic diimide semiconductors applied to flexible transistors. *Chem. Mater.*, **21** (1), 94–101.

161 See, K.C., Landis, C., Sarjeant, A., and Katz, H.E. (2008) Easily synthesized naphthalene tetracarboxylic diimide semiconductors with high electron mobility in air. *Chem. Mater.*, **20** (11), 3609–3616.

162 Chesterfield, R.J., McKeen, J.C., Newman, C.R., Frisbie, C.D., Ewbank, P.C., Mann, K.R., and Miller, L.L. (2004) Variable temperature film and contact resistance measurements on operating *n*-channel organic thin film transistors. *J. Appl. Phys.*, **95** (11), 6396–6405.

163 Chesterfield, R.J., McKeen, J.C., Newman, C.R., Ewbank, P.C., da Silva, D.A., Bredas, J.L., Miller, L.L., Mann, K.R., and Frisbie, C.D. (2004) Organic thin film transistors based on *N*-alkyl perylene diimides: charge transport kinetics as a function of gate voltage and temperature. *J. Phys. Chem. B*, **108** (50), 19281–19292.

164 Tatemichi, S., Ichikawa, M., Koyama, T., and Taniguchi, Y. (2006) High-mobility *n*-type thin-film transistors based on *N,N'*-ditridecyl perylene diimide with thermal treatments. *Appl. Phys. Lett.*, **89** (11), 112108.

165 Schmidt, R., Oh, J.H., Sun, Y.-S., Deppisch, M., Krause, A.-M., Radacki, K., Braunschweig, H., Könemann, M., Erk, P., Bao, Z., and Würthner, F. (2009) High-performance air-stable *n*-channel organic thin film transistors based on halogenated perylene bisimide semiconductors. *J. Am. Chem. Soc.*, **131** (17), 6215–6228.

166 Jones, B.A., Facchetti, A., Marks, T.J., and Wasielewski, M.R. (2007) Cyanonaphthalene diimide semiconductors for air-stable, flexible, and optically transparent *n*-channel field-effect transistors. *Chem. Mater.*, **19** (11), 2703–2705.

167 Yoo, B., Jung, T., Basu, D., Dodabalapur, A., Jones, B.A., Facchetti, A., Wasielewski, M.R., and Marks, T.J. (2006) High-mobility bottom-contact *n*-channel organic transistors and their use in complementary ring oscillators. *Appl. Phys. Lett.*, **88** (8), 082104.

168 Ling, M.M., Erk, P., Gomez, M., Koenemann, M., Locklin, J., and Bao, Z. (2007) Air-stable *n*-channel organic semiconductors based on perylene diimide derivatives without strong electron withdrawing groups. *Adv. Mater.*, **19** (8), 1123–1127.

169 Gao, X.K., Di, C.A., Hu, Y.B., Yang, X.D., Fan, H.Y., Zhang, F., Liu, Y.Q., Li, H.X., and Zhu, D.B. (2010) Core-expanded naphthalene diimides fused with 2-(1,3-dithiol-2-ylidene)malonitrile groups for high-performance, ambient-stable, solution-processed *n*-channel

organic thin film transistors. *J. Am. Chem. Soc.*, **132** (11), 3697–3699.

170 Zhao, Y., Di, C.-A., Gao, X., Hu, Y., Guo, Y., Zhang, L., Liu, Y., Wang, J., Hu, W., and Zhu, D. (2011) All-solution-processed, high-performance *n*-channel organic transistors and circuits: toward low-cost ambient electronics. *Adv. Mater.*, **23** (21), 2448–2453.

171 Zheng, Q., Huang, J., Sarjeant, A., and Katz, H.E. (2008) Pyromellitic diimides: minimal cores for high mobility *n*-channel transistor semiconductors. *J. Am. Chem. Soc.*, **130** (44), 14410–14411.

172 Wang, Z., Kim, C., Facchetti, A., and Marks, T.J. (2007) Anthracenedicarboximides as air-stable *n*-channel semiconductors for thin-film transistors with remarkable current on–off ratios. *J. Am. Chem. Soc.*, **129** (44), 13362–13363.

173 Itaka, K., Yamashiro, M., Yamaguchi, J., Haemori, M., Yaginuma, S., Matsumoto, Y., Kondo, M., and Koinuma, H. (2006) High-mobility C_{60} field-effect molecular-wetting controlled transistors fabricated on substrates. *Adv. Mater.*, **18** (13), 1713–1716.

174 Anthopoulos, T.D., Singh, B., Marjanovic, N., Sariciftci, N.S., Montaigne Ramil, A., Sitter, H., Cölle, M., and de Leeuw, D.M. (2006) High performance *n*-channel organic field-effect transistors and ring oscillators based on C_{60} fullerene films. *Appl. Phys. Lett.*, **89** (21), 213504.

175 Singh, T.B., Marjanovic, N., Stadler, P., Auinger, M., Matt, G.J., Gunes, S., Sariciftci, N.S., Schwodiauer, R., and Bauer, S. (2005) Fabrication and characterization of solution-processed methanofullerene-based organic field-effect transistors. *J. Appl. Phys.*, **97** (8), 083714.

176 Lee, T.W., Byun, Y., Koo, B.W., Kang, I.N., Lyu, Y.Y., Lee, C.H., Pu, L., and Lee, S.Y. (2005) All-solution-processed *n*-type organic transistors using a spinning metal process. *Adv. Mater.*, **17** (18), 2180–2184.

177 Chikamatsu, M., Nagamatsu, S., Yoshida, Y., Saito, K., Yase, K., and Kikuchi, K. (2005) Solution-processed *n*-type organic thin-film transistors with high field-effect mobility. *Appl. Phys. Lett.*, **87** (20), 203504.

178 Chikamatsu, M., Itakura, A., Yoshida, Y., Azumi, R., and Yase, K. (2008) High-performance *n*-type organic thin-film transistors based on solution-processable perfluoroalkyl-substituted C_{60} derivatives. *Chem. Mater.*, **20** (24), 7365–7367.

179 Bao, Z., Dodabalapur, A., and Lovinger, A.J. (1996) Soluble and processable regioregular poly(3-hexylthiophene) for thin film field-effect transistor applications with high mobility. *Appl. Phys. Lett.*, **69** (26), 4108–4110.

180 Assadi, A., Svensson, C., Willander, M., and Inganas, O. (1988) Field-effect mobility of poly(3-hexylthiophene). *Appl. Phys. Lett.*, **53** (3), 195–197.

181 Sirringhaus, H., Brown, P.J., Friend, R.H., Nielsen, M.M., Bechgaard, K., Langeveld- Voss, B.M.W., Spiering, A.J.H., Janssen, R.A.J., Meijer, E.W., Herwig, P., and de Leeuw, D.M. (1999) Two-dimensional charge transport in self-organized, high-mobility conjugated polymers. *Nature*, **401** (6754), 685–688.

182 Sirringhaus, H., Tessler, N., and Friend, R.H. (1998) Integrated optoelectronic devices based on conjugated polymers. *Science*, **280** (5370), 1741–1744.

183 Li, J., Sun, Z., and Yan, F. (2012) Solution processable low-voltage organic thin film transistors with high-*k* relaxor ferroelectric polymer as gate insulator. *Adv. Mater.*, **24** (1), 88–93.

184 McCulloch, I., Bailey, C., Giles, M., Heeney, M., Love, I., Shkunov, M., Sparrowe, D., and Tierney, S. (2005) Influence of molecular design on the field-effect transistor characteristics of terthiophene polymers. *Chem. Mater.*, **17** (6), 1381–1385.

185 Ong, B.S., Wu, Y.L., Liu, P., and Gardner, S. (2004) High-performance semiconducting polythiophenes for organic thin-film transistors. *J. Am. Chem. Soc.*, **126** (11), 3378–3379.

186 McCulloch, I., Heeney, M., Bailey, C., Genevicius, K., Macdonald, I., Shkunov, M., Sparrowe, D., Tierney, S., Wagner, R., Zhang, W.M., Chabinyc, M.L., Kline, R.J., McGehee, M.D., and Toney, M.F.

(2006) Liquid-crystalline semiconducting polymers with high charge-carrier mobility. *Nat. Mater.*, **5** (4), 328–333.

187 Heeney, M., Bailey, C., Genevicius, K., Shkunov, M., Sparrowe, D., Tierney, S., and McCulloch, I. (2005) Stable polythiophene semiconductors incorporating thieno[2,3-*b*]thiophene. *J. Am. Chem. Soc.*, **127** (4), 1078–1079.

188 Li, J., Qin, F., Li, C.M., Bao, Q.L., Chan-Park, M.B., Zhang, W., Qin, J.G., and Ong, B.S. (2008) High-performance thin-film transistors from solution-processed dithienothiophene polymer semiconductor nanoparticles. *Chem. Mater.*, **20** (6), 2057–2059.

189 Fong, H.H., Pozdin, V.A., Amassian, A., Malliaras, G.G., Smilgies, D.-M., He, M., Gasper, S., Zhang, F., and Sorensen, M. (2008) Tetrathienoacene copolymers as high mobility, soluble organic semiconductors. *J. Am. Chem. Soc.*, **130** (40), 13202–13203.

190 Pan, H., Wu, Y., Li, Y., Liu, P., Ong, B.S., Zhu, S., and Xu, G. (2007) Benzodithiophene copolymer – a low-temperature, solution-processed high-performance semiconductor for thin-film transistors. *Adv. Funct. Mater.*, **17** (17), 3574–3579.

191 Osaka, I., Sauvé, G., Zhang, R., Kowalewski, T., and McCullough, R.D. (2007) Novel thiophene-thiazolothiazole copolymers for organic field-effect transistors. *Adv. Mater.*, **19** (23), 4160–4165.

192 Osaka, I., Zhang, R., Sauvé, G.V., Smilgies, D.-M., Kowalewski, T., and McCullough, R.D. (2009) High-lamellar ordering and amorphous-like π-network in short-chain thiazolothiazole–thiophene copolymers lead to high mobilities. *J. Am. Chem. Soc.*, **131** (7), 2521–2529.

193 Sirringhaus, H., Wilson, R.J., Friend, R.H., Inbasekaran, M., Wu, W., Woo, E.P., Grell, M., and Bradley, D.D.C. (2000) Mobility enhancement in conjugated polymer field-effect transistors through chain alignment in a liquid-crystalline phase. *Appl. Phys. Lett.*, **77** (3), 406–408.

194 Kim, Y.M., Lim, E., Kang, I.-N., Jung, B.-J., Lee, J., Koo, B.W., Do, L.-M., and Shim, H.-K. (2006) Solution-processable field-effect transistor using a fluorene- and selenophene-based copolymer as an active layer. *Macromolecules*, **39** (12), 4081–4085.

195 Chena, M., Crispin, X., Perzon, E., Andersson, M.R., Pullerits, T., Andersson, M., Inganäs, O., and Berggren, M. (2005) High carrier mobility in low band gap polymer-based field-effect transistors. *Appl. Phys. Lett.*, **87**, 252105.

196 Usta, H., Lu, G., Facchetti, A., and Marks, T.J. (2006) Dithienosilole– and dibenzosilole–thiophene copolymers as semiconductors for organic thin-film transistors. *J. Am. Chem. Soc.*, **128** (28), 9034–9035.

197 Zhang, M., Tsao, H.N., Pisula, W., Yang, C.D., Mishra, A.K., and Müllen, K. (2007) Field-effect transistors based on a benzothiadiazole-cyclopentadithiophene copolymer. *J. Am. Chem. Soc.*, **129** (12), 3472–3473.

198 Tsao, H.N., Cho, D., Andreasen, J.W., Rouhanipour, A., Breiby, D.W., Pisula, W., and Müllen, K. (2009) The influence of morphology on high-performance polymer field-effect transistors. *Adv. Mater.*, **21** (2), 209–212.

199 Wang, S., Kappl, M., Liebewirth, I., Müller, M., Kirchhoff, K., Pisula, W., and Müllen, K. (2012) Organic field-effect transistors based on highly ordered single polymer fibers. *Adv. Mater.*, **24** (3), 417–420.

200 Lin, H.-W., Lee, W.-Y., and Chen, W.-C. (2012) Selenophene-DPP donor-acceptor conjugated polymer for high performance ambipolar field effect transistor and nonvolatile memory applications. *J. Mater. Chem.*, **22** (5), 2120–2128.

201 Babel, A. and Jenekhe, S.A. (2002) Electron transport in thin-film transistors from an *n*-type conjugated polymer. *Adv. Mater.*, **14** (5), 371–374.

202 Babel, A. and Jenekhe, S.A. (2003) High electron mobility in ladder polymer field-effect transistors. *J. Am. Chem. Soc.*, **125** (45), 13656–13657.

203 Chen, Z., Zheng, Y., Yan, H., and Facchetti, A. (2008) Naphthalenedicarboximide- vs. perylenedicarboximide-based copolymers. Synthesis and semiconducting properties in bottom-gate *n*-channel organic transistors. *J. Am. Chem. Soc.*, **131** (1), 8–9.

204 Yan, H., Chen, Z.H., Zheng, Y., Newman, C., Quinn, J.R., Dotz, F., Kastler, M., and Facchetti, A. (2009) A high-mobility electron-transporting polymer for printed transistors. *Nature*, **457** (7230), 679–687.

205 Zhan, X., Tan, Z.A., Domercq, B., An, Z., Zhang, X., Barlow, S., Li, Y., Zhu, D., Kippelen, B., and Marder, S.R. (2007) A high-mobility electron-transport polymer with broad absorption and its use in field-effect transistors and all-polymer solar cells. *J. Am. Chem. Soc.*, **129** (23), 7246–7247.

206 Dodge, J.A., Bain, J.D., and Chamberlin, A.R. (1990) Regioselective synthesis of substituted rubrenes. *J. Org. Chem.*, **55** (13), 4190–4198.

207 Valiyev, F., Hu, W.-S., Chen, H.-Y., Kuo, M.-Y., Chao, I., and Tao, Y.-T. (2007) Synthesis and characterization of anthra[2,3-*b*]thiophene and tetraceno[2,3-*b*]thiophenes for organic field-effect transistor applications. *Chem. Mater.*, **19** (12), 3018–3026.

208 Wang, C.L., Wei, Z.M., Meng, Q., Zhao, H.P., Xu, W., Li, H.X., and Hu, W.P. (2010) Dibenzo[*b,d*]thiophene based oligomers with carbon-carbon unsaturated bonds for high performance field-effect transistors. *Org. Electron.*, **11** (4), 544–551.

3
Organic/Polymeric Field-Effect Transistors

Chengliang Wang, Lang Jiang, Wenping Hu

Organic/polymeric field-effect transistors (OFETs) are not only essential building blocks for the next generation of cheap and flexible organic circuits, but also provide an important insight into the charge transport of π-conjugated systems. Therefore, they may serve as strong tools to explore the structure–property relationships of π-conjugated systems, such as parameters of field-effect mobility, current on/off ratio, and threshold voltage. OFETs have attracted much attention since the 1980s, with investigations having included the discovery, design, and synthesis of organic/polymeric conjugated systems for OFETs, device optimization, the development of applications in radiofrequency identification (RFID) tags, flexible displays, electronic papers, and sensors.

In this chapter, OFETs based on organic and polymeric semiconductors are introduced, including the device configurations, their working principles, and technologies for the fabrication of such devices. The factors that dominate the performance of these transistors will be also analyzed, such as the grain boundaries, morphology, interfaces and crystallization. The way in which organic semiconductor molecules are ordered and the alignment of energy levels of organic semiconductors and electrodes for the fabrication of high-performance transistors are also discussed. Finally, organic single-crystal transistors, including the controllable growth of organic single crystals, new fabrication techniques, transistor applications, structure–properties dependence and the transport anisotropy of organic single crystals, are described.

3.1
Introduction

Although various reports have referred to the semiconducting properties of organic materials [1] since the first reports of inorganic transistors were made during the 1940s, it was not until 1977 that organic materials began to attract interest in electronic devices, with the discovery of conducting polyacetylene [2]. Subsequently, organic/polymeric conjugated systems have attracted much attention as futuristic materials for the development and production of the next generation of

Organic Optoelectronics, First Edition. Edited by Wenping Hu.

electronic devices. These advances have been based on the benefits of these materials when compared to their inorganic counterparts, including their flexibility, light weight, ease of functionalization, and an ability to be fabricated in large areas at low cost.

The first OFET was created by Tsumura and coworkers in 1986, using polythiophene as a semiconducting layer [3] that would conduct charge, thus eliminating the need to use expensive metal oxide semiconductors. Since that time, much progress has been made in this area, with FET performance having been observed for thousands of organic semiconductors and their mobilities being improved from $10^{-5}\,cm^2/(V \cdot s)$ [3] to $10\,cm^2/(V \cdot s)$ [4, 5] for thin-film transistors, and to $15\sim40\,cm^2/(V \cdot s)$ [6–8] for single-crystal transistors.

3.1.1
Configurations of Organic Field-Effect Transistors

A typical OFET consists of a semiconducting layer, an insulating layer (gate dielectric), and three terminals (gate, drain, and source) (see Figure 3.1). According to the position of the electrodes, OFETs can be classified as four types [9]: bottom-gate top-contact (BGTC, Figure 3.1a); bottom-gate bottom-contact (BGBC, Figure 3.1b); top-gate top-contact (TGTC, Figure 3.1c); and top-gate bottom-contact (TGBC, Figure 3.1d). Because the organic semiconductors are sensitive to thermal irradiation and solvents, the BGBC configuration is the most convenient for organic materials, taking into consideration the influence of the deposition sequence on the organic semiconductors. However, BGTC usually showed a higher performance than BGBC, due to the larger contact region (the contact resistance does not dominate the performance) and a better contact quality between the semiconductor and the drain/source electrodes in BGTC mode [9, 10]. Consequently, the BGTC configuration is also widely used to construct OFETs in order to achieve an intimate contact between the organic semiconductor and the electrodes. However,

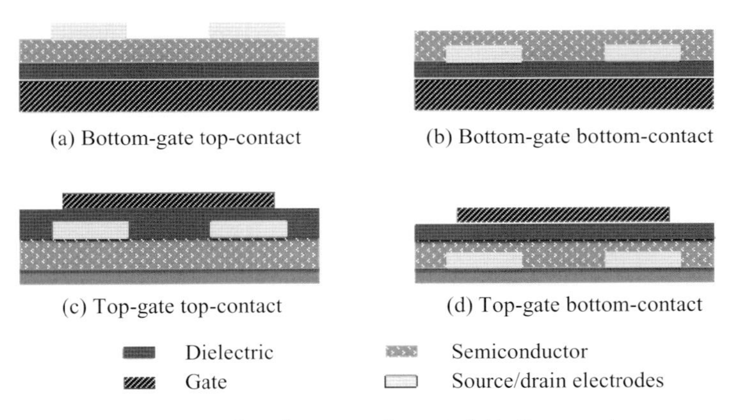

Figure 3.1 Four types of configuration of organic field-effect transistors.

when considering the practical applications of OFETs, top-gate modes are most likely more convenient for device fabrication.

3.1.2
Working Principle of Organic Field-Effect Transistors

An OFET can be regarded as a capacitor that consists of the conducting channel (between source and drain electrodes) and the gate electrode. If a voltage is applied to the gate electrode, charges are induced at the conducting channel, which serves as the other plate of the capacitor. In this case, if a bias is applied to the drain/source electrode, a current will be generated from the source electrode to the drain electrode. The conducting area of the semiconductor layer is termed the conducting channel; in OFETs, it is well recognized that the conducting channel is located close to the semiconductor/insulator interface, with a thickness of one to several molecular layers.

Compared to inorganic semiconductors (which are highly conductive through controllable and efficient doping), organic semiconductors often show a low conductivity, and consequently OFETs usually operate in accumulation mode. In the case of *p*-channel OFETs, holes should be accumulated in the channel by injecting holes from the source electrode to the semiconductor, which requires the semiconductor molecules to be stable as cations. In contrast, in the case of *n*-channel OFETs, the electrons should be accumulated in the conducting channel, and the semiconductor molecules should be stable as anions. Hence, *p*-channel OFETs will operate with negative gate voltages, whereas *n*-channel transistors operate with positive gate voltages (see Figure 3.2). It is difficult to obtain a defect-free semiconductor/insulator interface, as some of the induced charge carriers may become trapped at the interface. Moreover, charge transport may be affected by other factors such as the morphology (e.g., thin film, grain boundaries as traps) or the environmental conditions (oxygen or moisture). Hence, only when the gate voltage V_G is larger than a certain value can the channel be turned "on"; this gate voltage is termed the threshold voltage (V_T).

The carrier concentration profile of a FET during its operation is shown in Figure 3.3. When no drain-source bias is applied, the charge carriers are distributed uniformly in the channel (Figure 3.3a); however, as the drain-source bias V_D is increased, and if the bias is very small (much less than $V_G - V_T$), the charge concentration will become distributed as a linear gradient in the channel (Figure 3.3b). In this situation, the current will be increased linearly with the gate voltage. The current can then be described by the following equation:

$$I_D = \mu C_i W/L([V_G - V_T]V_D - V_D^2/2)$$

where I_D is the drain current, W and L are the channel width and length respectively, μ is the mobility (charge carrier velocity per unit electric field), C_i is the electric capacitance of the insulator per unit area, and V_G, V_T, and V_D are the gate, threshold, and drain voltages, respectively. If $V_D << V_G - V_T$, the term $V_D^2/2$ can be dropped, under which circumstances the transistor will operate in the *linear*

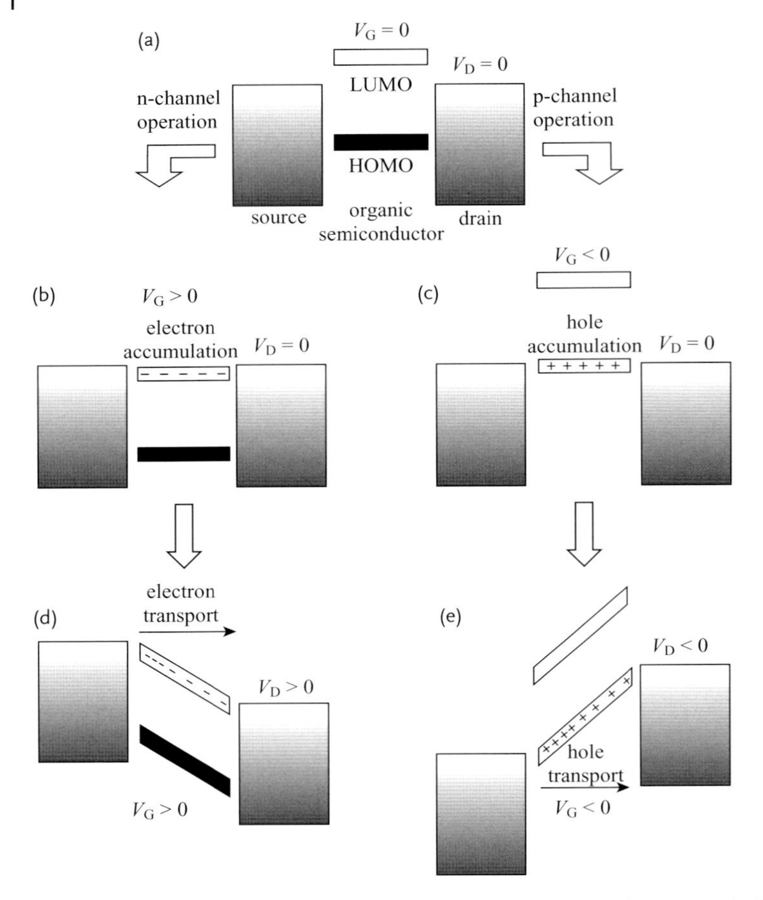

Figure 3.2 Ideal energy level diagram of an OFET. (a) When no voltage is applied; (b–e) When operated in electron accumulation (b and d) and hole accumulation (c and e) mode. Reprinted with permission from Ref. [11]; © 2004, American Chemical Society.

regime. In this region the drain current is also increased linearly by increasing the drain voltage, which can be described by the following equation:

$$I_D = \mu C_i W / L(V_G - V_T)V_D.$$

As the drain voltage further increases, $V_D \approx V_G - V_T$, the carrier concentration profile becomes nonuniform in the channel. The term $V_D^2/2$ cannot be omitted from the equation, so that the current is no longer proportional to the drain voltage. When $V_D = V_G - V_T$, there is no longer any potential difference between the gate and the drain electrode in the channel near the drain electrode; this point is regarded as "pinched off" voltage. A depletion area is also formed near the drain electrode, in which there are no charge carriers left. By using

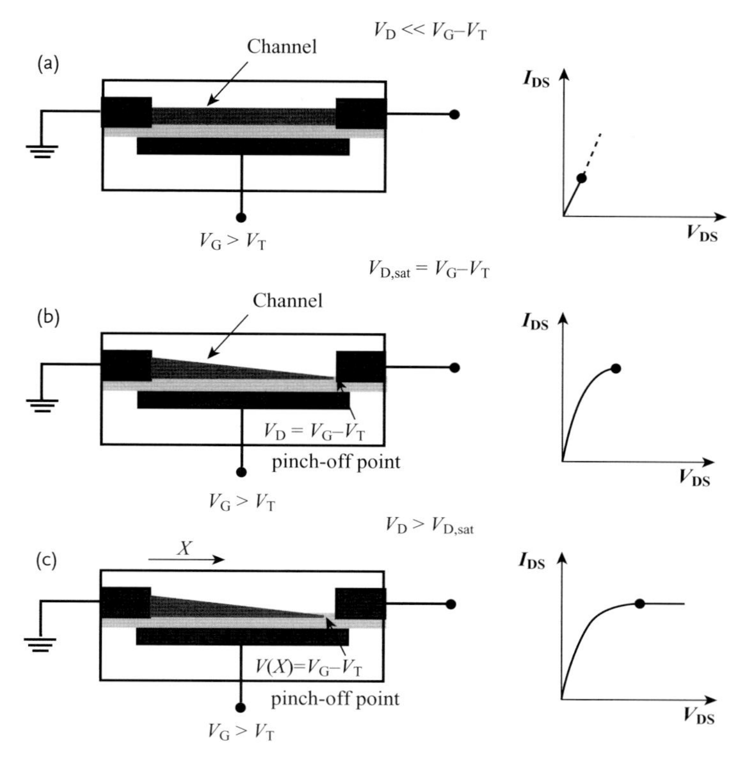

Figure 3.3 Carrier concentration profile of a FET. (a) In the linear regime; (b) When pinch-off occurs ($V_D \approx V_G - V_T$); (c) In the saturation regime. Reprinted with permission from Ref. [12]; © 2007, American Chemical Society.

$V_D = V_G - V_T$, it is possible to obtain the relationship between the drain current and the gate voltage:

$$\begin{aligned} I_D &= \mu C_i W/L([V_G - V_T]V_D - V_D^2/2) = \mu C_i W/L([V_G - V_T][V_G - V_T] - [V_G - V_T]^2/2) \\ &= \mu C_i (V_G - V_T)^2 \end{aligned}$$

Further increasing the drain voltage only results in the "pinched-off" point inclining towards the source electrode; that is, the drain current is no longer increased. Under this circumstance, the transistor will operate in the *saturation regime*.

All of the above discussions relate to a general condition in which the source electrode is grounded and the drain and gate voltages are positive for *n*-channel transistors and negative for *p*-channel transistors. In other words, if the gate voltage is negative as for the *n*-channel transistors, then no electrons will be accumulated in the channel and the channel will be "off". In fact, this is not an exact situation, as a region also exists where the channel can be turned "on," even if the drain and gate voltage are negative for *n*-type semiconductors. If the drain electrode (*bias*) is regarded as the "source" electrode and the source electrode (*grounded*) as the "drain" electrode, then under these circumstances the "V_{DS}" and the "V_{GS}" (which are equal to the V_{SD} and V_{GD}, respectively) are equivalent to positive values. Because it is difficult to accumulate holes in organic semiconductors, the device is operated in depletion mode when a negative gate voltage is applied for *n*-channel transistors. This means that an efficient "gate" voltage (absolute value) must be equal to or lower than the "source" voltage (absolute value), in order to guarantee that the electrons would not be extensively depleted ($|V_D| \geq |V_G - V_T|$). In other words, if – in the general condition – the gate voltage is positive and greater than the threshold voltage, then the channel will be "on" (including the *linear regime* and *saturation regime*). However, if the drain and gate voltage are negative, then only if the efficient gate voltage is equal to or lower than the drain voltage can the channel also be turned "on." Under these conditions, the device will operate in the "*saturation regime*," although this "*saturation regime*" is slightly different from that mentioned above, which is usually termed a *super-linear regime*. As "V_{GS}" should be V_{GD}, and "I_D" should be $-I_D$, the drain current (the grounded terminal is regarded as the source electrode) can be described by the following equation:

$$I_D = -\mu C_i(\text{“}V_{GS}\text{”} - V_T)^2 = -\mu C_i(V_G - V_D - V_T)^2$$

The three types of operation region and off-states can be described diagrammatically, as shown in Figure 3.4 [13, 14].

According to the scanning condition, two types of *I–V* curve are usually presented, namely *output* (constant gate voltage, Figure 3.5a) and *transfer* (constant drain voltage, Figure 3.5b) curves. Important parameters such as mobility and threshold voltage can be derived from the transfer curves, by using the slope and

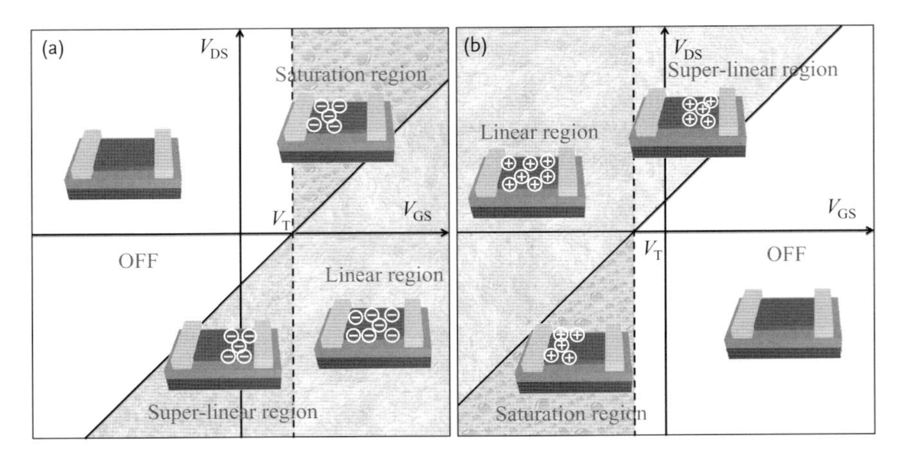

Figure 3.4 Four operation regimes of (a) *n*-channel and (b) *p*-channel FET: off-state, saturation, linear and super-linear regimes. Adapted with permission from Refs [13, 14]; © 2010 and 2011, Elsevier B.V.

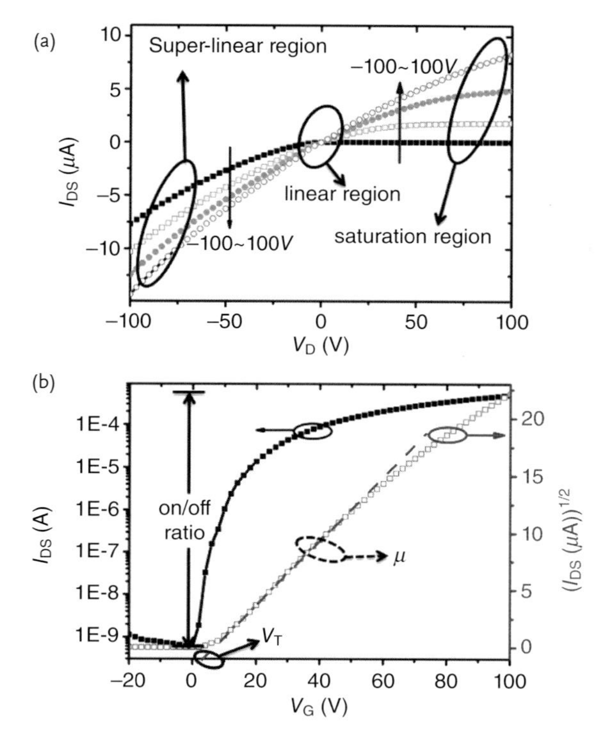

Figure 3.5 Typical (a) output and (b) transfer (e.g., obtained in saturation region) curves of an *n*-type FET.

intercept respectively of the drain current (linear regime) or the square-root of the drain current (saturation regime) as a function of gate voltage. The on/off current ratio is calculated from the ratio of maximum and minimum drain currents. Because the super-linear region is equivalent to the saturation region, the relevant *I–V* curves and detailed analyses will not be further described at this point.

3.2 Carriers Transport in Organic Field-Effect Transistors

3.2.1 Molecular Arrangement in Organic Semiconductors

It is well acknowledged that charge transport in OFETs is much different from that in inorganic FETs, the reasons for which are quite clear. In the case of inorganic semiconductors (e.g., silicon) the charge carriers are transported in "one molecule" (i.e., the whole molecule constitutes a domain or a single crystal through strong covalent bonding, and so can be regarded as one molecule). However, in the case of organic semiconductors the charge carriers are transported through many molecules (especially for small-molecule semiconductors, the channel is constituted by molecules via weak van der Waals' force), unless a single-molecule channel can be obtained (which is very difficult to achieve).

Consequently, charge transport in OFETs is highly dependent on the molecular arrangement, because the charge carriers transport much more slowly through intermolecular routes (which will serve as bottleneck and hence crucially determine the rate of charge transport) than through intramolecular paths. Hence, at this point it is important to consider the typical molecular packing motif, as simplified in Figure 3.6. As the specific arrangements for any certain material are quite different, only those arrangements that are efficient for charge transport are shown here [16]:

- Herringbone packing without π–π overlap (face-to-face) between adjacent molecules (e.g., pentacene [17], one of the benchmark molecules for OFETs).

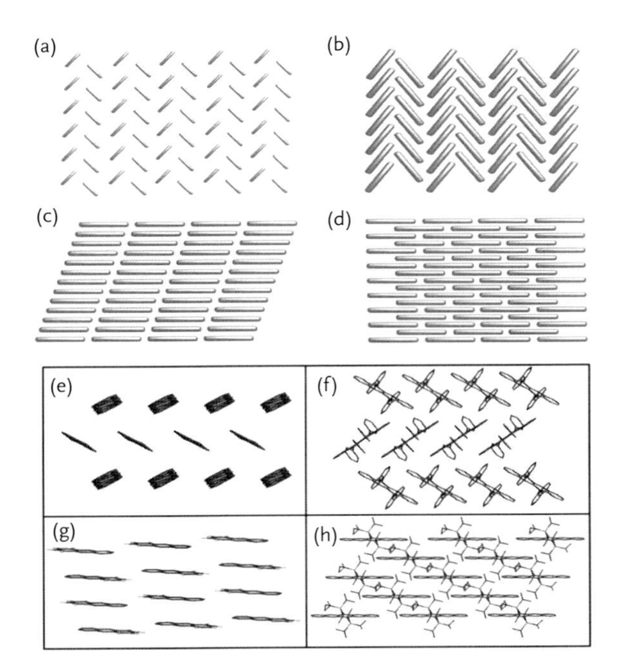

Figure 3.6 Typical efficient molecular packing motifs in crystals. (a–d) Schematic diagrams; (e–h) Example packing diagrams in their single crystals. (a, e) Herringbone packing without overlap between adjacent molecules; (b, f) Herringbone packing with overlap between adjacent molecules; (c, g) Lamellar motif, 1-D stacking; (d, h) Lamellar motif, 2-D stacking. Reprinted with permission from Refs [15, 16]; © 2012, American Chemical Society; © 2010, American Chemical Society, respectively.

- Herringbone packing with π–π overlap between adjacent molecules (e.g., rubrene, one of the highest mobilities reported to date, by using its single crystals as a semiconductor [6]).
- Lamellar motif, one-dimensional π-stacking (e.g., C_8-PTCDI, one of the highest electron mobilities reported to date [18]).
- Lamellar motif, two-dimensional (2-D) π-stacking (e.g., TiOPc, one of the highest mobilities reported to date, by using its thin films as a semiconductor [4]).

The sandwich–herringbone packing motif (which is unsuitable for charge transport in long range, e.g., perylene [19] and BTEA [20]) and 2-D π–π stacking motif (e.g., C_{18}-PTCDI [21]) have been omitted.

Several types of charge transport route via neighbor molecules are shown in Figure 3.7. Besides edge-to-edge and face-to-face packing (e.g., π–π stacking), in some semiconductors an edge-to-face packing (e.g., herringbone packing) also exist (see Figures 3.6 and 3.7). Among the types of intermolecular transport identified, charge transport along the π–π stacking direction is believed to be the most efficient, as the π–π overlap along this direction would facilitate charge carrier transport from one molecule to the next. Consequently, much effort has been focused on how to obtain π–π stacking in the solid states. The most efficient approach is to use chemical modification; for example, substitutions on the side-positions of acenes can alter the herringbone packing (the acenes) into π–π stacking (the acene derivatives) (see Chapter 2 for details).

As with device physics, there are today many methods available to improve molecular arrangements. For example, by using different solvents, α- and β-phase tetrathiafulvalene (TTF) could be obtained respectively, which showed quite

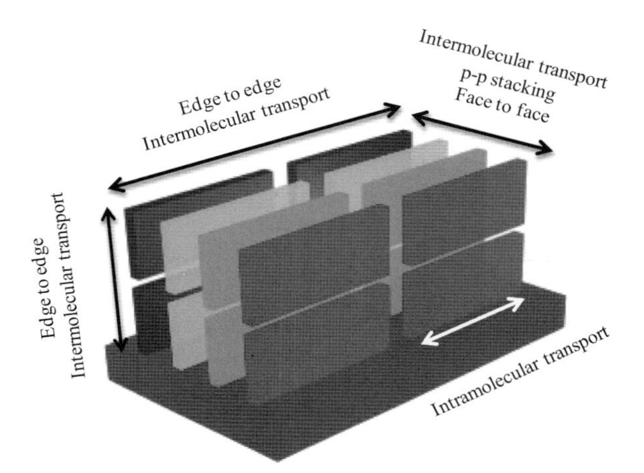

Figure 3.7 Intermolecular and intramolecular charge transport (take 1-D π–π stacking for example).

different packing motifs and different performances [22]. The same approach is quite commonly adopted to create crystalline thin films with a large grain size, by using different substrate temperatures or different self-assembled monolayers (SAMs) to modify the substrate [23–25], in order to obtain a high performance.

3.2.2
Charge Transport Models in Organic Semiconductors

In inorganic semiconductors, in which the atoms are linked by strong covalent bonds (e.g., Si, 76 kcal mol^{-1}), the charges are transported "... as highly delocalized plane waves in wide bands" [26], which in turn provides a very high mobility at room temperature. Under such circumstances, transport is limited by the vibrations of the crystal lattice (phonons) that scatter the carriers, and hence mobility is reduced as the temperature is increased (a high temperature means a strong vibration of the crystal lattice). However, as with organic semiconductors, the situation may be much more complex. For example, if the channel is constituted by more than one molecule, then the charge carriers must transport through either intramolecular transport, intermolecular transport, and/or charge transport through chemical impurities, grain boundaries, or other defects. Consequently, the interpretation of a certain condition may often differ, depending on the crucial role(s) of these three types of route.

Due to the polarization, the charge carriers may have a definite effect on these circumstances. It has been suggested that the charge carriers transport as polarons, which are quasi-particles composed of a charge and an accompanying polarization field. The most powerful evidence verifying the existence of the polaron is the dependence of mobility on the dielectric constant. It has been suggested that mobility should increase in line with the dielectric constant (ε), since a high value of ε means a that a high electric field will be induced at the same voltage. However, the results obtained were the opposite. If the charge carriers exist as polarons, then the paradox could be solved; because the polaron has an efficient mass, it will move under the influence of the dielectric constant. A large dielectric constant would hinder movement of the polaron, which would result in a negative coefficient-dependence of mobility on the dielectric constant [27, 28].

In the case of ultrapure single crystals at low temperature [26, 29] and highly ordered crystalline polymers (within large-sized domain), a band-like transport model has been prevalent to explain the mechanism of charge transport. The reason for this is that in such a situation the grain boundaries and defects would be negligible for the charge transport, and that the charge carriers would move through intramolecular and intermolecular transport. It has been supposed that carriers can "delocalize" (extended-state band conduction [30]) in the whole channel, because the molecules in the channel are perfectly ordered. As a result, the vibration of the molecules (or crystal lattice) becomes responsible for limiting the mobility. If the temperature is increased, the vibration would become violent, leading to a reduced mobility. In other words, the mobility is coincident with a negative temperature coefficient, following a power law: $\mu \propto T^{-n}$. Band-like charge

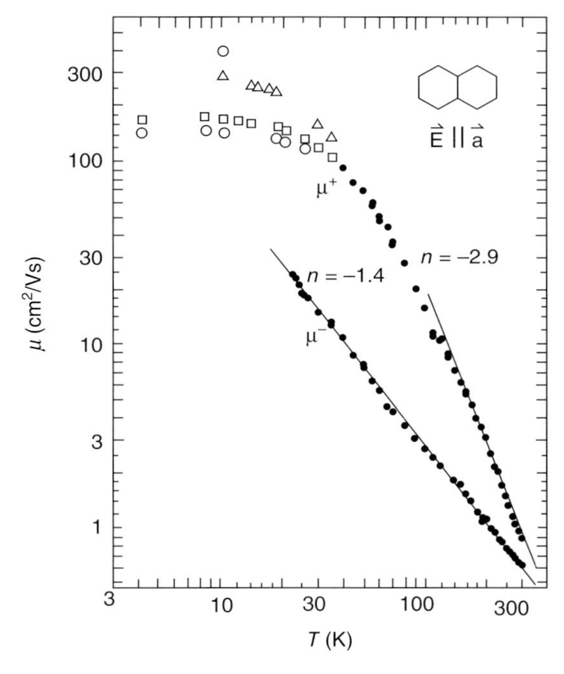

Figure 3.8 The power-law temperature dependence of electron and hole mobilities in ultrapure naphthalene. Reprinted with permission from Ref. [31]; © 1985, American Physical Society.

transport has succeeded in an interpretation of the temperature-dependent mobility for types of pure single crystals, such as ultrapure naphthalene [31, 32], pentacene [33], and DPA [34] (see curves simulated for naphthalene between 150~300 K in Figure 3.8). These results are often obtained by first detrapping the single crystals by using ultrapure single crystals, and employing a time-of-flight (TOF) method (i.e., a large field). By utilizing an air-gap technique, Podzorov *et al.* [35] also obtained an intrinsic polaronic band-like transport in rubrene single crystals at low field (see Figure 3.9, at 150 to 300 K). A similar band-like transport was also observed in solution-processed single crystals of C8-BTBT [36]. Reasonably, two strategies exist to achieve this form of transport: (i) strengthen the intermolecular interactions; and (ii) to use polymer molecules as a single-molecular channel [26]., although unfortunately both are difficult to achieve.

Normally, it is generally recognized that the charge carriers are delocalized in band-like transport (extended-state band conduction). However, by using optical charge modulation spectroscopy (CMS) of gate-induced charge carriers (Figure 3.10), based on single crystalline TIPS-TAP thin films (where the grain size is much larger than the channel dimension), Sakanoue and Sirringhaus [30] suggested that the charge carrier in a band-like transport model might also be localized and its transport limited by thermal lattice fluctuations.

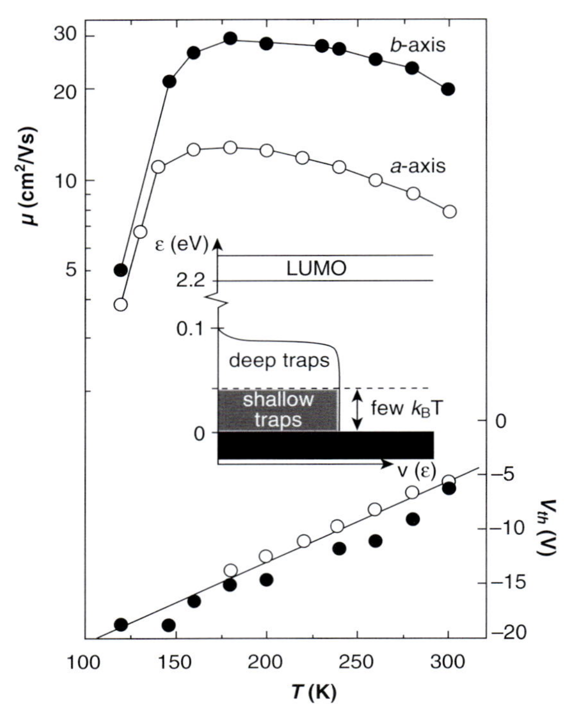

Figure 3.9 Temperature dependence of the mobility and threshold voltage along the a and b axes of rubrene single crystals. 150~300 K: band-like transport; <150 K: MTR mechanism. Inset: Schematic diagram of the electronic states. Reprinted with permission from Ref. [35]; © 2004, American Physical Society.

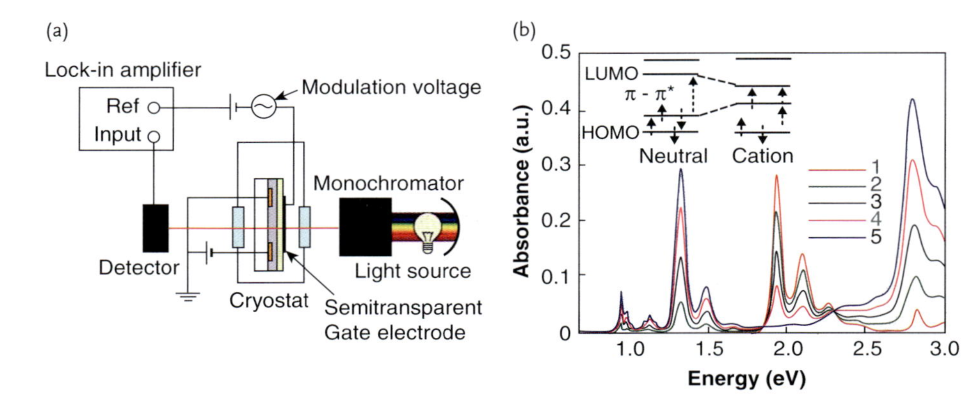

Figure 3.10 Charge modulation spectra of isolated, neutral and radical-cation molecules. (a) Schematic diagram of the instrument; (b) Spectra obtained. This method provides a direct and visible molecular nature changes of/without charge carriers. Reprinted with permission from Ref. [30]; © 2010, Macmillan Publishers Limited.

Incompatibly, the intermolecular interaction between neighbor organic molecules is much weaker than that of covalent bonds, and hence the charge carriers usually cannot delocalize in the whole channel, nor even between adjacent molecules. In other words, the mean free path of an organic semiconductor is lower than the molecular size, and in this sense the band-like transport is not appropriate for organic semiconductors. Various investigations suggested, however, that a "hopping model" could resolve this problem, and consequently this has emerged as the most prevalent mechanism for organic semiconductors. This means that the charge carriers can transport from one localized state to another one by hopping. Under these circumstances, it is not difficult to appreciate that hopping is assisted by phonons, and that the mobility is increased in line with the temperature increasing and the activation energy decreasing: $\mu \propto \exp(-E_a/kT)$, where E_a is the activation energy and k is the Boltzmann constant. In the above-mentioned case, a band-like transport is usually observed at low temperature; however, on increasing the temperature to close to ambient the mobility is, in turn, increased with temperature. Previously, it has been suggested that thermally activated hopping transport is responsible for this phenomenon [37–41] and, based on the relationship between mobility and temperature, the activation energy could be approximated (ca. 0.2 eV) [42, 43].

The types of transport model referred to above only take account of the transport in the ordered domain, without mention of the grain boundaries and traps. Yet, because the traps and defects are unavoidable, a multiple trapping and release (MTR) model–which typically is used to interpret amorphous silicon–has been introduced to solve this problem. It has been suggested that two types of trap exist: (i) a deep trap (more than a few kT from HOMO or LUMO energy level); and (ii) a shallow trap (less than a few kT). When the gate voltage is lower than the threshold voltage (V_T), the charge carriers are first localized in the deep traps (the latter do not affect the mobility), shifting the Fermi level to the HOMO or LUMO energy level. Only when the gate voltage is higher than V_T can the charge carriers become localization in the shallow traps; the conductivity is then dramatically increased [35]. Because the trap and release of the traps are dominant in this case, the mobility is also highly dependent on the temperature, which can have a major effect on the rate of trap and release. At a high temperature, the time for which a polaron traps (τ_{tr}) in the shallow trap is much less than the time taken to move between neighbor traps (τ). Hence, the effective mobility should follow the equation: $\mu = \mu_0(\tau/[\tau + \tau_{tr}]) \approx \mu_0$. However, at a low temperature τ_{tr} is much larger than τ, such that the mobility should obey the following equation: $\mu = \mu_0(\tau/[\tau + \tau_{tr}]) \propto \exp(-E_{tr}/kT)$. Hence, the mobility increases in line with any temperature increase. Because, in this situation, the polarons traps in the shallow traps for most of the time, the mobility will be dominated by the trap and this will lead to a disappearance of the anisotropy properties (see Figure 3.9).

Until now, the typical method used to assign the mechanism of charge transport has been to analyze the relationship between mobility and temperature. However, the mobility can be affected by many other factors, including impurities, channel dimensions, the interface between the semiconductor and the insulator, and the contact resistance between the semiconductor and the electrodes. All of these factors make the situation more complex, and will influence any estimation of the mechanism. Moreover, in some cases the mobility is almost independent of the temperature, which suggests that different mechanisms of charge transport might also exist. Notably, the "real" description of charge transport in organic semiconductors remains unclear.

3.2.3 Factors Influencing Charge Transport in the Conducting Channel of Organic Transistors

Although many factors are known to influence charge transport in the conducting channel of organic transistors, undoubtedly the most important factor is the semiconductor itself, the mobility of which is usually termed intrinsic mobility. It is well known that two important parameters can affect the intrinsic mobility, namely the transfer integral and the reorganization energy [44–46]. As noted above, as the charge carriers need to transport from one molecule to a neighbor molecule, the mobility is highly dependent on a splitting of the frontier molecular orbital between adjacent molecules (the HOMO and LUMO energy levels) – that is, the electronic coupling. This type of electronic coupling, which is referred to as the transfer integral, is highly dependent on the arrangement. There is no doubt that increasing the transfer integral will help to increase the mobility and, indeed, calculations have suggested that a decrease in the intermolecular distance can exponentially increase splitting of the HOMO and LUMO energy levels (Figure 3.11a). However, such splitting does not increase with the lateral increase in π–π overlap; with the shift of adjacent stacking molecules to form an larger π overlap, the splitting of HOMO and LUMO energy present as oscillations in the values (Figure 3.11b) [44]. In other words, although it is usually pursued to enlarge the π–π overlap, a large overlap does not necessarily lead to a large transfer integral and a high mobility.

Both, the transfer integral and the intrinsic mobility are determined by the reorganization energy, which represents the energy loss when a charge carrier transports through a molecule [46]. It is not difficult to appreciate that the reorganization energy must be sufficiently low to facilitate charge transport, and normally the larger the π-conjugated system, the lower will be the reorganization energy. The reasons for this are that a larger π-conjugated system can stabilize the cation (+1 charged state), and also that stable *n*-type semiconductors require electron-withdrawing groups in their chemical structures (to stabilize the anion's −1 charged state). Consequently, the reorganization energy is highly dependent on the conjugated length and degree, the electron-donating or electron-withdrawing group, and the arrangement.

Besides the transfer integral and reorganization energy, both of which determine the intrinsic mobility of the material, many factors can influence device performance. These include: the adopted device configuration (e.g., bottom-contact or top-contact; see above), the materials of the dielectric layer and the electrode; the interface between the semiconductor and the insulator or the interface between

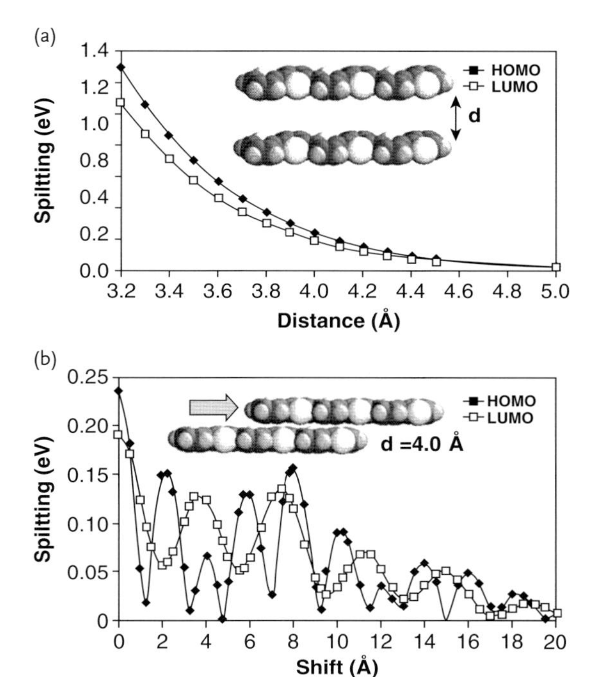

Figure 3.11 Evolution of the electronic splitting of the HOMO and LUMO energy levels as a function of (a) the intermolecular distance and (b) the lateral shift distance. Reprinted with permission from Ref. [44]; © 2002, National Academy of Sciences.

the semiconductor and the electrode; the dimension of the channel (especially the thickness of the semiconducting layer, the length and width of the channel); the morphology; the anisotropic property (different conducting direction); the drain and gate voltage [47]; and the fabrication and testing conditions. These factors will be described more fully in the following sections.

3.3 Electrodes, Insulators, and Interfaces of Organic Field-Effect Transistors

3.3.1 Electrodes

During the operation of OFETs, which have three electrodes (source, drain, and gate) in their device structure [12], the current between the source and drain (I_{DS}) is tuned by the electric field which is in turn induced by the gate voltage (V_G). Two basic requirements must be upheld when selecting an electrode material: (i) the material should be much more conductive than the semiconductor; and

(ii) the material should form homogeneous films, so as to provide a uniform electric field.

Usually, silicon wafers with SiO_2 on top of doped silicon are used as substrates for the fabrication of OFETs. In this case, the highly doped *p*- or *n*-type silicon is highly conductive and can function as the gate electrode. In order to obtain transparent devices [48], indium-tin oxide (ITO), carbon nanotubes (CNTs) and graphene [49–52] have also been used as materials for gate electrodes. Besides the two basic requirements noted above, the selection of a gate electrode material is also dependent on the procedure of device fabrication.

In contrast, the selection of the source and drain electrode materials is much more important than that of the gate electrode for a good device performance, because these are in direct contact with the active semiconductor layers and thus determine the charge carrier injection. Consequently, the energy alignment between the work function of an electrode and the HOMO (*p*-type) or LUMO (*n*-type) of an organic semiconductor is especially important for lowering the charge injection barrier. The work-functions of the metallic materials most frequently used as electrodes are listed in Table 3.1.

The materials used for the source and drain electrodes can be varied from metals to organic charge transfer complexes, or even conductive polymers. The most commonly used metals for OFET electrodes are gold and platinum. Most organic semiconductors have HOMO levels in the range of -5.1 ± 0.3 eV, which matches well with the work-functions of Au (5.1 eV). Consequently, it is easy to achieve an efficient charge carrier injection for *p*-type materials (hole transport) by using gold as the drain/source electrodes [53]. For *n*-type semiconductors, however, the charge carriers are electrons and the related LUMO levels are usually higher than 4.0 eV. Thus, low-work-function metals such as Ca [54], Mg [55], Al [56], and Ag [57] were chosen for the *n*-type and ambipolar OFETs. An example of a light-emitting OFET with ambipolar charge injection and transport, using an asymmetrical device structure, is shown in Figure 3.12 [55]. Here, the source and drain electrodes were tailored for efficient carrier injection by choosing a high-work-

Table 3.1 The work functions (W_F, eV) of the metallic materials most frequently used as electrodes.

Metal	Ag	Al	As	Au	B	Ba	Be	Bi	C	Ca	Cd	Ce
W_F (eV)	4.2	4.2	3.7	5.2	4.5	2.7	5.0	4.2	5.0	2.8	4.2	2.9
Metal	Co	Cr	Cs	Cu	Eu	Fe	Ga	Gd	Hf	Hg	In	Ir
W_F (eV)	5.0	4.5	2.1	4.7	2.5	4.5	4.2	3.1	3.9	4.5	4.2	5.2
Metal	K	La	Li	Lu	Mg	Mn	Mo	Na	Nb	Nd	Ni	Os
W_F (eV)	2.3	3.5	2.9	3.3	3.7	4.1	4.6	2.8	4.3	3.2	5.2	4.8
Metal	Pb	Pt	Rb	Re	Rh	Ru	Sb	Sc	Se	Si	Sm	Sn
W_F (eV)	4.2	5.6	2.2	5.0	5.0	4.7	4.6	3.5	5.9	4.8	2.7	4.4
Metal	Sr	Ta	Tb	Te	Th	Ti	Tl	U	V	W	Y	Zn
W_F (eV)	2.6	4.3	3.0	5.0	3.4	4.3	3.8	3.6	4.3	4.6	3.1	4.3

(a)

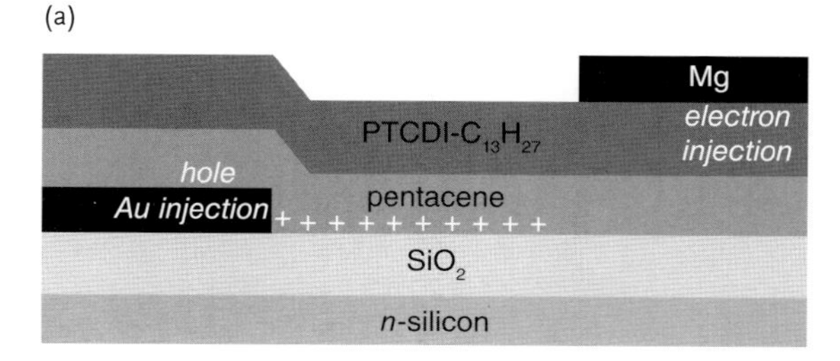

(b)

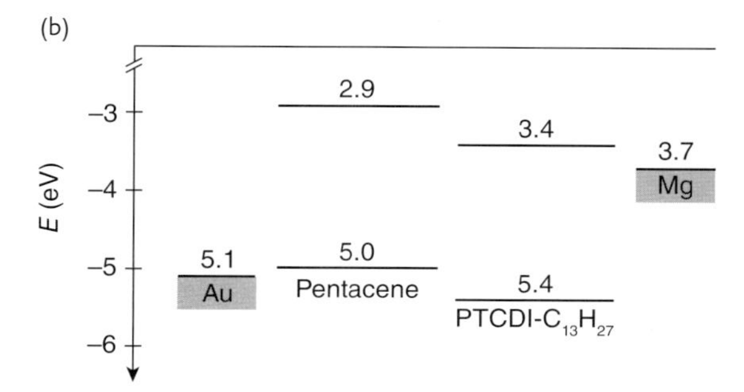

Figure 3.12 (a) Device structure of the ambipolar OFET based on an organic heterostructure; (b) Energy levels (HOMO and LUMO) of pentacene and PTCDI-$C_{13}H_{27}$ and work-function of two metal contacts (Au and Mg). Reprinted with permission from Ref. [55]; © 2004, Elsevier B.V.

function metal, Au, for hole injection into pentacene (*p*-type), and a low-work-function metal, Mg, for electron injection into PTCDI-$C_{13}H_{27}$ (*n*-type). A similar asymmetric device configuration [58] was also reported for high-performance air-stable *n*-type single crystal devices used to build stepwise energy levels between the drain/source electrodes and the LUMO energy levels of the semiconductor $F_{16}CuPc$, and which proved beneficial to both electron injection and flow out of the semiconductor. In order to achieve the goal of low cost, Cu was also used as a drain/source electrode material [59]; the results obtained confirmed that OFETs based on several *p*-type organic semiconductors with Cu top-contact source/drain electrodes could exhibit excellent field-effect properties. The reason for this might be attributed to the obscure diffusion of Cu atoms into the organic semiconducting layer, as had occurred in the Au and Ag electrodes. Moreover, electrode performance would be further improved by the oxidation of Cu, as the Cu_xO layer facilitated hole injection.

Besides the conventional metals, many organic materials have recently been used as electrodes in OFETs, mainly because they are superior to their inorganic counterparts in several aspects:

- Organic materials are flexible, which makes them intriguing for applications to flexible electronics.

- An organic electrode can form a more intimate contact with an organic semiconductor than can an inorganic electrode. This is especially valuable when growing thin films of organic semiconductors (in bottom-contact devices), and may also facilitate charge injection and transport.
- A printing technique could be used to deposit the organic electrodes, with a dramatic decrease in production costs.

Among the materials used for electrode construction, CNTs (and also other nanowires of polymer composites [60]) and graphene [48] have attracted much attention over the years. The main benefit is that the conductivity (σ) of a single CNT could reach 200 000 S cm^{-1}, and even 6600 S cm^{-1} for undoped and randomly oriented CNT films. The main challenge, however, is to obtain CNTs for sufficiently long periods at a high yield, at low cost, and with high purity, and to couple this with a fabrication process that could yield an homogeneous film while maintaining a high conductivity (the resistance between different tubes may range from 200 kΩ up to 20 MΩ).

A similar problem of achieving homogeneous films has also been encountered with other nanowires. For example, graphene–which has a similar chemical structure (large π-conjugated system) to organic semiconductors–has attracted much attention as a potential electrode material. Another benefit of graphene is its ultrathin and flexible nature, which would reduce not only the device size but also the edge effect of the electrodes. The main challenge when using graphene is to obtain sufficiently large amounts of material, but with a low level of defects. Nonetheless, during the past year the production of 75-cm graphene films with resistances as low as ~125 $\Omega\,m^2$ and an optical transmittance as high as 97.4% has been reported [61], using the method of chemical vapor deposition (CVD). Clearly, the application of graphene as transparent electrodes has a bright future!

The second type of organic electrode is the organic conductive polymer. The compatibility between organic electrodes and organic semiconductors is helpful when forming a high-quality interface (better morphology, ordered arrangement and low contact resistance) between the two. For example, highly conducting electrodes (102 S cm^{-1}) obtained from pyrolyzed poly(l,3,4-oxadiazole) (PPOD) thin films were utilized in transistors by using a low-cost, ink-jet printing method [62]. In this case, the crystalline size of pentacene in the conducting channel close to the electrode/semiconductor contact edge with the organic electrodes was clearly larger than that in a device with an Au electrode.

Other materials that might be used as top-contact electrodes in OFETs include poly(3,4-ethylenedioxythiophene) : poly(styrenesulfonate) (PEDOT : PSS) [63, 64] and the charge-transfer complex tetrathiafulvalene : tetracyanoquinodimethane (TTF : TCNQ) [65]. These organic electrodes possess similar advantages to the conducting polymers; for example, although the work-function of Au electrodes matched the HOMO levels of BNT better than the TTF : TCNQ electrode, a bottom-contact device with TTF : TCNQ electrodes exhibited a much better performance than that with Au electrodes [66]. The reason for this might be a better compatibility between organic–organic materials which would, by preference, form a better interface.

3.3.2 Insulators

The performance of OFETs is sensitive to the characteristics of the insulator surface. A good dielectric layer should possess the following characteristics:

- A compact structure and a low leakage current at minimal thickness.
- A smooth surface and a better compatibility with organic semiconductors, so that fewer defects, a better morphology and an ordered arrangement can be obtained.
- A high-*k* insulation, as this could increase the electric field at a certain voltage and successively scale down the device dimensions, increase the integrated degree, and reduce both the operating voltage and energy consumption. However, a high value of *k* is usually linked to a high dielectric constant (ε), which is harmful to performance (Figure 3.13) [27, 67].

3.3.2.1 Oxides

Among the metal oxides used most frequently as insulators can be included silicon oxide (SiO_2) which, conventionally, in used in some inorganic FETs. SiO_2 is very

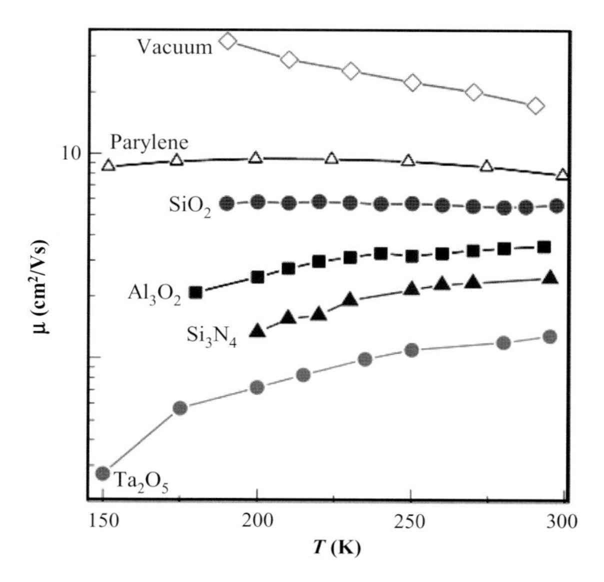

Figure 3.13 Temperature-dependence of mobility with different insulators (ε: Ta_2O_5 [25] > Al_2O_3 [9.4] > Si_3N_4 [7.5] > SiO_2 [3.9] > parylene [2.9] > vacuum [1]). Reprinted with permission from Ref. [28]; © 2006, Macmillan Publishers Limited.

popular in this role because it can be facially oxidized directly on top of a silicon semiconductor, with high quality (a compact structure with a low leakage current) to form a perfect interface, thus simplifying the fabrication process and allowing the material to be polished to produce a very smooth surface. Based on its commercial availability, smooth surface, low leakage current and low cost, SiO_2 is widely used in OFETs. Other metal oxide insulators have been developed with sthe aim of scaling down the device dimensions, decreasing the drive voltage and increasing the integrated degree by utilizing their high-k characteristic and achieving a solution-processed technique to reduce production costs.

Among the most commonly used insulators are included Ta_2O_5, TiO_2, ZrO_2, and Al_2O_3. The most commonly adopted deposition methods for these materials include thermal evaporation, sputtering, anodization, atomic layer deposition, and sol–gel deposition [68]. Of these techniques, the solution-processing and sol–gel methods have attracted the most attention. The application of inorganic oxides in OFETs is mainly limited by their flexibility and solution-processibility, despite inorganic oxides usually possessing a high k and a high environmental stability compared to organic insulators. Moreover, the –OH groups on the material's surface may harm the growth of the semiconductor, trap the electron carriers, and also hinder charge transport. As a consequence, many methods have been developed to overcome this problem, one of the most popular being to modify the insulator surface by applying a buffer layer that might include SAMs. *N*- octyltrichlorosilane (OTS), *n*-octadecyltrichlorosilane (ODTS), *n*-octadecyltrimethoxysilane (OTMS), and *n*-octadecylphosphonic acid (ODPA) have each been used to modify the substrate by anchoring the monolayer through Si–O–Si bonds. These monolayers are typically about 1~2 nm thick and have a compact structure, but have no any significant influence on capacitance. However, they may form a smooth surface and a compact structure, so that the organic end-groups can facilitate the growth of organic films. Their subsequent growth can be controlled by tailoring either the polarity of the end group or the coverage by SAMs. For example, by controlling the rinsing time after assembly of the buffer layer molecules, both monolayer and bilayer ODPAs (in fact, these were islands of bilayer ODPA on top of a whole monolayer) could be obtained [69]. Similar results were also reported whereby the use of different self-assembled processes led to the production of amorphous and crystalline monolayers [70].

Occasionally, in order to combine the merits of two insulators, two or more dielectric layers may be adopted. For example, a double-layer dielectric of TiO_2/Al_2O_3 would combine the merits of TiO_2 (high k) and Al_2O_3 (large band gap), while avoiding the disadvantages of TiO_2 (low band gap, high leakage current) and Al_2O_3 (moderate k value) [71, 72].

3.3.2.2 **Polymers**

Insulating polymers are one of the most suitable candidates to serve as dielectric layers, due to their flexibility, low cost, solution-processability, compatibility to organic semiconductors, and tunable surface chemistry for the control of device-critical interfacial trap state densities. Indeed, to search for and use the correct polymer as a dielectric layer represents one of the most important issues in accomplishing fully organic (i.e., flexible) devices.

The first insulating polymer-based OFETs, which were reported in 1990 [73], used various types of polymer for fabricating devices, including polymethylmethacrylate (PMMA), polystyrene (PS), polyvinylphenol (PVP), and cyanoethylpullulan (CYEPL). The results obtained suggested that a high *k* would be helpful to achieve a high performance. A variety of insulating polymers has been applied in OFETs, including BCB (a divinyltetramethylsiloxane-bis(benzocyclobutene) derivative) [74], polyimide (PI) [75], and cytop [76]. Of these materials, cytop attracted the greatest attention due to its low defect density which leads to negligible hysteresis [77].

Most of these polymer insulators must be solution-processed, but this may lead to pinhole or high leakage currents [76], especially after annealing at high temperature. The reason for this is that the materials are sensitive to heat, which in turn causes a phase transition or a morphology change. One way to overcome this problem would be to increase the thickness of the polymer insulator which, in most cases, may be between 500 nm and 1 μm. By comparison, the thicknesses of most inorganic insulators are less than 500 nm, and less than even 100 nm is quite common [68].

An alternative, and also very effective, approach involves the use of parylene as a dielectric layer. In its normal state parylene is a dimer, but it can be evaporated at ~100 °C and then pyrolyzed at ~700 °C to produce the monomer. The latter can move to a destination zone (at room temperature) and then polymerize into thin films (see Figure 3.14) [78]. Because the process does not require a solvent, and the film-forming procedure is a chemical reaction, the film obtained is very compact and free from pinholes and defects. This type of insulator possesses a low leakage, even when the thickness is as low as 100~200 nm. Moreover, the film does not need to be further annealed and, because the process is free from solvents, the sample can be placed in the zone at room temperature while the predeposited layers will suffer negligible influence due to the parylene deposition.

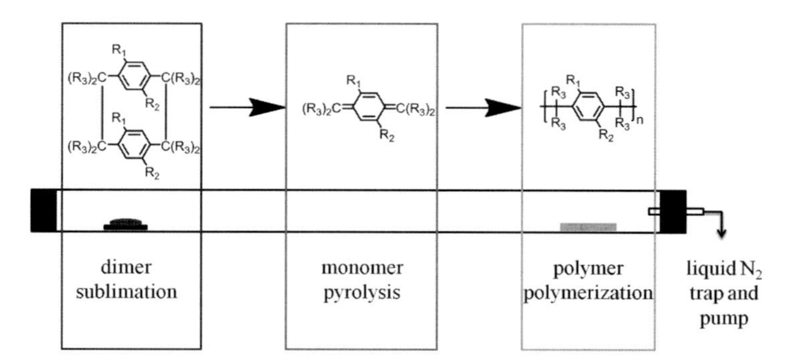

Figure 3.14 Chemical structure and deposition process of parylene.

Consequently, it is incomparably superior in devices using a top-gate mode. In addition, as the polymer is insoluble in most common solvents it shows promise for solution-processed semiconductors. Depending on the substituents (R), the forms of parylene include parylene N ($R_1 = R_2 = R_3 = H$), parylene C ($R_1 = R_3 = H$, $R_2 = Cl$), parylene D ($R_1 = R_2 = Cl$, $R_3 = H$), and parylene HT ($R_1 = R_2 = H$, $R_3 = F$).

The biggest disadvantages of polymer insulators is that they are unsuitable for application in low-voltage operations, due to their low dielectric constant [68]. The intentional increase in thickness of the dielectric layer intensifies this problem because the increased thickness would result in a lower electric field at the same voltage. (An exception to this situation involved devices with a sub-20 nm cytop dielectric, which had a low leakage current and exhibited an outstanding performance at low voltage [79].) Composites that include inorganic nanoparticles with a high *k*, or double-layer insulators containing one high-*k* dielectric layer, may solve this problem. Indeed, the double-layer method [80–82] is much more intriguing, as the top layer of the polymer might benefit from the alignment of organic semiconductors, similar to SAMs as a type of buffer layer.

Biomaterials can also be utilized as insulators in OFETs. For example, the use of chicken albumin (proteins) as a gate insulator in OFETs [83] showed much promise for the fabrication of both *p*-type pentacene and *n*-type C_{60} OFETs. Moreover, no hysteresis was detected in either pentacene or C_{60} OFETs that could be attributed to the natural protein's properties.

3.3.2.3 **Self-Assembled Layers**

It is clear that SAMs themselves might also serve as dielectric layers, their main advantage being that they are only several nanometers thick (typically ca. 2 nm), which would lead to a significant lowering of the drive voltage. Another major advantage is that the SAMs always possess a smooth surface and a good interface compatibility with organic semiconductors, so that a good morphology and arrangement can be expected. Moreover, highly ordered SAMs have a compact self-assembly structure which could reduce the leakage current, although highly ordered SAMs are not easily achieved. Most SAMs are fabricated on the substrate with a native or intentionally oxidized layer, in order to anchor the SAM molecules. For example, Halik and coworkers reported low-voltage organic transistors with an amorphous molecular gate dielectric [84]. Based on a 2.5 nm-thick SAM of (18-phenoxyoctadecyl) trichlorosilane (PhO-OTS) grown on heavily doped Si with native silicon oxide, pentacene thin-film OFETs showed a high mobility of $1\ cm^2/(V \cdot s)$ at a low operation voltage of -2.5 V with an on/off current ratio of 10^6.

3.3.2.4 **Air Dielectric**

Air or vacuum may also be used as dielectric layers in OFETs, since they each possess a very low dielectric constant ($\varepsilon = 1$), such that a high mobility would be expected (see Figure 3.13 [28]). There may be two reasons for this: (i) the low polarity of the dielectric layer facilitates charge transport (a high polarity would result in a larger size of polaron, which would hinder the later transport of polaron); and (ii) the defect of the surface of the dielectric layer no longer exists, which would eliminate or minimize the charge traps. Therefore, it would not be surprising if a mobility of up to 200 000 $cm^2/(V \cdot s)$ were to be obtained from suspended graphene devices when using a vacuum dielectric [85]. Clearly, due to its processing limits this type of dielectric would be applicable only to single-crystal devices [28, 41, 43], making it an unlikely technique for practical applications.

3.3.3 Interfaces

As interfaces are deemed crucial to the performance of devices, extensive investigations have been conducted on the interface engineering of OFETs [86–88]. The transport properties of OFETs are greatly determined by two types of interfaces: (i) those between semiconductors and electrodes, where a charge injection occurs from the electrodes into the semiconductors; and (ii) those between semiconductors and insulators, where charge transport takes place in the semiconductor layer. Hence, the defect density in the interface can play an important role in the performance. Moreover, the interface (i.e., the surface of the pre-deposited layers) might also affect the morphology and arrangement of the semiconductors, which in turn could have a dramatic effect on performance. In fact, the effect that an interface might have on the performance can be separated into two facets: (i) the energy level alignment, which refers to the interface between the semiconductor and the electrode; and (ii) interface compatibility, which refers to the interface between the semiconductor and the electrodes and the interface between the semiconductor and the dielectric layer, because the compatibility not only affects the growth of the semiconductor but also determines the contact.

3.3.3.1 Energy Level Alignment

Although the exact charge injection and transport mechanism remain unclear, it is well acknowledged that the Fermi level of the drain/source electrodes should be better matched with the HOMO or LUMO energy levels of the semiconductors, so that the charge injection can occur. A good match between these energy levels would facilitate the hole/electron carriers injecting from source electrodes to semiconductors, and then flowing out of the semiconductors into the drain electrodes. The results of various investigations [15] have suggested that most *p*-type semiconductors possess HOMO energy levels within -5.1 ± 0.3 eV, and this matches well with high-work-function metals such as Au (5.1 eV) and Pt. On the other hand, most *n*-type semiconductors have LUMO energy levels higher than −4.0 eV, which only matches well with low-work-function metals such as Al, Ca (2.87 eV), and Mg (3.66 eV). It is important to design molecules by taking into consideration the molecular energy gap and levels. The HOMO, LUMO, and energy gap of some organic materials with distinguished properties are summarized in Figure 3.15 [89]. For example, rubrene is a typical *p*-type semiconductor although, with Ag as a drain/source electrode, an ambipolar property was observed [57]. Due to the large energy barrier between the LUMO energy of rubrene (2.09 eV [57], or 3.15 eV) and the Fermi level of a silver electrode (−4.26 eV),

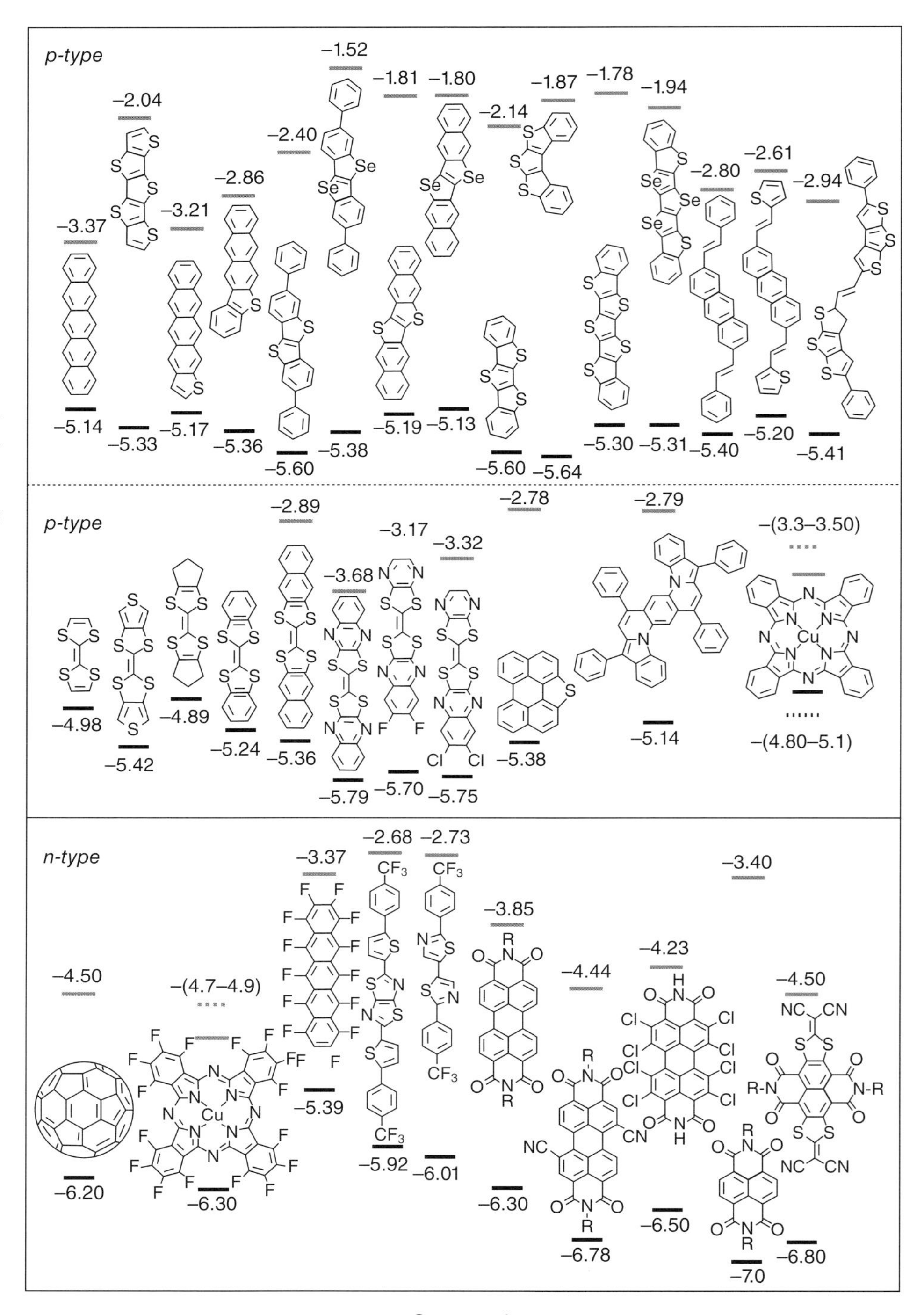

Figure 3.15 HOMO, LUMO energy levels of some organic materials with distinguished properties (black, HOMO level; gray, LUMO level). Reprinted from Ref. [89]; © 2010, Royal Society of Chemistry.

a large electric field should be applied so that a large threshold voltage is observed. With regards to typical *n*-type semiconductors, however, a low work-function will lead to a significant increase in performance. For example, electron mobility based on devices with Ag electrodes was several-fold higher than for those with Au electrodes for TIPS-TAP [90]. When using metals with an even lower work-function, such as Ca [55, 74] or Mg–Ag alloy [91], most typical *p*-type semiconductors exhibited an electron mobility. In addition, by using SAMs to modify the electrodes, the work-function of the electrodes could be significantly altered [92]. From this point of view it is not difficult to appreciate that a stepwise configuration from the source electrode, LUMO energy level of the semiconductor to the drain electrode (Au/Ag asymmetrical electrodes; Figure 3.15), would result in a higher performance than the symmetrical configuration devices–that is, Au/Au or Ag/Ag electrodes [58].

It should be noted that materials with a low work-function usually are not air-stable, and that metals with a low work-function should be protected when used. On the other hand, in order to obtain high-performance *n*-type transistors, electron-withdrawing groups are usually introduced into the chemical structures to lower the LUMO energy levels, allowing materials with a high work-function to be used as electrodes (this point is discussed in Chapter 2).

3.3.3.2 **Interface Compatibility**

It has been reported that, in OFETS, the conducting channel constitutes the first few molecular layers located exactly at the interface between the semiconductor and the insulator [87]. Hence, the quality of the semiconductor/insulator interface (e.g., the charge distribution, trap density, etc.) plays a very important role in charge transport. Moreover, the properties of this interface will have a major effect on the molecular alignment and morphology of the organic semiconductor. Thus, the semiconductor/insulator interface engineering is important for enhancing the performance of OFETs.

SAMs are widely used to modify the gate insulators, and can obviously alter the semiconductor morphology and either decrease or eliminate the defects on the surface of the dielectric layer (e.g., hydroxyl group on silicon oxide, which can trap charge carriers), and thereby efficiently improving device performance [88, 93]. Long alkyl chains with trichlorosilane [94], $-Si(OC_2H_5)_3$ [95], or phosphonic acid [96] as the anchoring groups can be utilized as SAMs on SiO_2/Si or AlO_x/Al gate insulators. With such SAMs, contact between an organic semiconductor and the insulator can be improved, followed by a significant improvement in OFET performance. ODTS, as a typical organosilane, is frequently used for the surface modification of SiO_2 substrates to fabricate silicon-based OFETs. For example, a SAM of ODTS exhibited remarkable effects on the molecular alignment of vacuum-deposited TiOPc films [4]. In this case, a highly ordered α-phase TiOPc film with an "edge on" molecular orientation was obtained on the ODTS-treated SiO_2 substrate with a mobility as high as $10\,cm^2/(V\cdot s)$ (this is one of the best results for thin-film-based OFETs). In contrast, TiOPc devices on pure SiO_2 substrates showed only a very low mobility of 10^{-4}~$10^{-5}\,cm^2/(V\cdot s)$. SAMs were also able to significantly affect the charge distribution close to the semiconductor/insulator interface. For example, due to a dipole effect of the F-containing SAM (tridecafluoro-1,1,2,2-tetrahydrooctyl)trichlorosilane [FTS; $(CF_3)(CF_2)_7$ $(CH_2)_2Si(OC_2H_5)_3$], holes were induced in the semiconductor near the semiconductor/insulator interface, which shifted the threshold voltage positively as a result [95].

Semiconductors that are easy to crystallize may also be used as a buffer layer to improve the morphology and arrangement of other semiconductors. For example, pentacene has been used widely to improve the growth of rubrene [97] and C_{60} [98], while similar results were also observed in the weak epitaxy growth of types of semiconductor on a p-6P buffer layer [99]. The contact intimacy of the interface can be judged simply by the wettability of the heterogeneous surface [100]. Wettability, which is an important property of a solid surface that is governed by the chemical composition and geometric structure of that surface, is a simple and useful parameter to value surface contact intimacy/surface compatibility. A surface with a water contact angle (CA) >90° is regarded as a hydrophobic surface, whereas a CA of <90° is regarded as hydrophilic. Only surfaces with a similar hydrophobicity or hydrophilicity can adhere intimately to one another, according to the surface "compatibility and similitude principle." Interface compatibility is also an important benefit for polymer insulators, for which a better morphology could be expected with an intriguingly improved performance [80, 101].

Besides the effects of interface compatibility on morphology and arrangements, the interface also affects device performance from the point of its defect density (e.g., hydroxyl group). This situation is particularly important for *n*-channel devices, where the electron is the main charge carrier and is highly sensitive to the defect density of the interface. Clearly, a defect-free dielectric is very important for *n*-channel devices, and SAMs or polymers without hydroxyl groups are the favorite choices for this purpose.

The interface between a semiconductor and the drain/source electrodes also plays an important role in the charge injection. Besides the energy level alignment (as noted above), interface compatibility also has a significant influence on the contact between the semiconductor and electrodes [87, 102, 103]. Of course, this type of effect only reacts on the bottom-contact, when the semiconductor is deposited after the drain/source electrodes. For example, Ag (Cu) electrodes could form a charge-transfer complex with TCNQ, whereby the pentacene films deposited on the Ag-TCNQs showed a strip crystal morphology with a length of 400~800 nm; in contrast, films deposited on the SiO_2 or Ag surfaces exhibited much smaller crystal grains with average diameters of 100~200 nm [102]. After modification, those devices with the Ag-TCNQs as electrodes showed a mobility of $0.18\,cm^2/(V\cdot s)$, which was much higher than devices with untreated Ag electrodes. Similar results were also observed in the graphene-modified Ag or Cu source/drain electrodes [103], where graphene showed an excellent electrode/organic interface contact and a low injection barrier. In some cases, the interface compatibility is even much more important than the energy level alignment. For example, although the Fermi level of Au electrode matches with the HOMO energy levels of BNT better than the TTF–TCNQ complex, those devices using a TTF–TCNQ complex as drain/source electrodes exhibited much a higher mobility than devices with Au electrodes, due to the much better surface contact [66].

3.4 Organic/Polymeric Thin Film Field-Effect Transistors

3.4.1 Techniques for Thin Film Preparation

During the 20-year period since the first OFET was reported, a wide diversity of methods has been adopted to prepare organic thin films for the fabrication of transistors. The first OFET [3] was obtained directly by using an electrochemically deposited thin film of polythiophene. Subsequently, solution-processed techniques were widely used to deposit organic thin films, due mainly to the low cost of such processes and the facility of obtaining large-area films by a variety of printing techniques. In contrast, vacuum deposition hand – due mainly to an absence of solvents – may result in high-quality thin films, and has been widely adopted for the preparation of small-molecule semiconductors.

The working mechanism for vacuum deposition originates from the different sublimation pressures of various semiconductors. In order to achieve sublimated thin films, the melting point, the decomposition point, and the degradation point must each be higher than the sublimation point, and for this purpose a high-vacuum environment is adopted, typically lower than 10^{-4} Pa. The sublimation rate and the substrate temperature can be tuned to improve the morphology and the arrangement of the semiconductors. A gradual increase in the sublimation rate is helpful first to form large grains near the interface (at a low sublimation rate), and subsequently to fill the interspaces between grains, leading to high-quality thin films [104]. Occasionally, sublimation at a different temperature is also adopted; this involves a first sublimation at a higher substrate temperature to obtain a large grain size close to the interface, followed by a lower substrate temperature to fill the interspaces [105]. Due to its freedom from solvents, the vacuum deposition method usually results in high-quality thin films, in addition to a high performance; an example is TiOPc [4], with a mobility as high as $10\,cm^2/(V\cdot s)$.

In the case of semiconductors where sublimation cannot be used to prepare thin films, due to either decomposition or degradation, it is necessary to adopt solution-processed techniques. These have shown great promise for practical application, with OFETs being superior to their inorganic counterparts, especially after employing various printing techniques (see Section 3.4.5). Many other parameters can also affect the quality of the thin films, including the solvent, the concentration, the evaporation temperature, and the annealing process (temperature or solvent annealing) after thin-film deposition. Various other solution-processed techniques are available for the preparation of organic thin films, including spin-coating, dip-coating, drop-casting, zone-casting, and the Langmuir–Blodgett (LB) technique:

- Spin coating: This is one of the most traditional methods, being used especially for polymers so as to provide the most uniform thin films in-coating [106]. In addition to the parameters noted above, the rotation speed can also be controlled to achieve a desired thickness and morphology of the thin film.
- Dip-coating and drop-casting: These are both very convenient methods, as no complex and/or expensive equipment is required. Dip-coating may also be controlled by adjusting the polling speed [107], to provide thin films with a high orientation. Drop-casting is usually regarded as a self-assembled technique (see Section 3.4.5) (Figure 3.16).
- Zone-casting: This can be used to provide thin films with a high orientation that, in turn, will result in a high anisotropy with a strong π–π interaction [108] along the casting direction.
- LB technique: This provides the ability to create amphiphilic molecules [109], but has also been found to be efficient for π-conjugated organic semiconductors [110]. The LB technique is especially meaningful for its layer-by-layer growth, which can result in compact and ordered arrangements in every semiconducting layer [111].

3.4.2 Effect of Thin-Film Microstructure on the Performance of Transistors

So far, a clear image has emerged regarding the importance of the morphology of a thin film, and how the arrangements of the molecules in that film can affect the performance of the devices created. Without doubt, ordered arrangements–especially π–π stacking arrangements–are more favorable for charge transport than amorphous thin films, and the existence of boundaries between crystalline grains in the channel should be minimized as much as possible. In other words, the crystalline grain should be sufficiently large, and the thin film should be continuous. An image has also been created that the layer deposited before the semiconducting layer should have a good interface compatibility with the semiconductor, regardless of whether that layer is the substrate (top-gate devices), the dielectric layer (bottom-gate devices), or the drain/source electrode (bottom-contact devices). Details of the techniques used most often for investigating microstructure will be presented in the following subsections, together with some examples.

One of the most useful techniques for detecting the microstructure of a thin film is atomic force microscopy (AFM). Usually, "tapping mode" AFM is employed due to the flexibility of the organic thin films, which might be damaged if a contact mode is utilized. Typically, AFM has provided the most widely visualized images of microstructures, with the ability to determine whether a film is continuous, or not, by direct observation. Notably, through this analysis it has been possible to achieve further device optimization to: (i) increase the substrate temperature for a larger crystal size, or to lower the substrate temperature for continuous films; (ii) accelerate or reduce the sublimation rate; and (iii) introduce or change the

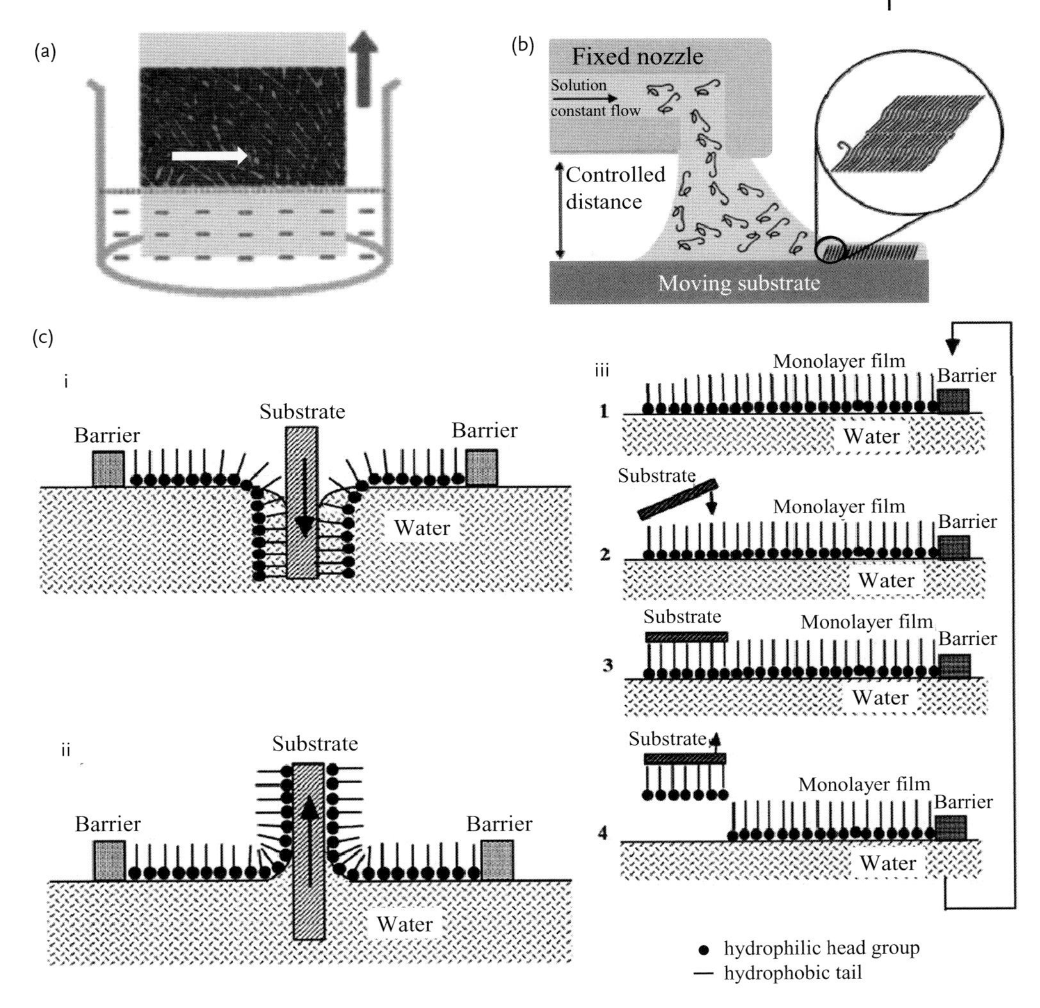

Figure 3.16 (a) Dip-coating. Reprinted with permission from Ref. [107]; © 2010, American Chemical Society; (b) Zone-casting. Reprinted with permission from Ref. [108]; © 2006, American Chemical Society; (c) LB technique: (i–iii) three alternative methods to transfer LB films onto desired substrates. Reprinted with permission from Ref. [112]; © 2008, Royal Society of Chemistry.

SAM layers. Following device optimization, the most suitable fabrication processes and conditions were obtained. For example, AFM images showed that films of 2,6-bis(phenylethynylene)benzo[1,2-*b*:4,5-*b*′]dithiophene (BPEBDT), when deposited at different substrate temperatures, tended to have a crystalline grain size that increased in line with temperature along with boundaries (Figure 3.17a–d) [105]. Hence, a two-step technique was adopted to obtain a large crystalline grain size in the channel close to the semiconductor/insulator interface, after which the

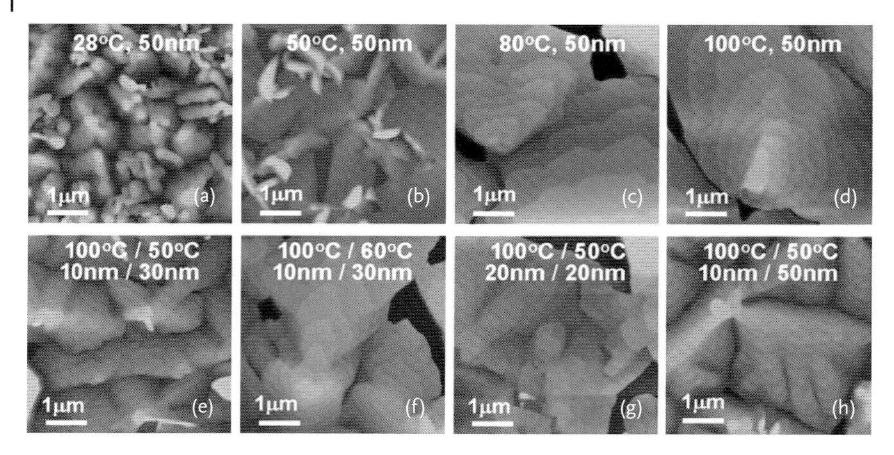

Figure 3.17 BPEBDT thin films deposited at (a–d) one-step substrate temperature and (e–h) two-step substrate temperatures. Reprinted with permission from Ref. [105]; © 2010, Royal Society of Chemistry.

domain interspaces were filled by lowering the substrate temperature (Figure 3.17e–h). When using this method, the highest mobility obtained for BPEBDT [1.17 $cm^2/(V \cdot s)$] was much higher than for the device fabricated via a one-step substrate temperature [0.72 $cm^2/(V \cdot s)$] [105]. Subsequently, the height distance between the different layers (for layer-by-layer growth) was obtained from cross-sections of the AFM images. In addition, by comparing the height distance and the crystal lattice parameters (or molecular size), the growth mode could be roughly assigned, for example, growing with the molecular long axis normal to the substrate [113]. Occasionally, if grains were seen to be growing upwards, such that the thin film had excessive surface roughness, scanning electron microscopy (SEM) was used to investigate the surface microstructure, without height details [105].

In order to determine the molecular arrangements in the microstructures, X-ray diffraction (XRD) patterns have proved to be the most powerful tool. XRD analyses are valuable for determining whether the thin films are crystalline or amorphous, and also to confirm a definite growth mode that usually can be verified with AFM images. Without doubt, the more ordered the molecular arrangement, the higher will be the performance obtained. The results of these analyses also allowed the optimum fabrication conditions to be determined. For example, when a TiOPc thin film was deposited at substrate temperature of 150 °C, the intensity of the XRD was much enhanced compared to a lower temperature, indicating the higher ordered structure of the film (Figure 3.18a). In addition, by analyzing the spacing distance (Bragg's law), an α-phase arrangement with an "edge-on" orientation (62° with respect to the substrate) was adopted when the TiOPc molecules were deposited (Figure 3.18b) [4].

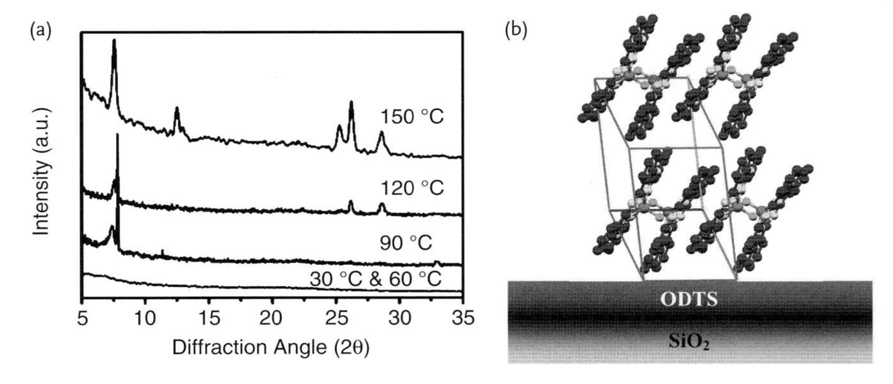

Figure 3.18 (a) XRD patterns of TiOPc thin films deposited at different substrate temperatures; (b) Schematic diagram of the arrangement of TiOPc molecules standing on the substrate. ODTS, octadecyltrichlorosilane. Reprinted with permission from Ref. [4]; © 2007, John Wiley & Sons, Inc.

It has been noted above that charge transport most likely occurs in the region close to the semiconductor/insulator interface and, indeed, the results of various investigations have suggested that this region generally consists of only several molecular layers or monolayers [114]. Thus, the investigation and analyses of the molecular arrangement in these few molecular layers have proved more important than those of bulk films. Most importantly, the arrangement in the interface region is occasionally different from that in the bulk films [114], although the more traditional technique of powder XRD is unsuitable for detecting the structure of such thin films. This led to the use of a grazing incidence X-ray diffraction (GIXD) technique, which employs small incident angles to detect diffraction information that might scatter the degree of order, both out-of-plane (i.e., vertical to the substrate, as traditional XRD can only detect the signal in this direction) and in-plane (parallel to the substrate). Hence, GIXD can be used not only to study the surface or even a monolayer [114, 115], but is also very efficient for determining "edge-on" or "face-on" orientations (Figure 3.19) [39, 116]. Normally, GIXD analyses are performed using a synchrotron.

Near-edge X-ray absorption fine structure (NEXAFS) analyses [117] are also typically provided by a synchrotron. This method has recently become widespread for the characterization of organic materials, mainly because the C–C 1s→π^* resonance is strong and well-resolved. In NEXAFS, a fluorescent photon, Auger electron, or an inelastically scattered photoelectron can be detected; consequently, NEXAFS is also a surface-analysis technique, and can also be used not only to probe the chemical structure of organic compounds but also to track the orientation of the resonant bonds [118].

Transmission electron microscopy (TEM) images and selected area electron diffraction (SAED) patterns have also been used to characterize the microstructure

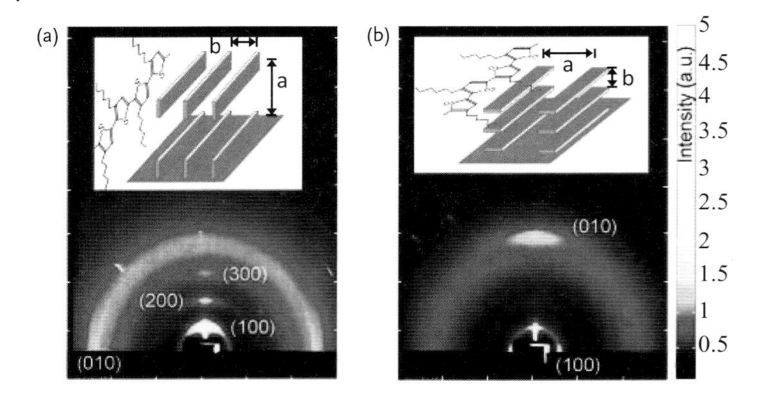

Figure 3.19 (a) "Edge-on" and (b) "face-on" orientation revealed by the GIXD technique. Reprinted with permission from Ref. [39]; © 1999, Macmillan Publishers Limited.

of organic thin films. Unfortunately, the technique of depositing a thin film onto the electron microscope grid directly [119] failed to reflect the actual structure of the thin film in the devices, because the difference between the grid surface and the dielectric layer in the device would most likely result in different arrangements. In contrast, the traditional method of preparing inorganic TEM samples–that is, to mechanically polish the device until it is thin enough–would be unsuitable for organic thin films. In order to resolve this problem, carbon-films were used as an efficient supporting layer through either electron-beam evaporation or sputtering to deposit the supporting films directly onto the organic thin film. Subsequently, the dielectric layer, the gate electrode, and the substrate could be wiped off chemically, in direct fashion. Notably, the only major limiting factor of this technique is that the organic semiconductor layer must remain stable during the deposition and polishing process.

3.4.3 High-Performance Transistors of Small Molecules

As a summary of the materials employed in high-performance transistors was provided in Chapter 2, the aim here is rather to summarize the properties of some representative high-performance materials, and to highlight their chemical structures, arrangements and deposition conditions. In this way, it is hoped that the relationship between structures and their properties can be more clearly defined.

The details of some representative small-molecule materials that have been reported with thin film mobilities of over 1 $cm^2/(V \cdot s)$, together with their chemical structures, arrangements, deposition conditions and properties, are listed in Table 3.2. Based on these data, some preliminary conclusions can be drawn as to *why* these devices exhibit a high performance in transistors [15].

Table 3.2 Representative small-molecule materials with high thin-film mobilities [>1 $cm^2/(V·s)$] and their chemical structures, arrangements, deposition conditions, and properties.

	El	E_g	P	Insulator	C, *W*/*L*, Electrode	T	μ[a]	Reference(s)
P-type molecules								
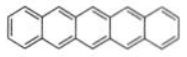	−5.0	1.8	H	AMS/Al_2O_3 (150 nm, or SiO_2 [100 nm])	BGTC, 1000/110, Au	—	5	[121, 123]
	−5.5	3.3	H	HMDS/SiO_2	BGTC, 3000/30, Au	—	3.2	[122, 124]
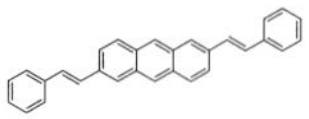	−5.4	2.6	H	ODTS/SiO_2	BGTC, 100/20, Au	80	1.3	[123]
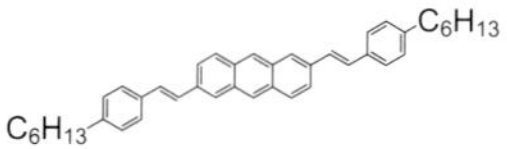 C_6H_{13} C_6H_{13}	−5.5	2.6	H	OTS/SiO_2	BGTC, 600/60, Au	60	1.28	[125]
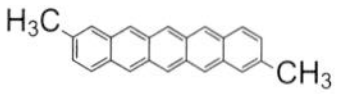 H_3C CH_3	—	—	—	SAM_1/Al_2O_3	BGTC, 1000/40~100, Au	—	2.5	[126]
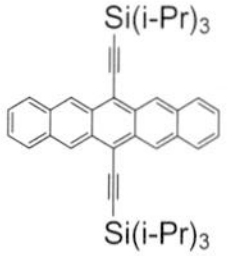 Si(i-Pr)$_3$ Si(i-Pr)$_3$	−5.11	1.69	2-D	HMDS/SiO_2 (370 nm)	BGBC, 220/25, PFBT/Au	Drop-casting	1.8	[127, 128]
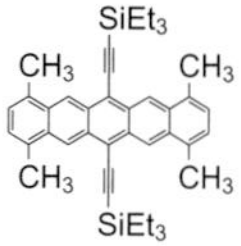 $SiEt_3$ CH_3 CH_3 CH_3 CH_3 $SiEt_3$	−5.19	1.87	1-D π	OTS/SiO_2 (300 nm)	BGTC, 2000/60, Au	Drop-casting	2.5	[129]

(*Continued*)

Table 3.2 (Continued)

	El	E_g	P	Insulator	C, *W/L*, Electrode	T	μ[a]	Reference(s)
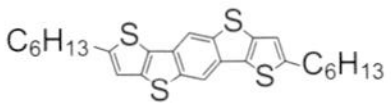 C_6H_{13}, S, S, S, S, C_6H_{13}	−5.56	3.36	π	Untreated SiO_2 (150 nm)	BGTC, 290/25, Au	Dip-coating	1.7	[130]
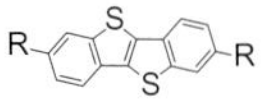 R, S, S, R	−5.5	3.5	H	Untreated SiO_2 (200 nm)	BGTC, 1500/50, Au	Spin-coating	1.71~2.75 (C_8H_{17}, $C_{11}H_{23}$, $C_{12}H_{25}$, $C_{13}H_{27}$)	[131]
				Untreated, OTS (or ODTS) treated SiO_2 (200 nm)	BGTC, 1500/50, Au	rt	1.1~3.9 (C_8H_{17}, $C_{10}H_{21}$, $C_{12}H_{25}$)	[132]
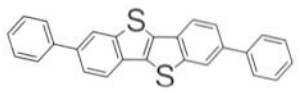 S, S	−5.6	3.2	—	OTS/SiO_2 (200 nm)	BGTC, 1500/50, Au	100	2.0	[133]
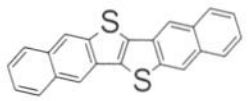 S, S	−5.44	3.0	H	OTS/SiO_2 (200 nm)	BGTC, 1500/50, Au	60	2.9	[134]
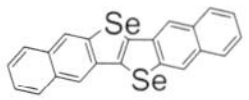 Se, Se	−5.38	2.9	—	OTS/SiO_2 (200 nm)	BGTC, 1500/50, Au	rt	1.9	[134]
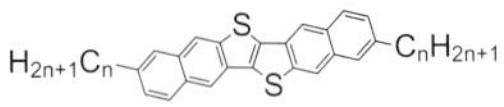 $H_{2n+1}C_n$, S, S, C_nH_{2n+1}	−5.3~−5.4	3.0	—	ODTS/SiO_2 (200 nm)	BGTC, 1500/190, Au	60	7.9 ($C_{10}H_{21}$)	[120]
				Untreated SiO_2 (200 nm)	BGTC, 1500/50, Au	60 or 100	2.3 (C_8H_{17}) 3.7 ($C_{10}H_{21}$) 3.1 ($C_{12}H_{25}$)	[120]
				SAM_2/SiO_2 (500 nm)	BGTC, Au	Solution-process	11 ($C_{10}H_{21}$)	[5]
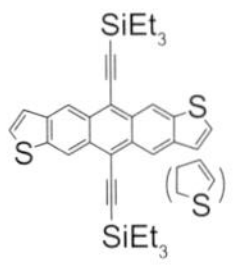 $SiEt_3$, S, S, S, $SiEt_3$	−5.31	—	2-D π	SiO_2 (370 nm)	BGBC, 340/22, PFBT/Au	Solution-process	1.0	[135, 136]

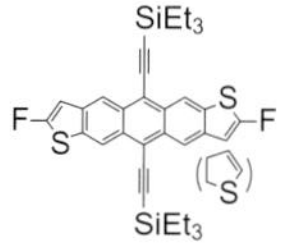	—	—	2-D π	SiO_2 (200 nm)	BGBC, 220/10, PFBT/Au	Spin-coating	1.5	[137]
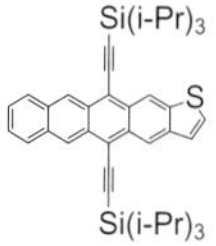	−5.21	2.03	2-D π	Untreated SiO_2 (300 nm)	BGTC, 1000/50, Au	60	1.25	[138]
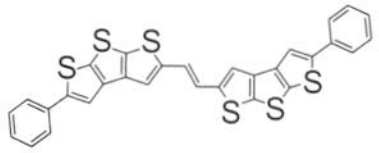	−5.41	2.90	H	ODTS/SiO_2	BGTC, 5300/110, Au	100/27[b)]	2.0	[139]
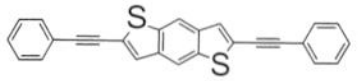	−5.66	3.07	—	OTS/SiO_2 (300 nm)	BGTC, Au	100/50[b)]	1.17	[105]
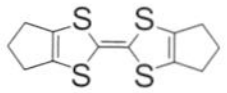	−4.61	—	2-D π	OTS/SiO_2 (300 nm)	BGTC, 1000/100, Au	40	3.6	[140]
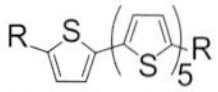 R=C_2H_5, C_6H_{13}	—	—	—	PVP (270 nm)	BGTC, 170/130, Au	—	1.1, 1.0	[10]
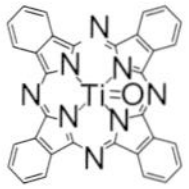	−5.7	1.7	2-D π	ODTS/SiO_2 (300 nm)	BGTC, 5300/110, Au	150	10	[4, 141]
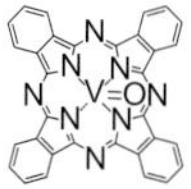	−5.3	1.4	2-D π	ODTS/SiO_2 (300 nm)	BGTC, 5300/110, Au	90	1.0	[100, 142, 143]

(Continued)

Table 3.2 (*Continued*)

	El	E_g	P	Insulator	C, *W*/*L*, Electrode	T	$\mu^{a)}$	Reference(s)
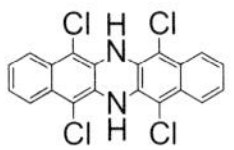	—	—	Slipped π	NTS/SiO_2 (300 nm)	BGTC, 500/50, Au	55	1.4	[144]
N-type molecules								
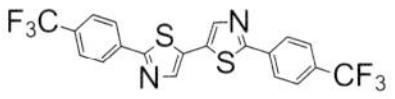	−2.73	2.9	2-D π	ODTS/SiO_2 (200 nm)	BGTC, 1000/50 or 100, Au	25	1.83	[145]
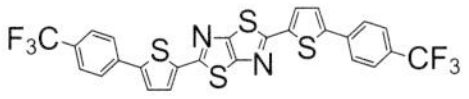	−2.68/2.82	2.76	—	ODTS/SiO_2 (200 nm)	BGTC, 1000/50, Au	—	1.2	[146–148]
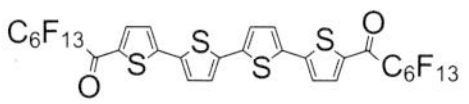	−3.96	2.4	—	PS (24 nm)/SiO_2 (300 nm)	BGTC, 5000/100, Au	~50	1.7	[149, 150]
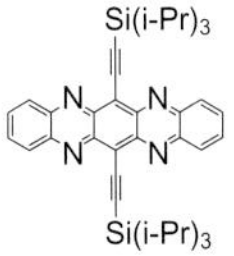	−4.01	1.74	2-D	OTMS/SiO_2 (300 nm)	BGTC, 1000 or 2000/50~150, Au	25	3.3	[151]
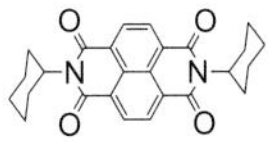	−3.71	3.26	2-D	ODTS/SiO_2 (185 nm)	BGTC, 650/50~150, Au	22	7.5	[47, 152]
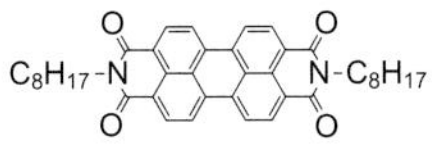	−3.9	2.4	1-D	PMS/SiO_2 (300 nm)	BGTC, 1000/100, Ag	75	1.7	[18, 148, 153]

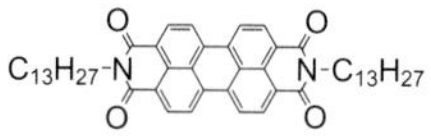	−3.4	2.0	—	SiO_2 (200 nm)	BGTC, 2000/100, Au	rt	2.1	[154, 155]
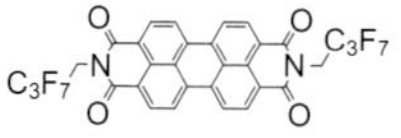	−4.1	2.3	1-D	OTMS/SiO_2 (300 nm)	BGTC, 1000/50, Au	125	1.44	[148, 156]
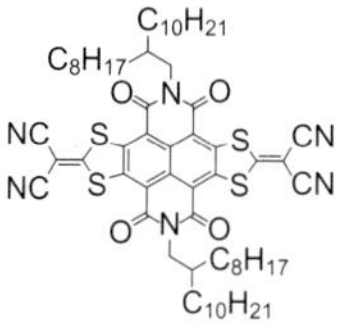	−4.3	2.0	—	ODTS/SiO_2 (300 nm)	BGBC, 1400/5~50, PFBT/Au	Spin-coating	1.2	[157, 158]
	−4.5/3.8	1.7/2.4	NA	BCB	BGTC, 1500/30, LiF/Al	250	6	[159–162]

a) The unit of μ is $cm^2/(V \cdot s)$.
b) Two-step deposition.
El: the energy level (eV), HOMO energy levels for *p*-type semiconductors and LUMO energy levels for *n*-type semiconductors; E_g: the energy gap (eV) between HOMO and LUMO energy levels;
P: packing mode; H: herringbone packing; π: π-stacking; C: contact mode; BG: bottom gate;
TC: top contact; BC: bottom contact; *W*/*L*: the width and length of the channel; T: temperature of substrate (°C) when deposited or techniques of solution-process; rt: room temperature; AMS: poly α-methylstyrene; SAM_1: 1-phosphonohexadecane; PFBT: pentafluorobenzenethiol; SAM_2: decyltriethoxysilane or *N*,*N*-dimethyl-10-aminodecyltriethoxysilane; NTS: *n*-nonyltrichlorosilane; PS: polystyrene; PMS: poly (α-methylstyrene); NA: not applicable; —: not found.

- Among these materials, only four are oligomers; this suggests the importance of extending π-systems for high performance.
- Most of the materials adopted a π-stacking mode, indicating that a strong π–π intermolecular interaction is favorable to achieve a high performance.
- Among these materials, all acenes with substitutions at peri-positions adopted π–π stacking, which suggests that peri-substitution is an efficient way of obtaining π–π stacking. Likewise, all the *n*-type materials adopted π–π stacking, which can support the assumption that electron-withdrawing groups (or polar groups) are helpful to achieve π–π stacking.
- Among these materials, only C_{60} have a nonplanar structure. All others have a planar or planar-like structure, which suggests that a planar π-system is important for high performance. One reason for this is that a planar structure is helpful in achieving crystalline thin films.
- Among these materials, only two do not possess C_2 symmetry, indicating the importance of such symmetry for molecular packing.

There is no doubt that the semiconductor itself – that is, its chemical structure and its arrangement – will determine the properties detected. On the other hand, many other external parameters may also affect device performance, such as the dielectric layer (the dielectric constant, the thickness, the roughness of the surface, the interface compatibility of the surface with organic semiconductors, etc.), the electrodes (the energy level alignment with the semiconductor, the interface compatibility with the semiconductor, etc.), and the channel dimensions. All of these could affect device performance through affecting the morphology and arrangement, the charge injection barrier, or the induced electric field. The following are some representative rules on these points:

1) Most *p*-type materials possess HOMO energy levels in the range -5.1 ± 0.3 eV, assuming that the drain/source electrodes are gold, which match well with each other between the HOMO energy levels and the work-function of the electrodes. For *n*-type materials, the LUMO energy levels should also match with the work-function of the electrodes adopted. Because the commonly used materials (e.g., Au, Ag, Cu, Al, Zn) all have a work-function higher than 4.0 eV, it is not difficult to understand why *n*-type materials with a high performance possess LUMO energy levels close to -4.0 eV. All of these points confirm the importance of energy level alignments between the HOMO/LUMO energy levels and the work-function of the electrodes.
2) Almost all of the insulators are modified by SAMs or buffer layers (e.g., polymer). If the device configuration is bottom-contact, then all of the electrodes will also be modified. This highlights the importance of an interface compatibility for molecular arrangements and high performance.
3) Almost all of the mobilities were obtained with a channel length greater than 20~30 μm. Studies have suggested that a short-channel effect would appear

when the channel length was scaled down to 5 μm in some cases. A short-channel effect means that, when the channel length is decreased to a certain value, the contact resistance cannot be omitted. Hence, a short-channel effect is heavily dependent on the contact quality and the conductivity of the channel (i.e., it relies on the charge transport in the domain and through the grain boundaries). If the contact quality is better and the conductivity lower, the short-channel effect will appear in a lower channel length. Therefore, as with C10-DNTT (dinaphtho[2,3-*b*:2′,3′-*f*]thieno[3,2-*b*]thiophene), which possesses a high conductivity in the channel, a short-channel effect will appear even when the channel length is as large as 190 μm [120]. In other words, the channel length should be large enough in some devices to achieve a high performance.

4) Due to the anisotropic properties, the highest performances usually were obtained along the direction with strong intermolecular interactions (C–H···π, or π–π interactions), which should be the orientation channel. In the meantime, the thickness of the semiconductor (the channel) should be low enough (typically 30~60 nm), to guarantee the charge carrier injection from the source electrodes through the semiconducting layer into the interface of semiconductor/insulator (for BGTC and TGBC devices).

Besides the factors mentioned above, many other factors have also been identified that may influence the performance, including the deposition conditions (rate, vacuum pressure) and the testing conditions (environment, pressure, temperature) [121] (for discussions of these points, see above). For example, the results of various investigations have suggested that exposure in oxygen may increase the performance of *p*-type devices [122].

3.4.4 High-Performance Transistors of Conjugated Polymers

Clearly, polymeric semiconductors have lagged far behind the small-molecule semiconductors, the main reason being that it is very difficult to obtain both monodisperse polymers and crystalline polymeric thin films. Nonetheless, it is interesting to examine the history of OFETs. The reason why organic electronics have attracted so much attention has been the discovery of high-conductive polymers [2], and the fact that the first OFET device was based on polymers [3]. Moreover, polymers are much more attractive than small-molecule semiconductors for practical applications, due to their greater suitability for solution-processed techniques, which is an intriguing benefit of organic electronics over inorganic electronics. Hence, printing techniques may be used that could, at least potentially, be applied to large-area products so as to reduce the production costs. In addition, due to its facile application in printing electronics, top-gate devices have occasionally been adopted in polymeric semiconductors, in which the insulators also serve as the protection layer.

In addition to the "rules" for achieving high-performance devices listed above (e.g., C_2 symmetry, energy level alignment and interface compatibility with insulator and electrodes), two very important points affect the performance of polymeric semiconductors, namely their molecular weight and annealing treatment. The details of some representative polymeric semiconductors that have been reported with thin-film mobilities >1 $cm^2/(V \cdot s)$, as well as their chemical structures, deposition conditions and properties, are listed in Table 3.3. It has been suggested that increasing the molecular weight can improve the molecular order, which in turn results in a higher FET performance [163]. Likewise, annealing (whether thermally or by solvent) can induce a self-assembly of the molecules (especially in the case of materials with a glass transition temperature), which also leads to ordered molecular arrangements.

Table 3.3 Representative polymeric semiconductors with high thin-film mobilities [>1 $cm^2/(V \cdot s)$] and their chemical structures, deposition conditions, and properties.

P-type	*El*	E_g	Mn	Insulator	C, *W*/*L* Electrode	Method	μ[a]	Reference(s)
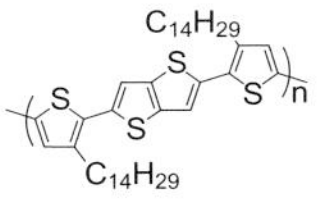 $C_{14}H_{29}$, S, S, S, n, $C_{14}H_{29}$	—	—	28K	ODTS/ SiO_2 (200 nm)	BGBC, 1 000/5, Au or Pt	Spin-coating/185 °C	1	[164, 165]
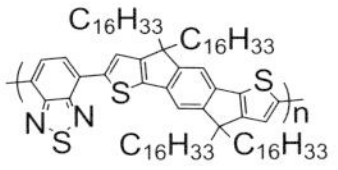 $C_{16}H_{33}$, $C_{16}H_{33}$, S, S, n, N, S, N, $C_{16}H_{33}$, $C_{16}H_{33}$	−3.7	1.7	38K	Cytop	TGBC, 1 000/30, PFBT/Au	Spin-coating/100 °C	1.2	[166]
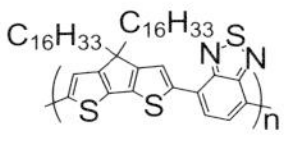 $C_{16}H_{33}$, $C_{16}H_{33}$, S, N, N, S, S, n	−3.57	1.73	50K	PTES/ SiO_2 (150 nm)	BGTC, 290/25, Au	Dip-coating/200 °C	1.4	[163, 167]
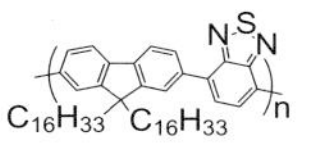 N, S, N, $C_{16}H_{33}$, $C_{16}H_{33}$, n	—	—	35K	HMDS/ SiO_2 (200 nm)	BGTC, 1 400/20, Au	Drop-casting/200 °C	3.3	[168]
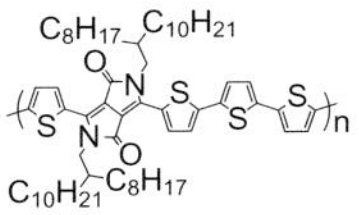 C_8H_{17}, $C_{10}H_{21}$, O, N, S, S, S, S, N, O, n, $C_{10}H_{21}$, C_8H_{17}	—	1.39	58K	OTS/SiO_2 (300 nm)	BGTC, 1 500/100	Spin-coating/150 °C	1	[169]
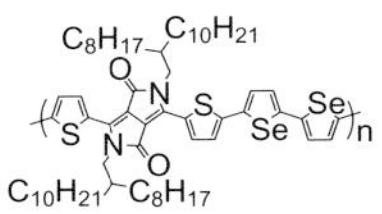 C_8H_{17}, $C_{10}H_{21}$, O, N, S, Se, S, Se, N, O, n, $C_{10}H_{21}$, C_8H_{17}	—	1.32	19K	OTS/SiO_2 (300 nm)	BGTC, 1 500/100	Spin-coating/180 °C	1.5	[169]
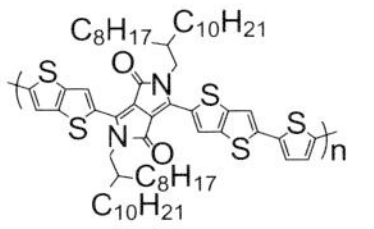 C_8H_{17}, $C_{10}H_{21}$, O, N, S, S, S, S, N, O, n, C_8H_{17}, $C_{10}H_{21}$	−5.06	1.38	14K	PMMA (550 nm)	TGBC, 1 000/20, Au	Spin-coating/100 °C	1.95	[170]
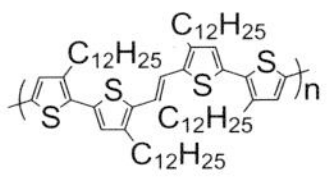 $C_{12}H_{25}$, $C_{12}H_{25}$, S, S, S, S, n, $C_{12}H_{25}$, $C_{12}H_{25}$	−5.02	1.79	25K	PMMA (520 nm)	TGBC, 1 000/20, Au	Spin-coating/200 °C	1.05	[171]

Table 3.3 (*Continued*)

P-type	*El*	E_g	Mn	Insulator	C, *W/L* Electrode	Method	$\mu^{a)}$	Reference(s)
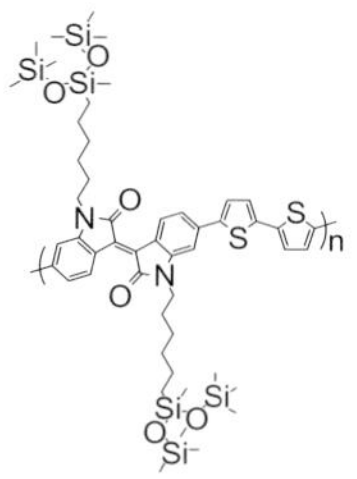	−5.2	1.61	138K	OTMS/ SiO_2 (300 nm)	BGTC, 1 000/50, Au	Spin-coating/170 °C	2.48	[172]
N-type	*El*	E_g	Mn	Insulator	C, *W/L* Electrode	T	$\mu^{a)}$	Reference(s)
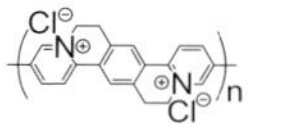	−4.0	2.56		HMDS/ SiO_2 (230 nm)	BGBC, 10 000/2.5∼20, ITO/Au	Spin-coating	3.4	[173]

a) The unit of μ is $cm^2/(V \cdot s)$.
El: the energy level (eV), HOMO energy levels for *p*-type semiconductors and LUMO energy levels for *n*-type semiconductors; E_g: the energy gap (eV) between HOMO and LUMO energy levels; Mn: molecular mass ($g\,mol^{-1}$); C: contact mode; BG: bottom gate; TG, top gate; TC: top contact; BC: bottom contact; *W/L*: the width and length of the channel; Top: top contact; TG: top gate; T: techniques of solution-process and annealing temperature; PFBT: pentafluorobenzenethiol; PTES: phenyltriethoxysilane; —: not found.

3.4.5 New Techniques for Organic/Polymeric Thin Film Field-Effect Transistors

3.4.5.1 Self-Assembly

The most common techniques used for the deposition of organic/polymeric thin films, including vacuum sublimation, spin-coating, dip-coating, drop-casting, zone-casting, and the LB technique, have been detailed above. In theory, each of these methods can refer to self-assembled techniques, with or without external induction. For example, an increase in substrate temperature during vacuum sublimation is favorable in order to form better molecular arrangements. On the other hand, a fast speed of spin-coating will result in a rapid evaporation, which would harm the self-assembly. A similar situation also exists for dip-coating, where a lower pulling speed would help to provide better arrangements.

Clearly, there is no doubt that drop-casting–which is almost free from external disturbance–will lead to a higher performance than either spin-coating and/or dip-coating techniques [127, 130]. It should be noted that the temperature, the use of a closed or opened system, and the atmosphere might also affect the self-assembled process. Occasionally, a liquid-holding structure may be used

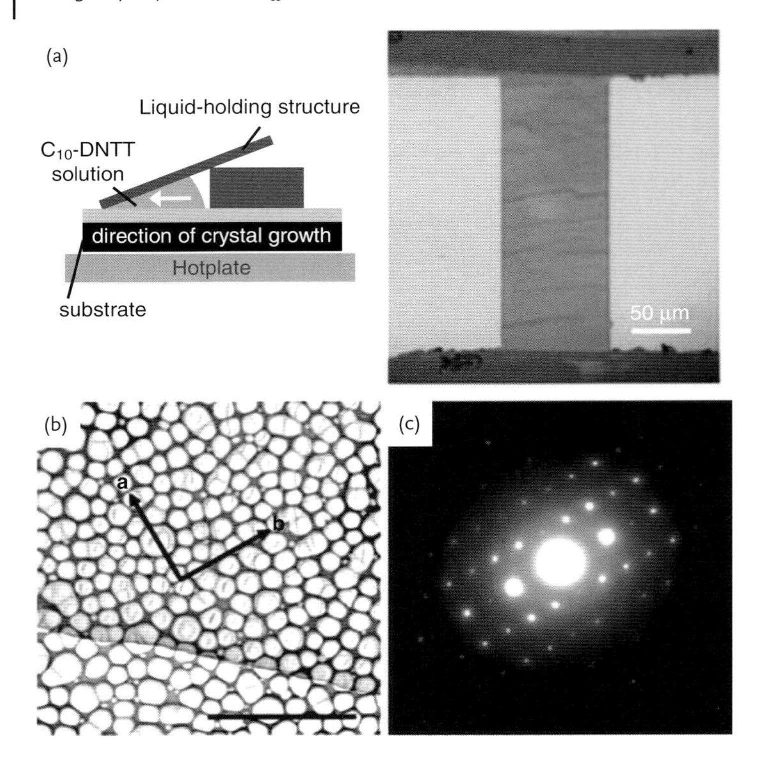

Figure 3.20 (a) Liquid-holding structure for promotion of crystal growth (left) and thin films of C10-DNTT obtained by using this method (right). Reprinted with permission from Ref. [5]; © 2011, Wiley-VCH Verlag GmbH & Co. KGaA; (b) TEM image (scale bar = 10 μm); (c) Corresponding SAED patterns of drop-cast large-area crystalline films, based on HTEB. Reprinted with permission from Ref. [174]; © 2011, Wiley-VCH Verlag GmbH & Co. KGaA.

(Figure 3.20a, left) to regulate the gradual shift direction of the liquid–atmosphere boundary, which would promote crystal growth into large-area crystalline films (see Figure 3.20a, right) [5]. On the other hand, without any such induction, large-area crystalline thin films could also be obtained for some materials, even with millimeter dimensions. The reason for this might be attributed to a self-assembled layer-by-layer growth of HTEB, originating from its intermolecular interactions [174].

The same mechanism applies to thermal or solvent annealing, with a thermal or solvent atmosphere being used to induce the self-assembly process following deposition of the thin film. In some cases, rubbed substrates may be used to induce the alignment of thin films (Figure 3.21) [175, 177]. Zone-casting represents another technique that can induce a long-range ordered structure. There is no doubt that the basis of the LB technique is also self-assembly when, due to the

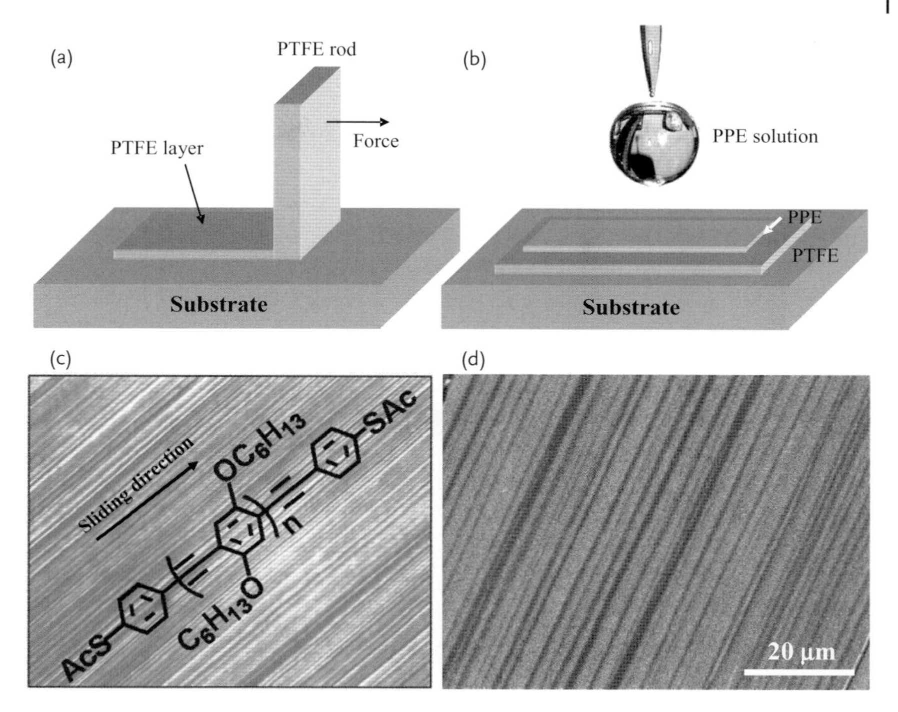

Figure 3.21 (a) A PTFE rod is used to obtain a rubbed aligned substrate; (b) Application of the substrate in the spin-casting technique; (c,d) Aligned thin films obtained using this induction method. Reprinted with permission from Refs [175, 176]; © 2008, American Chemical Society; © 2011, Royal Society of Chemistry, respectively.

unrestrained nature when floating on the surface of water or dissolved in organic solutions and the strong intermolecular interactions, the molecules can be assembled into an ordered monolayer that can be transferred onto the substrates (see Figure 3.16). The LB technique has also been employed to create dense and ordered SAM monolayers [178].

3.4.5.2 **Printing**

Among the diverse methods used to fabricate organic thin films, the most exciting is that of printing, because it can provide large-area integrated circuits by solution-processing, at low cost. Although, in theory, each of the methods used to deposit organic films (vacuum deposition, spin-coating, drop-casting, dip-coating, LB technique) can be used to provide large-area devices, it is often unrealistic from a practical viewpoint due to: (i) the size limitations of the equipment (e.g., vacuum deposition, spin-coating, dip-coating, LB technique); (ii) the limit of forming homogeneous films (drop-casting); and (iii) the waste of materials. Among the types of printing technique available, microcontact printing (μCP) and ink-jet printing are considered to show the most promise.

Microcontact printing is a type of soft lithography which uses a stamp (usually polydimethylsiloxane; PDMS) to transfer the objects (usually as a patterned stamp) to a desired substrate, or simply to reverse the stamp as substrate [179]. The patterned stamp is obtained using traditional photolithographic techniques, but the objects may be SAM monolayers [180], semiconductors [181], electrodes [182] or insulators [179], or they may exist as solids or liquids. The stamp, when applied directly to the subject, undergoes a selective adhesion according to its features. If the stamp is to be printed onto a desired substrate (the receiver), the objects can be transferred to the substrate by removing the stamp, either by heating [182], dissolving [183], or simply peeling back [184] (Figure 3.22a).

In ink-jet printing, the main challenge when applied to organic electronics is the size limit (the droplet size is usually >20 μm). Sirringhaus and coworkers have used patterned surfaces with different surface energies to dewet the inkjet-printed droplets (Figure 3.22b), such that devices with a channel length of 500 nm were obtained. The same group also used a self-aligned inkjet printing technique to provide sub-100 nm channel lengths. The inkjet-printed droplets can also serve as semiconductors [187], electrodes [185] or insulators. The use of an ink-jet printing technique led to one of the highest mobilities [31.3 $cm^2/(V \cdot s)$] being obtained from a solution-processed device [8].

In the past, the roll-to-roll (R2R) process has also attracted much attention, as it can be used to print flexible products over large areas, especially in the case of multilayer devices. The main components of the R2R process include a series of

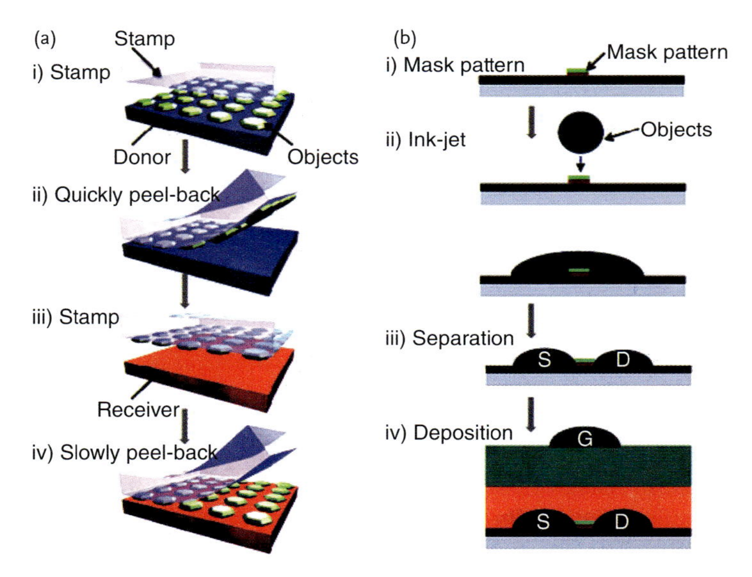

Figure 3.22 Schematic diagrams. (a) Microcontact printing; (b) ink-jet printing technique. Reprinted with permission from Refs [15, 185, 186]; © 2004, 2006, Macmillan Publishers Limited; © 2012, American Chemical Society, respectively.

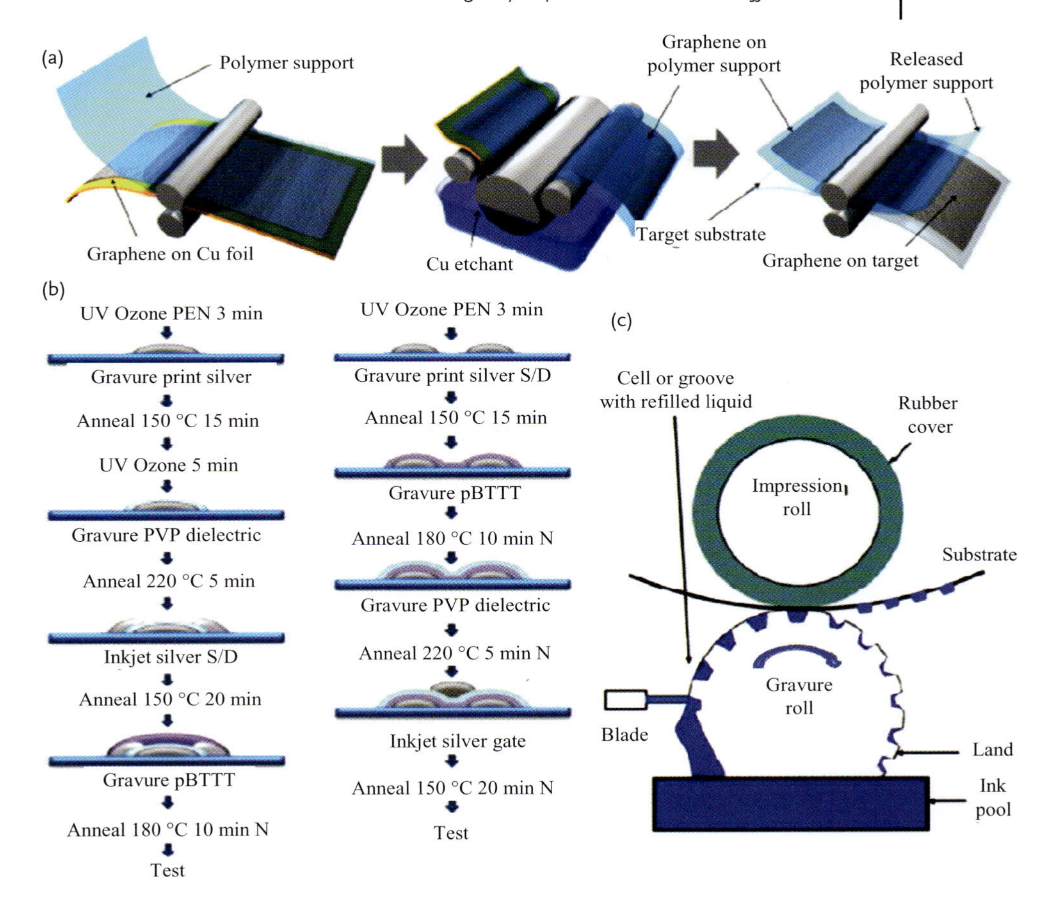

Figure 3.23 Schematic diagrams. (a) The roll-based production of graphene films. Reprinted with permission from Ref. [61]; © 2010, Macmillan Publishers Limited; (b) Hybrid gravure and ink-jet printing process. Reprinted with permission from Ref. [188]; © 2011, Elsevier B.V.; (c) The gravure R2R process. Reprinted with permission from Ref. [189]; © 2010 IEEE.

gravure rolls that provide μCP, and/or rolls that provide ink-jet printing. Such use of successive rolls, which provides the opportunity to fabricate multilayer devices on an assembly line, is especially exciting for practical applications, as the R2R process can integrate μCP and ink-jet printing into a single scheme. An example of R2R printing is shown in Figure 3.23a, where graphene is transferred first from a Cu foil to the polymer support, and then to the target by μCP [61]. The combination of hybrid gravure and ink-jet printing [188], where the gravure printing acts as μCP, is shown in Figure 3.23b (fine details of the gravure printing process are shown in Figure 3.23c [189]).

3.5
Organic/Polymeric Single Crystal Field-Effect Transistors

3.5.1
Organic/Polymeric Single Crystals

As mentioned above, the molecular arrangement takes an important role in OFETs. There is also no doubt that a long-range order of molecules is beneficial to the charge transport, because charge carriers injected from the source electrode must be transported from one molecule to neighbor molecules until they reach the drain electrode. Hence, the disordered structure (e.g., boundaries in thin films) may have a major effect on performance. On the other hand, organic single crystals are free from grain boundaries with minimized defects, and can provide devices with a high performance. Moreover, due to the absence of many external affects, organic single crystals can reveal the intrinsic properties of materials and demonstrate the charge transport mechanism of OFETs. In theory, polymers are difficult to create from single crystals, due to problems in obtaining monodisperse polymers. Nonetheless, micro/nanometer-sized wires of polymers can be created with high-order, low-defect characteristics and freedom from grain boundaries; however, these have also been regarded as single crystals in some cases [190].

3.5.2
Growth of Organic/Polymeric Crystals

The low thermal stability of organic/polymeric materials (decomposition) and the harsh conditions associated with growing inorganic crystals (e.g., melting method) means that these traditional techniques cannot be used to grow organic crystals. However, in order to obtain large crystals, especially for testing in TOF methods, the vapor-Bridgman growth process can be used. In this case, by controlling the temperature of the source and deposition zone, and also the movement of the reflector (i.e., irradiation of heat), it is possible to achieve a controlled growth of large, organic single crystals, albeit of low quality (Figure 3.24) [191].

3.5.2.1 Vapor Process for the Growth of Organic Crystals

In the physical vapor transport (PVT) technique, which can be used to provide high-quality organic single crystals, the equipment is very similar to the tube furnace used to deposit parylene thin films (see above). The main differences are that, in PVT, the tube furnace requires a carrier gas to achieve an improved removal of impurities, but not an ultrahigh temperature zone to pyrolyze the source (see Figure 3.25a). It has been suggested that the use of a carrier gas (e.g., Ar, N_2, or H_2) might lead to higher-quality crystals than are prepared with the vapor-Bridgman method. In PVT, the source is placed in the high-temperature zone, where it undergoes sublimation; the sublimated molecules are then transported by the carrier gas to the low-temperature zone, where they crystallize. By

Figure 3.24 Large-sized crystal of tetracene, obtained using the vapor-Bridgman growth method. Reprinted with permission from Ref. [191]; © 2005, American Institute of Physics.

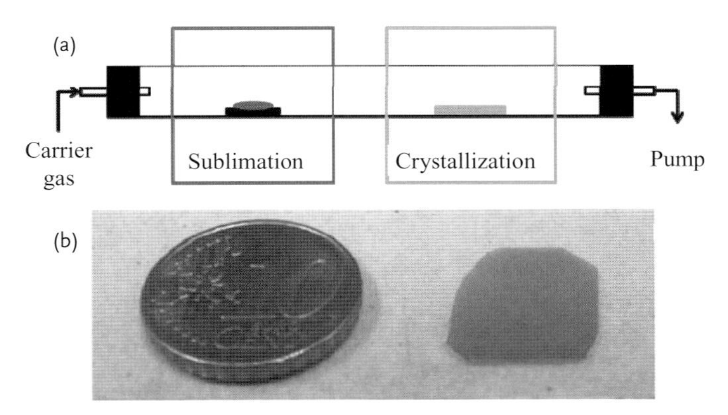

Figure 3.25 (a) Schematic diagram of tube furnace used in the PVT method; (b) Crystal of tetracene obtained using the PVT technique. Reprinted with permission from Ref. [192]; © 2004, Wiley-VCH Verlag GmbH & Co. KGaA.

controlling the amount of source material, the temperatures of sublimation (the closer to the sublimation point, the larger the crystal size and the higher the quality) and of the crystallization zone, the temperature gradient (the shallower the gradient, the greater and higher quality of the crystals; usually 2~5 °C cm^{-1}) and the flow velocity of the carrier gas (a slower velocity is better, usually 100~300 standard cc per min; sccm), it is possible to control not only the size of the crystals but also their quality. Three representative morphologies of dibenzo[*d*,*d*′]thieno[3,2-*b*;4,5-*b*′]dithiophene (DBTDT) obtained by controlling the growth temperature, but

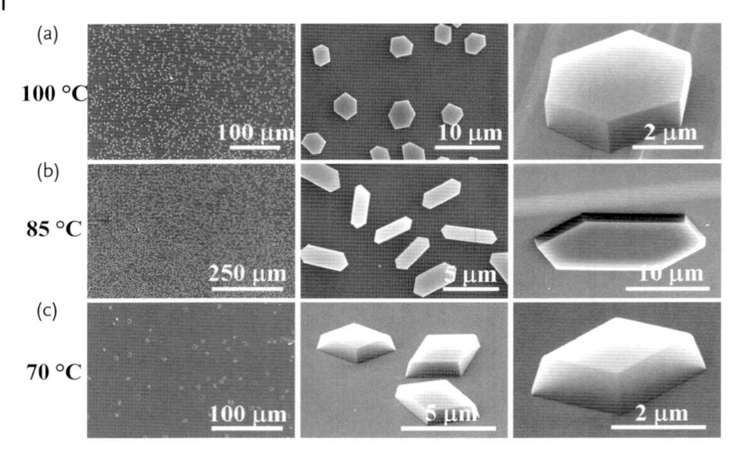

Figure 3.26 Three representative morphologies of DBTDT obtained by the using PVT technique, while controlling the growth temperature with other factors kept constant. Reprinted with permission from Ref. [193]; © 2010, American Chemical Society.

with other factors held constant, are shown in Figure 3.26. By using the PVT method, large crystal sizes could also be achieved (Figure 3.25b) of up to several centimeters [7, 192], which were almost comparable with those produced via the vapor-Bridgman method, based on the same material (tetracene). When compared to the solution-processed techniques, the PVT method can provide high-quality single crystals due to an absence of solvents that might serve as a source of impurities and defects.

3.5.2.2 **Solution Process for the Growth of Organic/Polymeric Crystals**

The solution process has already been discussed with regards to its convenient fabrication procedure, suitability for organic materials, and low cost. When solution-processed techniques (e.g., dip-coating and drop-casting) are used to prepare organic single crystals, the solubility and intermolecular interactions are *internal* factors that determine the crystal growth, while the solvent, the concentration, the deposition temperature, and the atmosphere are the major *external* factors. Three different morphologies obtained under the same external conditions, based on three materials and with minimal modification of the substituents, are shown in Figure 3.27 [20]. Similar results were also reported for another series of materials by adjusting the intermolecular hydrophilic interactions, π-stacking, and hydrogen bonds through a modification of the chemical structures [194, 195].

On occasion, a solvent-exchange method (Figure 3.28a) [196] or a solvent atmosphere (Figure 3.28b) [90] can be used to grow single crystals. Usually, these methods are suitable for materials that are highly soluble in common solvents but almost insoluble in other solvents. As noted above, solution-processed techniques usually result in low-quality crystals compared to the PVT method, but are very effective for materials with a low thermal stability. Notably, solution-processed

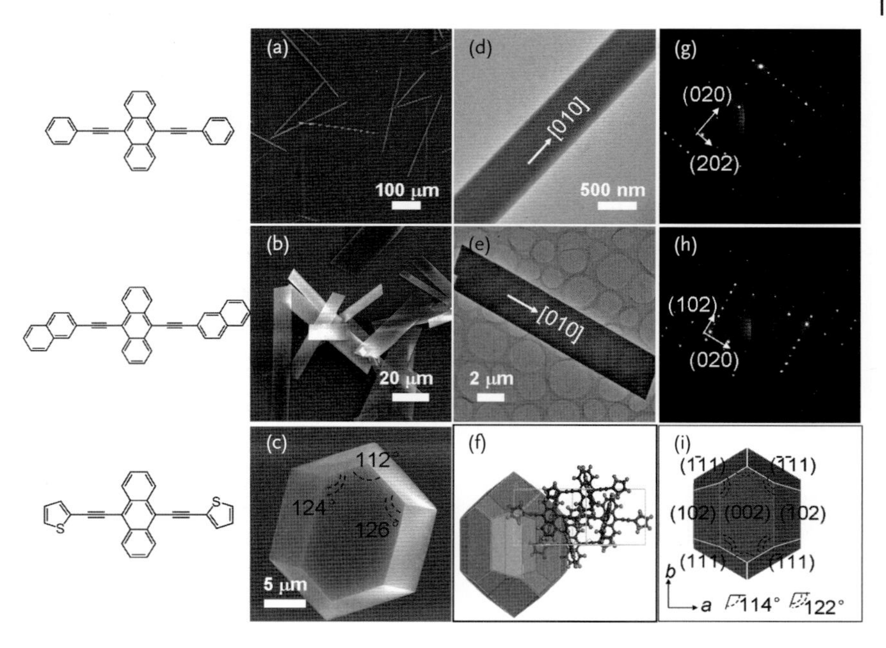

Figure 3.27 Different morphologies obtained under the same external conditions, based on three materials with minimal modification of the chemical structures. Reprinted with permission from Ref. [20]; © 2009, American Chemical Society.

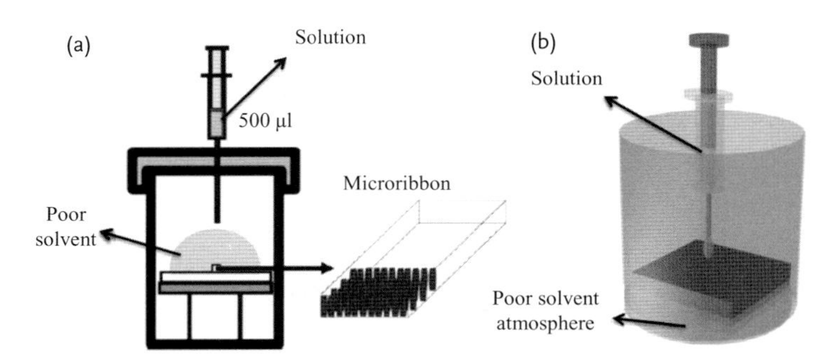

Figure 3.28 (a) Schematic diagram of the solvent-exchange method. Reprinted with permission from Ref. [196]; © 2007, Wiley-VCH Verlag GmbH & Co. KGaA; (b) Schematic diagram of the drop-casting technique, assisted by a solvent atmosphere. Reprinted with permission from Ref. [90]; © 2011, The Royal Society of Chemistry.

techniques are the only methods capable of achieving polymeric crystals, with up to millimeter-sized crystals being obtained [174].

3.5.3
Fabrication Techniques for Organic Field-Effect Transistors of Single Crystals

Due to their sensitivity to heat and electron beams, and their general fragility, techniques that are used to fabricate inorganic FETs (e.g., photolithography, ultrasonics, electron beam, focused ion beam) are not suitable for preparing organic crystals. Moreover, since the intermolecular interactions between organic molecules (van der Waals forces) are much weaker than covalent bonds, as occur in inorganic semiconductors, organic crystals are usually very small (micro/nanometer-sized). Thus, the adoption of traditional methods for inorganic semiconductors in organic crystals–that is, to fabricate devices via electron beam or focused ion beam deposition–represents a major challenge. Consequently, the development of new methods to fabricate organic single crystals is a topic of great urgency.

3.5.3.1 Electrostatic-Bonding Technique

In electrostatic bonding, organic single crystals are placed onto pre-deposited electrodes to form a contact via an electrostatic force. As shown in Figure 3.29, the organic crystals are deposited during the last stage of the method, so that they are free from any damage that might have originated from thermal irradiation, ultrasonic, lithography, and so on. Hence, this method can be used to eliminate any potential damage to crystals during device fabrication, and is also convenient if there is a need to change the position of the crystals, in order that the anisotropic properties can be facially investigated. Unfortunately, electrostatic bonding always leads to a poor contact between the crystal and the drain/source electrodes. A possible solution to this problem would be to use a flexible substrate (e.g., PDMS, as shown in Figure 3.29 [6]), but this would also require the contact mode to be changed into a top-contact mode (or other method) to further improve the contact.

3.5.3.2 Drop-Casting Technique

The use of drop-casting [22], dip-coating [197] or PVT [180] to grow single crystals directly onto pre-patterned drain/source electrode pairs represents just one of the methods available to improve contact between the semiconductor and the drain/source electrodes (as occurs in electrostatic bonding). The reason for this is that the *in situ* growth of single crystals may result in a better contact with the electrodes than would electrostatic-bonding contacts, especially when the electrodes have been modified by SAMs, or organic electrodes (e.g., TTF-TCNQ) are used, due to interface compatibility.

Drop-casting techniques are widely used for the deposition of organic thin films and single crystals (as noted above). However, in order to avoid the adverse effects brought about by subsequent deposition on single crystals, the most efficient approach is to use a BGBC mode to fabricate these devices. All other layers, as

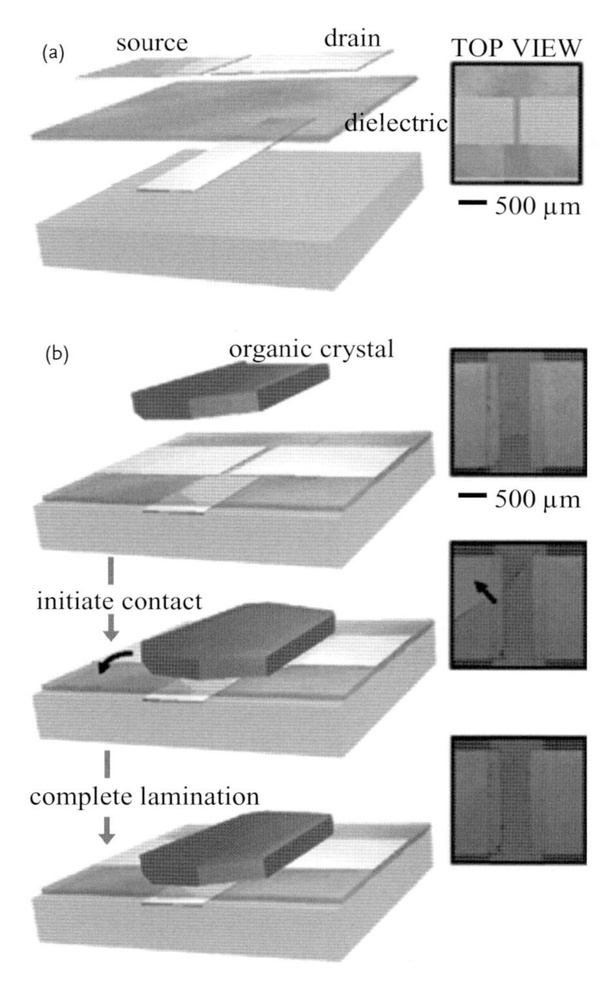

Figure 3.29 Schematic diagrams of electrostatic bonding technique for the fabrication of organic single-crystal transistors (BGBC). Reprinted with permission from Ref. [6]; © 2004, American Association for the Advancement of Science.

well as the insulator, gate, drain and source electrodes, are pre-deposited, which is similar to the electrostatic-bonding technique. The crystals may then be grown onto the drain/source electrodes to form conductive channels by using a drop-casting technique.

Likewise, dip-coating and PVT for the growth of organic single crystals could also be used to deposit the crystals directly onto the electrodes, so as to form the channels *in situ* (Figure 3.30a). The major disadvantage of these methods is that a poor contact still exists between the organic single crystals and the drain/source electrodes, due to the bottom-contact configuration.

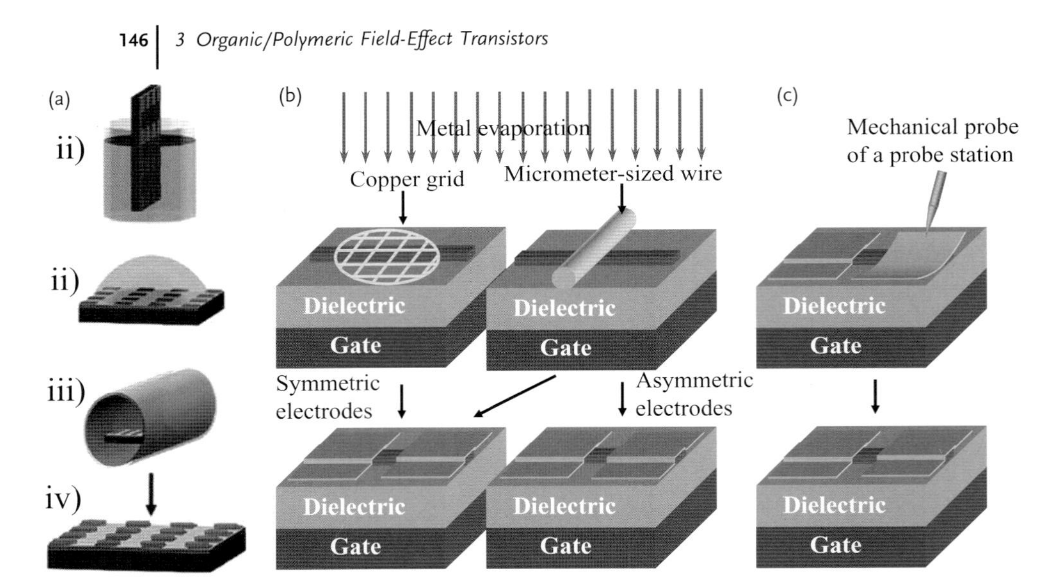

Figure 3.30 (a) The use of (i) dip-coating, (ii) drop-casting, and (iii) PVT to deposit organic single crystals onto the patterned electrodes, with channels formed simultaneously. Adapted with permission from Ref. [198]; © 2007, Elsevier B.V.; (b) Using a copper grid, gold wire, fiber, or organic ribbon as a shadow mask to deposit drain/source electrodes; (c) Using the probe of a probe station to manipulate gold films onto the single crystals as drain/source electrodes. Adapted with permission from Refs [15, 193, 199]; © 2008 Wiley-VCH Verlag GmbH & Co. KGaA; © 2010 and 2012, American Chemical Society, respectively.

3.5.3.3 **Deposition Parylene Dielectric Technique**

In addition to the above-mentioned reasons (i.e., small size, sensitivity to electron beam, focused ion beam, lithography, etc.), there is one critical reason that makes top-gate devices unsuitable for organic single crystals, namely that the layers deposited onto single crystals (especially the dielectric layer) will cause either thermal- or solvent-based damage to the organic single crystals. However, parylene as a polymer can avoid these potential damages, because its deposition does not require the use of solvents and it can also be placed in the zone at room temperature. Moreover, parylene can form a dense structure, leading to a low leakage current even at very low thickness. The detailed deposition of parylene was discussed in Section 3.3.2.2. By using parylene as the dielectric layer, one of the highest mobilities [$31.3\,cm^2/(V\cdot s)$] [8] has been obtained from solution-processed devices.

Although it is difficult to fabricate top-gate devices for organic single crystals by using other insulators (the main reason being that the deposition of after-deposited layers will harm or damage the semiconducting layer), the possibility is not totally excluded. When Palstra and coworkers used 6,13-pentacenequinone (PQ), which could be thermally evaporated onto the semiconductor, as the dielectric layers, a high performance was achieved for pentacene crystals [7]. Later, Frisbie and colleagues [200] inserted a buffer layer (5 nm thickness) between the organic single crystals and the subsequently deposited dielectric layer in order to reduce the solvent effects.

3.5.3.4 Shadow Mask Technique

There is no doubt that, in order to obtain a better contact between the organic single crystals and the drain/source electrodes, and between the crystals and the dielectric layer, a top-contact configuration represents one of the most efficient approaches. Because the organic single crystals are usually micro/nanometer-sized, due to the weak intermolecular interactions (in order to minimize defects in single crystals, a smaller size is preferred to achieve a higher performance), the greatest challenge for the BGTC mode is to identify a suitable shadow mask. The results of various investigations have suggested that a copper grid (~40 μm) [190], gold wire (~20 μm) [58], polyethylene fiber (~20 μm) [201] and organic micro/nanoribbon (ranging from hundreds of nanometers to several microns in size) [202] can be used as a shadow mask (Figure 3.30b). These methods could permit a significant scaling down of the channel length in order to adapt the crystal size and to form a good contact between the single crystals and the drain/source electrodes. Moreover, when the size of the mask is reduced to several microns, it is possible to use crossed masks on the organic single crystals, which in turn makes it possible to investigate the anisotropy on an individual single crystal (requirement: plate or disk crystals; Figure 3.31). Unfortunately, however, these methods suffer from problems of thermal irradiation when the electrodes are deposited, which in turn affects the contact between the semiconductor and the electrodes.

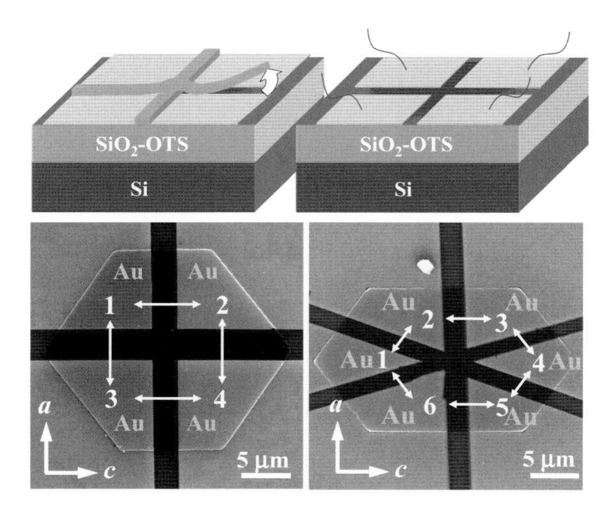

Figure 3.31 Using crossed organic micro/nanoribbons as a shadow mask to fabricate drain/source electrodes. Reprinted with permission from Ref. [193, 203]; © 2007, Wiley-VCH Verlag GmbH & Co. KGaA; © 2010, American Chemical Society, respectively.

3.5.3.5 Gold Layer Glue Technique

The gold layer glue technique, the aim of which is to eliminate the thermal irradiation and to maintain a good contact of the top-contact mode, employs a stamping technique to place a gold film directly onto the single crystals as drain/source electrodes. To achieve this, a small piece of gold film is first glued to the probe of probe station, and then transferred (by manipulation) onto the single crystals, such that a conducting channel is formed simultaneously (Figure 3.30c). Notably, other types of conducting material (e.g., Ag, Cu, Zn, Al) can also used as electrodes [90] when using this method. When using the gold layer glue technique, asymmetric electrodes can be obtained with relative ease using only two different electrode films, and much more easily than with the two-step shadow mask technique (Figure 3.30b) [58].

3.5.4
Performance of Organic/Polymeric Single Crystals in Field-Effect Transistors

3.5.4.1 Organic/Polymeric Crystals

During the past two decades, much progress has been made in the creation of organic/polymeric crystals, with the highest single-crystal mobilities reaching $40\,cm^2/(V \cdot s)$. Not all of the materials adopted in single crystal devices will be described at this point; rather, the details of some representative high-performance materials will be provided, together with their chemical structures, arrangements, and deposition conditions, the aim being to identify any clear relationship between structures and properties.

The details of some representative semiconductors with high single-crystal mobilities over $1\,cm^2/(V \cdot s)$, as well as their chemical structures, arrangements, preparation conditions, and properties, are listed in Table 3.4. Because the performance of these devices depends heavily on the adopted direction along the drain to the source, the directions investigated are also included. The data in Table 3.4 show clearly that the highest performance is usually obtained from the direction of strongest intermolecular interactions. Because polymeric crystals are rarely reported, they have been included in this table without the limit of mobility. From the data in Table 3.4, some conclusions can also be drawn as to why these devices exhibit a high performance [15]. As with thin-film transistors, there is no doubt that the semiconductor itself will determine the possible performance, while various rules – that is, the factors that are influenced by the semiconductors – are also known to exist, such as the importance of extending π-systems, the planar π-system, C_2 symmetry, and the π-stacking mode for high performance. In the meantime, other external parameters that might affect device performance include the dielectric layer (e.g., dielectric constant, thickness, etc.), the electrodes (e.g., energy level alignment), the channel dimensions (length and width), and crystal thickness. Although such factors mainly affect device performance by affecting the charge injection barrier or the induced electric field, the growth conditions of the single crystals (PVT or solution-processed techniques) might also affect performance by influencing crystal morphology and arrangement. In addition, other

Table 3.4 Representative semiconductors with high single-crystal mobilities [>1 $cm^2/(V{\cdot}s)$], polymers is not limited in this value) and their chemical structures, arrangements, preparation conditions, and properties.

	El	E_g	P	T	Insulator	C, E	D	$\mu^{a)}$	Reference(s)
P-type									
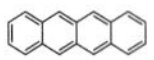	−5.21	2.57	H	PVT	PDMS	BGBC, Au	—	2.4	[218, 219]
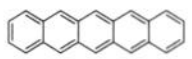	−5.0	1.8	H	PVT	PQ[b)]	TGTC, Ag-epoxy	—	40	[7, 123]
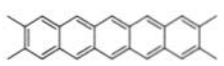	−4.41	1.95/2.03	H	PVT	SiO_2	BGBC, Au[c)]	—	1	[220]
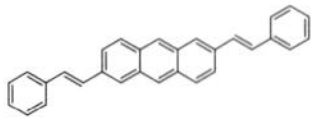	−5.4	2.6	H	PVT	ODTS/SiO_2	BGTC, Au	[100]	4.3	[123, 217]
							[010]	2.2	
	−5.36	2.21	Slipped π	PVT	Air	BGTC, Au	[010], π	24.5	[89, 205]
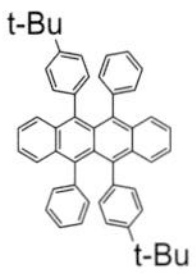 t-Bu, t-Bu	—	—	Slipped π	PVT	ODTS/SiO_2	BGBC, Au	[010], π	12	[221]
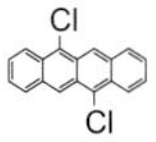 Cl, Cl	—	—	Slipped π	PVT	Parylene	TGTC, graphite ink	[100], π	1.6	[222]
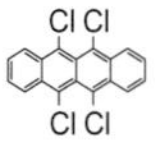 Cl Cl, Cl Cl	−5.7a	—	Slipped π	PVT	Parylene-N	TGTC, graphite ink	[100], π	1.7	[223]

(*Continued*)

Table 3.4 (*Continued*)

	El	E_g	P	T	Insulator	C, E	D	$\mu^{a)}$	Reference(s)
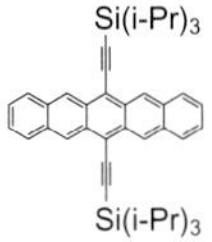	−5.11	1.69	2-D π	Solvent-exchange	SiO_2	BGTC, Au	[010]	1.42	[128, 196]
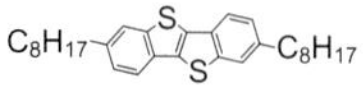	−5.5	3.5	H	Solvent vapor annealing	PMMA/SiO_2	BGTC, Au	—	9.1	[8, 36, 131]
				Ink-jet printing	Parylene-C	TGTC, Au	—	31.3	
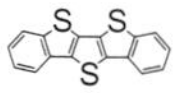	−5.6	3.46	H	PVT	ODTS/SiO_2	BGTC, Au	[001]	1.8	[104, 203]
							[100]	0.8	
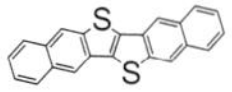	−5.44	3.0	H	PVT	Cytop/SiO_2	BGTC, Au/ TTF-TCNQ	—	8.3	[134, 224]
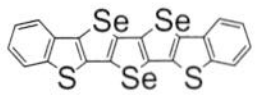	−5.31	3.37	Slipped π	PVT	Parylene	TGTC, Au paste	[001], π	1.1	[225]
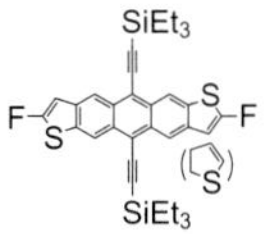	—	—	2-D π	PVT	OTS/SiO_2	BGBC, Au	—	6	[137]
	−5.39	2.90	Slipped π	PVT	ODTS/SiO_2	BGTC, Au	[010], π	2.66	[212]
	−5.3	3.48/3.12	Slipped π	Solution-process	ODTS/SiO_2	BGTC, Au	[010], π	2.13	[226]

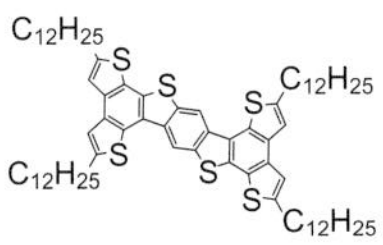	−5.7	3.1	—	Solution-process	$PMMA/SiO_2$	BGTC, Au	—	2.1	[201, 227]
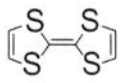	−5.09	2.76	Slipped π	Drop-casting	$ODTS/SiO_2$	BGBC, Au	[010], π	1.2	[22, 228]
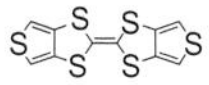	−5.14	2.51	Slipped π	Drop-casting	SiO_2	BGBC, Au	[010], π	3.6	[209, 229]
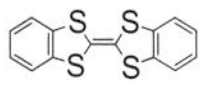	−4.96	2.4	Slipped π	Drop-casting	SiO_2	BGBC, Au	[010], π	1	[230, 231]
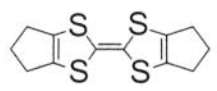	−4.61	—	2-D π	PVT	Parylene-C	TGTC, TTF-TCNQ	[100]	11.2	[140, 232]
				Solution-process				10.4	
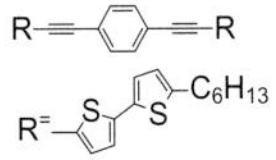	−5.45	2.78	—	Drop-casting	SiO_2	BGTC, Au	—	1.0	[174, 233]
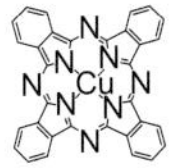	−5.2	1.7	π	PVT	Parylene-N	TGTC, graphite	[010], π	1.0	[160, 234]
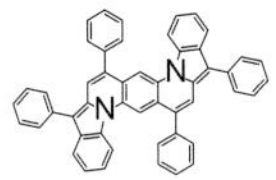	−5.19	2.59	1-D π	PVT	SiO_2	BGBC, Au	[001], π	1.0	[235]
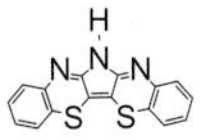	−5.44	2.41	Slipped π	PVT	$ODTS/SiO_2$	BGTC, Au	[100], π	3.6	[207, 236]
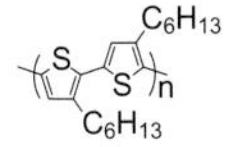	−5.1	1.8	NA	Drop-casting	$ODTS/SiO_2$	BGTC, Au	π	—	[190, 216]

(Continued)

Table 3.4 (*Continued*)

	El	E_g	P	T	Insulator	C, E	D	μ[a)]	Reference(s)
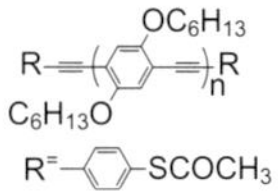 OC_6H_{13}, R, n, $C_6H_{13}O$, R= $SCOCH_3$	—	—	NA	Drop-casting	SiO_2	BGTC, Au	Intra-molecular	0.1	[237]
N-type									
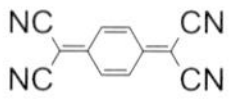 NC, CN, NC, CN	−4.63	2.25	Slipped π	PVT	Air/vacuum	BGBC, Au	[100]	1.6	[238–240]
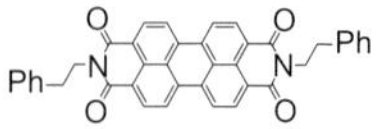 O, O, Ph, N, N, Ph, O, O	−4.1	2.0	Slipped π	Drop-casting	SiO_2	BGBC, Au	[100], π	1.4	[214]
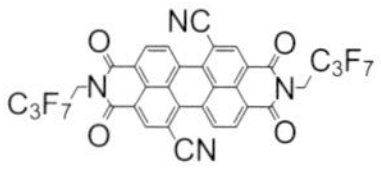 NC, O, O, C_3F_7, N, N, C_3F_7, O, CN, O	−4.5	2.3	2-D π	PVT	PMMA/SiO_2	BGBC, Au	—	6	[148, 241, 242]
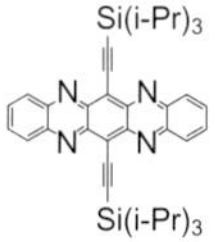 Si(i-Pr)$_3$, N, N, N, N, Si(i-Pr)$_3$	−4.01	1.74	2-D π	Drop-casting	OTMS/SiO_2	BGTC, Ag	[010], π	1.77	[90, 151]

a) The unit of μ is $cm^2/(V \cdot s)$.
b) 6,13-pentacenequinone.
c) 4-nitrobenzenethiol.
El: the energy level (eV), HOMO energy levels for *p*-type semiconductors and LUMO energy levels for *n*-type semiconductors; E_g: the energy gap (eV) between HOMO and LUMO energy levels; P: packing mode; H: herringbone packing; π: π-stacking; T: techniques for growth of organic single crystals; C: contact mode; E: the electrode material of drain/source electrodes; TC: top contact; TG: top gate; BC: bottom contact; BG: bottom gate; D: direction of the performance; NA: not applicable; —: not found.

factors such as the testing conditions (e.g., environment, pressure, temperature) may influence performance. Due to their freedom from grain boundaries, minimized defects and perfect order, single crystals are superior to thin films, the main advantage being that they can reveal structure–property relationships (as defined in the following section).

3.5.4.2 Structure–Property Relationship of Organic/Polymeric Single Crystals

Based on the above analysis, there is no doubt that chemical structure and molecular arrangements play important roles when defining the properties of single crystals. Due to the various intermolecular interactions that occur along different directions, the crystals are formed with different shapes. Typically, the stronger the interaction the larger will be the surface energy in this direction, and the smaller facet will appear in this direction. As a result, crystals will prefer to grow along this direction, which forms the basis of the Bravais–Friedel–Donnay–Harker (BFDH) method [20, 204].

From this point of view, it is not difficult to understand that the materials with π–π stacking in their single crystals usually showed one-dimensional (1-D) rods or wires, or 2-D ribbons or plates, because the intermolecular interactions along the π–π stacking orientation were usually stronger than the other one or two orientations (e.g., C–H···π, or simply van der Waals forces between alkyl groups). Many examples of this have been reported, including 9,10-bis(phenylethynyl) anthracene (BPEA) [20], rubrene [205], 9,10-diphenylanthracene (DPA) [206], 6*H*-pyrrolo-[3,2-*b*:4,5-*b′*]bis[1,4]benzothiazine (PBBTZ) [207], *syn*-bis(benzothieno) thiophene (*syn*-BBTT) [208], dithiophene-tetrathiafulvalene (DT-TTF) [209], triisopropylsilylethynyl-pentacene (TIPS-pentacene) [196], CuPc [210], F_{16}CuPc [58], 6,13-bis((triisopropylsilyl)ethynyl)-5,7,12,14-tetraazapentacene (TIPS-TAP) [90], perylo[1,12-*b,c,d*]thiophene (PET) [211], perylo[1,12-*b,c,d*]selenophene (PESE) [212], hexathiapentacene (HTP) [213], perylene-3,4:9,10-tetracarboxylic diimide (PTCDI) [214, 215], and poly(3-hexylthiophene) (P3HT) [216]. On the other hand, those materials without long-range π–π-stacking in their single crystals usually showed 2-D ribbons or plates, or three-dimensional (3-D) plates or particles, because the C–H···π intermolecular interactions usually were comparable along two or three orientations. Extensive examples of this type of material have also been described, including pentacene [180], tetracene [192], di(phenylvinyl) anthracene (DPV-ant) [217], *anti*-BBTT [208], and 9,10-bis(thiophen-2-ylethynyl)anthracene (BTEA) [20]. An example of two isomers showing 1-D and 2-D self-assembled morphologies, due respectively to their different packing modes in their single crystals (i.e., with or without π–π-stacking), is provided in Figure 3.32.

Clearly, in addition to intrinsic factors, the external conditions also could affect the intermolecular interactions, which in turn would result in different morphologies. Many of the factors which affect the growth of single crystals were discussed in Sections 3.5.2.1 and 3.5.2.2. For example, for solution-processed techniques there is no doubt that the intermolecular interactions between neighboring solvent molecules, and interactions between the semiconductor and solvent molecules, will affect the precipitation of organic single crystals. For example, weak

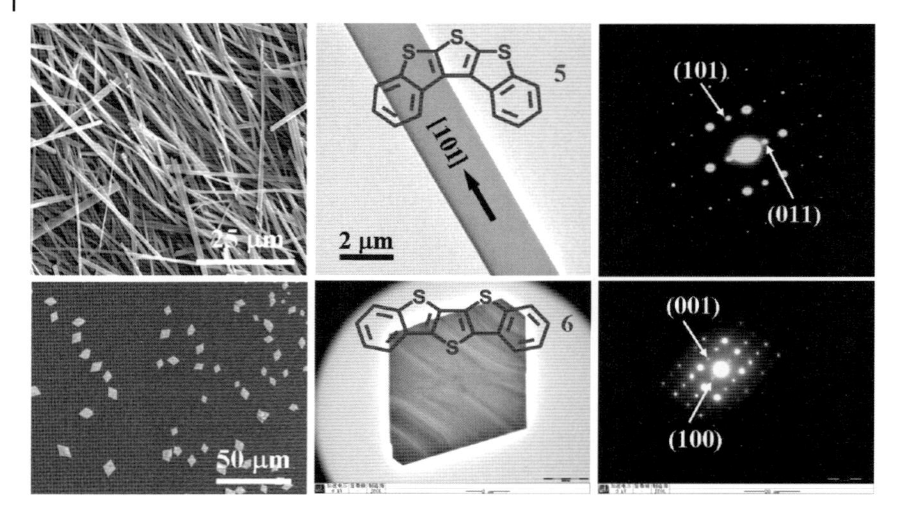

Figure 3.32 One- and two-dimensional morphologies due to different intermolecular interactions with or without π-π stacking. Reprinted with permission from Ref. [193]; © 2010, American Chemical Society.

interactions between the semiconductor and solvent molecules would result in an aggregation of semiconductor molecules and a rapid growth of single crystals. Hence, both the solvent and concentration would affect the morphology and arrangement of the semiconductor. Likewise, growth conditions such as temperature and atmosphere could also affect single-crystal growth.

The concept that different morphologies could be obtained due to different arrangements by controlling the solvents, concentration, and growth conditions is illustrated in Figure 3.33. Clearly, different interactions would result in property anisotropy among the crystals, and the stronger the intermolecular interactions the higher the performance in this direction. The reason for this is quite clear:

- In theory, if the intermolecular interaction is stronger, then the transfer integral along this direction will be larger, which would result in a higher performance.
- Experimentally, if the intermolecular interaction is stronger (especially for π–π interactions), then the charge carriers would transport along this direction much more easily, due to the stronger interactions. This also is the reason why inorganic materials always showed a much higher performance than organic semiconductors, due to their strong covalent bonds.

Hence, it is not difficult to understand that most of the semiconductors with π–π stacking showed the highest performance along the π–π stacking orientation (see Table 3.3), but a lower performance in other directions. An example of the property anisotropy of rubrene along different directions, in which the highest performance was obtained along the *b*-direction (the π–π-stacking direction) is shown in Figure 3.34.

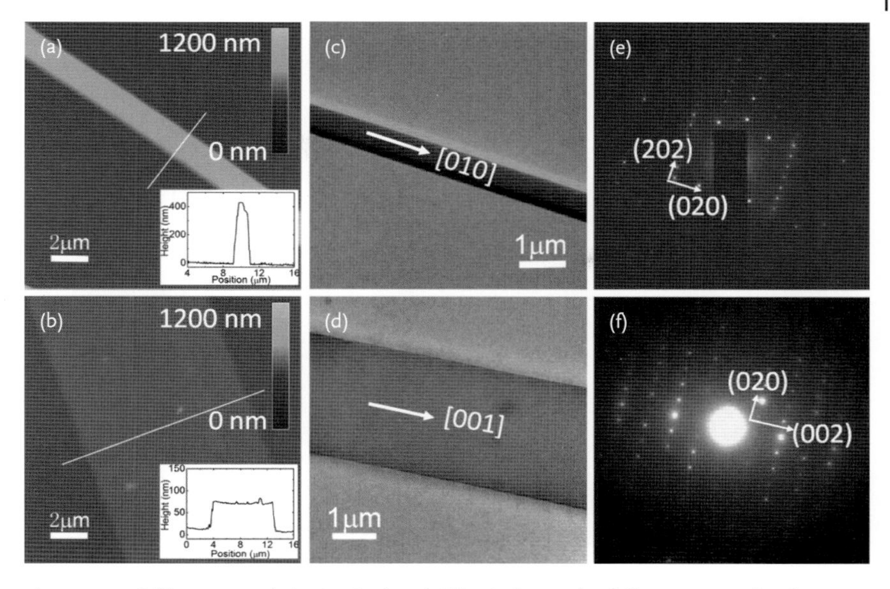

Figure 3.33 Different morphologies (rod and ribbon) due to the different intermolecular interactions induced by different growth conditions. Adapted with permission from Ref. [243]; © 2010, American Institute of Physics.

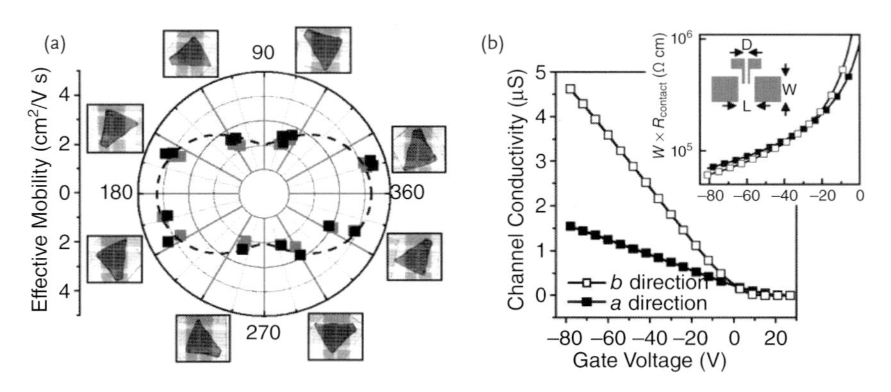

Figure 3.34 Property anisotropy of rubrene. Reprinted with permission from Ref. [6]; © 2004, American Association for the Advancement of Science.

3.6 Outlook

During the past two decades, vast improvements have been made in the area of OFETs, with the highest mobilities reaching 40 $cm^2/(V \cdot s)$, and with demonstration systems and products based on OFETs becoming available on a commercial basis. Nonetheless, there are still problems to be resolved, notably that the performance of the OFETs continues to lag far behind that of their inorganic counterparts. Likewise, the printing techniques used for the large-area production of OFETs – which is a major strongpoint of organic semiconductors – have still to be finalized. Although various types of charge transport mechanism have been proposed, these can explain only limited phenomena, and require further development. It is clear that, with many challenges and opportunities continuing to coexist in the area of OFETs, much effort remains to be expended before their practical application becomes truly achievable.

References

1 Eley, D.D. (1948) Phthalocyanines as semiconductors. *Nature*, **162** (4125), 819–819.

2 Chiang, C.K., Fincher, C.R., Park, Y.W., Heeger, A.J., Shirakawa, H., Louis, E.J., Gau, S.C., and MacDiarmid, A.G. (1977) Electrical conductivity in doped polyacetylene. *Phys. Rev. Lett.*, **39** (17), 1098–1101.

3 Tsumura, A., Koezuka, H., and Ando, T. (1986) Macromolecular electronic device: field-effect transistor with a polythiophene thin film. *Appl. Phys. Lett.*, **49** (18), 1210–1212.

4 Li, L.Q., Tang, Q.X., Li, H.X., Yang, X.D., Hu, W.P., Song, Y.B., Shuai, Z.G., Xu, W., Liu, Y.Q., and Zhu, D.B. (2007) An ultra closely pi-stacked organic semiconductor for high performance field-effect transistors. *Adv. Mater.*, **19** (18), 2613–2617.

5 Nakayama, K., Hirose, Y., Soeda, J., Yoshizumi, M., Uemura, T., Uno, M., Li, W., Kang, M.J., Yamagishi, M., Okada, Y., Miyazaki, E., Nakazawa, Y., Nakao, A., Takimiya, K., and Takeya, J. (2011) Patternable solution-crystallized organic transistors with high charge carrier mobility. *Adv. Mater.*, **23** (14), 1626–1629.

6 Sundar, V.C., Zaumseil, J., Podzorov, V., Menard, E., Willett, R.L., Someya, T., Gershenson, M.E., and Rogers, J.A. (2004) Elastomeric transistor stamps: reversible probing of charge transport in organic crystals. *Science*, **303** (5664), 1644–1646.

7 Jurchescu, O.D., Popinciuc, M., Wees, B.J.V., and Palstra, T.T.M. (2007) Interface-controlled, high-mobility organic transistors. *Adv. Mater.*, **19** (5), 688–692.

8 Minemawari, H., Yamada, T., Matsui, H., Tsutsumi, J.Y., Haas, S., Chiba, R., Kumai, R., and Hasegawa, T. (2011) Inkjet printing of single-crystal films. *Nature*, **475** (7356), 364–367.

9 Di, C.A., Liu, Y.Q., Yu, G., and Zhu, D.B. (2009) Interface engineering: an effective approach toward high-performance organic field-effect transistors. *Acc. Chem. Res.*, **42** (10), 1573–1583.

10 Halik, M., Klauk, H., Zschieschang, U., Schmid, G., Ponomarenko, S., Kirchmeyer, S., and Weber, W. (2003) Relationship between molecular structure and electrical performance of oligothiophene organic thin film transistors. *Adv. Mater.*, **15** (11), 917–922.

11 Newman, C.R., Frisbie, C.D., da Silva Filho, D.A., Bredas, J.-L., Ewbank, P.C., and Mann, K.R. (2004) Introduction to organic thin film transistors and design of *n*-channel organic semiconductors. *Chem. Mater.*, **16** (23), 4436–4451.

12 Zaumseil, J. and Sirringhaus, H. (2007) Electron and ambipolar transport in organic field-effect transistors. *Chem. Rev.*, **107** (4), 1296–1323.

13 Kim, J.B., Fuentes-Hernandez, C., Kim, S.J., Potscavage, W.J., Jr, Choi, S., and Kippelen, B. (2010) Ambipolar thin-film transistors with a co-planar channel geometry. *Org. Electron.*, **11** (8), 1351–1356.

14 Wang, C., Wang, X., Min, J., Zhao, N., and Xu, J. (2011) Super-linear rectifying property of rubrene single crystal devices. *Org. Electron.*, **12** (10), 1731–1735.

15 Wang, C., Dong, H., Hu, W., Liu, Y., and Zhu, D. (2012) Semiconducting pi-conjugated systems in field-effect transistors: a material odyssey of organic electronics. *Chem. Rev.*, **112**, 2208–2267.

16 Wang, C., Dong, H., Li, H., Zhao, H., Meng, Q., and Hu, W. (2010) Dibenzothiophene derivatives: from herringbone to lamellar packing motif. *Cryst. Growth Des.*, **10** (9), 4155–4160.

17 Mattheus, C.C., de Wijs, G.A., de Groot, R.A., and Palstra, T.T.M. (2003) Modeling the polymorphism of pentacene. *J. Am. Chem. Soc.*, **125** (20), 6323–6330.

18 Chesterfield, R.J., McKeen, J.C., Newman, C.R., Ewbank, P.C., da Silva Filho, D.A., Bredas, J.-L., Miller, L.L., Mann, K.R., and Frisbie, C.D. (2004) Organic thin film transistors based on N-alkyl perylene diimides: charge transport kinetics as a function of gate voltage and temperature. *J. Phys. Chem. B*, **108** (50), 19281–19292.

19 Gavezzotti, A., and Desiraju, G.R. (1988) A systematic analysis of packing energies and other packing parameters for fused-ring aromatic-hydrocarbons. *Acta Crystallogr.*, **B44**, 427–434.

20 Wang, C., Liu, Y., Ji, Z., Wang, E., Li, R., Jiang, H., Tang, Q., Li, H., and Hu, W. (2009) Cruciforms: assembling single crystal micro- and nanostructures from one to three dimensions and their applications in organic field-effect transistors. *Chem. Mater.*, **21** (13), 2840–2845.

21 Gsanger, M., Oh, J.H., Konemann, M., Hoffken, H.W., Krause, A.-M., Bao, Z., and Wurthner, F. (2010) A crystal-engineered hydrogen-bonded octachloroperylene diimide with a twisted core: an *n*-channel organic semiconductor. *Angew. Chem. Int. Ed.*, **49** (4), 740–743.

22 Jiang, H., Yang, X., Cui, Z., Liu, Y., Li, H., Hu, W., Liu, Y., and Zhu, D. (2007) Phase dependence of single crystalline transistors of tetrathiafulvalene. *Appl. Phys. Lett.*, **91** (12), 123505.

23 Wen, Y., Liu, Y., Guo, Y., Yu, G., and Hu, W. (2011) Experimental techniques for the fabrication and characterization of organic thin films for field-effect transistors. *Chem. Rev.*, **111** (5), 3358–3406.

24 Virkar, A.A., Mannsfeld, S., Bao, Z., and Stingelin, N. (2010) Organic semiconductor growth and morphology considerations for organic thin-film transistors. *Adv. Mater.*, **22** (34), 3857–3875.

25 Miozzo, L., Yassar, A., and Horowitz, G. (2010) Surface engineering for high performance organic electronic devices: the chemical approach. *J. Mater. Chem.*, **20** (13), 2513–2538.

26 Dimitrakopoulos, C.D. and Malenfant, P.R.L. (2002) Organic thin film transistors for large area electronics. *Adv. Mater.*, **14** (2), 99–117.

27 Veres, J., Ogier, S.D., Leeming, S.W., Cupertino, D.C., and Khaffaf, S.M. (2003) Low-k insulators as the choice of dielectrics in organic field-effect transistors. *Adv. Funct. Mater.*, **13** (3), 199–204.

28 Hulea, I.N., Fratini, S., Xie, H., Mulder, C.L., Iossad, N.N., Rastelli, G., Ciuchi, S., and Morpurgo, A.F. (2006) Tunable Frohlich polarons in organic single-crystal transistors. *Nat. Mater.*, **5** (12), 982–986.

29 Kim, E.G., Coropceanu, V., Gruhn, N.E., Sanchez-Carrera, R.S., Snoeberger, R., Matzger, A.J., and Bredas, J.L. (2007) Charge transport parameters of the pentathienoacene crystal. *J. Am. Chem. Soc.*, **129** (43), 13072–13081.

30 Sakanoue, T. and Sirringhaus, H. (2010) Band-like temperature dependence of mobility in a solution-processed organic semiconductor. *Nat. Mater.*, **9** (9), 736–740.

31 Warta, W. and Karl, N. (1985) Hot holes in naphthalene: high, electric-field-dependent mobilities. *Phys. Rev. B*, **32** (2), 1172.

32 Warta, W., Stehle, R., and Karl, N. (1985) Ultrapure, high mobility organic photoconductors. *Appl. Phys. A*, **36** (3), 163–170.

33 Jurchescu, O.D., Baas, J., and Palstra, T.T.M. (2004) Effect of impurities on the mobility of single crystal pentacene. *Appl. Phys. Lett.*, **84** (16), 3061–3063.
34 Tripathi, A.K., Heinrich, M., Siegrist, T., and Pflaum, J. (2007) Growth and electronic transport in 9,10-diphenylanthracene single crystals – an organic semiconductor of high electron and hole mobility. *Adv. Mater.*, **19** (16), 2097–2101.
35 Podzorov, V., Menard, E., Borissov, A., Kiryukhin, V., Rogers, J.A., and Gershenson, M.E. (2004) Intrinsic charge transport on the surface of organic semiconductors. *Phys. Rev. Lett.*, **93** (8), 086602.
36 Liu, C., Minari, T., Lu, X., Kumatani, A., Takimiya, K., and Tsukagoshi, K. (2011) Solution- processable organic single crystals with bandlike transport in field-effect transistors. *Adv. Mater.*, **23** (4), 523–526.
37 Brown, A.R., Jarrett, C.P., de Leeuw, D.M., and Matters, M. (1997) Field-effect transistors made from solution-processed organic semiconductors. *Synth. Met.*, **88** (1), 37–55.
38 Sirringhaus, H., Tessler, N., and Friend, R.H. (1998) Integrated optoelectronic devices based on conjugated polymers. *Science*, **280** (5370), 1741–1744.
39 Sirringhaus, H., Brown, P.J., Friend, R.H., Nielsen, M.M., Bechgaard, K., Langeveld-Voss, B.M.W., Spiering, A.J.H., Janssen, R.A.J., Meijer, E.W., Herwig, P., and de Leeuw, D.M. (1999) Two-dimensional charge transport in self-organized, high-mobility conjugated polymers. *Nature*, **401** (6754), 685–688.
40 Horowitz, G., Hajlaoui, R., Bourguiga, R., and Hajlaoui, M. (1999) Theory of the organic field-effect transistor. *Synth. Met.*, **101** (1–3), 401–404.
41 Tang, Q.X., Tong, Y.H., Li, H.X., and Hu, W.P. (2008) Air/vacuum dielectric organic single crystalline transistors of copper- hexadecafluorophthlaocyanine ribbons. *Appl. Phys. Lett.*, **92** (8), 083309.
42 Suglyama, H., Nagano, T., Nouchi, R., Kawasaki, N., Ohta, Y., Imai, K., Tsutsui, M., Kubozono, Y., and Fujiwara, A. (2007) Transport properties of field-effect transistors with thin films of C-76 and its electronic structure. *Chem. Phys. Lett.*, **449** (1–3), 160–164.
43 Xia, Y., Kalihari, V., Frisbie, C.D., Oh, N.K., and Rogers, J.A. (2007) Tetracene air-gap single- crystal field-effect transistors. *Appl. Phys. Lett.*, **90** (16), 162106.
44 Bredas, J.L., Calbert, J.P., da Silva, D.A., and Cornil, J. (2002) Organic semiconductors: a theoretical characterization of the basic parameters governing charge transport. *Proc. Natl Acad. Sci. USA*, **99** (9), 5804–5809.
45 Brédas, J.-L., Beljonne, D., Coropceanu, V., and Cornil, J. (2004) Charge-transfer and energy-transfer processes in π-conjugated oligomers and polymers: a molecular picture. *Chem. Rev.*, **104** (11), 4971–5004.
46 Bromley, S.T., Mas-Torrent, M., Hadley, P., and Rovira, C. (2004) Importance of intermolecular interactions in assessing hopping mobilities in organic field effect transistors: pentacene versus dithiophene-tetrathiafulvalene. *J. Am. Chem. Soc.*, **126** (21), 6544–6545.
47 Shukla, D., Nelson, S.F., Freeman, D.C., Rajeswaran, M., Ahearn, W.G., Meyer, D.M., and Carey, J.T. (2008) Thin-film morphology control in naphthalene-diimide-based semiconductors: high mobility *n*-type semiconductor for organic thin-film transistors. *Chem. Mater.*, **20** (24), 7486–7491.
48 Hecht, D.S., Hu, L., and Irvin, G. (2011) Emerging transparent electrodes based on thin films of carbon nanotubes, graphene, and metallic nanostructures. *Adv. Mater.*, **23** (13), 1482–1513.
49 Liu, W., Jackson, B.L., Zhu, J., Miao, C.-Q., Chung, C.-H., Park, Y.J., Sun, K., Woo, J., and Xie, Y.-H. (2010) Large scale pattern graphene electrode for high performance in transparent organic single crystal field-effect transistors. *ACS Nano*, **4** (7), 3927–3932.
50 Pal, B.N., Dhar, B.M., See, K.C., and Katz, H.E. (2009) Solution-deposited sodium beta-alumina gate dielectrics for low-voltage and transparent field-effect transistors. *Nat. Mater.*, **8** (11), 898–903.
51 Artukovic, E., Kaempgen, M., Hecht, D.S., Roth, S., and Grüner, G. (2005)

Transparent and flexible carbon nanotube transistors. *Nano Lett.*, **5** (4), 757–760.

52 Chen, Y.S., Xu, Y.F., Zhao, K., Wan, X.J., Deng, J.C., and Yan, W.B. (2010) Towards flexible all-carbon electronics: flexible organic field-effect transistors and inverter circuits using solution-processed all-graphene source/drain/gate electrodes. *Nano Res.*, **3** (10), 714–721.

53 Murphy, A.R. and Fréchet, J.M.J. (2007) Organic semiconducting oligomers for use in thin film transistors. *Chem. Rev.*, **107** (4), 1066–1096.

54 Yamada, K., Takeya, J., Takenobu, T., and Iwasa, Y. (2008) Effects of gate dielectrics and metal electrodes on air-stable *n*-channel perylene tetracarboxylic dianhydride single-crystal field- effect transistors. *Appl. Phys. Lett.*, **92** (25), 253311.

55 Rost, C., Karg, S., Riess, W., Loi, M.A., Murgia, M., and Muccini, M. (2004) Light-emitting ambipolar organic heterostructure field-effect transistor. *Synth. Met.*, **146** (3), 237–241.

56 Lee, T.-W., Byun, Y., Koo, B.-W., Kang, I.-N., Lyu, Y.-Y., Lee, C.H., Pu, L., and Lee, S.Y. (2005) All-solution-processed *n*-type organic transistors using a spinning metal process. *Adv. Mater.*, **17** (18), 2180–2184.

57 Takahashi, T., Takenobu, T., Takeya, J., and Iwasa, Y. (2006) Ambipolar organic field-effect transistors based on rubrene single crystals. *Appl. Phys. Lett.*, **88** (3), 033505.

58 Tang, Q., Li, H., Liu, Y., and Hu, W. (2006) High-performance air-stable *n*-type transistors with an asymmetrical device configuration based on organic single-crystalline submicrometer/nanometer ribbons. *J. Am. Chem. Soc.*, **128** (45), 14634–14639.

59 Di, C., Yu, G., Liu, Y., Guo, Y., Wang, Y., Wu, W., and Zhu, D. (2008) High-performance organic field-effect transistors with low-cost copper electrodes. *Adv. Mater.*, **20** (7), 1286–1290.

60 Gaynor, W., Burkhard, G.F., McGehee, M.D., and Peumans, P. (2011) Smooth nanowire/polymer composite transparent electrodes. *Adv. Mater.*, **23** (26), 2905–2910.

61 Bae, S., Kim, H., Lee, Y., Xu, X., Park, J.-S., Zheng, Y., Balakrishnan, J., Lei, T., Ri Kim, H., Song, Y.I., Kim, Y.-J., Kim, K.S., Ozyilmaz, B., Ahn, J.-H., Hong, B.H., and Iijima, S. (2010) Roll-to-roll production of 30-inch graphene films for transparent electrodes. *Nat. Nano*, **5** (8), 574–578.

62 Zhang, J., Zhao, Y., Wei, Z., Sun, Y., He, Y., Di, C.-A., Xu, W., Hu, W., Liu, Y., and Zhu, D. (2011) Inkjet-printed organic electrodes for bottom- contact organic field-effect transistors. *Adv. Funct. Mater.*, **21** (4), 786–791.

63 Sirringhaus, H., Kawase, T., Friend, R.H., Shimoda, T., Inbasekaran, M., Wu, W., and Woo, E.P. (2000) High-resolution inkjet printing of all-polymer transistor circuits. *Science*, **290** (5499), 2123–2126.

64 Li, L., Hirtz, M., Wang, W., Du, C., Fuchs, H., and Chi, L. (2010) Patterning of polymer electrodes by nanoscratching. *Adv. Mater.*, **22** (12), 1374–1378.

65 Yamada, T., Kumai, R., Takahashi, Y., and Hasegawa, T. (2010) High performance organic thin-film transistors based on hexamethylenetetrathiafulvalene lying flat-on-surface with non-layered packing motif. *J. Mater. Chem.*, **20** (28), 5810–5812.

66 Mamada, M., Nishida, J.I., Kumaki, D., Tokito, S., and Yamashita, Y. (2008) High performance organic field-effect transistors based on [2,2′]bi[naphtho[2,3-*b*]thiophenyl] with a simple structure. *J. Mater. Chem.*, **18** (29), 3442–3447.

67 Stassen, A.F., de Boer, R.W.I., Iosad, N.N., and Morpurgo, A.F. (2004) Influence of the gate dielectric on the mobility of rubrene single-crystal field-effect transistors. *Appl. Phys. Lett.*, **85** (17), 3899–3901.

68 Ortiz, R.P., Facchetti, A., and Marks, T.J. (2010) High-k organic, inorganic, and hybrid dielectrics for low-voltage organic field-effect transistors. *Chem. Rev.*, **110** (1), 205–239.

69 Li, Z., Du, J., Tang, Q., Wang, F., Xu, J.-B., Yu, J.C., and Miao, Q. (2010)

Induced crystallization of rubrene in thin-film transistors. *Adv. Mater.*, **22** (30), 3242–3246.
70 Virkar, A.A., Mannsfeld, S.C.B., and Bao, Z. (2010) Energetics and stability of pentacene thin films on amorphous and crystalline octadecylsilane modified surfaces. *J. Mater. Chem.*, **20** (13), 2664–2671.
71 Robertson, J. (2004) High dielectric constant oxides. *Eur. Phys. J.–Appl. Phys.*, **28** (03), 265–291.
72 Shi, L., Xia, Y.D., Xu, B., Yin, J., and Liu, Z.G. (2007) Thermal stability and electrical properties of titanium-aluminum oxide ultrathin films as high-k gate dielectric materials. *Appl. Phys. Lett.*, **101** (3), 034102.
73 Peng, X., Horowitz, G., Fichou, D., and Garnier, F. (1990) All organic thin film transistors made of alpha-sexithienyl semiconducting and various polymeric insulating layers. *Appl. Phys. Lett.*, **57** (19), 2013–2015.
74 Chua, L.L., Zaumseil, J., Chang, J.F., Ou, E.C.W., Ho, P.K.H., Sirringhaus, H., and Friend, R.H. (2005) General observation of *n*-type field- effect behaviour in organic semiconductors. *Nature*, **434** (7030), 194–199.
75 Diallo, K., Erouel, M., Tardy, J., Andre, E., and Garden, J.L. (2007) Stability of pentacene top gated thin film transistors. *Appl. Phys. Lett.*, **91** (18), 183508.
76 Cheng, X., Caironi, M., Noh, Y.-Y., Wang, J., Newman, C., Yan, H., Facchetti, A., and Sirringhaus, H. (2010) Air stable cross-linked cytop ultrathin gate dielectric for high-yield low-voltage top-gate organic field-effect transistors. *Chem. Mater.*, **22** (4), 1559–1566.
77 Ismail, A.G. and Hill, I.G. (2011) Stability of *n*-channel organic thin-film transistors using oxide, SAM-modified oxide and polymeric gate dielectrics. *Org. Electron.*, **12** (6), 1033–1042.
78 Podzorov, V., Pudalov, V.M., and Gershenson, M.E. (2003) Field-effect transistors on rubrene single crystals with parylene gate insulator. *Appl. Phys. Lett.*, **82** (11), 1739–1741.
79 Walser, M.P., Kalb, W.L., Mathis, T., and Batlogg, B. (2009) Low-voltage organic transistors and inverters with ultrathin fluoropolymer gate dielectric. *Appl. Phys. Lett.*, **95** (23), 233301.
80 Kim, C., Facchetti, A., and Marks, T.J. (2007) Polymer gate dielectric surface viscoelasticity modulates pentacene transistor performance. *Science*, **318** (5847), 76–80.
81 Kim, J., Lim, S.H., and Kim, Y.S. (2010) Solution-based TiO_2-polymer composite dielectric for low operating voltage OTFTs. *J. Am. Chem. Soc.*, **132** (42), 14721–14723.
82 Onoue, T., Nakamura, I., Sakabe, Y., Yasuda, T., and Tsutsui, T. (2006) Low-operating-voltage organic field-effect transistors with poly-*p*- xylylene/high-k polymer bilayer gate dielectric. *Jpn. J. Appl. Phys. Part 2*, **45** (29–32), L770–L772.
83 Chang, J.-W., Wang, C.-G., Huang, C.-Y., Tsai, T.-D., Guo, T.-F., and Wen, T.-C. (2011) Chicken albumen dielectrics in organic field-effect transistors. *Adv. Mater.*, **23** (35), 4077–4081.
84 Halik, M., Klauk, H., Zschieschang, U., Schmid, G., Dehm, C., Schutz, M., Maisch, S., Effenberger, F., Brunnbauer, M., and Stellacci, F. (2004) Low-voltage organic transistors with an amorphous molecular gate dielectric. *Nature*, **431** (7011), 963–966.
85 Bolotin, K.I., Sikes, K.J., Hone, J., Stormer, H.L., and Kim, P. (2008) Temperature-dependent transport in suspended graphene. *Phys. Rev. Lett.*, **101** (9), 096802.
86 Don Park, Y., Lim, J.A., Lee, H.S., and Cho, K. (2007) Interface engineering in organic transistors. *Mater. Today*, **10** (3), 46–54.
87 Di, C., Liu, Y., Yu, G., and Zhu, D. (2009) Interface engineering: an effective approach toward high-performance organic field-effect transistors. *Acc. Chem. Res.*, **42** (10), 1573–1583.
88 DiBenedetto, S.A., Facchetti, A., Ratner, M.A., and Marks, T.J. (2009) Molecular self-assembled monolayers and multilayers for organic and unconventional inorganic thin-film transistor applications. *Adv. Mater.*, **21** (14–15), 1407–1433.

89 Dong, H., Wang, C., and Hu, W. (2010) High-performance organic semiconductors for field-effect transistors. *Chem. Commun.*, **46**, 5211–5222.

90 Wang, C., Liang, Z., Liu, Y., Wang, X., Zhao, N., Miao, Q., Hu, W., and Xu, J. (2011) Single crystal *n*-channel field effect transistors from solution-processed silylethynylated tetraazapentacene. *J. Mater. Chem.*, **21** (39), 15201–15204.

91 Yamao, T., Shimuzu, Y., Kuriki, H., Katagiri, T., and Hotta, S. (2010) Improved field-effect transistor characteristics of an *n*-type semiconducting thiophene/phenylene co-oligomer. *Jpn. J. Appl. Phys.*, **49**, 01A–001.

92 Campbell, I.H., Rubin, S., Zawodzinski, T.A., Kress, J.D., Martin, R.L., Smith, D.L., Barashkov, N.N., and Ferraris, J.P. (1996) Controlling Schottky energy barriers in organic electronic devices using self-assembled monolayers. *Phys. Rev. B*, **54** (20), R14321–R14324.

93 Kim, D.H., Lee, H.S., Yang, H.C., Yang, L., and Cho, K. (2008) Tunable crystal nanostructures of pentacene thin films on gate dielectrics possessing surface-order control. *Adv. Funct. Mater.*, **18** (9), 1363–1370.

94 Pernstich, K.P., Haas, S., Oberhoff, D., Goldmann, C., Gundlach, D.J., Batlogg, B., Rashid, A.N., and Schitter, G. (2004) Threshold voltage shift in organic field effect transistors by dipole monolayers on the gate insulator. *J. Appl. Phys.*, **96** (11), 6431–6438.

95 Kobayashi, S., Nishikawa, T., Takenobu, T., Mori, S., Shimoda, T., Mitani, T., Shimotani, H., Yoshimoto, N., Ogawa, S., and Iwasa, Y. (2004) Control of carrier density by self-assembled monolayers in organic field-effect transistors. *Nat. Mater.*, **3** (5), 317–322.

96 Ma, H., Acton, O., Ting, G., Ka, J.W., Yip, H.-L., Tucker, N., Schofield, R., and Jen, A.K.-Y. (2008) Low-voltage organic thin-film transistors with pi-sigma-phosphonic acid molecular dielectric monolayers. *Appl. Phys. Lett.*, **92** (11), 113303.

97 Seo, J.H., Park, D.S., Cho, S.W., Kim, C.Y., Jang, W.C., Whang, C.N., Yoo, K.H., Chang, G.S., Pedersen, T., Moewes, A., Chae, K.H., and Cho, S.J. (2006) Buffer layer effect on the structural and electrical properties of rubrene- based organic thin-film transistors. *Appl. Phys. Lett.*, **89** (16), 163505.

98 Itaka, K., Yamashiro, M., Yamaguchi, J., Haemori, M., Yaginuma, S., Matsumoto, Y., Kondo, M., and Koinuma, H. (2006) High-mobility C60 field-effect transistors fabricated on molecular-wetting controlled substrates. *Adv. Mater.*, **18** (13), 1713–1716.

99 Wang, H.B., Zhu, F., Yang, J.L., Geng, Y.H., and Yan, D.H. (2007) Weak epitaxy growth affording high-mobility thin films of disk-like organic semiconductors. *Adv. Mater.*, **19** (16), 2168–2171.

100 Li, L., Tang, Q., Li, H., and Hu, W. (2008) Molecular orientation and interface compatibility for high-performance organic thin film transistor based on vanadyl phthalocyanine. *J. Phys. Chem. B*, **112** (34), 10405–10410.

101 Sun, X., Liu, Y., Di, C.-A., Wen, Y., Guo, Y., Zhang, L., Zhao, Y., and Yu, G. (2011) Interfacial heterogeneity of surface energy in organic field-effect transistors. *Adv. Mater.*, **23** (8), 1009–1014.

102 Di, C., Yu, G., Liu, Y., Xu, X., Wei, D., Song, Y., Sun, Y., Wang, Y., Zhu, D., Liu, J., Liu, X., and Wu, D. (2006) High-performance low-cost organic field-effect transistors with chemically modified bottom electrodes. *J. Am. Chem. Soc.*, **128** (51), 16418–16419.

103 Di, C.-A., Wei, D., Yu, G., Liu, Y., Guo, Y., and Zhu, D. (2008) Patterned graphene as source/drain electrodes for bottom-contact organic field- effect transistors. *Adv. Mater.*, **20** (17), 3289–3293.

104 Gao, J., Li, R., Li, L., Meng, Q., Jiang, H., Li, H., and Hu, W. (2007) High-performance field- effect transistor based on dibenzo [d,d′]thieno [3,2-*b*;4,5*b*′]dithiophene, an easily synthesized semiconductor with high

ionization potential. *Adv. Mater.*, **19** (19), 3008–3011.

105 Meng, Q., Jiang, L., Wei, Z., Wang, C., Zhao, H., Li, H., Xu, W., and Hu, W. (2010) Development of organic field-effect properties by introducing aryl-acetylene into benzodithiophene. *J. Mater. Chem.*, **20** (48), 10931–10935.

106 Chang, J.F., Sun, B.Q., Breiby, D.W., Nielsen, M.M., Solling, T.I., Giles, M., McCulloch, I., and Sirringhaus, H. (2004) Enhanced mobility of poly(3-hexylthiophene) transistors by spin-coating from high-boiling-point solvents. *Chem. Mater.*, **16** (23), 4772–4776.

107 Li, L.Q., Gao, P., Schuermann, K.C., Ostendorp, S., Wang, W.C., Du, C.A., Lei, Y., Fuchs, H., De Cola, L., Mullen, K., and Chi, L.F. (2010) Controllable growth and field-effect property of monolayer to multilayer microstripes of an organic semiconductor. *J. Am. Chem. Soc.*, **132** (26), 8807–8809.

108 Miskiewicz, P., Mas-Torrent, M., Jung, J., Kotarba, S., Glowacki, I., Gomar-Nadal, E., Amabilino, D.B., Veciana, J., Krause, B., Carbone, D., Rovira, C., and Ulanski, J. (2006) Efficient high area OFETs by solution based processing of a *p*-electron rich donor. *Chem. Mater.*, **18** (20), 4724–4729.

109 Wang, Y., Chen, Y., Li, R., Wang, S., Su, W., Ma, P., Wasielewski, M.R., Li, X., and Jiang, J. (2007) Amphiphilic perylenetretracarboxyl diimide dimer and its application in field effect transistor. *Langmuir*, **23** (10), 5836–5842.

110 Wei, Z.M., Cao, Y., Ma, W.Z., Wang, C.L., Xu, W., Guo, X.F., Hu, W.P., and Zhu, D.B. (2009) Langmuir–Blodgett monolayer transistors of copper phthalocyanine. *Appl. Phys. Lett.*, **95** (3), 033304.

111 Su, W., Jiang, J.Z., Xiao, K., Chen, Y.L., Zhao, Q.Q., Yu, G., and Liu, Y.Q. (2005) Thin-film transistors based on Langmuir–Blodgett films of heteroleptic bis(phthalocyaninato) rare earth complexes. *Langmuir*, **21** (14), 6527–6531.

112 Metzger, R.M. (2008) Unimolecular electronics. *J. Mater. Chem.*, **18** (37), 4364–4396.

113 Wang, C., Wei, Z., Meng, Q., Zhao, H., Xu, W., Li, H., and Hu, W. (2010) Dibenzothiophene derivatives with carbon-carbon unsaturated bonds for high performance field-effect transistors. *Org. Electron.*, **11** (4), 544–551.

114 Fritz, S.E., Martin, S.M., Frisbie, C.D., Ward, M.D., and Toney, M.F. (2004) Structural characterization of a pentacene monolayer on an amorphous SiO_2 substrate with grazing incidence X-ray diffraction. *J. Am. Chem. Soc.*, **126** (13), 4084–4085.

115 Yang, H.C., Shin, T.J., Ling, M.M., Cho, K., Ryu, C.Y., and Bao, Z.N. (2005) Conducting AFM and 2D GIXD studies on pentacene thin films. *J. Am. Chem. Soc.*, **127** (33), 11542–11543.

116 Dong, H.L., Li, H.X., Wang, E., Yan, S., Zhang, J.M., Yang, C.M., Takahashi, I., Nakashima, H., Torimitsu, K., and Hu, W.P. (2009) Molecular orientation and field-effect transistors of a rigid rod conjugated polymer thin films. *J. Phys. Chem. B*, **113** (13), 4176–4180.

117 DeLongchamp, D.M., Kline, R.J., Fischer, D.A., Richter, L.J., and Toney, M.F. (2011) Molecular characterization of organic electronic films. *Adv. Mater.*, **23** (3), 319–337.

118 Yuan, Q., Mannsfeld, S.C.B., Tang, M.L., Toney, M.F., Luning, J., and Bao, Z. (2008) Thin film structure of tetraceno[2,3-*b*]thiophene characterized by grazing incidence X-ray scattering and near-edge X-ray absorption fine structure analysis. *J. Am. Chem. Soc.*, **130** (11), 3502–3508.

119 Torsi, L., Lovinger, A.J., Crone, B., Someya, T., Dodabalapur, A., Katz, H.E., and Gelperin, A. (2002) Correlation between oligothiophene thin film transistor morphology and vapor responses. *J. Phys. Chem. B*, **106** (48), 12563–12568.

120 Kang, M.J., Doi, I., Mori, H., Miyazaki, E., Takimiya, K., Ikeda, M., and Kuwabara, H. (2011) Alkylated dinaphtho[2,3-*b*:2′,3′-*f*]thieno[3,2-*b*] thiophenes (Cn-DNTTs): organic semiconductors for high-performance thin-film transistors. *Adv. Mater.*, **23** (10), 1222–1225.

121 Kelley, T.W., Muyres, D.V., Baude, P.F., Smith, T.P., and Jones, T.D. (2003) High performance organic thin film transistors. *Mater. Res. Soc. Symp. Proc.*, **771**, 169–179.

122 Okamoto, H., Kawasaki, N., Kaji, Y., Kubozono, Y., Fujiwara, A., and Yamaji, M. (2008) Air-assisted high-performance field-effect transistor with thin films of picene. *J. Am. Chem. Soc.*, **130** (32), 10470–10471.

123 Klauk, H., Zschieschang, U., Weitz, R.T., Meng, H., Sun, F., Nunes, G., Keys, D.E., Fincher, C.R., and Xiang, Z. (2007) Organic transistors based on di(phenylvinyl)anthracene: performance and stability. *Adv. Mater.*, **19** (22), 3882–3887.

124 Kawasaki, N., Kubozono, Y., Okamoto, H., Fujiwara, A., and Yamaji, M. (2009) Trap states and transport characteristics in picene thin film field-effect transistor. *Appl. Phys. Lett.*, **94** (4), 043310.

125 Meng, H., Sun, F.P., Goldfinger, M.B., Gao, F., Londono, D.J., Marshal, W.J., Blackman, G.S., Dobbs, K.D., and Keys, D.E. (2006) 2,6-bis[2-(4- pentylphenyl) vinyl]anthracene: a stable and high charge mobility organic semiconductor with densely packed crystal structure. *J. Am. Chem. Soc.*, **128** (29), 9304–9305.

126 Kelley, T.W., Boardman, L.D., Dunbar, T.D., Muyres, D.V., Pellerite, M.J., and Smith, T.Y.P. (2003) High-performance OTFTs using surface- modified alumina dielectrics. *J. Phys. Chem. B*, **107** (24), 5877–5881.

127 Park, S.K., Jackson, T.N., Anthony, J.E., and Mourey, D.A. (2007) High mobility solution processed 6,13-bis(triisopropyl-silylethynyl) pentacene organic thin film transistors. *Appl. Phys. Lett.*, **91** (6), 063514.

128 Kaur, I., Jia, W.L., Kopreski, R.P., Selvarasah, S., Dokmeci, M.R., Pramanik, C., McGruer, N.E., and Miller, G.P. (2008) Substituent effects in pentacenes: gaining control over HOMO-LUMO gaps and photooxidative resistances. *J. Am. Chem. Soc.*, **130** (48), 16274–16286.

129 Llorente, G.R., Dufourg-Madec, M.-B., Crouch, D.J., Pritchard, R.G., Ogier, S., and Yeates, S.G. (2009) High performance, acene-based organic thin film transistors. *Chem. Commun.*, (21), 3059–3061.

130 Gao, P., Beckmann, D., Tsao, H.N., Feng, X., Enkelmann, V., Baumgarten, M., Pisula, W., and Mullen, K. (2009) Dithieno[2,3-*d*;2′,3′-*d*′]benzo [1,2-*b*;4,5-*b*′] dithiophene (DTBDT) as semiconductor for high-performance, solution-processed organic field-effect transistors. *Adv. Mater.*, **21** (2), 213–216.

131 Ebata, H., Izawa, T., Miyazaki, E., Takimiya, K., Ikeda, M., Kuwabara, H., and Yui, T. (2007) Highly soluble [1] benzothieno[3,2-b]benzothiophene (BTBT) derivatives for high-performance, solution-processed organic field-effect transistors. *J. Am. Chem. Soc.*, **129** (51), 15732–15733.

132 Izawa, T., Miyazaki, E., and Takimiya, K. (2008) Molecular ordering of high-performance soluble molecular semiconductors and re-evaluation of their field-effect transistor characteristics. *Adv. Mater.*, **20** (18), 3388–3392.

133 Takimiya, K., Ebata, H., Sakamoto, K., Izawa, T., Otsubo, T., and Kunugi, Y. (2006) 2,7-Diphenyl [1]benzothieno[3,2-b] benzothiophene, a new organic semiconductor for air-stable organic field-effect transistors with mobilities up to $2.0\,cm^2\,V^{-1}\,s^{-1}$. *J. Am. Chem. Soc.*, **128** (39), 12604–12605.

134 Yamamoto, T. and Takimiya, K. (2007) Facile synthesis of highly pi-extended heteroarenes, dinaphtho[2,3-*b* : 2′,3′-*f*] chalcogenopheno [3,2- *b*] chalcogenophenes, and their application to field-effect transistors. *J. Am. Chem. Soc.*, **129** (8), 2224–2225.

135 Payne, M.M., Parkin, S.R., Anthony, J.E., Kuo, C.-C., and Jackson, T.N. (2005) Organic field-effect transistors from solution-deposited functionalized acenes with mobilities as high as $1\,cm^2/Vs$. *J. Am. Chem. Soc.*, **127** (14), 4986–4987.

136 Anthony, J.E. (2006) Functionalized acenes and heteroacenes for organic electronics. *Chem. Rev.*, **106** (12), 5028–5048.

137 Subramanian, S., Park, S.K., Parkin, S.R., Podzorov, V., Jackson, T.N., and Anthony, J.E. (2008) Chromophore fluorination enhances crystallization and stability of soluble anthradithiophene semiconductors. *J. Am. Chem. Soc.*, **130** (9), 2706–2707.

138 Tang, M.L., Reichardt, A.D., Siegrist, T., Mannsfeld, S.C.B., and Bao, Z. (2008) Trialkylsilylethynyl-functionalized tetraceno [2, 3-*b*]thiophene and anthra[2,3-*b*]thiophene organic transistors. *Chem. Mater.*, **20** (14), 4669–4676.

139 Zhang, L., Tan, L., Wang, Z., Hu, W., and Zhu, D. (2009) High-performance, stable organic field-effect transistors based on *trans*-1,2-(dithieno [2,3-*b*:3′,2′-*d*] thiophene)ethene. *Chem. Mater.*, **21** (9), 1993–1999.

140 Kanno, M., Bando, Y., Shirahata, T., Inoue, J.I., Wada, H., and Mori, T. (2009) Stabilization of organic field-effect transistors in hexamethylenetetrathiafulvalene derivatives substituted by bulky alkyl groups. *J. Mater. Chem.*, **19** (36), 6548–6555.

141 Yanagi, H., Chen, S., Lee, P.A., Nebesny, K.W., Armstrong, N.R., and Fujishima, A. (1996) Dye-sensitizing effect of TiOPc thin film on *n*-TiO_2 (001) surface. *J. Phys. Chem.*, **100** (13), 5447–5451.

142 Barlow, D.E. and Hipps, K.W. (2000) A scanning tunneling microscopy and spectroscopy study of vanadyl phthalocyanine on Au(111): the effect of oxygen binding and orbital mediated tunneling on the apparent corrugation. *J. Phys. Chem. B*, **104** (25), 5993–6000.

143 Nollau, A., Pfeiffer, M., Fritz, T., and Leo, K. (2000) Controlled *n*-type doping of a molecular organic semiconductor: naphthalenetetracarboxylic dianhydride (NTCDA) doped with bis(ethylenedithio)-tetrathiafulvalene (BEDT-TTF). *J. Appl. Phys.*, **87** (9), 4340–4343.

144 Weng, S.-Z., Shukla, P., Kuo, M.-Y., Chang, Y.-C., Sheu, H.-S., Chao, I., and Tao, Y.-T. (2009) Diazapentacene derivatives as thin-film transistor materials: morphology control in realizing high-field-effect mobility. *ACS Appl. Mater. Interfaces*, **1** (9), 2071–2079.

145 Ando, S., Murakami, R., Nishida, J., Tada, H., Inoue, Y., Tokito, S., and Yamashita, Y. (2005) *n*-Type organic field-effect transistors with very high electron mobility based on thiazole oligomers with trifluoromethylphenyl groups. *J. Am. Chem. Soc.*, **127** (43), 14996–14997.

146 Kumaki, D., Ando, S., Shimono, S., Yamashita, Y., Umeda, T., and Tokito, S. (2007) Significant improvement of electron mobility in organic thin-film transistors based on thiazolothiazole derivative by employing self-assembled monolayer. *Appl. Phys. Lett.*, **90** (5), 053506.

147 Ando, S., Nishida, J.-I., Tada, H., Inoue, Y., Tokito, S., and Yamashita, Y. (2005) High performance *n*-type organic field-effect transistors based on *p*-electronic systems with trifluoromethylphenyl groups. *J. Am. Chem. Soc.*, **127** (15), 5336–5337.

148 Jones, B.A., Facchetti, A., Wasielewski, M.R., and Marks, T.J. (2007) Tuning orbital energetics in arylene diimide semiconductors. Materials design for ambient stability of *n*-type charge transport. *J. Am. Chem. Soc.*, **129** (49), 15259–15278.

149 Yoon, M.-H., Kim, C., Facchetti, A., and Marks, T.J. (2006) Gate Dielectric chemical structure-organic field-effect transistor performance correlations for electron, hole, and ambipolar organic semiconductors. *J. Am. Chem. Soc.*, **128** (39), 12851–12869.

150 Yoon, M.-H., DiBenedetto, S.A., Facchetti, A., and Marks, T.J. (2005) Organic thin-film transistors based on carbonyl-functionalized quaterthiophenes: high mobility *n*-channel semiconductors and ambipolar transport. *J. Am. Chem. Soc.*, **127** (5), 1348–1349.

151 Liang, Z., Tang, Q., Xu, J., and Miao, Q. (2011) Soluble and stable N-heteropentacenes with high field-effect mobility. *Adv. Mater.*, **23** (13), 1535–1539.

152 Adiga, S.P. and Shukla, D. (2010) Electronic structure and charge-transport

properties of *N,N'*-bis(cyclohexyl) naphthalene diimide. *J. Phys. Chem. C*, **114** (6), 2751–2755.

153 Malenfant, P.R.L., Dimitrakopoulos, C.D., Gelorme, J.D., Kosbar, L.L., Graham, T.O., Curioni, A., and Andreoni, W. (2002) N-type organic thin-film transistor with high field-effect mobility based on a *N,N'*-dialkyl-3,4,9,10-perylene tetracarboxylic diimide derivative. *Appl. Phys. Lett.*, **80** (14), 2517–2519.

154 Tatemichi, S., Ichikawa, M., Koyama, T., and Taniguchi, Y. (2006) High mobility *n*-type thin-film transistors based on *N,N'*-ditridecyl perylene diimide with thermal treatments. *Appl. Phys. Lett.*, **89** (11), 112108.

155 Seo, H.-S., An, M.-J., Zhang, Y., and Choi, J.-H. (2010) Characterization of perylene and tetracene-based ambipolar light-emitting field-effect transistors. *J. Phys. Chem. C*, **114** (13), 6141–6147.

156 Schmidt, R., Oh, J.H., Sun, Y.-S., Deppisch, M., Krause, A.-M., Radacki, K., Braunschweig, H., Konemann, M., Erk, P., Bao, Z., and Wurthner, F. (2009) High-performance air-stable *n*-channel organic thin film transistors based on halogenated perylene bisimide semiconductors. *J. Am. Chem. Soc.*, **131** (17), 6215–6228.

157 Gao, X., Di, C.-A., Hu, Y., Yang, X., Fan, H., Zhang, F., Liu, Y., Li, H., and Zhu, D. (2010) Core-expanded naphthalene diimides fused with 2-(1,3-dithiol-2-ylidene)malonitrile groups for high-performance, ambient-stable, solution-processed *n*-channel organic thin film transistors. *J. Am. Chem. Soc.*, **132** (11), 3697–3699.

158 Zhao, Y., Di, C.-A., Gao, X., Hu, Y., Guo, Y., Zhang, L., Liu, Y., Wang, J., Hu, W., and Zhu, D. (2011) All-solution-processed, high- performance *n*-channel organic transistors and circuits: toward low-cost ambient electronics. *Adv. Mater.*, **23** (21), 2448–2453.

159 Anthopoulos, T.D., Singh, B., Marjanovic, N., Sariciftci, N.S., Ramil, A.M., Sitter, H., Colle, M., and de Leeuw, D.M. (2006) High-performance *n*-channel organic field-effect transistors and ring oscillators based on C_{60} fullerene films. *Appl. Phys. Lett.*, **89** (21), 213504.

160 Peumans, P. and Forrest, S.R. (2001) Very-high-efficiency double-heterostructure copper phthalocyanine/C_{60} photovoltaic cells. *Appl. Phys. Lett.*, **79** (1), 126–128.

161 Schulze, K., Uhrich, C., Schuppel, R., Leo, K., Pfeiffer, M., Brier, E., Reinold, E., and Bauerle, P. (2006) Efficient vacuum-deposited organic solar cells based on a new low-bandgap oligothiophene and fullerene C_{60}. *Adv. Mater.*, **18** (21), 2872–2875.

162 Song, Q.L., Yang, H.B., Gan, Y., Gong, C., and Ming Li, C. (2010) Evidence of harvesting electricity by exciton recombination in an n–*n*-type solar cell. *J. Am. Chem. Soc.*, **132** (13), 4554–4555.

163 Tsao, H.N., Cho, D., Andreasen, J.W., Rouhanipour, A., Breiby, D.W., Pisula, W., and Mullen, K. (2009) The influence of morphology on high-performance polymer field-effect transistors. *Adv. Mater.*, **21** (2), 209–212.

164 McCulloch, I., Heeney, M., Bailey, C., Genevicius, K., MacDonald, I., Shkunov, M., Sparrowe, D., Tierney, S., Wagner, R., Zhang, W., Chabinyc, M.L., Kline, R.J., McGehee, M.D., and Toney, M.F. (2006) Liquid-crystalline semiconducting polymers with high charge-carrier mobility. *Nat. Mater.*, **5** (4), 328–333.

165 Hamadani, B.H., Gundlach, D.J., McCulloch, I., and Heeney, M. (2007) Undoped polythiophene field-effect transistors with mobility of $1\,cm^2\,V^{-1}\,s^{-1}$. *Appl. Phys. Lett.*, **91** (24), 243512.

166 Zhang, W., Smith, J., Watkins, S.E., Gysel, R., McGehee, M., Salleo, A., Kirkpatrick, J., Ashraf, S., Anthopoulos, T., Heeney, M., and McCulloch, I. (2010) Indacenodithiophene semiconducting polymers for high-performance, air-stable transistors. *J. Am. Chem. Soc.*, **132** (33), 11437–11439.

167 Mühlbacher, D., Scharber, M., Morana, M., Zhu, Z., Waller, D., Gaudiana, R., and Brabec, C. (2006) High photovoltaic performance of a low-bandgap polymer. *Adv. Mater.*, **18** (21), 2884–2889.

168 Tsao, H.N., Cho, D.M., Park, I., Hansen, M.R., Mavrinskiy, A., Yoon, D.Y., Graf,

R., Pisula, W., Spiess, H.W., and Müllen, K. (2011) Ultrahigh mobility in polymer field-effect transistors by design. *J. Am. Chem. Soc.*, **133** (8), 2605–2612.

169 Ha, J.S., Kim, K.H., and Choi, D.H. (2011) 2,5-Bis(2-octyldodecyl) pyrrolo[3,4-c] pyrrole-1,4-(2*H*,5*H*)-dione-based donor–acceptor alternating copolymer bearing 5,5′-di(thiophen-2-yl)-2,2′-biselenophene exhibiting 1.5 $cm^2 V^{-1} s^{-1}$ hole mobility in thin-film transistors. *J. Am. Chem. Soc.*, **133** (27), 10364–10367.

170 Bronstein, H., Chen, Z., Ashraf, R.S., Zhang, W., Du, J., Durrant, J.R., Shakya Tuladhar, P., Song, K., Watkins, S.E., Geerts, Y., Wienk, M.M., Janssen, R.A.J., Anthopoulos, T., Sirringhaus, H., Heeney, M., and McCulloch, I. (2011) Thieno[3,2-*b*]thiophene-diketopyrrolopyrrole-containing polymers for high-performance organic field-effect transistors and organic photovoltaic devices. *J. Am. Chem. Soc.*, **133** (10), 3272–3275.

171 Kim, J., Lim, B., Baeg, K.-J., Noh, Y.-Y., Khim, D., Jeong, H.-G., Yun, J.-M., and Kim, D.-Y. (2011) Highly soluble poly(thienylenevinylene) derivatives with charge-carrier mobility exceeding 1 $cm^2 V^{-1} s^{-1}$. *Chem. Mater.*, **23** (21), 4663–4665.

172 Mei, J., Kim, D.H., Ayzner, A.L., Toney, M.F., and Bao, Z. (2011) Siloxane-terminated solubilizing side chains: bringing conjugated polymer backbones closer and boosting hole mobilities in thin-film transistors. *J. Am. Chem. Soc.*, **133** (50), 20130–20133.

173 Izuhara, D. and Swager, T.M. (2009) Poly(pyridinium phenylene)s: water-soluble N-type polymers. *J. Am. Chem. Soc.*, **131** (49), 17724–17725.

174 Jiang, L., Dong, H.L., Meng, Q., Li, H.X., He, M., Wei, Z.M., He, Y.D., and Hu, W.P. (2011) Millimeter-sized molecular monolayer two- dimensional crystals. *Adv. Mater.*, **23** (18), 2059–2063.

175 Dong, H., Li, H., Wang, E., Wei, Z., Xu, W., Hu, W., and Yan, S. (2008) Ordering rigid rod conjugated polymer molecules for high performance photoswitchers. *Langmuir*, **24** (23), 13241–13244.

176 Shao, W., Dong, H., Jiang, L., and Hu, W. (2011) Morphology control for high performance organic thin film transistors. *Chem. Sci.*, **2** (4), 590–600.

177 Sirringhaus, H., Wilson, R.J., Friend, R.H., Inbasekaran, M., Wu, W., Woo, E.P., Grell, M., and Bradley, D.D.C. (2000) Mobility enhancement in conjugated polymer field-effect transistors through chain alignment in a liquid-crystalline phase. *Appl. Phys. Lett.*, **77** (3), 406–408.

178 Virkar, A., Mannsfeld, S., Oh, J.H., Toney, M.F., Tan, Y.H., Liu, G.-Y., Scott, J.C., Miller, R., and Bao, Z. (2009) The role of OTS density on pentacene and C_{60} nucleation, thin film growth, and transistor performance. *Adv. Funct. Mater.*, **19** (12), 1962–1970.

179 Mannsfeld, S.C.B., Tee, B.C.K., Stoltenberg, R.M., Chen, C.V.H.H., Barman, S., Muir, B.V.O., Sokolov, A.N., Reese, C., and Bao, Z. (2010) Highly sensitive flexible pressure sensors with microstructured rubber dielectric layers. *Nat. Mater.*, **9** (10), 859–864.

180 Briseno, A.L., Mannsfeld, S.C.B., Ling, M.M., Liu, S., Tseng, R.J., Reese, C., Roberts, M.E., Yang, Y., Wudl, F., and Bao, Z. (2006) Patterning organic single-crystal transistor arrays. *Nature*, **444** (7121), 913–917.

181 Chen, L., Degenaar, P., and Bradley, D.D.C. (2008) Polymer transfer printing: application to layer coating, pattern definition, and diode dark current blocking. *Adv. Mater.*, **20** (9), 1679–1683.

182 Hur, S.-H., Khang, D.-Y., Kocabas, C., and Rogers, J.A. (2004) Nanotransfer printing by use of noncovalent surface forces: applications to thin-film transistors that use single-walled carbon nanotube networks and semiconducting polymers. *Appl. Phys. Lett.*, **85** (23), 5730–5732.

183 Wei, Q., Tajima, K., and Hashimoto, K. (2009) Bilayer ambipolar organic thin-film transistors and inverters prepared by the contact-film-transfer method. *ACS Appl. Mater. Interfaces*, **1** (9), 1865–1868.

184 Rogers, J.A., Bao, Z., Baldwin, K., Dodabalapur, A., Crone, B., Raju, V.R., Kuck, V., Katz, H., Amundson, K.,

Ewing, J., and Drzaic, P. (2001) Paper-like electronic displays: large-area rubber-stamped plastic sheets of electronics and microencapsulated electrophoretic inks. *Proc. Natl Acad. Sci. USA*, **98** (9), 4835–4840.

185 Wang, J.Z., Zheng, Z.H., Li, H.W., Huck, W.T.S., and Sirringhaus, H. (2004) Dewetting of conducting polymer inkjet droplets on patterned surfaces. *Nat. Mater.*, **3** (3), 171–176.

186 Meitl, M.A., Zhu, Z.-T., Kumar, V., Lee, K.J., Feng, X., Huang, Y.Y., Adesida, I., Nuzzo, R.G., and Rogers, J.A. (2006) Transfer printing by kinetic control of adhesion to an elastomeric stamp. *Nat. Mater.*, **5** (1), 33–38.

187 Yan, H., Chen, Z.H., Zheng, Y., Newman, C., Quinn, J.R., Dotz, F., Kastler, M., and Facchetti, A. (2009) A high-mobility electron-transporting polymer for printed transistors. *Nature*, **457** (7230), 679–687.

188 de la Fuente Vornbrock, A., Sung, D., Kang, H., Kitsomboonloha, R., and Subramanian, V. (2010) Fully gravure and ink-jet printed high speed pBTTT organic thin film transistors. *Org. Electron.*, **11** (12), 2037–2044.

189 Jinsoo, N., Dongsun, Y., Chaemin, L., Hwajin, C., Jukyung, H., Junseok, K., Yongsu, P., Subramanian, V., and Gyoujin, C. (2010) Scalability of roll-to-roll gravure-printed electrodes on plastic foils. *IEEE Trans. Electron. Packag. Manuf.*, **33** (4), 275–283.

190 Kim, D.H., Han, J.T., Park, Y.D., Jang, Y., Cho, J.H., Hwang, M., and Cho, K. (2006) Single-crystal polythiophene microwires grown by self-assembly. *Adv. Mater.*, **18** (6), 719–723.

191 Niemax, J., Tripathi, A.K., and Pflaum, J. (2005) Comparison of the electronic properties of sublimation- and vapor-Bridgman-grown crystals of tetracene. *Appl. Phys. Lett.*, **86** (12), 122105.

192 de Boer, R.W.I., Gershenson, M.E., Morpurgo, A.F., and Podzorov, V. (2004) Organic single-crystal field-effect transistors. *Phys. Stat. Sol. (a)*, **201** (6), 1302–1331.

193 Li, R., Hu, W., Liu, Y., and Zhu, D. (2010) Micro- and nanocrystals of organic semiconductors. *Acc. Chem. Res.*, **43** (4), 529–540.

194 Jiang, L., Fu, Y., Li, H., and Hu, W. (2008) Single-crystalline, size, and orientation controllable nanowires and ultralong microwires of organic semiconductor with strong photoswitching property. *J. Am. Chem. Soc.*, **130** (12), 3937–3941.

195 Jiang, L., Gao, J., Fu, Y., Dong, H., Zhao, H., Li, H., Tang, Q., Chen, K., and Hu, W. (2010) Tuning intermolecular non-covalent interactions for nanowires of organic semiconductors. *Nanoscale*, **2** (12), 2652–2656.

196 Kim, D.H., Lee, D.Y., Lee, H.S., Lee, W.H., Kim, Y.H., Han, J.I., and Cho, K. (2007) High-mobility organic transistors based on single-crystalline microribbons of triisopropylisilylethynl pentacene via solution- phase self-assembly. *Adv. Mater.*, **19** (5), 678–682.

197 Briseno, A.L., Aizenberg, J., Han, Y.J., Penkala, R.A., Moon, H., Lovinger, A.J., Kloc, C., and Bao, Z.N. (2005) Patterned growth of large oriented organic semiconductor single crystals on self-assembled monolayer templates. *J. Am. Chem. Soc.*, **127** (35), 12164–12165.

198 Reese, C. and Bao, Z. (2007) Organic single-crystal field-effect transistors. *Mater. Today*, **10** (3), 20–27.

199 Tang, Q., Jiang, L., Tong, Y., Li, H., Liu, Y., Wang, Z., Hu, W., Liu, Y., and Zhu, D. (2008) Micrometer- and nanometer-sized organic single-crystalline transistors. *Adv. Mater.*, **20** (15), 2947–2951.

200 Panzer, M.J. and Frisbie, C.D. (2006) High charge carrier densities and conductance maxima in single-crystal organic field-effect transistors with a polymer electrolyte gate dielectric. *Appl. Phys. Lett.*, **88** (20), 203504.

201 Zhou, Y., Wang, L., Wang, J., Pei, J., and Cao, Y. (2008) Highly sensitive, air-stable photodetectors based on single organic sub-micrometer ribbons self-assembled through solution processing. *Adv. Mater.*, **20** (19), 3745–3749.

202 Jiang, L., Gao, J.H., Wang, E.J., Li, H.X., Wang, Z.H., and Hu, W.P. (2008)

Organic single-crystalline ribbons of a rigid "H"-type anthracene derivative and high-performance, short-channel field-effect transistors of individual micro/nanometer-sized ribbons fabricated by an "organic ribbon mask" technique. *Adv. Mater.*, **20** (14), 2735–2740.

203 Li, R., Jiang, L., Meng, Q., Gao, J., Li, H., Tang, Q., He, M., Hu, W., Liu, Y., and Zhu, D. (2009) Micrometer-sized organic single crystals, anisotropic transport, and field-effect transistors of a fused-ring thienoacene. *Adv. Mater.*, **21** (44), 4492–4495.

204 Donnay, J.D.H. and Harker, D. (1937) A new law of crystal morphology extending the Law of Bravais. *Am. Mineral.*, **22** (5), 446–467.

205 Zhang, Y., Dong, H., Tang, Q., He, Y., and Hu, W. (2010) Mobility dependence on the conducting channel dimension of organic field-effect transistors based on single-crystalline nanoribbons. *J. Mater. Chem.*, **20** (33), 7029–7033.

206 Zhang, X., Yuan, G., Li, Q., Wang, B., Zhang, X., Zhang, R., Chang, J.C., Lee, C.-S., and Lee, S.-T. (2008) Single-crystal 9,10- diphenylanthracene nanoribbons and nanorods. *Chem. Mater.*, **20** (22), 6945–6950.

207 Wei, Z., Hong, W., Geng, H., Wang, C., Liu, Y., Li, R., Xu, W., Shuai, Z., Hu, W., Wang, Q., and Zhu, D. (2010) Organic single crystal field-effect transistors based on 6*H*-pyrrolo [3,2-*b*:4,5-*b*′] bis [1,4]benzothiazine and its derivatives. *Adv. Mater.*, **22** (22), 2458–2462.

208 Li, R., Dong, H., Zhan, X., Li, H., Wen, S.-H., Deng, W.-Q., Han, K.-L., and Hu, W. (2011) Physicochemical, self-assembly and field-effect transistor properties of *anti*- and *syn*-thienoacene isomers. *J. Mater. Chem.*, **21** (30), 11335–11339.

209 Mas-Torrent, M., Durkut, M., Hadley, P., Ribas, X., and Rovira, C. (2004) High mobility of dithiophene-tetrathiafulvalene single-crystal organic field effect transistors. *J. Am. Chem. Soc.*, **126** (4), 984–985.

210 Tang, Q.X., Li, H.X., He, M., Hu, W.P., Liu, C.M., Chen, K.Q., Wang, C., Liu, Y.Q., and Zhu, D.B. (2006) Low threshold voltage transistors based on individual single-crystalline submicrometer-sized ribbons of copper phthalocyanine. *Adv. Mater.*, **18** (1), 65–68.

211 Sun, Y.M., Tan, L., Jiang, S.D., Qian, H.L., Wang, Z.H., Yan, D.W., Di, C.G., Wang, Y., Wu, W.P., Yu, G., Yan, S.K., Wang, C.R., Hu, W.P., Liu, Y.Q., and Zhu, D.B. (2007) High- performance transistor based on individual single-crystalline micrometer wire of perylo[1,12-*b*,*c*,*d*]thiophene. *J. Am. Chem. Soc.*, **129** (7), 1882–1883.

212 Tan, L., Jiang, W., Jiang, L., Jiang, S., Wang, Z., Yan, S., and Hu, W. (2009) Single crystalline microribbons of perylo[1,12-*b*,*c*,*d*]selenophene for high performance transistors. *Appl. Phys. Lett.*, **94** (15), 153306.

213 Briseno, A.L., Mannsfeld, S.C.B., Lu, X., Xiong, Y., Jenekhe, S.A., Bao, Z., and Xia, Y. (2007) Fabrication of field-effect transistors from hexathiapentacene single-crystal nanowires. *Nano Lett.*, **7** (3), 668–675.

214 Stoltenberg, R.M., Jung, E., Jin, Y.W., Kim, J.M., Yoo, J.B., and Bao, Z.N. (2009) Solution-processed, high-performance *n*-channel organic microwire transistors. *Proc. Natl Acad. Sci. USA*, **106** (15), 6065–6070.

215 Briseno, A.L., Mannsfeld, S.C.B., Reese, C., Hancock, J.M., Xiong, Y., Jenekhe, S.A., Bao, Z., and Xia, Y. (2007) Perylenediimide nanowires and their use in fabricating field-effect transistors and complementary inverters. *Nano Lett.*, **7** (9), 2847–2853.

216 Merlo, J.A. and Frisbie, C.D. (2004) Field effect transport and trapping in regioregular polythiophene nanofibers. *J. Phys. Chem. B*, **108** (50), 19169–19179.

217 Jiang, L., Hu, W., Wei, Z., Xu, W., and Meng, H. (2009) High-performance organic single-crystal transistors and digital inverters of an anthracene derivative. *Adv. Mater.*, **21** (36), 3649–3653.

218 Chen, Z., Muller, P., and Swager, T.M. (2006) Syntheses of soluble, pi-stacking tetracene derivatives. *Org. Lett.*, **8** (2), 273–276.

219 Reese, C., Chung, W.-J., Ling, M.-M., Roberts, M., and Bao, Z. (2006) High-performance microscale single-crystal transistors by lithography on an elastomer dielectric. *Appl. Phys. Lett.*, **89** (20), 202108.

220 Briseno, A.L., Tseng, R.J., Li, S.-H., Chu, C.-W., Yang, Y., Falcao, E.H.L., Wudl, F., Ling, M.-M., Chen, H.Z., Bao, Z., Meng, H., and Kloc, C. (2006) Organic single-crystal complementary inverter. *Appl. Phys. Lett.*, **89** (22), 222111.

221 Haas, S., Stassen, A.F., Schuck, G., Pernstich, K.P., Gundlach, D.J., Batlogg, B., Berens, U., and Kirner, H.J. (2007) High charge-carrier mobility and low trap density in a rubrene derivative. *Phys. Rev. B*, **76** (11), 115203.

222 Moon, H., Zeis, R., Borkent, E.J., Besnard, C., Lovinger, A.J., Siegrist, T., Kloc, C., and Bao, Z.N. (2004) Synthesis, crystal structure, and transistor performance of tetracene derivatives. *J. Am. Chem. Soc.*, **126** (47), 15322–15323.

223 Chi, X.L., Li, D.W., Zhang, H.Q., Chen, Y.S., Garcia, V., Garcia, C., and Siegrist, T. (2008) 5,6,11,12-Tetrachlorotetracene, a tetracene derivative with pi-stacking structure: the synthesis, crystal structure and transistor properties. *Org. Electron.*, **9** (2), 234–240.

224 Haas, S., Takahashi, Y., Takimiya, K., and Hasegawa, T. (2009) High-performance dinaphtho-thieno-thiophene single crystal field-effect transistors. *Appl. Phys. Lett.*, **95** (2), 022111.

225 Yamada, K., Okamoto, T., Kudoh, K., Wakamiya, A., Yamaguchi, S., and Takeya, J. (2007) Single-crystal field-effect transistors of benzoannulated fused oligothiophenes and oligoselenophenes. *Appl. Phys. Lett.*, **90** (7), 072102.

226 Jiang, W., Zhou, Y., Geng, H., Jiang, S., Yan, S., Hu, W., Wang, Z., Shuai, Z., and Pei, J. (2010) Solution-processed, high-performance nanoribbon transistors based on dithioperylene. *J. Am. Chem. Soc.*, **133** (1), 1–3.

227 Zhou, Y., Lei, T., Wang, L., Pei, J., Cao, Y., and Wang, J. (2010) High-performance organic field-effect transistors from organic single-crystal microribbons formed by a solution process. *Adv. Mater.*, **22** (13), 1484–1487.

228 Chu, C.W., Ouyang, J., Tseng, J.-H., and Yang, Y. (2005) Organic donor-acceptor system exhibiting electrical bistability for use in memory devices. *Adv. Mater.*, **17** (11), 1440–1443.

229 Mas-Torrent, M., Hadley, P., Crivillers, N., Veciana, J., and Rovira, C. (2006) Large photoresponsivity in high-mobility single-crystal organic field-effect phototransistors. *ChemPhysChem*, **7** (1), 86–88.

230 Mas-Torrent, M., Hadley, P., Bromley, S.T., Crivillers, N., Veciana, J., and Rovira, C. (2005) Single-crystal organic field-effect transistors based on dibenzo-tetrathiafulvalene. *Appl. Phys. Lett.*, **86** (1), 012110.

231 Naraso, Nishida, J.-I., Ando, S., Yamaguchi, J., Itaka, K., Koinuma, H., Tada, H., Tokito, S., and Yamashita, Y. (2005) High-performance organic field-effect transistors based on *p*-extended tetrathiafulvalene derivatives. *J. Am. Chem. Soc.*, **127** (29), 10142–10143.

232 Takahashi, Y., Hasegawa, T., Horiuchi, S., Kumai, R., Tokura, Y., and Saito, G. (2007) High mobility organic field-effect transistor based on hexamethylenetetrathiafulvalene with organic metal electrodes. *Chem. Mater.*, **19** (26), 6382–6384.

233 Meng, Q., Gao, J.H., Li, R.J., Jiang, L., Wang, C.L., Zhao, H.P., Liu, C.M., Li, H.X., and Hu, W.P. (2009) New type of organic semiconductors for field-effect transistors with carbon-carbon triple bonds. *J. Mater. Chem.*, **19** (10), 1477–1482.

234 Zeis, R., Siegrist, T., and Kloc, C. (2005) Single-crystal field-effect transistors based on copper phthalocyanine. *Appl. Phys. Lett.*, **86** (2), 022103.

235 Ahmed, E., Briseno, A.L., Xia, Y., and Jenekhe, S.A. (2008) High mobility single-crystal field-effect transistors from bisindoloquinoline semiconductors. *J. Am. Chem. Soc.*, **130** (4), 1118–1119.

236 Hong, W., Wei, Z.M., Xi, H.X., Xu, W., Hu, W.P., Wang, Q.R., and Zhu, D.B. (2008) 6H-pyrrolo[3,2-*b*:4,5-*b'*]bis[1,4]

benzothiazines: facilely synthesized semiconductors for organic field-effect transistors. *J. Mater. Chem.*, **18** (40), 4814–4820.

237 Dong, H., Jiang, S., Jiang, L., Liu, Y., Li, H., Hu, W., Wang, E., Yan, S., Wei, Z., Xu, W., and Gong, X. (2009) Nanowire crystals of a rigid rod conjugated polymer. *J. Am. Chem. Soc.*, **131** (47), 17315–17320.

238 Menard, E., Podzorov, V., Hur, S.H., Gaur, A., Gershenson, M.E., and Rogers, J.A. (2004) High-performance *n*- and *p*-type single-crystal organic transistors with free-space gate dielectrics. *Adv. Mater.*, **16** (23–24), 2097–2051.

239 Jiang, H., Yang, X., Cui, Z., Liu, Y., Li, H., and Hu, W. (2009) Micro-organic single crystalline phototransistors of 7,7,8,8- tetracyanoquinodimethane and tetrathiafulvalene. *Appl. Phys. Lett.*, **94** (12), 123308.

240 Kozaki, M., Isoyama, A., Akita, K., and Okada, K. (2004) Preparation, properties, and reduction of heteroaromatic quinoids with 1,4- diazacyclopentadien-2-ylidene terminals. *Org. Lett.*, **7** (1), 115–118.

241 Molinari, A.S., Alves, H., Chen, Z., Facchetti, A., and Morpurgo, A.F. (2009) High electron mobility in vacuum and ambient for PDIF-CN_2 single-crystal transistors. *J. Am. Chem. Soc.*, **131** (7), 2462–2463.

242 Jones, B.A., Ahrens, M.J., Yoon, M.-H., Facchetti, A., Marks, T.J., and Wasielewski, M.R. (2004) High-mobility air-stable *n*-type semiconductors with processing versatility: dicyanoperylene-3,4 : 9,10-bis(dicarboximides). *Angew. Chem. Int. Ed.*, **43** (46), 6363–6366.

243 Wang, C., Liu, Y., Wei, Z., Li, H., Xu, W., and Hu, W. (2010) Biphase micro/nanometer-sized single crystals of organic semiconductors: control syntheses and their strong phase dependent optoelectronic properties. *Appl. Phys. Lett.*, **96** (14), 143302.

4
Organic Circuits and Organic Single-Molecule Transistors

Qinqxin Tang, Yanhong Tong, Wenping Hu

The greatest and ultimate scientific goal of organic electronics is to fabricate circuits based on organic semiconductors. In this chapter, attention is first focused on ambipolar transistors, complementary inverters, logic gates and ring oscillators, after which circuits based on organic thin films, self-assembled and printed circuits, and circuits of organic single crystals are described.

4.1
Introduction

4.1.1
Ambipolar Transistors

An ambipolar transistor is one in which both electrons and holes are accumulated, depending on the applied voltages [1]. During recent years, ambipolar organic field-effect transistors (OFETs) have attracted considerable interest for realizing organic integrated circuits, which operate with a low power dissipation, wide noise margins, and great operational stability [1–4]. Moreover, these circuits can be fabricated more easily than those based on *p*- and *n*-channel semiconductors, as the ambipolar material can be deposited without the need for micro-patterning of the individual *p*- and *n*-type FETs [4].

Using a *n*-type organic semiconductor transistor as an example, the transistor at a given positive drain voltage V_d (the source potential is held at ground, $V_s = 0$) will start working with a positive gate voltage of $V_g = V_d$. Just as in a unipolar transistor, the gate is more positive than the source electrode, and thus electrons are injected from the source into the accumulation layer and then drift towards the drain, given that $V_g > V_{Th,e}$ ($V_{Th,e}$ = threshold for electron accumulation). Only one polarity of charge carriers is present; this is termed the unipolar regime. However, when V_g is smaller than V_d, the gate potential is more negative than that of the drain electrode by $(V_g - V_d)$. Whilst, for $V_g < V_{Th,e}$, the source will not inject electrons, the drain electrode in an ambipolar transistor will inject holes into the channel if $(V_g - V_d) < V_{Th,h}$ ($V_{Th,h}$ = threshold for hole accumulation). Thus, the

Organic Optoelectronics, First Edition. Edited by Wenping Hu.

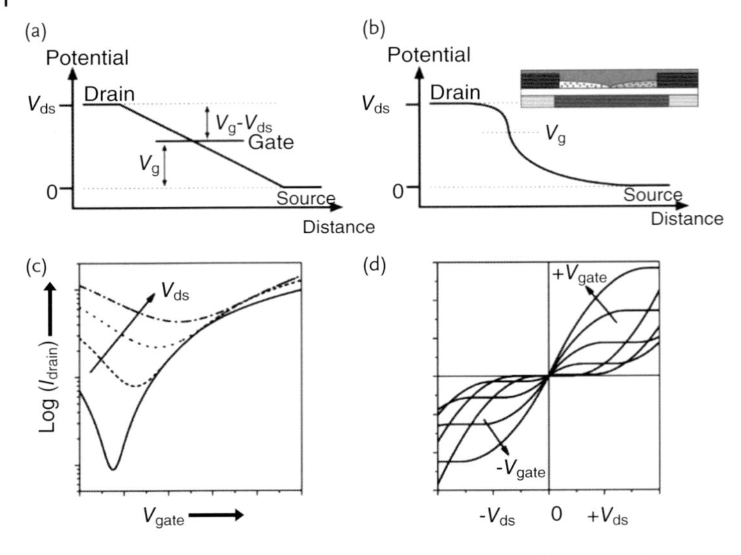

Figure 4.1 (a) Illustration of the source, drain and gate potentials with respect to each other in a field-effect transistor (FET); (b) Channel potential in a FET in the ambipolar regime with two separate channels of holes and electrons that meet within the transistor channel, where opposite charge carriers recombine (inset); (c) Calculated transfer characteristics for an ambipolar transistor with equal hole and electron mobilities and slightly different threshold voltages in a semilog plot for positive gate voltages and different positive source drain voltages; (d) Calculated ambipolar output characteristics for the same transistor for positive (first quadrant) and for negative (third quadrant) V_g and V_{ds}, respectively [1].

drain electrode should actually now be spoken of as a "hole source." A hole current will flow, and thus, the measured drain current will be high, unlike in a unipolar, *n*-channel transistor, which would now be in an off-state. If the gate potential is somewhere between V_d and V_s, so that it is bigger than $V_{Th,e}$ but also $(V_g - V_d) < V_{Th,h}$ (see Figure 4.1a), then both the source and the drain electrode will inject the respective charge carriers such that both electrons and holes will be present in the channel. This regime is termed ambipolar, in contrast to the unipolar regime, where only one polarity of charges is present in the channel for any particular biasing condition [1, 3].

In an ideal ambipolar transistor with just one semiconducting layer, the ambipolar regime is characterized by a hole and an electron accumulation layer next to the respective electrode, which meet at some point within the transistor channel (see inset of Figure 4.1b). There, oppositely charged carriers recombine, and in electroluminescent materials this leads to light emission from within the channel [5]. The length of each channel – and thus the meeting point and position of the recombination zone – depend on the applied gate and source–drain voltage and mobility ratio. The potential of the transistor channel in the ambipolar regime can roughly be imagined as that of a saturated hole and electron channel in series, resulting in an S-shaped transition region (Figure 4.1b), as shown in theoretical studies [1].

The transfer curves of ambipolar transistors exhibit a characteristic V-shape, with one arm indicating electron transport and the other indicating hole transport (see Figure 4.1c). For positive (negative) applied voltages, the effective gate voltage for holes (electrons) depends on the applied source–drain voltage, which in turn gives rise to the characteristic dependence of transfer characteristics on the source–drain voltage. The output curves are characterized by a superposition of standard saturated behavior for one carrier at high V_g, and a superlinear current increase at low V_g and high V_{ds} due to the injection of the opposite carrier (Figure 4.1) [1].

Ambipolar transistors can be achieved by using a single semiconductor [6–12], two parallel semiconductors [1, 13], a semiconductor blend [14–16], or a multilayer semiconductor film [17–20]. The main challenge to create high-performance ambipolar transistors is an efficient injection of both charge carriers [1, 16, 21]. For most organic semiconductors, the ionization potential [the highest occupied molecular orbital (HOMO) level] has the same order of magnitude as the work-functions, ϕ, of the normally used metal electrodes (e.g., for Au, $\phi \sim 5.2\,\text{eV}$). As a result, while the barrier for hole injection is small, the barrier for electron injection is large, such that it will be difficult for the device to exhibit n-type operation, or the n-type performance is poor. However, the use of narrow band-gap semiconductors, low-work-function metals, or asymmetric electrodes can reduce the injection barriers, and facilitate the observation of electron injection and matched electron and hole mobilities [11, 12, 15, 22–25]. The morphology of the semiconductor interface at the organic semiconductor/organic dielectric is also a critical parameter for ambipolar transport [26]. However, the number of the narrow-band-gap organic semiconductors is very limited. Moreover, very few stable metals are available of which the work-functions match with the electron affinity [the lowest unoccupied molecular orbital (LUMO) level] of organic semiconductors. The deposition of the asymmetric electrodes is relatively complex. Another challenge for the realization of bipolar transport is the trapping of one or both carriers [12]. For example, electrons are likely to be trapped by impurities, moisture, oxygen, or the hydroxyl groups of the dielectric [27, 28]. As a result, the electron mobilities are generally much smaller than the hole mobilities in ambipolar devices, and air-stable ambipolar OFETs are an important issue in organic electronics. In addition, the high charge mobility that is required in all organic transistors in order to build electronic circuits with high-frequency operation [3] represents another challenge for the creation of ambipolar devices.

4.1.2 Inverter Circuits

In order to develop electronic circuits based on organic semiconductors, logic elements are required that go beyond single transistors. The most basic element is the voltage inverter, which inverts the incoming signal V_{in} into the outgoing signal

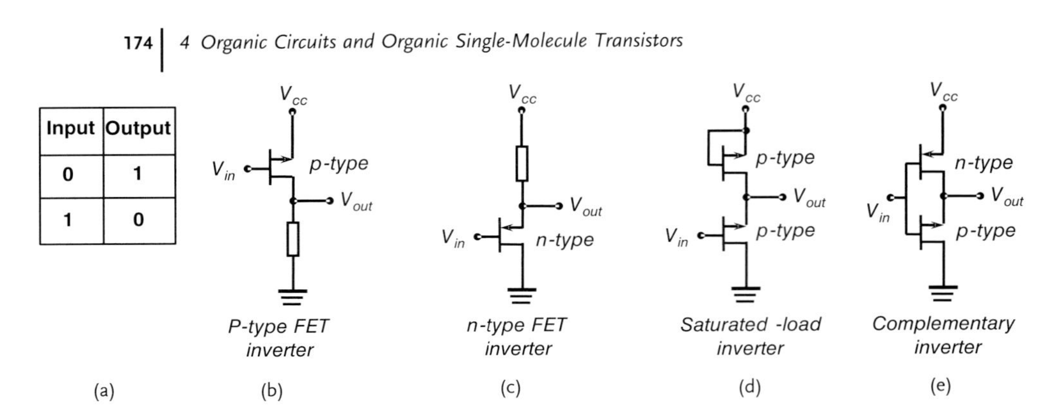

Figure 4.2 Schematic of inverter. (a) Truth table; (b) Inverter with *p*-type FET and resistor; (c) Inverter with *n*-type FET and resistor; (d) Inverter with two FETs of same type; (e) Inverter with *n*-type and *p*-type.

V_{out} [1]. The truth table for this is shown in Figure 4.2a. This represents a perfect switching behavior, which is the defining assumption in digital electronics. Inverters are basic circuit elements for logical functions such as NAND, NOR, multiplexers, decoders, state machines, and other sophisticated digital devices. They are also used in ring oscillators, which consist of a series of inverters, with the output of each inverter stage connected to the input of the following stage. The output of the last stage is then connected to the input of the first stage. Inverters are useful to demonstrate the basic parameters of organic transistors, given by their oscillation frequency and propagation delay per stage [1].

Inverters can be constructed using a single *n*-type FET or a single *p*-type FET coupled with a resistor, or two FETs of the same type (see Figure 4.2b,c). Inverters can also be implemented with two *p*-type (or *n*-type) transistors in either an enhancement-load or a depletion-load (see Figure 4.2d). As this "resistive-drain" approach uses only a single type of transistor, it can be fabricated at low cost. However, because current flows through the resistor in one of the two states, the resistive-drain configuration is at a disadvantage in terms of its power consumption and processing speed. Unipolar architectures consist of one-channel FETs (see Figure 4.2d), which are simple to fabricate as they require single metal electrodes and a single semiconductor material, which can be either evaporated or solution-processed. Despite this great advantage, unipolar circuits generally have a poor performance, exhibiting a narrow noise margin, a low yield, and a high power consumption. In order to improve their performance, more sophisticated architectures are generally but, although beneficial, such an approach increases the circuit complexity by almost 100% [1, 12, 29]. Alternatively, inverters can be constructed using a two complementary FETs configuration (see Figure 4.2e); this involves using a *p*- and an *n*-channel transistor where the gates are connected and serve as an input node (V_{in}). The drains of the two transistors are also connected and serve as an output node (V_{out}) [1]. The complementary configuration leads to a considerable reduction in power consumption, as one of the transistors is always

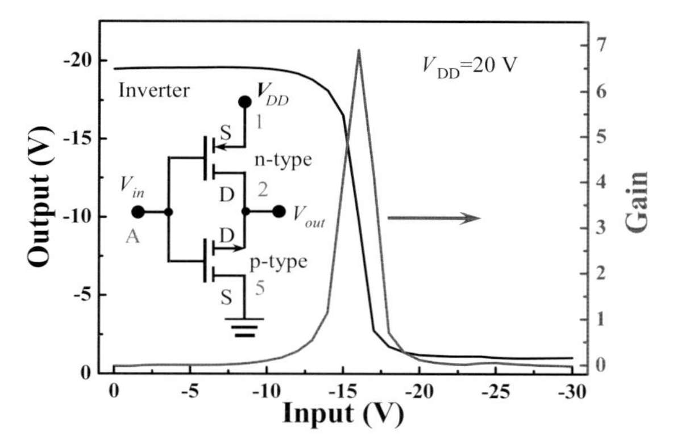

Figure 4.3 Inverter circuit configurations; the *p*-channel FET is the driver and the *n*-channel is the load [32].

off in both logic states. Only during switching are both transistors on, but for only a very short time. The processing speed can also be improved due to the relatively low resistance compared to *n*-type-only or *p*-type-only devices. Inverters can also be constructed with an ambipolar FET in either a resistor-transistor logic (Figure 4.2b–d) or a transistor-transistor logic (Figure 4.2e) configuration.

The very first results on complementary OFET-based circuits were reported by the research team at Bell Labs [30, 31], who subsequently demonstrated the feasibility of fabricating large-scale, organic integrated circuits [2, 32]. The transfer characteristics of the inverter with a *n*-channel FET load is shown in Figure 4.3. The value of the output voltage (V_{out}) for small values of input voltage (V_{in}) is approximately: $V_{our} = V_{\sup ply}\{R_{off}(n)/[R_{on}(p)+R_{off}(n)]\}$. Under these bias conditions, the *p*-channel load is on, and the *n*-channel switch transistor is off. R_{on} and R_{off} are the on-resistance and off-resistance, respectively, of the transistor-type indicated in parentheses. As the input voltage is increased, the *n*-channel TFT is gradually turned on, while the *p-ch* device is gradually turned off. Ultimately, at a large V_{in}, the output voltage is given by: $V_{our} = V_{\sup ply}\{R_{on}(n)/[R_{on}(n)+R_{off}(p)]\}$. The above discussion illustrates the importance of low on/off resistance ratios for the transistors. Low on-resistances will also enhance the speeds at which such circuits can operate while, for the off-resistance, departures from the ideal will increase the static power dissipation and adversely affect the noise margins. The ability to control the channel dimensions of the *n*- and *p*-channel TFTs offers some flexibility in controlling the shape of the inverter transfer characteristics, for a given combination of active materials [4, 30].

In order to fabricate a complementary inverter [2], the *n*- and *p*-type transistor active areas must be spatially separated by depositing the semiconductors through two shadow masks [33]. With this approach, however, the density of the layout is limited. Klauk *et al.* reported on a truly complementary organic technology [34], whereby the *p*-type semiconductor is deposited and patterned using a water-based photoresist and an oxygen plasma etch [35]. Next, the *n*-type material is deposited, but this is not subsequently patterned. Although this approach offers a relatively easy means of creating truly complementary organic FETs and circuits, it requires an additional processing step between the two semiconductor depositions. A third method of creating complementary organic logic circuits has been proposed, using organic semiconductors with ambipolar transport properties [15]. In addition, Vusser *et al.* used the integrated shadow mask method and deposited semiconductors at a 45° angle [36].

4.1.3 Ring Oscillator Circuits

A ring oscillator is a device that is composed of an odd number of inverters attached in a chain and arranged such that the output of the last inverter is fed back into the first (Figure 4.4a). Because a single inverter computes the logical

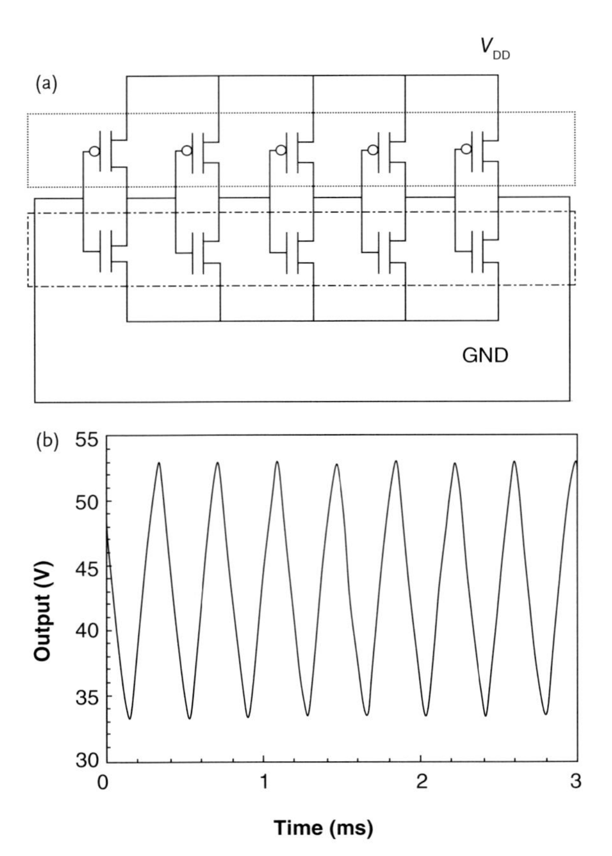

Figure 4.4 (a) Schematic diagram of the layout of an organic complementary five-stage ring oscillator; (b) Output characteristics of an OFET complementary five-stage ring oscillator [31].

NOT of its input, it can be shown that the last output of a chain of an odd number of inverters is the logical NOT of the first input. This final output is asserted a finite amount of time after the first input is asserted by the gate delay, and the feedback of this last output to the input causes oscillation (see Figure 4.4b). In a physical device, no gate can switch instantaneously. In a FET, the gate capacitance must be charged before a current can flow between the source and the drain, and consequently the output of every inverter of a ring oscillator changes by a finite amount of time after the input has changed. Under these conditions, the delay is given by $t = C \times V_{osc}/I$, where C is the total capacitance of the transistor, V_{osc} is the oscillation amplitude, and I is the current that is switched [31]. It can be easily seen that adding more inverters to the chain will increase the total gate delay, thus reducing the frequency of oscillation.

A real ring oscillator only requires power to operate since, above a certain threshold voltage, oscillations will begin spontaneously. In order to increase the frequency of oscillation, two methods may be used:

- The applied voltage can be increased, which in turn increases both the frequency of oscillation and the power consumed, which is dissipated as heat. Unfortunately, the heat dissipated limits the speed of a given oscillator.
- A smaller ring oscillator can be fabricated, but this results in a higher frequency of oscillation, given a certain power consumption [37].

A circular chain composed of an even number of inverters cannot be used as a ring oscillator, as the last output in this case will be the same as the input. However, this configuration of inverter feedback can be used as a storage element, and serves as the basic building block of static random access memory (SRAM).

The ring oscillator, which is a distributed version of the delay oscillator, employs an odd number of inverters to produce the effect of a single inverting amplifier with a gain of greater than one. Rather than having a single delay element, each inverter contributes to the delay of the signal around the ring of inverters – hence the name "ring oscillator." Adding pairs of inverters to the ring increases the total delay, and thereby decreases the oscillator frequency; changing the supply voltage alters the delay through each inverter, with higher voltages typically decreasing the delay and increasing the oscillator frequency [37].

To measure the ring oscillator frequency, the channel of the output transistor is connected in series with a fast current amplifier, and a common voltage is applied. The ring oscillator begins to oscillate if the supply voltage reaches an onset value, which depends on the circuit design, OFET threshold voltage, and on/off ratio. The ring oscillator frequency is limited by the charging and discharging time of the capacitive load of a stage output, which is the sum of the input capacitance of the following stage and a parasitic capacitance. The charge and discharge currents are limited by the channel conductivity of the load OFET and drive OFET, respectively, and also by series resistances, such as the contact resistance between electrodes and semiconductor. Thus, not only is the mobility crucial, but the parasitic capacitances and series resistances which result from the OFET design and the circuit layout are also of great importance [37].

4.2
Circuits of Organic Thin Films

4.2.1
Circuits of Organic Thin Films Based on Ambipolar Transistors

Two ambipolar devices can be used to build inverters with the configuration shown in Figure 4.2e. The complementary inverter needs *n*-channel and *p*-channel FETs, while a controlled deposition of two different organic materials on a substrate requires several additional process steps, thus significantly increasing the process complexity and manufacturing costs [1]. However, ambipolar device architectures or materials can provide both *n*- and *p*-channel performance, which enables complementary-like inverters to be created. Another advantage of ambipolar inverters is that they function for both positive and negative V_{in} and V_{out}, depending on the supply voltage V_{supply}; this is unlike unipolar inverters, which only function for one polarity [1].

Until now, several research groups have used ambipolar OFETs in inverters and ring oscillators. Although pentacene has been typically used as a *p*-type organic semiconductor in most reports, Singh *et al.* showed that, with a correct choice of different surface energy organic dielectrics [e.g., poly(vinylalcohol); PVA], pentacene can exhibit an ambipolar operation [26]. The results of Singh *et al.* confirmed that PVA would indeed affect the growth of the semiconductor, where smaller grain sizes accompany the ambipolar charge transport properties with a high electron mobility of $0.05\,cm^2/(V \cdot s)$ and a hole mobility of $0.4\,cm^2/(V \cdot s)$. This device can be used for inverter circuits [38]. The transport characteristics of a pentacence FET based on deferent dielectrics, and the transfer characteristics of the inverter base on the devices, are shown in Figure 4.5.

Meijer *et al.*, at Philips Research Laboratories, used an OC_1C_{10}-PPV:PCBM (see Figure 4.6a) blend and PIF (see Figure 4.6b) as semiconductors to produce ambipolar FETs and demonstrate a CMOS-like inverter operation (Figure 4.6) [15]. A high gain of 10 for the OC_1C_{10}-PPV:PCBM blend inverter, and of 11 for the PIF inverter, was easily achieved (as indicated by the steepness of the inverter characteristic), in combination with a good noise margin (the position of the voltage switch in the inverter characteristic). Depending on the polarity of the supply voltage, V_{DD}, the inverter functions in the first or third quadrant of Figure 4.6, which is a particular feature of the ambipolar transistor-based inverter.

Narrow band-gap semiconductors are important for ambipolar OFETs. Chen *et al.*, at the University of Cambridge, used polyselenophenes with small band gaps and low LUMO levels to fabricate an ambipolar device, and observed matched electron and hole mobilities on the order of $0.03\,cm^2/(V \cdot s)$. These values were among the highest reported for solution-processed ambipolar organic semiconductors at the time [11]. These inverters, based on ambipolar devices, had a very high gain in switching (the absolute value was as high as 86) [11]. Subsequently,

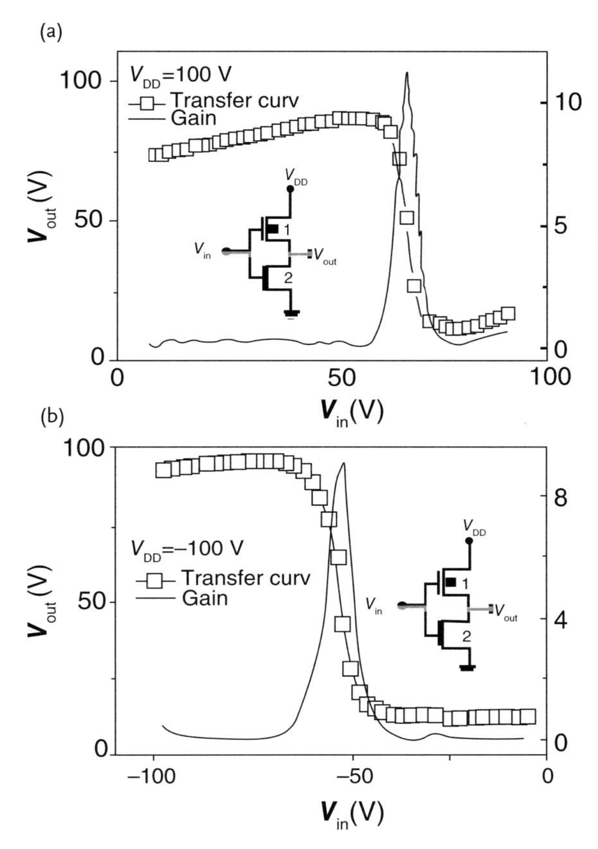

Figure 4.5 Transfer characteristics of complementary-like inverter using two identical ambipolar pentacene OFETs [38].

Anthopoulos *et al.*, at Imperial College London, investigated nickel dithiolene for application in ambipolar FET. This semiconductor had a narrow band-gap and was soluble. More importantly, the transistors proved to be very stable, and showed no noticeable degradation even after storage in ambient air (without encapsulation) for several months [12]. Although the air stability of the *p*-channel did not come as a surprise, the high stability of the *n*-channel was surprising, as it has been observed only in a very few organic materials [12]. The quasi-static transfer curves of inverters based on nickel dithiolene ambipolar devices are shown in Figure 4.7. Here, a sharp inversion of the input signal is observed with a maximum voltage gain, or a signal amplification of six, to test whether the complementary-like inverter is suitable for the fabrication of more complex circuits. Subsequently, Anthopoulos *et al.* created ring oscillators that were

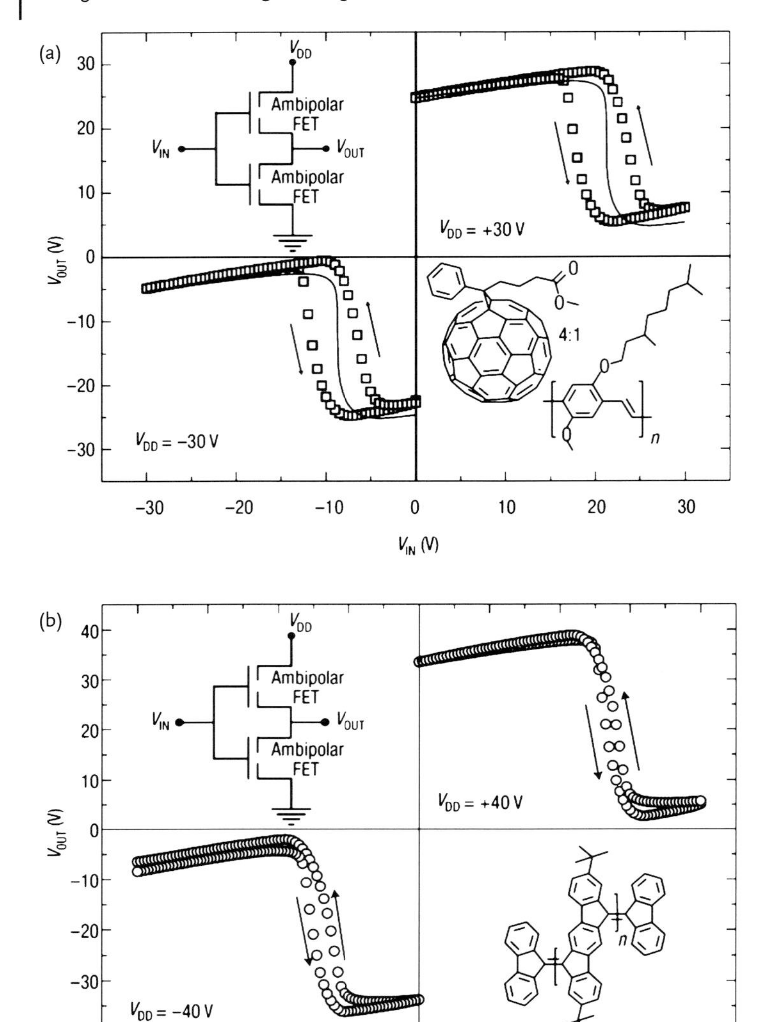

Figure 4.6 Transfer characteristics of CMOS-like inverters based on two identical ambipolar transistors. The insets show the schematic representation of the electrical connections in the inverter. (a) Transport characteristics of inverter based on two identical OC1C10-PPV:PCBM FETs; (b) Transport characteristics of inverter based on two identical PIF FETs [15].

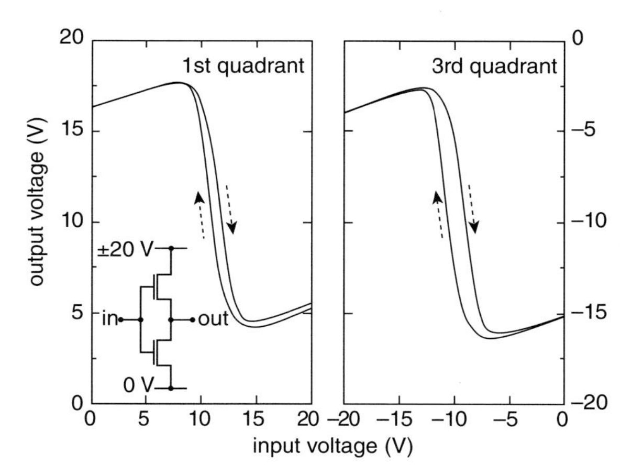

Figure 4.7 Quasi-static transfer characteristics of a complementary-like voltage inverter fabricated and measured in ambient conditions. Inset: Complementary circuit of the voltage inverter employed [12].

comprised of several inverter stages (Figure 4.8a) [12]. The output characteristics of an integrated five-stage ring oscillator, which was fabricated and tested in air, are shown in Figure 4.8b. While the maximum oscillation frequency measured was 710 Hz, more importantly the circuits exhibited a very high yield of >95% (i.e., the fraction of functioning circuits that emerged from batch fabrication), with excellent environmental stability under ambient light and air conditions.

Anthopoulos *et al.*, at the University of Groningen, investigated FETs based on solution-processible methanofullerene 6-phenyl-C_{71}-butyric acid methyl ester (PCBM). Depending on the gate bias, these devices showed either unipolar (*n*-type) or ambipolar characteristics [39]. Anthopoulos and colleagues first fabricated unipolar (*n*-type) ring oscillators, for which the maximum oscillation frequency (f_{OSC}) measured was 2.1 kHz at $V_{DD} = +130\,V$. By taking advantage of the ambipolar nature of PCBM transistors, complementary-like inverters were also fabricated that comprised two ambipolar OFETs, and for which the maximum gain was approximately 6 [39].

However, even in such an ambipolar inverter based on perfectly ambipolar device with balanced μ_e and μ_h and both $V_{Th,e}$ and $V_{Th,h}$ close to 0 V, neither transistor was ever fully switched off [4]. The presence of a parallel, parasitic *p*-channel (*n*-channel) that switches on in the pull-down (pull-up) FET when the *n*-channel (*p*-channel) is switched off, reduces the impedance of the transistor acting as the load, thus strongly reducing the output voltage swing. This effect produces the typical Z-shaped voltage transfer characteristics (see Figures 4.6 and 4.7), resulting in a high static power consumption and very limited noise-margins compared to conventional, complementary logic inverters [1, 4, 12].

In order to solve these problems, Baeg *et al.*, at the Electronics and Telecommunications Research Institute, Republic of Korea, demonstrated a method by

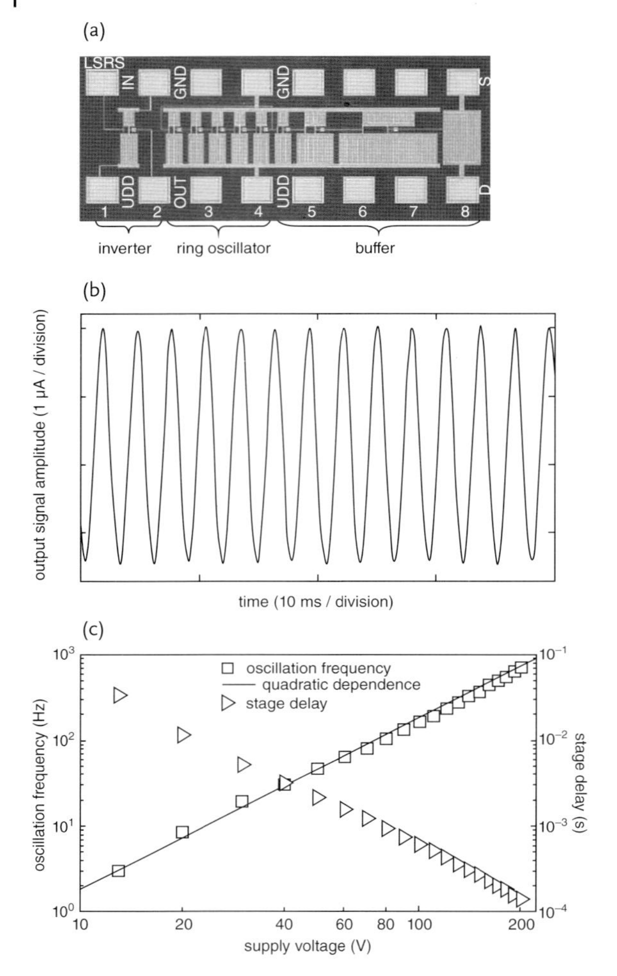

Figure 4.8 (a) Microphotograph of a five-stage complementary-like ring oscillator fabricated on a polymer substrate. The ring oscillator consists of five inverter stages and a four-stage buffer; (b) Output waveform (350 Hz) of a five-stage complementary-like ring oscillator at a supply voltage of 140 V; (c) Dependence of oscillation frequency on supply voltage and corresponding stage delay. The solid line represents a quadratic dependence plot for comparison [12].

which the charge injection and transport could be controlled in ambipolar OFETs, by engineering the electrical contacts [4]. It was shown that, solution-processed cesium (Cs) salts (Figure 4.9a), employed as electron-injection and hole-blocking layers at the interface between the semiconductor [e.g., poly(thienlylenevinylene-*co*-phthalimide)s functionalized at the imide nitrogen with 2-ethylhexyl (PTVPhI-Eh)] and the charge injection electrodes, would decrease the gold (Au) work-function (~4.1 eV) when compared to that of a pristine Au electrode (~4.7 eV) [4]. By controlling the electrode surface chemistry, *p*-channel [hole mobility ~0.1–0.6 $cm^2/(V \cdot s)$]

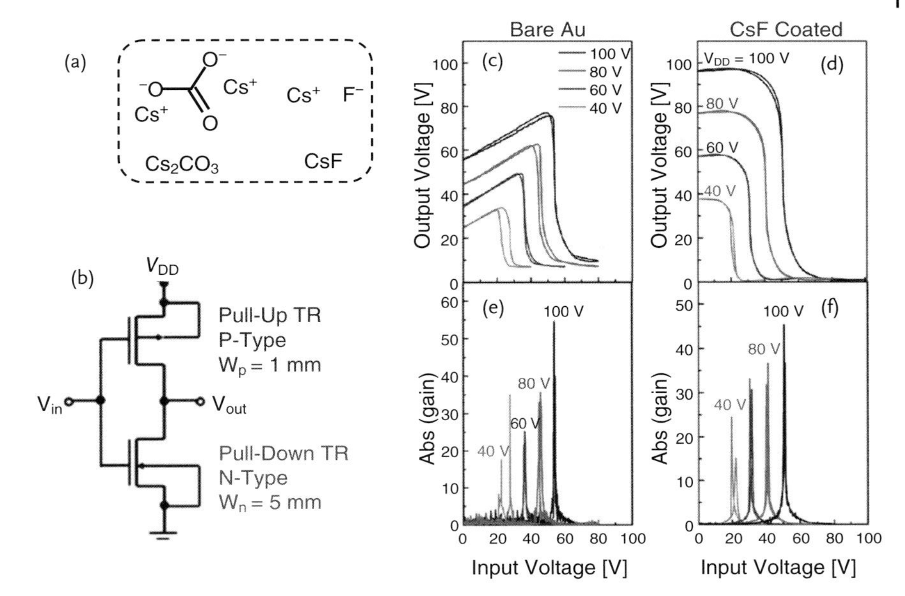

Figure 4.9 Complementary inverter circuits based on ambipolar PTVPhI-Eh OFETs. (a) Molecular structure of Cs_2CO_3 and CsF; (b) Circuit configuration of the complementary inverter; (c–f) Voltage transfer characteristics and corresponding output voltage gains of the inverters: (c, e) bare Au S/D electrode; (d, f) after selective spray-deposited CsF layer; (g) Noise margins of the corresponding inverter with or without the CsF layer only in the *n*-channel region [4].

and *n*-channel [electron mobility ~0.1–0.3 $cm^2/(V \cdot s)$] OFET characteristics with the same semiconductor could be demonstrated. Moreover, in these OFETs the counterpart charge carrier currents were highly suppressed for depletion mode operation (I_{off} < 70 nA when I_{on} > 0.1 mA). Based on the solution-processed ambipolar polymer and modified electrodes, high-performance, truly complementary inverters could be demonstrated (Figure 4.9b). As shown in Figure 4.9c, the complementary inverter based on the ambipolar PTVPhI-Eh FETs and bare Au contacts exhibited Z-shape-like voltage transfer characteristic curves for smaller V_{in}, and very narrow noise margins of ~33% and ~20% of ½ V_{DD} at low and high levels, respectively. Following a selective deposition of the CsF layer onto the *n*-channel region, the complementary inverter based on PTVPhI-Eh exhibited good voltage transfer characteristics curves of the inverter (Figure 4.9d). In this case, the V_{out} loss was markedly decreased to <+4 V in the static off-state (V_{in} = 0 V), and was negligible in the on-state (V_{in} = +80 V). These complementary inverters exhibited a very high voltage gain of >50 (at V_{DD} = +100 V), a negligible bias hysteresis, and good noise margins as high as ~75% of ½ V_{DD} (see Figure 4.9d,f).

Complementary ring oscillators were also demonstrated based on the solution-processed ambipolar polymer and modified electrodes (Figure 4.10) [4]. The bare

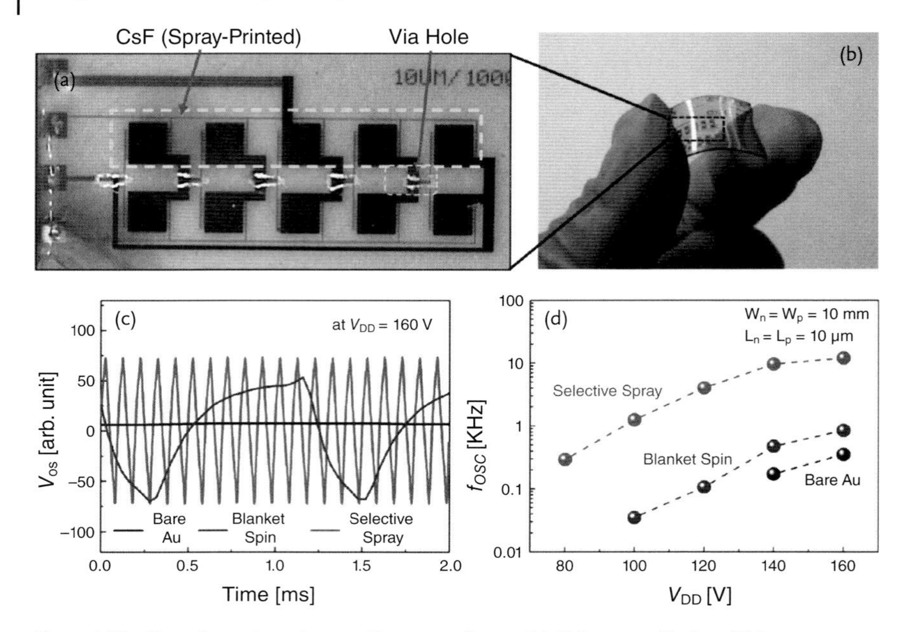

Figure 4.10 Complementary ring oscillators based on ambipolar PTVPhI-Eh OFETs. (a) Optical microscope image of the ring oscillators circuit; (b) Digital camera image of flexible ambipolar polymer ring oscillators fabricated onto a plastic substrate; selective spray deposition area of the CsF (*n*-channel region) and via-hole are indicated in the figure; (c) Voltage oscillation (V_{os}) at $V_{DD} = 160\,V$ of the RO under various Au S/D electrode conditions; black line = bare Au; blue line = blanket spin-coated CsF; red line = selective spray-printed CsF; (d) Dependence of the oscillation frequencies (f_{osc}) on the supply voltage (V_{DD}) in the 80 to 160 V range [4].

Au-based ring oscillators did not oscillate for $|V_{DD}| < 140\,V$, and very low frequencies (~0.2 kHz) were measured for larger V_{DD} (see Figure 4.10c). Whereas, a value of f_{osc} of up to 12 kHz was achieved by a selective deposition of the CsF layer onto the *n*-channel transistor region by spray-printing, the V_{DD} to induce a ring oscillator voltage was significantly reduced, from +140 V (bare Au) to +80 V (spray CsF). These circuits can be fabricated on a flexible substrate [4].

4.2.2
Circuits of Organic Thin Films Based on Unipolar Transistors

As shown in Figure 4.2d, inverters with unipolar architectures consist of one-channel FETs, which are easily fabricated as they require only a single metal electrode and a single semiconductor material. Moreover, this type of inverter can also be used to create ring oscillation.

Flexible OFETs and inverter circuits using solution-processable functionalized graphene for all of the electrodes (source, drain, and gate) were first fabricated by Chen *et al.*, at Nanjing University [40]. Likewise, by using soluble polymers for the active layer and insulating layer, Fix *et al.* created fast integrated circuits based on *p*-type organic transistors only; moreover, their ring oscillators had a stable performance and presented with frequencies of 106 kHz which, at the time, was the fastest yet devised [37]. Subsequently, Shang *et al.* at the Institute of Microelectronics, Chinese Academy of Sciences, used atomic layer deposition to deposit an aluminum oxide thin film (30 nm) as the dielectric layer so as to create low-voltage (5 V) unipolar OFETs, inverters (gain of 12), and five-stage ring oscillations (frequency of 0.8 kHz) [41]. Later, Graz *et al.*, at the University of Cambridge, fabricated flexible pentacene FETs and inverters on 1 mm-thick transparent polydimethylsiloxane (PDMS) [42]. Anthopoulos *et al.*, at Imperial College London, used *n*-type C_{60} FETs with a polymer or SiO_2 dielectric to fabricate a seven-stage ring oscillator and a ring oscillator that operated at a frequency (f_{osc}) of 30.5 kHz [43]. Low-voltage-operating OFETs and inverters (5 V) with a photopatternable ultrathin gate dielectric (70 nm) were fabricated by Jang *et al.*, at Pohang University of Science and Technology [44]; the gain of the inverter thus created was 13.5. In addition, Park *et al.*, at Penn State University, used a solution-processed 6,13-bis(triisopropylsilylethynyl) (TIPS)-pentacene to fabricate FETs on glass substrates; the unipolar seven-stage ring oscillators thus created had an oscillation frequency >10 kHz at a bias of –80 V [45]. When Smith *et al.*, at Imperial College London, fabricated short-channel, bottom-gate, bottom-contact transistors based on solution-processed polymer/small molecule blends, their unipolar seven-stage ring oscillators had an oscillation frequency >100 kHz at a bias of −120 V [46].

Attempts were also made to improve the performance of the unipolar circuits. For example, Stingelin-Stutzmann *et al.*, at Philips Research Laboratories, advanced a general route for a facile fabrication of OFETs and circuits from solution [47]. The main feature of this method was the incorporation of a glass-inducing diluent that enabled a controlled crystallization from an initial vitreous state of the organic semiconductor, formed in a selected area of the phase diagram of the two constituents. Because of the high crystallinity of the solution-processed material, the environmentally stable, discrete rubrene-based transistors obtained had mobilities of up to $0.7\,cm^2/(V \cdot s)$ and ON/OFF ratios $\geq 10^6$. The inverter and ring-oscillator structures are shown in Figure 4.11 [47].

For dual-gate OFETs, one gate controls the output of device while the other can be used to tune the threshold voltage. Hence, the performance of inverters based on dual-gate transistors can be tuned and have an increased noise margin. Spijkman *et al.* used a thermally oxidized SiO_2 layer as the bottom-gate dielectric, and polyisobutylmethacrylate as the top-gate insulator, to fabricate a solution-processed poly(triarylamine) dual-gate FETS [48]. In this case, the top gate could be used to independently tune the threshold voltage, the shift being described quantitatively by $\Delta V_{th} = (C_t/C_b)\,V_{top}$ gate, where C_t and C_b are the top- and bottom-gate capacitances. It could be shown that, by adjusting the top-gate biases, the noise margin of the dual-gate inverters could be significantly improved, from 1 V to about 5 V (see Figure 4.12) [48].

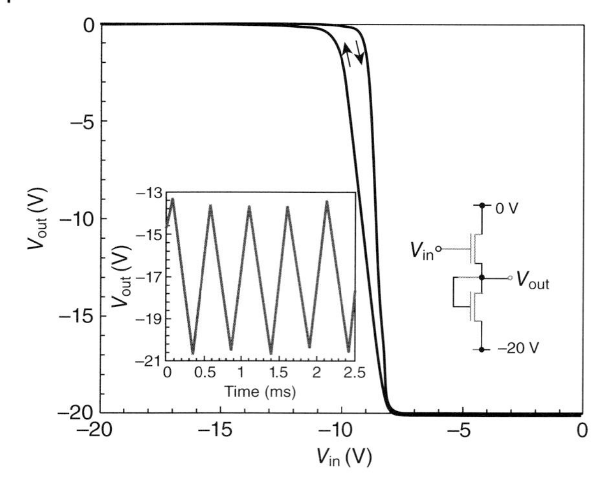

Figure 4.11 Transfer characteristics of a digital inverter gate consisting of two integrated FETs. The output characteristics of the integrated seven-stage ring-oscillator circuit are shown in the left inset (the switching frequency of ~1.5 kHz at a rail voltage of −45 V) [47].

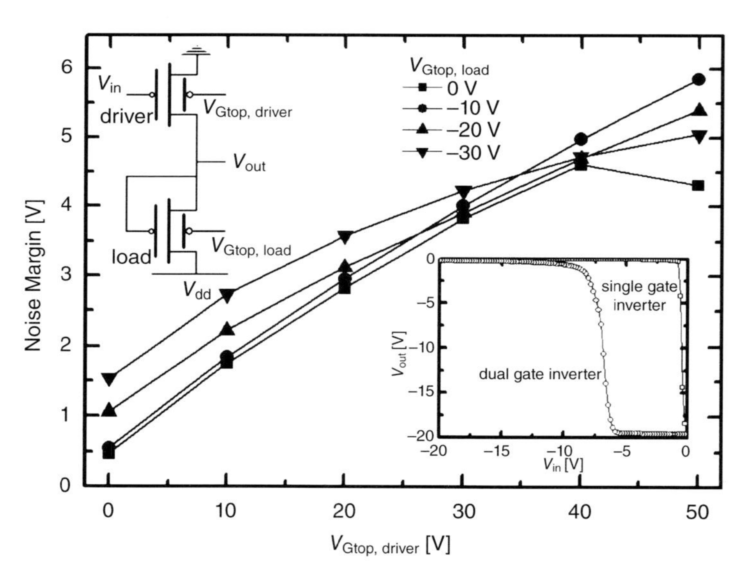

Figure 4.12 The noise margin of dual gate inverters as a function of the top gate bias on the driver transistor for several top gate biases on the load transistor. The supply voltage V_{DD}, was set at −20 V. The insets show the schematic of the dual-gate inverter and the improvement gained in input–output characteristics by using a dual-gate inverter [48].

4.2.3
Complementary Circuits of Organic Thin Films

Complementary circuits demonstrate the best performance for logical circuit applications because of their low power dissipation, high noise immunity, and operational stability [15]. Hence, although these circuits are more complex to fabricate they occupy a very important position in organic electronics.

For complementary organic circuits, the lack of a high-performance *n*-type semiconductor – and especially of an air-stable *n*-type semiconductor – has limited developments to date. $F_{16}CuPc$ is a typical air-stable *n*-type semiconductor which has been used by several research groups in circuits [49–52]. The synthesis of new, high-mobility *n*-type semiconductors that can be used in circuits represents one of the most important aspects of current organic electronics [53–56]. For example, Walser *et al.* at ETH Zurich used *N,N′*-ditridecylperylene- 3,4,9,10-tetracarboxylicdiimide (PTCDI-C_{13}) as an *n*-type semiconductor [saturation mobility of ~$0.2\,cm^2/(V \cdot s)$] to build complementary inverters, with a gain that was as high as 250 [56].

Organic circuits which operate at a low voltage should have great prospects for a variety of applications, and this represents another developmental direction of organic electronics. The gate-source voltage V_{GS} required to induce a charge density Q in the channel of a FET is determined by the gate dielectric capacitance C: $V_{GS} = Q/C = Qt/\varepsilon\varepsilon_0$, where t and ε are the thickness and permittivity of the dielectric, respectively. Thus, the operating voltage of a transistor scales with the dielectric thickness and permittivity. Yoon *et al.*, at Northwestern University, reported the creation of new, spin-coatable, ultrathin (<20 nm) crosslinked polymer blend (CPB) which exhibited excellent insulating properties (leakage current densities ~$10^{-8}\,A\,cm^{-2}$), large capacitances (up to ~$300\,nF\,cm^{-2}$), that enabled low-voltage OFETs to function [49]. These dielectrics, which exhibited a good uniformity over areas of ~$150\,cm^2$, were shown to be soluble in common solvents, could be patterned using standard microelectronic etching methodologies, and would adhere to/be compatible with n^+-Si, ITO, and Al gates [49]. Moreover, the CPB dielectric is robust and compatible with both *p*- and *n*-type semiconductors, such that the fabrication of complementary logic devices becomes possible. Complementary invertors based both on pentacene (*p*-type) and $F_{16}CuPc$ (*n*-type) transistors have been fabricated which function at 2 V (see Figure 4.13a), with a voltage gain of ≈3.5 being achieved. These inverters could be switched at frequencies of up to ~100 Hz, with a $\tau \approx 1.5$ ms fall time and a $\tau \approx 2.3$ ms rise time (Figure 4.13b). Such speeds would be more than enough to operate an electrophoretic ink-based display. Walser *et al.*, at ETH Zurich, prepared a single-layer fluoropolymer by spin-coating an ultrathin (<20 nm) and highly insulating gate dielectric, such that an inverter which operated voltages <2 V could be fabricated [57]. Na, at the University of Tokyo, employed a high-*k*, triple-layer dielectric structure $SiO_2/TiSiO_2/SiO_2$ (the $TiSiO_2$ is the high-*k* material) as a high capacitance gate insulator to fabricate low-voltage-operating organic complementary inverters and ring oscillators based on pentacene and C_{60} FETs [58]. The complementary inverters

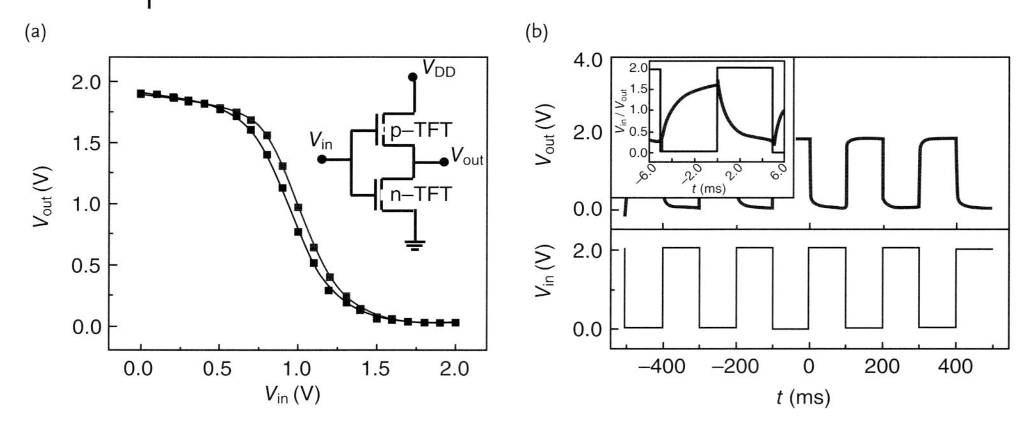

Figure 4.13 Performance of inverter based on pentacene and $F_{16}CuPc$ transistors with ultrathin CPB dielectric. (a) Output characteristics of inverter. Inset: Schematic electrical connections of the inverter; (b) Dynamic switching characteristics of the inverter at 5 Hz. Inset: Dynamic switching characteristics of the inverter at 100 Hz [49].

composed of these FETs operated over the voltage range of 2 to 10 V, with large gain values of up to 65. The inverter yielded five-stage ring oscillators which had an oscillation frequency of 80 Hz at 10 V.

Molecular self-assembled monolayers (SAMs) and aluminum oxide on aluminum each represent excellent ultrathin dielectric layer with a small leakage current [59–63]. Klauk *et al.*, at the Max Planck Institute for Solid State Research, selected phosphonic acid anchor groups for self-assembly on aluminum oxide to obtain high-quality, ultrathin dielectric layers [50]. In this case, the SAM was 2.1 nm thick, and the aluminum oxide 3.8 nm thick. Notably, although the SAM added only 2.1 nm to the total dielectric thickness, the leakage current density of the whole dielectric layer was reduced by three orders of magnitude, from ca. $5 \times 10^{-5}\,A\,cm^{-2}$ to $(5 \pm 1) \times 0^{-8}\,A\,cm^{-2}$ at an applied voltage of 2 V. Moreover, the capacitance of the SAM-covered aluminum oxide dielectric was $0.7 \pm 0.05\,\mu F\,cm^{-2}$, as determined using impedance spectroscopy for frequencies of between 1 Hz and 10 kHz. Circuits based on pentacene (*p*-type) and $F_{16}CuPc$ (*n*-type) transistors with this dielectric demonstrated the best performance, while the inverters with a SAM-based gate dielectric showed sharp switching with rail-to-rail output swings, a large signal gain (~100), and negligible hysteresis for supply voltages as low as 1.5 V (Figure 4.14). Complementary two-input NAND gates also showed the correct logic function (Figure 4.15). The static currents (when $V_{in} = 0$ or $V_{in} = V_{DD}$) in the complementary gates were very small, typically <100 pA. Thus, the static power dissipation was less than 1 nW per logic gate, which meant that these organic circuits were truly low-power [50]. Ball *et al.*, at Imperial College London, demonstrates both *p*- and *n*-channel solution-processed OFETs that operated at voltages below 1.5 V with ultrathin gate dielectrics based on solution-processed SAMs. These investiga-

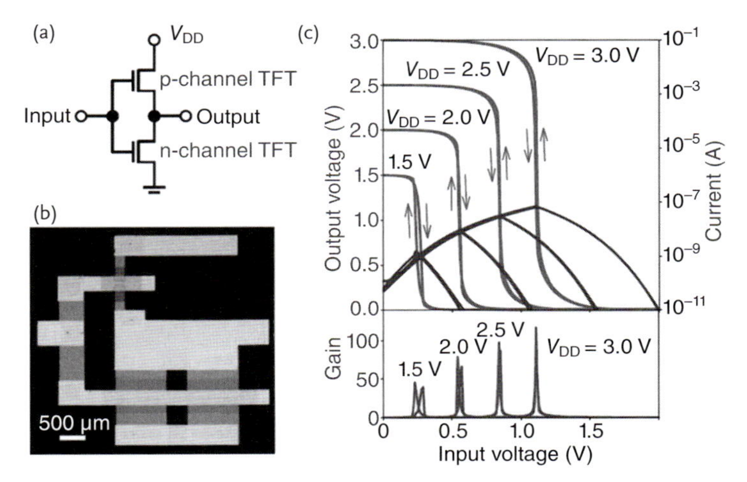

Figure 4.14 Complementary inverter with SAM gate dielectric. (a) Circuit schematic of the inverter; (b) Photograph of the inverter; (c) Output voltage, current, and small-signal gain as a function of input voltage for supply voltages between 1.5 and 3.0 V. The inverter shows rail-to-rail output switching, a maximum static current of 100 pA, and a small-signal gain as large as 100 [50].

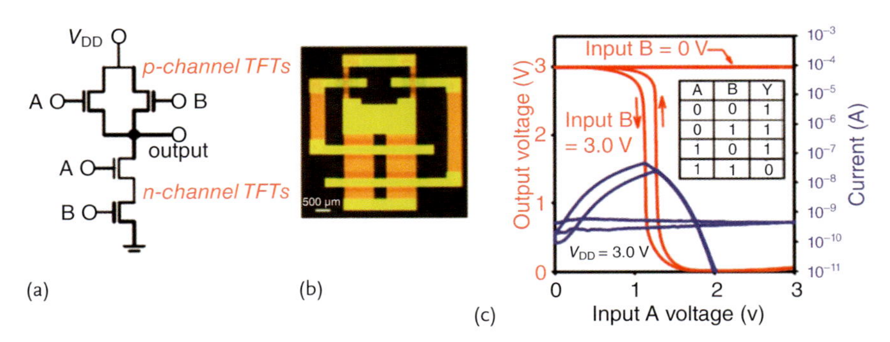

A	B	Y
0	0	1
0	1	1
1	0	1
1	1	0

Figure 4.15 Complementary two-input NAND gate. (a) Circuit schematic; (b) Photograph of the circuit; (c) Transfer characteristics (inset: truth table of the two-input NAND gate) [50].

tions confirmed the feasibility of fabricating low-voltage complementary organic circuits by means of solution processing [64].

Complementary five-stage ring oscillators were also fabricated based on pentacene and F_{16}CuPc transistor, using the SAM-based gate dielectric, by Klauk *et al.* These circuits showed stable oscillations for supply voltages as low as 1.5 V, which was the lowest operating voltage reported for an organic circuit. The circuits oscillated with a rail-to-rail output voltage and the signal was delayed by as little as 1.4 ms per stage (Figure 4.16). Notably, all of these measurements were

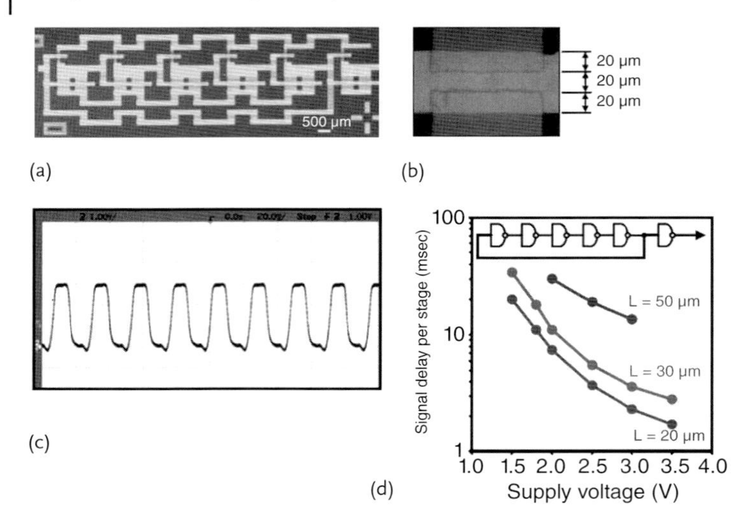

Figure 4.16 Complementary five-stage ring oscillators. (a) Photograph of the circuit; (b) Close-up photograph of a pentacene FET with channel length and contact-to-gate overlap of 20 µm; (c) Output voltage signal of a ring oscillator with critical dimensions of 20 µm for a supply voltage of 3 V, showing an amplitude of 3 V and a period of 23 ms (corresponding to a signal delay of 2.3 ms per stage); (d) Signal propagation delay as a function of channel length and supply voltage (inset: circuit schematic) [50].

performed in air, without the circuits being protected against ambient oxygen and humidity [50].

Flexible electronic circuits are an essential prerequisite for the development of rollable displays, conformable sensors, biodegradable electronics, and other applications with unconventional form factors [51]. During recent years, several groups worldwide have achieved excellent results in this area [52, 55, 65, 66]. For example, in 2005 Klauk, at Infineon Technologies, fabricated inverters and ring oscillators on thin, transparent sheets of polyethylene naphthalate [52]. In this case, a 50 nm-thick layer of solution-processed polyvinylphenol was used as the gate dielectric, while the transistors and circuits were operated at supply voltages as low as 8 V, and the ring oscillators had a signal propagation delay as low as 8 µs per stage [52]. However, the characteristics of the transistor and circuits when the substrates were bent were not reported. Graz *et al.*, at the University of Cambridge, reported the creation OFETs and inverters on 1 mm-thick transparent PDMS membranes [55], while Jang *et al.* at Pohang University of Science and Technology, described the production of a solution-processed, ultrathin cyclic olefin copolymer (COC)/Al_2O_3 bilayer gate dielectric for low-voltage and flexible PTCDI-C_{13}-based *n*-type OFETs, and their complementary circuits [65]. These inverters performed a voltage inversion satisfactorily even at a low voltage of 5 V, and with negligible hysteresis in ambient air (Figure 4.17). As the current levels of both OFETs were similar, the voltage inversion occurred at the center of the voltage swing, which resulted in

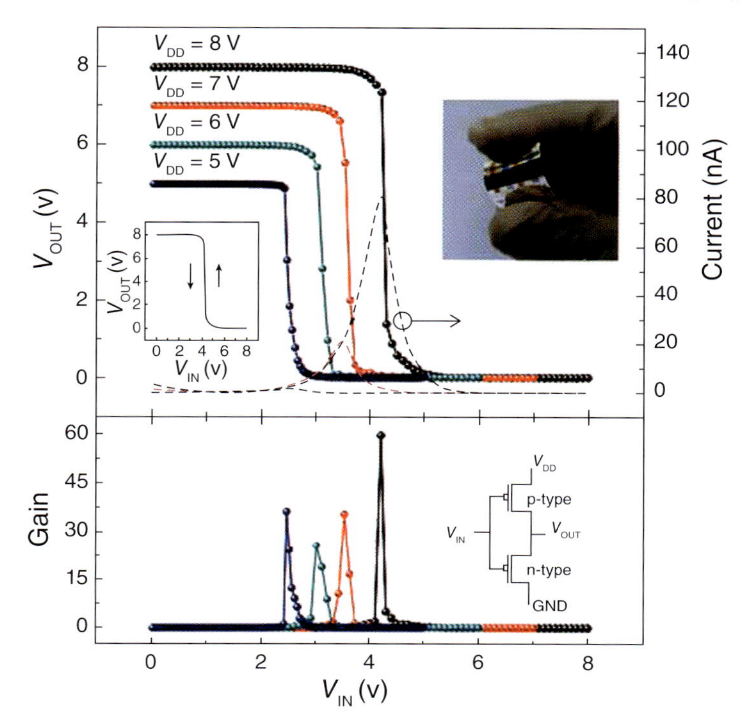

Figure 4.17 Upper panel: Voltage transfer characteristics and corresponding currents (top) for the complementary inverter with various V_{DD} (5 V to 8 V); Lower panel: Corresponding gains. Insets top and left: Voltage transfer characteristics of the inverter in dual V_{IN} sweeps (V_{DD}: 8 V); top and right: A photograph of the flexible inverter devices on a polyarylate substrate; Bottom: A schematic diagram of the complementary inverter. All electrical measurements were carried out in ambient air [65].

good noise margins. As shown in the lower part of Figure 4.16, these inverters also exhibited a high gain (defined as the absolute value of the slope of the inverter voltage-transfer curve) ,with an average value of 48 and a maximum of 63 [65]. Zhang *et al.*, at the Georgia Institute of Technology, reported the creation of flexible organic complementary inverters, using pentacene and C_{60} as active semiconductors fabricated on a plastic substrate. In this case, a high DC gain of 180 was achieved at $V_{DD} = 5\,V$ [66].

The smallest radius into which a circuit can be bent is a typical parameter for flexible device and circuits, and is limited by strain-induced damage to the active circuit elements. Sekitani *et al.*, at the University of Tokyo, demonstrated the preparation of organic pentacene and $F_{16}CuPc$ FETs and complementary circuits that continued to operate, without degradation, despite being folded into a radius of 100 μm. This enormous flexibility and bending stability was made possible by the existence of a very thin plastic substrate (12.5 μm), an atomically smooth planarization coating, and a hybrid encapsulation stack that placed the transistors in the neutral strain position [51]. Very few reports have been made of flexible organic complementary ring oscillators, and in most of these cases the circuits either required large operating voltages [34, 54] or had to be operated in an inert environment owing to the use of semiconductors which degraded in air [58]. The thickness of the gate dielectric produced was about 6 nm (the aluminum oxide had a thickness of about 4 nm, while the SAM on aluminum oxide had a thickness of about 2 nm), which ensured that the transistors and circuits would function on a voltage lower than 3 V [51]. The gain of the inverter was about 40. The bending stability of the organic complementary inverters and ring oscillators fabricated on flexible polyimide substrates was also evaluated (see Figure 4.18). When the transfer characteristics of the inverter (as shown in Figure 4.18b) were measured in the flat state, and again with the inverter bent into a radius of 300 μm, there was no discernible change in characteristics during the bending process. Neither were the output characteristics of the five-stage complementary ring oscillator altered during tight bending, when compared to the flat state [51].

4.2.4 Complex Circuits of Organic Thin Films

The inverter forms the basis of complex circuits, the fabrication of which is considered to be one of the ultimate goals of organic electronics.

In 1998, Drury *et al.*, at Philips Research Laboratories, developed a technology to fabricate all polymer-integrated circuits, and demonstrated a real logic functionality with the 15-bit mechanically programmable code generator [67], regarded as perhaps the first of the complex organic circuits. In the code generator, the layout of which is shown in Figure 4.19, the integrated circuit incorporated 326 transistors and over 300 vias, and also contained an onboard clock generator, a 5-bit counter, decoder logic, and 15 programming pads. As shown in Figure 4.20, the circuit produced a mechanically user-friendly serial data stream of 15 bits. In operation, a DC voltage was applied and the current through an output transistor measured; the default setting was with all programming pads connected to the power rail. The output (Figure 4.20a) was a start bit, 15 bits on, and 16 stopped bits for pattern recognition. Cutting the interconnection tracks between the programming pads and the power rail allowed remeasurement with none of the bits programmed; the output (Figure 4.20b) then yielded only start bits, and the code generator could then be programmed. A *via* in a program pad set that bit to 1, with 0 represented by the absence of a *via*. The output from a selectively programmed generator is shown in Figure 4.20c, where the bit rate obtained was 30 bits per second. Moreover, the circuits remained operational when the foils were sharply bent.

In 2004, Gelinck *et al.*, at Philips Research Laboratories demonstrated flexible active-matrix monochrome electrophoretic displays that were based on solution-processed *p*-type pentacene FETs on 25 μm-thick polyimide substrates. The displays could be bent to a radius of 1 cm, without any significant loss in performance. Using the same process flow, the same group prepared 32-stage shift registers

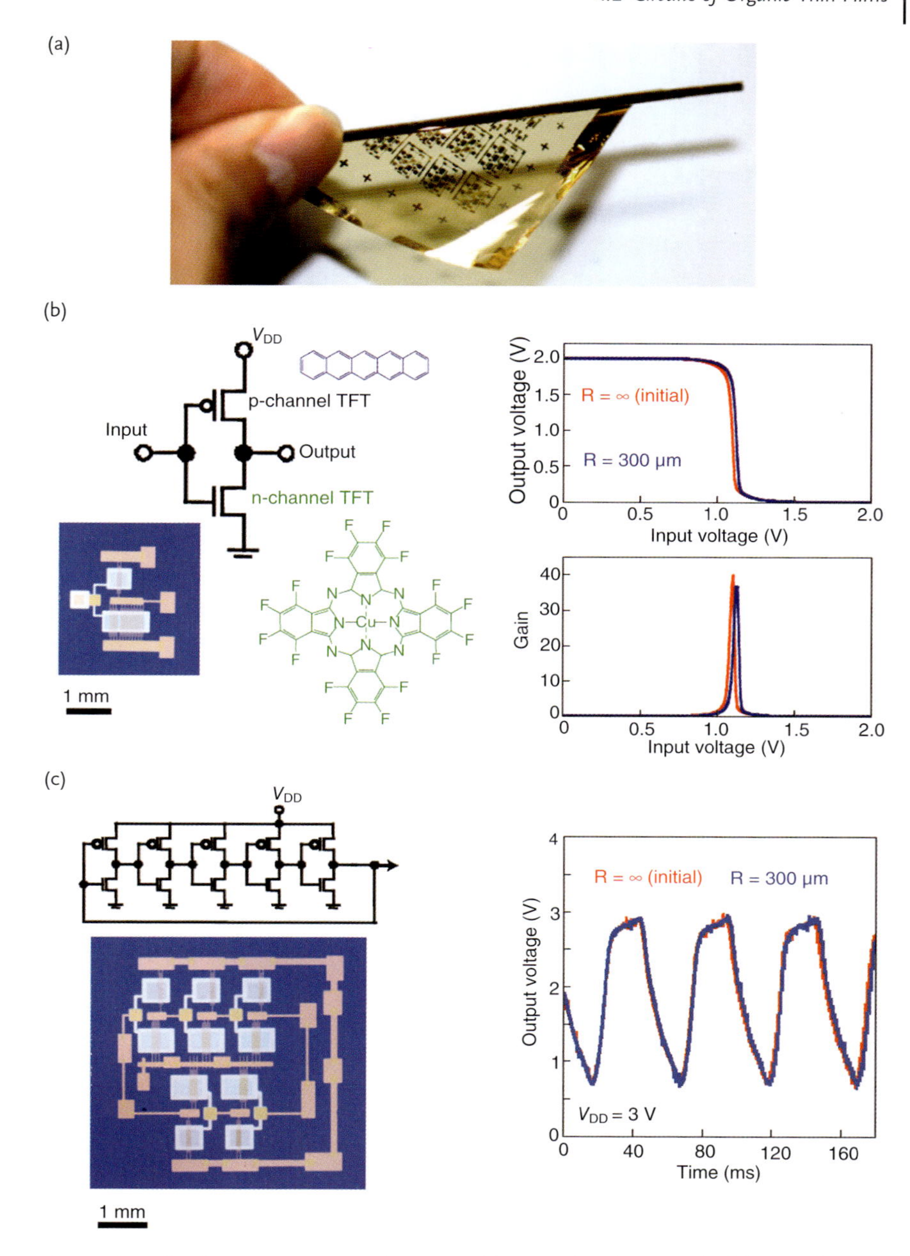

Figure 4.18 Ultraflexible organic circuits. (a) Photograph of a polyimide substrate with functional organic FETs and circuits wrapped around a cylinder with a radius of 300 µm; (b) Circuit diagram, photograph and electrical transfer characteristics of a complementary inverter (composed of a pentacene *p*-channel FET and a $F_{16}CuPc$ *n*-channel FET) operated with a supply voltage (V_{DD}) of 2 V; (c) Circuit diagram, photograph and output signals of a five-stage complementary ring oscillator operated with a supply voltage (V_{DD}) of 3 V in the flat state and while bent into a radius of 300 µm. The circuit oscillates with a signal delay per stage of 4.5 ms, both in the flat state and during tight bending [51].

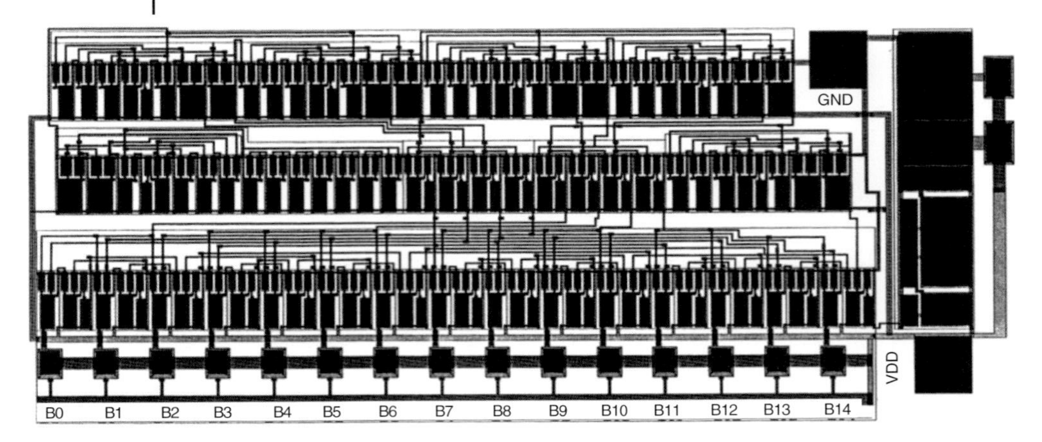

Figure 4.19 Layout of the 15-bit mechanically programmable code generator [67].

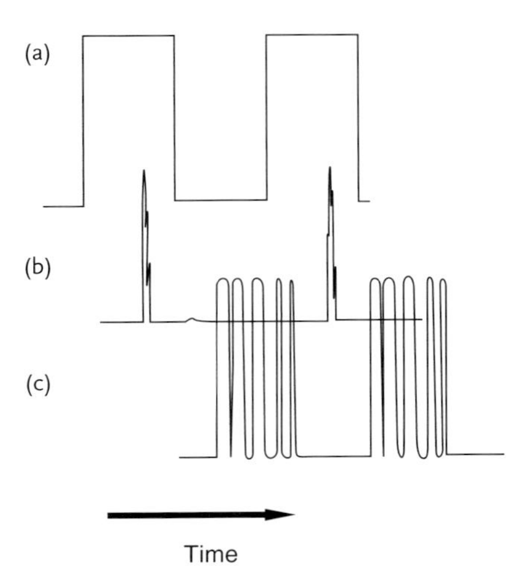

Figure 4.20 Output from the 15-bit code generator. (a) All programming pads connected to the power rail (default); (b) Programming pads disconnected without programming; (c) Selectively programmed [67].

with 1888 transistors, which were the largest organic integrated circuits reported at that time [68]. A schematic of the basic building block of shift registers, a standard logic inverter with two transistors of the same type, is shown in Figure 4.21a. An external clock pulse was used to transfer the row-select signal from one stage to the next, with each stage containing 29 transistors. The clock, the input signal, and the output signals of selected stages were then plotted as a function of time (see Figure 4.21a). One input data pulse was applied at the first clock cycle, and

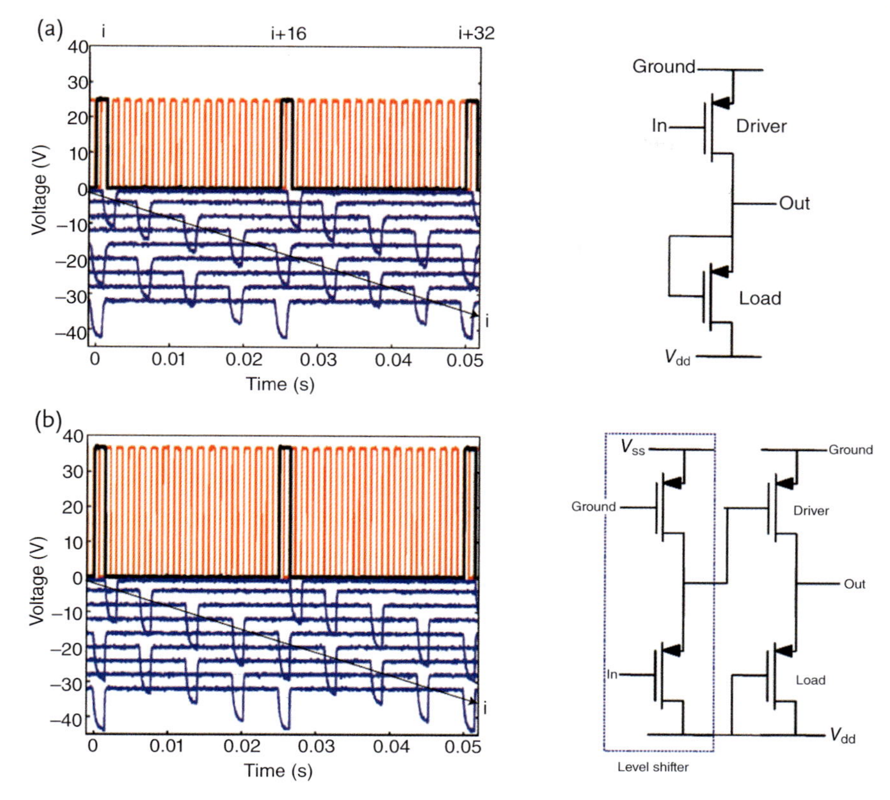

Figure 4.21 Characteristics of organics-based 32-stage shift registers. (a) Input data pulses i, i + 16 and i + 32 (black), clock frequency of 640 Hz (red), and buffered output (blue) at different stages of a 32-stage shift register based on standard logic; (b) Diode-connected logic. The arrows guide the eye through the stages displayed: 1, 4, 8, 12, 16, 20, 24, 28, and 32. Schematics of the inverters used in the two shift registers are also shown. The introduction of the level shifter, needed for a sufficient performance of the diode-connected inverter, requires a second voltage rail, the control voltage, V_{ss} [68].

the second input data pulse applied after 16 clock cycles. The circuits operated at a supply voltage, V_{dd}, of 25 V and a clock frequency of up to 0.7 kHz. The low operating frequency was related to the relatively low current of the load transistors that were always "off," that is, at $V_g = 0$ V. One way to increase the operating frequency was to use a more complex logic. A schematic of the inverter used is shown in Figure 4.21b, where the load transistor was always "on." The 32-stage shift registers based on this logic consisted of 1888 transistors. The shift registers operated at a supply voltage V_{dd} of 35 V and a control voltage V_{ss} of 10 V (Figure 4.21b). A clock frequency of up to 5 kHz could be reached, which corresponded to driving a 64-row display at a refresh rate of 75 Hz.

In addition to research team at Philips Research Laboratories, investigators at the Bell Laboratories also reported an important development of complex organic

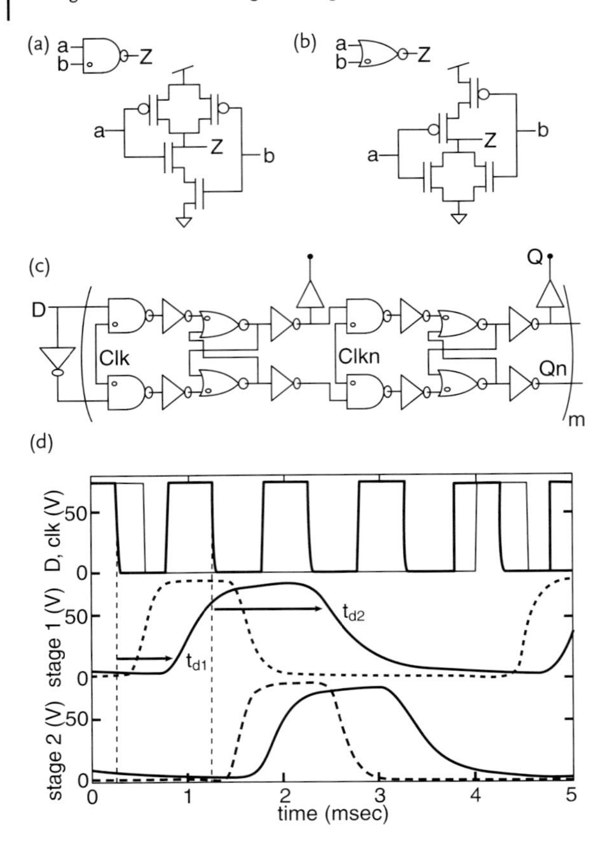

Figure 4.22 Schematics of a two-stage shift register. (a) NAND gate; (b) NOR gate; (c) m-stage NOR/NAND-based shift register; (d) Characteristics: the lower panel shows the input data D, clock (Clk) (bold line), and measured (solid line) and simulated (dashed line) outputs for the first and second stages of the shift register [29].

circuits. In 2000, Crone *et al.* described a clocked sequential complementary circuits with 864 transistors [33], followed by 48-stage shift registers and three-bit row decoders 2001 [29]. The measured (solid line) and simulated (dashed line) results of these systems for a two-stage shift register based on NOR and NAND gates are shown in Figure 4.22d [29]. The schematics for NAND and NOR gates are shown in Figure 4.22a,b, respectively, while the circuit schematic for a NOR, NAND gate-based D-flip-flop is shown in Figure 4.22c. These utilize a master-slave configuration which is composed of two D latches, a master with a clock input (Clk), and a slave with a negated clock input (Clkn). The circuit was measured with a 1 kHz clock frequency, and the data were then loaded into the master D latch when the clock was high, before being passed to the slave D latch when the clock was low. The data (solid line) and clock (long-dash line) and the buffered Q outputs of stages 1 and 2, plotted against time, are shown in Figure 4.22d. The flip-flop

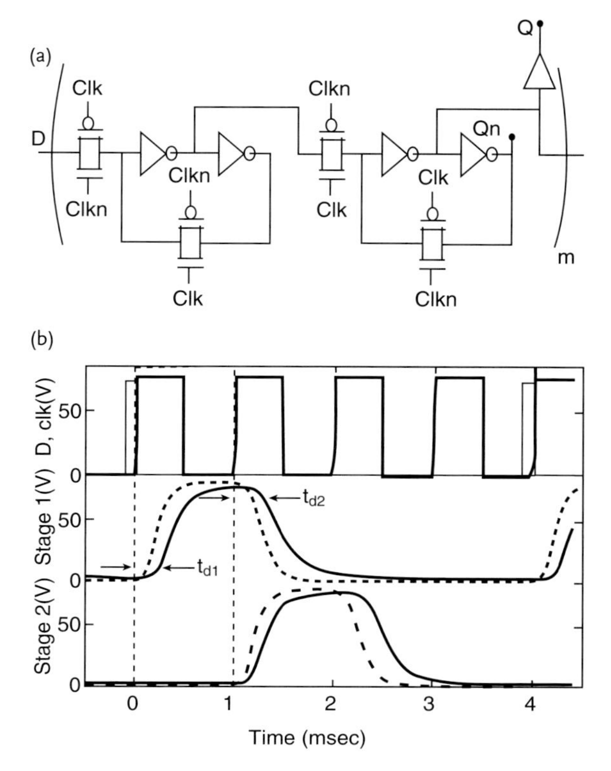

Figure 4.23 (a) Schematic of an m-stage pass transistor logic-based shift register; (b) Characteristics of a two-stage pass transistor logic-based shift register: the lower panel shows the input data D, clock (Clk) (bold line), and measured (solid line) and simulated (dashed line) outputs for the first and second stages of the shift register [29].

circuit consisted of 48 transistors (neglecting output buffers), and the propagation delay for the slave D latch was about 400 ms. The static current drawn by the two-stage NOR/NAND shift register was 20 μA for a static power of 0.16 mW. The circuit functions took the input data and passed them from stage to stage with the clock frequency [29].

Pass transistor logic requires fewer transistors than NOR/NAND-based logic. The circuit schematic for a pass transistor logic-based two-stage shift register is shown in Figure 4.23a [29]. Crone *et al.* again utilized a master-slave configuration composed of two D latches, a master with clock input (Clk), and a slave with a negated clock input (Clkn). The pass transistor logic-based D latch utilized fewer transistors (16 for flip flop, neglecting output buffers). The measured characteristics for a two-stage pass transistor logic-based shift register are shown in Figure 4.23b. In this case, the data were loaded into the master D latch when the clock was low, and passed to the slave D latch when the clock was high. For the data shown, the clock frequency was set to 1 kHz. The propagation delay for the slave

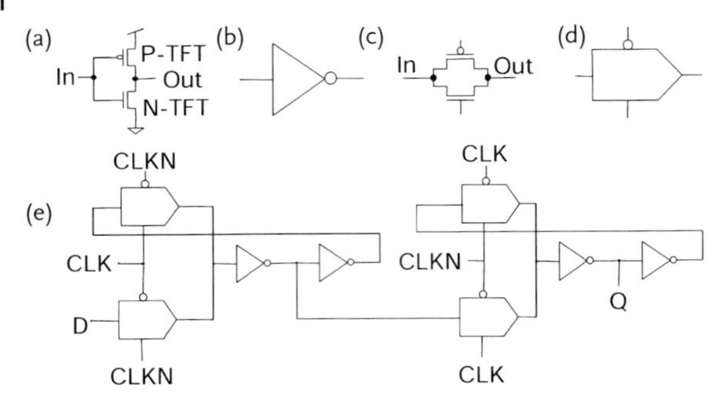

Figure 4.24 Representation of circuits and their constituents. (a) A complementary inverter; (b) Symbol of an inverter; (c) Layout of a complementary transmission gate; (d) Symbol of a transmission gate; (e) Schematic of a D flip-flop showing the data input (D) and the connections of the clock (CLK) and its complement (CLKN). Also shown is the output (Q) [33].

D latch was about 130 μs, less than for the NOR/NAND-based logic. The static current drawn by the two-stage pass transistor logic shift register was 7.6 μA, for a static power of 0.06 mW, about one-third of that drawn for the NOR/NAND-based register. Again, the simulation showed reasonable agreement, given the simplicity of the model [29].

Crone *et al.* designed logic circuits based on a particular type of complementary logic known as pass transistor logic [33]. The essential circuit building blocks were an inverter and a transmission gate (see Figure 4.24), where the building blocks had been combined to form latches and D flip-flops (Figure 4.24e). The use of pass transistor logic results in both space and power-dissipation reduction, as fewer transistors were required to implement most functions. The relatively simple layouts of pass transistors logic were also convenient for the implementation on flexible plastic substrates. Subsequently, Crone *et al.* also fabricated the complementary circuits of row decoders and shift registers [33]. The largest circuit to be evaluated was a 48-stage shift register with 24 output buffers, that incorporated 864 transistors. Each stage of the shift register was a D flip-flop, with the output of one stage connected to the input of the next. The clock and its complement drove all the stages. Every second stage had an output buffer, which consisted of two inverters with large transistors to facilitate probing, but without loading the circuit. Shift registers were ubiquitous in digital systems and could perform many functions; one function was to shift a "bit" in an orderly and predictable manner from one stage to the next every clock cycle. The operation of the shift register described above is illustrated schematically in Figure 4.25, where the clock, the data, and the output of the 24 output buffers are plotted as a function of time. The data consisted of a single bit, which was sequentially shifted over all stages of the register. Thus, the two-stage shift registers had been operated at clock rates of up to 1 kHz, with an operating voltage of 80 V [33].

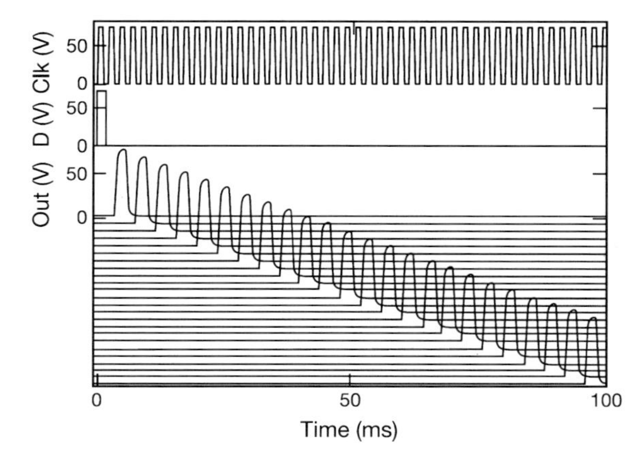

Figure 4.25 Characteristics of the 48-stage shift register. The clock (500 Hz) and data were shown along with the output voltages of the 24 output buffers as a function of time. The 24 outputs had been vertically offset for clarity [33].

A schematic representation of a three-bit row decoder, as fabricated by Crone *et al.*, is shown in Figure 4.26a [33]. This decoder, which was designed so that a single serial input activated one of eight outputs, consisted of three D flip-flops and a NOR array (Figure 4.26b). The output of the three flip-flops and their complements drove the NOR array, which was configured in such a way that for any given set of inputs only one output was "high." The characteristics of the decoder are shown in Figure 4.27.

The circuit used for the NOR array was significant for one other reason, namely that four transistors had been serially connected between supply and ground. Although these "four-deep" connections were often employed in silicon arithmetic and logical unit (ALU) circuits, the fact that such a configuration had been shown to function for organic semiconductors suggested that complex logic gates (such as those used in ALUs and simple microprocessors) could be constructed from OFETs.

Myny *et al.* [69], at the Katholieke Universiteit Leuven, designed a 64-bit organic RFID transponder chip that was based on *p*-type double-gate OFETs; the output signal, together with a photographic image of the chip, are shown in Figure 4.28.

4.2.5
Performance Modulation of Organic Thin-Film Circuits

Compared to devices based on Si, modulation of the performance of organic devices is more difficult, since doping cannot be used; rather, the performance is determined by the semiconductor and the fabrication process. As a result, it is difficult to optimize the performance of organic circuits based on devices with

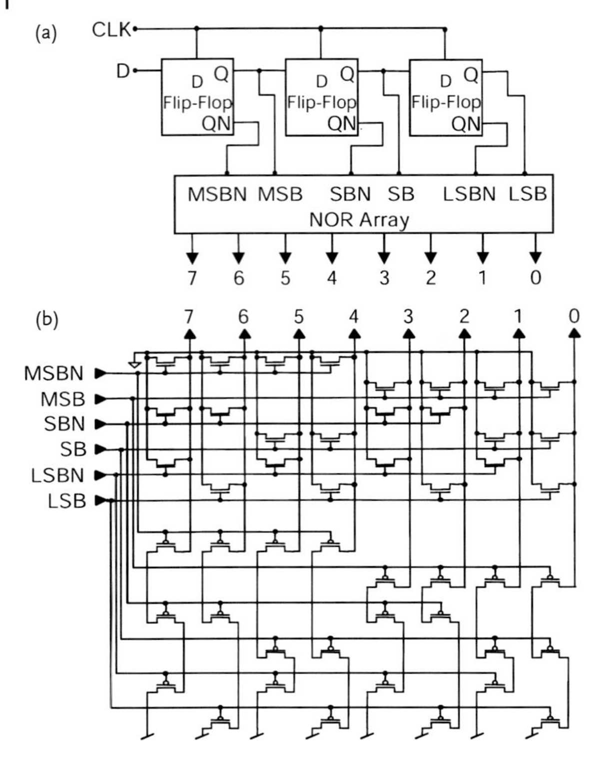

Figure 4.26 Row decoder design. (a) Schematic of a three-bit row decoder with eight outputs (0–7). Each square is a D flip-flop (see Figure 4.24e). The bits are designated as most significant bit (MSB), second bit (SB) and least significant bit (LSB). Complements are labeled with a N; (b) Details of the NOR array, showing the manner in which individual transistors are connected [33].

fixed characteristics. For example, in the case of inverters, if the threshold voltage of the transistors cannot be controlled during manufacture, then the switching voltage may be too close to either the ground potential or the supply voltage potential; in this case, the circuits may not operate reliably when exposed to electrical noise. Consequently, it is highly desirable that the threshold voltage of the transistors can be set to a specific value determined by the circuit designer.

Double-gate structures can independently control the threshold voltage of the *p*- or *n*-type OFETs, and this could be used to control the switch voltage of inverters. For example, in 2007 Hizu *et al.*, at the University of Tokyo, fabricated organic inverters that comprised *p*-type pentacene and *n*-type fluoroalkyl naphthalenetetracarboxylic di-imide FETs with double-gate structures [70]. As shown in Figure 4.29, the threshold voltage of both the *p*-type and *n*-type devices could be controlled by the top gate. For the inverters based on double-gate transistors, the output characteristics could be controlled when the top-gate voltages were applied to the

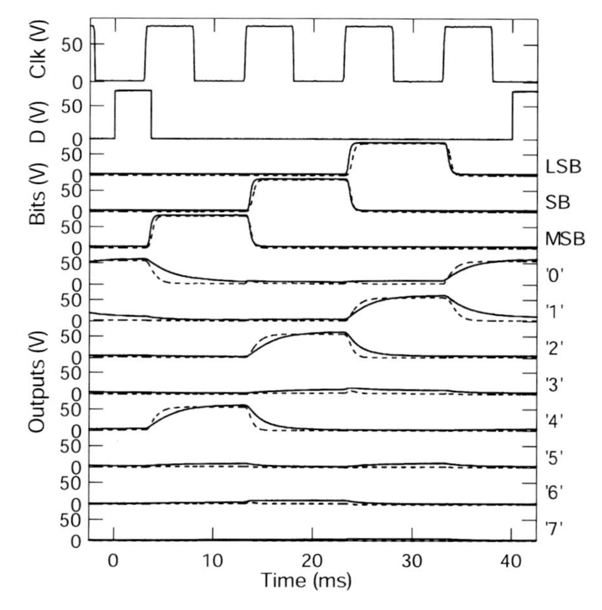

Figure 4.27 Characteristics of the three-bit decoder. In descending order, the traces are as follows. Clock, data, outputs of the three D flip-flops (labeled LSB, SB, MSB on the right-hand vertical axis). The eight remaining traces, labeled 0–7, show the output voltages of the eight outputs: at any time, one output at most is high. Also shown (dashed lines) are the simulated responses of the flip-flops and the outputs of the decoder. The good agreement between simulation and experiment may be noted. The simulations were performed with a set of tools developed for organic/polymer transistor circuit design [33].

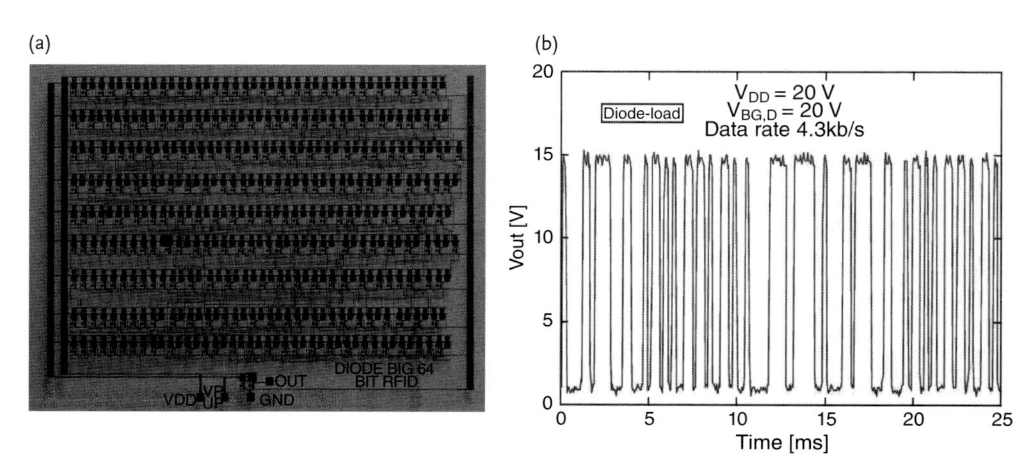

Figure 4.28 (a) Photographic image and (b) output signal of a 64-bit organic RFID transponder chip based on double-gate OFETs for a supply voltage of 20 V and a back-gate voltage of 45 V. The corresponding data rate is 4.3 kb s^{-1} [69].

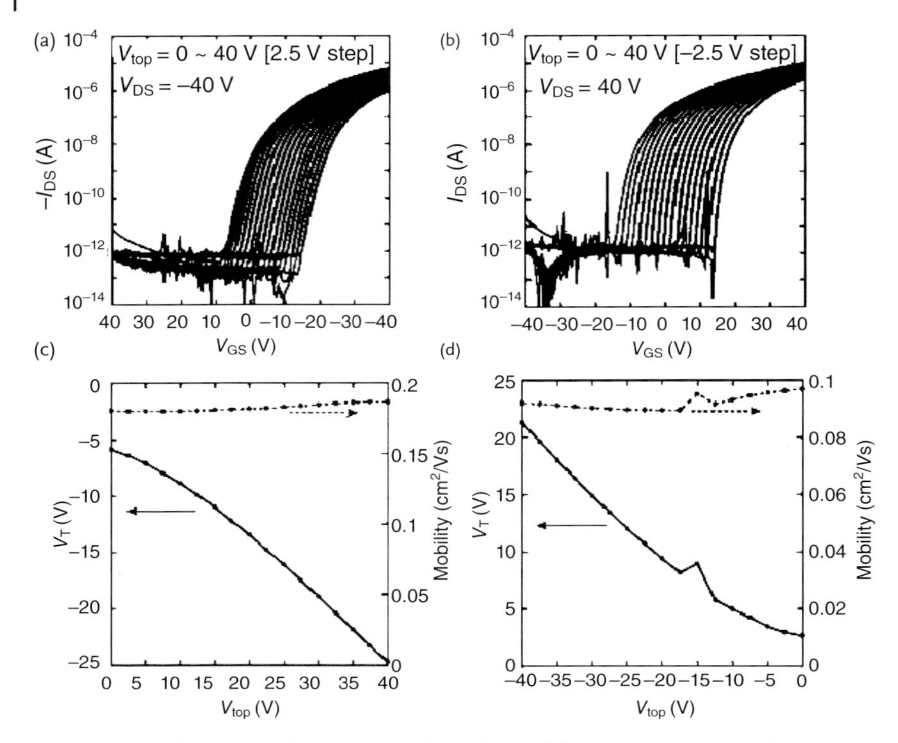

Figure 4.29 (a, b) Transfer characteristics of (a) *p*-type and (b) *n*-type organic transistors with double-gate structures. As the top-gate voltage applied is increased, the threshold voltage of the organic transistors decreases; (c,d) Mobility and threshold voltage changes in (c) *p*-type and (d) *n*-type organic transistors with double-gate structures [70].

n-type transistors, and the top-gate voltages of the *p*-type transistors were set at 0 V (see Figure 4.30) [70].

Myny *et al.* also designed double-gate OFETs with a similar structure to control the threshold voltage of transistors [69, 71], in which the circuits were based only on *p*-type transistors (see Figure 4.31). Again, the output characteristics could be controlled when the top-gate voltages were applied to the drive transistors, and the back-gate of the load was connected with the output node (see Figure 4.31) [69].

In 2011, Yokota *et al.*, in collaboration with Hizu, designed OFETs with a floating-gate structure and SAM-based gate dielectrics to demonstrate the threshold-voltage control [72]. A schematic cross-section of the floating-gate transistors and inverters is shown in Figure 4.32a; the transfer characteristics of the *p*- and *n*-type OFETs, recorded after program operations performed on the floating-gate, are shown in Figure 4.32b,c. The threshold voltages of both transistors were tuned by the program voltages of the floating gate [72].

These devices were applied to fabricate inverters and a five-stage complementary ring oscillator, while the floating-gate was used to tune the characteristics (see

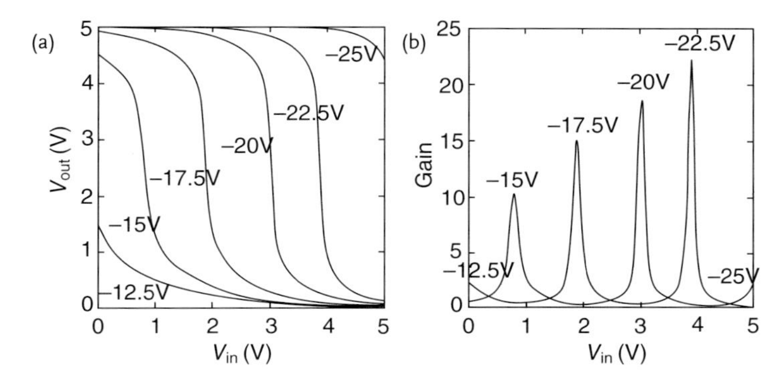

Figure 4.30 (a) Input–output characteristics and (b) gain of the organic CMOS inverter. V_{top} of *n*-type transistors changed from −12.5 to −25 V, while V_{top} of *n*-type transistors was 0 V [70].

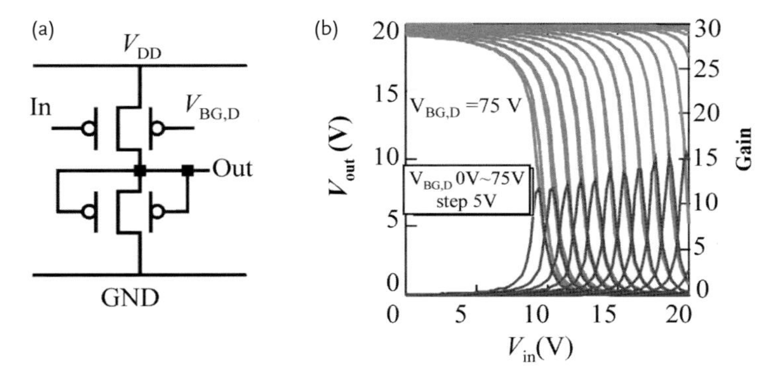

Figure 4.31 (a) Architecture of inverter architectures in dual-gate technology for zero-V_{GS}-load inverter; (b) Transfer curves of the inverter plotted for a supply voltage of 20 V, as a function of the voltage on the -control gate of the drive transistor [69].

Figure 4.33) [72]. The transfer characteristics of the inverter at a supply voltage of 1.5 V, measured after a program voltage had been applied to the input node (i.e., to the floating control gates of the OFETs), are shown in Figure 4.33a. The application of a positive (negative) program voltage caused the threshold voltages of the *p*- and *n*-channel TFTs, and hence the switching voltage of the inverter to shift systematically toward more positive (negative) voltages. In this way, the switching voltage of the inverter could be shifted across the entire range of input voltages (from 0 to 1.5 V) in a deterministic manner. The small-signal gain was almost constant for program voltages between −4 and +4 V. However, in the extreme case – that is, for program voltages of −6 and +6 V – the switching voltage was moved outside the input voltage range, so that the output voltages of the inverter were almost constant at 0 V and 1.5 V, respectively [72].

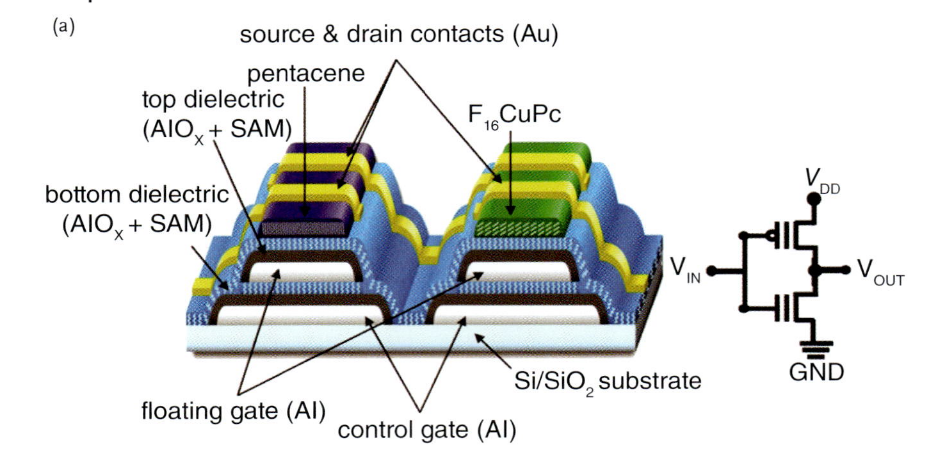

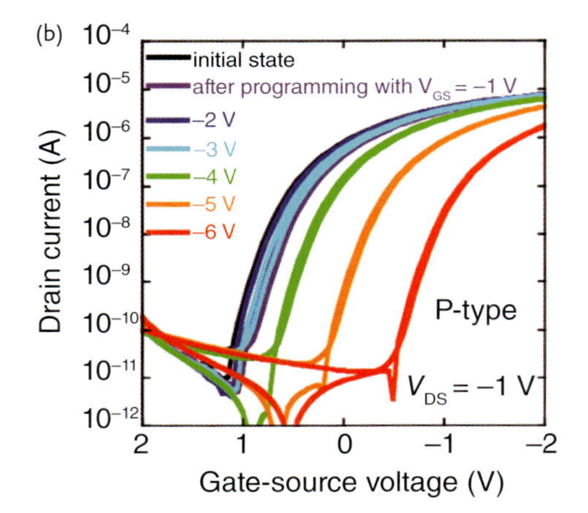

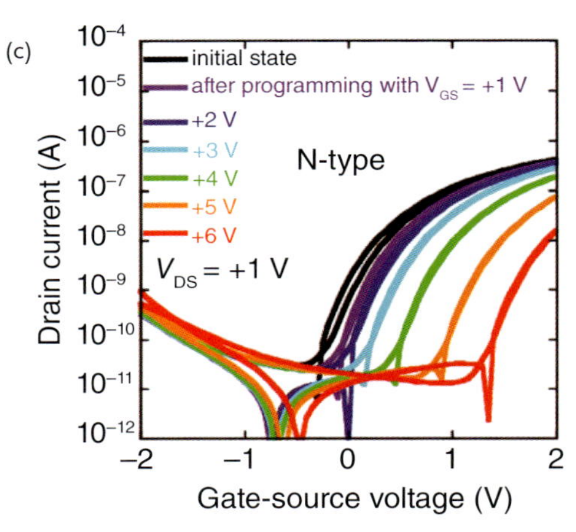

Figure 4.32 (a) Schematic cross-section and circuit diagram of the floating-gate complementary inverters. The substrate is silicon covered by thermally grown silicon dioxide; (b) Transfer characteristics of pentacene FETs recorded after program operations performed on the floating-gate with program voltages ranging from −1 to −6 V. The duration of each program pulse was 1 s; (c) Transfer characteristics of $F_{16}CuPc$ FETs recorded after program operations performed on the floating-gate with program voltages ranging from 1 to 6 V. The duration of each program pulse was 1 s [72].

The oscillation of the fabricated ring oscillator, with an oscillation frequency of 17.9 Hz at a supply voltage V_{DD} of 3 V, is shown in Figure 4.33c [72]. Surprisingly, even at a supply voltage as low as 1 V the ring oscillator remained functional, with a frequency of 0.38 Hz. The signal propagation delay as a function of V_{DD}. is shown in Figure 4.33d, while the output signal of the ring oscillator at a supply voltage of 1.5 V, measured after a program voltage had been applied to the control gates of all OFETs in the ring oscillator, is shown in Figure 4.33e. Whereas, the oscillator

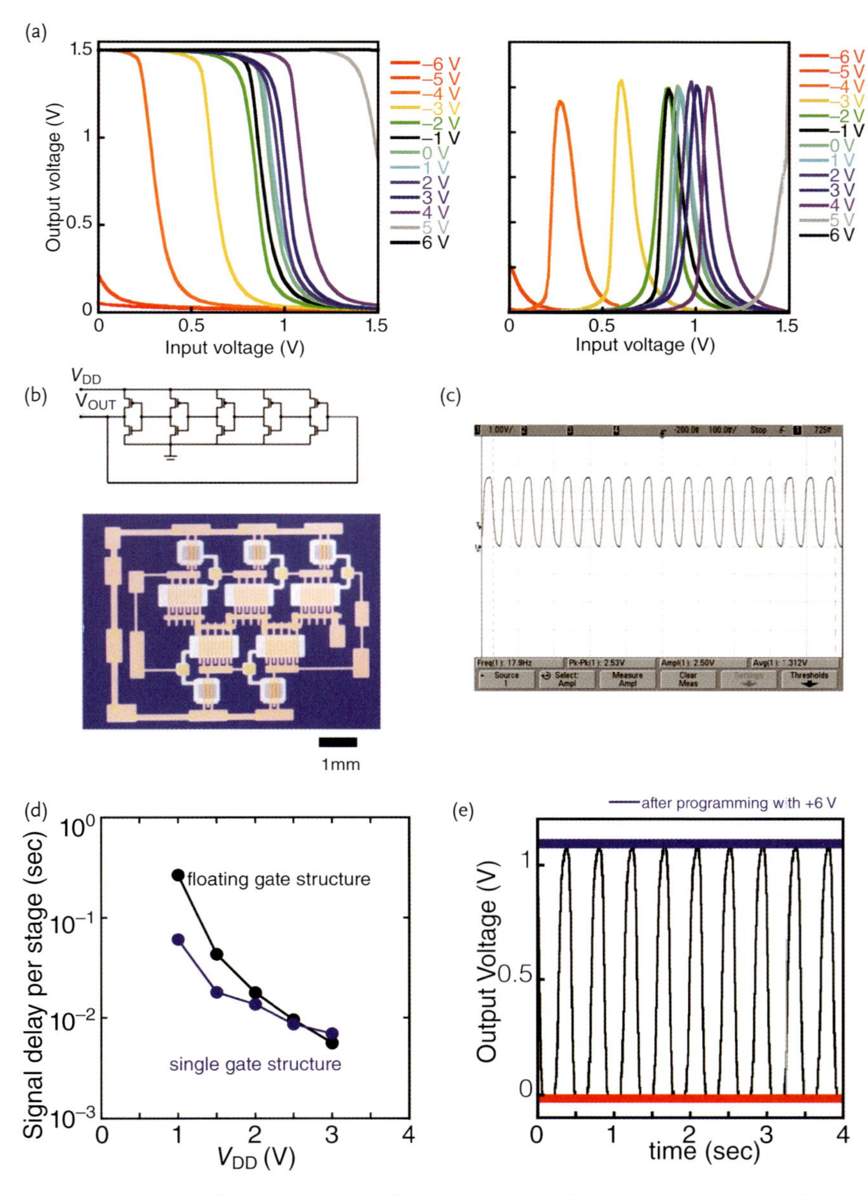

Figure 4.33 (a) Transfer characteristics of the inverter at a supply voltage of 1.5 V, measured after program voltages between −6 and +6 V had been applied to the input node for a duration of 1 s. The switching voltage was systematically controlled by the program pulses; (b) Circuit diagram and optical microscope image of a five-stage complementary ring oscillator; (c) Output signal of the ring oscillator at $V_{DD} = 3$ V; (d) Signal propagation delay as a function of supply voltage; (e) Output signals recorded at V_{DD} of 1.5 V after the application of program voltages of −6, 0, and 6 V to the control gates of all TFTs in the ring oscillator for a duration of 1 s [72].

frequency was 2.3 Hz for a program voltage of 0 V, when program voltages of +6 V and −6 V were applied, the ring oscillator no longer oscillated but rather exhibited constant output voltages of 1.1 V and 0 V, respectively. Hence, the oscillation frequency could be controlled systematically by the program voltage [72].

Zschieschang *et al.*, at the Max Planck Institute for Solid State Research, demonstrated a novel approach that allowed the threshold voltage of low-voltage (≤3 V) organic *p*-channel and *n*-channel FETs and the switching voltage and noise margin of low-power organic complementary circuits to be tuned continuously over a wide range during manufacture [73]. A continuous threshold voltage tuning was facilitated by employing gate dielectrics that were based on mixed alkyl/fluoroalkyl phosphonic acid SAMs, prepared by a simple solution process (Figure 4.34a–c). The threshold voltage of the OFETs was a linear function of the atomic fluorine concentration in the mixed SAM dielectric, with a modulation coefficient of 40 mV %$^{-1}$. This was because, compared to the alkyl SAMs, the electronegative substituents in fluoroalkyl SAMs reduced the electron density in the carrier channel of the transistors, causing a change in threshold voltage towards more positive values [73].

The transfer characteristics of complementary inverters, based on the pentacene *p*-channel and $F_{16}CuPc$ *n*-channel FETs from Figure 4.34d–f, are shown in Figure 4.34g–i [73]. The effect of the SAM composition on the inverter switching voltage can be clearly seen since, when both FET threshold voltages were negative (100% alkyl SAM), the inverter switching voltage was very close to the ground potential, so that the inverter was very susceptible to electrical noise. However, when both threshold voltages were positive (100% fluoroalkyl SAM), the switching voltage was actually greater than the supply voltage, so the inverter could not switch unless the input voltage exceeded the supply voltage. Consequently, by selecting a mixing ratio that provided symmetric FET threshold voltages (50% alkyl + 50% fluoroalkyl SAM), an inverter switching voltage of exactly half the supply voltage (1 V) was obtained (Figure 4.34h). One benefit of placing the switching voltage at half the supply voltage was to acquire a greater noise immunity: Even if noise were to cause the input voltage to deviate by close to half the supply voltage in either direction, the inverter would still produce the correct output signal. The inverter in Figure 4.34h (50% alkyl + 50% fluoroalkyl SAM) had a noise margin of 0.8 V (80% of ½ V_{DD}; this was the largest noise margin reported to date for an organic circuit).

Another important benefit of a symmetric switching voltage for inverters is that the circuits can be operated with lower supply voltages [73]. This can be seen in Figure 4.35, where the signal delay of two complementary ring oscillators is plotted as a function of the supply voltage. The ring oscillator that uses FETs with a 100% alkyl SAM requires a minimum supply voltage of 1.5 V, while the ring oscillator with the mixed SAM operates with supply voltages as low as 1.0 V, which is the smallest supply voltage reported for organic complementary circuits to date.

The rational design can also improve the performance of circuits based on OFETs. Crone *et al.*, at Bell Laboratories, designed an output buffer to tune the characteristics of inverters and other circuits [29]. The inverter and buffered inverter are shown schematically in Figure 4.36a,b, respectively. The measured

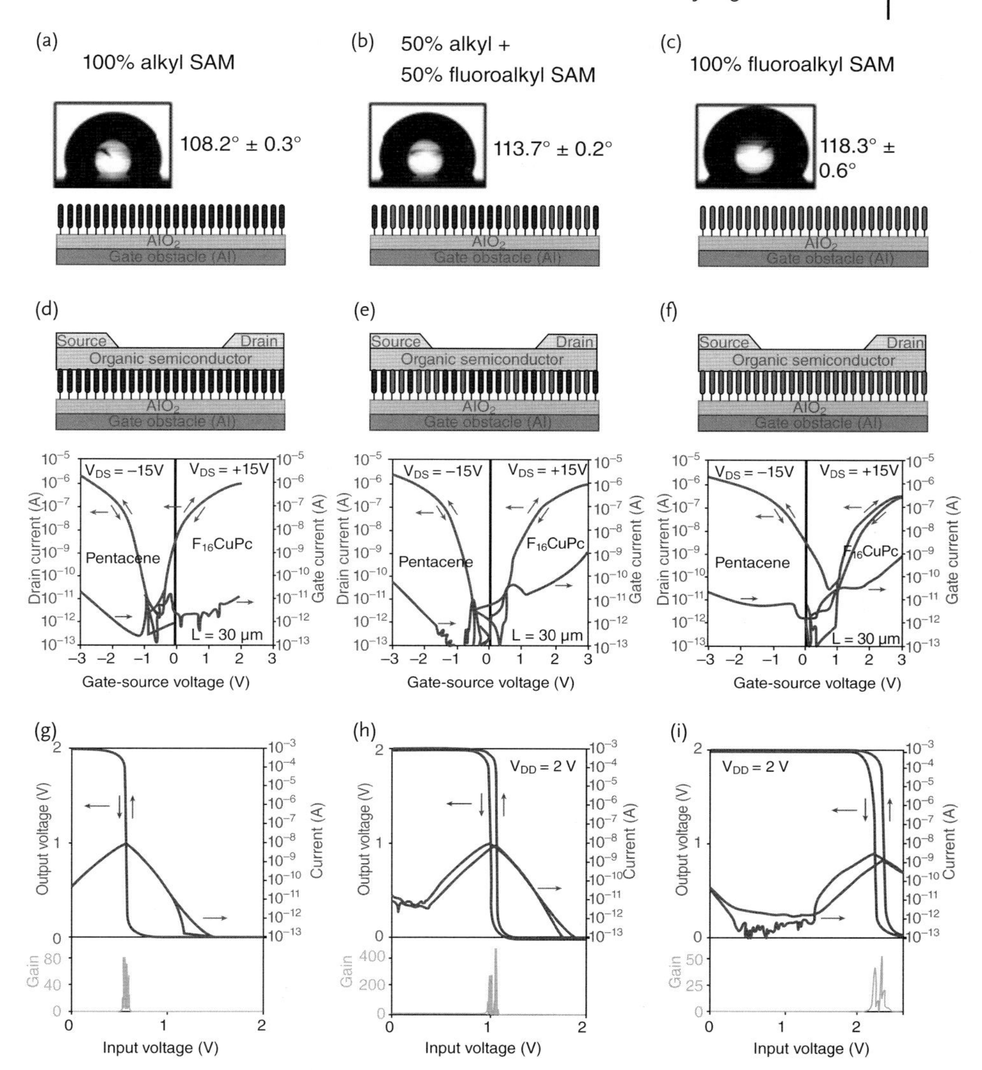

Figure 4.34 (a–c) Water contact angles measured on a self-assembled monolayer (SAM) of 100% alkyl phosphonic acid (a), on a mixed SAM of 50% alkyl and 50% fluoroalkyl phosphonic acid (b), and on a SAM of 100% fluoroalkyl phosphonic acid (c); (d–f) Schematic cross-sections and transfer characteristics of pentacene *p*-channel FETs and $F_{16}CuPc$ *n*-channel FETs with three different SAMs as the gate dielectric: SAM of 100% alkyl phosphonic acid, mixed SAM of 50% alkyl and 50% fluoroalkyl phosphonic acid, and SAM of 100% fluoroalkyl phosphonic acid; (g, h) Transfer characteristics of complementary inverters (each composed on one pentacene and one $F_{16}CuPc$ FET) with the same SAMs. Using the mixed SAM of 50% alkyl and 50% fluoroalkyl phosphonic acid, the inverter had a switching voltage of exactly half the supply voltage (1 V) and a low power consumption of 20 pW (the product of the supply voltage, 2 V, and the maximum supply current, 10 pA) [73].

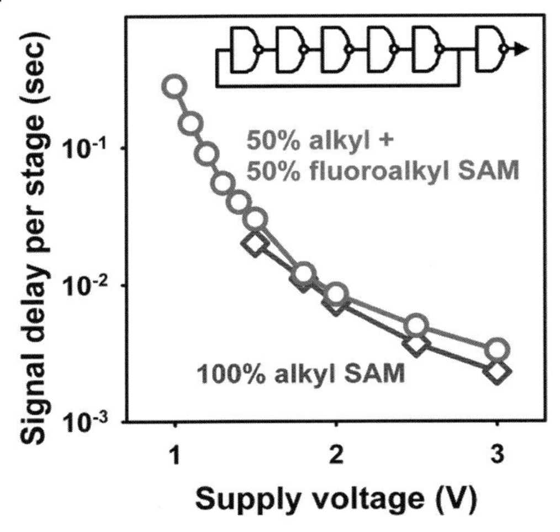

Figure 4.35 Signal delay per stage as a function of supply voltage for two five-stage organic complementary ring oscillators. One oscillator was based on FETs with a mixed SAM of 50% alkyl and 50% fluoroalkyl phosphonic acid as the gate dielectric (red curve); the other oscillator was based on FETs with a 100% alkyl phosphonic acid SAM dielectric (blue curve). The ring oscillator with the optimized mixed SAM gate dielectric operated with a supply voltage as low as 1 V [73].

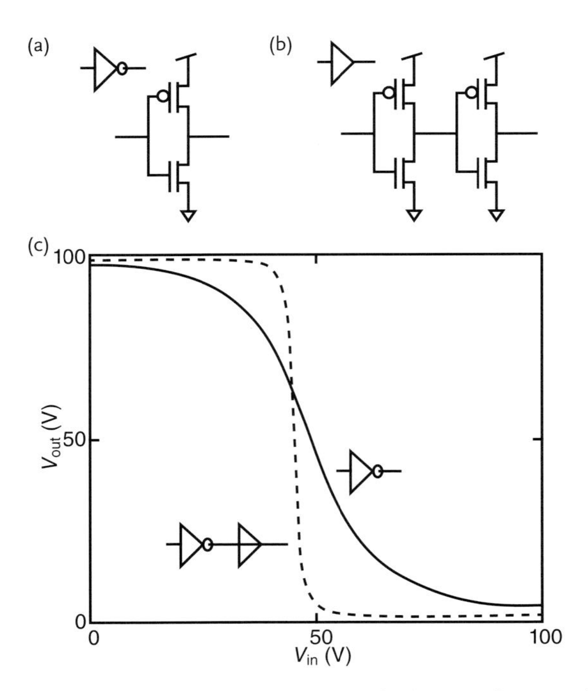

Figure 4.36 Symbolic representation and schematics for a complementary inverter (a) and buffer (b); (c) Measured DC characteristics for a complementary inverter and buffered inverter [29].

characteristics of an inverter with (dashed line) and without (solid line) an output buffer are shown in Figure 4.36c, where the unbuffered output showed a good inversion, although the curve was not very steep. The buffered output was much steeper, not only because it loaded the measurement less but also (and mostly) because the addition of inverters caused a natural sharpening of the output characteristics, leading eventually to a step function from the stable high to the stable low voltages of the inverter [29].

4.2.6 Analog Circuit Based on Organic Thin-Film Transistors

Before the invention of digital logic circuits, transistors were used extensively in analog circuits, in a host of applications. Previously, several research groups have investigated the feasibility of fabricating analog circuits by OFETs. For example, Marien *et al.* at the Katholieke Universiteit Leuven, reported the design and measurement of one of the first dual DC–DC upconverters implemented in a double-gate organic thin-film technology on foil, generating bias voltages of around +50 and −40 V from a 20 V power supply (Figure 4.37) [74]. The DC–DC upconverters circuit consisted of two Dickson cores: a two-stage core for the high-output voltage (V_h), and a three-stage core for the low-output voltage (V_l), as shown in Figure 4.37.

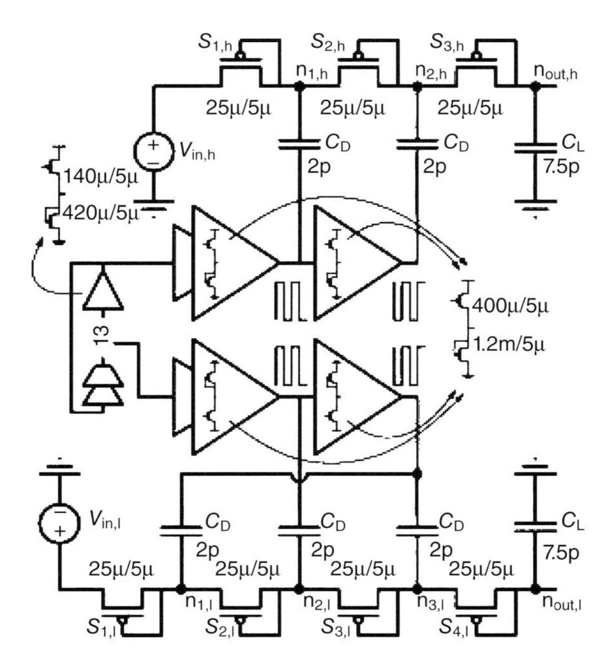

Figure 4.37 Schematic view of presented dual DC–DC upconverter [74].

Each Dickson core consisted of a 2 pF capacitor and a diode-connected transistor with a 25 μm/5 μm width/length (W/L) ratio. A 7.5 pF capacitor was applied to both output voltages. Two differential clock signals were required to make the converters operate; these were generated by a 13-stage ring oscillator, consisting of zero-voltage load inverters. The oscillator was followed by two buffer chains that drove the Dickson cores. According to simulations, the converter oscillated at 500 Hz, and the output voltage–which was generated with the purpose of improving the noise margin of digital circuitry–was applied directly as a backgate bias to the ring oscillator and the buffers.

4.3
Self-Assembled and Printed Organic Circuits

4.3.1
Self-Assembled Organic Circuits

The creation of the first chemisorbed SAMs during the 1980s inspired many to not only study the formation of the SAM but also to characterize its properties [75, 76]. Subsequently, SAMs have been used to change the wetting properties of materials [77], to adapt the work function of metals [78, 79], to build large-area molecular junctions [80], and/or to fabricate ultrathin dielectrics [50, 81]. SAMs can also be applied as transport channels in FETs [82–85], which is advantageous from a technological viewpoint because this represents one of the few practical strategies for producing ensembles of nanostructures [86]. The use of SAMs as semiconductors to fabricate FETs, and their combination into integrated circuits, was reported only by de Leeuw and colleagues. The demonstration of a real logic functionality leads to self-assembly becoming the ultimate technology for the bottom-up mass production of organic electronics [87, 88].

For this purpose, liquid-crystalline molecules were used that consisted of a π-conjugated semiconducting core, separated by a long aliphatic chain from a mono-functionalized anchor group to form a SAM on an atomically flat amorphous SiO_2 gate dielectric [87]. The molecule structure and schematic of the transistor are shown in Figure 4.38, where the trichlorosilanes or trialkoxysilanes are used as anchoring groups. SAMs can be formed via a condensation reaction with hydroxyl groups on the hydrolyzed SiO_2 (200 nm) surface. Defects may be formed as the result of any uncontrolled self-condensation, thus prohibiting any long-range order [89]. If these defects are to be prevented, then mono-functional anchoring groups are crucial. Any dimers formed upon self-condensation should not interfere with SAM formation on the gate dielectric. An α-substituted quinquethiophene is used as the core of the semiconducting molecule, which is a good compromise between solubility and charge carrier mobility; the semiconducting core is then functionalized with aliphatic chains. This spacer concept originates from the field of liquid crystals, where it is well known that the aliphatic chain assists in the anisotropic ordering of rod-like mesogenic groups [90–92]. In this

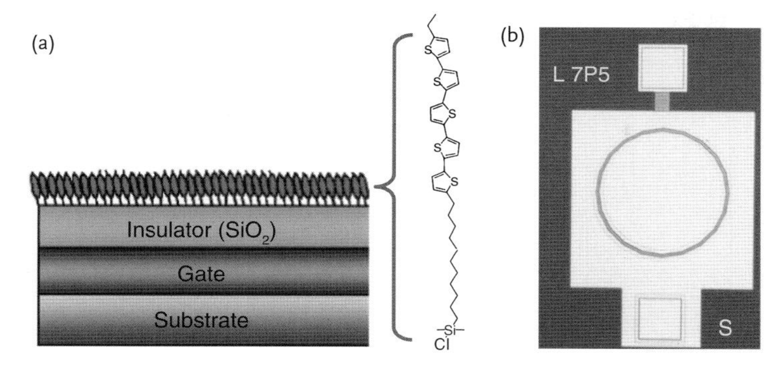

Figure 4.38 (a) Schematic cross-section of the molecules self-assembled on substrate and the molecule structure of the SAM molecules. The substrate is silicon covered by thermally grown silicon dioxide; (b) Optical photograph of a ring transistor based on the SAM semiconductor [87].

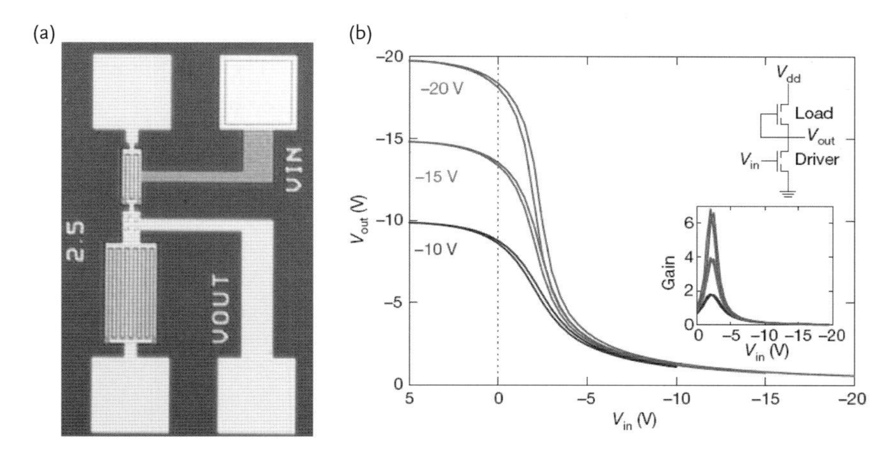

Figure 4.39 (a) Optical photograph of an inverter based on unipolar SAM FETs; (b) Static input–output characteristics of an inverter. The inverter was measured with supply voltages, V_{dd}, of −10, −15, and −20 V. The inset shows a diagram of the logic gate and a plot of the measured gain (color-coded as for main panel) [87].

case, an undecane spacer between the thiophene core and the anchoring group was used. The conformational degree of freedom allowed the molecule to self-assemble and to optimize its π-π stacking. For reasons of stability and solubility, an ethane chain is attached to the other side; the chemical structure is shown in Figure 4.38a [87].

An optical image of an inverter, together with a circuit diagram and the input–output characteristics, are presented in Figure 4.39. Typically, inverters show a voltage amplification with a gain that increases with the DC supply voltage. The

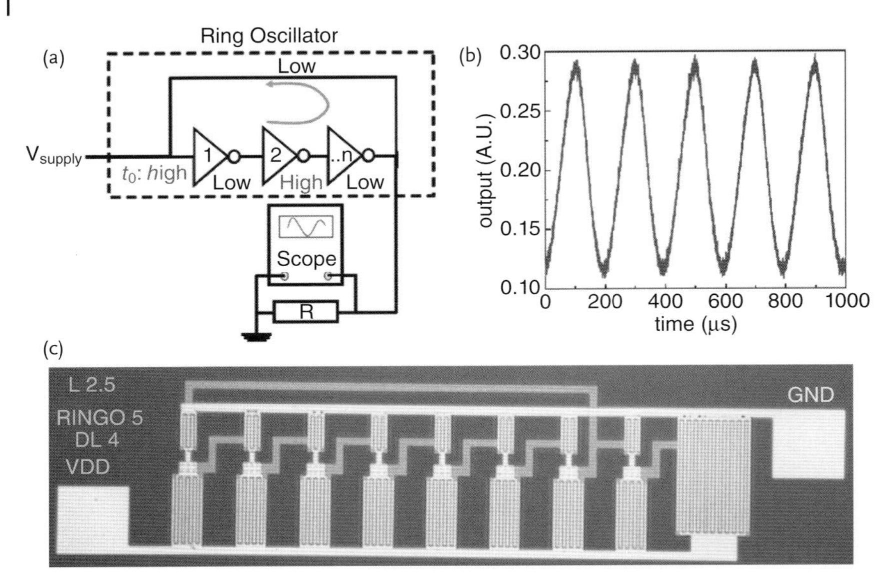

Figure 4.40 (a) Schematic overview of the measurement set-up for the ring oscillators; (b) A SAM FET-based integrated ring oscillator oscillating at a frequency of 5 kHz with a supply voltage of −10 V; (c) Optical photograph of a seven-stage ring oscillator based on SAM FETs [87].

noise margin is about 1 V. The small parameter spread in mobility and threshold voltage has allowed investigators to combine SAM FET inverters into seven-stage ring oscillators, which have a switching frequency of 5 kHz at a supply voltage of −10 V (see Figure 4.40). A buffer stage was added to determine the switching frequency.

Real logic functionality has been demonstrated with the realization of 15-bit code generators [87]. In this case, the integrated circuits combine over 300 SAM FETs, and contain an onboard clock generator, a hard-wired memory, a four-bit counter, decoder logic, and a load modulator. An optical image and a plot of the output of a 15-bit code generator are shown in Figure 4.41, where the bit rate is about 1 kbit s^{-1} at a supply voltage of −40 V. The circuit performance is similar to that of state-of-the-art organic integrated circuits developed for organic radiofrequency identification transponders [93]. As the SAM FETs and circuits are fabricated on an Si/SiO_2 substrate which cannot be applied in flexible electronics [87], the gate dielectric group was changed from SiO_2 into a polymeric [88]. The subsequent polymer gate dielectric was activated using a oxygen plasma to impose a driving force between the dielectric and molecules toward self-assembly. Seven-stage ring oscillators were also fabricated from the SAM FETs with a polymer gate dielectric. An oscillation frequency, *f*, of about 2 kHz was obtained at a supply voltage of −32 V. Four-bit code generators were fabricated that contained a clock generator (the ring oscillators), a hard-wired memory, a four-bit counter, decoder logic, and

(a)

(b)

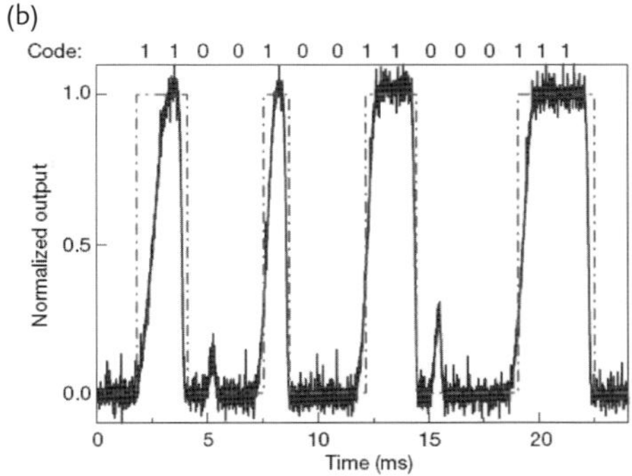

Figure 4.41 (a) Optical photograph of a functional 15-bit SAM FET code generator. The circuit combines over 300 SAM FETs; (b) Output of a 15-bit code generator based on SAM FETs. The bit rate is about 1 kbit s^{-1} at a supply voltage of −40 V. The outputted code is indicated at the top, and by the red line [87].

a load modulator. In this case, the bit rate was about 0.5 kbit s^{-1} at a supply voltage of −42 V [88].

4.3.2 Printed Organic Circuits

A major target in the creation of low-cost, flexible electronics has been the development of solution-processable, functional materials that can afford both the direct printing of electronic components (including resistors, diodes, capacitors, transistors, and interconnects) and high performance in devices [94–100]. The motivation to utilize printed electronics has include the potential economy of high-throughput printing methods, and the compatibility of printing with roll-to-roll processing of large-area plastic or paper substrates appropriate for signage, displays, or distributed sensing [100].

In 2000, Sirringhaus *et al.*, at the Cavendish Laboratory, described the direct inkjet printing of complete organic circuits, including *via*–hole interconnections, that was based on solution-processed polymer conductors, insulators, and self-organizing semiconductors [100]. These were the first organic circuits for which all components could be continuously fabricated by successive solution-deposition and printing steps, within the same environment. The source/drain electrodes gap was fabricated by confining the spreading of water-based conducting polymer ink droplets on a hydrophilic substrate with a pattern of narrow, repelling, hydrophobic surface regions, which define the critical device dimensions (Figure 4.42a). Following the inkjet printing of source-drain electrodes, FET devices are fabricated in a top-gate configuration (Figure 4.42b) by spin-coating a continuous film of the active semiconducting polymer, poly(9,9-dioctylfluoreneco-bithiophene). Subsequently, a 400 to 500 nm-thick film of the gate dielectric polymer, polyvinylphenol

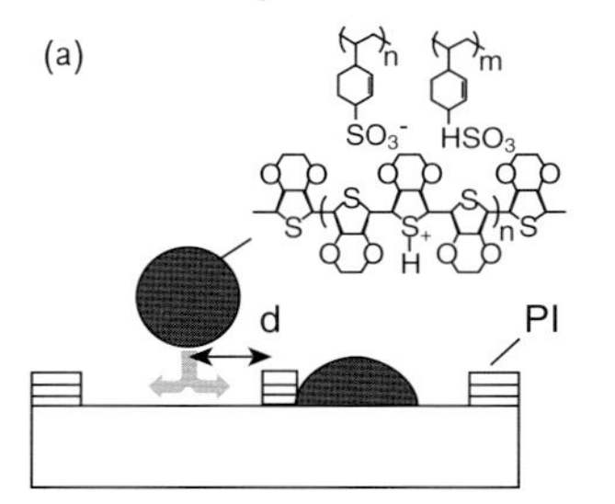

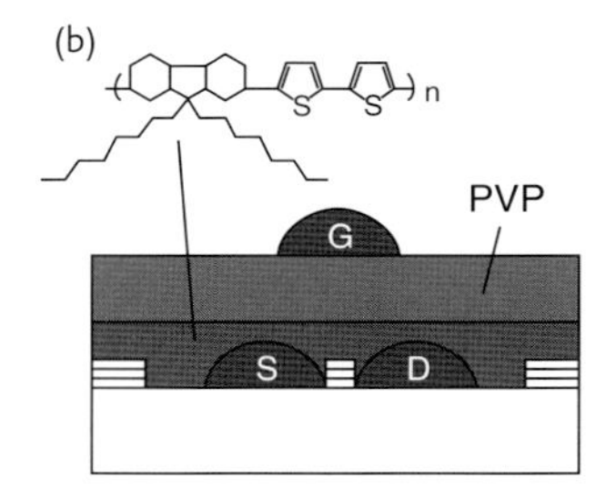

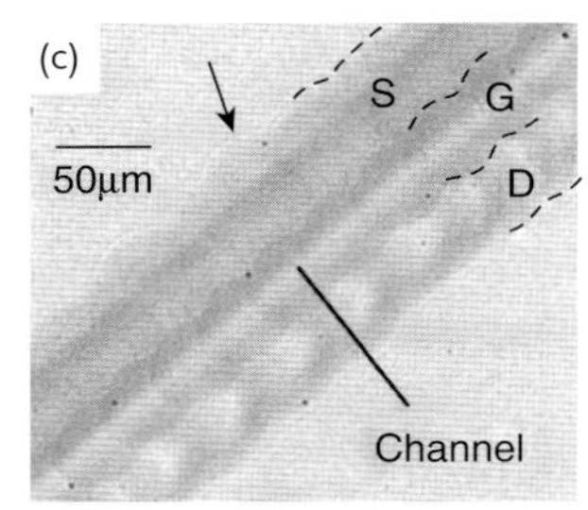

Figure 4.42 (a) Schematic diagram of high-resolution inkjet printing onto a prepatterned substrate; (b) Schematic diagram of the top-gate inkjet printing FET configuration (S, source; D, drain; G, gate); (c) Optical micrograph of an inkjet printing FET. The image was taken under crossed polarizers [100].

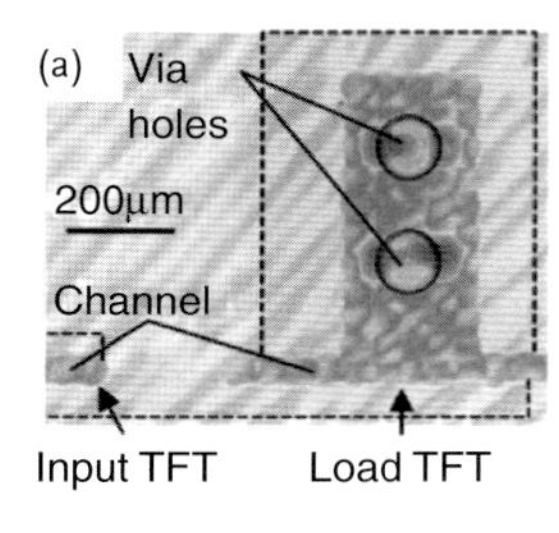

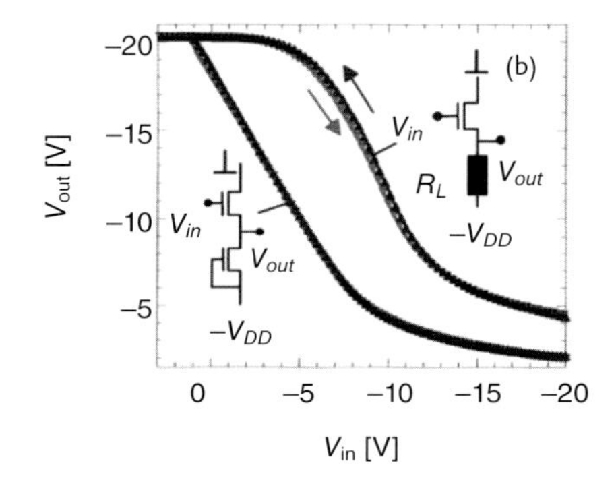

Figure 4.43 (a) Optical micrograph of two via-holes connecting the drain and gate electrodes of the load FET in an enhancement-load inkjet printing inverter. Some reddish interference contrast is seen due to small spin-coating-induced thickness variations of the PVP dielectric; (b) Static characteristics of an enhancement-load inverter and of a resistance-load inverter measured with increasing (red) and decreasing (blue) input voltage V_{in} (negative supply voltage $V_{DD} = 20\,V$) [100].

(PVP), is spin-coated on the semiconductor, and finally the gate electrode is inkjet-printed [100]. The inverter of an enhancement–load configuration, in which the drain and gate of the load transistor are connected together through a *via*-hole, is appropriate for inkjet-printed transistors (Figure 4.43a,b). Alternatively, inverters with a printed resistor as the load element have been used. In both configurations, a clean inverter action is observed for switching between logic "1" (−20 V) and logic "0" (0 V). The hysteresis of the characteristics was small, which reflected the stability of the transistor threshold voltage. The performance of the inkjet-printed, all-polymer FET circuits is believed to be adequate for applications such as active-matrix displays or identification tags [100].

Most printable organic circuits are fabricated via polymers, mainly because they are easier to dissolve. In addition to the developments of Sirringhaus *et al.* (as noted above), Cho *et al.* at the University of Minnesota used printable polymer electrolytes known as "ion-gels" as high-capacitance gate insulators in organic FETs. Ion-gel dielectrics hold substantial promise for flexible electronics applications by virtue of their extremely large capacitances, printability, and suitable frequency response [99]. The semiconductor, gate dielectric and gate electrode of transistors are printable, and the work voltage of device is lower than 5 V. Consequently, Cho and colleagues fabricated resistor-loaded inverters on a flexible polyimide substrate based on the printed devices, and an inverter action was demonstrated. The output voltage of the printed inverter responded well to a 1 kHz square-wave input voltage signal [99]. In 2004, Knobloch *et al.*, at PolyIC GmbH & Co. KG, Germany, designed the structural and electrical properties of printed polymeric thin films and multilayers to set up *p*-type FETs and circuits [101]. In this case, the source, drain and gate electrodes of devices were printed, while the semiconductor and gate dielectric of the devices were spin-coated. However, the seven-stage ring-oscillator showed only a maximum oscillation frequency of 0.86 Hz at −90 V. In contrast to the above reports, Tan *et al.*, at the Nanyang Technological University, fabricated complementary organic circuits by evaporating *n*-type $F_{16}CuPc$ and inkjet printing *p*-type polymer semiconductors instead of unipolar devices [102]. Although, in this case, a complementary circuit structure was used, the five-stage ring oscillator shown a oscillation frequency of 0.73 Hz at 100 V.

Although small-molecule semiconductors are easy to purify and have better field-effect characteristics than polymers, most of them are not printable. However, in 2011, Di and Gao *et al.* at the Institute of Chemistry and Shanghai Institute of Organic Chemistry, Chinese Academy of Sciences, synthesized a series of new printable air-stable *n*-type small molecules and used them to fabricate high-performance FETs and circuits [103]. The devices and circuits could be fabricated on different substrates, including Si, glass, and polyethylene terephthalate (PET). The electrodes, semiconductor and gate dielectric could all be printed, and the printed devices also showed a high performance. An example of an all-solution-processed flexible *n*-channel transistors array with an inkjet-printed active layer is shown in Figure 4.44a. The devices showed good saturation, sharp turn-on characteristics, and a high mobility of up to $0.45\,cm^2/(V\cdot s)$ (Figure 4.44b,c), confirming the good quality of the active films and the excellent interface between the organic film and the polyacrylonitrile (PAN)/polymethylsilsesquioxane (PMSQ) dielectric layer. Moreover, the all-solution-processed devices fabricated under ambient conditions showed outstanding operating and environmental stabilities. By virtue of the excellent device performance and remarkable stability of all-solution-processed *n*-channel devices, it is possible to create all solution-processed organic circuits based on a single *n*-type semiconductor, with all fabrication processes carried out in air. Organic inverters with single NDI2OD-DTYM2-based transistors were built on both glass and PET substrates (the inverter structure is shown in the inset of Figure 4.44e). An inverter switch response was clearly observed from logic "1"

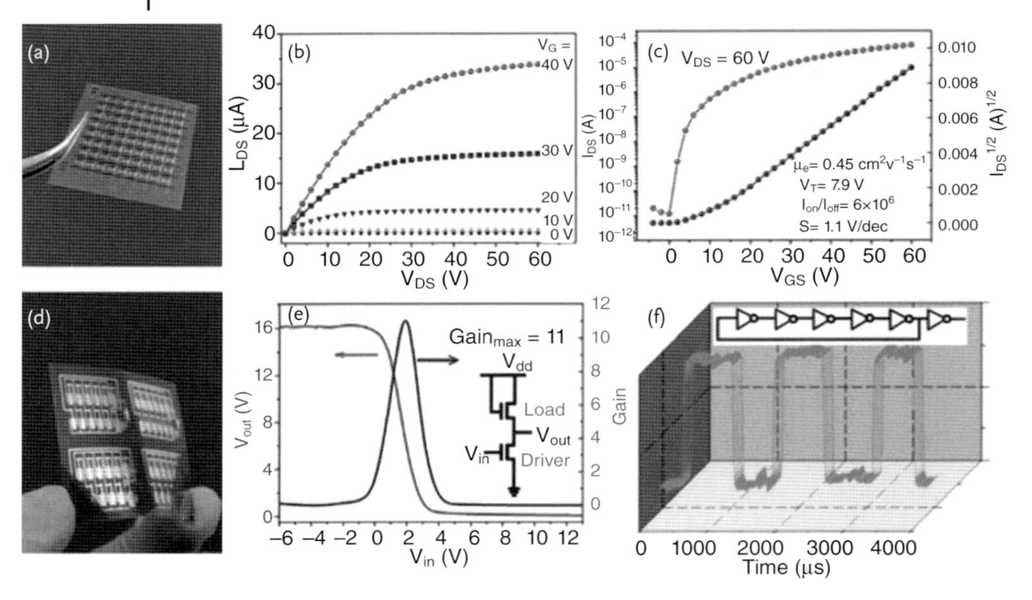

Figure 4.44 (a) Optical image, (b) output characteristics, and (c) transfer characteristics of all-solution-processed flexible OFETs; (d) Optical images of a flexible five-stage organic oscillator. Electrical characteristics of (e) an organic inverter and (f) an organic oscillator [103].

(16 V) to logic "0" (0 V) (Figure 4.44e). Voltage gains higher than 10 were obtained routinely for all of these flexible devices, implying their potential application in more complex logic circuits. Further integration was also successful, yielding a five-stage flexible ring oscillator with a switching frequency >1.2 kHz at a rail voltage of 30 V (the oscillator structure is shown in the inset of Figure 4.44f).

4.4
Circuits of Organic Crystals

Compared to organic thin films, single crystals can demonstrate intrinsic charge-transport properties and a high mobility because of: (i) the perfect order of molecules; and (ii) the absence of any grain boundaries and the minimized concentration of charge traps in single crystals, which provides the opportunity to reveal the relationship between the microscopic molecular packing and macroscopic charge transport of materials [104]. Moreover, the study of organic single crystals is also crucial for first, the fabrication of high-performance devices, and then to obtain high-performance organic circuits comparable to those based on silicon.

During recent years, enormous efforts have been expended to bring about significant progress in the growth of organic single crystals, the fabrication of single crystal FETs, and investigations into the structure–property relationships of organic semiconductors based on the crystals. However, having been limited by the device fabrication technologies, very few groups have fabricated single-crystal circuits successfully. The first organic circuit based on single-crystal transistors of *p*-type tetramethylpentacene (TMPC) and *n*-type *N,N′*-di[2,4-difluorophenyl]-3,4,9,10-perylenetetracarboxylic diimide (PTCDI) was reported by Briseno *et al.* at the University of California-Los Angeles in 2006 [105]. In this case, the single crystals were grown by horizontal physical vapor transport (PVT); the TMPC single crystals were typically 3×3 mm in size, with thicknesses ranging from 400 nm to 2 μm, while the PTCDI crystals were typically 0.5×2 mm in size and usually about 1–3 μm thick. Organic single crystals were electrostatically bonded across source-drain electrodes (bottom-contact configuration). The organic single-crystal inverter device configuration is shown in Figure 4.45a. The *p*-type TMPC single-crystal device showed a field-effect mobility of $1.03\,cm^2/(V \cdot s)$, while the *n*-type PTCDI showed a field-effect mobility of $0.006\,cm^2/(V \cdot s)$. The transfer characteristics of the single-crystal inverter are shown in Figure 4.45b. Despite the fact that there was a large mismatch in charge carrier mobilities between the two

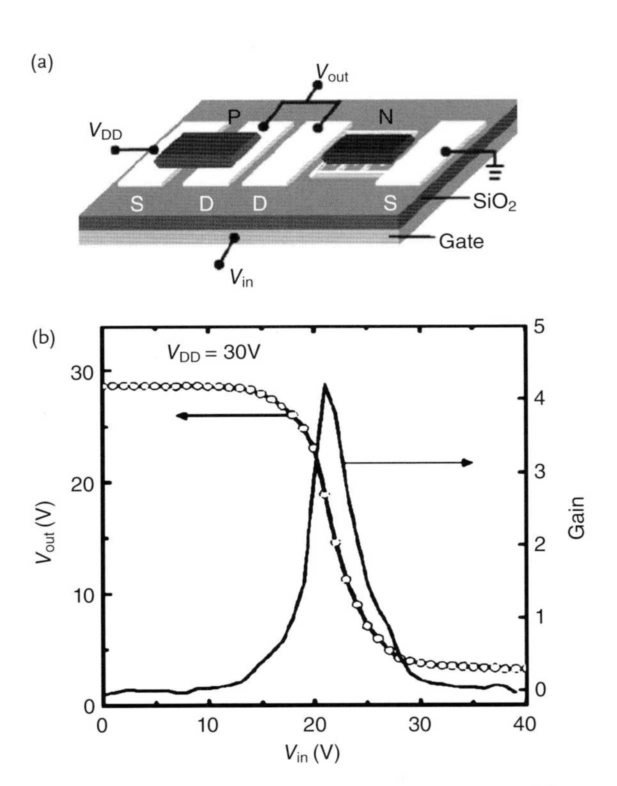

Figure 4.45 (a) A schematic complementary inverter circuit devised from a *p*-type TMPC and a *n*-type PTCDI single-crystal transistor; (b) The inverter and gain characteristics acquired from TMPC and PTCDI transistors, which were measured at room atmosphere. The supply voltage is $V_{DD} = 30$ V [105].

semiconducting materials, a well-defined transfer curve could still be observed. The gain of the inverter was 4.2 [105].

Uemura and colleagues, at Osaka University, also applied the bulk organic single crystal to build an inverter [106]. For this, rubrene was chosen to build both transistors and inverters, and the mobility of the single crystals was reported to be as high as $20\,cm^2/(V\cdot s)$ or even higher [107, 108]; consequently, a 100 MHz clock frequency could be predicted with micrometer-channel devices, assuming a standard charging model. Usually, rubrene single-crystal FETs are *p*-type when combined with the source electrode of a noble metal such as gold; however, they may be *n*-type if a low-work-function metal such as calcium is used for the electrode [109]. The monolithic complementary circuit presented by Uemura *et al.* was fabricated on the rubrene single crystal, using both gold and calcium electrodes to inject the holes and electrons, respectively. The configuration of the monolithic complementary inverter is illustrated schematically in Figure 4.46a, while

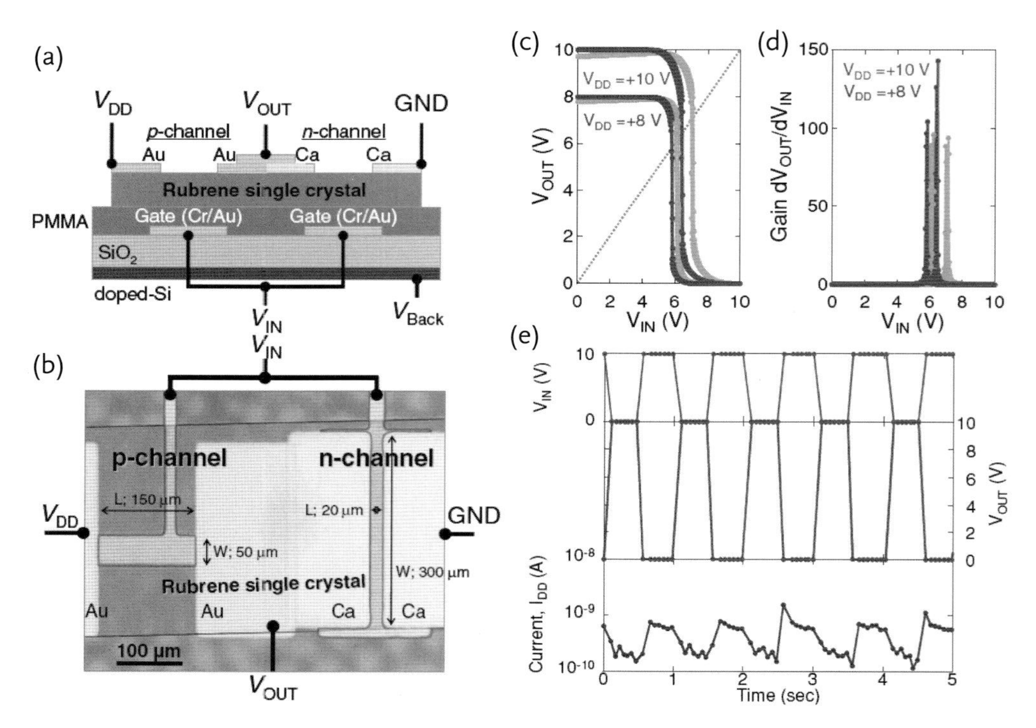

Figure 4.46 (a) Schematic illustration of a monolithic complementary inverter using a rubrene single crystal; (b) Optical view of the typical device. The *p*- and *n*-channels were separately fabricated on the same rubrene crystal; (c) Voltage transfer characteristics of a monolithic complementary inverter, applying a back-gate voltage of +10 V; (d) Corresponding DC values of inverter switching; (e) Demonstration of repeat switching of a monolithic complementary inverter applying a back-gate voltage of +10 V. Square waves of input voltages to V_{IN} terminal (red curve). Output voltages detected at V_{OUT} terminal (blue curve). Penetration current through the complementary inverter (green curve) [106].

an optical image of a typical device is shown in Figure 4.46b [106]. Both, *p*- and *n*-channels were fabricated on the same rubrene single crystal simply by patterning the electrodes with gold and calcium, and the doped-Si was used as a back gate to tune the characteristics of the transistors and circuit. The hole and electron mobilities were estimated at 1.2 and 0.2 $cm^2/(V \cdot s)$, respectively, which represented standard values for the top-contact rubrene single-crystal transistors [109]. The characteristics of the inverter switching with the application of the back-gate voltage of +10 V are shown in Figure 4.46c,d, where the characteristics without back-gating are plotted in gray. The voltage transfer characteristics with a supply voltage V_{DD} of +8 V and +10 V are presented in Figure 4.46c; these data indicate that the monolithic crystal inverter does indeed operate correctly, with minimal hysteresis. A DC gain of up to 140 at V_{DD} = +10 V, which was comparable to values reported for high-performance complementary inverters composed of organic semiconductors, is shown in Figure 4.46d [50, 56, 57, 66]. The switching performance of the inverter, repeated at 1 Hz with the application of V_{Back}, is demonstrated in Figure 4.46e. In this case, the inverter could be seen to consume only a few nW in the static state and, indeed, to realize a logic inversion with minimum power consumption [106].

Micro/nanocrystals of organic semiconductor possess all the advantages of bulk single crystal, while their small size is conducive to the miniaturization and integration of devices and circuits. The first inverter based on organic nanowires [*n*-type PTCDI and *p*-type hexathiapentacene (HTP)] was reported by Briseno *et al.* at the University of Washington in 2007 [110]. Here, the nanowires were grown via solution-phase self-assembly, after which the FETs were built by drop-casting nanowire/methanol solutions onto gold interdigitated electrodes that had been fabricated by conventional photolithography. The inverters exhibited near-symmetrical voltage-transfer curves and a gate switching with a gain of 8 [110]. The group of Hu, at the Institute of Chemistry, Chinese Academy of Science, completed a series of investigations on FETs based on organic micro/nano single crystals, and also investigated the circuit characteristics of these micro/nano crystals [32, 111, 112]. Tang *et al.* used conducting SnO_2:Sb nanowires as electrodes, CuPc nanowires as *p*-channel, and $F_{16}CuPc$ nanowires as *n*-channel to build the pure nano complementary inverter, NAND and NOR circuits [32]. The organic single-crystal nanowires were grown by PVT, and the circuits were built by means of nanomechanical manipulation to transfer the nanowires to the substrate; the gain of inverter was 7. Subsequently, by using nine nanowires of CuPc, $F_{16}CuPc$, and SnO_2:Sb as building blocks, logic NOR and NAND circuits were assembled (Figure 4.47a) that each demonstrated a low operational voltage (<10 V), low current (static current <4 pA) and an ultralow power consumption (<40 pW). The fact that the power consumption of the single-crystal logic circuits was even one order of magnitude lower than that of organic thin-film circuits (1 nW per logic gate) [33] was of great interest for possible future applications [32].

Jiang *et al.* grew di(phenylvinyl) anthracene (DPV-Ant) single-crystal disks by applying a PVT technique in a two-zone, horizontal-tube furnace [111]. The typical size of the crystalline disks ranged from several micrometers to tens of

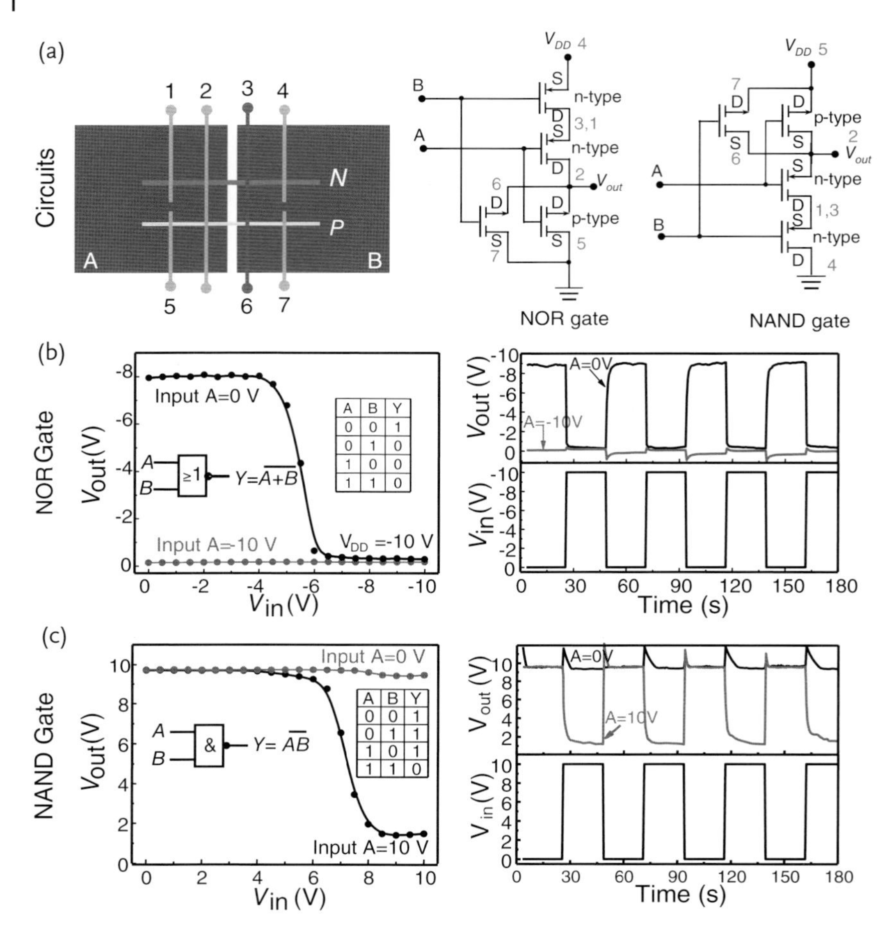

A	B	Y
0	0	1
0	1	0
1	0	0
1	1	0

A	B	Y
0	0	1
0	1	1
1	0	1
1	1	0

Figure 4.47 Organic single-crystal logic circuits of NOR and NAND gates fabricated by using nanowires of CuPc, $F_{16}CuPc$, and SnO_2:Sb as building blocks. (a) Circuits of NOR and NAND logic gates; (b) Transfer and dynamic switching characteristics of a NOR gate at $V_{DD} = -10\,V$; (c) Transfer and dynamic switching characteristics of a NAND gate at $V_{DD} = 10\,V$ [32].

micrometers, with the thickness varying from 18 nm to several hundred nanometers. DPV-Ant is a *p*-type semiconductor, and the highest mobility of DPV-Ant single-crystal FETs was observed as $4.3\,cm^2/(V\cdot s)$. Digital inverters consisting of two top-contact *p*-type single crystal transistors based on two micro-disk were fabricated in this way. The circuit diagram of the inverter is shown in Figure 4.48a, and a schematic of the corresponding device in Figure 4.48b. The inverter operated well over a wide voltage range, even at a low V_{DD} of −2.5 V, and showed ideal transfer curves with excellent logic-level conservation (as shown in Figure 4.48c).

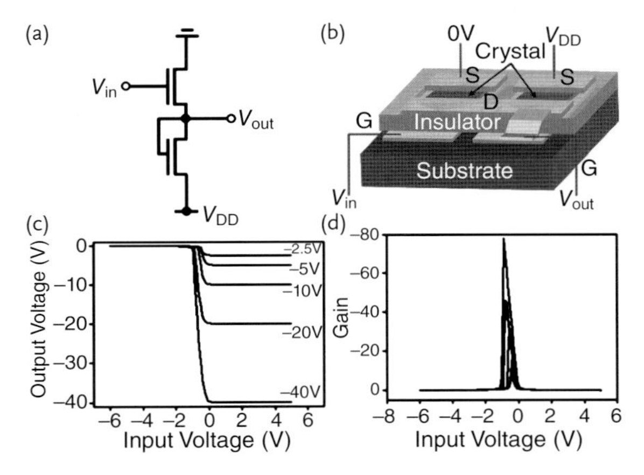

Figure 4.48 (a) Circuit diagram of the inverter; (b) Device schematic of the inverter; (c) Transfer characteristics of the inverter operated at supply voltages between −2.5 V and −40 V; (d) Signal gain of the inverter [111].

The maximum gain was about 80 for $V_{DD} = -40\,V$ [111]. Because only a few air stable *n*-type organic nanocrystals were available, Zhang *et al.* introduced a "hybrid" structure based on organic (*p*-type) and inorganic (*n*-type) single crystalline nanoribbons for the fabrication of organic bipolar FETs and inverters. In this case, the maximum gain of the inverters reached 29 [112]. Taken together, these results indicate that organic nano single crystals may serve as good candidates in organic electronics.

4.5 Single-Molecule Transistors

As early as in 1974, Aviram and Ratner had predicted that, one day, individual molecules would be used as circuit elements in devices and, indeed, the dream of utilizing molecules as functional units in electronic circuits has motivated research groups for many years [113]. If this suggestion were to come true, it would be possible to integrate transistors in billions. Molecular-based devices possess unique advantages for electronic applications [114], such as lower cost, lower power dissipation, higher efficiency, ability of self-assembly and recognition, distinct optical and electronic properties, and a synthetic tailoring ability via an elaborate choice of geometry and composition. A range of specific electronic functions performed by single molecules, including rectifiers [115, 116], switches [117, 118] and transistors [119–121], has been accordingly designed and reported. All of the above aspects render molecules as ideal candidates for the next generation of electronics [122].

4.5.1
Fabrication of Single-Molecule Transistors

There are two main approaches for wiring molecules between electrodes in order to fabricate devices. The first method is to create top-contact junctions, including scanning probe microscopy (SPM), scanning tunneling microscopy (STM) and conducting atomic force microscopy (C-AFM) [123–134], cross-wire junctions [135–137], mercury drop electrodes [138, 139] and thermally deposited metal films [117]. All devices manufactured via these methods can be categorized as "prototype devices," and are valuable in fundamental investigations, having already provided many important results [115–118, 123–143]. However, at present these devices are far from practical application, as a nanometer-sized device carrying a huge SPM system or other systems cannot be imagined. The second method involves the use of nanogap electrodes to form metal/molecule/metal devices [144–148].

4.5.1.1 Fabrication of Single-Molecule Prototype Devices

4.5.1.1.1 Single-Molecule Devices Based on Thermally Deposited Metal Film

As a typical example, in 2000 Collier and his coworkers from the University of California at Los Angeles fabricated the device from a single monolayer of a bistable [113] catenane that was anchored with amphiphilic phospholipid counterions and sandwiched between an *n*-type polycrystalline Si (poly-Si) electrode and a Ti/Al top electrode [117]. The bottom electrodes were 7 μm-wide *n*-type poly-Si (resistivity of $0.02\,\Omega\,cm^{-1}$). Typically, poly-Si films formed by direct chemical vapor deposition (CVD) growth onto SiO_2 are neither smooth nor defect-free. However, as amorphous Si films can be very smooth, it was used as a starting point for the fabrication of smooth poly-Si electrodes. The molecule layer was first deposited on the poly-Si electrodes using Langmuir–Blodgett (LB) technology. Following LB-monolayer deposition, the 10 mm-wide top electrodes (50 Ti followed by 1000 Al) were deposited onto the LB film using electron-beam evaporation through a shadow mask to complete the device fabrication. The device structure was stable and easy to fabricate, while the compactness of the single-molecule layer was crucial for the device yield. During the top electrode deposition, the metallic atoms easily diffuse through the pinhole of the molecular layer, leading to a connection with the bottom electrode and failure of the device.

4.5.1.1.2 Single-Molecule Devices with Scanning Probe Microscopy

STM observations performed under high-vacuum conditions can allow stable measurements of electron transport through a single-molecule layer. In 1996, Andres and coworkers at Purdue University, employed double-ended aryl dithiols (α,α′-xylyldithiol [XYL] and 4,4′-biphenyldithiol) to form SAMs on gold(111) substrates, and these were used to tether nanometer-sized gold clusters deposited from a cluster beam [123]. An ultra-high-vacuum STM system was then used to image these nanostructures and to measure their current–voltage characteristics as a function of the separation between the probe tip and the metal cluster (see

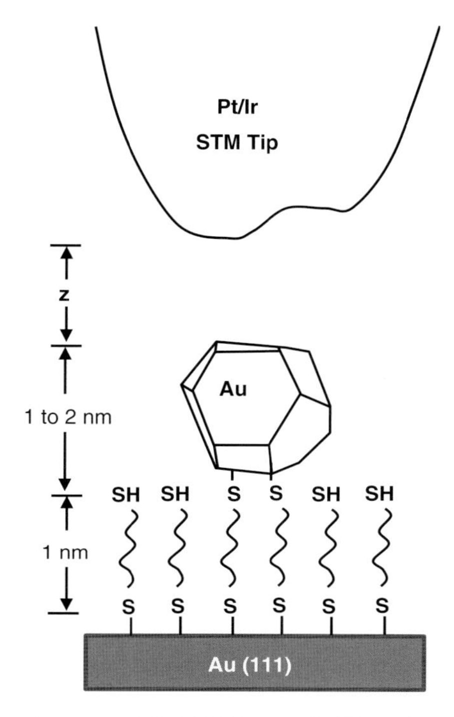

Figure 4.49 Dithiol SAM on an Au(111) substrate with an individual crystalline Au cluster tethered to the surface of the SAM. This self-assembled nanostructure allows the measurement of $I(V)$ data by a scanning transmission microscope (STM) tip [123].

Figure 4.49). The single-molecule double-barrier junction can be defined by positioning the microscope tip over an individual molecule adsorbed directly onto the bottom electrode surface [132]; alternatively, the individual molecular junctions can be built by repeatedly moving a gold scanning tunneling microscope tip into and out of contact with a gold substrate in a solution containing the sample molecules. The sample molecule with two anchored atoms on its two ends can bind strongly to gold electrodes to form a molecular junction [129].

4.5.1.1.3 Single-Molecule Devices with Crossed-Wire Junctions

Crossed-wire tunnel junctions were pioneered by Gregory in 1990 [140], and have been used previously to measure the inelastic electron tunneling spectra of molecular adsorbates [140, 141], as well as to study Coulomb blockade [142] and Kondo-type scattering [143]. Subsequent experiments conducted in 2002 by Kushmerick and coworkers at the Naval Research Laboratory constituted the first measurements of electron transport across organic monolayers with a crossed-wire tunnel [135]. A schematic representation of a crossed-wire tunnel junction is shown in Figure 4.50a, where the 10 μm-diameter wires, one of which has been modified

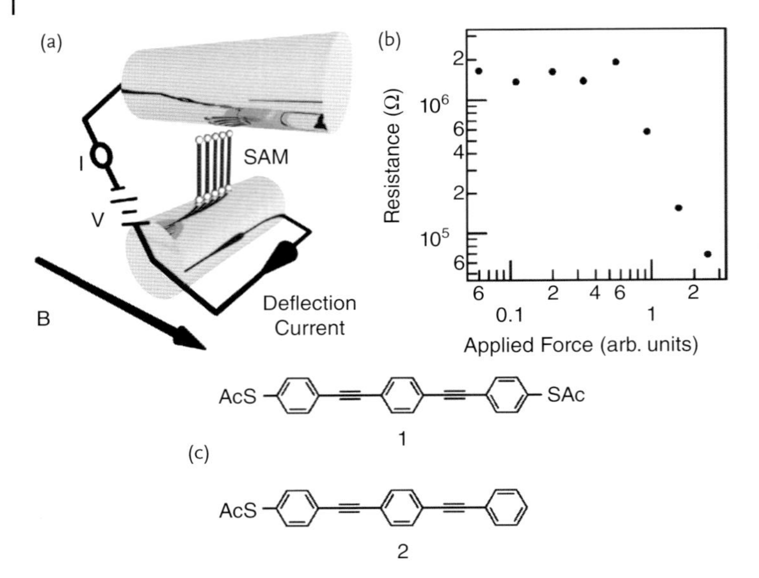

Figure 4.50 (a) Schematic representation of the experimental set-up (not to scale). The *I–V* characteristics of the junction are obtained by ramping the bias voltage (*V*) while monitoring the current flow (*I*) across the junction. In all cases, the bias voltage is applied to the wire that was initially modified with the SAM. All measurements were made in a nitrogen-purged Faraday cage at room temperature; (b) Junction resistance (calculated from the linear fits of the *I–V* characteristics at low bias voltage) as a function of applied force for a Au–1–Au junction; (c) Structure of the molecules investigated in this study [135].

with a SAM of the molecule of interest, are mounted onto a custom-built test stage in such a manner that the wires are in a crossed geometry with one wire perpendicular to the applied magnetic field (B). The junction separation is controlled by deflecting this wire with the Lorentz force generated from a small DC current (<5 mA). This deflection current is slowly increased to bring the wires gently together, forming a junction at the contact point. At low applied forces the junction resistance (calculated from linear fits of the *I–V* characteristics at low bias voltage) is constant (Figure 4.50b). Above a certain threshold force, the monolayer distorts, leading to a drop in resistance. This experimental approach enables the formation of metal–molecule–metal junctions with a high degree of control over the nature of the metal–molecule connection and the choice of metal used [135].

4.5.1.1.4 **Single-Molecule Devices with Mercury Drop Electrodes**

Mercury drop electrodes were first introduced in 1998 by Rampi and his coworkers at the University of Ferrara to form an Hg–SAM/SAM–Hg junction [138]. A micro-syringe body (2.5 mm inner diameter) was used to maintain the area of the two

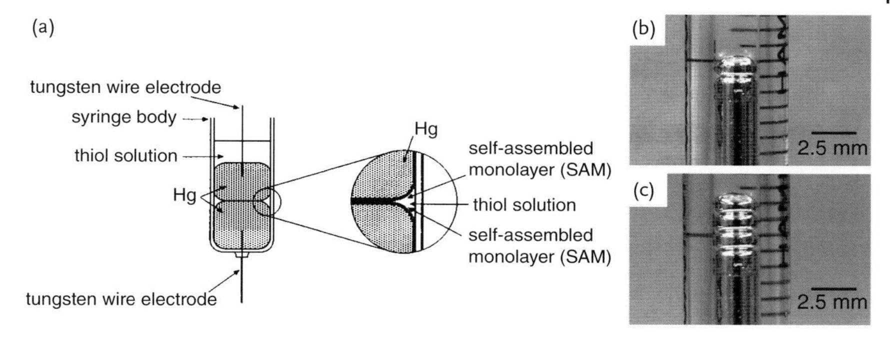

Figure 4.51 (a) Schematic representation of the geometry of the Hg–SAM/SAM–Hg junction; (b) Photograph of the system schematized in (a): one capacitor is formed by the Hg–SAM drop in contact with the underlying Hg–SAM surface; (c) Photograph of the system where three Hg–SAM drops are stacked and create three capacitors in series [138].

facing Hg surfaces constant; a thiol solution was then transferred to the syringe to form the first Hg–SAM interface. A calibrated drop of mercury (5 μl), previously stored in the thiol solution, was then rolled along the reclined syringe wall, bringing the second Hg–SAM surface gently into contact with the first surface. Electrodes (50 μm diameter tungsten wire) were inserted into the drops. A typical system is shown in Figure 4.51.

4.5.1.2 Fabrication of Single-Molecule Transistors by Nanogap Electrodes

Compared to the "prototype devices," the devices based on nanogap electrodes with conductive metal nanowire circuits and functional molecules inserted in the desired position have the potential ability to realize superintegrated circuits, and so are more likely to be used in practical applications. Moreover, because the nanogap electrodes are fabricated before the molecular components are subsequently inserted, and the junction can be characterized with and without molecules in place, this facilitates the distinction of the intrinsic molecule properties. In addition, as most nanogap electrodes present in a planar configuration it would be easier for higher-density integration and to take the underlying substrate as a gate contact to tune the electrical properties of the molecular components. Due to their superior characteristics and promising future, nanogap electrodes have attracted worldwide attention for almost 20 years, with many reports and significant progress having been made, ranging from fabrication techniques to device operations.

As the typical dimensions of target molecules are well below 5 nm, the fabrication of electrodes with separations that are suitable for specific molecules is a major challenge, because it extends beyond the capability of traditional microfabrication technologies. Also, if the gap were too small, it would be very likely to posit the molecule in a tense and distorted state, and result in unexpected performances. Consequently, a precise control of the spacing makes it more difficult to prepare these types of device. Despite these obstacles, several effective and creative methods for the fabrication of nanogap electrodes with controlled spacing have been reported during recent years, including mechanical break junctions [149], electron-beam lithography [150], electrochemical plating [151], electromigration [152], focused ion beam lithography [148], shadow mask evaporation [120], scanning probe and AFM lithography [153], on-wire lithography [154], and molecular rulers [155]. Whilst each of these methods has provided promising results and has its own characteristics, different methods are frequently combined to achieve a desired configuration. For example, electron-beam lithography is often used first to obtain nanogap electrodes with spaces at 10–20 nm, after which other techniques (e.g., shadow mask evaporation) may be implemented to further narrow the gap width to the 1–5 nm scale.

4.5.1.2.1 Mechanical Controllable Break Junctions

The mechanical controllable break (MCB) junction was first introduced by Moreland and coworkers at the US National Bureau of Standards, to form an electron tunneling junction [156]. Subsequently, this method was adopted creatively by Reed *et al.* at Yale University to fabricate nanogap electrodes, yielding electrodes with separations of several nanometers [149, 157]. A schematic of the sample mounting for the MCB technique is shown in Figure 4.52. Here, a notched metallic wire is first glued to an elastic substrate, which serves as the bending beam. The substrate is then bent by pushing at its center with a driving rod; as a consequence the notched wire is fractured, such that an adjustable tunneling gap can be established. The surfaces can then be brought together again, with the distance being controlled by a piezoelectric element. The breaking process is generally conducted under low-temperature and high-vacuum conditions, and this guarantees two atomically clean surfaces. Although, in principle, any elastic substrate can be used, a phosphor bronze substrate [144, 157, 159] is often chosen because it is more flexible, it is compatible with fine electron-beam lithography, and it allows a much wider range of metals to be studied. In this case, an insulating layer is deposited on the substrate before transfer of the metallic nanostructure, using

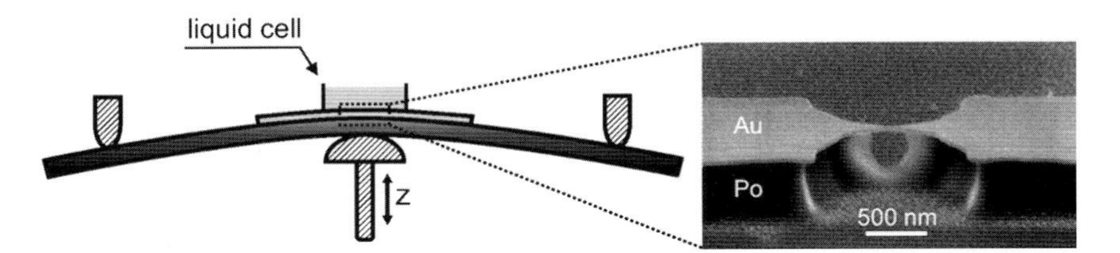

Figure 4.52 Schematics of the MCBJ principle with a liquid cell and a scanning electron microscopy (SEM) image of the central part of the microfabricated Au junction. Po is a polymer insulating layer [158].

standard electron-beam lithography techniques. The notched section is usually made with a scalpel, by rolling it over the filament, although other methods may also be applied depending on certain properties of different metals; for example, a spark method was used when dealing with the semimetal Sb [160]. To ensure that the notched section was unglued, a selective etching (e.g., isotropic reactive-ion etching) [144] of the insulating layer was often made to produce a metallic bridge that was suspended between the anchoring pads. The MCB technique is very stable (down to $0.2\,\mathrm{pm\,h^{-1}}$) [159], and the contact size can be continuously adjusted under the control of an ultra-fine piezoelectric, without polluting the junction. When the target molecule is anchored between the gap, the mechanical stress is also under control.

Currently, MCB junctions are widely used for single-molecule, even single-atom devices, including one-atom metallic point contacts [158–167] and metal–molecule–metal junctions [144, 149, 157, 168–174]. Many of the pioneering studies in this field were conducted by van Ruitenbeek *et al.* [159–166], especially regarding the creation of atomic-size contacts and tunnel junctions. In this case, the metallic point contacts were made by bringing two freshly prepared electrodes back into contact under precise adjustment of the piezo element. As with single-molecule devices, the molecules are often self-assembled to integrate between the metal leads. When Reed and Tour *et al.* [149] successfully prepared benzene-1,4-dithiol molecules self-assembled on a gold wire before the MCB process, a mechanical breakage of the wire in solution produced two opposing gold contacts that were SAM-covered. The tips were then moved slowly together until an onset of conductance was achieved, indicating that a single-molecule junction had been created (Figure 4.53). Subsequently, by using the MCB technique, Riel *et al.* [173] investigated the conductance switching behavior of single bipyridyl-dinitro-oligophenylene-ethynylene dithiol (BPDN-DT) molecules contacted by two symmetric leads. Here, the metal–single molecule–metal system could be controlled and reversibly switched between two distinct states, even after performing more than 500 positive and negative sweeps iteratively. On comparing these data with the experimental results obtained with bipyridyl oligophenylene-ethynylene dithiol (BP-DT) molecules, Riel *et al.* concluded that the switching behavior was caused by the nitro groups of BPDN-DT, thereby eliminating the effects of electrode properties or molecule–metal interfaces. More recently, Tian and Tao *et al.* [174] notably combined MCB with a surface-enhanced Raman scattering (SERS) technique to characterize molecules in a nanogap with a continuously adjustable gap width. In this case, the SERS intensity depended critically on the gap width and the incident light polarization, which in turn indicated a successful detection of the sample molecules inside the gap.

Although the MCB method is very useful for fundamental investigations, such as electronic transport at the molecular scale, it is more difficult to fabricate highly integrated molecular devices because of the constraint of the piezoelectric components; neither is it easy to build devices with three or even more electrodes. The MCB approach also appears to be difficult to control when fabricating relatively large gaps.

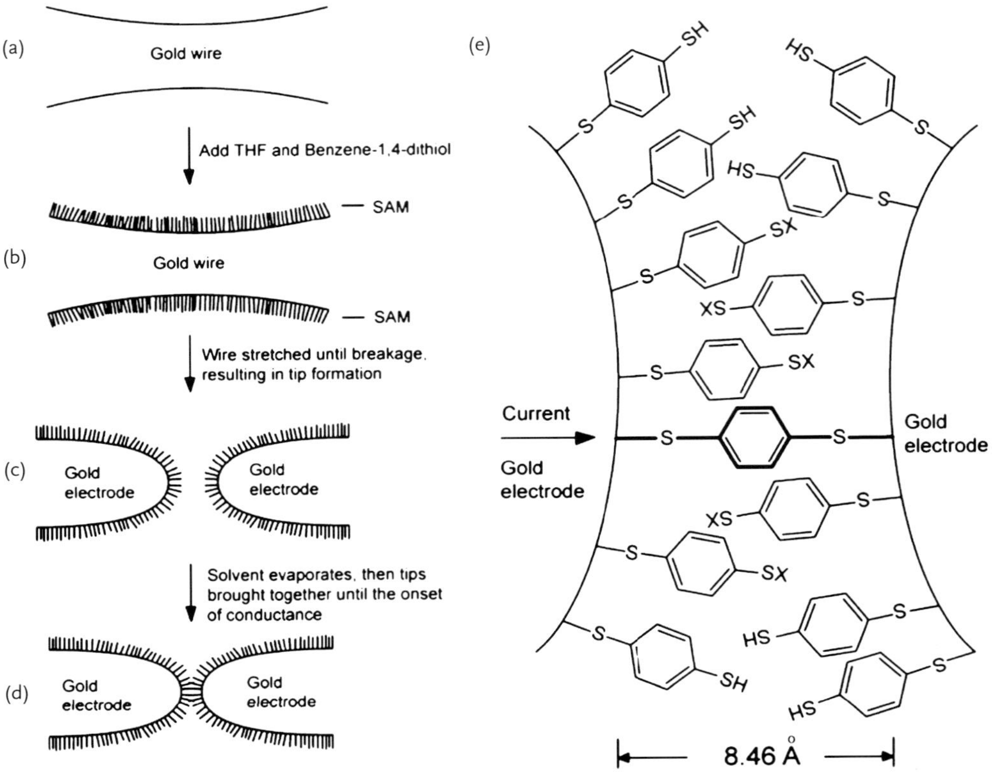

Figure 4.53 Schematic of the measurement process. (a) The gold wire of the break junction before breaking and tip formation; (b) After addition of benzene-1,4-dithiol, SAMs form on the gold wire surfaces; (c) Mechanical breakage of the wire in solution produces two opposing gold contacts that are SAM-covered; (d) After the solvent is evaporated, the gold contacts are slowly moved together until the onset of conductance is achieved. Steps (c) and (d) (without solution) can be repeated numerous times to test for reproducibility; (e) A schematic of a benzene-1,4-dithiolate SAM between proximal gold electrodes formed in an MCB. The thiolate is normally H-terminated after deposition; end groups denoted as X can be either H or Au, with the Au potentially arising from a previous contact/retraction event. These molecules remain nearly perpendicular to the Au surface, making other molecular orientations unlikely [149].

4.5.1.2.2 **Electrochemical and Chemical Deposition for Nanogap Electrodes**

Electrochemical and chemical deposition methods, when combined with standard lithography techniques, provide a simple, accurate, and reproducible means of fabricating nanogap electrodes. The initial electrodes with a relatively large gap are created by conventional lithography technique on substrates (e.g., Si/SiO_2), after which the gap is narrowed down to nanometer or even atomic scale by depositing specific atoms onto the lithographically defined electrodes. The process can also be reversed in dissolution mode to provide a controlled etching of atoms from the junction to solution, thus widening the gap back to the submicrometer scale.

Electrochemical and chemical deposition methods require no special technique or system [175], and can be used to conveniently prepare gaps ranging from several Angstroms to 10 nm, which is the best scale for fitting desired molecules or nanocrystals for nanodevices. The simplicity and robustness of the technique also provides a promising means of fabricating large-scale and highly integrated nanodevices. Furthermore, the gap dimension can be simultaneously monitored and precisely controlled by using a feedback system which, when taking electroplating method as an example, is frequently the current that is flowing through the gap electrodes and is monitored during the deposition process [151, 176–183]. When the electrodes are very close but not yet touching, the monitor current is extremely sensitive to the electrode distance; consequently, it is easy to control the separation on an atomic scale by stopping the electrodeposition process at predefined conductance values. Unlike the current-feedback mode, which uses both facing electrodes as working electrodes (WEs), Tian *et al.* [175] developed a method for the controllable electrochemical fabrication of electrodes with a nanometer/Angstrom-sized gap by using the potential distribution in the electric double layer as feedback, wherein a design that two facing electrodes served as the WE and the reference electrode (RE), respectively. The potential difference between the WE and the RE (V_{gap}) was monitored continuously during the electrodeposition process. The value of V_{gap} was maintained constant until the gap narrowed below a certain value, after which it decreased with time and finally approached zero, indicating a conducting state. This result suggested that V_{gap} could also be used as a feedback signal to control the gap width. Liu *et al.* [147, 184] fine-tuned the electrode gap width by utilizing high-frequency impedance in feedback. In this case, the dependency of the AC voltage frequency on the final gap size was determined, and gaps of 30 nm were obtained by using high-frequency feedback signal, with high accuracy and reproducibility. The gap size was affected by the frequency and amplitude of the AC signal, and could be fine-tuned to about 1 nm simply by controlling the subsequent deposition time (Figure 4.54).

Tao *et al.* [185] achieved excellent atom-size gaps and contacts between electrodes fabricated with a self-terminated electrochemical method, based on a built-in self-termination mechanism. In this case, an external resistor (R_{ext}) was connected to one of the electrodes, and the final gap width could be predetermined by elaborately choosing $1/R_{ext}$ compared to the conductance quantum ($G_0 = 2e^2h^{-1}$). The gap resistance (R_{gap}) was shown to decrease with time, and resistances in series (R_{ext} and R_{gap}) shared the total applied bias voltage (V_0). Finally, when R_{gap} was much smaller than R_{ext}, V_{gap} approached 0, which meant that almost no applied voltage was used for etching and deposition, and the fabrication process self-terminated. Another advantage of electrochemical deposition method is the ability to fabricate asymmetric nanogap electrodes (e.g., a gold and a platinum finger) and nanogap electrodes with three or more fingers [178], as shown by Kashimura

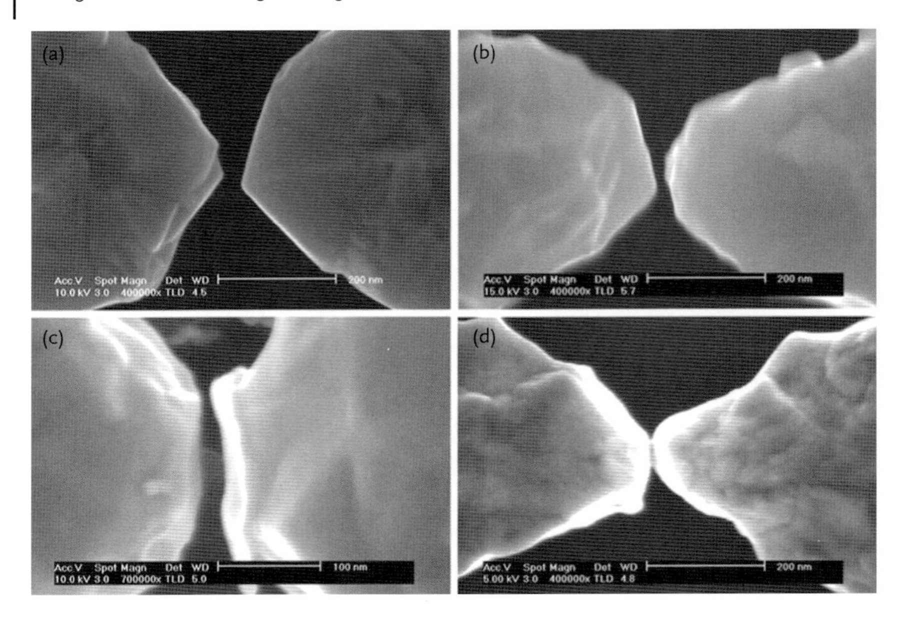

Figure 4.54 SEM images of the samples prepared at 3 KHz with different Δt values: (a) $\Delta t = 9$ s, d = 26 nm; (b) $\Delta t = 25$ s, d = 16 nm; (c) $\Delta t = 42$ s, d = 7 nm; (d) $\Delta t = 62$ s, d ≈ 1 nm [147].

and Torimitsu *et al.* Although the electrodes were difficult to achieve using conventional lithographic techniques, they were superior to those produced via the MCB or SPM techniques, as discussed above.

When performed with gold, the main drawback of electroplating lies in the difficulty of obtaining a simple gold electrolyte, while special care should also be taken to avoid generating highly toxic hydrocyanic acid gases. Accordingly, Umeno and Hirakawa [186] introduced a simple and cost-effective method in which "iodine tincture" was used as an electrolyte for gold electroplating, and was free from toxic compounds or strong acids. Kervennic *et al.* [176] showed that, compared to gold deposition, Pt promised a higher reproducibility and stability of the electrodes, and pairs of platinum electrodes with separations of between 20 and 3.5 nm were obtained with relative ease.

Compared to the electroplating method, surface-catalyzed chemical deposition represents a simpler method for narrowing gap electrodes that have been prepatterned by conventional lithography, as it does not even have to carry an external circuit. By localizing a catalyst on the initial electrodes, the metal atoms can be selectively deposited onto the electrode surface, assisted by reductive agents. The only requirement is to immerse the chips in a stock solution containing metal ions and a mild reducing agent; the gap distance will depend on the reaction time and reactant concentrations, such that this method has controllability and mass-producibility. Gu and coworkers [187] first modified the gold leads with

Figure 4.55 (a) Optical and (b) field emission scanning electron microscopy (FESEM) images of an array of the nanogaps with sub-5 nm separations after the surface-catalyzed chemical deposition. FESEM images of two-fingered nanogap electrodes with magnifications of (c) ×40 000 and (d) ×320 000 [188].

2-mercaptoethylamine, the amino groups of which bind Pd(II) particles for catalyzing the deposition of metallic copper, and this resulted in an electrode nanogap of 45 nm. Later, Yun *et al.* [188] reported the creative fabrication of integrated nanogap electrodes of a few nanometers separation, with a yield of over 90% (Figure 4.55). These authors proposed that the Au deposition rate of the gap edge was smaller than that of the other edges of the electrodes, mainly because of the mass-transport limitation in the narrow gap; consequently, the plating process was slowed down as the electrodes approached the tunneling distances. Multi-fingered nanogap electrodes can also be conveniently prepared via an electroless plating method [189]. If the fabrication process of chemical deposition can be monitored in real-time, and the main controlling factors (e.g., reaction time and reactant concentration) are deemed to be more theoretical than empirical, then a much better controllability of this method can be expected.

4.5.1.2.3 **Oblique Angle Shadow Evaporation for Nanogap Electrodes**

The oblique angle shadow evaporation method was pioneered by Dolan [190] in 1977, but subsequently adopted and refined for the fabrication of nanogap

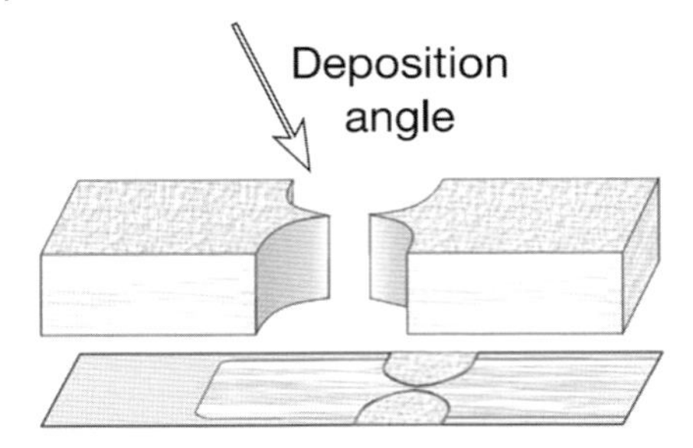

Figure 4.56 Oblique angle deposition with an elevated mask for preparation of two metal electrodes spaced to a few nanometers [120].

electrodes. Shadow evaporation is often combined with optical and electron-beam lithography to define the metal leads [120, 191–193]. By suspending the mask above the substrate and controlling the direction of the deposition angle to be oblique against the surface normal (Figure 4.56), feature sizes smaller than those of the masks can be obtained perfectly, as reported by Bjornholm and colleagues [120]. Gap lengths between the electrodes of less than 10 nm can be reproducibly fabricated in this way. Moreover, the gap size can be adjusted by changing the tilt angle stepwise until a desired space is obtained [193], and *in situ* sample conductance measurements can help to find the correct angle [194], which endows this method with a high degree of controllability.

If the substrate is uneven, it can provide a similar effect as an elevated mask, and can simplify the fabrication process; accordingly, the double-angle evaporation [195, 196] of thin metallic films was introduced. In this case, the separation of the electrodes can be adjusted by simply changing the film thickness. Furthermore, even without using conventional lithography methods, nanogap electrodes can also be accurately fabricated. Kawai *et al.* [197] developed a method based on oblique angle shadow evaporation to construct nanogap electrodes without a wet process, in order to measure the electrical characteristics of molecules after they had been positioned on the substrate (Figure 4.57). The gap length between the electrodes was easily controlled under 100 nm, while a top-contact configuration prevented any structural deformation of the molecules. This method proved to be especially suitable for the reliable fabrication of nanogap arrays for molecular electronics. For example, Sun *et al.* [198] achieved the fabrication of gap electrodes arrays as small as 3 nm in large quantities, and applied these successfully to electrical studies of nanocrystals. The oblique angle shadow evaporation method is frequently conducted at a low temperature [120] (e.g., in liquid helium), since under such conditions the grain sizes of the metal deposits will be relatively small, and this will result in the gap width remaining closer to the designed value. In this sense, other improvements – such as the use of metals with smaller granules,

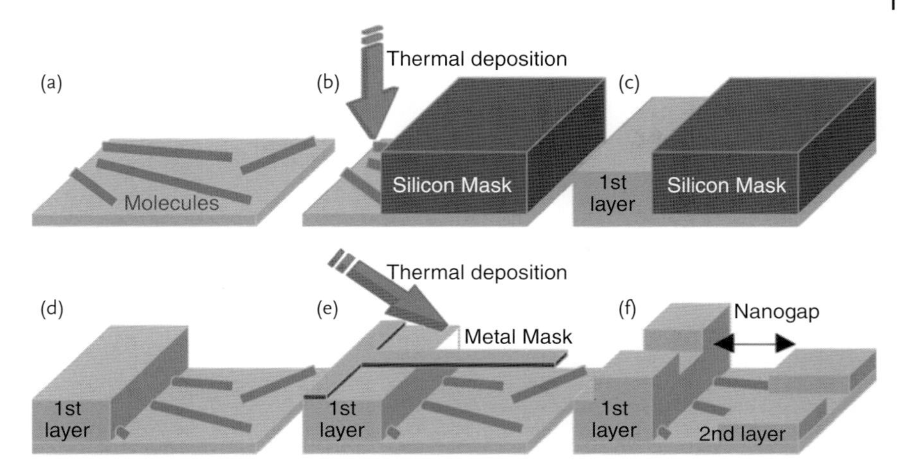

Figure 4.57 Fabrication procedure of angle-controlled shadow-masking method. The width of the metal mask shown in (e) is actually much wider than shown in the illustration [197].

decreasing the film thickness, and adjusting the evaporation rate – can be made to enable a narrower gap-width distribution [195].

4.5.1.2.4 **Electromigration and Electrical Breakdown Method for Nanogap Electrodes**

Electromigration for Nanogap Electrodes The phenomenon of electromigration has been known for more than 100 years [199]. In an applied electric field, the momentum of the moving electrons can be transferred to the ions that make up the lattice of the interconnect material, and this results in a gradual movement of the ions. Although electromigration has long been considered as the major failure mode in microelectronic circuitry [200], it has more recently been widely used to fabricate nanogap electrodes for nanodevices [201–206]. The passage of a large density current or the application of a large DC voltage to the thin metal wire that had been predefined by electron-beam lithography would cause the electromigration of metal atoms and an eventual breakage of the nanowire. This process can yield a stable electrode separation of 1 nm with high efficiency [201], and can be monitored in real time by observing the current–voltage characteristics until only a tunneling signal is present [152, 201]. Electromigration is especially advantageous for the preparation of three-electrode devices. Typically, a gate electrode can be readily fabricated on the substrate before the electrical break is carried out (Figure 4.58) [146, 207].

Furthermore, by refining the electromigration procedure, nanogap electrode arrays can be reliably obtained. For example, Johnson and colleagues [208] developed a feedback-controlled electromigration technique for the parallel formation of nanogap electrodes (Figure 4.59). In this case, by mounting the

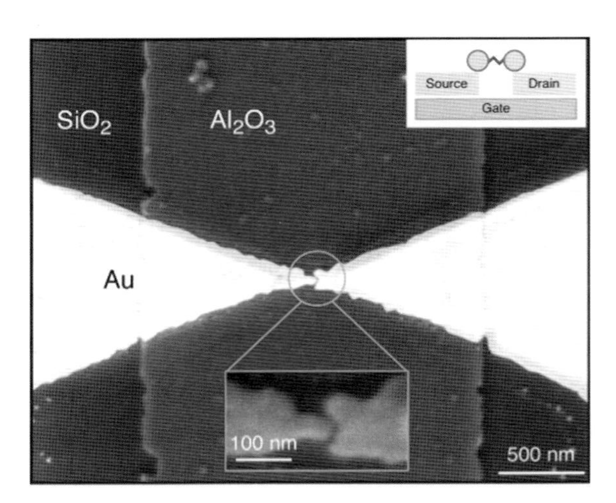

Figure 4.58 Scanning electron microscope images (false color) of the metallic electrodes fabricated by electron-beam lithography and the electromigration-induced break-junction technique. The image shows two gold electrodes separated by ~1 nm above an aluminum pad, which is covered with a ~3 nm-thick layer of aluminum oxide. The whole structure was defined on a silicon wafer. The bright yellow regions correspond to a gold bridge with a thickness of 15 nm and a minimum lateral size of ~100 nm. The paler yellow regions represent portions of the gold electrodes with a thickness of ~100 nm, Inset: schematic diagram of single-molecule transistors containing individual divanadium molecules [146].

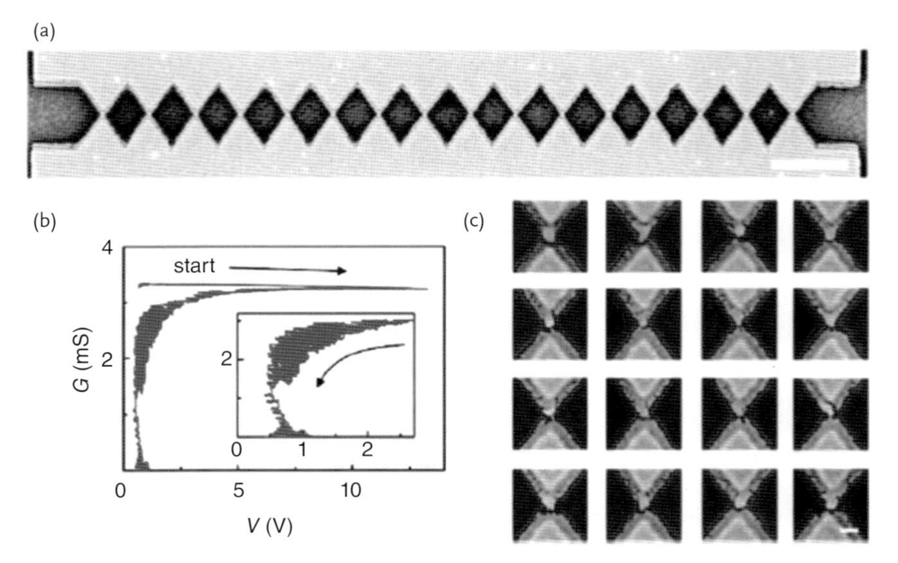

Figure 4.59 Feedback-controlled electromigration (FCE) of a 16-junction array. (a) Scanning electron microscope (SEM) image of an array of 16 junctions made by electron-beam lithography and shadow evaporation. Scale bar = 1 μm; (b) G–V data from FCE of the 16-junction array; (c) SEM images of nanogaps formed by parallel FCE of a 16-junction array, clearly showing the gold removed from the thin overlap junctions. Scale bar = 100 nm [208].

interjunction resistance smaller than the individual junction resistance, the problem was solved that uniform nanogaps could not be formed consistently due to an uneven power dissipation and, as a consequence, successfully achieved a parallel electromigration of nanogaps. This was considered highly promising for the construction of large-scale circuits of molecular devices.

As discussed above, electromigration provide an advantageous means of preparing three-electrode nanodevices, with the substrate as a gate. Subsequently, Park and coworkers made some great advances, including the creation of single-electron transistors based on colloidal cadmium selenide nanocrystals [201], and provided evidence of the nanomechanical oscillations of the C_{60} molecule against the gold electrode in a single-C_{60} transistor [119]. Coulomb blockade and the Kondo effect in single-atom or single-molecule transistors was also observed [146, 152]. Recently, van der Zant *et al.* reported some important findings in this field, such as electron measurements through single-molecule magnets [209], the observation of a Kondo effect in gold break junctions in the presence of magnetic impurities [210], in addition to studies of the temperature dependence of three-terminal molecular junctions, based on sulfur end-functionalized tercyclohexylidenes [211].

Nonetheless, two major issues remain to be addressed with regards to electromigration procedures:

- The first problem relates to the role of temperature in the breaking of a nanowire. It is generally accepted that electromigration is a thermally assisted process [205, 212], and some Joule heating is required in order for gap formation to begin. However, excessive heating should be avoided because it would cause an undesired melting of the metal [213]. Thus, by minimizing the total series resistance of the system, the temperature can be kept low during the electromigration process, and in this way a controllable formation of nanogaps smaller than 2 nm can be conveniently achieved [214].

- The second problem stems from the possible metal debris that might remain in the gap after fabrication, which would interfere with the insertion of molecules of interest and very probably mask the desired signals [210]. Usually, the residue is difficult to image because the resolution of scanning electron microscopy (SEM) is below 1 nm. Accordingly, Strachan *et al.* [215] prepared nanogaps on free-standing transparent SiN_x membranes which permitted the use of transmission electron microscopy (TEM) to image the gaps. As a consequence, the junctions were found to be clean and particle-free.

Electrical Breakdown of Carbon Nanotubes for Nanogap Electrodes Compared to thin metal wires (~20 nm) that can only be prepared using electron-beam lithography, carbon nanotubes (CNTs) possess sufficient conductivity, more favorable configurations, and better contacts to organic molecules via C–C bonding [216], for them to be considered as good substitutes for metal wires in the fabrication of nanogap electrodes. Moreover, due to their strong C–C bonds, CNTs have current-carrying capacities that are in excess of 10 μnA nm^{-2} [217], which is much greater than that of noble metals (~10 nA nm^{-2}). Initially, Avouris and coworkers [218, 219] found that, when applying a high-density current through multiwalled carbon nanotubes (MWNTs), failure did not occur due to electromigration, but that a layer-by-layer breakdown of the individual carbon shells eventually resulted in the formation of nanogaps within the nanotubes. Subsequently, this phenomenon was adopted to fabricate nanogap electrodes, and referred to as the "electrical breakdown" method. Later, Liu *et al.* [220] utilized the same method to cut CNTs successfully to form an initial gap of 10–60 nm, after which organic molecules were deposited onto the CNTs. Finally, an electron beam-induced decomposition (EBID) of the absorbed organic molecules resulted in an ingenious broadening of the diameter and thus a reduction in the gap size (Figure 4.60). When the EBID was operated in scanning electron microscopy, the *in-situ* characterization improved the accuracy of the fabrication of nanogap electrodes.

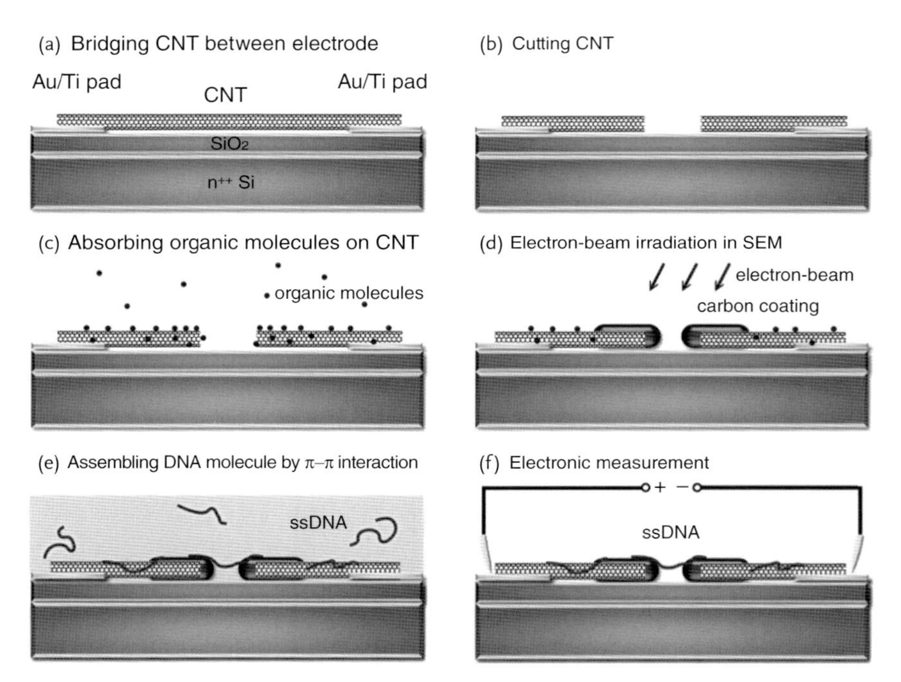

Figure 4.60 Schematic diagram of the process of the fabrication of a CNT electrode with a controlled nanogap (a–d) and the fabrication of a DNA electronic device (e, f). (a) Bridging a CNT between Au/Ti electrodes; (b) Cutting the CNT by current breakdown method; (c) Adsorbing organic molecules on or in the CNT; (d) Irradiating the gap of the CNT by electron beam with *in situ* observation in SEM; (e) Assembling DNA molecules between the CNT electrodes by immerging in DNA buffer solution; (f) Measuring the electronic properties of the bridged DNA molecules [218, 220].

By using the electrical breakdown method, OFETs with effective channel lengths down to 1–3 nm were fabricated by Dai *et al.* [221], and significantly improved switching characteristics were observed with single-walled carbon nanotube (SWNT) electrodes over conventional metal electrodes. Tsukagoshi *et al.* [222, 223] prepared organic nanotransistors using pentacene nanocrystals and MWNT nanogap electrodes, thus circumventing the problem of film discontinuity between gold electrodes and SiO_2 substrates in organic thin-film transistors.

4.5.1.2.5 Molecular Ruler and Nanostructure Template for Nanogap Electrodes

The main purpose of fabricating nanogap electrodes is to reliably configure metal electrodes with suitable separations that are capable of linking molecules or other nanometer-sized components. In contrast, molecules and nanostructures can in turn serve as templates for constructing the nanospaced electrodes. In this way, the gap dimensions are determined and easily tuned by the length of a certain molecule, the thickness of some atomic layer, or the width of a nanowire.

Weiss [155] innovated the fabrication of nanostructures (e.g., thin parallel nanowires) with mercaptoalkanoic acids as molecular templates (Figure 4.61).

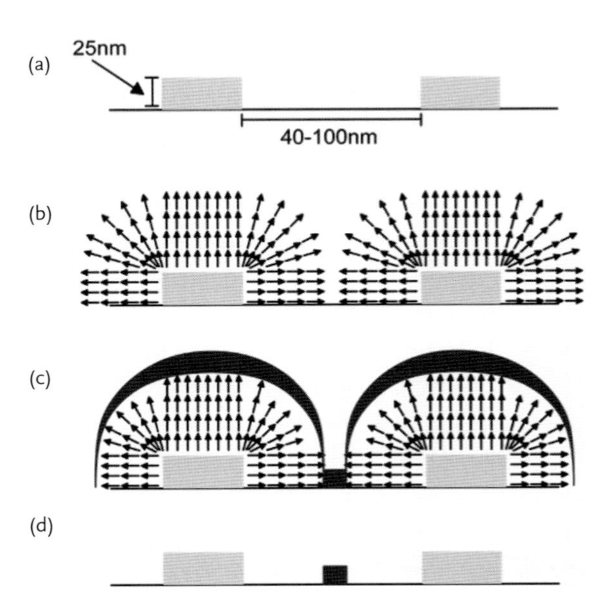

Figure 4.61 Schematic of the scaling-down process. (a) Electron-beam lithography (EBL) is used to form the parent structures (yellow rectangles); (b) Layer-by-layer construction of metal–organic resist (2 nm per layer). The arrows represent the mercaptoalkanoic acid (the tail is the SH group); Cu^{2+} ions are not depicted; (c) Metal (blue arcs and rectangle) evaporation into the gap (the gap size is determined by the number of organic layers deposited); (d) Lift-off of the organic layer in HCl/DMF solution. The thin Ti layer was omitted for clarity [155].

This method was later refined for the construction of nanogap electrodes [224–226]. Following metal deposition by electron-beam lithography and lift-off of the mercaptoalkanoic acid resist by chemical etching, nanogaps were formed with separations that were determined by the thickness of the mercaptoalkanoic acid layer. As molecular lithography provides a simple, highly reproducible and economical way of fabricating metallic electrodes separated on the molecular scale, it is especially promising for constructing arrays of molecular circuits [227]. Very recently, Bjornholm *et al.* [228] described an innovative method for the direct synthesis and growth of end-to-end linked gold nanorods using gold nanoparticle seeds that were linked with a dithiol-functionalized polyethylene glycol (SH-PEG-SH) linker, and which resulted in a nanogap with a size of 1–2 nm between two rods (Figure 4.62). The results of this study highlighted the possibility of fabricating nanostructures with a single or few molecules connected to two nanoelectrodes, by employing a bottom-up chemical assembly.

Alternatively, inorganic films [229–232] with precise atomic thickness separating two metal leads can also serve as templates (these are referred to as "atomic rulers"). In this case, the inorganic spacing layer should also be selectively removed in some way, much like the organic molecule rulers, while at the same time leaving the metal leads unaffected. By using molecular beam epitaxy, Tornow *et al.* [229, 230] were able to embed a thin gallium arsenide (GaAs) layer in between two aluminum gallium arsenide (AlGaAs) layers, with monolayer precision. The substrate was then cleaved to obtain an atomically flat surface, exposing the AlGaAs–GaAs–AlGaAs sandwich structure. Finally, after selective etching of the GaAs layer

Figure 4.62 High-resolution TEM images and schematic view of the linkage between two gold nanorods. The nanogap is 1–2 nm [228].

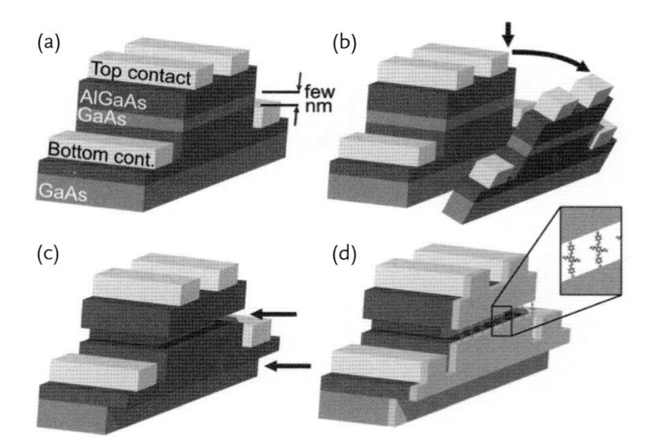

Figure 4.63 Schematic representation of the fabrication process for high-aspect-ratio nanogap electrodes (not to scale). (a) Molecular-beam epitaxy (MBE)-grown AlGaAs/GaAs heterostructure after conventional microstructuring with lithographically defined metallic leads; (b) Atomically smooth cleavage along the crystal plane; (c) Selective recess etching of the GaAs layers; (d) Complete device after thin-film metallization and subsequent deposition of molecules. The magnified section shows how several molecules bridge the gap. Molecules adsorbed on only one of the electrodes are omitted for clarity. Duplicate metallic leads connecting the bottom and top thin-film electrodes allow for a device operational test [230].

with citric acid and H_2O_2, a thin-film metal layer was deposited perpendicularly onto the cleaved surface (Figure 4.63). This high-aspect ratio configuration was potentially suitable for averaging molecular conductance measurements, provided that the target molecules were in a desired distribution, without interacting. Silicon nanogaps [233–235] for the connection of organic molecules into standard silicon technology (complementary metal oxide-semiconductor; CMOS) have also been exploited by selectively etching the intermediate layer of silicon dioxide. It should be noted at this point that, when the molecular or atomic templates have been removed, the final gap is unlikely to be the exact size of the ruler element due to a relaxing effect of the metal structures.

Recently, a template nanolithography method was developed for the processing of one-dimensional (1-D) wires with sacrificial materials located strategically along the long axis of the wire. Subsequent removal of the sacrificial segment resulted in gap electrodes ranging from a few nanometers to several hundred nanometers (Figure 4.64). In this technique – termed "on-wire lithography" [154, 237] – the 1-D wires (ideally of micrometer length to permit their easy manipulation and positioning) are prepared by template synthesis (e.g., anodic aluminum oxide template), using either electroplating or chemical polymerization. One well-studied configuration for on-wire lithography has been Au wires with Ag stripes [236, 238], whereby the thickness of the Ag segment dictates the size of the nanogap, which

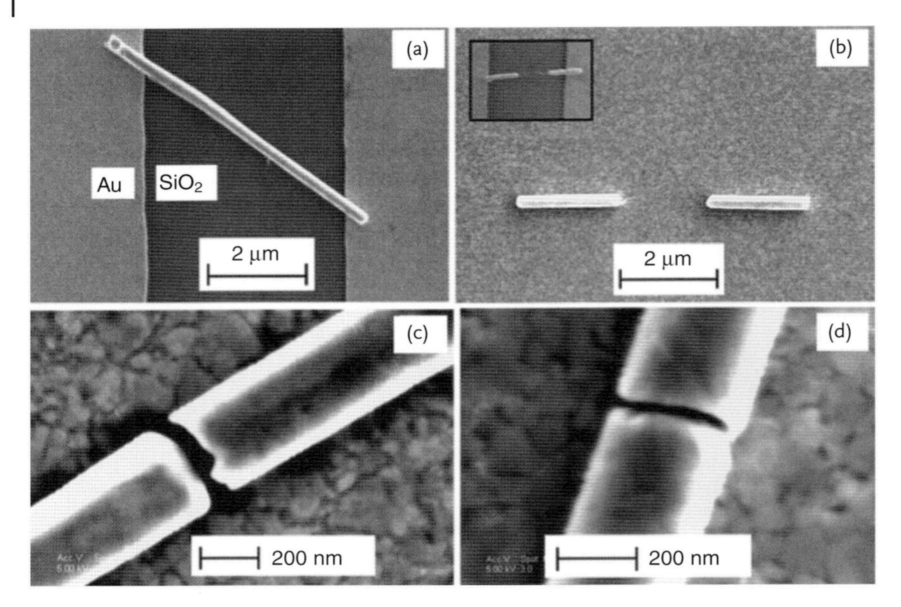

Figure 4.64 SEM images of (a) a nanowire lying across contact electrode pads and (b–d) gaps with different designed separations of 2 μm, 100 nm, and 20 nm, respectively. The inset in panel (b) shows nanowire electrodes connected with contact electrode pads [236].

can be easily controlled during the nanowire synthesis process [239]. The Ag spacer layer can be removed using high-temperature or chemical etching, and this results in gaps as small as 20 nm [236]. This method offers a precisely controlled and scalable means of preparing nanogap electrodes, of which the smallest gap reported was 5 nm [237]. The electrical properties of the nanogap junctions were detected [237] and OFETs of conjugated polymer molecules based on these nanogap electrodes were fabricated [236].

As CNTs with nanometer-size diameters and micrometer lengths demonstrate excellent mechanical stiffness and chemical stability, they have been considered a good choice for shadow masks when fabricating nanogap electrodes; moreover, CNTs can be easily combined with other micromachining processes for the batch production of nanogap electrodes. The diameter and spatial placement of the CNTs represent two key factors that contribute to a successful CNT shadow mask fabrication. The CNTs can be randomly dispersed via ultrasonication [240, 241], elaborately manipulated by electric field alignment [242–244], or positioned by direct mechanical transfer technique [245, 246]. After metal deposition and a successive lift-off process, nanogaps consistent with the diameters of the CNTs can be readily obtained. Subsequently, this method was refined by suspending the CNTs above the substrate [241, 246, 247], as this not only makes the gap dimensions more controllable but also allows a thicker deposition of metal films and cleaner nanojunctions. Johnson and coworkers [241] described a highly controllable and scalable method for fabricating nanogap arrays using suspended CNTs as the shadow mask, combined with a trilayer electron-beam lithography process. It should be noted that the incident angle and surface migration of the deposited atoms may greatly influence the ultimate dimension of the gap, however [248].

In addition to CNTs, other organic or inorganic nanowires (e.g., V_2O_5 nanowire) [249, 250] can also be utilized as shadow masks. Recently, the present author's group has developed an "organic ribbon mask" method [251] by which single-crystalline nanometer-sized ribbons were utilized as the template to determine the channel length of OFETs. This method offered a cheap, facile, and controllable way of fabricating single-crystal devices with short conducting channels. Moreover, the gaps between the metal leads can be easily scaled down to submicrometer level, depending on the width of the organic ribbons. More innovatively, Ongaro *et al.* [252] presented a thiol–DNA–biotin template method to assemble a protein-functionalized 10 nm gap gold electrode onto a silicon wafer substrate. In this case, the modified gold nanoparticles were selectively adsorbed onto the DNA backbone to form the electrode; the size of the formed gap was approximately 10 nm, and consistent with the dimensions of the proteins located between the DNA backbones. This approach opens up the possibility of building electronic nanodevices directly onto the molecular frameworks.

4.5.1.2.6 **Focused Ion Beam and Oxidative Plasma Ablation for Nanogap Electrodes**

Focused ion beam (FIB) lithography is a commonly used technique for fabricating nanostructures. Because the minimum diameter of a FIB is relatively small (ca. 5–20 nm), and the proximity effects of FIB are much less than that of electron-beam lithography, FIB might prove to be a reliable means of preparing nanogap electrodes, but with advantages of maskless nanostructuring, good reproducibility, and a high speed of nanogap cutting. Ochiai *et al.* [253] reported that, by cutting a wire of MWNT anchored by metal pads on a Si wafer with a focused Ga^{2+} ion beam, nanogap electrodes with a separation of approximately 50 nm could be obtained. Subsequently, by integrating the nanogap electrodes as source-drain electrodes with C_{60} as semiconductor, excellent FETs could be achieved. Leifer and coworkers [254] obtained large numbers of high-resistance and highly stable nanogaps electrodes on the order of 20 nm by using FIB combined with photolithography and electron-beam lithography. However, when considering the dimensions of the molecular device, the beam diameter appeared to be too large to fabricate nanogaps smaller than 10 nm. Later, Nagase *et al.* [148, 255, 256] introduced a Ti protective layer, which had a much lower etching rate than Au, and monitored the etching steps *in situ* by measuring a current fed to the Au electrodes, to form nanogaps that were much narrower than a FIB spot size (typically, nanogaps down to 3 nm were produced).

Similar to the FIB technique, oxidative plasma can also be used for the elaborate ablation of pre-patterned nanowires to form nanogap electrodes. In this field,

Nuckolls and Guo *et al.* [82, 257–261] have conducted many impressive studies, including the ingenious introduction of a precise oxidative cutting of SWNTs through a lithographic mask to produce ultra-small point contacts separated by only a few nanometers. Moreover, these point contacts were robust and could be utilized to fabricate devices with a variety of molecules, acting as pH sensors [257], chemoresponsive transistors [82], photogated switches [258], scaffolding for the assembly of biological macromolecules [259], and DNA hybridization sensors [260] (Figure 4.65). Notably, the technique of oxidative cutting may represent a new approach to creating devices on the molecular scale.

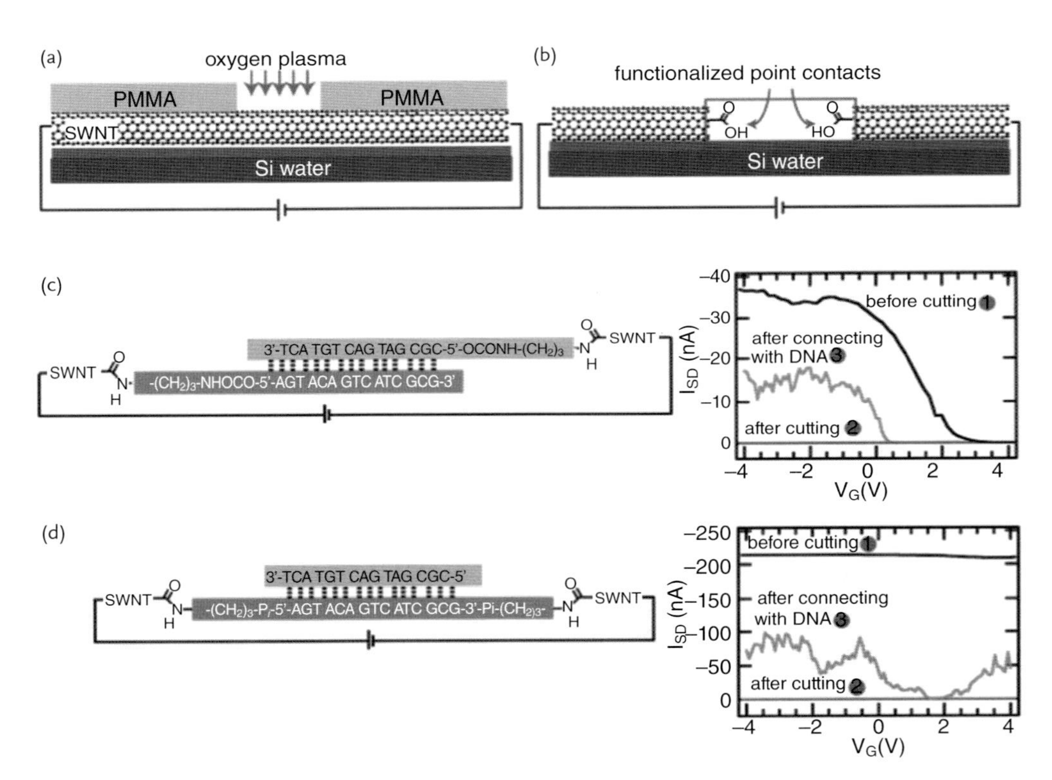

Figure 4.65 (a) Precise cutting of SWNTs with an oxygen plasma introduced through an opening in a window of PMMA defined with electron-beam lithography; (b) Oxidative opening of a tube produces two point-contacts functionalized on their ends with carboxylic acids and separated by as little as 2 nm; (c, d) Device characteristics for individual SWNTs connected with DNA. Source–drain current versus V_G at a constant source–drain voltage (50 mV) before cutting (black curve: 1), after cutting (red curve: 2), and after connection with the DNA sequence shown (green curve: 3), for a semiconducting SWNT device (a) and a metallic SWNT device (b). G, guanine; C, cytosine; A, adenine; T, thymine [257].

4.5.1.2.7 **Scanning Probe Lithography and Manipulation of Single Atoms**

Scanning probe lithography (SPL) describes a set of lithographic methods in which a microscopic or nanoscopic stylus is moved mechanically across a surface to form a pattern. Two main types of SPL have been developed:

- A destructive approach, in which the substrate surface is physically, chemically, or electronically deformed to generate the pattern by providing the substrate with energy [153, 262–264].
- A constructive approach, in which the patterning is achieved by directly transferring chemical species to the surface [264].

Because the sharp tips of the scanning probe microscope, with a radius smaller than 10 nm, can be neatly and controllably manipulated across the substrate, SPL may be utilized to construct nanogap electrodes.

The simplest method is to scratch the pre-patterned metal stripe with the scanning probe, so as to form a gap on the order of a few nanometers, as described by Notargiacomo *et al.* [153], with the atomic force microscope. Unfortunately, this method is not widely used due to the possibility of damaging the probe and also the unsatisfactory removal of material. In a second method, SPL was combined with electron-beam lithography and wet etching. Notargiacomo *et al.* have also reported a more complex set-up based on AFM-assisted mask patterning to achieve the smallest possible separation between electrode pairs [153]. Miyazaki *et al.* [265] also utilized the SPL technique to pattern the electron-beam negative resist, such that resist patterns were transferred to the underlying titanium film by wet etching. In this way, Ti nanowires with desired dimensions were easily obtained. Moreover, nanogaps can be fabricated on this line pattern by switching off the voltage bias between the AFM tip and the sample for a short period during the scanning process; this would result in an unexposed section that, eventually, would form the gap. A third method, termed dip-pen nanolithography (DPN), is of the "constructive" type. As an example of this approach, Mirkin and colleagues [266] successfully prepared sub-50 nm solid-state nanostructures, among which nanowires with deliberately designed gaps were fabricated by initially using DPN to pattern the etch resist (16-mercaptohexadecanoic acid) on $Au/Ti/SiO_x/Si$ substrates, and then removing the exposed gold by wet etching. In this way, nanogaps as small as 12 nm could be obtained (Figure 4.66).

Other than lithography, the fine tips of the scanning probe microscope can also be utilized to manipulate single atoms or molecules on specific substrates to form desired configurations [267, 268]. Ho *et al.* [267] reported the ingenious assembly of a copper(II) phthalocyanine molecule between two Au atomic chains on a NiAl(110) substrate by manipulating single atoms with scanning tunneling microscope tips. This innovative metal–molecule–metal junction provided a new method for the electronic characterization of single molecules and their contacts with metal electrodes. The advantages of SPM-related techniques stems from the technique's agility and controllability in the range below 100 nm, which is well-suited

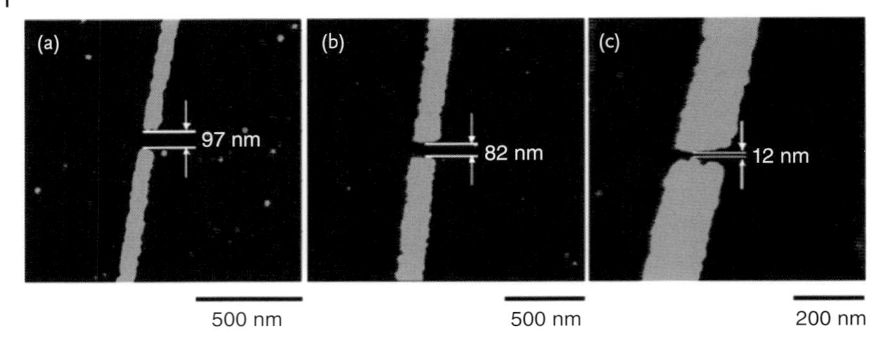

Figure 4.66 Trapping-mode AFM (TMAFM) topographic images of etched MHA/Au/Ti/SiO_x/Si nanogaps [266].

to investigations in the laboratory. However, as the scanning area and rate of the probe are limited, this presents a challenge for large-scale fabrication.

To date, many studies have focused on nanodevices with the integration of nanogap electrodes and organic molecules (small molecule [269], oligomer [270, 271], polymer [272], fullerene [119, 273], and biomolecule [274–277], etc.) or other nanometer-sized components (CNT [278], nanocrystal [201, 279], etc.) functionalizing effectively as switches, rectifiers, memories, and transistors. The elementary molecular digital logic circuits with fundamental logic gates (AND, OR, and XOR gates) and even more complicated molecular computing systems have received intensively theoretical and experimental consideration. As a consequence, significant progress has achieved due to efforts of talented and diligent scientists worldwide.

4.5.2 Behavior of Single-Molecule Transistors

Whilst a full understanding of the transport properties of a molecular junction represents a key step towards the realization of single-molecule electronic devices, a detailed microscopic characterization of the active region of the junction is also clearly required. Indeed, a major hurdle in the development of most single-molecule electronic devices is the unambiguous demonstration that charge transport occurs only through a single molecule of interest. For these reasons, the analysis of the transport properties of molecular junctions has attracted much attention in the field, and a wide variety of experimental techniques have been established during recent years [158]. Although many molecules have been investigated using single-molecule devices, it is impossible to describe all of these molecules in comprehensive fashion. Hence, the transport characteristics of only simple molecules (e.g., alkyl chains and π-conjugated molecules) will be introduced at this point.

4.5.2.1 Temperature- and Length-Variable Transport of Single Molecules

The charge transport mechanism of a molecular junction can be revealed by the characteristic temperature [280–282] and length dependences [282–284]. Therefore, measurements of the temperature- and length-variable transport of a molecular junction must be made in order to examine the charge transport mechanism. In particular, two distinct transport mechanisms have been extensively discussed in the literature [280–289]: (i) coherent transport via tunneling or superexchange; and (ii) incoherent thermally activated hopping.

Coherent tunneling or superexchange dominates through relatively short molecules, and the conductance value (G) decreases exponentially as the molecular length increases, according to Eq. (4.1):

$$G \propto \exp(-\beta d) \tag{4.1}$$

where d is the molecular length and β is the tunneling decay coefficient (which varies between 0.7 and 0.9 $Å^{-1}$ for alkyl chains and between 0.2 and 0.5 $Å^{-1}$ for π-conjugated molecules). In addition to the exponential decay of the conductance with molecular length, this coherent tunneling process is characterized by temperature-independent transport. On the other hand, incoherent hopping is known to be responsible for charge transport along long-conjugated molecular wires, and the conductance follows an Arrhenius relation given by:

$$G \propto \exp(-E_a / k_B T) \tag{4.2}$$

where k_B is the Boltzmann constant, T is the temperature, and E_a represents the hopping activation energy. The incoherent charge hopping is also characterized by a weak length-dependent transport that results in a conductance that scales linearly with the inverse of the molecular length.

A representative temperature-variable current (I)–voltage (V) of 1,8-octanedithiol bridging the Au nanogap electrodes broken by electromigration is shown in Figure 4.67a [282]. The $I(V)$ curves were measured at between 4.2 and 90 K, and no temperature dependence was observed. The temperature-independent $I(V)$ characteristic is a clear manifestation of coherent tunneling transport, and eliminates many other potential mechanisms. The conductance of five different alkanedithiols having between eight (DC8) and 12 (DC12) carbon atoms was also measured to examine the length-dependent conductance (Figure 4.67b) [282]. In accordance with Eq. (4.1), a semilog plot of the conductance versus the molecular length was linear. From the linear fit (the solid line across data points) in Figure 4.67b, the β-value was found to be 0.88 (=0.71 $Å^{-1}$) per carbon atom, assuming a through-bond tunneling [282], and was in good agreement with previously reported values for alkyl chains [281, 290]. Thus far, a consistent picture has emerged for the coherent tunneling mechanism of saturated alkyl chains and short-length conjugated molecules [126, 281, 291]. A coherent tunneling transport can be reasonably expected when the Fermi energy of the electrode lies within the large energy gap between HOMO and LUMO of the short molecules. Collectively, the correct exponential decrease of conductance upon a molecular length increase, the temperature-independent $I(V)$ characteristics, and

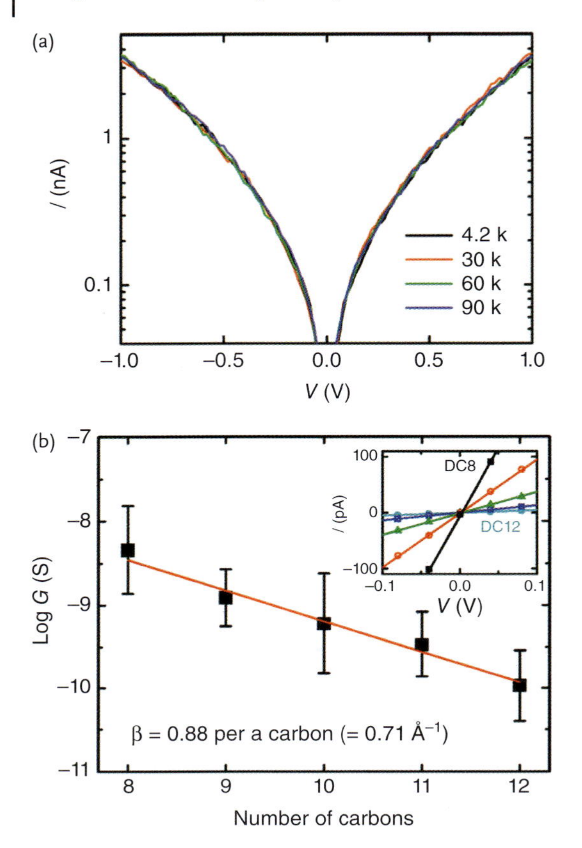

Figure 4.67 (a) Semilog plot of temperature-variable $I(V)$ characteristics for Au-1,8-octanedithiol-Au junctions at selected temperatures (4.2, 30, 60, and 90 K); (b) Semilog plot of the conductance versus the number of carbon atoms for five different length alkanedithiol nanogap junctions. The decay coefficient (β) can be determined from the linear fit (the solid line), yielding a β-value of 0.88 (=0.71 Å^{-1}) per carbon atom. The inset shows length-dependent $I(V)$ curves in the low-bias linear regime, where a conductance value is obtained from linear fits to the data [282].

the agreement with decay coefficients all point to the formation of a valid molecular junction.

Conjugated molecules made from repeating units with a modulated molecular length are ideal for understanding charge transport mechanisms, because such molecular systems permit the investigation not only of coherent tunneling and incoherent hopping but also of the transition between two distinct transport mechanisms by systematically changing the molecular length [283, 285]. This transition from tunneling to hopping has been observed by Frisbie and colleagues [283], who synthesized oligophenyleneimine (OPI) molecules of various lengths (ranging from 1.5 to 7.3 nm) bonded to Au through a thiolate linkage. The OPI molecular wires were grown on the Au substrate by a step-wise imination, with alternating addition of benzene-1,4-dicarboxaldehyde and benzene-1,4-diamine (see Figure 4.68a). The transport characteristics of the OPI wires were then measured using C-AFM. In the semilog plot of resistance versus molecular length (Figure 4.68b), a clear transition of the length dependence of the resistance was observed close to 4 nm (OPI 5), which indicated that the transport mechanism was different in short (OPI 1 to 4) and long (OPI 6 to 10) wires. In the short wires, the linear fit in Figure 4.68b indicated that the data were well described by Eq. (4.1) for coherent, nonresonant tunneling. The β-value was found to be $3\,\text{nm}^{-1}$, which was within the range of β-values for typical conjugated molecules [292]. For long OPI wires, a much flatter resistance versus molecular length relation ($\beta \approx 0.9\,\text{nm}^{-1}$) was shown. Such an extremely small β-value suggested that the principal transport mechanism was incoherent hopping [283]. A plot of resistance versus molecular length for long OPI wires was linear (see Figure 4.68b, inset), which was consistent with hopping as described above, and indicated that Eq. (4.1) would not apply for the long wires [283]. The change in transport mechanism apparent in the length-dependent measurements was verified by the temperature dependence. In Figure 4.68c, it can be seen that the resistance for OPI 4 was independent of temperature from 246 to 333 K, as expected for tunneling. In contrast, both OPI 6 and OPI 10 displayed the strongly thermally activated transport that is characteristic of hopping. The activation energies determined from the slopes of the data were identical at 0.28 eV for both OPI 6 and OPI 10 [283].

4.5.2.2 Inelastic Electron Tunneling Spectroscopy of Single Molecules

Inelastic electron tunneling spectroscopy (IETS), an all-electronic spectroscopy due to localized molecular vibrational modes, was discovered in 1966 by Jaklevic and Lambe [293]. These pioneering studies clearly showed an ability to detect the vibrational features of molecules buried in the interface of a metal insulator– metal (MIM) device. To explain the principles of IETS (see also Ref [294]), Figure 4.69 shows the energy-band diagrams (for elastic and inelastic tunneling paths) of a tunnel junction and the corresponding $I(V)$, $\mathrm{d}I/\mathrm{d}V$, and $\mathrm{d}^2I/\mathrm{d}V^2$ plots. When a negative bias (small with respect to the tunnel barrier) is applied to the left metal electrode, the left Fermi level is lifted. An electron from an occupied state on the left side then tunnels into an empty state on the right side, and its energy is conserved (process a in Figure 4.69); this process is elastic tunneling. During this process, the current increases linearly with the applied small bias (less than the vibrational energy) (Figure 4.69b); however, if there is a vibrational mode with a frequency of ω localized inside this barrier, then the electron can lose a quantum of energy, $\hbar\omega$, where $\hbar$ is Planck's constant divided by 2π, to excite the vibrational mode and tunnel into another empty state when the applied bias is large enough such that $eV \geq \hbar\omega$ (process b in Figure 4.69) [295, 296]. This process opens an inelastic tunneling channel for the electron, and its overall tunneling probability is increased. Thus, the total tunneling current has a kink that is a function of the applied bias (Figure 4.69b), and this kink becomes a step in the differential

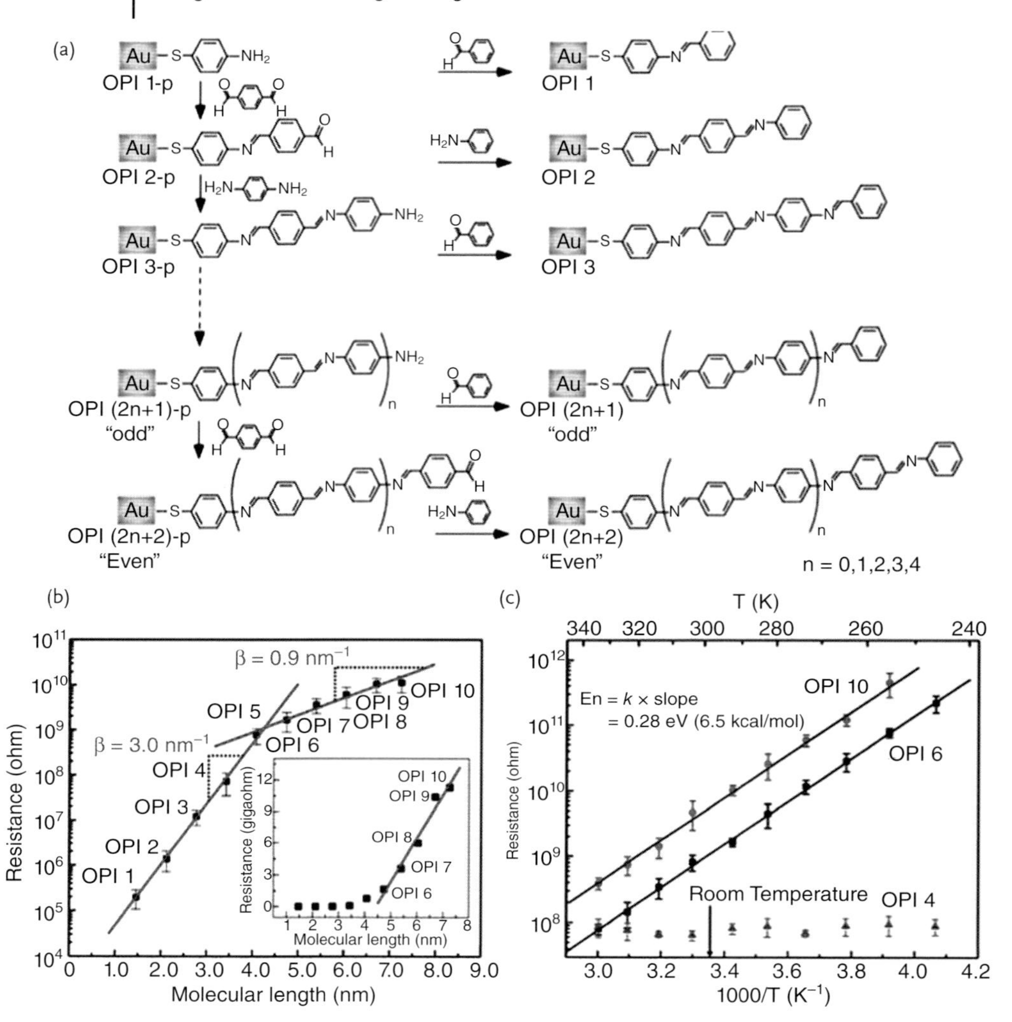

Figure 4.68 (a) Molecular structure and synthetic route to oligophenyleneimine wire precursors (OPI-p) and OPI monolayers on gold substrates; (b) Measurements of molecular wire resistance with CP-AFM. A gold-coated tip was brought into contact with an OPI monolayer on a gold substrate. The $I(V)$ traces were obtained over ±1.5 V for OPI 3 to 10, and ±1.0 V for OPI 1 and 2, at a load of 2 nN on the tip contact, and semilog plot of resistance versus molecular length for the gold–wire–gold junctions. Each datum point is the average differential resistance obtained from 10 $I(V)$ traces in the range −0.3 to +0.3 V. Error bars indicate 1 standard deviation (SD). Straight lines are linear fits to the data according to Eq. (4.1). Inset: A linear plot of resistance versus molecular length, demonstrating linear scaling of resistance with length for the long OPI wires; (c) Arrhenius plot for OPI 4, OPI 6, and OPI 10. Each datum point is the average differential resistance obtained at six different locations on samples in the range −0.2 to +0.2 V. Error bars indicate 1 SD. Straight lines are linear fits to the data [283].

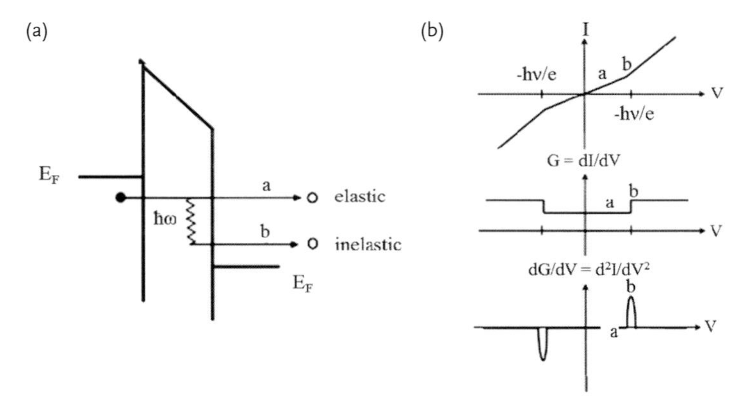

Figure 4.69 (a) Energy band diagram of a tunnel junction with a vibrational mode of frequency ω localized inside: "a" is the elastic tunneling process; "b" is the inelastic tunneling process; (b) Corresponding $I(V)$, $\mathrm{d}I/\mathrm{d}V$, and $\mathrm{d}^2I/\mathrm{d}V^2$ characteristics [294].

conductance ($\mathrm{d}I/\mathrm{d}V$) plot and a peak in the $\mathrm{d}^2I/\mathrm{d}V^2$ plot. Typically, only a very small fraction of electrons tunnel inelastically (the cross-section for such an excitation is very small because the electron traversal time is much smaller than the oscillator period), and thus the IETS conductance step is often too small to be conveniently detected. In practice, investigators use a phase-sensitive ("lock-in") detection technique to directly measure the peaks of the second derivative of $I(V)$. The IETS signal, which is proportional to the second derivative of $I(V)$, is usually measured by an AC modulation method. In theory, the signal can also be determined by a mathematical differential approach that computes the numerical derivatives of the directly measured $I(V)$ characteristics [297]. However, this method is generally not feasible in practice due to insufficient signal-to-noise ratios or bit resolutions of the instrumentation used to acquire the data. IETS has an ultra-high sensitivity, which makes it possible to obtain single-molecule vibronic information. Additionally, IETS is not subject to the selection rules of infrared or Raman spectroscopy. There is an orientation preference in IETS, but there are no rigorous selection rules. Both, infrared- and Raman-active vibrational modes are able to appear in IETS spectra. Collectively, IETS is a powerful and informative spectroscopic tool for probing molecular-scale charge transport.

IETS has recently become a primary characterization technique to identify the component molecules present in molecular junctions (not an adlayer or impurity, but molecules forming the active region of a junction) [280, 298–300], analogous to infrared and Raman spectroscopy for macroscopic samples, for the unambiguous determination of the molecular species in the junction. An example of experimental IETS measurements is shown in Figure 4.70 [300], which shows the $I(V)$ curve, the differential conductance ($\mathrm{d}I/\mathrm{d}V$), and the IETS ($\mathrm{d}^2I/\mathrm{d}V^2$) spectrum of Au- octanedithiol (ODT)–Au and Au–benzenedithiol (BDT)–Au junctions

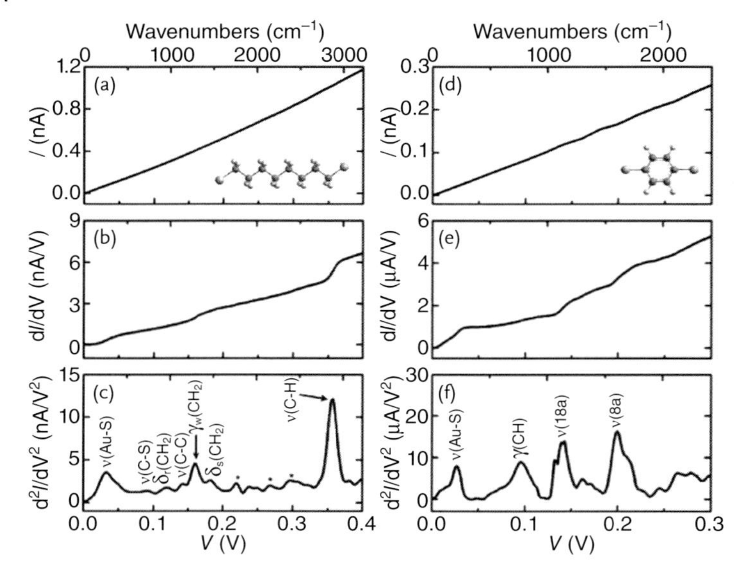

Figure 4.70 Transport properties of Au–ODT–Au (a–c) and Au–BDT–Au (d–f) junctions measured at 4.2 K. (a,d) $I(V)$ characteristics. The insets display the chemical structure of each molecule; (b,e) Differential conductance (dI/dV) obtained from lock-in first harmonic signal; (c,f) IETS spectrum (d^2I/dV^2) obtained from lock-in second harmonic signal. The peaks are labeled with the assigned vibrational modes [300].

measured at 4.2 K using an electromigrated break junction. Although the $I(V)$ characteristics seem to be linear over the bias range measured, the plots of $\mathrm{d}I/\mathrm{d}V$ and $\mathrm{d}^2I/\mathrm{d}V^2$ exhibit significant features corresponding to vibrational modes of the molecules under investigation. Standard AC modulation techniques with a lock-in amplifier are used to directly obtain the first and second harmonic signals proportional to $\mathrm{d}I/\mathrm{d}V$ and $\mathrm{d}^2I/\mathrm{d}V^2$, respectively [293, 301]. As explained above, a molecular vibration coupled to tunneling charge carriers gives rise to an increase in slope of the $\mathrm{d}I/\mathrm{d}V$ curve owing to an inelastic tunneling process, which then appears as a step and peak in the first ($\mathrm{d}I/\mathrm{d}V$) and second ($\mathrm{d}^2I/\mathrm{d}V^2$) derivatives, respectively. The plot of $\mathrm{d}^2I/\mathrm{d}V^2$ versus V is referred to as the IETS spectrum. The observed spectral features were assigned to specific molecular vibrations by comparison with previously reported infrared, Raman, and IETS measurements and by density functional theory (DFT) calculations. For the ODT junction (Figure 4.70c), peaks were reproducibly observed at 92, 119, 143, 161, 181, and 355 mV, which correspond to ν(C–S) stretching, δ_r (CH_2) rocking, ν(C–C) stretching, $\gamma_w(CH_2)$ wagging, $\delta_s(CH_2)$ scissoring, and ν(C–H) stretching modes, respectively. The absence of a prominent peak corresponding to the ν(S–H) stretching mode at 319 mV (2575 cm^{-1}) suggests that the thiol (–SH) anchoring group reacts with the Au electrode pairs broken during the electromigration. In the IETS spectrum of the

BDT junction (Figure 4.70f), three prominent peaks reproducibly appeared at 96, 142, and 201 mV, corresponding to γ(C–H) aryl out-of-plane bending, ν(18a) stretching, and ν(8a) stretching modes, respectively. These modes originate from vibrations of the phenyl ring. A theoretical study predicted that the ν(18a) and ν(8a) ring modes should have a strong vibronic coupling in phenylene molecules [301], and was consistent with these results. The dominance of aromatic ring modes in IETS spectra has also been observed experimentally for various conjugated molecules [302, 303]. The fully assigned IETS spectrum provides unambiguous experimental evidence of the existence of the desired molecules in the region of the junction and, in conjunction with the other characteristics of the junction transport, leaves the IETS-identified molecule as the only element in the junction through which tunneling occurs.

Over the past few years, IETS has evolved into an essential tool in the field of molecular electronics. Although IETS requires cryogenic temperatures, it is the only available method that provides both structural and electronic information about a single-molecule electronic device for a particular conformation and contact geometry of the molecular junction at a low temperature [289]. From sophisticated comparisons between experiments and theoretical computations, IETS can be more useful for characterizing numerous aspects of molecular junctions, such as an identification of the molecule, information on the nature of the interfaces, the orientation of the molecule, and even electronic pathways [286].

4.5.2.3 **Transition Voltage Spectroscopy of Single Molecules**

Transition voltage spectroscopy (TVS) is becoming an increasingly popular spectroscopic tool for molecular junctions [304–308] and other diverse nanoelectronic systems [309]. Specifically, TVS is used to provide insight into the energy offset between the contact Fermi level and the nearest molecular level responsible for charge transport in molecular junctions by measuring the transition voltage (V_{trans}) required to generate the inflection behavior of a Fowler–Nordheim (F–N) plot; that is, the corresponding analysis of $\ln(I/V^2)$ against $1/V$ for $I(V)$ characteristics [304, 305]. By combining TVS with ultraviolet photoelectron spectroscopy (UPS), Beebe *et al.* [304] correlated the charge transport properties of π-conjugated molecules with their effective band lineup. In this study, CP-AFM and crossed-wire tunnel junction measurements on molecular junctions revealed a characteristic minimum in the F–N plot at a bias voltage, V_{trans} (Figure 4.71a), which scaled linearly with the HOMO energy (which is the nearest molecular level for the measured molecules) obtained from UPS (Figure 4.71b) [304]. These results show that the magnitude of V_{trans} is molecule-specific (as a form of spectroscopy), and depends directly on the manner in which the conjugation path is extended. In general, the HOMO–LUMO gap of π-conjugated molecules decreases with an increase in conjugation length [312]. It is reasonable, therefore, to expect that longer conjugated molecules would exhibit a smaller value of V_{trans} than shorter conjugated molecules within a given molecular series [305]. Recently, TVS has also facilitated the calibration of orbital energy positions in molecular transistors [308, 313].

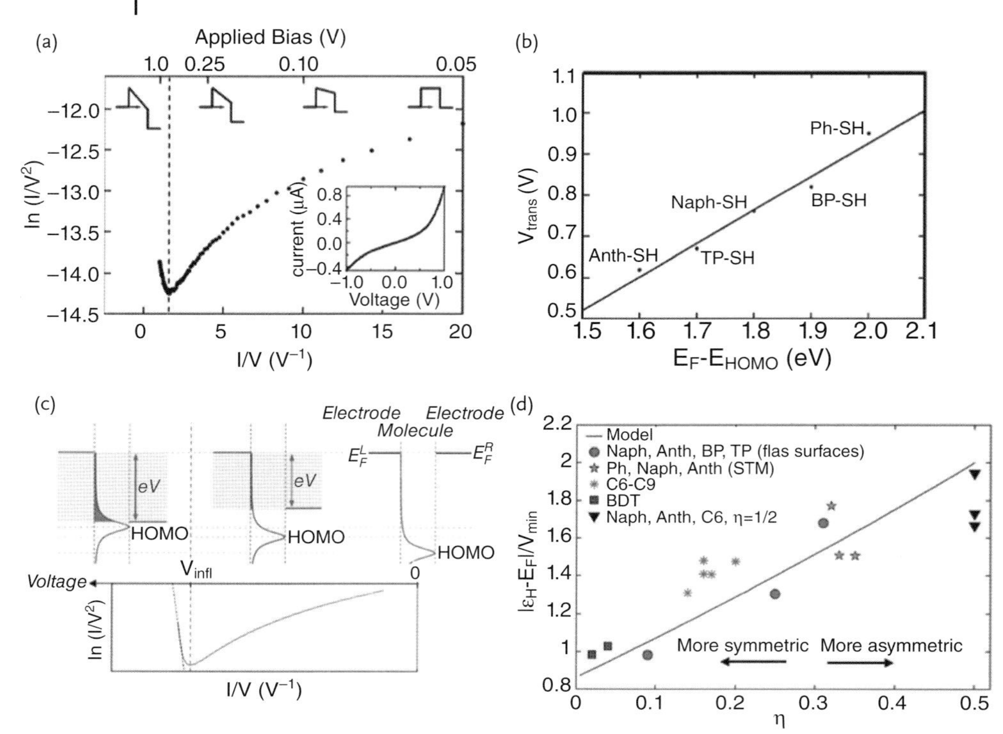

Figure 4.71 (a) Solid circles represent the average of 100 $I(V)$ curves for a Au–anthracenethiol–Au junction measured by CP-AFM. The dashed line corresponds to the voltage at which the tunneling barrier transitions from trapezoidal to triangular (V_{trans}). Also shown are representations of the barrier shape at various values of applied bias. The inset shows current–voltage data on standard axes; (b) V_{trans} (CP-AFM) versus E_F-E_{HOMO} energy difference (UPS) [304]; (c) Schematic of the theoretical model [310] to qualitatively explain the inflection of the F–N curve. Also shown are representations of the barrier shape at various values of applied bias. The inset shows current–voltage data on standard axes [310]; (d) Ratio between the HOMO energy (at zero bias) and the transition voltage (denoted as V_{min} in the figure), versus asymmetry parameter, η. A solid line is obtained from a Lorentzian transmission function; symbols are results of *ab initio* finite bias calculations (see Ref. [311] for details).

TVS was initially interpreted by a simple barrier picture for charge tunneling in a junction [304]. Within this interpretation, the transition behavior in the F–N plots corresponds to a change in the tunneling mechanism from direct tunneling through a trapezoidal barrier to F–N tunneling (or field emission) through a triangular barrier (see the barrier shapes in Figure 4.71a) [304]. The transition voltage equals the barrier height, which is interpreted as the energy gap from the metal electrode's Fermi level to the nearest molecular level. However, as pointed out by

Huisman *et al.* [307], the naive tunnel barrier model is inconsistent with experimental data. On the other hand, the TVS experiments on molecular junctions are more appropriately described by the coherent Landauer approach with a single transport level [307]. Charge transport through such a junction is described by a transmission function, which is assumed to have a Lorentzian shape. Within the coherent Landauer transport picture, V_{trans} can be directly scaled with the barrier height (Φ_B) in molecular junctions, thus providing valid information on molecular energy levels. Araidai *et al.* [310] undertook a theoretical investigation of the origin of an inflection behavior appearing in the F–N plot of $I(V)$ characteristics for molecular junctions. The results showed that the inflection does not necessarily indicate the transition between the two regimes of direct tunneling and F–N tunneling. In fact, a close examination of the relationship between the behavior of the F–N curve and the transmission function showed that the inflection takes place when the molecular level responsible for the charge transport approaches the edge of the electrode-bias window (Figure 4.71c) [310]. Although the origin of the inflection behavior drastically differed from the conventional model, the F–N plots obtained from calculations showed a very similar behavior to those from the recent experiments [310]. Recently, Chen *et al.* [311] described extensive *ab initio* calculations to simulate TVS for a broad class of molecular junctions. The numerical data closely followed the trend expected from an analytical model, with a Lorentzian-shaped transmission function. Interestingly, the ratio of V_{trans} to the HOMO level position was found to vary between 0.8 and 2.0, depending on the junction asymmetry (as shown in Figure 4.71d). This means that it is necessary to consider the asymmetry of the molecular junction in order to use TVS as a quantitative spectroscopic tool to probe the molecular levels [311].

4.5.3
Quanta and Theories of Single-Molecule Transistors

Limited by the capacity, only the simple theory of electron transport through molecular junction is introduced at this point [314]. Electron transport through a short channel of less than a few nanometers in the molecular junction is a complicated quantum scattering phenomenon [315–319]. A schematic picture of a molecular junction is shown in Figure 4.72. When a bias voltage is applied across the junction, charge carriers flow through the molecule from the source (left) to the drain (right) electrode, unless the molecules are short enough to allow a direct tunneling between both electrodes. In that case, molecular orbitals (MOs) should serve as conduits for electron transport. Therefore, the alignment of molecular energy levels with respect to the Fermi energy of both electrodes is a key parameter for conductance.

As a molecule is linked to metal electrodes, its MOs are changed. First, there can be a significant charge transfer from the metal to the molecule or *vice versa*, because of the difference between the work-function of the metal and the ionization potential of the molecule. This leads to a new equilibrium of the combined system, resulting in the MO energy level shift (Δ). Second, the molecular states of

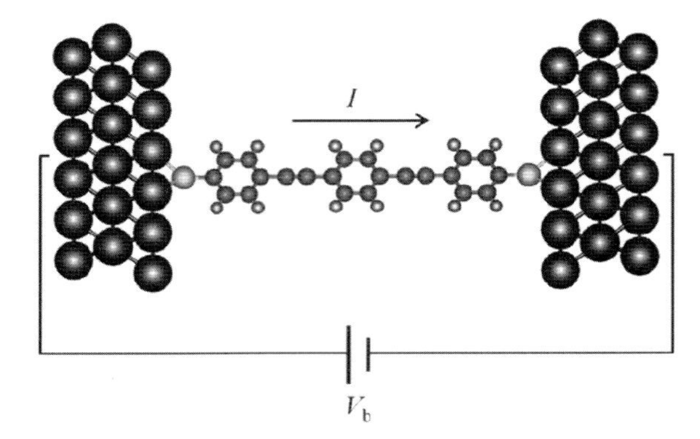

Figure 4.72 Schematic representation of a molecular junction. A *p*-conjugated organic molecule is contacted with the metal electrodes by a sulfur linkage [314].

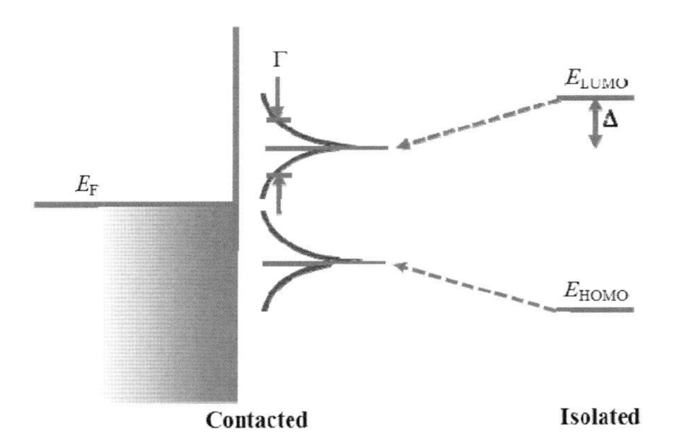

Figure 4.73 Readjustment (or re-normalization) of the molecular energy levels in the metal–molecule contact [319].

the isolated molecule are no longer eigenstates of the molecule contacted to metal electrodes, because the molecular states are coupled to the continuum states of the electrodes and this coupling results in energy level broadening (Γ) corresponding to the finite life time of occupied electrons in the molecular states. Consequently, the MO energy levels are readjusted or re-normalized by the contact effects raised from the metal–molecule bonding, as depicted in Figure 4.73 [320].

In theoretical calculations, the contact effects depend heavily on the microscopic geometry of a molecular junction. Hence, an accurate description of the contact geometry is essential to obtain reliable theoretical results. Although various types of metal–molecule contact models have been used, the common idea is that some atoms belonging to the electrode region should be included into the molecular part during the geometry optimization, to effectively take into account the contact effects. In this model, most studies have used the electrode part with a few atoms or metal clusters [124, 321–330]. Although a few atoms may be taken into account for the charge transfer effect, it may not be appropriate to describe the periodic nature of the semi-infinite metal electrode and to screen the long-range Coulomb potential induced from the metal–molecule contact. To date, very few studies have included several atomic layers with the periodic boundary condition of the electrodes, which provides a correct description of the infinite surface effect of the metallic bulk electrodes and the effective screening of Coulomb potential [324, 325, 329, 330].

In order to describe the molecular junction from a theoretical standpoint, there is a need to set up an appropriate Hamiltonian for a given system. For this, the total Hamiltonian is divided into three parts of: (i) a molecule (i.e., extended molecule); (ii) the left and right leads; and (iii) the interaction between the molecule and leads:

$$H = H_{\mathrm{M}} + H_{\mathrm{L}} + H_{\mathrm{R}} + V_{\mathrm{int}} \tag{4.3}$$

The extended molecule, including a few atoms or metal clusters, is a finite system. However, as each electrode is a semi-infinite system along the direction of current, it cannot be treated just by the k-point sampling along the surface normal direction. Thus, the Hamiltonian ($H_{\mathrm{L/R}}$) for each lead has the infinite dimension. In practical calculations, a screening approximation is useful to deal with the infinite system. Each electrode is a bulk system with well-defined periodic conditions, except near the surface region around the metal–molecule contact. Electrons close to the surface are influenced by the potential induced from the molecule, and thereby they are redistributed to screen the potential. Typical metal electrodes such as gold are good conductors, and a few atomic layers may be sufficient to fully screen the induced potential. Therefore, if a sufficient part of electrode is included into an extended molecule region, the remainder of the electrode parts can be described as a bulk system. In exploiting the screening approximation, the infinite dimension of the Hamiltonian is drastically reduced to the dimension of the molecular part, while all semi-infinite properties of the leads are effectively treated by using self-energy. Finally, the effective Hamiltonian (H_{eff}) which is now tractable is given by the following equation:

$$H_{\mathrm{eff}} = H_{\mathrm{M}} + \Sigma_{\mathrm{L}} + \Sigma_{\mathrm{R}} \tag{4.4}$$

It is now clear that the self-energy terms ($\Sigma_{\mathrm{L/R}}$) in the Hamiltonian should involve the re-normalization effect; the real part of the self-energy gives rise to the energy level shift [$\Delta_{\mathrm{L/R}} = \mathrm{Re}(\Sigma_{\mathrm{L/R}})$], while the imaginary part of the self-energy results in the energy level broadening [$\Gamma_{\mathrm{L/R}} = -\mathrm{Im}(\Sigma_{\mathrm{L/R}})$], as shown graphically in Figure 4.73.

If the interaction between electrons and molecular vibrations is neglected, the study of electron transport would require a numerically reliable calculation method to deal with a system that is composed of hundreds of atoms for the elastic and coherent transport regime; thus, DFT has a great advantage over other methods.

In addition, the electron density – which is the fundamental variable in DFT – can be obtained directly from the non-equilibrium Green's function (NEGF) method. The NEGF method coupled to DFT (NEGF + DFT) can be easily implemented [330].

The first step is to calculate the Hamiltonian and overlap matrices partitioned in the left (L), molecular (M), and right (R) parts by using localized bases in DFT as follows:

$$H = \begin{bmatrix} H_L & H_{LM} & 0 \\ H_{ML} & H_M & H_{MR} \\ 0 & H_{RM} & H_R \end{bmatrix}$$

and

$$S = \begin{bmatrix} S_L & S_{LM} & 0 \\ S_{ML} & S_M & S_{MR} \\ 0 & S_{RM} & S_R \end{bmatrix}$$

The self-energy is a main ingredient to determine renormalization of molecular energy levels. The self-energy for the left/right metal–molecule contact is given by:

$$\Sigma_{L/R}(E) = \nu_{L/R}(E) g^r_{L/R}(E) \nu_{L/R}(E) \tag{4.5}$$

where $v_L(E) = (H_{ML} - ES_{ML})$ and $v_R(E) = (H_{MR} - ES_{MR})$ are the interaction terms of the left and right electrode with the molecule, respectively and $g^r_{L/R}$ is the surface Green's function for the left/right electrode, which can be calculated by various ways such as transfer matrices method [331]. Once the self-energy matrix has been obtained, the effective Hamiltonian will be given by Eq. (4.4). The retarded Green's function matrix can then be calculated as:

$$G^r(E) = (ES - H_{eff})^{-1} \tag{4.6}$$

The retarded Green's function matrix is directly related to the electron density by the following two equations:

$$\rho(r) = \sum_{\alpha\beta} D_{\alpha\beta} \phi_\alpha(r) \phi_\beta(r) \tag{4.7}$$

and

$$D = -\frac{1}{\pi} \mathrm{Im} \int_{-\infty}^{\infty} [G^r(E) f(E - \mu)] dE \tag{4.8}$$

where $\phi_{\alpha/\beta}(r)$ is a localized atomic basis orbital and $D_{\alpha\beta} = \sum_{i=1}^{\mathrm{occ}} f_i c^*_{\alpha i} c_{\beta i}$ is the density matrix element where f_i is the occupation number of the i-th state and

The retarded Green's function G^r, based on the effective Hamiltonian, gives the electron density, but the Hamiltonian itself is a functional of the electron density. Therefore, the final electron density should be calculated in a self-consistent manner, as shown in Figure 4.74.

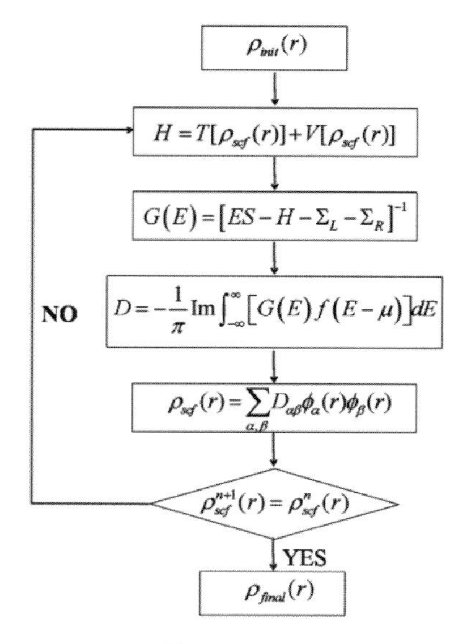

Figure 4.74 Self-consistent loop of the NEGF + DFT method [314].

By using the converged electron density, it is possible to calculate the transmission coefficients as a function of energy and current (I) at the given bias voltage (V_b) through the Landauer–Büttiker formula:

$$T(E, V_b) = T_r\left[\Gamma_L(E, V_b) G^a(E, V_b) \Gamma_R(E, V_b) G^r(E, V_b)\right] \tag{4.9}$$

and

$$I = \frac{2e}{h} \int [T(E, V_b)\{f_L(E, V_b) - f_R(E, V_b)\}] dE \tag{4.10}$$

where $f_{L/R}$ is the Fermi–Dirac distribution function for the left/right electrodes, and G^a is the advanced Green's function.

In the presence of a finite bias, a system becomes nonequilibrium because of the difference in chemical potential between the left and right electrodes. The present NEGF + DFT method gives a steady-state current under the nonequilibrium conditions.

In a molecular device, the interaction of electrons with nuclear vibrations plays a significant role in thermal relaxation or transition of transport mechanism. Electrons may either lose or gain their energies through interaction with nucleus motions (i.e., vibration or phonon) of the molecule. In this case, the Hamiltonian for the molecular part has additional terms:

$$H = H_e + H_{ph} + H_{e-ph} \tag{4.11}$$

The electron–phonon coupling (e–ph) term in the Hamiltonian is given by:

$$H_{\text{e-ph}} = \sum_{\alpha}\sum_{\text{ij}} M_{\text{ij}}^{\alpha} c_{\text{i}}^{+} c_{\text{j}} \left(b_{\alpha}^{+} + b_{\alpha}\right) \tag{4.12}$$

where $c_{\text{i}}^{+}c_{\text{ij}}$ and $(b_{\alpha}{}^{+}b_{\alpha})$ are electron and phonon creation (annihilation) operator, respectively, and M_{ij}^{α} is a e–ph coupling matrix element for the α phonon mode. It is difficult to take into account the nucleus motion explicitly in the Hamiltonian. Fortunately, the electron transport for a small molecule is much faster than the nucleus motion, so that the e–ph coupling term can be obtained by Taylor expansion from the adiabatic Born–Oppenheimer approximation with respect to the given molecular coordinate [332–335]. From the lowest order expansion, the coupling matrix element is defined by:

$$M_{\text{ij}}^{\alpha} = \sum_{\text{Ix}} \left\langle i \left| \frac{\partial H_{\text{e}}}{\partial R_{\text{Ix}}} \right| j \right\rangle \Bigg|_{RO} v_{\text{Ix}}^{\alpha} \sqrt{\frac{\hbar}{2m_{\text{I}}\omega_{\alpha}}} \tag{4.13}$$

where v_{Ix}^{α} is the nucleus displacement vector of the α phonon mode, ω_{α} is the corresponding phonon frequency, m_{I} is the mass of the i-th nucleus, and R_{Ix} is relative coordinate along the x spatial direction of the i-th nucleus from its equilibrium position (R_{o}). Each component of the matrix element can be obtained from a series of DFT calculations [335].

The inelastic transport can be calculated based on the NEGF + DFT as an extension of the method for the elastic transport. The Keldysh NEGF formalism provides a straightforward way, which adds the self-energy term due to the e–ph interaction [335]:

$$\sum_{\text{e-ph}}^{<(>)}(E) = i\int_{-\infty}^{\infty} \frac{d\varepsilon}{2\pi} M^{\alpha} d^{<(>)}(\alpha, E-\varepsilon) G^{<(>)}(E) M^{\alpha} \tag{4.14}$$

where $G^{<(>)}$ represents the lesser (greater) Green's function as the electron (hole) propagator and $d^{<(>)}$ is the phonon propagator for the α mode. For the weak e–ph interaction regime, the Born approximation gives a systematic perturbation series of the phonon self-energy up to the infinite order. In practical terms, the lowest-order Born approximation has been quite successful for the description of the weak interaction in small molecules [334, 335].

The total lesser (greater) self-energy due to the metal-molecule contact and e–ph interaction is given by:

$$\sum_{tot}^{<(>)}(E) = \sum_{L}^{<(>)}(E) + \sum_{R}^{<(>)}(E) + \sum_{e\text{-}ph}^{<(>)}(E) \tag{4.15}$$

The current in the left/right contact is given by the Landauer-type equation:

$$I_{L/R} = \frac{2e}{h}\int_{-\infty}^{\infty} T_r\left[\sum_{L/R}^{>}(E)G^{<}(E) - \sum_{L/R}^{<}(E)G^{>}(E)\right]dE \tag{4.16}$$

Finally, the total current in the left (or right) contact is just the sum of elastic and inelastic contributions:

$$I_{\text{L}} = I_{\text{elastic}} + I_{\text{inelastic}} \tag{4.17}$$

where

$$I_{(\text{in})\text{elastic}} = \frac{2e}{h}\int_{-\infty}^{\infty} T_{\text{r}}\left[\sum\nolimits_{\text{L/R}}^{>}(E)G_{(\text{in})\text{elastic}}^{<}(E) - \sum\nolimits_{L}^{<} G_{(\text{in})\text{elastic}}^{>}(E)\right]dE \tag{4.18}$$

4.6 Challenges and Outlooks

The greatest and final scientific target of organic electronics is to fabricate circuits based on organic semiconductors. In this chapter ambipolar transistors, complementary inverters, logic gates and ring oscillators have first been introduced, after which attention was focused on circuits based on organic thin films, self-assembled and printed circuits, and circuits of organic single crystals. Although research investigations on this topic are at their earliest stages, the prospects for, and high performance of, these circuits indicate a bright future for organic electronics.

The single-molecule transistor provides an ideal model to examine the properties of organic semiconductors at the molecular level, as well as to identify the potential of these devices for higher integrated circuits. Hence, a review of the transistors of single molecules has been provided, together with details of techniques used to fabricate nanogap electrodes and to prepare self-assembled single-molecule devices. The potential application of these devices for the analysis of materials at the molecular level is also introduced. Quanta and theories for single-molecule transistors are also addressed, and suggestions made as to the future challenges in this field.

References

1 Zaumseil, J. and Sirringhaus, H. (2007) Electron and ambipolar transport in organic field-effect transistors. *Chem. Rev.*, **107** (4), 1296–1323.

2 Tang, Q., Tong, Y., Li, H., Ji, Z., Li, L., Hu, W., Liu, Y., and Zhu, D. (2008) High-performance air-stable bipolar field-effect transistors of organic single-crystalline ribbons with an air-gap dielectric. *Adv. Mater.*, **20** (8), 1511–1515.

3 Cornil, J., Brédas, J., Zaumseil, J., and Sirringhaus, H. (2007) Ambipolar transport in organic conjugated materials. *Adv. Mater.*, **19** (14), 1791–1799.

4 Baeg, K., Kim, J., Khim, D., Caironi, M., Kim, D., You, I., Quinn, J., Facchetti, A., and Noh, Y. (2011) Charge injection engineering of ambipolar field-effect transistors for high-performance organic complementary circuits. *ACS Appl. Mater.*, **3** (8), 3205–3214.

5 Capelli, R., Dinelli, F., Toffanin, S., Todescato, F., Murgia, M., Muccini, M., Facchetti, A., and Marks, T. (2008) Investigation of the optoelectronic properties of organic light-emitting transistors based on an intrinsically ambipolar material. *J. Phys. Chem. C.*, **112** (33), 12993–12999.

6 Takahashi, T., Takenobu, T., Takeya, J., and Iwasa, Y. (2007) Ambipolar light-emitting transistors of a tetracene single crystal. *Adv. Funct. Mater.*, **17** (10), 1623–1628.

7 Zaumseil, J., Donley, C., Kim, J., Friend, R., and Sirringhaus, H. (2006) Efficient top-gate, ambipolar, light-emitting field-effect transistors based on a green-light-emitting polyfluorene. *Adv. Mater.*, **18** (20), 2708–2712.

8 de Boer, R., Stassen, A., Craciun, M., Mulder, C., Molinari, A., Rogge, S., and Morpurgo, A. (2005) Ambipolar Cu- and Fe-phthalocyanine single- crystal field-effect transistors. *Appl. Phys. Lett.*, **86** (26), 262109.
9 Yasuda, T., Goto, T., Fujita, K., and Tsutsui, T. (2004) Ambipolar pentacene field-effect transistors with calcium source-drain electrodes. *Appl. Phys. Lett.*, **85** (11), 2098–2100.
10 Zaumseil, J., Friend, R.H., and Sirringhaus, H. (2006) Spatial control of the recombination zone in an ambipolar light-emitting organic transistor. *Nat. Mat.*, **5** (1), 69–74.
11 Chen, Z., Lemke, H., Albert-Seifried, S., Caironi, M., Nielsen, M., Heeney, M., Zhang, W., McCulloch, I., and Sirringhaus, H. (2010) High mobility ambipolar charge transport in polyselenophene conjugated polymers. *Adv. Mater.*, **22** (21), 2371–2375.
12 Anthopoulos, T., Setayesh, S., Smits, E., Cölle, M., Cantatore, E., de Boer, B., Blom, P., and de Leeuw, D. (2006) Air-stable complementary-like circuits based on organic ambipolar transistors. *Adv. Mater.*, **18** (14), 1900–1904.
13 Zhang, Y., Dong, H., Tang, Q., Ferdous, S., Liu, F., Mannsfeld, S., Hu, W., and Briseno, A. (2010) Organic single-crystalline p-n junction nanoribbons. *J. Am. Chem. Soc.*, **132** (33), 11580–11584.
14 Kreouzisa, T., Baldwin, R.J., Shkunov, M., McCulloch, I., Heeney, M., and Zhang, W. (2005) High mobility ambipolar charge transport in a cross-linked reactive mesogen at room temperature. *Appl. Phys. Lett.*, **87** (17), 172110.
15 Meijer, E., de Leeuw, D., Setayesh, S., Van Veenendaal, E., Huisman, B., Blom, P., Hummelen, J., Scherf, U., and Klapwijk, T. (2003) Solution-processed ambipolar organic field-effect transistors and inverters. *Nat. Mater.*, **2** (10), 678–682.
16 Facchetti, A. (2011) π-conjugated polymers for organic electronics and photovoltaic cell applications. *Chem. Mater.*, **23** (3), 733–758.
17 Wang, J., Wang, H., Yan, X., Huang, H., Jin, D., Shi, J., Tang, Y., and Yan, D. (2006) Heterojunction ambipolar organic transistors fabricated by a two-step vacuum-deposition process. *Adv. Funct. Mater.*, **16** (6), 824–830.
18 Ye, R., Baba, M., Oishi, Y., Mori, K., and Suzuki, K. (2005) Air-stable ambipolar organic thin-film transistors based on an organic homostructure. *Appl. Phys. Lett.*, **86** (25), 253505.
19 Kang, S.J., Yi, Y., Kim, C.Y., Cho, K., Seo, J.H., Noh, M., Jeong, K., Yoo, K.H., and Whang, C.N. (2005) Ambipolar organic thin-film transistors using C_{60}/pentacene structure: characterization of electronic structure and device property. *Appl. Phys. Lett.*, **87** (23), 233502.
20 Dinelli, F., Capelli, R., Loi, M., Murgia, M., Muccini, M., Facchetti, A., and Marks, T. (2006) High-mobility ambipolar transport in organic light-emitting transistors. *Adv. Mater.*, **18** (11), 1416–1420.
21 Hu, Y., Tsubasa, K., and Hidenori, O. (2009) Ambipolar pentacene/C_{60}-based field-effect transistors with high hole and electron mobilities in ambient atmosphere. *Appl. Phys. Lett.*, **94** (2), 023305.
22 Blanco, R., Gomez, R., Seoane, C., Segura, J.L., Mena-Osteritz, E., and Ba1uerle, P. (2007) An ambipolar peryleneamidine monoimide-fused polythiophene with narrow band gap. *Org. Lett.*, **9** (11), 2171–2174.
23 Smits, E.C.P., Anthopoulos, T.D., Setayesh, S., van Veenendaal, E., Coehoorn, R., Blom, P.W.M., de Boer, B., and de Leeuw, D.M. (2006) Ambipolar charge transport in organic field-effect transistors. *Phys. Rev. B*, **73** (20), 205316.
24 Rost, C., Gundlach, D.J., Karg, S., and Riess, W. (2004) Ambipolar organic field-effect transistor based on an organic heterostructure. *J. Appl. Phys.*, **95** (10), 5782–5787.
25 Schmechel, R., Ahles, M., and von Seggern, H. (2005) A pentacene ambipolar transistor: experiment and theory. *J. Appl. Phys.*, **98** (8), 084511.
26 Singh, T., Meghdadi, F., Günes, S., Marjanovic, N., Horowitz, G., Lang, P., Bauer, S., and Sariciftci, N. (2005) High-performance ambipolar pentacene

organic field-effect transistors on poly(vinyl alcohol) organic gate dielectric. *Adv. Mater.*, **17** (14), 2315–2320.

27 Chua, L., Zaumseil, J., Chang, J., Ou, E.C.W., Ho, P.K.H., Sirringhaus, H., and Friend, R.H. (2005) General observation of *n*-type field-effect behaviour in organic semiconductors. *Nature*, **434** (7030), 194–199.

28 Kepler, R.G. (1960) Charge carrier production and mobility in anthracene crystals. *Phys. Rev.*, **119** (4), 1226–1229.

29 Crone, B., Dodabalapur, A., Sarpeshkar, R., Filas, R., Lin, Y., Bao, Z., O'Neill, J., Li, W., and Katz, H. (2001) Design and fabrication of organic complementary circuits. *J. Appl. Phys.*, **89** (9), 5125–5132.

30 Dodabalapur, A., Laquindanum, J., Katz, H., and Bao, Z. (1996) Complementary circuits with organic transistors. *Appl. Phys. Lett.*, **69** (27), 4227–4229.

31 Lin, Y., Dodabalapur, A., Sarpeshkar, R., Bao, Z., Li, W., Baldwin, K., Raju, V., and Katz, H. (1999) Organic complementary ring oscillators. *Appl. Phys. Lett.*, **74** (18), 2714–2716.

32 Tang, Q., Tong, Y., Hu, W., Wan, Q., and Bjørnholm, T. (2009) Assembly of nanoscale organic single-crystal cross-wire circuits. *Adv. Mater.*, **21** (42), 4234–4237.

33 Crone, B., Dodabalapur, A., Lin, Y., Filas, R., Bao, Z., LaDuca, A., Sarpeshkar, R., Katz, H., and Li, W. (2000) Large-scale complementary integrated circuits based on organic transistors. *Nature*, **403** (6769), 521–523.

34 Klauk, H., Halik, M., Zschieschang, U., Eder, F., Rohde, D., Schmid, G., and Dehm, C. (2005) Flexible organic complementary circuits. *IEEE Trans. Electron Devices*, **52** (4), 618–622.

35 Sheraw, C., Zhou, L., Huang, J., Gundlach, D., Jackson, T., Kane, M., Hill, I., Hammond, M., Campi, J., Greening, B., Francl, J., and West, J. (2002) Organic thin-film transistor-driven polymer-dispersed liquid crystal displays on flexible polymeric substrates. *Appl. Phys. Lett.*, **80** (6), 1088–1090.

36 Vusser, S., Steudel, S., Myny, K., Genoe, J., and Heremans, P. (2006) Low voltage complementary organic inverters. *Appl. Phys.*, **88** (16), 162116.

37 Fix, W., Ullmann, A., Ficker, J., and Clemens, W. (2002) Fast polymer integrated circuits. *Appl. Phys. Lett.*, **81** (9), 1735–1737.

38 Singh, T., Senkarabacak, P., Sariciftci, N., Tanda, A., Lackner, C., Hagelauer, R., and Horowitz, G. (2006) Organic inverter circuits employing ambipolar pentacene field-effect transistors. *Appl. Phys. Lett.*, **89** (3), 033512.

39 Anthopoulos, T., de Leeuw, D., Cantatore, E., Hof, P., Alma, J., and Hummelen, J. (2005) Solution processible organic transistors and circuits based on a C_{70} methanofullerene. *J. Appl. Phys.*, **98** (5), 054503.

40 Chen, Y., Xu, Y., Zhao, K., Wan, X., Deng, J., and Yan, W. (2010) Towards flexible all-carbon electronics: flexible organic field-effect transistors and inverter circuits using solution-processed all-graphene source/drain/gate electrodes. *Nano Res.*, **3** (10), 714–721.

41 Shang, L., Ji, Z., Chen, Y., Wang, H., Liu, X., Han, M., and Liu, M. (2011) Low voltage organic devices and circuits with aluminum oxide thin film dielectric layer. *Sci. China Technol. Sci.*, **54** (1), 95–98.

42 Graza, I. and Lacour, S. (2009) Flexible pentacene organic thin film transistor circuits fabricated directly onto elastic silicone membranes. *Appl. Phys. Lett.*, **95** (24), 243305.

43 Anthopoulos, T., Singh, B., Marjanovic, N., Sariciftci, N., Ramil, A., Sitter, H., Cölle, M., and de Leeuw, D. (2006) High performance *n*-channel organic field-effect transistors and ring oscillators based on C_{60} fullerene films. *Appl. Phys. Lett.*, **89** (21), 213504.

44 Jang, J., Kim, S., Hwang, J., Nam, S., Yang, C., Chung, D., and Park, C. (2009) Photopatternable ultrathin gate dielectrics for low-voltage- operating organic circuits. *Appl. Phys. Lett.*, **95** (7), 073302.

45 Park, S., Anthony, J., and Jackson, T. (2007) Solution-processed TIPS-pentacene organic thin-film-transistor

circuits. *IEEE Electron Device Lett.*, **28** (10), 877–879.

46 Smith, J., Hamilton, R., Heeney, M., de Leeuw, D., Cantatore, E., Anthony, J., McCulloch, I., Bradley, D., and Anthopoulos, T. (2008) High-performance organic integrated circuits based on solution processable polymer-small molecule blends. *Appl. Phys. Lett.*, **93** (25), 253301.

47 Stingelin-Stutzmann, N., Smits, E., Gem, H., Tanase, C., Blom, P., Smith, P., and de Leeuw, D. (2005) Organic thin-film electronics from vitreous solution-processed rubrene hypereutectics. *Nat. Mater.*, **4** (8), 601–606.

48 Spijkman, M., Smits, E., Blom, P., de Leeuw, D., Bon Saint Côme, Y., Setayesh, S., and Cantatore, E. (2008) High-performance organic integrated circuits based on solution processable polymer-small molecule blends. *Appl. Phys. Lett.*, **92** (14), 143304.

49 Yoon, M., Yan, H., Facchetti, A., and Marks, T. (2005) Low-voltage organic field-effect transistors and inverters enabled by ultrathin cross-linked polymers as gate dielectrics. *J. Am. Chem. Soc.*, **127** (29), 10388–10395.

50 Klauk, H., Zschieschang, U., Pflaum, J., and Halik, M. (2007) Ultralow-power organic complementary circuits. *Nature*, **445** (7129), 745–748.

51 Sekitani1, T., Zschieschang, U., Klauk, H., and Someya, T. (2010) Flexible organic transistors and circuits with extreme bending stability. *Nat. Mater.*, **9** (12), 1015–1022.

52 Klauk, H., Halik, M., Zschieschang, U., Eder, F., Rohde, D., Schmid, G., and Dehm, C. (2005) Flexible organic complementary circuits. *IEEE Electron Device Lett.*, **52** (4), 618–622.

53 Usta, H., Facchetti, A., and Marks, T.J. (2011) *n*-channel semiconductor materials design for organic complementary circuits. *Acc. Chem. Res.*, **44** (7), 501–510.

54 Yan, H., Zheng, Y., Blache, R., Newman, C., Lu, S., Woerle, J., and Facchett, A. (2008) Solution processed top-gate *n*-channel transistors and complementary circuits on plastics operating in ambient conditions. *Adv. Mater.*, **20** (18), 3393–3398.

55 Graz, I. and Lacour, S. (2010) Complementary organic thin film transistor circuits fabricated directly on silicone substrates. *Org. Electron.*, **11** (11), 1815–1820.

56 Walser, M.P., Kalb, W.L., Mathis, T., Brenner, T.J., and Batlogg, B. (2009) Stable complementary inverters with organic field-effect transistors on Cytop fluoropolymer gate dielectric. *Appl. Phys. Lett.*, **94** (5), 053303.

57 Walser, M.P., Kalb, W.L., Mathis, T., and Batlogg, B. (2009) Low-voltage organic transistors and inverters with ultrathin fluoropolymer gate dielectric. *Appl. Phys. Lett.*, **95** (23), 233301.

58 Na, J., Kitamura, M., and Arakawa, Y. (2009) Low- voltage-operating organic complementary circuits based on pentacene and C60 transistors. *Thin Solid Films*, **517** (6), 2079–2082.

59 Collet, J., Tharaud, O., and Chapoton, A. (2000) Vuillaume, D. Low-voltage, 30 nm channel length, organic transistors with a self- assembled monolayer as gate insulating films. *Appl. Phys. Lett.*, **76** (14), 1941–1943.

60 Halik, M., Klauk, H., Zschieschang, U., *et al.* (2004) Low-voltage organic transistors with an amorphous molecular gate dielectric. *Nature*, **431** (7011), 963–966.

61 Yoon, M.H., Facchetti, A., and Marks, T.J. (2005) σ-π molecular dielectric multilayers for low-voltage organic thin-film transistors. *Proc. Natl Acad. Sci. USA*, **102** (13), 4678–4682.

62 Park, Y.D., Kim, D.H., Jang, Y., *et al.* (2005) Low-voltage polymer thin-film transistors with a self-assembled monolayer as the gate dielectric. *Appl. Phys. Lett.*, **87** (24), 243509.

63 Kim, K.D. and Song, C.K. (2006) Low voltage pentacene thin film transistors employing a self-grown metal-oxide as a gate dielectric. *Appl. Phys. Lett.*, **88** (23), 233508.

64 Ball, J., Wöbkenberga, P., Kooistrab, F., Hummelenb, J., de Leeuwc, D., Bradleya, D., and Anthopoulosa, T. (2009) Complementary circuits based on solution processed low-voltage organic

field-effect transistors. *Synth. Met.*, **159** (21), 2368–2370.

65 Jang, J., Nam, S., Yun, W., Yang, C., Hwang, J., An, T., Chung, D., and Park, C. (2011) High Tg cyclic olefin copolymer/Al_2O_3 bilayer gate dielectrics for flexible organic complementary circuits with low-voltage and air-stable operation. *J. Mater. Chem.*, **21** (33), 12542–12546.

66 Zhang, X., Potscavage, W.J., Choi, S., and Kippelen, B. (2009) Low-voltage flexible organic complementary inverters with high noise margin and high dc gain. *Appl. Phys. Lett.*, **94** (4), 043312.

67 Drury, C., Mutsaers, C., Hart, C., Matters, M., and de Leeuw, D. (1998) Low-cost all-polymer integrated circuits. *Appl. Phys. Lett.*, **73** (1), 108–110.

68 Gelinck, G., Huiema, H., Veenendaal, E., Cantatore, E., Schrijnemakers, L., van der Putten, J., Geuns, T., Beenhakkers, M., Giesbers, J., Huisman, B., Meijr, E., Benito, E., Touwslager, F., Marsman, A., van Rens, B., and de Leeuw, D. (2004) Flexible active-matrix displays and shift registers based on solution-processed organic transistors. *Nat. Mater.*, **3** (2), 106–110.

69 Myny, K., Beenhakkers, M., van Aerle, N., Gelinck, G., and Genoe, J. (2011) Unipolar organic transistor circuits made robust by dual-gate technology. *IEEE J. Solid-State Circuits*, **46** (5), 1223–1230.

70 Hizu, K., Sekitani, T., and Someya, T. (2007) Reduction in operation voltage of complementary organic thin-film transistor inverter circuits using double-gate structures. *Appl. Phys. Lett.*, **90** (9), 093504.

71 Spijkman, M.J., Myny, K., Smits, E.C.P., Heremans, P., Blom, P.W.M., and de Leeuw, D.M. (2011) Dual-gate thin-film transistors, integrated circuits and sensors. *Adv. Mater.*, **23** (29), 3231–3242.

72 Yokota, T., Nakagawa, T., Sekitani, T., Noguchi, Y., Fukuda, K., Zschieschang, U., Klauk, H., Takeuchi, K., Takamiya, M., Sakurai, T., and Someya, T. (2011) Control of threshold voltage in low-voltage organic complementary inverter circuits with floating gate structures. *Appl. Phys. Lett.*, **98** (19), 193302.

73 Zschieschang, U., Ante, F., Schlörholz, M., Schmidt, M., Kern, K., and Klauk, H. (2010) Mixed self-assembled monolayer gate dielectrics for continuous threshold voltage control in organic transistors and circuits. *Adv. Mater.*, **22** (40), 4489–4493.

74 Marien, H., Steyaert, M., van Veenendaal, E., and Heremans, P. (2011) Organic dual DC-DC upconverter on foil for improved circuit reliability. *Electron. Lett.*, **47** (4), 278–280.

75 Nuzzo, R.G. and Allara, D.L. (1983) Adsorption of bifunctional organic disulfides on gold surfaces. *J. Am. Chem. Soc.*, **105** (13), 4481–4483.

76 Sagiv, J. (1980) Organized monolayers by adsorption. 1. Formation and structure of oleophobic mixed monolayers on solid surfaces. *J. Am. Chem. Soc.*, **102** (1), 92–98.

77 Ulman, A. (1998) *Self-Assembled Monolayers of Thiols*, vol. 24, Academic Press, San Diego, CA.

78 de Boer, B., Hadipour, A., Mandoc, M.M., van Woudenbergh, T., and Blom, P.W.M. (2005) Tuning of metal work functions with self-assembled monolayers. *Adv. Mater.*, **17** (5), 621–625.

79 Mathijssen, S.G.J., van Hal, P.A., van den Biggelaar, T.J.M., Smits, E.C.P., de Boer, B., Kemerink, M., Janssen, R.A.J., and de Leeuw, D.M. (2008) Manipulating the local light emission in organic light-emitting diodes by using patterned self-assembled monolayers. *Adv. Mater.*, **20** (14), 2703–2706.

80 Akkerman, H.B., Blom, P.W.M., de Leeuw, D.M., and de Boer, B. (2006) Towards molecular electronics with large-area molecular junctions. *Nature*, **441** (7089), 69–72.

81 DiBenedetto, S.A., Facchetti, A., Ratner, M.A., and Marks, T.J. (2009) Molecular self-assembled monolayers and multilayers for organic and unconventional inorganic thin-film transistor applications. *Adv. Mater.*, **21** (14–15), 1407–1433.

82 Guo, X., Myers, M., Xiao, S., Lefenfeld, M., Steiner, R., Tulevski, G.S., Tang, J.,

Baumert, J., Leibfarth, F., Yardley, J.T., Steigerwald, M.L., Kim, P., and Nuckolls, C. (2006) Chemoresponsive monolayer transistors. *Proc. Natl Acad. Sci. USA*, **103** (31), 11452–11456.

83 Tulevski, G.S., Miao, Q., Fukuto, M., *et al.* (2004) Attaching organic semiconductors to gate oxides: *in situ* assembly of monolayer field effect transistors. *J. Am. Chem. Soc.*, **126** (46), 15048–15050.

84 Mottaghi, M., Lang, P., Rodriguez, F., *et al.* (2007) Low-operating-voltage organic transistors made of bifunctional self-assembled monolayers. *Adv. Funct. Mater.*, **17** (4), 597–604.

85 Gholamrezaie, F., Andringa, A.M., Roelofs, W.S.C., Neuhold, A., Kemerink, M., Kemerink, M., Blom, P.W.M., and de Leeuw, D.M. (2012) Charge trapping by self-assembled monolayers as the origin of the threshold voltage shift in organic field-effect transistors. *Small*, **8** (2), 241–245.

86 Whitesides, G.M. and Grzybowski, B. (2002) Self-assembly at all scales. *Science*, **295** (5564), 2418–2421.

87 Smits, E., Mathijssen, S., Hal, P., Setayesh, S., Geuns, T., Mutsaers, K., Cantatore, E., Wondergem, H., Werzer, O., Resel, R., Kemerink, M., Kirchmeyer, S., Muzafarov, A., Ponomarenko, S., de Boer, B., and de Leeuw, P. (2008) Bottom-up organic integrated circuits. *Nature*, **455** (7215), 956–959.

88 Gholamrezaie, F., Mathijssen, S.G.J., Smits, E.C.P., Geuns, T.C.T., van Hal, P.A., Ponomarenko, S.A., Flesch, H.G., Resel, R., Cantatore, E., Blom, P.W.M., and de Leeuw, D.M. (2010) Ordered semiconducting self- assembled monolayers on polymeric surfaces utilized in organic integrated circuits. *Nano Lett.*, **10** (6), 1998–2002.

89 Onclin, S., Ravoo, B.J., and Reinhoudt, D.N. (2005) Engineering silicon oxide surfaces using self-assembled monolayers. *Angew. Chem. Int. Ed.*, **44** (39), 6282–6304.

90 van Breemen, A.J.J.M., Herwig, P.T., Chlon, C.H.T., *et al.* (2006) Large area liquid crystal monodomain field-effect transistors. *J. Am. Chem. Soc.*, **128** (7), 2336–2345.

91 Ponomarenko, S.A., Tatarinova, E.A., Muzafarov, A.M., *et al.* (2006) Star-shaped oligothiophenes for solution-processible organic electronics: flexible aliphatic spacers approach. *Chem. Mater.*, **18** (17), 4101–4108.

92 McCulloch, I., Heeney, M., Bailey, C., *et al.* (2006) Liquid-crystalline semiconducting polymers with high charge carrier mobility. *Nat. Mater.*, **5** (4), 328–333.

93 Cantatore, E., Geuns, T.C.T., Gelinck, G.H., *et al.* (2007) A 13.56-MHz RFID system based on organic transponders. *IEEE J. Solid State Circuits*, **42** (1), 84–92.

94 Noh, Y.Y., Zhao, N., Caironi, M., and Sirringhaus, H. (2007) Downscaling of self-aligned, all printed polymer thin-film transistors. *Nat. Nanotechnol.*, **2** (12), 784–789.

95 Liu, Y., Cui, T., and Varahramyan, K. (2003) All-polymer capacitor fabricated with inkjet printing technique. *Solid State Electron.*, **47** (9), 1543–1548.

96 Comiskey, B., Albert, J.D., Yoshizawa, H., and Jacobson, J. (1998) An electrophoretic ink for all-printed reflective electronic displays. *Nature*, **394**, 253–255.

97 Bharathan, J. and Yang, Y. (1998) Polymer electroluminescent devices processed by inkjet printing: I. Polymer light-emitting logo. *Appl. Phys. Lett.*, **72** (21), 2660–2662.

98 Garnier, F., Hajlaoui, R., Yassar, A., and Srivastava, P. (1994) All-polymer field-effect transistor realized by printing techniques. *Science*, **265** (5179), 1684–1686.

99 Cho, J.H., Lee, J., Xia, Y., Kim, B.S., He, Y., Renn, M.J., Lodge, T.P., and Daniel Frisbie, C. (2008) Printable ion-gel gate dielectrics for low-voltage polymer thin-film transistors on plastic. *Nat. Mater.*, **7** (11), 900–906.

100 Sirringhaus, H., Kawase, T., Friend, R.H., *et al.* (2000) High-resolution inkjet printing of all-polymer transistor circuits. *Science*, **290** (5499), 2123–2126.

101 Knobloch, A., Manuelli, A., Bernds, A., and Clemens, W. (2004) Fully printed integrated circuits from solution processable polymers. *J. Appl. Phys.*, **96** (4), 2286–2291.

102 Tan, H.S., Wang, B.C., Kamath, S., Chua, J., Shojaei-Baghini, M., Rao, V.R., Mathews, N., and Mhaisalkar, S.G. (2010) Complementary organic circuits using evaporated F16CuPc and inkjet printing of PQT. *IEEE Electron Devices Lett.*, **31** (11), 1311–1313.

103 Zhao, Y., Di, C., Gao, X., Hu, Y., Guo, Y., Zhang, L., Liu, Y., Wang, J., Hu, W., and Zhu, D. (2011) All-solution-processed, high-performance n-channel organic transistors and circuits: toward low-cost ambient electronics. *Adv. Mater.*, **23** (21), 2448–2453.

104 Jiang, L., Dong, H., and Hu, W. (2010) Organic single crystal field-effect transistors: advances and perspectives. *J. Mater. Chem.*, **20** (24), 4994–5007.

105 Briseno, A.L., Tseng, R.J., Li, S., Chu, C., Yang, Y., Falcao, E.H.L., Wudl, F., Ling, M., Chen, H., Bao, Z., Meng, H., and Kloc, C. (2006) Organic single-crystal complementary inverter. *Appl. Phys. Lett.*, **89** (22), 222111.

106 Uemura, T., Yamagishi, M., Okada, Y., Nakayama, K., Yoshizumi, M., Uno, M., and Takeya, J. (2010) Monolithic complementary inverters based on organic single crystals. *Adv. Mater.*, **22** (35), 3938–3941.

107 Menard, E., Podzorov, V., Hur, S.H., Gaur, A., Gershenson, M.E., and Rogers, J.A. (2004) High-performance *n*- and *p*-type single-crystal organic transistors with free-space gate dielectrics. *Adv. Mater.*, **16** (23–24), 2097–2101.

108 Takeya, J., Yamagishi, M., Tominari, Y., Hirahara, R., Nakazawa, Y., Nishikawa, T., Kawase, T., and Shimoda, T. (2007) Very high- mobility organic single-crystal transistors with in-crystal conduction channels. *Appl. Phys. Lett.*, **90** (10), 102120.

109 de Boer, R.W.I., Stassen, A.F., Craciun, M.F., Mulder, C.L., Molinari, A., Molinari, A., Molinari, A., Rogge, S., and Morpurgo, A.F. (2005) Ambipolar Cu- and Fe-phthalocyanine single-crystal field-effect transistors. *Appl. Phys. Lett.*, **86** (26), 262109.

110 Briseno, A.L., Mannsfeld, S.C.B., Reese, C., Hancock, J.M., Xiong, Y., Jenekhe, S.A., Bao, Z., and Xia, Y. (2007) Perylenediimide nanowires and their use in fabricating field-effect transistors and complementary inverters. *Nano Lett.*, **7** (9), 2847–2853.

111 Jiang, L., Hu, W., Wei, Z., Xu, W., and Hong, M. (2009) High-performance organic single-crystal transistors and digital inverters of an anthracene derivative. *Adv. Mater.*, **21** (36), 3649–3653.

112 Zhang, Y., Tang, Q., Li, H., and Hu, W. (2009) Hybrid bipolar transistors and inverters of nanoribbon crystals. *Appl. Phys. Lett.*, **94** (20), 203304.

113 Aviram, A. and Ratner, M.A. (1974) Molecular rectifiers. *Chem. Phys. Lett.*, **29** (2), 277–283.

114 Heath, J.R. and Ratner, M.A. (2003) Molecular electronics. *Phys. Today*, **56** (5), 43–49.

115 Kornyshev, A.A., Kuznetsov, A.M., and Ulstrup, J. (2006) *In situ* superexchange electron transfer through a single molecule: a rectifying effect. *Proc. Natl Acad. Sci. USA*, **103** (18), 6799–6804.

116 Metzger, R.M., Chen, B., Hopfner, U., Lakshmikantham, M.V., Vuillaume, D., Kawai, T., Wu, X.L., Tachibana, H., Hughes, T.V., Sakurai, H., Baldwin, J.W., Hosch, C., Cava, M.P., Brehmer, L., and Ashwell, G.J. (1997) Unimolecular electrical rectification in hexadecylquinolinium tricyanoquinodimethanide. *J. Am. Chem. Soc.*, **119** (43), 10455–10466.

117 Collier, C.P., Mattersteig, G., Wong, E.W., Luo, Y., Beverly, K., Sampaio, J., Raymo, F.M., Stoddart, F., and Heath, J.R. (2000) A [2]catenane-based solid state electronically reconfigurable switch. *Science*, **289** (5482), 1172–1175.

118 Blum, A.S., Kushmerick, J.G., Long, D.P., Patterson, C.H., Yang, J.C., Henderson, J.C., Yao, Y.X., Tour, J.M., Shashidhar, R., and Ratna, B.R. (2005) Molecularly inherent voltage-controlled

conductance switching. *Nat. Mater.*, **4** (2), 167–172.

119 Park, H., Park, J., Lim, A., Anderson, E., Alivisatos, A., and McEuen, P. (2000) Nanomechanical oscillations in a single-C_{60} transistor. *Nature*, **407** (6800), 57–60.

120 Kubatkin, S., Danilov, A., Hjort, M., Cornil, J., Bredas, J.L., Stuhr-Hansen, N., Hedegard, P., and Bjornholm, T. (2003) Single-electron transistor of a single organic molecule with access to several redox states. *Nature*, **425** (6959), 698–701.

121 Yu, L.H. and Natelson, D. (2004) The kondo effect in C_{60} single-molecule transistors. *Nano Lett.*, **4** (1), 79–83.

122 Li, T., Hu, W., and Zhu, D. (2010) Nanogap electrodes. *Adv. Mater.*, **22** (2), 286–300.

123 Andres, R.P., Bein, T., Dorogi, M., Feng, S., Henderson, J.I., Kubiak, C.P., Mahoney, W., Osifchin, R.G., and Reifenberger, R. (1996) "Coulomb staircase" at room temperature in a self-assembled molecular nanostructure. *Science*, **272** (5266), 1323–1325.

124 Datta, S., Tian, W.D., Hong, S.H., Reifenberger, R., Henderson, J.I., and Kubiak, C.P. (1997) Current-voltage characteristics of self- assembled monolayers by scanning tunneling microscopy. *Phys. Rev. Lett.*, **79** (13), 2530–2533.

125 Dhirani, A., Lin, P.H., Guyot Sionnest, P., Zehner, R.W., and Sita, L.R. (1997) Self- assembled molecular rectifiers. *J. Chem. Phys.*, **106** (12), 5249–5253.

126 Venkataraman, L., Klare, J.E., Nuckolls, C., Hybertsen, M.S., and Steigerwald, M.L. (2006) Dependence of single-molecule junction conductance on molecular conformation. *Nature*, **442** (7105), 904–907.

127 Venkataraman, L., Klare, J.E., Tam, I.W., Nuckolls, C., Hybertsen, M.S., and Steigerwald, M.L. (2006) Single-molecule circuits with well-defined molecular conductance. *Nano Lett.*, **6** (3), 458–462.

128 Fan, F.R.F., Yang, J.P., Cai, L.T., Price, D.W., Dirk, S.M., Kosynkin, D.V., Yao, Y.X., Rawlett, A.M., Tour, J.M., and Bard, A.J. (2002) Charge transport through self-assembled monolayers of compounds of interest in molecular electronics. *J. Am. Chem. Soc.*, **124** (19), 5550–5560.

129 Xu, B.Q. and Tao, N.J. (2003) Measurement of single-molecule resistance by repeated formation of molecular junctions. *Science*, **301** (5637), 1221–1223.

130 Xiao, X.Y., Nagahara, L.A., Rawlett, A.M., and Tao, N.J. (2005) Electrochemical gate-controlled conductance of single oligo (phenylene ethynylene)s. *J. Am. Chem. Soc.*, **127** (25), 9235–9240.

131 Xiao, X.Y., Xu, B.Q., and Tao, N.J. (2004) Measurement of single molecule conductance: benzenedithiol and benzenedimethanethiol. *Nano Lett.*, **4** (2), 267–271.

132 Wu, S.W., Ogawa, N., and Ho, W. (2006) Atomic-scale coupling of photons to single-molecule junctions. *Science*, **312** (5778), 1362–1365.

133 Lewis, P.A., Inman, C.E., Maya, F., Tour, J.M., Hutchison, J.E., and Weiss, P.S. (2005) Molecular engineering of the polarity and interactions of molecular electronic switches. *J. Am. Chem. Soc.*, **127** (49), 17421–17426.

134 He, J., Fu, Q., Lindsay, S., Ciszek, J.W., and Tour, J.M. (2006) Electrochemical origin of voltage-controlled molecular conductance switching. *J. Am. Chem. Soc.*, **128** (46), 14828–14835.

135 Kushmerick, J.G., Holt, D.B., Yang, J.C., Naciri, J., Moore, M.H., and Shashidhar, R. (2002) Metal-molecule contacts and charge transport across monomolecular layers: measurement and theory. *Phys. Rev. Lett.*, **89** (8), 086802.

136 Kushmerick, J.G., Holt, D.B., Pollack, S.K., Ratner, M.A., Yang, J.C., Schull, T.L., Naciri, J., Moore, M.H., and Shashidhar, R. (2002) Effect of bond-length alternation in molecular wires. *J. Am. Chem. Soc.*, **124** (36), 10654–10655.

137 Kushmerick, J.G., Naciri, J., Yang, J.C., and Shashidhar, R. (2003) Conductance scaling of molecular wires in parallel. *Nano Lett.*, **3** (7), 897–900.

138 Rampi, M.A., Schueller, O.J.A., and Whitesides, G.M. (1998) Alkanethiol self-assembled monolayers as the

dielectric of capacitors with nanoscale thickness. *Appl. Phys. Lett.*, **72** (14), 1781–1783.

139 Rampi, M.A. and Whitesides, G.M. (2002) A versatile experimental approach for understanding electron transport through organic materials. *Chem. Phys.*, **281** (2), 373–391.

140 Gregory, S. (1990) Inelastic tunneling spectroscopy and single-electron tunneling in an adjustable microscopic tunnel junction. *Phys. Rev. Lett.*, **64** (6), 689–692.

141 Zimmerman, D.T., Weimer, M.B., and Agnolet, G. (1999) An adjustable oxide-free tunnel junction for vibrational spectroscopy of molecules. *Appl. Phys. Lett.*, **75** (16), 2500–2502.

142 Gregory, S. (1991) Coulomb blockade and the intrinsically one-dimensional character of microscopic tunnel junctions. *Phys. Rev. B*, **44** (23), 12868–12872.

143 Gregory, S. (1992) Experimental observation of scattering of tunneling electrons by a single magnetic moment. *Phys. Rev. Lett.*, **68** (13), 2070–2073.

144 Kergueris, C., Bourgoin, J.P., Palacin, S., Esteve, D., Urbina, C., Magoga, M., and Joachim, C. (1999) Electron transport through a metal- molecule-metal junction. *Phys. Rev. B*, **59** (19), 12505–12513.

145 Saifullah, M.S.M., Ondarcuhu, T., Koltsov, D.K., Joachim, C., and Welland, M.E. (2002) A reliable scheme for fabricating sub-5 nm co-planar junctions for single-molecule electronics. *Nanotechnology*, **13** (5), 659–662.

146 Liang, W.J., Shores, M.P., Bockrath, M., Long, J.R., and Park, H. (2002) Kondo resonance in a single-molecule transistor. *Nature*, **417** (6890), 725–729.

147 Qing, Q., Chen, F., Li, P.G., Tang, W.H., Wu, Z.Y., and Liu, Z.F. (2005) Finely tuning metallic nanogap size with electrodeposition by utilizing high-frequency impedance in feedback. *Angew. Chem. Int. Ed.*, **44** (47), 7771–7775.

148 Nagase, T., Kubota, T., and Mashiko, S. (2003) Fabrication of nano-gap electrodes for measuring electrical properties of organic molecules using a focused ion beam. *Thin Solid Films*, **438–439** (1–2), 374–377.

149 Reed, M.A., Zhou, C., Muller, C.J., Burgin, T.P., and Tour, J.M. (1997) Conductance of a molecular junction. *Science*, **278** (5336), 252–254.

150 Chen, W., Ahmed, H., and Nakazoto, K. (1995) Coulomb blockade at 77 K in nanoscale metallic islands in a lateral nanostructure. *Appl. Phys. Lett.*, **66** (24), 3383–3384.

151 Morpurgo, A.F., Marcus, C.M., and Robinson, D.B. (1999) Controlled fabrication of metallic electrodes with atomic separation. *Appl. Phys. Lett.*, **74** (14), 2084–2086.

152 Park, J., Pasupathy, A.N., Goldsmith, J.I., Chang, C., Yaish, Y., Petta, J.R., Rinkoski, M., Sethna, J.P., Abruna, H.D., McEuen, P.L., and Ralph, D.C. (2002) Coulomb blockade and the Kondo effect in single-atom transistors. *Nature*, **417** (6890), 722–725.

153 Notargiacomo, A., Foglietti, V., Cianci, E., Capellini, G., Adami, M., Faraci, P., Evangelisti, F., and Nicolini, C. (1999) Atomic force microscopy lithography as a nanodevice development technique. *Nanotechnology*, **10** (4), 458–463.

154 Qin, L.D., Park, S., Huang, L., and Mirkin, C.A. (2005) On-wire lithography. *Science*, **309** (5731), 113–115.

155 Hatzor, A. and Weiss, P.S. (2001) Molecular rulers for scaling down nanostructures. *Science*, **291** (5506), 1019–1020.

156 Moreland, J. and Ekin, J.W. (1985) Electron tunneling experiments using NbSn "break" junctions. *J. Appl. Phys.*, **58** (10), 3888–3895.

157 Muller, C.J., Vleeming, B.J., Reed, M.A., Lamba, J.J.S., Hara, R., Jones, L., and Tour, J.M. (1996) Atomic probes: a search for conduction through a single molecule. *Nanotechnology*, **7** (4), 409–411.

158 Song, H., Reed, M.A., and Lee, T. (2011) Single molecule electronic devices. *Adv. Mater.*, **23** (14), 1583–1608.

159 Krans, J.M., Muller, C.J., Yanson, I.K., Govaert, T.C.M., Hesper, R., and van Ruitenbeek, J.M. (1993) One-atom point contacts. *Phys. Rev. B*, **48** (19), 14721–14724.

160 Krans, J.M. and van Ruitenbeek, J.M. (1994) Subquantum conductance steps in atom-sized contacts of the semimetal Sb. *Phys. Rev. B*, **50** (23), 17659–17661.

161 van Ruitenbeek, J.M., Alvarez, A., Pineyro, I., Grahmann, C., Joyez, P., Devoret, M.H., Esteve, D., and Urbina, C. (1996) Adjustable nanofabricated atomic size contacts. *Rev. Sci. Instrum.*, **67** (1), 108–111.

162 Muller, C.J., van Ruitenbeek, J.M., and de Jongh, L.J. (1992) Conductance and supercurrent discontinuities in atomic-scale metallic constrictions of variable width. *Phys. Rev. Lett.*, **69** (1), 140–143.

163 Yanson, I.K., Fisun, V.V., Hesper, R., Khotkevich, A.V., Krans, J.M., Mydosh, J.A., and van Ruitenbeek, J.M. (1995) Size dependence of kondo scattering in point contacts. *Phys. Rev. Lett.*, **74** (2), 302–305.

164 Vanderpost, N., Peters, E.T., Yanson, I.K., and van Ruitenbeek, J.M. (1994) Subgap structure as function of the barrier in atom-size superconducting tunnel junctions. *Phys. Rev. Lett.*, **73** (19), 2611–2613.

165 Krans, J.M., van Ruitenbeek, J.M., Fisun, V.V., Yanson, I.K., and de Ongh, L.J. (1995) The signature of conductance quantization in metallic point contacts. *Nature*, **375** (6534), 767–769.

166 Yanson, A.I., Boolinger, G.R., van den Brom, H.E., Agrait, N., and van Ruitenbeek, J.M. (1998) Formation and manipulation of a metallic wire of single gold atoms. *Nature*, **395** (6704), 783–785.

167 Zhou, C., Muller, C.J., Deshpande, M.R., Sleight, J.W., and Reed, M.A. (1995) Microfabrication of a mechanically controllable break junction in silicon. *Appl. Phys. Lett.*, **67** (8), 1160–1162.

168 van Ruitenbeek, J. and Scheer, E. (eds) (2005) *Introducing Molecular Electronics (Lecture Notes in Physics)*, Springer, Heidelberg.

169 Reichert, J., Ochs, R., Beckmann, D., Weber, H.B., Mayor, M., and von Lohneysen, H. (2002) Driving current through single organic molecules. *Phys. Rev. Lett.*, **88** (17), 176804.

170 Weber, H.B., Reichert, J., Weigend, F., Ochs, R., Beckmann, D., Mayor, M., Ahlrichs, R., and von Lohneysen, H. (2002) Electronic transport through single conjugated molecules. *Chem. Phys.*, **281** (2–3), 113–125.

171 Reichert, J., Weber, H.B., Mayor, M., and von Lohneysen, H. (2003) Low-temperature conductance measurements on single molecules. *Appl. Phys. Lett.*, **82** (23), 4137–4139.

172 Djukic, D. and van Ruitenbeek, J.M. (2006) Shot noise measurements on a single molecule. *Nano Lett.*, **6** (4), 789–793.

173 Lortscher, E., Ciszek, J.W., Tour, J., and Riel, H. (2006) Reversible and controllable switching of a single-molecule junction. *Small*, **2** (8–9), 973–977.

174 Tian, J.H., Liu, B., Li, X.L., Yang, Z.L., Ren, B., Wu, S.T., Tao, N.J., and Tian, Z.Q. (2006) Study of molecular junctions with a combined surface-enhanced Raman and mechanically controllable break junction method. *J. Am. Chem. Soc.*, **128** (46), 14748–14749.

175 Xiang, J., Liu, B., Wu, S.T., Ren, B., Yang, F.Z., Mao, B.W., Chow, Y.L., and Tian, Z.Q. (2005) A controllable electrochemical fabrication of metallic electrodes with a nanometer/angstrom-sized gap using an electric double layer as feedback. *Angew. Chem. Int. Ed.*, **44** (8), 1265–1268.

176 Kervennic, Y.V., Van der Zant, H.S.J., Morpurgo, A.F., Gurevich, L., and Kouwenhoven, L.P. (2002) Nanometer-spaced electrodes with calibrated separation. *Appl. Phys. Lett.*, **80** (2), 321–323.

177 Li, C.Z., He, H.X., and Tao, N.J. (2000) Quantized tunneling current in the metallic nanogaps formed by electrodeposition and etching. *Appl. Phys. Lett.*, **77** (24), 3995–3997.

178 Kashimura, Y., Nakashima, H., Furukawa, K., and Torimitsu, K. (2003) Fabrication of nano-gap electrodes using electroplating technique. *Thin Solid Films*, **438**, 317–321.

179 Li, C.Z., Bogozi, A., Huang, W., and Tao, N.J. (1999) Fabrication of stable metallic nanowires with quantized conductance. *Nanotechnology*, **10** (2), 221–223.

180 Yu, L. and Natelson, D. (2003) Zero-bias anomalies in electrochemically fabricated nanojunctions. *Appl. Phys. Lett.*, **82** (14), 2332–2334.

181 Li, C., He, H., Bogozi, A., Bunch, J., and Tao, N. (2000) Molecular detection based on conductance quantization of nanowires. *Appl. Phys. Lett.*, **76** (10), 1333–1335.

182 He, H., Boussad, S., Xu, B., Li, C., and Tao, N. (2002) Electrochemical fabrication of atomically thin metallic wire and electrodes separated with molecular-scale gaps. *J. Electroanal. Chem.*, **522** (2), 167–172.

183 He, H., Zhu, J., Tao, N.J., Nagahara, L.A., Amlani, I., and Tsui, R. (2001) A conducting polymer nanojunction switch. *J. Am. Chem. Soc.*, **123** (31), 7730–7731.

184 Chen, F., Qing, Q., Ren, L., Wu, Z., and Liu, Z. (2005) Electrochemical approach for fabricating nanogap electrodes with well controllable separation. *Appl. Phys. Lett.*, **86** (12), 123105–123107.

185 Boussaad, S. and Tao, N. (2002) Atom-size gaps and contacts between electrodes fabricated with a self-terminated. *Appl. Phys. Lett.*, **80** (13), 2398–2400.

186 Umeno, A. and Hirakawa, K. (2005) Fabrication of atomic-scale gold junctions by electrochemical plating using a common medical liquid. *Appl. Phys. Lett.*, **86** (14), 143103–143105.

187 Huang, L., Xu, L., Zhang, H., and Gu, N. (2002) Fabrication of a nano-scale gap by selective chemical deposition. *Chem. Commun.*, **7** (1), 72–73.

188 Ah, C., Yun, Y., Lee, J., Park, H., Ha, D., and Yun, W. (2006) Fabrication of integrated nanogap electrodes by surface-catalyzed chemical deposition. *Appl. Phys. Lett.*, **88** (13), 133116–133118.

189 Yasutake, Y., Kono, K., Kanehara, M., Teranishi, T., Buitelaar, M., Smith, C., and Majima, Y. (2007) Simultaneous fabrication of nanogap gold electrodes by electroless gold plating using a common medical liquid. *Appl. Phys. Lett.*, **91** (20), 203107.

190 Dolan, G. (1977) Offset masks for lift-off photoprocessing. *Appl. Phys. Lett.*, **31** (5), 337–339.

191 Klein, D., McEuen, P., Katari, J., Roth, R., and Alivisatos, A. (1996) An approach to electrical studies of single nanocrystals. *Appl. Phys. Lett.*, **68** (18), 2574–2576.

192 Naitoh, Y., Liang, T., Azehara, H., and Mizutani, W. (2005) Measuring molecular conductivities using single molecular-sized gap junctions fabricated without using electron beam lithography. *Jpn. J. Appl. Phys. Part 2*, **44** (12–15), L472–L474.

193 Kubatkin, S., Danilov, A., Hjort, M., Cornil, J., Bredas, J., Stuhr-Hansen, N., Hedegard, P., and Bjornholm, T. (2004) Single electron transistor with a single conjugated molecule. *Curr. Appl. Phys.*, **4** (5), 554–558.

194 Kubatkin, S., Danilov, A., Olin, H., and Claeson, T. (2000) Tunneling through a single quench-condensed cluster. *J. Low. Temp. Phys.*, **118** (5–6), 307–316.

195 Kanda, A., Wada, M., Hamamoto, Y., and Ootuka, Y. (2005) Simple and controlled fabrication of nanoscale gaps using double-angle evaporation. *Physica E*, **29** (3–4), 707–711.

196 Philipp, G., Weimann, T., Hinze, P., Burghard, M., and Weis, J. (1999) Shadow evaporation method for fabrication of sub 10 nm gaps between metal electrodes. *Microelectron. Eng.*, **46** (1–4), 157–160.

197 Otsuka, Y., Naitoh, Y., Matsumoto, T., Mizutani, W., Tabata, H., and Kawai, T. (2004) A simple fabrication method of nanogap electrodes for top-contacted geometry: application to porphyrin nanorods and a DNA network. *Nanotechnology*, **15** (11), 1639–1644.

198 Sun, L., Chin, S., Marx, E., Curtis, K., Greenham, N., and Ford, C. (2005) Shadow-evaporated nanometre-sized gaps and their use in electrical studies of nanocrystals. *Nanotechnology*, **16** (6), 631–634.

199 Black, J. (1969) Electromigration – A brief survey and some recent results. *IEEE Trans. Electron. Devices*, **16** (4), 338–347.

200 Ho, P. and Kwok, T. (1989) Electromigration in metals. *Rep. Prog. Phys.*, **52** (3), 301–348.

201 Park, H., Lim, A., Alivisatos, A., Park, J., and McEuen, P. (1999) Fabrication of metallic electrodes with nanometer separation by electromigration. *Appl. Phys. Lett.*, **75** (2), 301–303.

202 Strachan, D., Smith, D., Johnston, D., Park, T., Therien, M., Bonnell, D., and Johnson, A. (2005) Controlled fabrication of nanogaps in ambient environment for molecular electronics. *Appl. Phys. Lett.*, **86** (4), 043109–043111.

203 Mahapatro, A., Ying, J., Ren, T., and Janes, D. (2008) Electronic transport through ruthenium-based redox-active molecules in metal-molecule- metal nanogap junctions. *Nano Lett.*, **8** (8), 2131–2136.

204 Ghosh, S., Halimun, H., Mahapatro, A., Choi, J., Lodha, S., and Janes, D. (2005) Device structure for electronic transport through individual molecules using nanoelectrodes. *Appl. Phys. Lett.*, **87** (23), 233509.

205 Lambert, M., Goffman, M., Bourgoin, J., and Hesto, P. (2003) Fabrication and characterization of sub-3 nm gaps for single-cluster and single-molecule experiments. *Nanotechnology*, **14** (7), 772–777.

206 Bolotin, K., Kuemmeth, F., Pasupathy, A., and Ralph, D. (2006) From ballistic transport to tunneling in electromigrated ferromagnetic breakjunctions. *Nano Lett.*, **6** (1), 123–127.

207 Yu, H., Luo, Y., Beverly, K., Stoddart, J., Tseng, H., and Heath, J. (2003) The molecule– electrode interface in single-molecule transistors. *Angew. Chem. Int. Ed.*, **42** (46), 5706–5711.

208 Johnston, D., Strachan, D., and Johnson, A. (2007) Parallel fabrication of nanogap electrodes. *Nano Lett.*, **7** (9), 2774–2777.

209 Heersche, H., de Groot, Z., Folk, J., van der Zant, H., Romeke, C., Wegewijs, M., Zobbi, L., Barreca, D., Tondello, E., and Cornia, A. (2006) Electron transport through single Mn_{12} molecular magnets. *Phys. Rev. Lett.*, **96** (20), 206801–206804.

210 Heersche, H., Groot, Z., Folk, J., Kouwenhoven, L., van der Zant, H., Houck, A., Labaziewicz, J., and Chuang, I. (2006) Kondo effect in the presence of magnetic impurities. *Phys. Rev. Lett.*, **96** (1), 017205–017208.

211 Poot, M., Osorio, E., O'Neill, K., Thijssen, J., Vanmaekelbergh, D., van Walree, C., Jenneskens, L., and van der Zant, H. (2006) Temperature dependence of three-terminal molecular junctions with sulfur end-functionalized tercyclohexylidenes. *Nano Lett.*, **6** (5), 1031–1035.

212 Esen, G. and Fuhrer, M. (2005) Temperature control of electromigration to form gold nanogap junctions. *Appl. Phys. Lett.*, **87** (26), 263101–263103.

213 Ramachandran, G., Edelstein, M., Blackburn, D., Suehle, J., Vogel, E., and Richter, C. (2005) Nanometre gaps in gold wires are formed by thermal migration. *Nanotechnology*, **16** (8), 1294–1299.

214 Trouwborst, M., van der Molen, S., and van Wees, B. (2006) The role of Joule heating in the formation of nanogaps by electromigration. *J. Appl. Phys.*, **99** (11), 114316–114322.

215 Strachan, D., Smith, D., Fischbein, M., Johnston, D., Guiton, B., Drndic, M., Bonnell, D., and Johnson, A. (2006) Clean electromigrated nanogaps imaged by transmission electron microscopy. *Nano Lett.*, **6** (3), 441–444.

216 Dai, H. (2002) Carbon nanotubes: synthesis, integration, and properties. *Accounts Chem. Res.*, **35** (12), 1035–1044.

217 Frank, S., Poncharal, P., Wang, Z., and de Heer, W. (1998) Carbon nanotube quantum resistors. *Science*, **280** (5370), 1744–1746.

218 Collins, P., Hersam, M., Arnold, M., Martel, R., and Avouris, P. (2001) Current saturation and electrical breakdown in multiwalled carbon nanotubes. *Phys. Rev. Lett.*, **86** (14), 3128–3131.

219 Collins, P., Arnold, M., and Avouris, P. (2001) Engineering carbon nanotubes and nanotube circuits using electrical breakdown. *Science*, **292** (5517), 706–709.

220 Wei, D., Liu, Y., Cao, L., Wang, Y., Zhang, H., and Yu, G. (2008) Real time and *in situ* control of the gap size of nanoelectrodes for molecular devices. *Nano Lett.*, **8** (6), 1625–1630.

221 Qi, P., Javey, A., Rolandi, M., Wang, Q., Yenilmez, E., and Dai, H. (2004) Miniature organic transistors with carbon nanotubes as quasi-one-dimensional electrodes. *J. Am. Chem. Soc.*, **126** (38), 11774–11775.

222 Tsukagoshi, K., Yagi, I., and Aoyagi, Y. (2004) Pentacene nanotransistor with carbon nanotube electrodes. *Appl. Phys. Lett.*, **85** (6), 1021–1023.

223 Yagi, I., Tsukagoshi, K., Watanabe, E., and Aoyagi, Y. (2004) Carbon nanotubes with a nanogap for nanoscale organic devices. *Microelectron. Eng.*, **73–74**, 675–678.

224 Anderson, M., Tan, L., Tanaka, H., Mihok, M., Lee, H., Horn, M., and Weiss, P. (2003) Advances in nanolithography using molecular rulers. *J. Vac. Sci. Technol. B*, **21** (6), 3116–3119.

225 Negishi, R., Hasegawa, T., Terabe, K., Aono, M., Ebihara, T., Tanaka, H., and Ogawa, T. (2006) Fabrication of nanoscale gaps using a combination of self-assembled molecular and electron beam lithographic techniques. *Appl. Phys. Lett.*, **88** (22), 223111.

226 Negishi, R., Hasegawa, T., Terabe, K., Aono, M., Tanaka, H., Ogawa, T., and Ozawa, H. (2007) *I–V* characteristics of single electron tunneling from symmetric and asymmetric double-barrier tunneling junctions. *Appl. Phys. Lett.*, **90** (22), 223112.

227 McCarty, G. (2004) Molecular lithography for wafer-scale fabrication of molecular junctions. *Nano Lett.*, **4** (8), 1391–1394.

228 Jain, T., Westerlund, F., Johnson, E., Moth-Poulsen, K., and Bjornholm, T. (2009) Self-assembled nanogaps via seed-mediated growth of end-to end linked gold nanorods. *ACS Nano*, **3** (4), 828–834.

229 Luber, S., Strobel, S., Tranitz, H., Wegscheider, W., Schuh, D., and Tornow, M. (2005) Nanometre spaced electrodes on a cleaved AlGaAs surface. *Nanotechnology*, **16** (8), 1182–1185.

230 Luber, S., Zhang, F., Lingitz, S., Hansen, A., Scheliga, F., Thorn-Csanyi, E., Bichler, M., and Tornow, M. (2007) High-aspect-ratio nanogap electrodes for averaging molecular conductance measurements. *Small*, **3** (2), 285–289.

231 Dirk, S., Howell, S., Zmuda, S., Childs, K., Blain, M., Simonson, R., and Wheeler, D. (2005) Novel one-dimensional nanogap created with standard optical lithography and evaporation procedures. *Nanotechnology*, **16** (10), 1983–1985.

232 Chen, C., Sheu, J., Chiang, S., and Sheu, M. (2006) A novel nanofabrication technique for the array of nanogap electrodes. *Jpn. J. Appl. Phys. Part 1*, **45** (6B), 5531–5534.

233 Berg, J., Che, F., Lundgren, P., Enoksson, P., and Bengtsson, S. (2005) Electrical properties of Si SiO_2 Si nanogaps. *Nanotechnology*, **16** (10), 2197–2202.

234 Howell, S., Dirk, S., Childs, K., Pang, H., Blain, M., Simonson, R., Tour, J., and Wheeler, D. (2005) Mass-fabricated one-dimensional silicon nanogaps for hybrid organic/nanoparticle arrays. *Nanotechnology*, **16** (6), 754–758.

235 Yi, M., Jeong, K., and Lee, L. (2005) Theoretical and experimental study towards a nanogap dielectric biosensor. *Biosens. Bioelectron.*, **20** (7), 1320–1326.

236 Liu, S., Tok, J., and Bao, Z. (2005) Nanowire lithography: fabricating controllable electrode gaps using Au-Ag-Au nanowires. *Nano Lett.*, **5** (6), 1071–1076.

237 Qin, L., Jang, J., Huang, L., and Mirkin, C. (2007) Sub-5-nm gaps prepared by on-wire lithography: correlating gap size with electrical transport. *Small*, **3** (1), 86–90.

238 Martin, C. and Baker, L. (2005) Expanding the molecular electronics toolbox. *Science*, **309** (5731), 67–68.

239 Nicewarner-Pena, S., Freeman, R., Reiss, B., He, L., Pena, D., Walton, I., Cromer, R., Keating, C., and Natan, M. (2001) Submicrometer metallic barcodes. *Science*, **294** (5540), 137–141.

240 Xu, T. and Metzger, R. (2002) Nanoditches fabricated using a carbon nanotube as a contact mask. *Nano Lett.*, **2** (10), 1061–1065.

241 Lefebvre, J., Radosavljevic, M., and Johnson, A. (2000) Fabrication of

nanometer size gaps in a metallic wire. *Appl. Phys. Lett.*, **76** (25), 3828–3830.
242 Chen, Z., Hu, W., Guo, J., and Saito, K. (2004) Fabrication of nanoelectrodes based on controlled placement of carbon nanotubes using alternating-current electric field. *J. Vac. Sci. Technol. B*, **22** (2), 776–780.
243 Chung, J. and Lee, J. (2003) Nanoscale gap fabrication and integration of carbon nanotubes by micromachining. *Sens. Actuators A*, **104** (3), 229–235.
244 Chung, J., Lee, K., and Lee, J. (2003) Nanoscale gap fabrication by carbon nanotube-extracted lithography (CEL). *Nano Lett.*, **3** (8), 1029–1031.
245 Huang, X., Caldwell, R., Huang, L., Jun, S., Huang, M., Sfeir, M., O'Brien, S., and Hone, J. (2005) Controlled placement of individual carbon nanotubes. *Nano Lett.*, **5** (7), 1515–1518.
246 De Poortere, E., Stormer, H., Huang, L., Wind, S., O'Brien, S., Huang, M., and Hone, J. (2006) Single-walled carbon nanotubes as shadow masks for nanogap fabrication. *Appl. Phys. Lett.*, **88** (14), 143124.
247 De Poortere, E., Stormer, H., Huang, L., Wind, S., O'Brien, S., Huang, M., and Hone, J. (2006) 1-to 2-nm-wide nanogaps fabricated with single-walled carbon nanotube shadow masks. *J. Vac. Sci. Technol. B*, **24** (6), 3213–3216.
248 Chopra, N., Xu, W., De Long, L., and Hinds, B. (2005) Incident angle dependence of nanogap size in suspended carbon nanotube shadow lithography. *Nanotechnology*, **16** (1), 133–136.
249 Sordan, R., Burghard, M., and Kern, K. (2001) Removable template route to metallic nanowires and nanogaps. *Appl. Phys. Lett.*, **79** (13), 2073–2075.
250 Ancona, M., Kooi, S., Kruppa, W., Snow, A., Foos, E., Whitman, L., Park, D., and Shirey, L. (2003) Patterning of narrow Au nanocluster lines using V_2O_5 nanowire masks and ion-beam milling. *Nano Lett.*, **3** (2), 135–138.
251 Jiang, L., Gao, J., Wang, E., Li, H., Wang, Z., Hu, W., and Jiang, L. (2008) Organic single-crystalline ribbons of a rigid "H"-type anthracene derivative and high-performance, short-channel field-effect transistors of individual micro/nanometer-sized ribbons fabricated by an "organic ribbon mask" technique. *Adv. Mater.*, **20** (14), 2735–2740.
252 Ongaro, A., Griffin, F., Nagle, L., Iacopino, D., Eritja, R., and Fitzmaurice, D. (2004) DNA-templated assembly of a protein-functionalized nanogap electrode. *Adv. Mater.*, **16** (20), 1799–1803.
253 Horiuchi, K., Kato, T., Hashii, S., Hashimoto, A., Sasaki, T., Aoki, N., and Ochiai, Y. (2005) Fabrication of nanoscale C_{60} field-effect transistors with carbon nanotubes. *Appl. Phys. Lett.*, **86** (15), 153108–153110.
254 Blom, T., Welch, K., Stromme, M., Coronel, E., and Leifer, K. (2007) Fabrication and characterization of highly reproducible, high resistance nanogaps made by focused ion beam milling. *Nanotechnology*, **18** (28), 285301.
255 Nagase, T., Gamo, K., Kubota, T., and Mashiko, S. (2005) Maskless fabrication of nanoelectrode structures with nanogaps by using Ga focused ion beams. *Microelectron. Eng.*, **78–79**, 253–259.
256 Nagase, T., Gamo, K., Kubota, T., and Mashiko, S. (2006) Direct fabrication of nano-gap electrodes by focused ion beam etching. *Thin Solid Films*, **499** (1–2), 279–284.
257 Guo, X., Small, J., Klare, J., Wang, Y., Purewal, M., Tam, I., Hong, B., Caldwell, R., Huang, L., O'Brien, S., Yan, J., Breslow, R., Wind, S., Hone, J., Kim, P., and Nuckolls, C. (2006) Covalently bridging gaps in single-walled carbon nanotubes with conducting molecules. *Science*, **311** (5759), 356–359.
258 Whalley, A., Steigerwald, M., Guo, X., and Nuckolls, C. (2007) Reversible switching in molecular electronic devices. *J. Am. Chem. Soc.*, **129** (42), 12590–12591.
259 Guo, X., Whalley, A., Klare, J., Huang, L., O'Brien, S., Steigerwald, M., and Nuckolls, C. (2007) Single-molecule devices as scaffolding for multicomponent nanostructure assembly. *Nano Lett.*, **7** (5), 1119–1122.

260 Guo, X., Gorodetsky, A., Hone, J., Barton, J., and Nuckolls, C. (2008) Conductivity of a single DNA duplex bridging a carbon nanotube gap. *Nat. Nanotechnol.*, **3** (3), 163–167.

261 Feldman, A., Steigerwald, M., Guo, X., and Nuckolls, C. (2008) Molecular electronic devices based on single-walled carbon nanotube electrodes. *Accounts Chem. Res.*, **41** (12), 1731–1741.

262 Venema, L., Wildoer, J., Temminck Tuinstra, H., Dekker, C., Rinzler, A., and Smalley, R. (1997) Length control of individual carbon nanotubes by nanostructuring with a scanning tunneling microscope. *Appl. Phys. Lett.*, **71** (8), 2629–2631.

263 Kim, D., Koo, J., and Kim, J. (2003) Cutting of multiwalled carbon nanotubes by a negative voltage tip of an atomic force microscope: a possible mechanism. *Phys. Rev. B*, **68** (11), 113406–113409.

264 Bouzehouane, K., Fusil, S., Bibes, M., Carrey, J., Blon, T., Du, M., Seneor, P., Cros, V., and Vila, L. (2003) Nanolithography based on real-time electrically controlled indentation with an atomic force microscope for nanocontact elaboration. *Nano Lett.*, **3** (11), 1599–1602.

265 Miyazaki, T., Kobayashi, K., Horiuchi, T., Yamada, H., and Matsushige, K. (2001) Fabrication of a nanogap on a metal nanowire using scanning probe lithography. *Jpn. J. Appl. Phys. Part 1*, **40** (6B), 4365–4367.

266 Zhang, H., Chung, S., and Mirkin, C. (2003) Fabrication of sub-50-nm solid-state nanostructures on the basis of dip-pen nanolithography. *Nano Lett.*, **3** (1), 43–45.

267 Nazin, G., Qiu, X., and Ho, W. (2003) Visualization and spectroscopy of a metal-molecule-metal bridge. *Science*, **302** (5642), 77–81.

268 Repp, J., Meyer, G., Paavilainen, S., Olsson, F., and Persson, M. (2006) Imaging bond formation between a gold atom and pentacene on an insulating surface. *Science*, **312** (5777), 1196–1199.

269 Keane, Z., Ciszek, J., Tour, J., and Natelson, D. (2006) Three-terminal devices to examine single-molecule conductance switching. *Nano Lett.*, **6** (7), 1518–1521.

270 Osorio, E., O'Neill, K., Stuhr-Hansen, N., Nielsen, O., Bjornholm, T., and van der Zant, H. (2007) Addition energies and vibrational fine structure measured in electromigrated single- molecule junctions based on an oligophenylenevinylene derivative. *Adv. Mater.*, **19** (2), 281–285.

271 Chen, J. and Reed, M. (2002) Electronic transport of molecular systems. *Chem. Phys.*, **281** (2–3), 127–145.

272 Hu, W., Jiang, J., Nakashima, H., Luo, Y., Kashimura, Y., Chen, K., Shuai, Z., Furukawa, K., Lu, W., Liu, Y., Zhu, D., and Torimitsu, K. (2006) Electron transport in self-assembled polymer molecular junctions. *Phys. Rev. Lett.*, **96** (2), 027801–027804.

273 Park, J., Pasupathy, A., Goldsmith, J., Soldatov, A., Chang, C., Yaish, Y., Sethna, J., Abruna, H., Ralph, D., and McEuen, P. (2003) Wiring up single molecules. *Thin Solid Films*, **438–439**, 457–461.

274 Maruccio, G., Visconti, P., Arima, V., D'Amico, S., Blasco, A., D'Amone, E., Cingolani, R., Rinaldi, R., Masiero, S., Giorgi, T., and Gottarelli, G. (2003) Field effect transistor based on a modified DNA base. *Nano Lett.*, **3** (4), 479–483.

275 Maruccio, G., Biasco, A., Visconti, P., Bramanti, A., Pompa, P., Calabi, F., Cingolani, R., Rinaldi, R., Corni, S., Di Felice, R., Molinari, E., Verbeet, M., and Canters, G. (2005) Towards protein field-effect transistors: report and model of a prototype. *Adv. Mater.*, **17** (7), 816–822.

276 Yoo, K.H., Ha, D.H., Lee, J.O., Park, J.W., Kim, J., Kim, J.J., Lee, H.Y., Kawai, T., and Choi, H.Y. (2001) Electrical conduction through poly(dA)-poly(dT) and poly(dG)-poly(dC) DNA molecules. *Phys. Rev. Lett.*, **87** (19), 198102–198105.

277 Porath, D., Bezryadin, A., de Vries, S., and Dekker, C. (2000) Direct measurement of electrical transport through DNA molecules. *Nature*, **403** (6770), 635–638.

278 Seidel, R., Graham, A., Kretz, J., Rajasekharan, B., Duesberg, G., Liebau, M., Unger, E., Kreupl, F., and Hoenlein,

W. (2005) Sub-20 nm short channel carbon nanotube transistors. *Nano Lett.*, **5** (1), 147–150.

279 Klein, D., Roth, R., Lim, A., Alivisatos, A., and McEuen, P. (1997) A single-electron transistor made from a cadmium selenide nanocrystal. *Nature*, **389** (6652), 699–701.

280 Galperin, M., Ratner, M.A., Nitzan, A., and Troisi, A. (2008) Nuclear coupling and polarization in molecular transport junctions: beyond tunneling to function. *Science*, **319** (5866), 1056–1060.

281 Wang, W., Lee, T., and Reed, M. (2003) Mechanism of electron conduction in self-assembled alkanethiol monolayer devices. *Phys. Rev. B*, **68** (3), 035416–035422.

282 Song, H., Kim, Y., Jeong, H., Reed, M., and Lee, T. (2010) Coherent tunneling transport in molecular junctions. *J. Phys. Chem. C*, **114** (48), 20431–20435.

283 Choi, S., Kim, B., and Frisbie, C. (2008) Electrical resistance of long conjugated molecular wires. *Science*, **320** (5882), 1482–1486.

284 Lafferentz, L., Ample, F., Yu, H., Hecht, S., Joachim, C., and Grill, L. (2009) Conductance of a single conjugated polymer as a continuous function of its length. *Science*, **323** (5918), 1193–1197.

285 Hines, T., Diez-Perez, I., Hihath, J., Liu, H., Wang, Z., Zhao, J., Zhou, G., Muellen, K., and Tao, N. (2010) Transition from tunneling to hopping in single molecular junctions by measuring length and temperature dependence. *J. Am. Chem. Soc.*, **132** (23), 11658–11664.

286 Cuevas, J. and Scheer, E. (2009) *Molecular Electronics: An Introduction to Theory and Experiment*, World Scientific, Singapore.

287 Tao, N. (2006) Electron transport in molecular junctions. *Nat. Nanotechnol.*, **1** (3), 173–181.

288 Akkerman, H. and de Boer, B. (2008) Electrical conduction through single molecules and self-assembled monolayers. *J. Phys. Condens. Matter.*, **20** (1), 013001.

289 McCreery, R. and Bergren, A. (2009) Progress with molecular electronic junctions: meeting experimental challenges in design and fabrication. *Adv. Mater.*, **21** (43), 4303–4322.

290 Wold, D. and Frisbie, C. (2001) Fabrication and characterization of metal-molecule-metal junctions by conducting probe atomic force microscopy. *J. Am. Chem. Soc.*, **123** (23), 5549–5556.

291 Wold, D., Haag, R., Rampi, M., and Frisbie, C. (2002) Distance dependence of electron tunneling through self-assembled monolayers measured by conducting probe atomic force microscopy: unsaturated versus saturated molecular junctions. *J. Phys. Chem. B*, **106** (11), 2813–2816.

292 Salomon, A., Cahen, D., Lindsay, S., Tomfohr, J., Engelkes, V., and Frisbie, C. (2003) Comparison of electronic transport measurements on organic molecules. *Adv. Mater.*, **15** (22), 1881–1890.

293 Jaklevic, R. and Lambe, J. (1966) Molecular vibration spectra by electron tunneling. *Phys. Rev. Lett.*, **17** (22), 1139–1140.

294 Reed, M. (2008) Inelastic electron tunneling spectroscopy. *Mater. Today*, **11** (11), 46–50.

295 Hansma, P. (1977) Inelastic electron-tunneling. *Phys. Rep.*, **30** (2), 145–206.

296 Adkins, C. and Phillips, W. (1985) Inelastic electron tunneling spectroscopy. *J. Phys. C*, **18** (7), 1313–1346.

297 Horiuchi, T., Ebisawa, F., and Tabei, H. (1989) New inelastic electron tunneling spectrometer with an absolute peak intensity. *Rev. Sci. Instrum.*, **60** (6), 993–996.

298 Wang, W., Lee, T., Kretzschmar, I., and Reed, M. (2004) Inelastic electron tunneling spectroscopy of an alkanedithiol self-assembled monolayer. *Nano Lett.*, **4** (4), 643646.

299 Kushmerick, J., Lazorcik, J., Patterson, C., Shashidhar, R., Seferos, D., and Bazan, G. (2004) Vibronic contributions to charge transport across molecular junctions. *Nano Lett.*, **4** (4), 639–642.

300 Song, H., Kim, Y., Ku, J., Jang, Y., Jeong, H., and Lee, T. (2009) Vibrational spectra of metal-molecule-metal junctions in electromigrated nanogap electrodes by inelastic electron tunneling. *Appl. Phys. Lett.*, **94** (10), 103110–103112.

301 Troisi, A., Ratner, M., and Nitzan, A. (2003) Vibronic effects in off-resonant molecular wire conduction. *J. Chem. Phys.*, **118** (13), 6072–6082.

302 Kushmerick, J.G., Lazorcik, J., Patterson, C., Shashidhar, R., Seferos, D., and Bazan, G. (2004) Vibronic contributions to charge transport across molecular junctions. *Nano Lett.*, **4** (4), 639–642.

303 Long, D., Lazorcik, J., Mantooth, B., Moore, M., Ratner, M., Troisi, A., Yao, Y., Ciszek, J., Tour, J., and Shashidhar, R. (2006) Effects of hydration on molecular junction transport. *Nat. Mater.*, **5** (11), 901–908.

304 Beebe, J., Kim, B., Gadzuk, J., Frisbie, C., and Kushmerick, J. (2006) Transition from direct tunneling to field emission in metal-molecule-metal junctions. *Phys. Rev. Lett.*, **97** (2), 026801–026804.

305 Beebe, J., Kim, B., Frisbie, C., and Kushmerick, J. (2008) Measuring relative barrier heights in molecular electronic junctions with transition voltage spectroscopy. *ACS Nano*, **2** (5), 827–832.

306 Yu, L., Gergel-Hackett, N., Zangmeister, C., Hacker, C., Richter, C., and Kushmerick, J. (2008) Molecule-induced interface states dominate charge transport in Si-alkyl-metal junctions. *J. Phys. Condens. Matter.*, **20** (37), 374114.

307 Huisman, E., Guédon, C.M., van Wees, B., and van der Molen, S. (2009) Interpretation of transition voltage spectroscopy. *Nano Lett.*, **9** (11), 3909–3913.

308 Song, H., Kim, Y., Jang, Y., Jeong, H., Reed, M., and Lee, T. (2009) Observation of molecular orbital gating. *Nature*, **462** (7276), 1039–1043.

309 Chiu, P. and Roth, S. (2008) Transition from direct tunneling to field emission in carbon nanotube intramolecular junctions. *Appl. Phys. Lett.*, **92** (4), 042107–042109.

310 Araidai, M. and Tsukada, M. (2010) Theoretical calculations of electron transport in molecular junctions: inflection behavior in Fowler-Nordheim plot and its origin. *Phys. Rev. B*, **81** (23), 235114–235120.

311 Chen, J., Markussen, T., and Thygesen, K. (2010) Quantifying transition voltage spectroscopy of molecular junctions: *Ab initio* calculations. *Phys. Rev. B*, **82** (12), 121412(R).

312 O'Neill, L. and Byrne, H. (2005) Structure-property relationships for electron-vibrational coupling in conjugated organic oligomeric systems. *J. Phys. Chem. B*, **109** (26), 12685–12690.

313 Baldea, I. (2010) Revealing molecular orbital gating by transition voltage spectroscopy. *Chem. Phys.*, **377** (1–3), 15–20.

314 Kim, W.Y., Choi, Y.C., Min, S.K., Cho, Y., and Kim, K.S. (2009) Application of quantum chemistry to nanotechnology: electron and spin transport in molecular devices. *Chem. Soc. Rev.*, **38** (8), 2319–2333.

315 Patrone, L., Palacin, S., Charlier, J., Armand, F., Bourgoin, J.P., Tang, H., and Gauthier, S. (2003) Evidence of the key role of metal-molecule bonding in metal-molecule-metal transport experiments. *Phys. Rev. Lett.*, **91** (9), 096802–096805.

316 Hu, Y., Zhu, Y., Gao, H., and Guo, H. (2005) Conductance of an ensemble of molecular wires: a statistical analysis. *Phys. Rev. Lett.*, **95** (15), 156803–156806.

317 Venkataraman, L., Park, Y.S., Whalley, A.C., Nuckolls, C., Hybertsen, M.S., and Steigerwald, M.L. (2007) Electronics and chemistry: varying single-molecule junction conductance using chemical substituents. *Nano Lett.*, **7** (2), 502–506.

318 Park, Y.S., Whalley, A.C., Kamenetska, M., Steigerwald, M.L., Hybertsen, M.S., Nuckolls, C., and Venkataraman, L. (2007) Contact chemistry and single-molecule conductance: a comparison of phosphines, methyl sulfides, and amines. *J. Am. Chem. Soc.*, **129** (51), 15768–15769.

319 Quek, S.Y., Venkataman, L., Choi, H.J., Louie, S.G., Hybertsen, M.S., and Neaton, J.B. (2007) Amine-gold linked single-molecule circuits: experiment and theory. *Nano Lett.*, **7** (11), 3477–3482.

320 Neaton, J.B., Hybertsen, M.S., and Louie, S.G. (2006) Renormalization of molecular electronic levels at metal-molecule interfaces. *Phys. Rev. Lett.*, **97** (21), 216405–216408.

321 Nardelli, M.B. (1999) Electronic transport in extended systems: application to carbon nanotubes. *Phys. Rev. B*, **60** (11), 7828–7833.

322 Ventra, M.D., Pantelides, S.T., and Lang, N.D. (2000) First-principles calculation of transport properties of a molecular device. *Phys. Rev. Lett.*, **84** (5), 979–982.

323 Derosa, P.A. and Seminario, J.M. (2001) Electron transport through single molecules: scattering treatment using density functional and green function theories. *J. Phys. Chem. B*, **105** (2), 471–481.

324 Taylor, J., Guo, H., and Wang, J. (2001) Ab initio modeling of quantum transport properties of molecular electronic devices. *Phys. Rev. B*, **63** (24), 245407–245419.

325 Brandbyge, M., Mozos, J.L., Ordejon, P., Taylor, J., and Stokbro, K. (2002) Density-functional method for nonequilibrium electron transport. *Phys. Rev. B*, **65** (16), 165401–165417.

326 Palacios, J.J., Perez-Jimenez, A.J., Louis, E., and Verges, J.A. (2001) Fullerene-based molecular nanobridges: a first-principles study. *Phys. Rev. B*, **64** (11), 115411–115414.

327 Ke, S.H., Baranger, H.U., and Wang, W. (2004) Electron transport through molecules: self-consistent and non-self-consistent approaches. *Phys. Rev. B*, **70** (8), 085410–085421.

328 Kim, Y.H., Tahir-Kheli, J., Schultz, P.A., and Goddard, W.A., III (2006) First-principles approach to the charge-transport characteristics of monolayer molecular-electronics devices: application to hexanedithiolate devices. *Phys. Rev. B*, **73** (23), 235419–235428.

329 Rocha, A.R., Garcia-Suarez, V.M., Bailey, S.W., Lambert, C.J., Ferrer, J., and Sanvito, S. (2006) Spin and molecular electronics in atomically generated orbital landscapes. *Phys. Rev. B*, **73** (8), 085414–085435.

330 Kim, W.Y. and Kim, K.S. (2008) Carbon nanotube, graphene, nanowire, and molecule-based electron and spin transport phenomena using the nonequilibrium Green's function method at the level of first principles theory. *J. Comput. Chem.*, **29** (7), 1073–1083.

331 Sancho, M.P.L., Sancho, J.M.L., and Rubio, J. (1984) Quick iterative scheme for the calculation of transfer matrices: applications to Mo (100). *J. Phys. F Met. Phys.*, **14** (5), 1205–1215.

332 Troisi, A. and Ratner, M.A. (2006) Molecular transport junctions: propensity rules for inelastic electron tunneling spectra. *Nano Lett.*, **6** (8), 1784–1788.

333 Jiang, J., Kula, M., and Luo, Y. (2006) A generalized quantum chemical approach for elastic and inelastic electron transport in molecular electronic devices. *J. Chem. Phys.*, **124** (3), 034708–034717.

334 Pecchia, A., Carlo, A.D., Gagliardi, A., Sanna, S., Frauenheim, T., and Gutierrez, R. (2004) Incoherent electron-phonon scattering in octanethiols. *Nano Lett.*, **4** (11), 2109–2114.

335 Frederiksen, T., Paulsson, M., Brandbyge, M., and Jauho, A.P. (2007) Inelastic transport theory from first principles: methodology and application to nanoscale devices. *Phys. Rev. B*, **75** (20), 205413–205434.

5
Polymer Light-Emitting Diodes (PLEDs): Devices and Materials

Xiong Gong

5.1
Introduction

Since the discovery of the metallic properties of doped polyacetylene in 1977, remarkable progress has been made in synthesizing conjugated polymers with unique properties: the electronic and optical properties of metals and semiconductors in combination with the processing advantages and mechanical properties of polymers. These materials are under development for use in electronic and optical applications, including polymeric light-emitting diodes (PLEDs), photodetectors, photovoltaic cells, sensors, field-effect transistors (FETs), and lasers [1–7].

Research investigations carried out over the past decade have demonstrated the commercial opportunities associated with the implementation of PLEDs in passive and active matrix displays. As a result, the development of PLEDs that show efficient, stable blue, green, and red emissions is an active ongoing research effort in laboratories in Europe, Asia, and the United States.

The aim of this chapter is to introduce concepts and progress in the field of PLEDs and polymer lasers, with details of PLEDs fabricated from conjugated polymers described in Section 5.2, with descriptions of the accurate measurement of PLED device parameters in Section 5.3. The device physics of PLEDs is then discussed in Section 5.4, and the materials used for PLEDs are detailed in Section 5.5. The properties of electrophosphorescent and white-light PLEDs are detailed in Sections 5.6 and Section 5.7, respectively. A summary and details of recommended reference and reading material are provided in Section 5.8 and the References.

Organic Optoelectronics, First Edition. Edited by Wenping Hu.

5.2
PLEDs Fabricated from Conjugated Polymers

5.2.1
Device Architecture

The basic and simplified architecture of a PLED, which is largely built up of conjugated polymers, is shown in Figure 5.1. Under the action of a driving voltage of a few and/or a couple of volts, electrons are injected from a metal cathode with a low work-function into the electronic state corresponding to the lowest unoccupied molecular orbital (LUMO) of a conjugated polymer; holes are injected from a bilayer anode [PEDOT:PSS (poly(3,4-ethylene dioxythiophene): poly(styrene sulfonic acid) on indium tin oxide (ITO)] into the electronic state corresponding to the highest occupied molecular orbital (HOMO) of a conjugated polymer. Both electrons and holes, deriving from the different electrodes, move from opposite directions towards the recombination zone where they can combine to form excitons. This leads to a population of excited states of the emissive polymers which subsequently emits light.

5.2.2
Device Fabrication

Single-layer PLEDs were prepared according to the following procedure. The ITO-coated glass substrate was first cleaned with detergent, ultrasonicated in acetone and isopropyl, and subsequently dried in an oven for at least 6 h. Medium-conducting PEDOT:PSS (e.g., Baytron P 4083) was spin-cast (4000 r.p.m.) with a thickness of about 40 nm from aqueous solution after being passed through a 0.45 μm filter. The substrate was then dried for 10 min at 160 °C in air and moved into a glove box for spin-casting of the emissive layer. Subsequently, the device was pumped down in vacuum ($<10^{-7}$ Torr), and a ~5 nm Ba and/or Ca with 200 nm Al and/or Ag encapsulation film was deposited on top of the active layer.

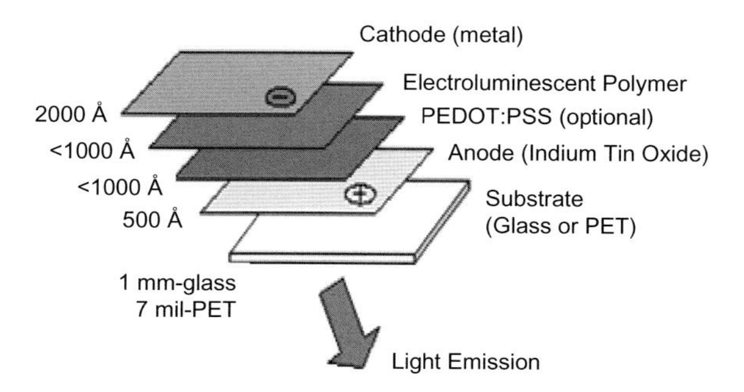

Figure 5.1 Device architecture of single-layer PLEDs.

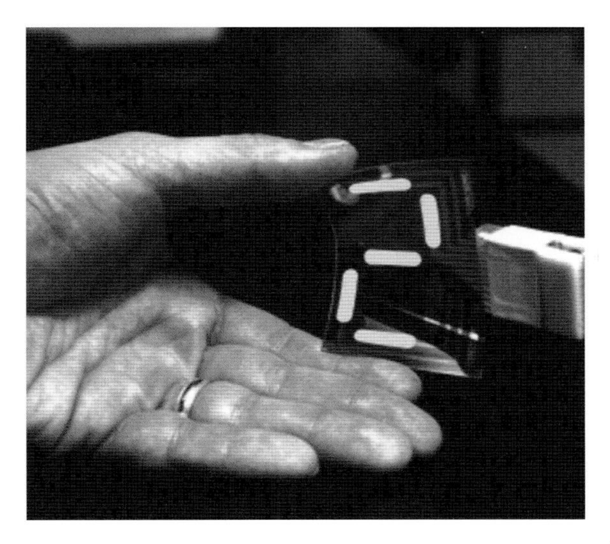

Figure 5.2 Photographic image of a thin-film, flexible, seven-segment display fabricated from conjugated polymers.

For the demonstration device shown in Figure 5.2, polyaniline (PANI) was chosen as the anode material because it is flexible, conducts current, and is transparent to visible light (there is no ITO layer in the device of Figure 5.2). More generally, however, a bilayer electrode comprising a thin layer of metallic polymer cast onto ITO is used for the anode; the ITO carries the current and the metallic polymer layer serves to planarize the ITO and to improve the injection of holes into the luminescent conjugated polymer. The emissive layer of the display shown in Figure 5.2 was formed by spin-casting a layer of polyphenylenevinylene (PPV) derivatives over the polyaniline. The top electrodes that define the seven-segment display were formed by evaporating calcium through a patterned shadow mask. Because of the low conductivity of undoped emissive polymers, it was not necessary to pattern the polymer or the bottom electrode to prevent current spreading between neighboring pixels.

5.3 Accurate Measurement of PLED Device Parameters

5.3.1 Photopic Luminosity

A prime concern of the PLED community is the level of uncertainty and inconsistency of results that have been reported in the scientific literature. Consequently, the details of an accurate method for the measurement of PLED optical properties are provided in the following subsections.

Table 5.1 The quantity and unit of measurement.

Technique of measurement	Type of measurement	Quantity	Unit of measurement
Photometry	Total flux	Total luminous flux	Lumen
Radiometry	Total flux	Total radiant flux	Watt
Spectroradiometry	Total Flux	Total spectral flux	Watt nm^{-1}
Photometry	Angular intensity	Luminous intensity	Candelas = lumen sr^{-1}
Radiometry	Angular intensity	Radiant intensity	Watts sr^{-1}
Spectroradiometry	Angular intensity	Spectroradiometric intensity	Watts $(sr \cdot nm)^{-1}$
Photometry	At a surface	Illuminance	Lux = lumen m^{-2}
Radiometry	At a surface	Irradiance	Watts m^{-2}
Spectroradiometry	At a surface	Spectral irradiance	Watts $(m^{-2} nm)^{-1}$
Photometry	At a surface	Luminance	Candela m^{-2}
Photometry	At a surface	Radiance	Watts $(sr \cdot m^2)^{-1}$
Spectroradiometry	At a surface	Spectral radiance	Watts $(sr \cdot m^2 \cdot nm)^{-1}$

In addition to transmittance, absorbance, and color temperature, light is described by the following characteristics:

- Flux: the total luminous power (measured in lumens)
- Intensity: the angular concentration of flux (candelas)
- Illuminance: the surface density of incident flux (lux or lumen m^{-2})
- Luminance: the intensity emitted per unit area (candelas m^{-2})

The various quantities, their units, and the related measurement techniques are summarized in Table 5.1.

Because the major application of the PLEDs is for displays, the response of the human eye – described by the photopic luminosity function – must be taken into account [8, 9]. By using the photopic luminosity function (as shown in Figure 5.3), the radiance ($W \cdot sr^{-1} \cdot m^{-2}$) is converted into the luminance [candela m^{-2} (cd m^{-2}) or lumen/($sr \cdot m^{-2}$) [10]. Therefore, photometry is used to measure the forward-viewing luminance at the surface of the PLEDs.

The luminous intensity (luminance) can be determined by measuring the flux in any given solid angle, Ω (the ratio of the size of aperture divided by the square of the distance between the light and the aperture). Consider a flat emitting surface, each point of which emits light equally in all directions; that is, a Lambertian source. The PLED is a Lambertian source if the luminous intensity follows the cosine law when measured with a small area detector placed far away from the surface (very small Ω, see below) [11]. It is assumed that, for the PLEDs, deviations from Lambertian emission are relatively small.

To measure the luminous intensity, a reference direction for the measurements must first be chosen, after which the solid angle to be used in the measurement is determined. For display applications, the reference direction should be chosen as the forward-viewing direction, (along the direction perpendicular to the surface

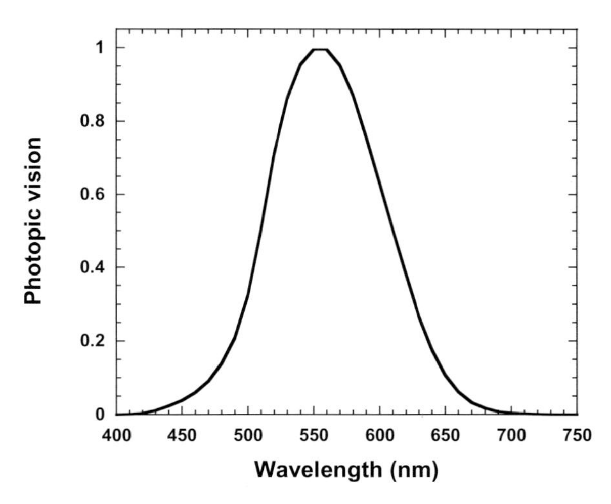

Figure 5.3 The photopic luminosity function.

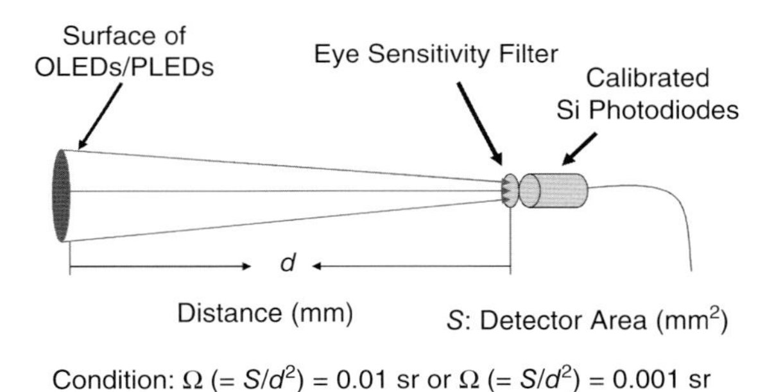

Figure 5.4 Configuration for measuring the luminous intensity of a PLED.

of the PLEDs). The luminous intensity is defined as the emission (in cd m^{-2}) from the emitting surface.

An accurate and convenient configuration for measuring luminous efficiency from the PLEDs is shown in Figure 5.4. Because the luminous efficiency depends heavily on the PLED's emission spectrum (even for constant quantum efficiency), an eye-sensitivity filter is mounted directly onto the surface of the calibrated photodiode (see Figure 5.4).

5.3.2 Measurement of PLEDs

Assuming a Lambertian intensity profile and a disc-shaped source with radius r, the light intensity on a point detector placed a distance d away (see Figure 5.4) can be expressed as follows [9]:

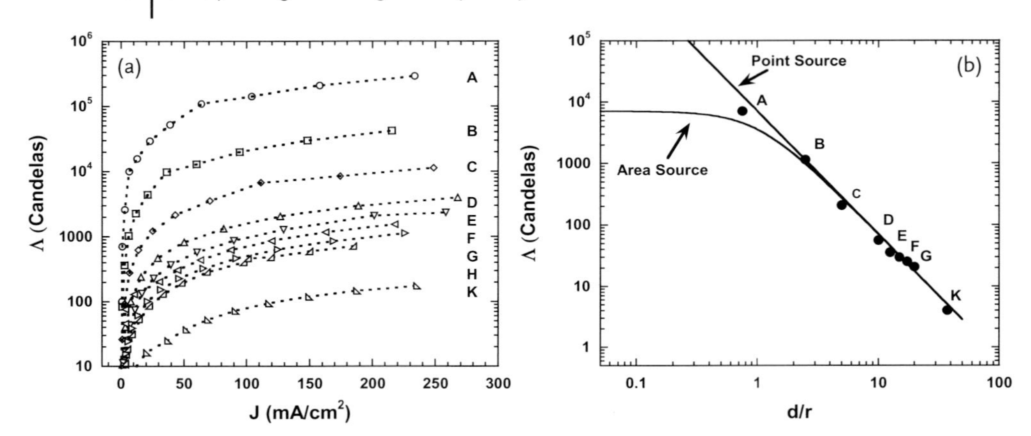

Figure 5.5 (a) The luminous intensity (Λ, cd) versus current density (j, mA cm^{-2}) measured at the different distances (d, mm) between the surface of the PLEDs and the calibrated photodiodes. A, $d = 3$ mm; B, $d = 10$ mm; C, $d = 20$ mm; D, $d = 40$ mm; E, $d = 50$ mm; F, $d = 60$ mm; G, $d = 70$ mm; H, $d = 80$ mm; K, $d = 150$ mm. The PLEDs were made from MEH-PPV. The surface area of calibrated photodiodes was 4×4 mm^2; (b) Inverse square law approximation error.

$$I = I_0(r^2/[r^2 + d^2]) \qquad 5.1$$

For a detector with a finite area, this expression can be used provided that the detector subtends a sufficiently small solid angle (see Figure 5.4). The condition recommended by CIE is that $\Omega \leq 0.01$ sr; under these conditions the luminous intensity (Λ, in candelas) can be obtained by comparing the signal from the PLEDs to that obtained from a calibrated lamp. The brightness is then given by $L = \Lambda/\pi r^2$.

The data from a set of such measurements are shown in Figure 5.5 (for these data, $r = 2$ mm). As expected, the signal falls off as d^{-2}. Note that even for $d/r = 1$, there is no serious error from the finite area of the detector.

Once the luminance, L (cd m^{-2}), is accurately measured, the luminous efficiency, LE (cd A^{-1}), luminous power efficiency, PE (lm W^{-1}) and external quantum efficiency, η_{ext} (the ratio of the number of photons emitted by the PLEDs into the viewing direction of the number of electrons injected) can be determined using the following expressions:

$$\begin{aligned} LE(cdA^{-1}) &= \frac{L}{j} \\ PE(lmW^{-1}) &= \frac{\pi * L}{jV} \\ \eta_{ext}(\%) &= \frac{5.0\times10^3}{(h\nu)\varphi(\lambda)} LE \end{aligned} \qquad (5.2)$$

where j is the current density, V is the applied voltage, $h\nu$ is the photon energy (in eV) of the emission, and $\Phi(\lambda)$ is the photopic luminosity function (Figure 5.4).

Another option is to place the PLEDs into an integrating sphere containing a calibrated detector, and then to measure the total output from the device. In an integrating sphere, all of the emitted photons are collected, including those that are guided to the edge of the substrate. It should be noted, however, that photons emitted from the edge of the substrate are not useful in display applications, and therefore the integrating sphere approach must be used with caution. If the edges of the PLEDs are sealed to absorb the waveguided and scattered light, the external quantum efficiency can be accurately obtained. Once the external quantum efficiency has been measured, *L*, *LE*, and *PE* can be calculated using the above equations. The values obtained by using the integrating sphere are consistent with corresponding values obtained from the configuration shown in Figure 5.5.

The conditions utilized for the measurement of PLEDs carried out in the present authors laboratories and also at UC Santa Barbara, are as follows:

- The surface area (*S*) of the calibrated photodiode with eye-sensitivity filter is $4 \times 4\,\mathrm{mm}^2$.
- The distance (*d*) between the photodiode and the PLEDs is 4 cm; that is, $\Omega = 0.01\,\mathrm{sr}$ ($\Omega = S\,d^{-2}$).

It should be noted that, as indicated by Optronic Laboratories, the response of a diffuser/fiber combination is far from uniform or ideal. Not only does the fiber/diffuser combination fail to provide the correct measurement area, but also the nonuniform spatial response makes the result highly sensitive to alignment. Therefore, the fiber/diffuser combination should be avoided if the intensity of the PLED is to be correctly measured.

5.4 Devices Physics of PLEDs

5.4.1 Elementary Microscopic Process of PLEDs

It is convenient to consider the overall operation of a PLED as a sequence of five steps: (1) charge injection at the electrodes; (2) transport through the structure; (3) recombination into a neutral excited state; (4) emission of a photon; and (5) transport of the photon out of the diode. These steps are shown in Figure 5.6 [12].

5.4.1.1 Injection

The equilibrium concentrations of carriers in the polymer materials used are very small; the forbidden gap is large, and shallow donors or acceptors are absent. The carriers present in the operating PLEDs must be injected from electrodes. Injection may either proceed over the interface barrier by thermoionic emission or through tunneling. For electrons, the injection barrier is given by the equation: $\Delta E_e = W_c - A$; and for holes: $\Delta E_h = I - W_a$, where W_a and W_c are the work-functions

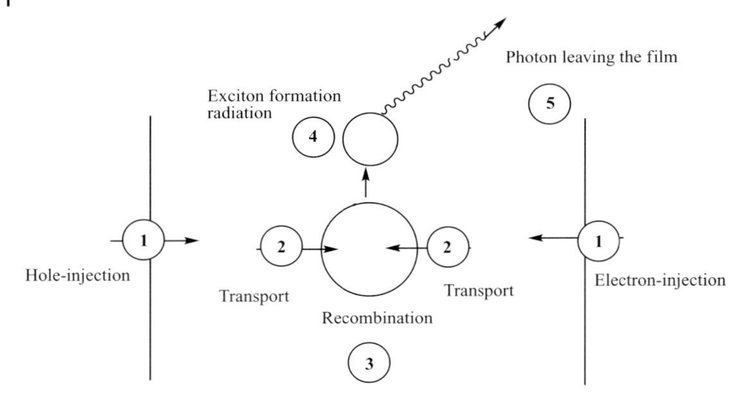

Figure 5.6 Schematic drawing of the five elementary functional steps of PLEDs.

of the anode and cathode, and I and A are the ionization potential and electron affinity, respectively, of conjugated polymers in the solid state.

5.4.1.2 Carrier Transport

Once an electron and a hole have been injected, they must be transported through the film to the point where they will recombine. In PLEDs, carrier transport is believed to occur by hopping between localized states that are usually loosely referred to as *polarons* and *bipolarons*. The details of carrier transport are described in Section 5.4.2.

5.4.1.3 Carrier Recombination

The dielectric constant of conjugated polymers is small, typically 3 to 4. Coulomb screening is inefficient, and the attractive interaction between an electron and a hole is $>KT$, large, and typically 15 nm at room temperature. Therefore, it is believed that the carrier recombination in PLEDs is "Langevin recombination," and that the recombination rate constants are given by the equation:

$$\gamma_r = \frac{e(\mu^+ + \mu^-)}{\varepsilon\varepsilon_0} \tag{5.3}$$

and is independent of any applied field. This has been well established experimentally in the case of molecular crystals, and is quite different from the case of semiconductors.

5.4.1.4 Photon Emission

Almost all conjugated polymers used in PLEDs are composed of light elements, mainly C and H; thus spin–orbit coupling effects are inefficient. Neutral excited states are well classified as singlet (S = 0) and triplets (S = 1), according to their spins and multiplicity. Higher spin states are generally at too-high energies to play any role here; the singlet–triplet energy splitting is typically 1 eV. If the quantum yield for singlet emission, the fluorescence, is η_s, and for triplet emission, the phosphorescence, is η_t, then the electroluminescence quantum yield is proportional to:

$$\frac{\eta_s + 3\eta_t}{4} \tag{5.4}$$

5.4.1.5 Photon Extraction

The emitting layer of conjugated polymers being generally disordered and non textured, the emission is therefore isotropic inside the film, leading to an approximately Lambertian emission pattern outside the diode. This is a luminance that is independent of the angle of view, and is indeed advantageous for display applications.

5.4.2 Carrier Transport in PLEDs

Most of the organic electroluminescence materials, small molecules and conjugated polymers are low-conductance materials; typically, the h^+ mobility in these materials is 10^{-7} to $10^{-3}\,cm^2/(V\cdot s)$, and the e-mobility is typically lower by a factor of 10 to 100. However, it is now clear that low-mobility is due to the disorder in the amorphous or polycrystalline materials. The application of an external field causes the injection of h^+ from the ITO and of an e^- from the cathode. The injection from the metallic injection is usually less efficient than from the ITO. The asymmetry in carrier injection leads to an imbalance in the concentrations of the injected carriers that reduces the device's efficiency.

Unlike inorganic semiconductors, the transport and the injection properties in PLEDs are determined by intersite hopping of charge carrier between localized states, as well as hopping from delocalized states in the metal to localized states in the organic layer. The actual transition rate from one site to another depends on their energy difference, and on the distance between them. The energy states involved in the hopping transport of h^+ and e^- from narrow bands are around the HOMO and LUMO levels. The widths of these bands are determined by the intermolecular interactions and by the level of disorder.

The transport in PLEDs has been extensively studied using time-of-flight (TOF), and by an analysis of the DC current–voltage characteristics. The universal dependence of charge carrier mobility on the electric field is:

$$\mu(E,T) = \mu(0,T)\exp(\gamma\sqrt{E}) \tag{5.5}$$

where $\mu(0, T)$ is the low-field mobility and γ, an empirically determined coefficient, is observed for the vast majority of materials.

Two models – polaron and disorder – were applied for an explanation of carrier hopping in PLEDs. Those models based on polaron formation assume that a localized carrier interacts strongly with molecular vibrations of the host and neighboring molecules, so that a significant relaxation of the local molecule structure occurs around the carrier. This carrier can then move to an adjacent molecule simply by carrying that relaxation along with it. Clearly, such a relaxation or stabilization will lower the energy of the negative carrier below the LUMO level, and the energy of the positive carrier above the HOMO level.

Although the experimental evidence for polaronic relaxation in PPV and other conjugated polymers is extensive, other experiments render the polaron models problematic due to: (i) using the Arrhenius relation to describe the temperature dependence of the mobility leads to pre-factor mobilities well in excess of unity; and (ii) the polaron models cannot account for the dispersive transport observed at low temperature.

In the disorder models, it is assumed that the coupling of a charge carrier to molecular modes is weak, and that the activation energy reflects the static disorder of the hopping sites. In polaron models, it is assumed that the energetic disorder energy is small compared to the deformation energy. The fundamental difference between disorder and polaron models is related to the difference in energy of hopping sites due to disorder and the change in molecular conformation upon the addition or removal of a charge at a given site. Generally, despite the better agreement between the disorder-based models and transport measurements, it is widely believed that the charge carrier exist as polarons rather than free e^-s and h^+s.

5.4.3
Electronic Characteristic of PLEDs

5.4.3.1 Current–Voltage Characteristics

The current–voltage characteristics of PLEDs, measured over many orders of magnitude in the applied field, exhibit several well-pronounced regimes. An example, as presented in Figure 5.7 shows the low-field ohmic conduction and space–charge-limited conduction (SCLC) in the presence of shallow traps, followed by the free-trap conduction in the upper limit of applied field.

Although having superficially similar device characteristics to conventional diodes made from inorganic semiconductors, polymeric devices differ in many details. For Schottky-barrier or *p-n* junction-type devices, the operational voltage is typically equal to, or less than, the inorganic semiconductor band gap. In contrast, polymer devices typically begin to generate light at voltage much greater than their band gap. Reducing the thickness lowers the device operating voltage. It can be seen clearly in Figure 5.8 that the *I–V* characteristics depend not on the voltage, but rather on the electric field strength. It can also be seen clearly that a tunneling model for carrier injection in which one (or both) of the carriers is field-emitted through a barrier nature at the electrode/polymer interface. The details of the tunneling model are described in Section 5.4.4.

5.4.3.2 Space–Charge-Limited Currents

Organic solids are usually insulators. However, unlike in inorganic semiconductors, impurities normally act as traps for charge carriers rather than as sources of them. Exceptions to this rule are the conjugated polymers; for example, PPV fab-

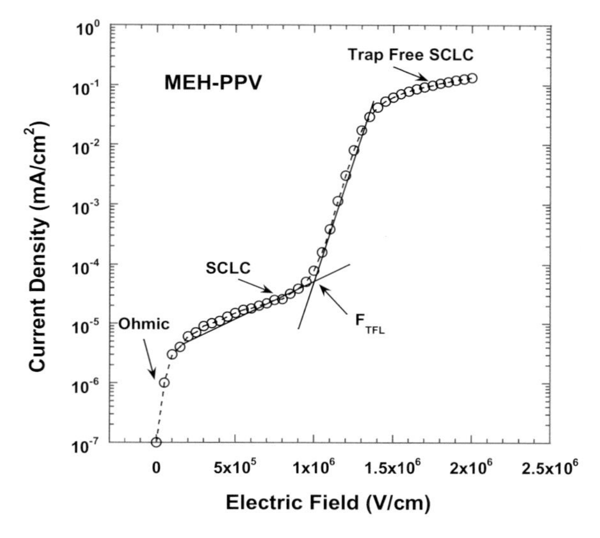

Figure 5.7 Unipolar current versus electric field for 80 nm-thickness MEH-PPV with device configuration of ITO/MEH-PPV/Ca.

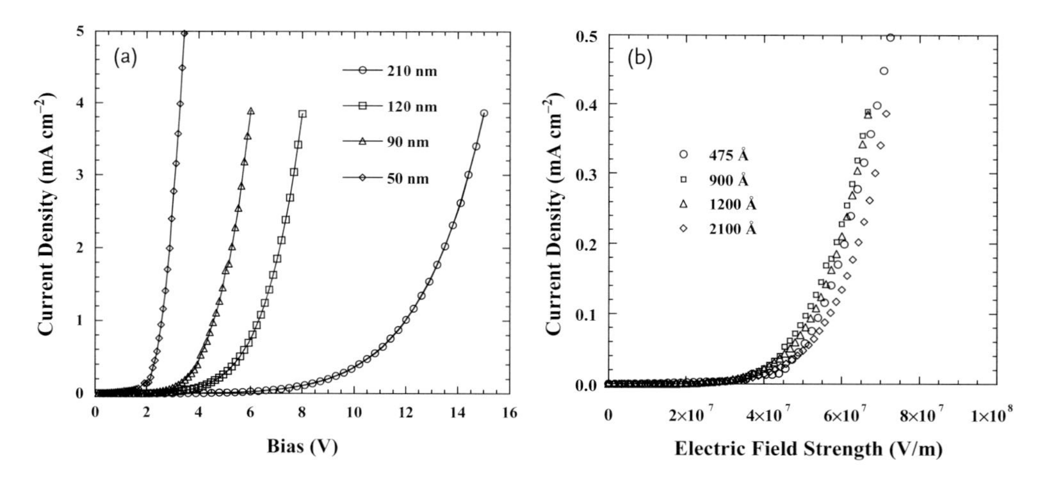

Figure 5.8 (a) Thickness-dependence of the *I–V* characteristics in an ITO/MEH-PPV/Ca device; (b) Electric field versus current for the above devices.

ricated via a special precursor route was shown to be *p*-doped with doping concentrations in the order of $10^{17}\,cm^{-3}$. In that case, a Schottky-type depletion zone can be established near a metal contact, although in the vast majority of cases the concentration of impurities is small enough so as not to perturb the electric field distribution inside a solid-state sample. In those cases, the dark electrical conduction is very low, and the solids are considered to be good insulators. Such solids can be made to conduct a relatively large current if the contacts permit the introduction in them of an excess of free carriers. If the carrier enters through a surface boundary, the process is referred to as "charge injection."

The charge-injected conduction is governed by charge injection barriers at the electrode contacts and the charge transport properties of the materials. Depending on the charge injection efficiency and mobility of charge carrier, the current is either SCLC or injection-limited current (ILC).

For a perfectly ordered or disordered insulating material, or those containing very shallow traps ($\Delta E << kT$), the SCL current in a sample of thickness d obeys Child's law:

$$j_{\mathrm{SCL}} = \frac{9}{8}\varepsilon_0 \varepsilon \mu \frac{F^2}{d} \tag{5.6}$$

In the presence of discrete traps

$$j_{\mathrm{SCL}} = \frac{9}{8}\varepsilon_0 \varepsilon \Theta \mu \frac{F^2}{d} \tag{5.7}$$

where μ is the microscopic mobility of the carriers, ε the dielectric constant, ε_0 is the permittivity in vacuum, and Θ is the fraction of free (nf) to trapped (nt) space charge.

If local traps are distributed in energy (E), they will be filled from bottom to top as the electric fields, F, increase. The quasi-Fermi level will scan the distribution shifting towards the transport band, and $\Theta \approx \mathrm{nf}/\mathrm{nt}$ will become a function of F.

5.4.3.3 Injection-Limited Currents

The current becomes limited by injection when the average charge density in the sample approaches the charge density at the injecting contact. At this point, the injecting contact can no longer act as a reservoir and thus ceases to be ohmic, and the current from such an electrode will saturate at a sufficiently high voltage. On the other hand, very high electric fields can make some contacts ohmic by causing a strong injection via tunneling or other mechanisms super-linear with the electric field. Although the average charge density in the samples is comparable with the charge density at the contact, both of these should be much smaller than the capacitor charge related to unit volume. ILC will be observed only for relatively low currents at high electric fields with high mobility and large-value dielectric permittivity materials formed into high chemical and structural perfection thin layers. These are often-met features of thin organic films sandwiched between metals or semiconductors with moderate work-functions.

5.4.3.4 Diffusion-Controlled Currents

Charge carrier injection from a metallic electrode is said to be diffusion-controlled if the space-charge effects can be neglected and the diffusion current is comparable or exceeds the drift current flow.

5.4.4 Fowler–Nordheim Tunneling in Conjugated Polymer MIM Diodes

The operating mechanism of PLEDs is quite different from conventional *p-n* junction LEDs. In a polymer LED, a pure undoped film of luminescent conjugated polymer is sandwiched between a high work-function metal anode and a low work-function metal cathode. The charge carrier concentration in such pure semiconducting films is sufficiently low ($\sim 10^{14}$–$10^{15}\,cm^{-3}$), that any residual carriers introduced by impurities are swept out by the built-in field that arises from any difference in the work-functions of the two electrodes. The depletion depth of pristine PPV is approximately 250 μm, which is much larger than the thickness of the polymer layer in an LED (typically <100 nm). Consequently, the electronic structure of the LED can be approximated by the rigid band model displayed in Figure 5.9. The built-in field causes the uniform slope in the energies of the states in the bulk of the conjugated polymer; there is negligible band bending.

- Zero bias with a common Fermi Level across the device (note that within the semiconducting polymer layer, the Fermi level moves across the energy gap on passing from the anode to the cathode) (Figure 5.9a).
- A flat-band condition occurs when the applied voltage equals the difference in the work-functions of the anode and the cathode (Figure 5.9b). This is the minimum voltage required for the injection of electrons and holes. Ideally, the electroluminescent emission should turn on at this voltage.
- Forward bias; carriers are injected through the triangular barriers at the anode (holes) and cathode (electrons) and meet within the polymer film where they radiatively recombine (electroluminescence) (Figure 5.9c).

When a positive bias is applied to the LED, the Fermi level of the cathode is raised relative to that of the anode (see Figure 5.9c). Thus, the thickness of the barrier is a function of the applied voltage; the barrier thickness decreases as the voltage is

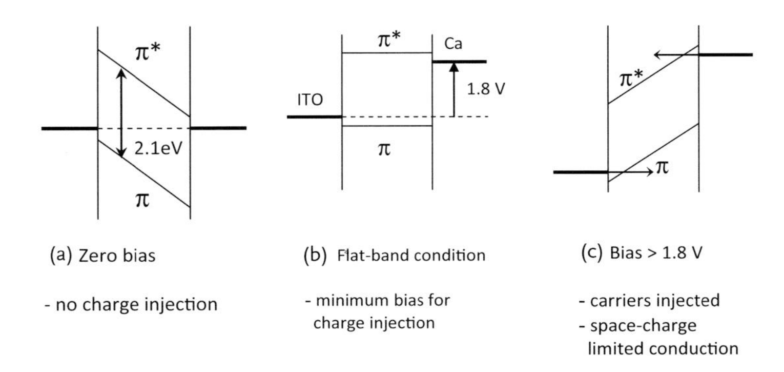

Figure 5.9 The electronic structure of the LED. See text for details.

increased. Carriers tunnel through the barrier primarily by Fowler–Nordheim field emission, tunneling from the anode and cathode into the π-band (holes from the anode), and the π*-band (electrons from the cathode) of the conjugated polymer. Thermionic emission over the barriers can also play a role if the barriers are small and the temperature is relatively high. As the rate of injection by Fowler–Nordheim tunneling is determined by the strength of the electric field, it is important for the polymer layer to be thin so that high electric fields can be obtained at low voltages.

To optimize the performance of PLEDs, it is important to minimize the barriers for charge injection by choosing electrodes with work-functions that are well matched to the bands of the polymer; for this, ITO, PANi, polypyrrole (PPy), and PEDOT are the most commonly used anode materials. They also have the important property of being transparent, and therefore allow the emitted light to escape from the device. Ca, Ba and Mg are commonly chosen as the cathode because of their low work-functions. Unfortunately, low-work-function metals are highly reactive, and PLEDs must therefore be hermetically sealed for long life. An improved electron injection from stable metals such as aluminum can be achieved by coating the electrode with a polar self-assembled monolayer (SAM). In this case, the dipole layer effectively shifts the electrode work-function.

If the electrodes are well matched to the bands of the polymer, then the barrier for charge injection is small and the current that passes through the LED is not limited by injection. Instead, the hole current is space–charge-limited and the electron current is trap-limited. Space–charge limiting arises because the space–charge which builds up near the anode due to the population of holes screens the field between the two electrodes, and thereby limits the current. The traps that limit electron transport originate from defects that have energy levels just below the conduction band (due to disorder in the polymer).

Once the electrons and holes have been injected into the polymer, they must encounter each other and recombine radiatively in order to give off light. In this context, the low mobility of the charge carriers (polarons) in semiconducting polymers is helpful, as the slow drift of the charge carriers across the thickness of the semiconducting polymer will allow enough time for the carriers to meet and recombine radiatively. There are several factors that determine the efficiency of an LED. Maximum efficiency can only be achieved through balanced electron and hole currents. If one carrier type is injected much more efficiently and drifts in the applied electric field with higher mobility than the other, then many of the majority carriers will traverse the entire polymer layer without recombining with a minority carrier. As shown in Figure 5.6, this problem can be minimized by carefully choosing appropriate electrodes, so that the Fermi level of the anode is close in energy to the top of the π-band, while the Fermi level of the cathode is close in energy to the bottom of the π*-band. With such well-matched electrodes, both carriers are injected efficiently.

As demonstrated in Figure 5.10, diodes fabricated as described above show excellent rectification ratios and strong electroluminescent light emission in forward bias. Light emission turns on close to the flat band condition; that is,

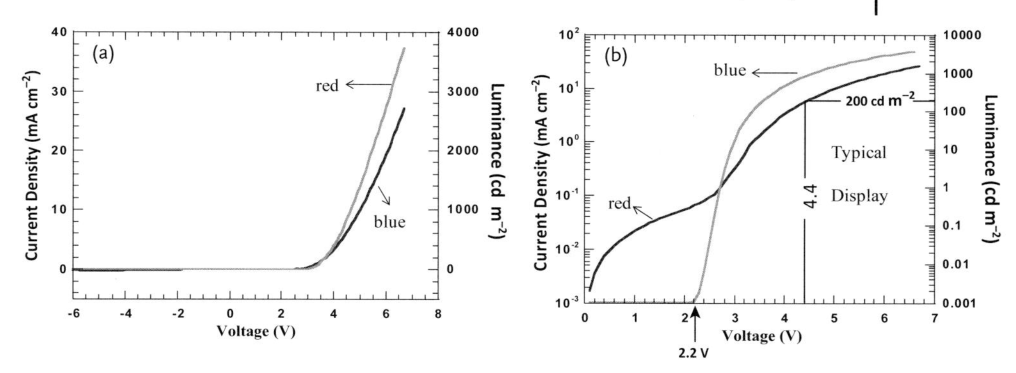

Figure 5.10 (a) Current density (red) and luminance (blue) versus voltage for devices fabricated with MEH-PPV using polyaniline as the anode and calcium as the cathode; (b) The current at voltages below 2 V due to the residual leakage current resulting from microshorts and imperfections.

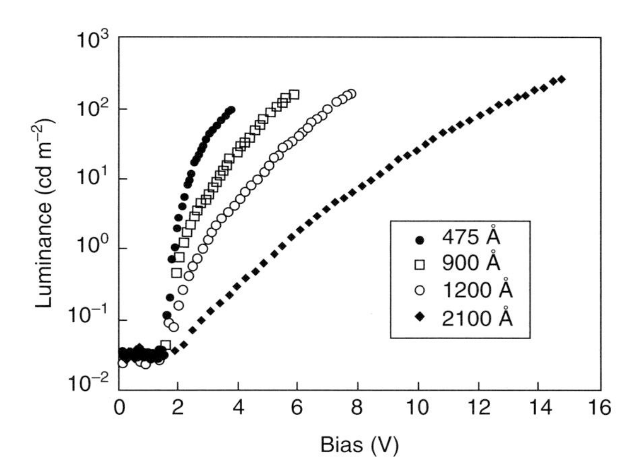

Figure 5.11 Luminance versus voltage for ITO/MEH-PPV/Ca LEDs with semiconducting polymer layers of different thicknesses. Note that the turn-on voltage is independent of thickness, and equal to that needed to reach the flat band condition.

when the applied voltage is greater than the difference between the work-functions of the two electrodes (see Figure 5.10a). The current at voltages below 2 V in Figure 5.10b is a residual leakage current resulting from microshorts and imperfections.

Parker [12] observed that the turn-on voltages for devices with different thicknesses of the conjugated polymer layer are roughly the same, and equal to the voltage required to reach the flat band condition. As shown in Figure 5.11, however, the current density is a strong function of the thickness.

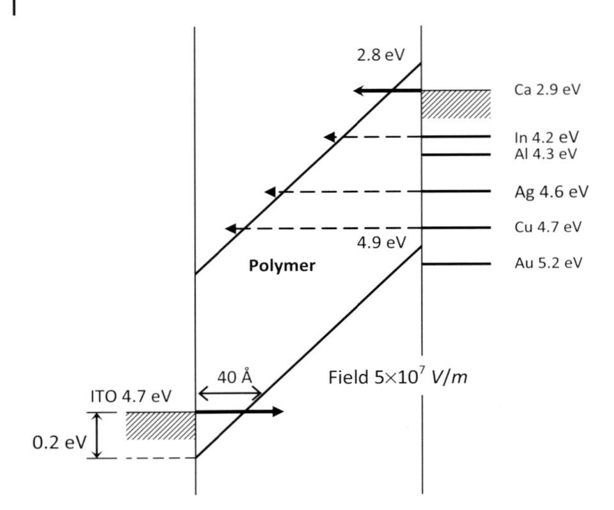

Figure 5.12 Band diagram (in forward bias) indicating the magnitudes of the triangular tunneling barrier for various cathode materials.

An understanding of the operating mechanism of PLEDs is complicated by the two-carrier nature of these diodes. In order to study the details of the carrier injection, single carrier devices were fabricated in which the current density of one of the carriers was reduced to negligible levels.

5.4.4.1 **Single Carrier Devices**

"Hole-only" devices were fabricated by replacing the low-work-function Ca cathode (2.9 eV) with higher-work-function metals such as In (4.2 eV), Al (4.4 eV), Ag (4.6 eV), Cu (4.7 eV), or Au (5.2 eV) (see Figure 5.12).

Increasing the work-function of the cathode increases the offset between the Fermi energy of the cathode and the bottom of the π^*-band of MEH-PPV at 2.8 eV. This in turn reduces the number of injected electrons to levels at which the injected holes dominate. Despite the fact that the work-function of the cathode increases by more than 1 eV (from In to Au), the *I*–*V* curves of the devices made with In, Al, Ag, Cu, and Au cathodes are almost identical, indicating that electron injection has been shut off.

Fowler–Nordheim tunneling theory predicts that the tunneling current is an exponential function of $1/F$:

$$I \,\alpha\, F^2 \exp(-\kappa/F) \tag{5.8}$$

where I is the current, F is the electric field strength, and κ is a parameter that depends on the barrier shape. For a triangular barrier,

$$k = 8\pi(2m^*)^{1/2}\varphi^{3/2}/3qh. \tag{5.9}$$

Here, φ is the barrier height, m^* is the effective mass of holes in the semiconducting polymer, q is the electron charge, and h is Planck's constant.

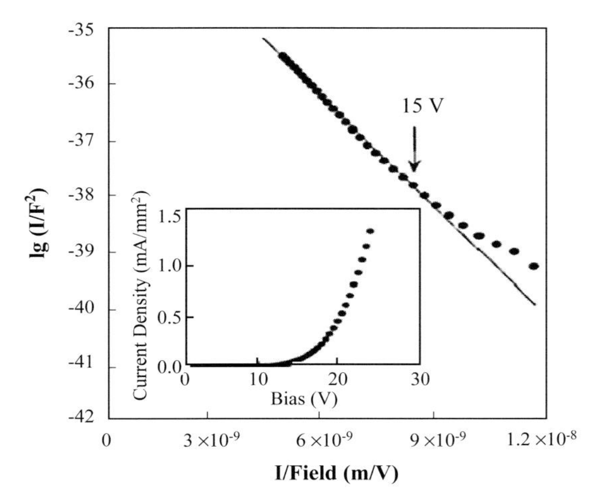

Figure 5.13 Fowler–Nordheim plot of $\ln[I/F^2]$ versus I/F for a hole-only device fabricated using MEH-PPV with ITO as the anode and Au as the cathode.

The *I–V* data from a hole-only device fabricated from MEH-PPV with Au as the cathode and ITO as the anode are plotted as I/F^2 versus I/F in Figure 5.13. As predicted, the plot is close to linear, particularly at high fields, and the literal assumption of tunneling through a triangular barrier appears to be an excellent approximation. The deviation from linearity at lower fields probably indicates an additional contribution to the current from thermionic emission.

Assuming that the electric field is constant across the semiconducting polymer, and that $m^* = m$ (the free electron mass), the calculated barrier heights are found to be 0.2–0.3 eV for all the hole-only devices indicated in Figure 5.12. The fact that the infrared barrier height is unchanged despite the large variation in cathode work-function indicates that the barrier must be at the ITO/polymer interface. Indeed, $\varphi = 0.2$–0.3 eV is in agreement with the energy diagram in Figure 5.12.

A similar approach can be taken to study electron injection. Replacing ITO with a lower-work-function metal (e.g., Nd or Mg) yields devices in which the carriers are almost exclusively electrons. A similar analysis of the data from a range of "electron-only" devices indicates that electrons tunnel into the π^*-band of MEH-PPV at 4.9 eV through a triangular barrier at the polymer/cathode interface. With Ca as the cathode, this barrier is 0.1 eV, again in agreement with the energy diagram in Figure 5.12.

5.4.4.2 **LED Operating Voltage and Efficiency**

The operating voltage is very sensitive to the barrier height, and such sensitivity is predicted by the Fowler–Nordheim tunneling model. Equations (5.8) and (5.9) indicate that, for the same tunneling current, the ratio $\varphi^{3/2}/V$ must be the same. Thus, increasing the barrier height from 0.1 eV (PANi) to 0.6 eV (Cr) should

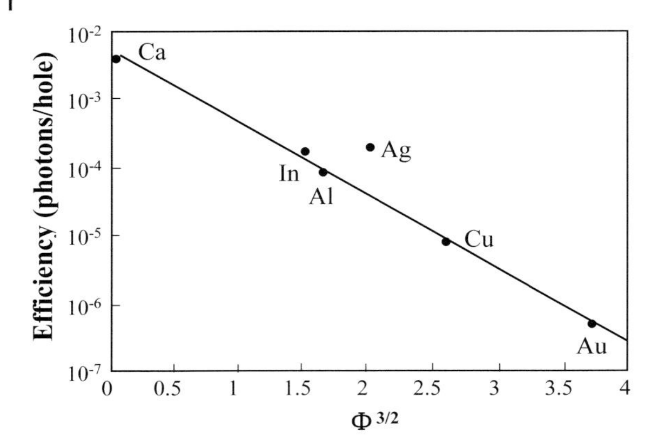

Figure 5.14 Semilog plot of quantum efficiency (photons per electron) versus $\varphi^{3/2}$.

increase the operating voltage by a factor of 11. The experimentally determined increase was a factor of 9.

It is clear that the device efficiency, η, must also be very sensitive to the barrier height, as the efficiency is limited by the minority carrier density. As suggested by Eqs (5.8) and (5.9), Figure 5.14 plots $\ln(\eta)$ versus $\varphi^{3/2}$. The excellent agreement between the theory and the data confirms the use of the Fowler–Nordheim tunneling model for describing the carrier injection into the band structure of the conjugated polymer.

There is evidence of interface modification when metal films are deposited onto conjugated polymers. However, as these interactions involve specific chemical reactions between the metal atoms and certain groups on the polymer, it is not clear how such chemical reactions could give rise to the systematic variation in the device efficiency with a work-function that is evident in Figure 5.14.

5.4.4.3 **Limits of the Model**

When the barrier heights are small, the simple comparison of work-functions for the prediction of barrier heights is inappropriate; thermionic emission becomes important for barriers less than a few tenths of eV. Moreover, the barrier will not be perfectly triangular as the model assumes, but must be somewhat ill-defined as a result of disorder-induced band-tailing near the polymer band edges. In addition, the evaporated electrode materials are likely to be full of defects (possibly even amorphous), and are therefore not expected to have precisely the well-defined work-function values listed in reference books. Other factors, such as the mobility of each charge species and space–charge-limiting effects, may also have an increasing influence over the performance for low barrier-height devices. Nevertheless, it is clear that an energy band picture for the conjugated polymer supplemented by Fowler–Nordheim tunneling theory provides an excellent starting point for understanding the operating mechanism of LEDs fabricated from conjugated polymers.

5.4.5
Approaches to Improved Carrier Injection

It is not always possible to find electrode materials that are well matched to the electronic structure of an electroluminescent polymer. For example, for large band-gap blue-emitting polymers such as the polyfluorenes, the energy barrier at the cathode is typically too large for an efficient injection of electrons via Fowler–Nordheim tunneling. Similarly, for more electronegative, stable conjugated polymers, the barrier at the anode will be too large for efficient hole injection. In such cases, hole and/or electron blocking layers can be added to improve the balance of electron and hole currents. The electronic structure of an LED with a hole-blocking layer is shown in Figure 5.15, where the blocking layer creates a barrier at the interface of two polymers that blocks the flow of the majority carrier. As the density of the majority carrier increases at the blocking interface, the electric field at the minority carrier injecting electrode increases, thereby enhancing the minority carrier injection. As a result, the electron and hole currents will be more nearly balanced. An analogous diagram can be drawn for a device with an electron-blocking layer.

An improved injection can also be achieved through the addition of defects and impurities. End-capping with electron-accepting groups and/or hole-accepting groups is a useful example and, as shown in Figure 5.16, these groups lead to the formation of localized states near the band edges. Thus, in forward bias an electron can tunnel into the π^*-band in two steps: first from the cathode into a localized state on an electron-accepting site; and then from that site into the π^*-band. Since the tunneling probability decreases exponentially as $\varphi^{3/2}$, the two-step injection can be more efficient than direct tunneling through the barrier into the π^*-band. As sketched in Figure 5.16 (dotted arrow), a two-step process involving tunneling

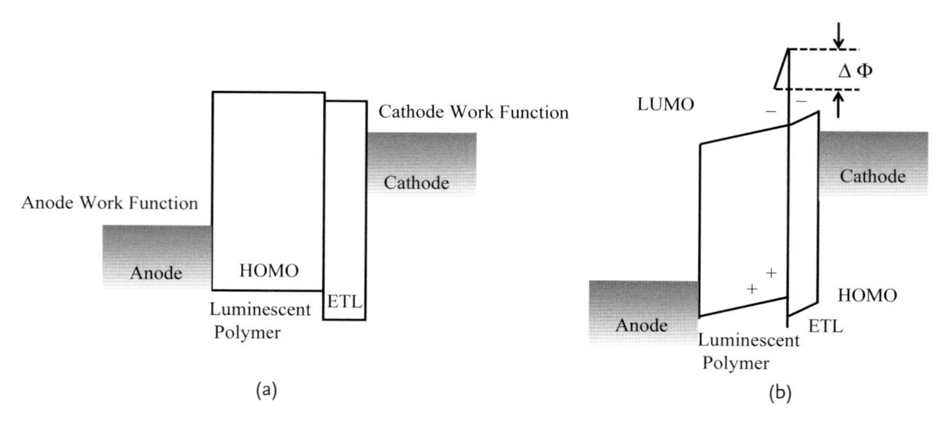

Figure 5.15 (a) Flat band diagram of the electronic structure of an LED with a hole-blocking layer; (b) Electronic structure of an LED with a hole-blocking layer in forward bias. The electric field is primarily across the ETL; the field in the luminescent polymer is nearly zero.

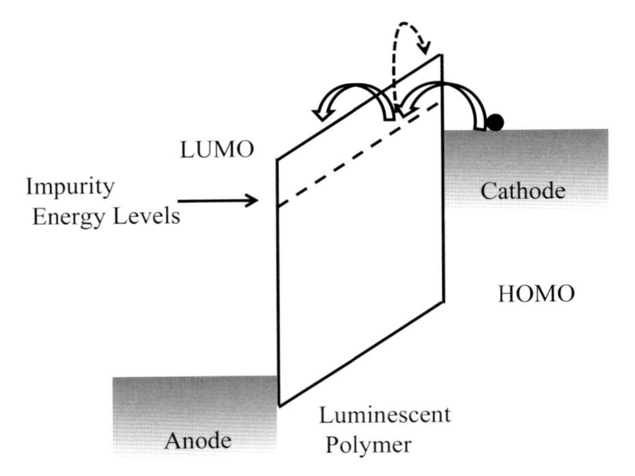

Figure 5.16 End-capping with electron-accepting groups and/or hole-accepting groups leads to the formation of localized states near the band edges; electron-accepting states are at energies just below the π^*-band, while hole-accepting states are at energies just above the π-band. Alternatively, such acceptor units can be inserted into the main chain as "dopants." Small molecule acceptors can also be used.

followed by thermal excitation from the impurity acceptor state into the band can also help.

Electron-accepting and/or hole-accepting groups can also be incorporated into the chain (copolymers) or, by the addition of appropriate small molecules, into the film. In the latter approach, however, clustering and phase separation are likely problems.

5.5 Materials for PLEDs

5.5.1 Conjugated Polymers for PLEDs

The design of luminescent materials for use in PLED devices is as critical to device performance as the process of constructing the device itself. Processability, purity, thermal and oxidative stability, color of emission, luminance efficiency, balance of charge carrier mobility, and others factors are among many important material properties required for a system to be viable in commercial PLED devices applications. During the past two decades, several types of conjugated polymer with different molecular structures have been developed and applied for the fabrication of PLEDs. Among conjugated polymers that have included poly(*p*-phenylenevinylene)s (PPVs), polyphenylenes (PPPs), polyfluorenes (PFs) and polythiophenes (PTs), the PPVs and the PFs have been the most widely studied and have emerged as the leading candidates for PLEDs applications, mainly because of their tunable emission frequencies and relatively high photoluminescence (PL) yields.

Semiconducting polymers have been synthesized with different molecular structures and with an associated wide range of energy gaps. Consequently, luminescent semiconducting polymers can be obtained with emission colors that span the full range of the visible spectrum. Side-chain functionalization of the same main chain (e.g., with alkoxy or alkyl groups) can be used to shift the color of the emitted light over a substantial portion of the visible spectrum. The use of synthesis to create homopolymers with different molecular structures and to create copolymers is the "band gap engineering" methodology for semiconducting polymers. Some examples of polymers demonstrating the range of different band gaps and emission colors that can be obtained through band gap engineering are shown in Figure 5.17. The use of copolymers is particularly interesting since, by using block-copolymers, well-defined quantum well structures can be created.

5.5.1.1 **Poly(p-phenylenevinylene)s (PPVs)**

Poly(*p*-phenylenevinylene) is an insoluble, intractable, and infusible conjugated polymers with a green–yellow emission. The core molecular structure of PPV is shown in Figure 5.18. The main problem encountered with PPV when making PLEDs is that any direct synthesis will produce a solid. An alternative solution is to use a solution-processable precursor polymer, which is first cast from solution and then converted into PPV by thermal treatment. The synthesis of PPV through a precursor polymer is shown in Figure 5.18. Here, the conversion temperature can be reduced to 100 °C, from over 200 °C, by using bromide salt, instead of the chloride counterpart, or by vacuum treatment.

Since 1991, Fred Wudl and his colleagues have demonstrated the synthesis of soluble MEH-PPV, and many PPVs derivatives have subsequently been reported. These include alkyl- and alkoxy-substituted, cyano-substituted, silyl-substituted, and phenyl-substituted PPVs. The representative molecular structures of PPV derivatives are shown in Figure 5.19.

5.5.1.2 **Polyphenylenes (PPPs)**

Polyphenylenes (PPPs) have received considerable attraction due to their blue color emission. Like PPV, the unsubstituted PPP is insoluble in any solvent; hence, soluble PPPs can be designed by the incorporation of alkyl-, aryl-, or alkoxy-solublizing groups into their side chains. The representative molecular structures of PPPs derivatives are shown in Figure 5.20.

5.5.1.3 **Polyfluorenes (PFs)**

Polyfluorenes with substitutes at C-9 are soluble in conventional organic solvents. The PFs and their copolymers have evolved as a major class of emitting materials for PLEDs. A solid-state PL yield of up to 80% has been reported, and PLEDs fabricated from PFs and their copolymers exhibit blue, green, and red emissions. The molecular structures of PFs and their copolymers are shown in Figure 5.21.

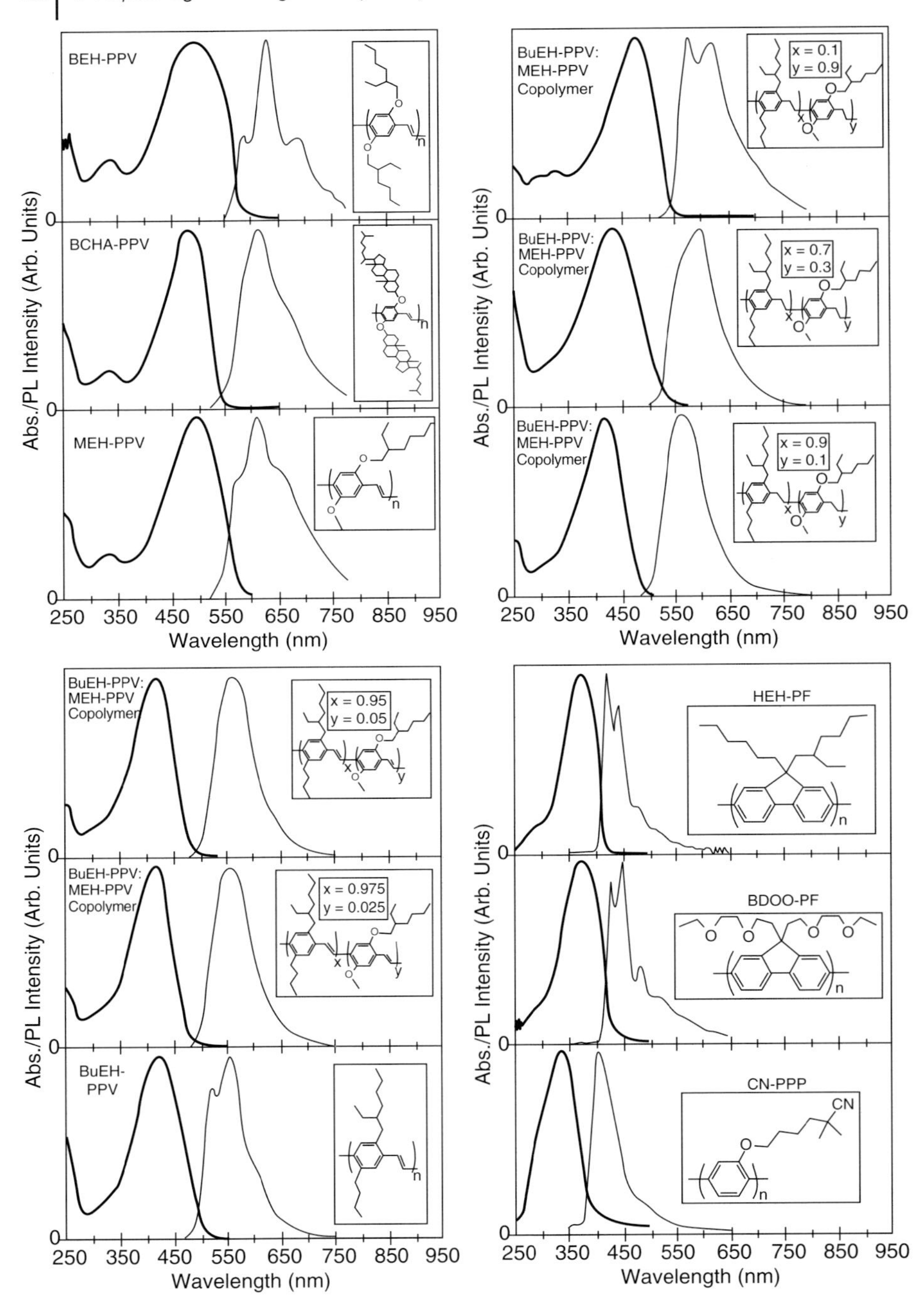

Figure 5.17 Absorption and emission spectra of a number of conjugated polymers with different molecular structures. The emission colors span the visible spectrum.

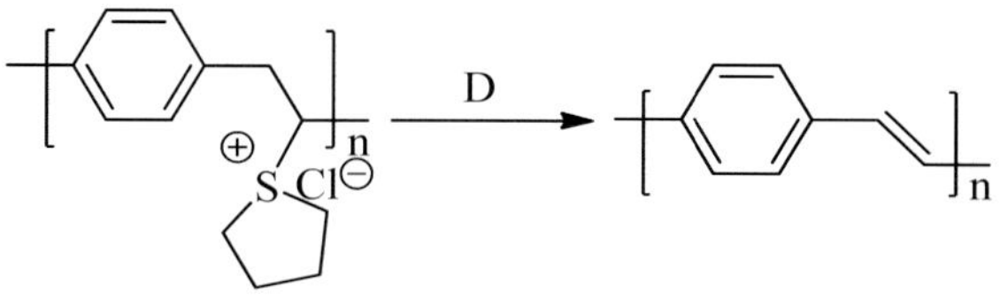

D ⊕ S Cl⊖ n n

Figure 5.18 Synthesis of PPV through a precursor polymer.

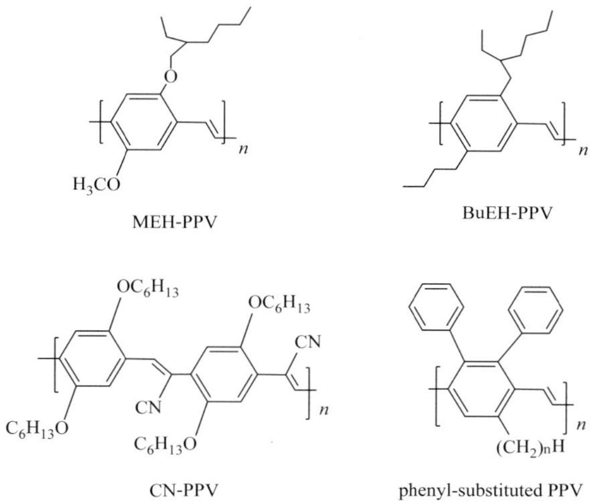

O H_3CO n MEH-PPV

n BuEH-PPV

OC_6H_{13} OC_6H_{13} CN CN $C_6H_{13}O$ $C_6H_{13}O$ n CN-PPV

n $(CH_2)_nH$ phenyl-substituted PPV

Figure 5.19 Molecular structures of representative PPV derivatives.

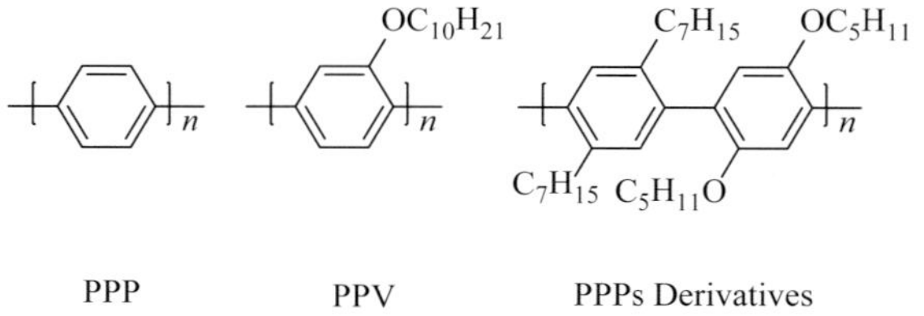

n PPP

$OC_{10}H_{21}$ n PPV

C_7H_{15} OC_5H_{11} C_7H_{15} $C_5H_{11}O$ n PPPs Derivatives

Figure 5.20 Representative molecular structures of PPPs derivatives.

5.5.1.4 **Polythiophenes (PTs)**

Soluble poly(3-alkythiophene)s with alkyl chain lengths varying from 6 to 22 carbon atoms have been reported. The emission intensity increases with increasing side-chain length, and emission wavelength is blue-shifted with cyclohexyl group substituents or alkyl groups substitutions in both the 3-and 4-positions. As the PTs have a low PL quantum yield, PT-based PLEDs are prepared from their

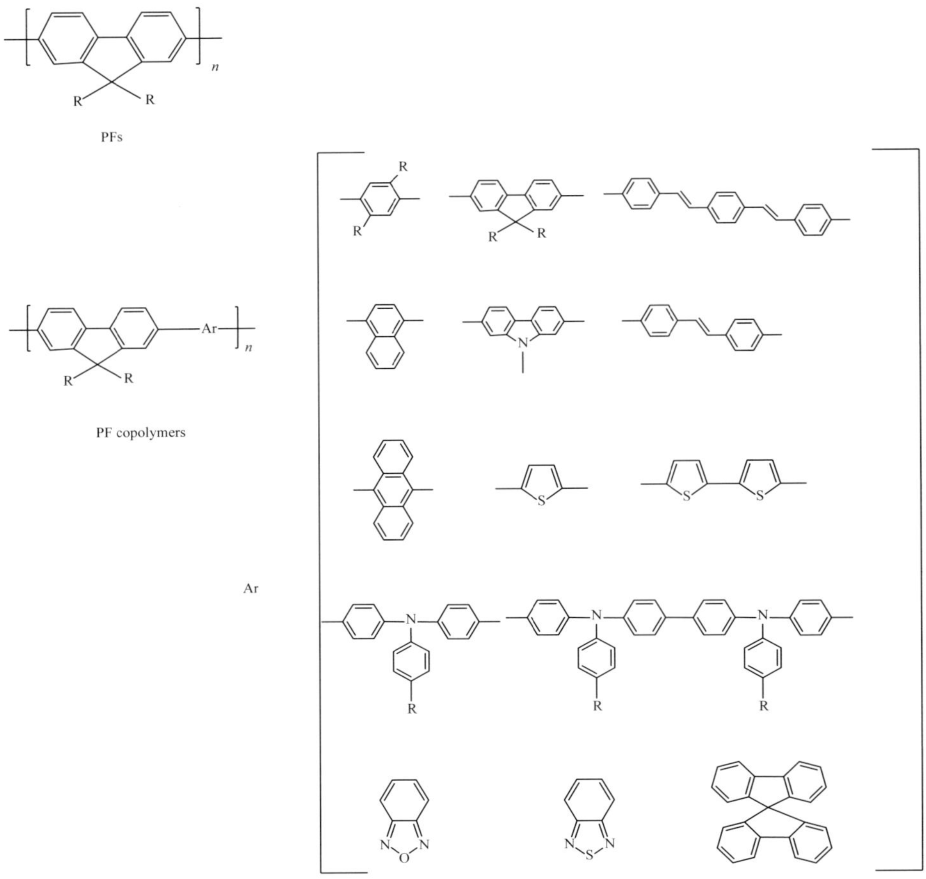

Figure 5.21 Molecular structures of PFs and examples of monomers, Ar, used with PF copolymers.

derivatives. The representative molecular structures of PTs and PT derivatives are shown in Figure 5.22.

5.5.2 Anode and Cathode

5.5.2.1 Anodes

Relatively few materials have been explored as anodes for PLEDs. Although an ITO is widely used as the anode in PLEDs, because it has the virtue of optical transparency, it is not a well-controlled material. Several alternative materials have been recently examined as anodes, including doped ITO [Ge-doped ITO and fluorine-doped tin oxide (FTO)], doped zinc oxide [Al-doped zinc oxide (AZO)] and transparent conductive oxides (TCOs).

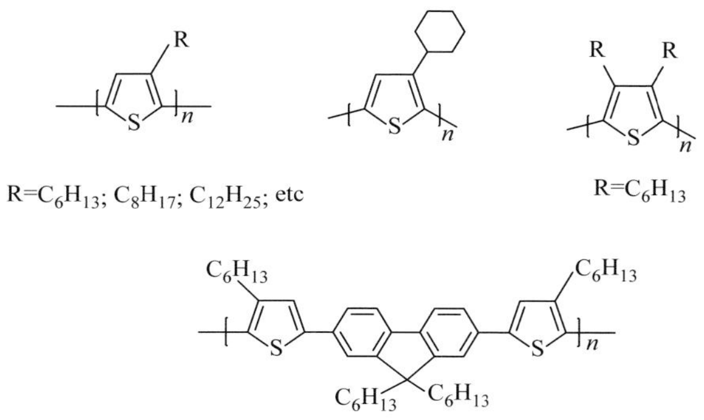

$R = C_6H_{13}$; C_8H_{17}; $C_{12}H_{25}$; etc

$R = C_6H_{13}$

Figure 5.22 Representative molecular structures of PT and PT derivatives.

Table 5.2 Electronic properties of typical electrode metals.

Element	Ionization potential (eV)	Preferred work-function (eV)
Cs	3.89	2.14
K	4.34	2.30
Ba	5.21	2.70
Na	5.14	2.75
Ca	6.11	2.87
Li	5.39	2.90
Mg	7.65	3.66
In	5.79	4.12
Ag	7.58	4.30
Al	5.99	4.28
Nb	6.88	4.30
Cr	6.77	4.50
Cu	7.73	4.65
Si	8.15	4.85
Au	9.23	5.10

5.5.2.2 **Cathodes**

Different cathode metals have been employed in device architecture, with varying degree of success. The most obvious change to the choice of electron-injecting media is to vary the electronic work-function and to observe performance. The various metallic electronic work-functions and ionization potentials are summarized in Table 5.2.

Attempts to use Ca, K, and Li as effective cathode materials have revealed that these exhibit a poor corrosion resistance and a high chemical reactivity with the organic medium. Thus, a variety of low-work-function metal alloys such as Mg–Ag and Al–Li, and metal compounds such as alkali metal compounds/Al, are used for cathodes.

LiF and CsF exhibit pronounced differences from those reactive alkali metal compounds, as they are thermodynamically stable with respect to either Al or organic/polymeric materials. Devices with a bilayer electrode showed significantly better *I–V* characteristics and higher electroluminescence efficiencies than those with a standard MgAg, Al cathode. Hence, LiF/Al and CsF/Al have been applied to both PLEDs and OLEDs to form an effective electron injector.

5.5.3
Hole-Injection/Transporting Materials

5.5.3.1 Hole-Injection Materials

Beside the surface treatment of ITO, the alternative approach to enhance hole-injection is to insert a nanometer-thick layer with its HOMO between the ITO Fermin level and the HOMO of the hole-transporting layer (HTL). This inserted layer, which is termed the "HIL," creates "a ladder-type" energy structure which has been shown to improve hole-injection. A variety of materials have been shown to act as a HIL, including copper phthalocyanine (CuPC), alkaline halogens such as LiF, amorphous carbon, platinum, and conductive polymers such as PEDOT:PSS, starburst polyamines, and polyaniline.

5.5.3.2 Hole-Transporting Materials

The HIL in PLEDs is to facilitate hole injection from the anode into the conjugated polymer layer, accepting holes, and transporting injected holes to the emitting layer. The HIL also functions to block electrons against escaping from the emitting layer to the anode. Consequently, hole-transporting materials should fulfill the requirements of energy level matching for the injection of holes from the anode. Moreover, they should possess electron-donating properties, and their anodic oxidation processes should be reversible to form stable cation radicals. They also should form homogeneous thin films with both morphological and thermal stability. Details of the most widely used hole-transporting materials, together with their physical properties, are listed in Table 5.3. Recently, water-soluble conjugated polymers have been reported to be used as HILs.

5.5.4
Electron-Transporting Materials

The electron-transport layer (ETL) in PLEDs facilitates electron injection from the cathode into the conjugated polymer layer, accepting electrons and transporting injected electrons to the emitting layer. The ETL also functions to block the holes from escaping from the emitting layer to the cathode. Consequently, electron-transporting materials should fulfill the requirements of energy level matching for the injection of electrons from the cathode. Their electron mobility should be desirably high, and they also should form homogeneous thin films with both morphological and thermal stability. Details of the most widely used hole-transporting materials, together with their physical properties, are listed in Table

Table 5.3 HTL/ETL materials and their glass-transition temperature (T_g) and oxidation/reduction potentials.

Material	T_g (°C)	$(E_{1/2}^{ox}/E_{1/2}^{red})^{a)}$	Function
m-MTDATA	75	0.06	HTL
1-TNATA	113	0.08	HTL
t-Bu-TBATA	203	0.09	HTL
o-PTDATA	93	0.06	HTL
TFATA	131	0.08	HTL
TCTA	151	0.69	HTL
p-MTDAPB	110	0.64	HTL
TFAPB	150	0.61	HTL
TPTE	140		HTL
p-BPD	102	0.50	HTL
TPD	60	0.48	HTL
α-NPD	100	0.51	HTL
CBP		0.72	HTL
OXD-7			ETL
TAZ			ETL
TPOB	137	−2.10 ($E_{1/2}^{red}$)	ETL
Spiro-PBD	163		ETL

a) Versus Ag/Ag^+.

5.3. Recently, water-soluble conjugated polymers have been reported to be used as ETLs.

5.6 Electrophosphorescent PLEDs

5.6.1 Energy Transfer

PLEDs are under active investigation because they have potential applications in flat panel displays [13, 14], and because they can be fabricated via inexpensive solution-processing methods at room temperature, for example, by spin-casting [2], screen-printing [15], or inkjet-printing [16].

During recent years the performance of PLEDs has improved dramatically. In these devices, electrons and holes are injected from opposite electrodes and combine to form either singlet or triplet excitons [17]. As the triplet radiative decay is typically forbidden in organic materials [18], and as spin statistics imply that the ratio of triplets to singlets is 3 : 1, a 100% photoluminescence efficiency from the singlet excitations results in a maximum internal quantum efficiency of 25% for electroluminescence. Experiments have shown, however, that in luminescent semiconducting polymers, the singlet cross-section is considerably larger than that of the triplet (by a factor of 3–4) [19]. As a result, electroluminescent quantum efficiencies as high as 50% have been reported [20].

Efficient triplet emitters have been introduced by several groups to improve the LED efficiency [21–24]. Recently, considerable progress has been demonstrated with electrophosphorescent organic light-emitting diodes (OLEDs; based on small molecules) as hosts, and with heavy-metal (Pt, Ir, Os) complexes and rare-earth metal complexes as guests [21–31]. For example, a green light emission with an external quantum efficiency (QE_{ext}) of 19.2% photons per electron (ph/el) and *LE* of $73\,cd\,A^{-1}$ at a current density (*j*) of $0.55\,mA\,cm^{-2}$ was reported for devices comprising "starburst" perfluorinated phenylenes as the hole- and exciton-blocking layer and the hole-transport material, 4,4′-4″-tri(*N*-carbazolyl)triphenylamine, as host with tris(2-phenylpyridine)iridium (III) ($Ir(ppy)_3$) as the phosphorescent guest [30].

Electrophosphorescent LEDs based on small molecules require the evaporation of multiple layers under high vacuum, which involves the use of a relatively expensive process [23, 24, 29–32]. Electrophosphorescent LEDs based on a polymer as host could, in principle, be fabricated using simpler techniques; for example, by spin-casting, screen-printing, or inkjet-printing [22, 25–27, 33]. Several groups have reported electrophosphorescent LEDs using poly(vinylcarbazole) (PVK) as the host polymer. Kim *et al.* [29] reported an QE_{ext} of 1.9% ph/el and a luminance (*L*) of $2500\,cd\,m^{-2}$ for PVK-doped $Ir(ppy)_3$. Yang and Tsutsui [34] reported a QE_{ext} of 7.5% ph/el from $Ir(ppy)_3$ in PVK in a triple-layer device configuration, while Thompson and Forrest and colleagues reported QE_{ext} values of 5.1% ph/el [35] and 3.5% ph/el [36] from two different Ir-complexes doped into PVK.

Electrophosphorescence based on conjugated polymer blends as hosts offers additional advantages. In principle, a balanced charge injection and efficient transport can be achieved when heavy-metal complexes (e.g., Pt, Ir, Os) are doped into conjugated polymers as hosts [22, 25, 28, 37, 38]. For example, O'Brien *et al.* [27] reported a QE_{ext} of 3.5% ph/el at $L = 10\,cd\,m^{-2}$ with a maximum *L* of $200\,cd\,m^{-2}$ by doping platinum octaethylporphyrin into polyfluorene. Zhu *et al.* [25] reported a QE_{ext} of 5.1% ph/el and a *LE* of $12\,cd\,A^{-1}$, by incorporating $Ir(Bu\text{-}ppy)_3$ into poly(2-(6′-cyano-6′-methylheptyl-oxy)-1,4-phenylene.

In iridium phenylpyridine complexes, spin–orbit coupling at the metal atom enables a radiative relaxation from the triplet excited states to the ground state. By using different ligands, this phosphorescence can be tuned to wavelengths across the visible spectrum [26, 27, 32–36, 39–41]. In contrast, in rare-earth metal complexes, the inner f-shell transitions are insensitive to the structures of the ligands and offer limited emission frequencies [22]. The Ir-complexes also have shorter radiative lifetimes [40, 41] than are observed with rare-earth metal complexes [21, 22].

Emission from the guest molecules in electrophosphorescent LEDs involves exciton localization on the organometallic emitter [42, 43]. This process can occur by various mechanisms, including excitation transfer via Förster and/or Dexter energy-transfer mechanisms from the host to the organometallic dopant. Alternatively, a direct, sequential trapping of electrons and holes can result in triplet excitations localized on the metal-organic center.

In the Förster mechanism [44, 45], the dipole–dipole interaction mediates transfer of the singlet excited-state from the host to the guest. Förster energy transfer can lead to lower self-absorption losses in blends because of the red-shift of the emission relative to the absorption edge [46, 47]. The rate (K_{FET}) of Förster energy transfer is given by:

$$K_{FET} = \tau_d^{-1}(R / R_0)^6 \quad (5.10)$$

where τ_d is the lifetime of the host in the absence of the guest, R is the distance between the host and the guest, and R_0 is the characteristic Förster distance which is given by

$$R_0^6 = \alpha \int_0^\infty F_d(\nu)\varepsilon_\alpha(\nu)\nu^{-4}d\nu \quad (5.11)$$

where α depends on the relative orientation of the host and the guest dipole moments, the fluorescence quantum yield of the host in the absence of the guest, and the refractive index of the medium; $F_d(\nu)$and $\varepsilon_\alpha(\nu)$ are the fluorescence and extinction spectra of the host and the guest, respectively. The efficiency of Förster energy transfer depends on the spectral overlap between the host emission spectrum and the guest absorption spectrum. Typically, the maximum distance over which Förster energy transfer can occur is 30~50 Å.

The Förster energy transfer efficiency from the host to the guest can be written as:

$$\eta = \frac{k_{FET}}{k_{FET} + 1/\tau_d} = \frac{1}{1+(R / R_0)^6} \quad (5.12)$$

where η is the Förster energy transfer efficiency and k_{FET} and R_0 are given in Eqs (5.10) and (5.11), respectively. The distance between the host and the guest, R, is given by [48]:

$$R = \left(N_G * \frac{4\pi}{3}\right)^{-1/3} \quad (5.13)$$

where N_G is the doping concentration of the guest. Combining Eqs (5.12) and (5.13), the energy transfer efficiency is given by:

$$\eta = \frac{1}{1+\left[R_0 \Big/ \left(N_G * \frac{4\pi}{3}\right)^{-1/3}\right]^6} \quad (5.14)$$

Dexter energy transfer requires direct quantum mechanical tunneling of electrons between the host and the guest. It is, therefore, a short-range process that requires a separation of no more than a few Ångstroms. In addition to singlet–singlet energy transfer, the Dexter mechanism can also allow triplet–triplet energy transfer.

Förster and Dexter energy transfer can occur simultaneously if the singlet (or triplet) in the host is resonant with the corresponding levels in the guest. Although the conditions for Förster energy transfer can be easily evaluated from the absorption and emission, an evaluation of the conditions for efficient Dexter transfer requires knowledge of the absolute energies of the excited states. These energies are not available for most light-emitting polymers [49].

In the electron- and hole-trapping mechanism [50, 51], an excited guest molecule is formed by the sequential trapping of first a hole (or electron) and then an electron (or hole) onto the organometallic dopant. The hole- and electron-trapping mechanism is most favorable if the HOMO level of the guest is above (closer to vacuum level) that of the host, and if the LUMO level of the guest is below (farther from vacuum level) that of the host. If, however, only the HOMO of the guest is above that of the host, holes will be readily trapped to form a cationic excited state of the guest that will then function as an electron trap. Charge trapping and localization of carriers onto the guest requires an overlap of the molecular orbitals of the host and guest molecules.

In the following subsection, the results obtained from three guest–host systems are reported. The first system comprises blends of PVK and 2-*tert*-butylphenyl-5-biphenyl-1,3,4-oxadiazole (PBD) as the host and the complexes, tris-[9,9-dihexyl-2-(pyridinyl-2′)fluorene] iridium(III) ($Ir(DPF)_3$), tris-[9,9-dihexyl-2-(phenyl-4′-(-pyridin-2″-yl))fluorene] iridium(III) ($Ir(DPPF)_3$), and tris-[2,5-bis-2′-(9,9′-dihexylfluorene) iridium] ($Ir(HFP)_3$), as the guests. The second guest–host system contains the conjugated polymer, poly(9,9-dioctylfluorenyl-2,7-diyl) (PFO) as the host and $Ir(HFP)_3$ as the guest. The third system uses the conjugated copolymer, poly(9,9-dihexylfluorene)-*co*-2,5-dicyanophenylene (PF_3CNP_1) as host and $Ir(HFP)_3$ as guest. The molecular structures of PVK, PBD, PFO, PF_3CNP_1, $Ir(DPF)_3$, Ir(DPPF)3 and $Ir(HFP)_3$ are shown in Figure 5.23. In the three systems, Förster energy transfer from polymer to iridium complex was investigated by steady-state photoluminescence. High-performance electrophosphorescent LEDs were demonstrated using these three host–guest systems.

5.6.2 Electrophosphorescent PLEDs

The synthesis and characterization of $Ir(DPF)_3$, $Ir(DPPF)_3$ and $Ir(HFP)_3$ have been reported. PVK and PBD were obtained from Aldrich and used without further purification, while PFO was provided by American Dye Source (Canada). The synthesis and characterization of PF_3CNP_1 has also been reported [52]. The HOMO and LUMO energy levels of PVK, PBD, PFO, PF_3CNP_1, $Ir(DPF)_3$, $Ir(DPPF)_3$ and $Ir(HFP)_3$ are shown in Figure 5.24. For spectroscopic measurements, thin films were prepared by spin-casting from dichloroethane onto quartz substrates. The absorption spectra were measured using a Shimadzu UV-2401PC UV-visible spectrophotometer. The photoluminescence (PL) spectra were measured using a Spex Fluoromax-2 spectrometer.

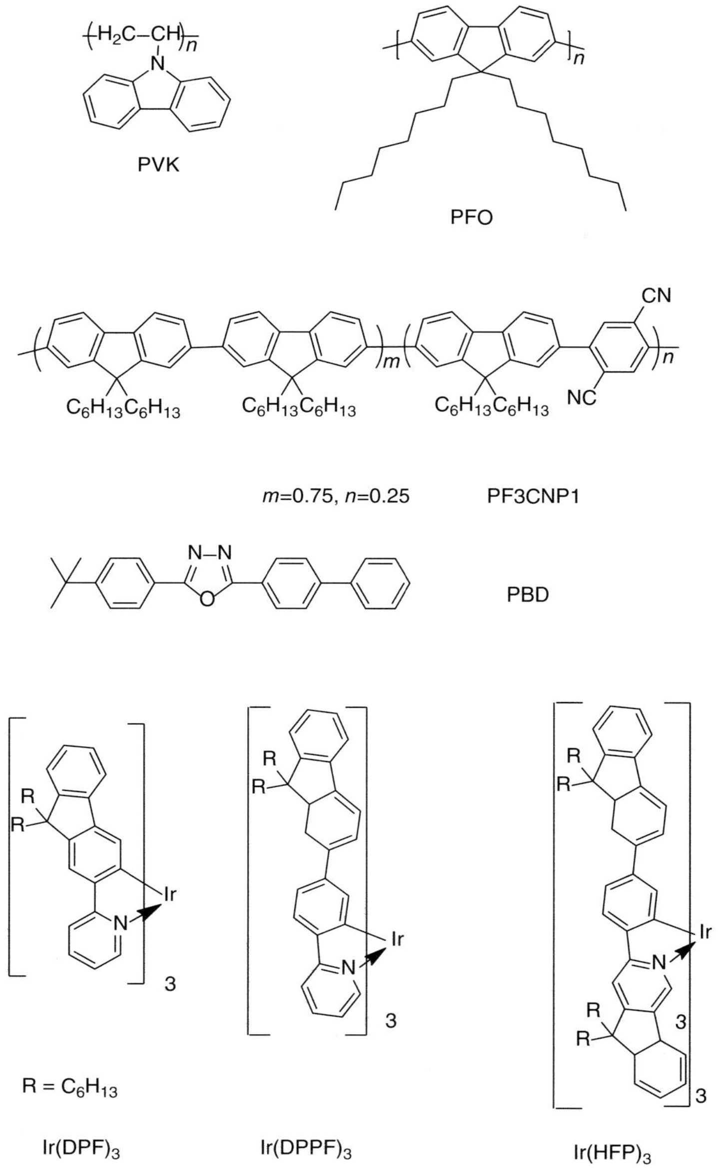

PVK

PFO

m=0.75, n=0.25 PF3CNP1

PBD

$R = C_6H_{13}$

$Ir(DPF)_3$ $Ir(DPPF)_3$ $Ir(HFP)_3$

Figure 5.23 Molecular structures of PVK, PFO, PF_3CNP1, PBD, $Ir(DPF)_3$, $Ir(DPPF)_3$, and $Ir(HFP)_3$.

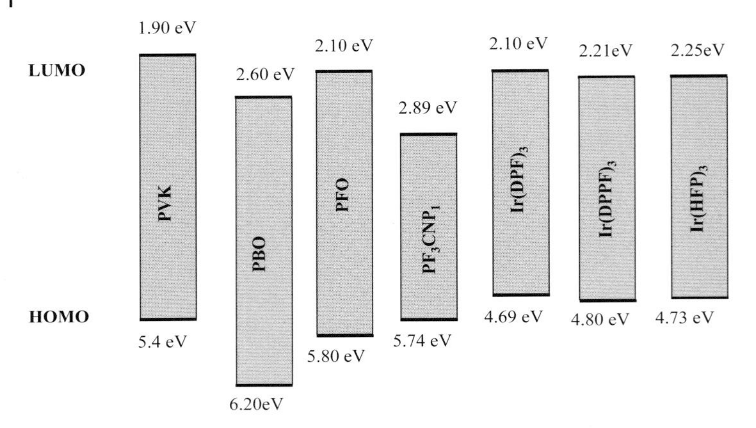

Figure 5.24 HOMO and LUMO energy levels of PVK, PBD, PFO, PF_3CNP1, $Ir(DPF)_3$, $Ir(DPPF)_3$, and $Ir(HFP)_3$.

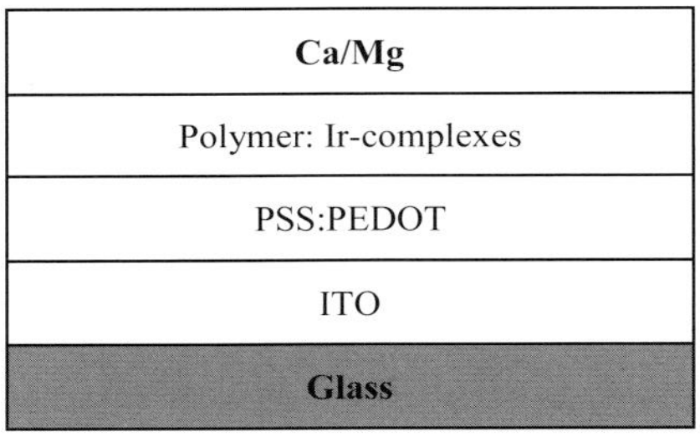

Ca/Mg
Polymer: Ir-complexes
PSS:PEDOT
ITO
Glass

Figure 5.25 Device configuration.

For device fabrication, only the single-active-layer configuration was employed, with PEDOT:PSS on ITO as the hole-injecting bilayer electrode. The device configuration was (ITO)/PEDOT:PSS/polymer:Ir-complex/Ca/Ag. PEDOT:PSS was spun-cast onto ITO-coated glass, after which the emitting layer, polymers with different concentrations of Ir-complexes, was spun-cast onto the PEDOT:PSS layer. The film thicknesses ranged from 100 to 150 nm. The Ca/Ag cathode was deposited through a shadow mask by thermal evaporation at 4×10^{-7} Torr. The device configuration is shown in Figure 5.25.

The current–voltage and brightness–voltage characteristics were measured using a Keithley 236 source measurement unit and a calibrated silicon photodiode [22, 53] (computer interfaced with LabView™ supplied by National Instruments). The emission spectra were measured with a single-grating monochromator equipped with a photometric charge-coupled device (CCD) camera as the detector. All device fabrication and testing were carried out under nitrogen.

5.6.3
Nonconjugated Polymer-Based Electrophosphorescent PLEDs

PVK [54–56] was selected as the host because its emission spectrum overlaps the absorption spectrum of $Ir(DPF)_3$, $Ir(DPPF)_3$, and $Ir(HFP)_3$. PVK is also known to be a good hole-transporting material, but a poor electron-transporting material [56]; thus, PBD was blended with PVK to enable a balanced transport in the host [56]. The normalized absorption and PL spectra of thin films of PVK-PBD (40 wt%) and neat $Ir(DPF)_3$, $Ir(DPPF)_3$ and $Ir(HFP)_3$ are shown in Figure 5.26a–c. There was a good overlap between the fluorescence spectrum of PVK-PBD (40 wt%) and the metal-ligand charge transfer (MLCT) band of these three Ir-complexes, and the overlap satisfied the conditions for Förster energy transfer. Hence, fast intersystem crossing in the Ir-complexes and subsequent emission from this state are built-in features [57].

The current density versus voltage and brightness versus voltage characteristics of devices made using PVK-PBD (40 wt%) with 1 wt% $Ir(DPF)_3$, $Ir(DPPF)_3$ and $Ir(HFP)_3$ are shown in Figure 5.27. The turn-on voltage of these devices was approximately 10 V, and the devices had maximum brightnesses in excess of 8300 cd m^{-2} (at 55 V) for $Ir(DPF)_3$, over 3500 cd m^{-2} (at 30 V) for $Ir(DPPF)_3$, and 2700 cd m^{-2} (at 26 V) for $Ir(HFP)_3$.

Typical *LE* and QE_{ext} curves versus current density for a device made from PVK-PBD (40 wt%) with 1 wt% $Ir(DPF)_3$ are shown in Figure 5.28. Devices made from PVK-PBD (40 wt%) doped with other Ir-complexes showed similar features. The *LE* and QE_{ext} were increased initially and then decreased with increasing current density, similar to the QE_{ext} behavior reported for devices containing $Ir(ppy)_3$ [30]. The inset in Figure 5.28 shows QE_{ext} versus $Ir(DPF)_3$ concentration at constant current density, $j = 4\,mA\,cm^{-2}$, whereby the highest QE_{ext}, of approximately 10% ph/el, was obtained from the device with 0.3 wt% $Ir(DPF)_3$ in PVK-PBD (40 wt%). The QE_{ext} decreased as the doping concentration increased, possibly due to the aggregation and self-quenching of $Ir(DPF)_3$ [58].

A maximum QE_{ext} was achieved at 0.3 wt% of $Ir(DPF)_3$, which was significantly lower than the doping concentration (8 wt%) reported for $Ir(ppy)_3$:PVK devices [29]. The maximum *LE* of 36 cd A^{-1} and QE_{ext} of 10% ph/el were achieved at a current density of 4 mA cm^{-2} and at a brightness of 403 cd m^{-2}, while the power efficiency of the devices was 3.14 lum W^{-1} at 36 V. It was possible to run the devices at much higher current densities (e.g., 4 mA cm^{-2}) than had been reported for devices made with $Ir(ppy)_3$ in PVK or for small-molecule devices (0.01 or 0.2 mA cm^{-2}) [26, 29]. Even at 50 mA cm^{-2}, these devices exhibited a relatively high QE_{ext} (5.1% ph/el), *LE* (24 cd A^{-1}), and power efficiency (1.40 lum W^{-1}).

The operating conditions and the characteristics of LEDs fabricated from $Ir(DPF)_3$ doped into PVK-PBD (40 wt%) are summarized in Table 5.4. The presence of PBD was seen to improve device performance.

The performance of devices made from different Ir-complexes is summarized in Table 5.5. The high performance of these devices was due, in part, to the advantageous molecular features of the $Ir(DPF)_3$, $Ir(DPPF)_3$, and $Ir(HFP)_3$ dopants,

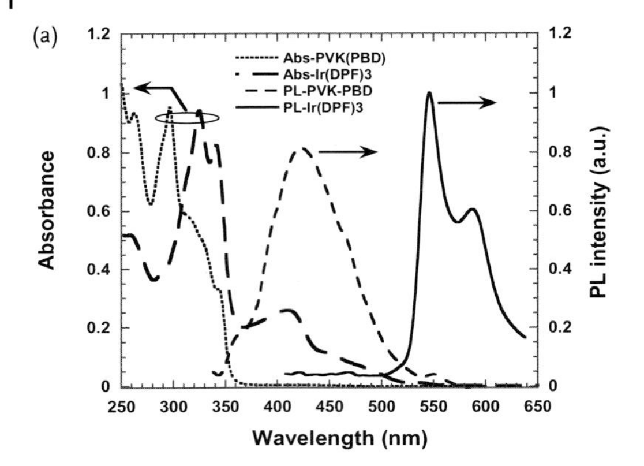

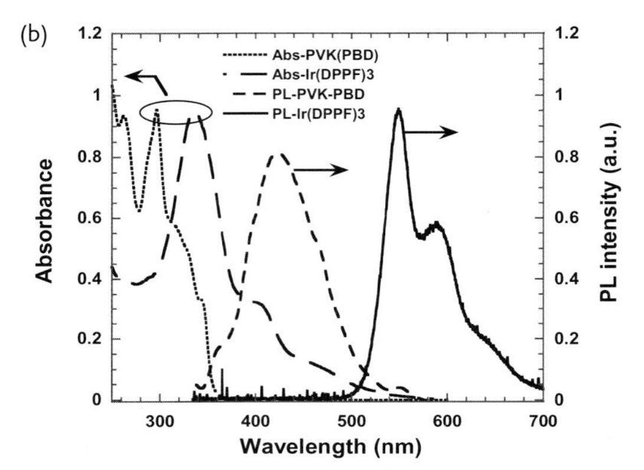

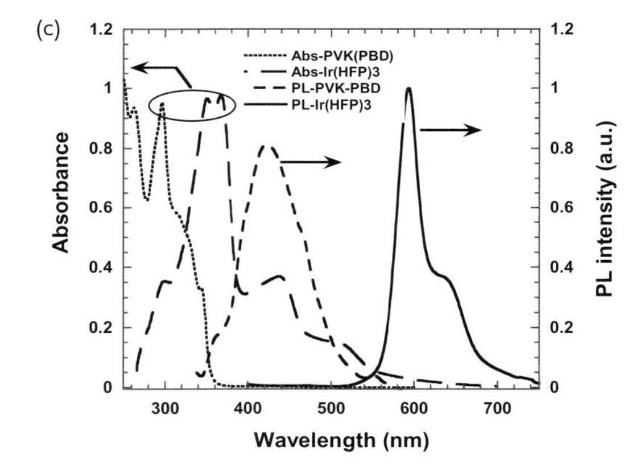

Figure 5.26 The absorption and photoluminescence spectra of thin films of PVB-PBD (40 wt%) and (a) $Ir(DPF)_3$, (b) $Ir(DPPF)_3$, and (c) $Ir(HFP)_3$.

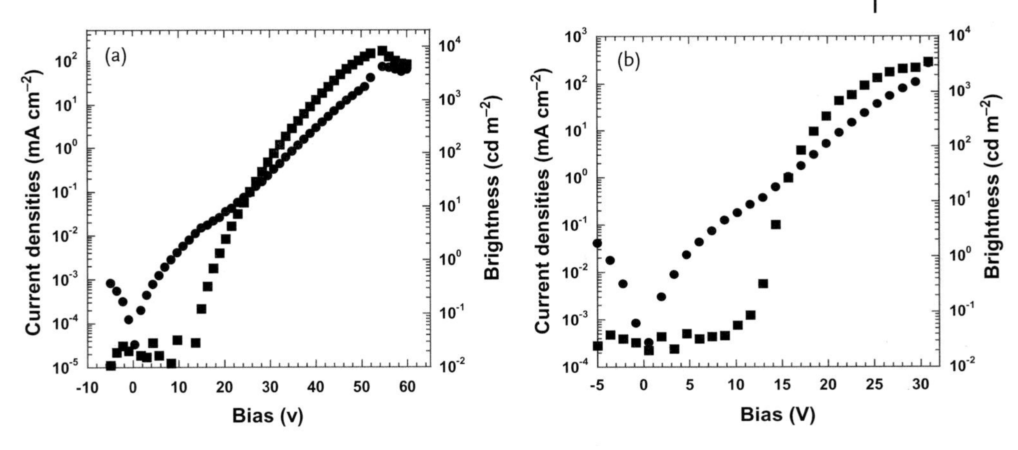

Figure 5.27 Brightness (solid squares) and current density (solid circles) versus applied voltage of devices made from Ir-complexes in PVK-PBD. (a) $Ir(DPF)_3$; (b) $Ir(DPPF)_3$.

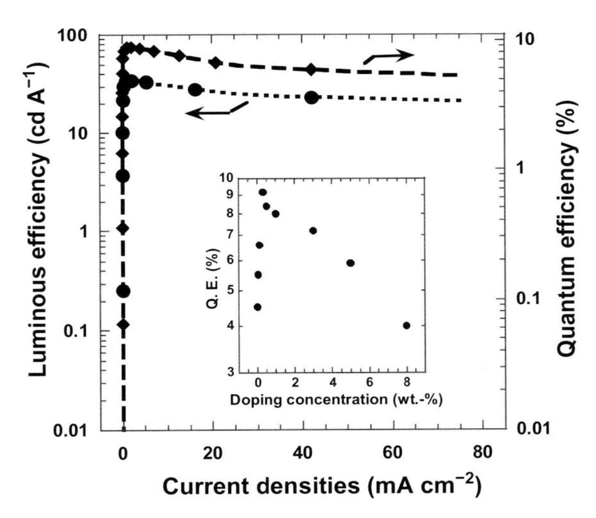

Figure 5.28 Luminous efficiency (solid circles) and quantum efficiency (solid diamonds) versus current density of $Ir(DPF)_3$:PVK-PBD (40 wt%) LEDs with 1 wt% doping concentration of $Ir(DPF)_3$. Inset: The external quantum efficiency versus doping concentration (wt%) of $Ir(DPF)_3$/PVK-PBD (40 wt%) LEDs at the constant current density of 4 mA cm^{-2}.

Table 5.4 Summary of operating conditions and characteristics of LEDs fabricated from $Ir(DPF)_3$ and PVK-PBD (40wt%).

Concentration of $Ir(DPF)_3$ (wt%)	Bias @1 mA cm^{-2} (V)	Bias@ 10 cd m^{-2} (V)	L_{max} [b] (cd m^{-2})
PVK only [a]	5	9	59 (@ ~12 V)
0.05	11	25	1638 (@ 48 V)
0.3	16	20	3054 (@ 53 V)
1	32	23	8321 (@ 55 V)
3	33	30	2853 (@ 54 V)
5	37	32	1238 (@ 55 V)
3 [c]	47	40	370 (@ 51 V)

a) Reference [56].
b) Maximum luminance.
c) Device structure is same as others, but without PBD.

Table 5.5 Comparison of the devices' performance.

Dopant	Doping concentration (wt%)	λ_{max} (nm)	L_{max} (cd m^{-2})	*LE* (cd A^{-1})	QE_{ext} (ph/el, %)
$Ir(DPF)_3$	1	546	8320	36	10
$Ir(DPPF)_3$	1	550	3500	29	8
$Ir(HFP)_3$	3	600	2700	7.2	5.6

λ_{max}: maximum emission wavelength; *L*: Luminance; *LE*: Luminous efficiency; QE_{ext}: external quantum efficiency.

which include a large ligand framework that reduces the tendency towards phase separation. Additionally, by controlling the ligand conjugation length, it was possible to obtain yellow-green to red emission. The large ligand size was also shown to minimize emission quenching in the solid-state [37], and to enhance electron-hole trapping on the Ir-complex [59].

In order to test energy-transfer efficiencies, thin films of PVK-PBD (40wt%) with different concentrations of $Ir(DPF)_3$, $Ir(DPPF)_3$ and $Ir(HFP)_3$ were prepared, and the PL spectra recorded; PL spectra (excited at 325 nm) are shown in Figure 5.29. The PL spectra of PVK-PBD (40wt%) with different $Ir(DPF)_3$ concentrations were similar [26], with all PL profiles containing two peaks: one peak centered at 425 nm, which resulted from the emission of the PVK-PBD host; and a second peak at 550 nm with a shoulder at 600 nm for $Ir(DPF)_3$ and $Ir(DPPF)_3$, and at 600 nm with a shoulder at 620 nm for $Ir(HFP)_3$ triplet emission. The emission intensity at 425 nm decreased significantly as the Ir-complex concentration was increased. At 8wt% of the Ir-complex, there was no virtually emission from the host polymer, and the Ir-complex emission dominated. Hence, an efficient Förster energy transfer from PVK-PBD to Ir-complexes was confirmed. The absorption

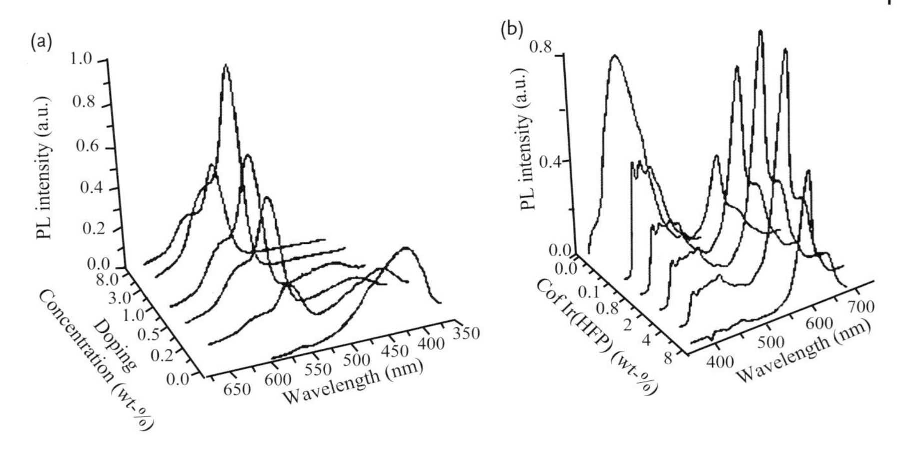

Figure 5.29 Photoluminescence (PL) spectra of thin films of PVK-PBD (40wt%) and PVK-PBD (40wt%) with different concentrations of (a) $Ir(DPPF)_3$ and (b) $Ir(HFP)_3$.

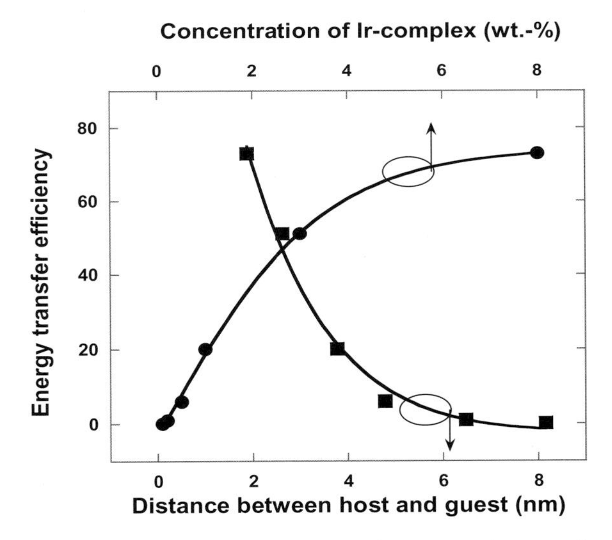

Figure 5.30 Relative energy transfer efficiency as a function of $Ir(DPPF)_3$ concentration and the distance between the host and the guest (datum points represent the calculated values, lines represent the fitted curves.

and PL spectra obtained from films made from PVK-PBD blended with different Ir-complexes exhibited similar trends. Thus, α, $F_d(\nu)$, and $\varepsilon_\alpha(\nu)$in Eq. (5.2) can be assumed to be constants, implying a constant R_0 in Eq. (5.5) for all the films with different Ir-complex concentrations. Thus, the relative energy-transfer efficiency can be predicted either as a function of Ir-complex concentration or, equivalently, the distance between the host and the guest. As an example, Figure 5.30 shows the data obtained from PVK-PBD doped with $Ir(DPPF)_3$.

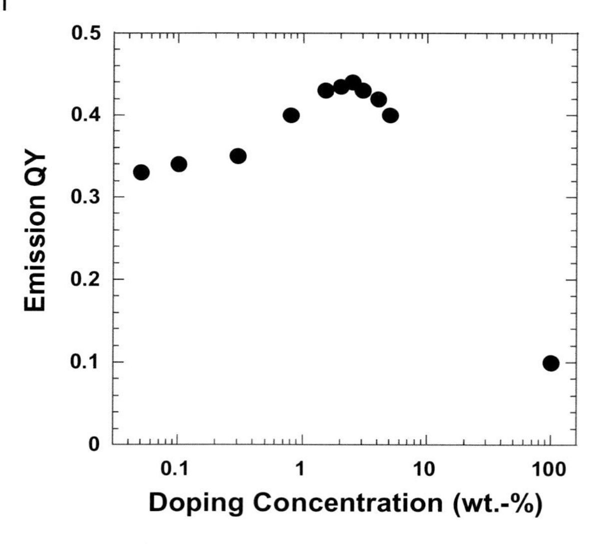

Figure 5.31 Photoluminescence quantum yield as a function of the doping concentrations of $Ir(HFP)_3$ in PVK-PBD.

The PL quantum yield as a function of the $Ir(HFP)_3$ concentration in PVK-PBD is shown in Figure 5.31. Both, Figures 5.30 and 5.31 demonstrate that at lower doping concentrations, energy transfer from the PVK-PBD is incomplete because the average distance from a photoexcited polymer chain to the nearest Ir-complexes is too large ($R < R_0$) [22]. Although, at higher concentrations ($R > R_0$), energy was transferred to the Ir-complexes, the quantum yield was reduced by concentration quenching and saturation of Ir-complexes emission. The PL from neat Ir-complex films was also found to be weak [40]. Taken together, these results indicated an efficient Förster energy transfer from PVK-PBD to Ir-complexes, even at low concentrations of the Ir-complexes, as might be expected for the relatively long-range (dipole–dipole) Förster coupling [23, 29].

The EL spectra of devices made with PVK-PBD with different Ir-complex concentrations showed only the emission from the Ir-complex [26, 33, 60], even at 0.01 wt% guest concentration [23]. The EL spectra of devices made from PVK-PBD (40 wt%) with $Ir(DPF)_3$ are shown in Figure 5.32, while the EL spectra of the device made from 0.01 wt% doping concentration of $Ir(DPF)_3$ at different applied voltages are shown in Figure 5.33. It should be recalled that the concentration required to completely quench the PVK-PBD (40 wt%) under optical excitation was 8 wt% of $Ir(DPF)_3$. The absence of PVK or PBD EL emission from devices containing $Ir(DPF)_3$, even at the lowest concentration of $Ir(DPF)_3$ (0.01 wt%), was consistent with charge trapping on the Ir-complex instead of Förster transfer as the dominant mechanism in the LEDs. It can be concluded, therefore, that the

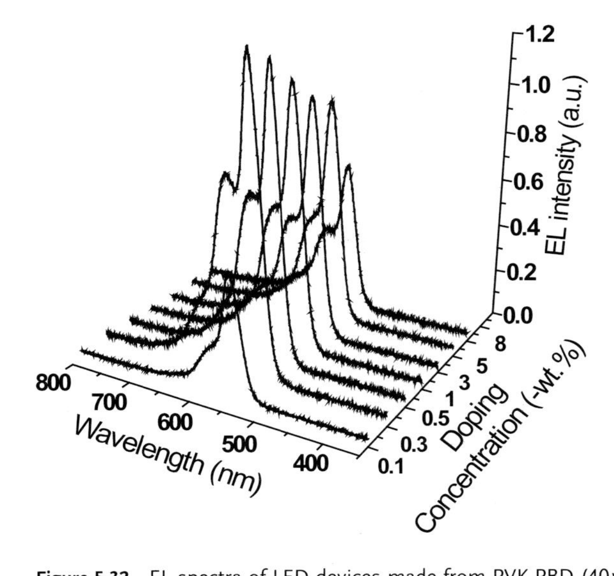

Figure 5.32 EL spectra of LED devices made from PVK-PBD (40 wt%) with different concentrations of $Ir(DPF)_3$.

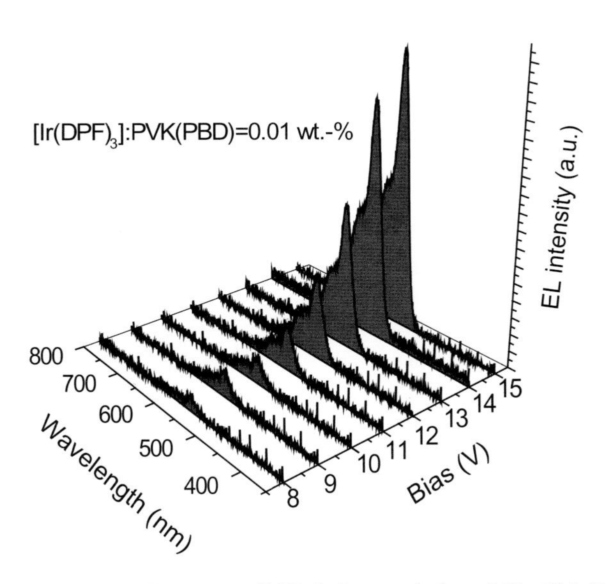

Figure 5.33 EL spectra of LED devices made from 0.01 wt% $Ir(DPF)_3$ doped into PVK-PBD (40 wt%) at different bias voltages.

Ir-complex traps electrons *and* holes, which enables direct recombination on the Ir-complex. If energy transfer were the dominant EL mechanism, then an EL emission from PVK or PVK-PBD would be expected to appear when the triplet of Ir-complexes became saturated – that is, at higher applied voltages and lower concentrations [28, 42]. The observed increase in the operating voltages with increasing $Ir(DPF)_3$ concentration (Table 5.4) provided additional evidence of charge trapping in this system.

Since the Ir-complexes function as both hole and electron traps, the HOMO and/or the LUMO of the Ir-complex must fall within the band gap of the polymer. The HOMO and LUMO of PVK are −5.4 eV and −1.9 eV, respectively [56], while the HOMO and LUMO of $Ir(DPF)_3$, $Ir(DPPF)_3$ and $Ir(HFP)_3$ are −4.69 eV, −4.73 eV, −4.82 eV, and −2.10 eV, −2.21 eV, −2.25 eV, respectively, as determined by cyclic voltammetry measurements. Hence, the HOMOs of Ir-complexes are 0.71 eV, 0.67 eV, and 0.58 eV above that of the PVK, which implies that holes are readily trapped on the Ir-complexes. Assuming, therefore, that the HOMO functions as the hole trap, the electron can subsequently hop and localize onto a positive Ir-complex to form the excited state. The results observed with time-resolved photoluminescence at different temperatures were also consistent with the charge trapping mechanism [43, 61, 62].

In summary, the results of steady-state photoluminescence studies have indicated an efficient Förster energy transfer from PVK-PBD to the Ir-complexes. However, Förster energy transfer plays only a minor role in electrophosphorescent PLEDs; rather, direct charge trapping on the Ir-complexes is the main operating mechanism.

5.6.4 Conjugated Polymer-Based Electrophosphorescent PLEDs

Although high-performance devices were demonstrated by using PVK-PBD, the operating voltages of these devices are relatively high. In order to demonstrate electrophosphorescent PLEDs with lower turn-on voltages, conjugated polymers were examined as hosts. The absorption and PL spectra of a PFO film, and of a neat film of $Ir(HFP)_3$, are shown in Figure 5.34. Again, the emission of PFO overlaps with the MLCT absorption band of $Ir(HFP)_3$, which meets the requirement for an efficient Förster transfer. The PL spectra from $Ir(HFP)_3$ doped into PFO (excitation at 375 nm) are shown in Figure 5.35; in this case it was surprising that the drop in PFO PL was not recovered by $Ir(HFP)_3$ PL. In fact, even at 8 wt% $Ir(HFP)_3$ concentration, the emission from PFO was not completely quenched.

The EL spectra of devices made with PFO and PFO doped with $Ir(HFP)_3$ at different concentrations are shown in Figure 5.36. The strong peak at 530 nm in the EL spectrum of devices made using only PFO is most likely due to fluorenone defects generated during device operation [63]. The EL spectra of the devices made by PFO with $Ir(HFP)_3$ show the characteristic spectrum of $Ir(HFP)_3$, with a peak at 600 nm and a secondary band at 620 nm. The devices prepared from

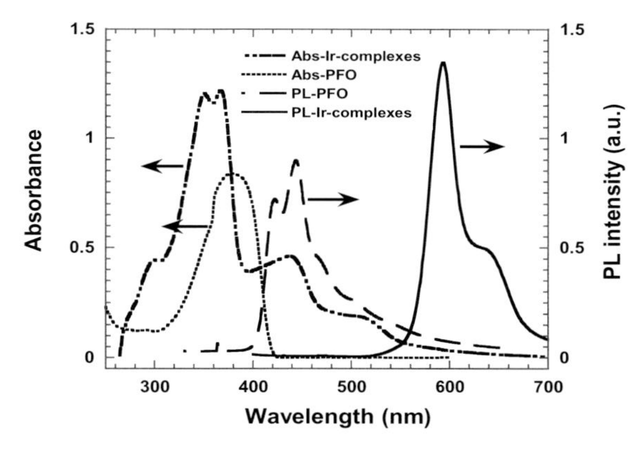

Figure 5.34 Absorption and PL spectra of thin films of PFO.

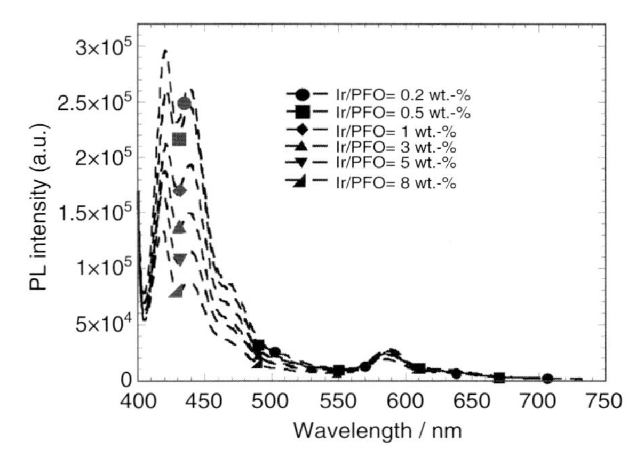

Figure 5.35 PL spectra of thin films of PFO with different concentrations of $Ir(HFP)_3$ excited at 375 nm.

PFO with 0.2 wt% $Ir(HFP)_3$ also showed a weak peak at 450 nm characteristic of PFO, and a broad emission peak at 530 nm which previously had been assigned to fluorenone defects in PFO [60]. There was no PFO emission from any other devices containing $Ir(HFP)_3$ and, again, the absence of a PFO emission peak from LEDs was consistent with charge trapping on the Ir-complex (rather than Förster transfer).

The current density versus voltage and brightness versus voltage characteristics of a device made with 1 wt% $Ir(HFP)_3$ in PFO are shown in Figure 5.37. This device turned on at approximately 5 V, which was significantly lower than that observed with devices made from PVK-PBD and $Ir(HFP)_3$ [33]. The devices had a maximum

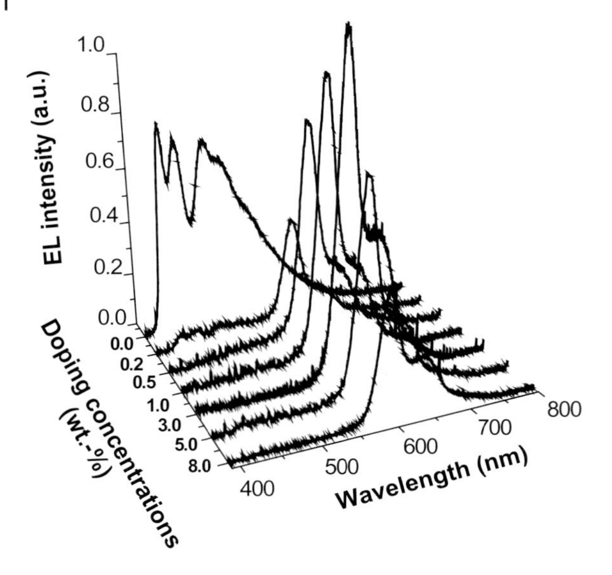

Figure 5.36 El spectra of devices made using neat PFO and PFO doped with different concentrations of $Ir(HFP)_3$.

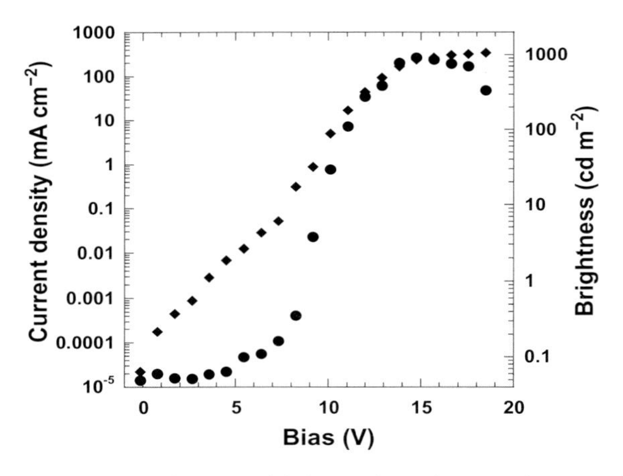

Figure 5.37 Brightness (solid diamonds) and current density (solid circles) versus applied voltage of a device made from PFO doped with $Ir(HFP)_3$.

L in excess of 1000 cd m^{-2} (at 17 V). The *LE* and QE_{ext} as a function of the current density for devices made with 1 wt% $Ir(HFP)_3$ in PFO are shown in Figure 5.38a. In this case, an initial increase was observed in *LE* and QE_{ext}, followed by a decrease with increasing current density, similar to the QE_{ext} dependence reported for devices made from PVK-PBD [26, 32, 43] and $Ir(ppy)_3$ [30]. The QE_{ext} and *LE* versus

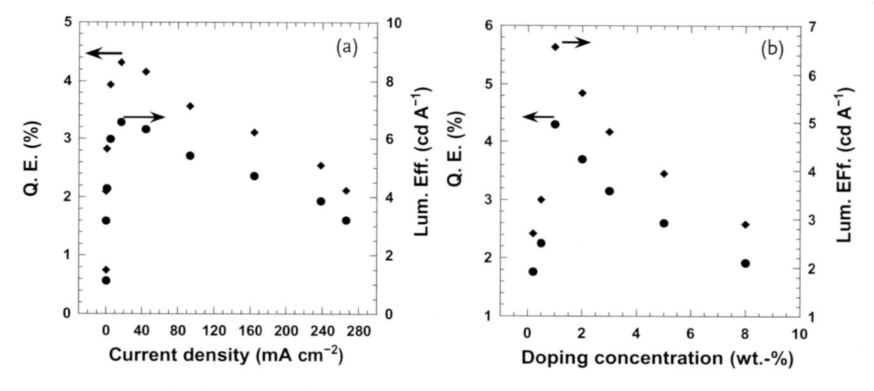

Figure 5.38 (a) The luminous efficiency (solid circles) and external quantum efficiency (solid diamonds) versus current density of $Ir(HFP)_3$/PFO LEDs with 1 wt% doping concentration of $Ir(HFP)_3$; (b) The external quantum efficiency and luminous efficiency versus doping concentration (wt%) of $Ir(HFP)_3$/PFO LEDs at the constant current density.

$Ir(HFP)_3$ concentration at constant current density is shown in Figure 5.38b; in this case, the highest QE_{ext} (4.5% ph/el) and LE (6.2 cd A^{-1}) were obtained from the device with 1 wt% $Ir(HFP)_3$ in PFO at 10 mA cm^{-2}. However, as the doping concentration increased, the QE_{ext} and LE were seen to decrease, possibly due to $Ir(HFP)_3$ aggregation [58].

In summary, red electrophosphorescent LEDs with QE_{ext} 4.5% ph/el, LE 6.2 cd A^{-1}, and L >1000 cd m^{-2} can be achieved despite the weak Förster energy transfer observed with PFO doped with $Ir(HFP)_3$.

The optoelectronic properties of a conjugated polymer are governed primarily by the chemical structure of the polymer backbone. Consequently, by incorporating suitable functional groups onto the polymer main chain, or by synthesizing copolymer structures, it is possible to fine-tune the HOMO and LUMO energy levels and the electronic properties that, eventually, will affect the charge transport characteristics of the bulk [51, 64–69].

The absorption and PL spectra of films of neat $Ir(HFP)_3$ and of PF_3CNP_1 are shown in Figure 5.39 where, once again, there is good overlap between the absorption bands of singlet MLCT (1MLCT, ~λ = 440 nm) and triplet MLCT (3MLCT, ~λ = 515 nm) and the fluorescence spectrum of PF_3CNP_1 (420–590 nm).

In an attempt to optimize the device performance, the concentration dependence ($Ir(HFP)_3/PF_3CNP_1$) was examined over a range between 0.05 wt% to 8 wt%. The luminance and current density versus bias voltage of devices with 1 wt% doping $Ir(HFP)_3$ are shown in Figure 5.40. Here, the turn-on voltage, arbitrarily defined as the voltage required for a luminance of 0.1 cd m^{-2}, was approximately 4.5 V. Moreover, increasing the doping concentrations also increased the turn-on voltage; for example, the latter was increased from 4.5 V for 1 wt% to 8 V for 8 wt% $Ir(HFP)_3$, while the maximum L was 2200 cd m^{-2} at 15 V at a 1 wt% doping concentration of $Ir(HFP)_3$.

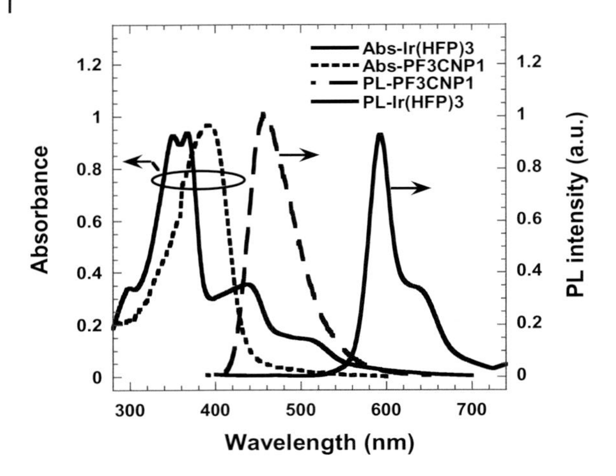

Figure 5.39 Absorption and PL spectra of thin films of $Ir(HFP)_3$ and PF_3CNP_1.

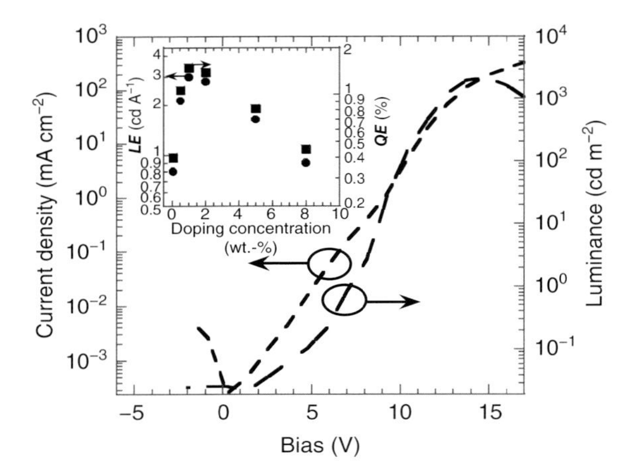

Figure 5.40 Luminance (L) and current density (j) versus voltage of a device with PF_3CNP_1 doped with 1 wt% $Ir(HFP)_3$. The inset shows the QE_{ext} and LE versus the concentration of $Ir(HFP)_3$.

The dependence of QE_{ext} and LE on $Ir(HFP)_3$ concentration is shown in the inset of Figure 5.40. At low concentrations, QE_{ext} and LE increased sharply as the $Ir(HFP)_3$ concentration was increased, with a maximum at approximately 1 wt%. However, above a level of 1 wt%, both QE_{ext} and LE decreased with increasing concentration. Although the reasons for this decrease were not understood in detail, it might have resulted from triplet–triplet or triplet polaron annihilation [42], as well as concentration quenching [58].

Devices with 1 wt% $Ir(HFP)_3$ exhibited QE_{ext} of $3\,cd\,A^{-1}$ and LE of 1.5% ph/el, respectively, at $L = 142\,cd\,m^{-2}$ ($j = 4.8\,mA\,cm^{-2}$ and $V = 10\,V$). This QE (LE) value was approximately at the maximum on a plot of QE_{ext} (LE) versus current density. The j-value of $4.8\,mA\,cm^{-2}$ was higher than that reported for devices made with $Ir(ppy)_3$ in PVK and in small-molecule devices (0.01 or $0.2\,mA\,cm^{-2}$) [23, 24, 27, 29, 30], but was far below that required to excite all of the dopant molecules [43]. It should be noted that, even at $j = 50\,mA\,cm^{-2}$, where $L = 1210\,cd\,m^{-2}$, the devices retained an LE of $2.3\,cd\,A^{-1}$ and a QE_{ext} of 1.2%.

When an initial study of the stability of devices prepared from PF_3CNP_1 doped with 1 wt% and 2 wt% $Ir(HFP)_3$ was carried out, the devices were not encapsulated, while all measurements were carried out under N_2 under a constant applied voltage with an initial L-value of $100\,cd\,m^{-2}$. The luminance decay curves (as a function of time) are shown in Figure 5.41. For 1 wt% $Ir(HFP)_3$, after an initial drop to 68% of the initial value, the luminance remained essentially constant for more than 15 h. By comparison, the 2 wt% $Ir(HFP)_3$ devices showed a similar stability, with an improved stability expected for correctly encapsulated devices. These initial results demonstrates that single-layer devices made from the copolymer, PF_3CNP_1, when doped with $Ir(HFP)_3$, have a promising operational stability.

The PL spectra of films of PF_3CNP_1 and PF_3CNP_1 doped with different concentrations of $Ir(HFP)_3$ (excited at 375 nm) are shown in Figure 5.42a. The EL spectra of a device made from 0.05 wt% $Ir(HFP)_3$ in PF_3CNP_1 at different applied voltages are shown in Figure 5.42b. The PL profile showed two peaks: one peak was centered at 475 nm, from PF_3CNP_1 emission, and the second peak at 600 nm with

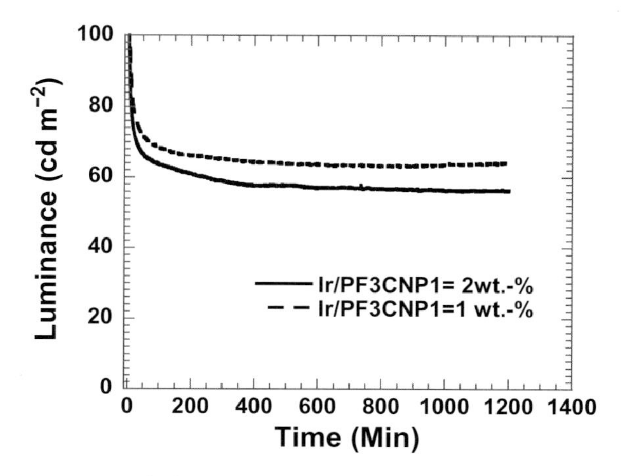

Figure 5.41 The luminance decay curves of the devices made from PF_3CNP_1 doped with $Ir(HFP)_3$ as a function of the time.

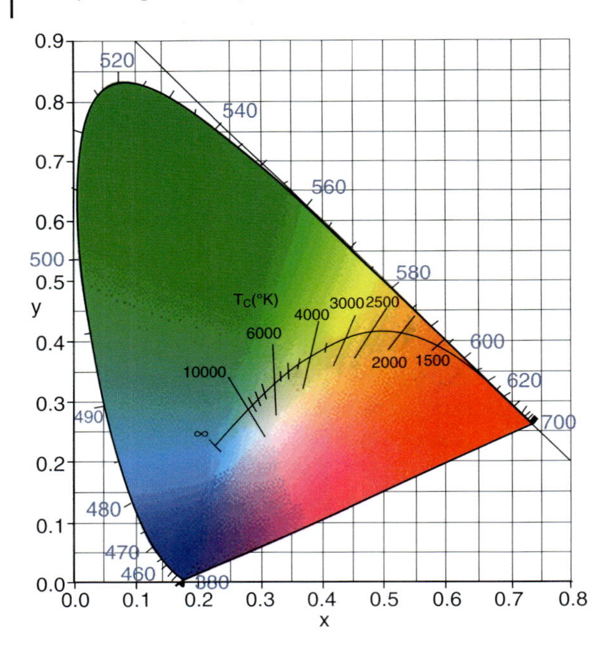

Figure 5.42 The CIE chromaticity diagram. The curve inside the diagram represents the colors of blackbody radiation corresponding to CCT values superior to 1000 K.

a shoulder at 620 nm, from the $Ir(HFP)_3$ emission. The host emission at 475 nm decreased with increasing doping concentration, with 8 wt% $Ir(HFP)_3$ being required to completely quench the host PL. The EL spectra were identical to the PL spectra of neat $Ir(HFP)_3$. There was no EL emission from the PF_3CNP_1 host of any of the PLEDs that contained $Ir(HFP)_3$; in fact, the host EL was completely quenched even at the lowest doping concentration (0.05 wt%).

The EL spectra of a device made with 5 wt% $Ir(HFP)_3$ in PF_3CNP_1, recorded at different current densities, are shown in Figure 5.43. Even at high current density, there was no EL emission from the PF_3CNP_1 host. These results again showed that charge trapping, rather than energy transfer, was the dominant EL mechanism.

Taken together, these data show that red electrophosphorescent LEDs with a high brightness and a promising operational stability can be fabricated from the conjugated copolymer, PF_3CNP_1, as the host, and with $Ir(HFP)_3$ as the guest.

In summary, electrophosphorescent PLEDs with high brightness, high external quantum efficiency, high luminous efficiency and promising operational stability, were demonstrated using PVK-PBD, PFO and PF_3CNP_1 as hosts, and $Ir(DPF)_3$, $Ir(DPPF)_3$ and $Ir(HFP)_3$ as guests.

Steady-state photoluminescence studies have indicated an efficient Förster energy transfer from the polymer hosts to the different Ir-complexes. However,

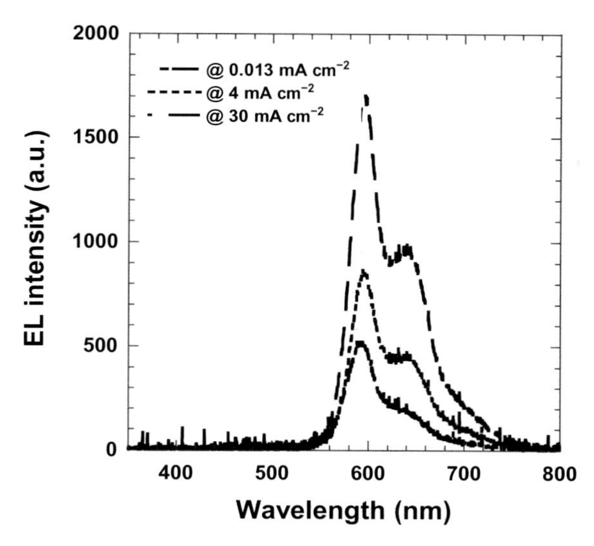

Figure 5.43 EL spectra of a device made with 5 wt% $Ir(HFP)_3$ in PF_3CNP_1 at different current densities.

data from the electrophosphorescent LEDs have demonstrated that Förster energy transfer plays a minor role in these devices. Rather, direct charge trapping on the Ir-complexes guest is the main operating mechanism for electrophosphorescence.

From a practical perspective, these results have demonstrated that high-performance electrophosphorescent PLEDs can be realized with polymers as hosts and heavy metal complexes as guests, and that these devices can be fabricated by processing the luminescent layer from solution. These results also demonstrate an opportunity to realize high-performance LEDs with low operating voltages, by using Ir-complexes in combination with conjugated semiconducting polymers as hosts.

5.7 White-Light PLEDs

5.7.1 Solid-State Lighting

The need to reduce energy consumption associated with the low efficiency of conventional lighting systems (e.g., incandescent bulbs) has prompted considerable research effort towards the development of white-light-emitting diodes (WLEDs) as high-performance alternative devices [70–72]. Actually, incandescent bulbs, which have for many years represented the most common lighting sources, are very inefficient and dissipate as heat the main part of the electrical energy absorbed, converting only 5% of this energy into light. Even energy-saving compact fluorescent lamps (CFLs), which are rapidly replacing incandescent bulbs, are only about 20% efficient. Moreover, CFLs represent a highly toxic cumulative product because they are composed of phosphor-coated gas-discharge tubes containing milligram quantities of mercury [72]. Efficient solid-state lighting (SSL) sources based on WLEDs are predicted to become the next generation of general illumination systems, with a significant impact on energy saving [73]. Indeed, a representative estimate indicates that, in the USA alone, the replacement of current white-lighting technologies with 50% efficient SSL would reduce the electricity used for lighting by 62%, and the total electrical energy consumption by 13% [74]. While common LEDs produce light by the electroluminescence of Group III–V mixed crystal inorganic semiconductors, PLEDs generate light from conjugated polymers and/or organometallic molecules [75]. Moreover, PLED technology offers the distinctive possibility to fabricate lightweight, large-area, and also flexible and transparent light-emitting devices by a low-cost processing that is based on various thin-film deposition techniques such as vacuum thermal evaporation, spin-coating, or casting from solution. Building on these attractive properties, sources for general lighting based on white organic light-emitting diodes (WOLEDs) are being actively investigated [76–78].

5.7.2
Characterization of White Light

The main parameters used to characterize the quality of a white light source are:

1) The Commission Internationale d'Eclairage (CIE) chromaticity coordinates (x, y) locate the emission color in the chromaticity diagram shown in Figure 5.42. The perfect white light has CIE coordinates (0.33, 0.33). However, there is a quite broad region of the diagram around this point that can be considered white light.

2) The Color Rendering Index (CRI) is a number that ranges from 0 to 100 and is used to measure the ability of a source lighting an object to reproduce the true color of that object. CRI values less than 70 are unacceptable for indoor lighting applications. CRI values of WOLEDs are excellent, being similar to those of incandescent bulbs (>90), and can be higher than those of fluorescent tubes and most common inorganic LEDs [77];

3) The Correlated Color Temperature (CCT) is the temperature of a blackbody radiator emitting the same color of the light source. In general, CCT values in the range 2500–6500 K are required for lighting: for example, incandescent lamps have CCT ≈ 2700 K (warm white), while fluorescent lamps can range from about 3000 K to more than 4000 K (cool white) [74]. Cool white light is less desired by consumers, particularly for inhouse lighting.

5.7.3 Fabrication of White-Light PLEDs

The fabrication of white-light PLEDs, by processing the active materials from solution, promises to be much less expensive than that of white OLEDs (based on small molecules), where the active layers require high-vacuum deposition. Although several approaches have been used to generate white-light PLEDs, a good example is the use of PFO as a host material containing 1% fluorenone [PFO-F(1%)] and the Ir complex, tris[2,5-bis-2′-(9′,9′-dihexylfluorene)pyridine-κ2NC3′] iridium(III) [$Ir(HFP)_3$]. In this case, the PFO provides the blue component, the PFO-F(1%) provides the green component [7], and the $Ir(HFP)_3$ [60] provides the red component. The molecular structures of the various components are shown in Figure 5.44.

Subsequently, PFs–as efficient blue emitters–have emerged as an important class of conjugated polymers, as they exhibit a high PL efficiency, good charge-transport, and thermal stability. As host materials, the PFs enable full color (blue, green, and red) emission via energy transfer to longer-wavelength emitters in blends with other conjugated polymers, phosphorescent dyes, and organometallic emitters. Consequently, the PFs represent interesting candidates as materials for use in the fabrication of PLEDs that emit white light. In such white-light-emitting PLEDs the PFs can function both as the host and as the blue emitter.

Typically, however, a low-energy (2.2~2.4 eV) green PL band appears in the emission from PFs over time, and this causes the blue emission to degrade. The green emission band originates from fluorenone defects (impurities) in the PF chain

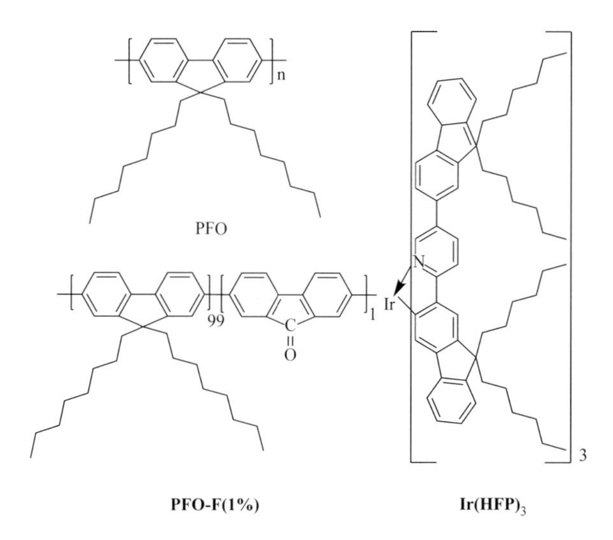

Figure 5.44 Molecular structures of PFO, PFO-F(1%), and $Ir(HFP)_3$.

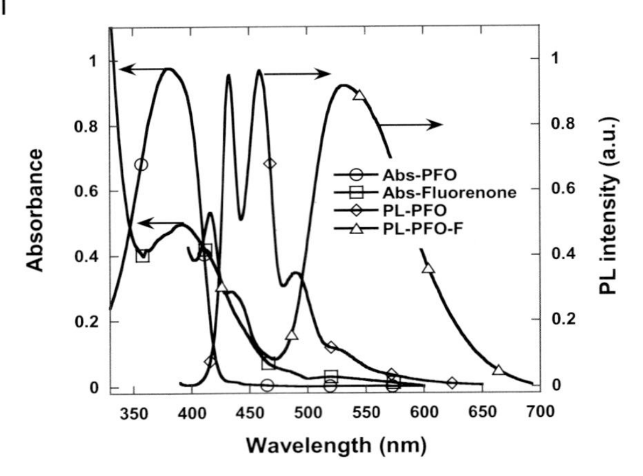

Figure 5.45 Normalized thin-film absorption spectra of PFO and fluorenone and photoluminescence spectra of PFO and PFO-F(1%).

[79, 80]. Hence, having identified the role of fluorenone defects in the PFs, it is possible to take advantage of the fluorenone defect to create materials with a stable green emission [81].

The absorption and PL spectra of PFO-F(1%) thin films are shown in Figure 5.45. In the case of PFO-F(1%), the absorption onsets at approximately 430 nm (2.88 eV), with a maximum absorption at 384 nm (3.23 eV), corresponding to the expected π–π^* transition from the fluorene backbone. When compared to PFO, however, an additional weak absorption band appears at ~420 nm, which is associated with the n-π^* transition observed in the spectroscopy of fluorenone molecules. Under irradiation with light of 384 nm, the PL from PFO-F(1%) is green with a weak blue component – in other words, it is the same as the PL from oxidized PFO films. The green emission from PFO-F(1%) films thus originates from fluorenone units within the copolymer.

5.7.4
Efficient Excitation Energy Transfer from PFO to the Fluorenone Defect

The absorption spectra of PFO and fluorenone thin films and the PL spectra of PFO and PFO-F(1%) thin films are shown in Figure 5.45. The absorption maxima for PFO and fluorenone are at 384 nm and 415 nm, respectively. Under irradiation with 380 nm light, the PL shows well-defined vibronic features, while a weaker long-wavelength emission (~530 nm) is also evident in the spectra. The spectra in Figure 5.45 demonstrate that there is good overlap between the absorption spectrum of fluorenone and the emission spectrum of PFO, implying an efficient Förster energy transfer with subsequent emission from the fluorenone.

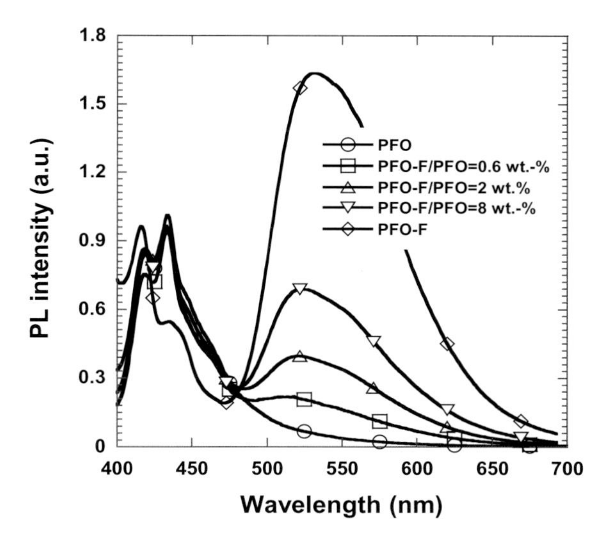

Figure 5.46 Normalized thin-film photoluminescence spectra of PFO, PFO-F(1%), and PFO blended with different concentrations of PFO-F(1%).

In order to test the efficiency of Förster energy transfer, thin films of pure PFO, PFO-F(1%) and PFO containing different concentrations of PFO-F(1%) were prepared and excited optically with 380 nm radiation. Normalized thin-film PL spectra of PFO and PFO blended with the different concentrations of PFO-F(1%) are shown in Figure 5.46. Here, the PL profile indicates two species: (i) the PL from PFO with maxima at 420 nm and 450 nm; and (ii) the PL from fluorenone with a maximum at 530 nm. The blue emission (at 420 nm and 450 nm) decreases and the green emission (at 530 nm) increases as the concentration of PFO-F(1%) is increased. Direct measurements of the optical absorption at 380 nm indicate that the absorption coefficient of PFO is 20-fold greater than that of the fluorenone component. Thus, for PFO-F(1%), essentially all of the incident photons are absorbed by the PFO. Taken together, these data indicate an efficient Förster energy transfer from PFO to the fluorenone defects.

The HOMO and LUMO energy levels of fluorenone are shown in Figure 5.47. Consistent with charge trapping, the fluorenone defects function as both a hole trap and an electron trap; the HOMO and LUMO of fluorenone fall within the π–π^* gap of PFO [82]. In addition, the hole (electron) can be injected from the PEDOT:PSS (Ca) electrode directly into the HOMO (LUMO) of fluorenone, because of the small energy barrier between PEDOT:PSS and the HOMO (or between Ca and the LUMO) of fluorenone.

Therefore, the more pronounced green emission from PFO containing fluorenone defects results from a combination of an efficient energy transfer, charge carrier trapping, and the relatively easy injection (from the electrodes) of carriers into the fluorenone traps.

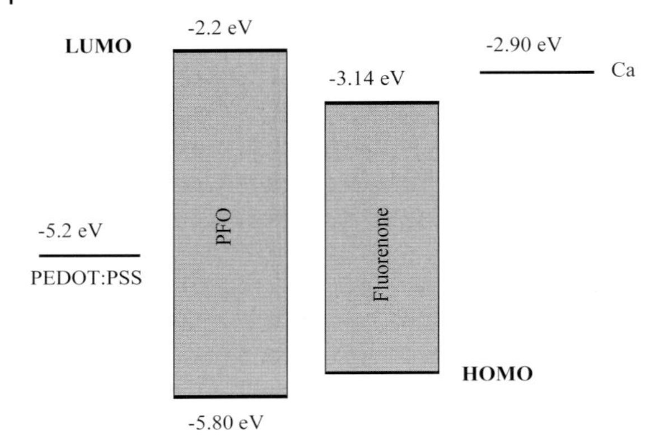

Figure 5.47 HOMO and LUMO energy levels of fluorenone are compared to the π- and π*-band edges of PFO.

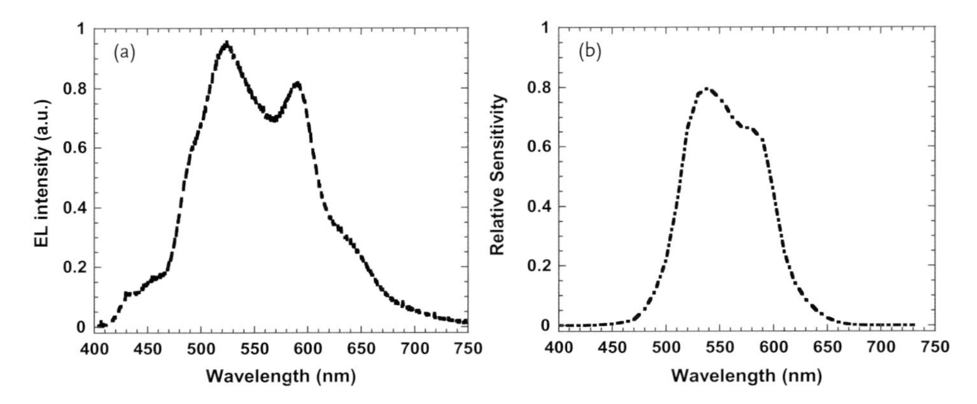

Figure 5.48 (a) EL spectra obtained from white-light-emitting electrophosphorescent PLEDs; (b) The same spectra as in panel (a), but renormalized to the sensitivity of the human eye.

5.7.5
White Electrophosphorescent PLEDs

White-light-emitting PLEDs were fabricated using PFO, PFO-F, and $Ir(HFP)_3$. The EL spectra obtained from devices with a configuration of (ITO)/PEDOT:PSS/emitting layer/Ba/Al are shown in Figure 5.48a, where the emitting layer comprised $Ir(HFP)_3$:PFO-F(1%):PFO. The same spectrum, renormalized to the sensitivity of the human eye, is shown in Figure 5.48b. In this case, the emission was shown to be well-matched to the response of the human eye.

In the electrophosphorescent PLEDs made from blends of $Ir(HFP)_3$:PFO-F(1%):PFO, the injected holes and electrons could recombine by two processes:

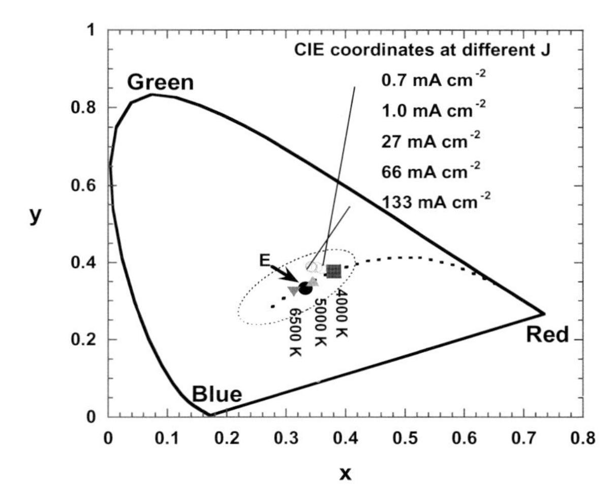

Figure 5.49 CIE (1931) chromaticity diagram, with coordinates corresponding to the emission from white-light-emitting devices (○ ○ ○) biased at different current densities. Also shown are the equi-energy point (E) for pure white light (0.333, 0.333) (●) and the coordinates corresponding to color temperatures of 4000 K (■), 5000 K (▲), and 6500 K (▼). The dotted line indicates different color temperatures; the dotted oval indicates the approximate area where the human eye perceives the color as white.

(i) by direct recombination on the main chain (PFO) to produce a blue emission in parallel with electron and hole trapping on the fluorenone units; and (ii) on the $Ir(HFP)_3$ followed by radiative recombination, with green light from PFO-F(1%) and red light from the triplet excited state of $Ir(HFP)_3$

The CIE coordinates, color temperature (CT), and CRI were evaluated quantitatively from the EL spectra [83, 84]. The 1931 CIE chromaticity diagram, with coordinates corresponding to the emission from electrophosphorescent PLEDs, is shown in Figure 5.49, where data points are indicated for devices (open circles) biased at different current densities. In Figure 5.49, the dotted line indicates different color temperatures, while the dotted oval indicates the approximate area where the human eye perceives the color as white. The CIE coordinates for the $Ir(HFP)_3$:PFO-F(1%):PFO devices were (0.352, 0.388) at $j = 1\,mA\,cm^{-2}$, which were very close to the CIE coordinates for pure white light (0.333, 0.333). The CIE coordinates showed only minor shifts at different j-values. The stability of the CIE coordinates as a function of the brightness and applied voltage was much better than had been reported previously for PLEDs/OLEDs [85–90].

The $Ir(HFP)_3$:PFO-F(1%):PFO devices had CT ~4600 K and CRI = 86; the CT ~4600 K was very close to that of sunlight at 20° solar altitude (4700 K) [91]. Notably, the CTs and CRIs are insensitive to brightness and j.

The white emission turns on at approximately 5 V, with $L = 6100\,cd\,m^{-2}$ at 17 V. A Lambertian intensity profile was assumed to calculate the LE ($cd\,A^{-1}$) and the power efficiency (lm W^{-1}) with the following results: $LE = 3\,cd\,A^{-1}$ and $L = 255\,cd\,m^{-2}$ at $j = 8.5\,mA\,cm^{-2}$. The power efficiency was approximately $1\,lm\,W^{-1}$.

Since the charge carriers are not balanced in the devices described in Figures 5.48 and 5.44 [92], the carrier injection and carrier balance can be improved through the use of hole injection and electron injection layers. By using this approach, a higher efficiency has been achieved by fabricating multilayer white electrophosphorescent PLEDs. For this, the multilayer white PLEDs were fabricated by using luminescent semiconducting polymers and organometallic complexes as the emission layer, water-soluble (or methanol-soluble) poly(vinylcarbazole) sulfonic lithium (PVK-SO_3Li) as the HTL, and water-soluble (or methanol-soluble)4-(5-(4-*tert*-butylphenyl)-1,3,4-oxadiazole-2-yl)-biphenyl-4′-yl sulfonic sodium (*t*-Bu-PBD-SO_3Na) as the ETL. All layers were spin-cast from solutions in the device configuration: ITO/PEDOT/HTL/emissive layer/ETL/Ba/Al. The results demonstrated that white-light-emitting PLEDs with both HTL and ETL had a significantly enhanced luminance, luminous efficiency, and power efficiency. The multilayer white electrophosphorescent PLEDs had $LE = 10.4\,cd\,A^{-1}$, $PE = 3\,lm\,W^{-1}$, and $L = 2391\,cd\,m^{-2}$ at a current density j of $23\,mA\,cm^{-2}$ (V = 11 V) with $L = 2.4 \times 104\,cd\,m^{-2}$ at 16 V. For solid-state lighting applications, where a wide-angle emission is used, the corresponding values were >20 cd A^{-1} and 5 lm W^{-1} [92].

5.7.6 Outlook of White PLEDs

The most recent progress on white OLEDs/PLEDs has been surveyed in a series of recent reviews [93–96].

Today, only a few niche PLED luminaries and panels exist, and these have been produced at a high price, in very limited quantities, and with significant performance limitations, mainly with the aim of introducing this new technology to the market. In the case of these prototype white PLED products, an efficiency of about 23 lm W^{-1} and a lifetime of 5000 h have been estimated, with an average cost of about US$ 25.0 per kilolumen (klm); this contrasts with reported values of US$ 128 klm^{-1} for LED lamps and US$ 2 klm^{-1} for compact fluorescent lamps. It should be noted, however, that cost estimates for white PLED panels contain a high degree of uncertainty, as they are based on only a few prototype/demonstration products that are available.

It is clear that significant performances improvements will be necessary before PLED technology can become competitive for general illumination applications. Currently, investigations are being conducted actively at both academic and industrial levels, prompted by the appealing features of this new class of devices. Nonetheless, a market for white PLEDs, operating at a lower efficiency and involving a higher cost than inorganic LEDs can be expected because of their unique form factors, which would be difficult to accomplish by using other lighting technologies. In fact, white PLEDs may lead to a profound change in the concept of indoor/outdoor illumination architecture, which is presently based on small and intense light sources, by developing wide surfaces that not only emit good quality white light but may also be envisaged as rollable/bendable "wallpaper-style" illumination devices. Thus, high-value-added innovative lighting products might represent a profitable niche market for the early developers of organic solid-state lighting sources and, perhaps, serve as a starting point to create PLEDs that are truly competitive for general illumination purposes.

The possibility of white PLEDs achieving a level of a broadly diffused lighting technology lies primarily with the development of new materials, which would be expected to achieve the main targets in terms of both their stability and efficiency.

5.8
Summary

Although investigations into PLEDs has achieved significant progress during the past decade, several challenges remain that continue to limit their widespread application as commercial products. Among the most important of these are the achievement of a high materials purity, low-cost, high brightness, and long operational lifetimes. For emissive devices, impurities may result in exciton quenching, thereby decreasing not only the emission efficiency but also operational stability. In addition to the materials' impurities, the main challenges that currently are preventing the widespread manufacture and use of white PLEDs for solid-state lighting include manufacturing costs, efficiency, low out-coupling efficiency, large-area devices, and low operational stability.

References

1 (a) Heeger, A.J. (1998) Light emission from semiconducting polymers: light-emitting diodes, light-emitting electrochemical cells, lasers and white light for the future. *Solid State Commun.*, **107** (11), 673–679;
(b) Heeger, A.J. (2001) Nobel lecture: semiconducting and metallic polymers: the fourth generation of polymeric materials. *Rev. Modern Phys.*, **73** (3), 681–700.

2 Braun, D. and Heeger, A.J. (1991) Visible light emission from semiconducting polymer diodes. *Appl. Phys. Lett.*, **58** (18), 1982–1984.

3 Yu, G., Wang, J., McElvain, J., and Heeger, A.J. (1998) Large-area, full-color image sensors made with semiconducting polymers. *Adv. Mater.*, **10** (17), 1431–1434.

4 Heeger, P.S. and Heeger, A.J. (1999) Making sense of polymer-based biosensors. *Proc. Natl Acad. Sci. USA*, **96** (22), 12219–12221.

5 Dimitrakopoulos, D. and Mascaro, D.J. (2001) Organic thin-film transistors: a review of recent advances. *IBM J. Res. Dev.*, **45** (1), 11–28.

6 McGehee, M.D. and Heeger, A.J. (2000) Semiconducting (conjugated) polymers as materials for solid-state lasers. *Adv. Mater.*, **12** (22), 1655–1668.

7 Scherf, U. and List, E.J.W. (2002) Semiconducting polyfluorenes–towards reliable structure–property relationships. *Adv. Mater.*, **14** (7), 477–487.

8 Whitaker, J.C. (2001) *Video Display Engineering*, McGraw-Hill Companies, Inc.

9 Ryer, D. (1998) *Light Measurement Handbook*, International Light Inc.

10 Keitz, H.A.E. (1971) *Light Calculations and Measurements*, 2nd edn, Macmillan and Co. Ltd.

11 Greenham, N.C., Friend, R.H., and Bradley, D.D.C. (1994) Angular dependence of the emission from a conjugated polymer light-emitting diode: implications for efficiency calculations. *Adv. Mater.*, **6** (6), 491–494.

12 Parker, D. (1994) Carrier tunneling and device characteristics in polymer light-emitting diodes. *J. Appl. Phys.*, **75** (3), 1656–1666.

13 Heeger, A.J. (2001) Semiconducting and metallic polymers: the fourth generation of polymeric materials (Nobel Lecture). *Angew. Chem. Int. Ed.*, **40** (14), 2591–2611.

14 Friend, R.H., Gymer, R.W., Holmes, A.B., Burroughes, J.H., Marks, R.N., Taliani, C., Bradley, D.D.C., Dos Santos, D.A., Brédas, J.L., Lögdlund, M., and Salaneck, W.R. (1999) Electroluminescence in conjugated polymers. *Nature*, **397** (6517), 121–128.

15 Pschenitzha, F. and Sturm, J.C. (1999) Three-color organic light-emitting diodes patterned by masked dye diffusion. *Appl. Phys. Lett.*, **74** (13), 1913–1915.

16 Sirringhaus, H., Kawase1, T., Friend, R.H., Shimoda, T., Inbasekaran, M., Wu, W., and Woo, E.P. (2000) High-resolution inkjet printing of all-polymer transistor circuits. *Science*, **290** (5499), 2123–2126.

17 Brown, A.R., Pichler, K.N., Greenham, C., Bradley, D.D.C., Friend, R.H., and Holmes, A.B. (1993) Optical spectroscopy of triplet excitons and charged excitations in poly(*p*-phenylene) light-emitting diodes. *Chem. Phys. Lett.*, **210** (1–3), 61–66.

18 Tero-Kubota, S., Katsuki, A., and Kobori, Y. (2001) Spin-orbit coupling induced electron spin polarization in photoinduced electron transfer reactions. *J. Photochem. Photobiol., C*, **2** (1), 17–33.

19 Wohlgenannt, M., Tandon, K., Mazumdar, S., Ramasesha, S., and Vardeny, Z.V. (2001) Formation cross-sections of singlet and triplet excitons in-conjugated polymers. *Nature*, **409** (6819), 494–497.

20 Cao, Y., Parker, I.D., Yu, G., Zhang, C., and Heeger, A.J. (1999) Improved quantum efficiency for electroluminescence in semiconducting polymers. *Nature*, **397** (6718), 414–417.

21 Kido, J., Hayase, H., Hongawa, K., Nagai, K., and Okuyama, K. (1994) Bright red light-emitting organic electroluminescent devices having a europium complex as an emitter. *Appl. Phys. Lett.*, **65** (17), 2124–2126.

22 McGehee, M.D., Bergstedt, T., Zhang, C., Saab, A.P., O'Regan, M.B., Bazan, G.C., Srdanov, V.I., and Heeger, A.J. (1999) Narrow bandwidth luminescence from blends with energy transfer from semiconducting conjugated polymers to europium complexes. *Adv. Mater.*, **11** (16), 1349–1354.

23 Baldo, M.A., Thompson, M.E., and Forrest, S.R. (2000) High-efficiency fluorescent organic light-emitting devices using a phosphorescent sensitizer. *Nature*, **403** (6771), 750–753.

24 Adachi1, C., Baldo1, M.A., Forrest, S.R., and Thompson, M.E. (2000) High-efficiency organic electrophosphorescent devices with tris(2-phenylpyridine)iridium doped into electron-transporting materials. *Appl. Phys. Lett.*, **77** (6), 904–906.

25 Zhu, W., Mo, Y., Yuan, M., Yang, W., and Cao, Y. (2002) Highly efficient electrophosphorescent devices based on conjugated polymers doped with iridium complexes. *Appl. Phys. Lett*, **80** (12), 2045–2047.

26 Gong, X., Robinson, M.R., Ostrowski, J.C., Moses, D., Bazan, G.C., and Heeger, A.J. (2002) High-efficiency polymer-based electro- phosphorescent devices. *Adv. Mater.*, **14** (8), 581–585.

27 Lamansky, S., Kwong, R.C., Nugent, M., Djurovich, P.I., and Thompson, M.E. (2001) Molecularly doped polymer light emitting diodes utilizing phosphorescent Pt(II) and Ir(III) dopants. *Org. Electron.*, **2** (1), 53–62.

28 O'Brien, F., Giebler, C., Fletcher, R.B., Cadby, A.J., Palilis, L.C., Lidzey, D.G., Lane, P.A., Bradley, D.D.C., and Blau, W. (2001) Tuning of emission color for blue dendrimer blend light-emitting diodes. *Synth. Met.*, **116** (1–3), 379–383.

29 Lee, C.-L., Lee, K.B., and Kim, J.-J. (2000) Polymer phosphorescent light-emitting devices doped with tris(2-phenylpyridine) iridium as a triplet emitter. *Appl. Phys. Lett.*, **77** (15), 2280–2282.

30 Ikai, M., Tokito, S., Sakamoto, Y., Suzuki, T., and Taga, Y. (2001) Highly efficient phosphorescence from organic light-emitting devices with an exciton-block layer. *Appl. Phys. Lett.*, **79** (2), 156–158.

31 D'Andrade, B.W., Thompson, M.E., and Forrest, S.R. (2002) Controlling exciton diffusion in multilayer white phosphorescent organic light emitting devices. *Adv. Mater.*, **14** (2), 147–151.

32 Lamansky, S.S., Djurovich, P., Murphy, D., Abdel-Razzaq, F., Lee, H.-E., Adachi, C., Burrows, P.E., Forrest, S.R., and Thompson, M.E. (2001) Highly phosphorescent bis-cyclometalated iridium complexes: synthesis, photophysical characterization, and use in organic light emitting diodes. *J. Am. Chem. Soc.*, **123** (18), 4304–4312.

33 Gong, X., Ostrowski, J.C., Bazan, G.C., Moses, D., and Heeger, A.J. (2002) Red electrophosphorescence from polymer doped with iridium complex. *Appl. Phys. Lett.*, **81** (20), 3711–3713.

34 Yang, M.-J. and Tsutsui, T. (2000) Use of poly(9-vinylcarbazole) as host material for iridium complexes in high-efficiency organic light-emitting devices. *Jpn. J. Appl. Phys.*, **39**, L828–L829.

35 Kawamura, Y., Yanagida, S., and Forrest, S.R. (2002) Energy transfer in polymer electrophosphorescent light emitting devices with single and multiple doped luminescent layers. *J. Appl. Phys.*, **92** (1), 87–93.

36 Lamansky, S., Djurovich, P.I., Abdel-Razzaq, F., Garon, S., Murphy, D.L., and Thompson, M.E. (2002) Cyclometalated Ir complexes in polymer organic light-emitting devices. *J. Appl. Phys.*, **92** (3), 1570–1575.

37 Köhler, A., Wilson, J.S., and Friend, R.H. (2002) Fluorescence and phosphorescence in organic materials. *Adv. Mater.*, **14** (10), 701–707.

38 Gong, X., Ostrowski, J.C., Bazan, G.C., Moses, D., Heeger, A.J., Liu, M., and Jen, A.K.-Y. (2003) Electrophosphorescence from a conjugated copolymer doped with an iridium complex: high brightness and improved operational stability. *Adv. Mater.*, **15** (1), 45–49.

39 Cleave, V., Yahioglu, G., Barny, P.L., Friend, R.H., and Tessler, N. (1999) Harvesting singlet and triplet energy in polymer LEDs. *Adv. Mater.*, **11** (4), 285–288.

40 Ostrowski, J.C., Robinson, M.R., Heeger, A.J., and Bazan, G.C. (2002) Amorphous iridium complexes for electrophosphorescent light emitting devices. *Chem. Commun.*, (7), 784–785.

41 Grushin, V.V., Herron, N., LeCloux, D.D., Marshall, W.J., Petrov, V.A., and Wang, Y. (2001) New, efficient electroluminescent materials based on organometallic Ir complexes. *Chem. Commun.*, (16), 1949–1495.

42 Baldo, M.A., and Forrest, S.R. (2000) Transient analysis of organic electrophosphorescence: I. Transient analysis of triplet energy transfer. *Phys. Rev. B*, **62** (16), 10958–10966.

43 Gong, X., Lim, S.-H., Ostrowski, J.C., Moses, D., Bardeen, C.J., and Bazan, G.C. (2004) Phosphorescence from iridium complexes doped into polymer blends. *J. Appl. Phys.*, **95** (3), 948–953.

44 Förster, T. (1959) 10th Spiers Memorial Lecture. Transfer mechanisms of electronic excitation. *Discuss. Faraday Soc.*, **27**, 7–17.

45 Dexter, D.L. (1953) A theory of sensitized luminescence in solids. *J. Chem. Phys.*, **21** (5), 836–850.

46 Gupta, R., Stevenson, M., Dogariu, A., McGehee, M.D., Park, J.Y., Srdanov, V., Heeger, A.J., and Wang, H. (1998) Low-threshold amplified spontaneous emission in blends of conjugated polymers. *Appl. Phys. Lett.*, **73** (24), 3492–3494.

47 Gupta, R., Stevenson, M., and Heeger, A.J. (2002) Low threshold distributed feedback lasers fabricated from blends of conjugated polymers: reduced losses through Förster transfer. *J. Appl. Phys.*, **92** (9), 4874–4877.

48 Dogariu, A., Gupta, R., Heeger, A.J., and Wang, H. (1999) Time-resolved Förster energy transfer in polymer blends. *Synth. Met.*, **100** (1), 95–100.

49 Monkman, A.P., Burrows, H.D., Hartwell, L.J., Horsburgh, L.E., Hamblett,

I., and Navaratnam, S. (2001) Triplet energies of π-conjugated polymers. *Phys. Rev. Lett.*, **86** (7), 1358–1361.
50 Utsugi, K. and Takano, S. (1992) Luminescent properties of doped organic EL diodes using naphthalimide derivative. *J. Electrochem. Soc.*, **139** (12), 3610–3615.
51 Suzuki, H. and Hoshino, A. (1996) Effects of doping dyes on the electroluminescent characteristics of multilayer organic light-emitting diodes. *J. Appl. Phys.*, **79** (11), 8816–8822.
52 Liu, M.S., Jiang, X., Herguth, P., and Jen, A.K.-Y. (2001) Efficient cyano-containing electron-transporting polymers for light-emitting diodes. *Chem. Mater.*, **13** (11), 3820–3822.
53 Gao, J., Li, Y., Yu, G., and Heeger, A.J. (1999) Polymer light-emitting electrochemical cells with frozen junctions. *J. Appl. Phys.*, **86** (8), 4594–4599.
54 Kido, J., Hongawa, K., Okuyama, K., and Nagai, K. (1993) Bright blue electroluminescence poly(N-vinylcarbazole). *Appl. Phys. Lett.*, **63** (19), 2627–2629.
55 Zhang, C., Seggern, H., Pakbaz, K., Schimidt, H.-W., and Heeger, A.J. (1994) Blue electroluminescence diodes utilizing blends of poly(*p*-phenylphenylenevinylene) in poly(9-vinyl-carbazole). *Synth. Met.*, **62** (1), 35–40.
56 Muyata, S. (1997) *Organic Electroluminescent Materials and Devices*, Gordon and Breach Publishers.
57 Colombo, M.G., Hauser, A., and Güdel, H.U. (1994) Competition between ligand centered and charge transfer lowest excited states in bis cyclometalated Rh^{3+} and Ir^{3+} complexes. *Top. Curr. Chem.*, **171**, 143–171.
58 Baldo, M.A., O'Brien, D.F., You, Y., Shoustikov, A., Sibley, S., Thompson, M.E., and Forrest, S.R. (1998) Highly efficient phosphorescent emission from organic electroluminescent devices. *Nature*, **395** (6698), 151–154.
59 Bao, Z., Lovinger, A.J., and Brown, J. (1998) New air-stable *n*-channel organic thin film transistors. *J. Am. Chem. Soc.*, **120** (1), 207–208.
60 Gong, X., Iyer, P.K., Moses, D., Bazan, G.C., Heeger, A.J., and Xiao, S. (2003) Stabilized blue emission from polyfluorene-based light-emitting diodes: elimination of fluorenone defects. *Adv. Funct. Mater.*, **13** (4), 325–330.
61 Negres, R.A., Gong, X., Ostrowski, J.C., Bazan, G.C., Moses, D., and Heeger, A.J. (2003) Origin of efficient light emission from a phosphorescent polymer/organometallic guest-host system. *Phys. Rev. B*, **68** (11), 115209–115216.
62 Lim, S.-H., Gong, X., Ostrowski, J., Bazan, G.C., Moses, D., and Bardeen, C.J. (2003) Temperature dependence of electronic energy transfer from a polymer host to a triplet emitter in light emitting diode materials. *Chem. Phys. Lett.*, **376** (1–2), 55–61.
63 Gong, X., Ostrowski, J.C., Moses, D., Bazan, G.C., and Heeger, A.J. (2003) Electrophosphorescence from a polymer guest-host system with an iridium complex as guest: Förster energy transfer and charge trapping. *Adv. Funct. Mater.*, **13** (6), 439–444.
64 Renak, M.L., Bartholomew, G.P., Wang, S., Ricatto, P.J., Lachicotte, R.J., and Bazan, G.C. (1999) Fluorinated distyrylbenzene chromophores: effect of fluorine regiochemistry on molecular properties and solid-state organization. *J. Am. Chem. Soc.*, **121** (34), 7787–7799.
65 Oelkrug, D., Tompert, A., Gierschner, J., Egelhaaf, H.-J., Hanack, M., Hohloch, M., and Steinhuber, E. (1998) Tuning of fluorescence in films and nanoparticles of oligophenylenevinylenes. *J. Phys. Chem. B*, **102** (11), 1902–1907.
66 Strehmel, B., Sarker, A.M., Malpert, J.H., Strehmel, V., Seifert, H., and Neckers, D.C. (1999) Effect of aromatic ring substitution on the optical properties, emission dynamics, and solid-state behavior of fluorinated oligophenylenevinylenes. *J. Am. Chem. Soc.*, **121** (6), 1226–1236.
67 Redecker, M., Bradley, D.D.C., Inbasekaran, M., and Woo, E.P. (1998) Nondispersive hole transport in an electroluminescent polyfluorene. *Appl. Phys. Lett.*, **73** (11), 1565–1567.
68 Janietz, S., Bradley, D.D.C., Grell, M., Giebeler, C., Inbasekaran, M., and Woo, E.P. (1998) Electrochemical

determination of the ionization potential and electron affinity of poly(9,9-dioctylfluorene). *Appl. Phys. Lett.*, **73** (17), 2453–2455.

69 Liu, M.S., Jiang, X.Z., Liu, S., Herguth, P., and Jen, A.K.-Y. (2002) Effect of cyano substituents on electron affinity and electron-transporting properties of conjugated polymers. *Macromolecules*, **35** (9), 3532–3538.

70 Dawson, T.L. (2010) Development of efficient and durable sources of white light. *Color Technol.*, **126** (1), 1–10.

71 Humphreys, J. (2008) Solid-state lighting. *MRS Bull.*, **33** (4), 459–470.

72 Takei, Y. (2009) Basic research needs for solid-state lighting. *Sci. Technol.- Q. Rev.*, **32**, 59.

73 Müllen, K. and Scherf, U. (eds) (2006) *Organic Light-Emitting Devices: Synthesis, Properties and Applications*, Wiley-VCH Verlag GmbH & Co. KGaA, Weinheim.

74 D'Andrade, B.W. and Forrest, S.R. (2004) White organic light-emitting devices for solid-state lighting. *Adv. Mater.*, **16** (18), 1585–1595.

75 Misra, P., Kumar, M., Kamalasanan, N., and Chandra, S. (2006) White organic LEDs and their recent advancements. *Semicond. Sci. Technol.*, **21**, R35–R47.

76 Service, R.F. (2005) Organic LEDs look forward to a bright, white future. *Science*, **310** (5755), 1762–1763.

77 Franky, S., Kido, J., and Burrows, P. (2008) Organic light-emitting devices for solid-state lighting. *MRS Bull.*, **33** (7), 663–669.

78 Gong, X., Moses, D., and Heeger, A.J. (2004) Excitation energy transfer from polyfluorene to fluorenone defects. *Synth. Met.*, **141** (1–2), 17–20.

79 Gong, X., Ma, W.L., Ostrowski, J.C., Bazan, G.C., Moses, D., and Heeger, A.J. (2004) White electrophosphorescence from semiconducting polymer blends. *Adv. Mater.*, **16** (7), 615–619.

80 Shaheen, S.E., Kippelen, B., Peyghambarian, N., Wang, J.F., Anderson, J.D., Mash, E.A., Lee, P.A., Armstrong, N.R., and Kawabe, Y. (1999) Energy and charge transfer in organic light-emitting diodes: a soluble quinacridone study. *J. Appl. Phys.*, **85** (11), 7939–7945.

81 Wyszecki, G. and Stiles, W.S. (1982) *Color Science*, 2nd edn, John Wiley & Sons, Inc., New York.

82 Judd, B. and Wyszecki, G. (1975) *Color in Business, Science and Industry*, 3th edn, John Wiley & Sons.

83 Zhang, C. and Heeger, A.J. (1998) Gallium nitride/conjugated polymer hybrid light-emitting diodes: performance and lifetime. *J. Appl. Phys.*, **84** (3), 1579–1583.

84 Kido, J., Shionoya, H., and Nagai, K. (1995) Single-layer white light-emitting organic electroluminescent devices based on dye-dispersed poly(*N*-vinylcarbazole). *Appl. Phys. Lett.*, **67** (16), 2281–2283.

85 Shen, Z., Burrows, P.E., Bulvić, V., Forrest, S.R., and Thompson, M.E. (1997) Three-color, tunable, organic light-emitting devices. *Science*, **276** (5321), 2009–2011.

86 Hamada, Y., Sano, T., Fujii, H., and Nishio, Y. (1996) Observation of upconversion lasing within a thulium-ion-doped glass powder film containing titanium dioxide particles. *Jpn. J. Appl. Phys.*, **35**, L1337–L1339.

87 Wang, Y.Z., Sun, R.G., Meghdadi, F., Leising, G., and Epstein, A.J. (1999) Multicolor multilayer light-emitting devices based on pyridine-containing conjugated polymers and para-sexiphenyl oligomer. *Appl. Phys. Lett.*, **74** (24), 3613–3615.

88 Strukelj, M., Jordan, R.H., and Dodabalapur, A. (1996) Organic multilayer white light emitting diodes. *J. Am. Chem. Soc.*, **118** (5), 1213–1214.

89 Yamamoto, T. (1992) Electrically conducting and thermally stable [pi]-conjugated poly(arylene)s prepared by organometallic processes. *Prog. Polym. Sci.*, **17**, 1153–1205.

90 Hunt, R.W.G. (1991) *Measuring Color*, 2nd edn, Ellis Horwood.

91 Gong, X., Ma, W.L., Ostrowski, J.C., Bechgaard, K., Bazan, G.C., Heeger, A.J., Xiao, S., and Moses, D. (2004) End-capping as a method for improving carrier injection in electrophosphorescent light-emitting diodes. *Adv. Funct. Mater.*, **14** (4), 393–397.

92 Gong, X., Wang, S., Bazan, G.C., and Heeger, A.J. (2005) Multilayer polymer

light-emitting diodes: white-light emission with high efficiency. *Adv. Mater.*, **17** (17), 2053–2058.

93 Reineke, S., Lindner, F., Schwartz, G., Seidler, N., Walzer, K., Lüssem, B., and Leo, K. (2009) White organic light-emitting diodes with fluorescent tube efficiency. *Nature*, **459** (7244), 234–238.

94 Kamtekar, K.T., Monkman, A.P., and Bryce, M.R. (2010) Recent advances in white organic light-emitting materials and devices. *Adv. Mater.*, **22** (5), 572–582.

95 Beaupré, S., Boudreault, P.-L.T., and Leclerc, M. (2010) Solar energy production and energy efficient lighting: photovoltaic devices and white-light-emitting diodes using poly(2,7-fluorene), poly(2,7-carbazole) and poly(2,7-dibenzosilole) derivatives. *Adv. Mater.*, **22** (8), E6–E27.

96 Raja, U.H., Lee, J.Y., Kim, I.T., and Lee, S.H. (2008) Recent progress in the development of polymers for white light-emitting polymer devices. *Monatsh. Chem.*, **7** (139), 725–737.

6
Organic Solids for Photonics

Hongbing Fu

6.1
Introduction

During the past decade, inorganic semiconductor or metallic materials, in particular semiconductors of Groups II–VI/III–V of the Periodic Table, have been extensively investigated and utilized in the creation of light-emitting devices [1], field-effect transistors, solar cells [2], lasers [3], and photodetectors [4]. Compared to their inorganic counterparts, organic semiconductors (which are also termed "organic solids") provide many advantages, including the tuning of properties by molecular design, high optical tenability [5], high luminescence efficiency [6], low cost of materials fabrication, ease of large-area processing and compatibility with flexible and lightweight plastic substrates [7], and thus may open broader applications for the next generation of photonic devices.

Organic solids are packed via the van der Waals interactions between adjacent molecules, but these are weaker than the covalent bonds of inorganic solids. Moreover, owing to the different nature of bonding in organic solids, they always exhibit different mechanical, thermodynamic, and optical properties compared to those of inorganic solids. For example, the weaker intermolecular interactions lead to a reduced hardness and lowered melting temperatures, thermal instability, as well as changes in mechanical properties. In addition, the charges confined to each molecule do not expand over the nanometer and micrometer domains because of the much smaller radius of the Frenkel excitons; hence, the size-dependent optical properties of organic solids cannot be explained on the basis of the so-called "quantum confinement effect". Consequently, there remain many challenges to acquire an in-depth understanding of organic solids. Fortunately, the emerging of nanostructures has provided an ideal platform for the study of organic solids, and especially of their intrinsic electronic and optical properties that are based on single crystalline nanostructures. For example, since Nakanishi and coworkers first described reprecipitation [8] as a facile method to prepare organic nanoparticles, increasing attention has been paid to the generation of organic nanostructures, but with the ability to control the morphology, crystallinity, and optoelectronic properties of these materials. It has been noted that many key

Organic Optoelectronics, First Edition. Edited by Wenping Hu.

properties depend on the size of organic aggregates; for example, Chernyak *et al.* confirmed the size-dependence of Stokes shift on organic aggregates [9]. However, the limitation of molecular exchange interactions was also found to result in an increase of the nonlinear optical properties in organic aggregates [10]. Notably, the melting point of organic nanocrystals is also size-dependent [11].

In this chapter the growth techniques applied to organic nanostructures will be introduced, and an outline provided of the size effects of nanostructures in terms their properties of absorption and luminescence. Finally, the details of composite nanostructures–including core/sheath and uniformly doped structures–will be described in an effort to investigate the interactions that occur between various compositions in nanomaterials. This will include fluorescence resonance energy transfer (FRET) and photo-induced electron transfer.

6.2 Size Effects on the Optical Properties of Organic Solids

It has been well documented that the optical properties of inorganic semiconductor nanoparticles change drastically, depending on the particle size, as the result of a quantum confinement of the excitons. The optical properties of organic nanostructures have not been well studied, however. It was shown recently that the properties of some organic nanostructures can also be modulated by size. Such size effects of organic nanostructures, which are similar to but cannot be explained by the so-called "quantum confinement effect" that has been observed in inorganic semiconductor quantum dots (QDs), were first observed in perylene nanocrystals and reported by Nakanishi and coworkers [12]. Subsequently, Horn and coworkers investigated the effect of both supramolecular structure and particle size on the absorption spectra of β-carotene nanoparticles [13]. Yao *et al.* also systematically investigated the size-tunable optical properties of organic nanoparticles prepared from a series of aromatically substituted pyrazoline compounds. The effect of size on the absorption and luminescence properties of these nanoparticles, based on pyrazoline compounds, will be outlined in the following sections.

6.2.1 Exciton Confinement Effect

During the past decade, nanostructures based on organic optical materials, serving as fundamental elements in modern photonic devices, have attracted a great deal of research interest, due to the unique advantages of organic compounds, including high reaction activity, good processability, and high photoluminescence (PL) efficiency. Based on the synthesis and modification of their molecular structures, organic nanomaterials are believed to exhibit diverse and tunable optical properties.

As part of these investigations, nanoparticles of 1-phenyl-3-((dimethylamino)-styryl)-5-((dimethylamino)phenyl)-2-pyrazoline (PDDP) were prepared by repre-

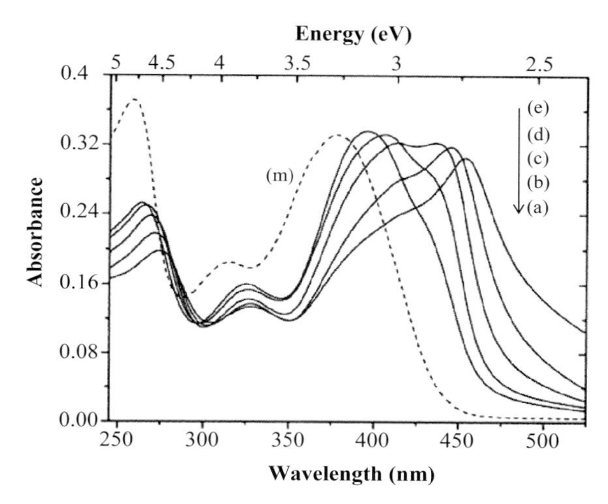

Figure 6.1 UV-visible absorption spectra of PDDP nanoparticle dispersions in water with different sizes: (a) 20 nm, (b) 50 nm, (c) 105 nm, (d) 190 nm, (e) 310 nm. Curve (m) shows the spectrum of the PDDP/ethanol solution (1.0×10^{-5} mol l^{-1}). Inset: Molecular structure of PDDP.

cipitation [14]. The absorption spectra of the PDDP nanoparticles and PDDP monomers in ethanol are shown in Figure 6.1, where three resolved absorption bands were seen to arise from the phenyl ring transition and the pyrazoline ring n-π^* and π-π^* transitions, labeled P_{phenyl}, $P_{n\text{-}\pi^*}$, and $P_{\pi\text{-}\pi^*}$, respectively. None of these three bands shifted when the concentration of the solution was changed from 1.0×10^{-5} to 1.0×10^{-3} mol l^{-1}. In the case of the nanoparticles, when the size increased from tens to hundreds of nanometers, the P_{phenyl} and $P_{\pi\text{-}\pi^*}$ bands were observed to shift to a longer wavelength, and concomitantly a new peak gradually appeared and also shifted to the lower energy side. This bathochromic shift was proposed to result from an increased overlap of the pyrazoline ring p-orbital and intermolecular interactions between PDDP molecules when the nanoparticle size increased. The newly emerged absorption peak was ascribed to the transition from an extended charge-transfer (CT) state of PDDP aggregates derived from PDDP molecules closely stacked in the nanoparticles. The red shift was considered to result from the CT exciton confinement effect.

6.2.2 Size-Tunable Emission

Organic nanoparticles with average diameters ranging from 40 to 160 nm were prepared from another pyrazoline compound, 1,3-diphenyl-5-(2-anthryl)-2-pyrazoline (DAP) [15]. Subsequently, the absorption transitions of the DAP nanoparticles at the lower-energy side were found to have experienced a bathochromic shift with an increase of the particle size as a result of increased intermolecular

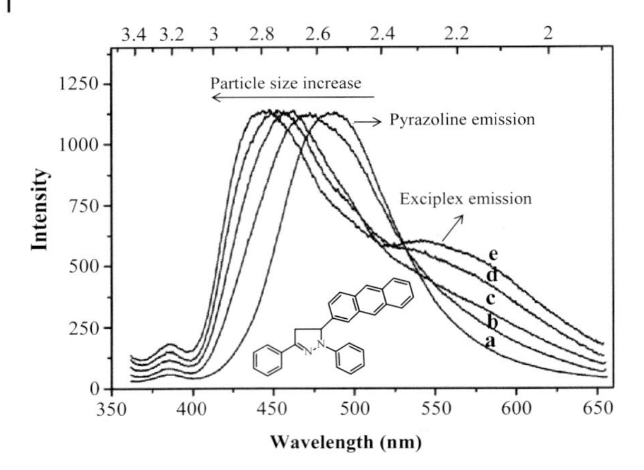

Figure 6.2 Fluorescence emission spectra of DAP nanoparticle dispersions with different sizes: Curve (a) 40 nm, (b) 60 nm, (c) 90 nm, (d) 120 nm, (e) 160 nm. The excitation wavelength for all the samples is 350 nm. Inset: Molecular structure of DAP.

interactions, whereas the higher-energy bands of anthracene split due to the electronic coupling between the pyrazoline ring of one molecule and the anthracene moiety of the neighboring molecule. It is worth noting from Figure 6.2 that the nanoparticle emission in the blue light region from the pyrazoline chromophore shifted to shorter wavelengths with an increase in particle size, accompanied by a relatively gradual dominance of the emission at about 540 nm from the exciplex formed by the pyrazoline ring of one molecule and the anthracene moiety of the neighboring molecule. The hypsochromic shift in the emission of DAP nanoparticles was identified as originating from the pronounced decrease in the Stokes shift, due to the restraint of vibronic relaxation and the configuration reorganization induced by increased intermolecular interaction.

In another series of studies examining fluorescence narrowing, the single-crystalline nanowires were first fabricated from a small organic functional molecule, 2,4,5-triphenylimidazole (TPI) (see inset of Figure 6.3 for chemical structure), by adsorbent-assisted photo-voltaic deposition [16], in which the introduction of the adsorbents proved to be indispensable for improving the uniformity of the TPI nanowires. Further data acquired with selected area electronic diffraction (SAED) and X-ray diffraction (XRD) indicated that the TPI nanowires were single crystalline and grew along the b-axis of the TPI crystal. Moreover, the degree of preferential orientation was obviously strengthened, with a decrease in diameter from 500 nm to 40 nm. The optical properties of the TPI nanowires are dependent primarily on the diameter, and are insensitive to the length. The fluorescence emission spectra of TPI nanowires of different diameters are displayed in Figure 6.3. In this case, the spectrum of the TPI monomer measured at ambient temperature showed an approximately symmetric peak centered at 383 nm, whereas the vibronic levels were well resolved in the low-temperature spectrum measured at 77 K. It is

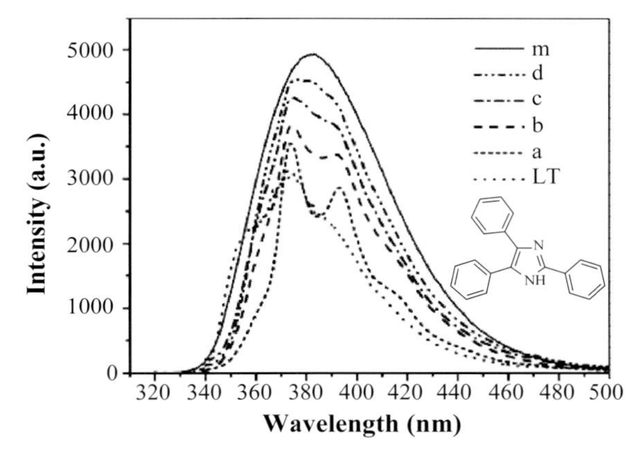

Figure 6.3 Fluorescence emission spectra of TPI nanowires with different diameters deposited onto quartz wafers: (a) 40 nm, (b) 120 nm, (c) 300 nm, (d) 500 nm. Curve (m) shows the spectrum of TPI monomers measured at room temperature. Curve LT is the low-temperature (77 K) fluorescence spectrum of the TPI monomer. All emission spectra were excited at 310 nm.

interesting to note, that a narrowing of the spectra was also observed in the nanowire samples even at room temperature, and that this became more pronounced with decreasing diameter. In an amorphous environment, the emission band broadening would result mainly from the fact that the molecules each have somewhat different energy levels. In the single-crystalline nanowires, however, the molecules are confined distinctly in the lattice with a specific geometric configuration and a high degree of orientation. In this case, many molecules would share the same energy level, so that the degree of freedom of their vibrations would be drastically reduced. The local environment of the molecules in the nanowires does not change in time during the emission process, and only those molecules that have energy levels corresponding to the excitation wavelength can be excited, which in turn results in the emergence of a vibrational fine structure. Moreover, the vibrational structure became increasingly resolved as the diameter decreased from 500 nm to 40 nm, but this was assigned to an increase in long-range order and the degree of orientation with a decreasing wire diameter.

6.2.3 Multiple Emissions

In the case of 1,3-diphenyl-5-pyrenyl-2-pyrazoline (DPP) [17], the nanoparticles showed a different emission behavior compared to either the solution or the bulk material. DPP contains two chromophores – the pyrene and pyrazoline groups – whose conjugated systems are separate from each other due to steric effects. Diluted solutions exhibit only the emission characteristics of the pyrene chromophore, while the bulk material shows only the emission of the pyrazoline chromophore [18]. However, the nanoparticles presented multiple emissions from

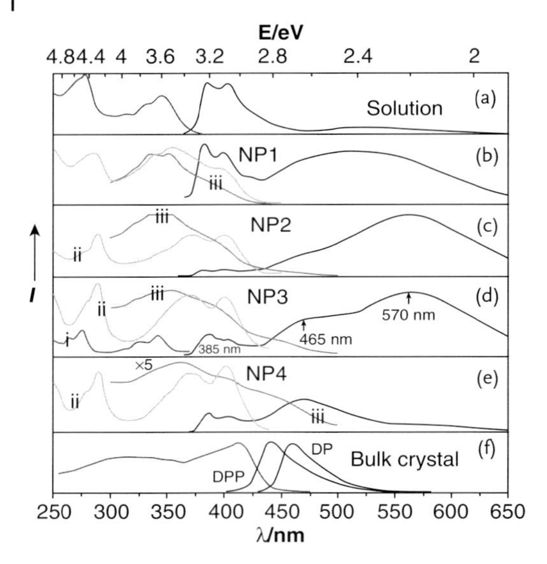

Figure 6.4 Fluorescence excitation and emission spectra of: (a) DPP solution (in acetonitrile, 1.0×10^{-5} mol l^{-1}); (b–e) DPP nanoparticles (NPs) with different sizes; (f) DPP and DP bulk crystals. The lines are the excitation spectra obtained by monitoring the emission at 385 (i), 465 nm (ii), and 570 nm (iii), except for the red line in panel (e), which was obtained by monitoring the emission of DPP at 445 nm. The black lines are emission spectra obtained by excitation at 345 nm.

pyrene, pyrazoline and, additionally, a CT complex between pyrene and pyrazoline. As shown in Figure 6.4, the relative intensity of the emissions was dependent on the size of the nanoparticles. Moreover, these emissions possessed an individual optical channel, and could be excited by their original excitation wavelength. This means that the emission could be tuned by an alteration of either the excitation wavelength or the particle size. The multiple emissions, and their evolution as a function of particle size, were considered to be caused by molecular aggregation and surface effects.

6.3
Aggregation-Induced Enhanced Emission

In general, the fluorescence efficiency of organic chromophores decreases in the solid state, as a result of concentration quenching, even though they demonstrate a high fluorescence efficiency in solution. However, a phenomenon termed aggregation-induced enhanced emission (AIEE) has been reported during recent years, in which an enhanced emission rather than fluorescence quenching was observed in the solid state for some fluorophores [18, 19]. These findings have provided a possible solution to the limitation caused by concentration quenching in applications such as ultrahigh-density optical memory and full color flat-panel displays. Recently, Park and coworkers reported AIEE from organic nanoparticle systems [6], and subsequently synthesized the organic molecule 1-cyano-*trans*-1,2-bis- (40-methylbiphenyl)ethylene (CN-MBE) and prepared the corresponding nanoparticles. It was interesting to note that, although the fluorescence emission from CN-MBE solution was very weak, the nanoparticles emitted a very strong photoluminescence with an intensity almost 700-fold that of the solution (Figure 6.5a). It was proposed that the aggregation induced the planarization of the

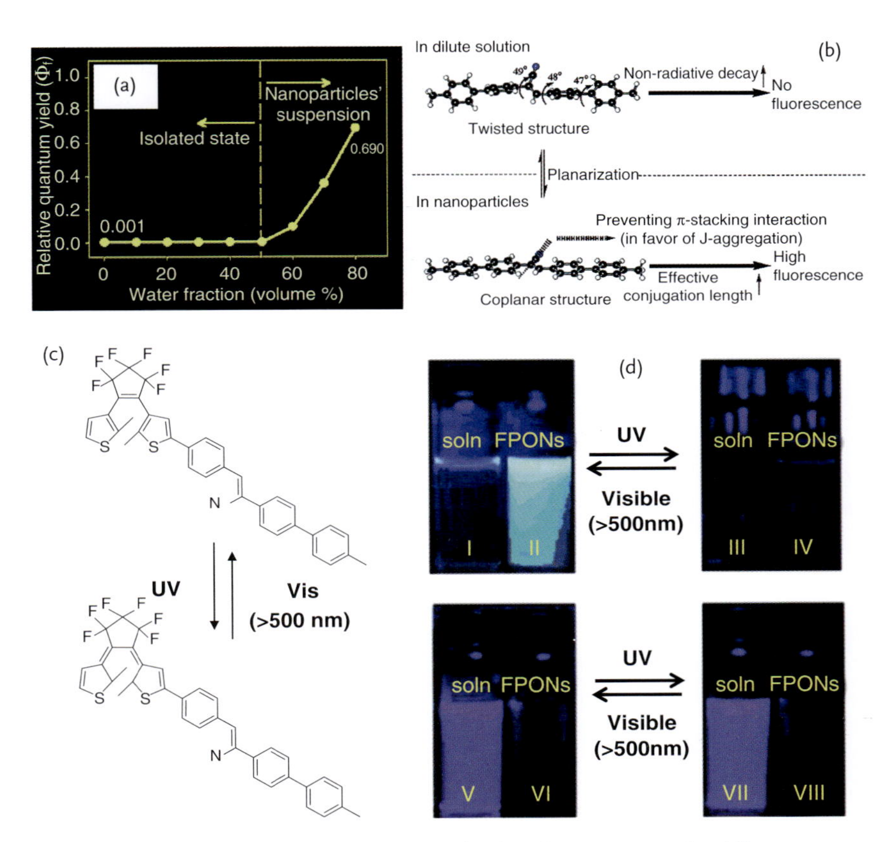

Figure 6.5 (a) Relative quantum yields of CN-MBE (2×10^{-5} mol l^{-1}) depending on water fractions in tetrahydrofuran. The addition of water induces the aggregation of CN-MBE molecules; (b) Proposed mechanism of enhanced emission in CN-MBE nanoparticles. Inset: The molecular structure of CN-MBE; (c) Upper: chemical structure of the open-ring BTE-CN-MBE and fluorescence images of its THF solution (I, 2×10^{-4} mol l^{-1}) and the colloidal suspension (II) of fluorescent photochromic organic nanoparticles (FPONs, 2×10^{-4} mol l^{-1}). Lower: chemical structure of the closed-ring compound and fluorescence images of its THF solution (III, 2×10^{-4} mol l^{-1}) and the colloidal suspenison of FPONs (IV, 2×10^{-4} mol l^{-1}) in the photostationary state; (d) Photo-rewritable fluorescence imaging on the polymer film loaded with 20 wt% of BTE-CN-MBE nanoparticles using UV (365 nm, hand-held lamp, 1.2 mW cm^{-2}) and visible light (>500 nm). The dark regions represent the parts irradiated with UV light; the real size of the photomasks is about 1 × 1 cm.

CN-MBE molecules in the nanoparticles, which in turn resulted in strong intermolecular interactions that caused a specific aggregation. In addition, the bulky and polar cyano group restricted the parallel face-to-face intermolecular interactions and prevented any parallel orientation of the conjugated chromophores. This favored the formation of J- instead of H-aggregation, and restricted excimer formation in the solid state (Figure 6.5b). The synergetic effect of intramolecular planarization and J-aggregate formation in the nanoparticles was considered to be responsible for the enhanced emission.

On the basis of these investigations, Park and coworkers designed a multifunctional fluorescent molecule for a photoswitchable memory media by replacing one of the end tolyl groups in CN-MBE with the photochromic 1,2-bisthienylethene (BTE) moiety [20]. As shown in Figure 6.5c, this compound provided an opportunity to combine the AIEE property of the CN-MBE unit with the bistable photochromism of the BTE unit, to solve the general problem of concentration quenching in the fluorescence switch system. With this novel compound, Park *et al.* prepared size-controllable nanoparticles, and further used the neat nanoparticles and a nanoparticle-loaded poly(methyl methacrylate) (PMMA) film as a photoswitchable memory media. This allowed them to achieve information storage with a high capability, high sensitivity, and a high-contrast on/off signal ratio. Photo-rewritable fluorescence imaging on the polymer film loaded with 20 wt% of BTE-CN-MBE nanoparticles is illustrated in Figure 6.5d.

6.4
Composite Solid

Doping based on FRET has been widely used in electroluminescent (EL) devices, because it may help to improve the luminescence efficiency and to tune the emission colors [21]. In a doped system, the slight variation in the content of the energy acceptor will result in a significant PL color change. For example, white-light emission – which is pivotal to full-color displays and the backlight of portable display devices – can be obtained by doping blue fluorescent dyes with green and red dyes. In addition, the energy transfer also significantly increases the PL quantum yield of the energy acceptor. For instance, the quantum yield of a pure rubrene (5,6,11,12-tetraphenylnaphthacene) thin film prepared by vacuum deposition is about 30%, whereas in films with rubrene as the acceptor the quantum yield may be as high as 100% [22]. Until now, most organic doping systems have been based on amorphous film materials, whereas doped organic nanomaterials (especially crystalline nanomaterials) have attracted much less attention. Recently, Yao and coworkers prepared doped organic nanoparticles and crystalline binary organic nanowires to investigate their tunable and switchable emissions, and also to study energy transfer in the doped nanostructures.

For this, doped organic nanoparticles were first prepared using reprecipitation, with 4-(dicyanomethylene)-2-methyl-6-(*p*-dimethyl-aminostyryl)-4*H*-pyran (DCM, see inset of Figure 6.6) as the energy acceptor and 1,3,5-triphenyl-2-pyrazoline

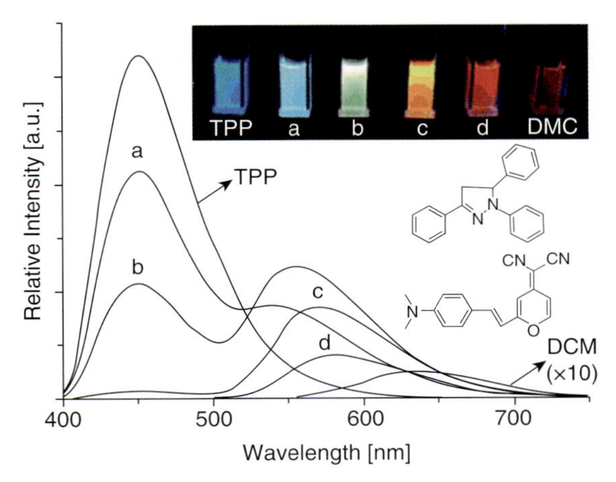

Figure 6.6 Fluorescence emission spectra of suspensions of DCM-doped TPP nanoparticles (intensity of the emission of the dispersion of pure DCM nanoparticles is multiplied by 10 and the excitation wavelength is 365 nm). Inset: Photographs of the fluorescence emissions from suspensions of the doped TPP nanoparticles taken under the UV lamp (365 nm). From left to right, the doping concentrations of DCM are 0, 0.1%, 0.2%, 2%, 10%, and 100% (molar ratio).

(TPP, see inset of Figure 6.6) as the energy donor [5]. For the preparation, a mixed solution of TPP with DCM was used in the reprecipitation process. As shown in Figure 6.6, the emission colors of the doped nanoparticles dispersions evolved from blue to red with increasing DCM concentration. The effective quenching of the emission from TPP indicated the occurrence of an efficient energy transfer in the nanoparticles. The fluorescence decay measurements proved that the nonradiative Förster resonance type was the dominant mechanism of energy transfer. In order to fulfill the requirements of practical applications, polymer films dispersed with DCM-doped TPP nanoparticles were fabricated, which also demonstrated tunable emissions. Recently, when Sun *et al.* also investigated this DCM/TPP-doped system from a theoretical standpoint [23], their results suggested that intermolecular charge transfer processes may also be involved in the fluorescence quenching mechanism.

Another application of organic composite solids is the emission switch. Photoswitches based on photochromic materials have attracted a great deal of interest in recent years, mainly because of their potential application as ultrahigh-density optical memory media [24]. In most switch systems, the fluorescent probe moieties are bound covalently to their photochromic counterparts, which can change between two distinct isomeric states, representing 0 and 1 of a digital mode, when irradiated at different wavelengths. The design and synthesis of these multifunctional fluorescent photochromic molecules is a difficult task, especially considering the tunability of the productive absorption bands corresponding to the "write," "erase," and "read" processes [25].

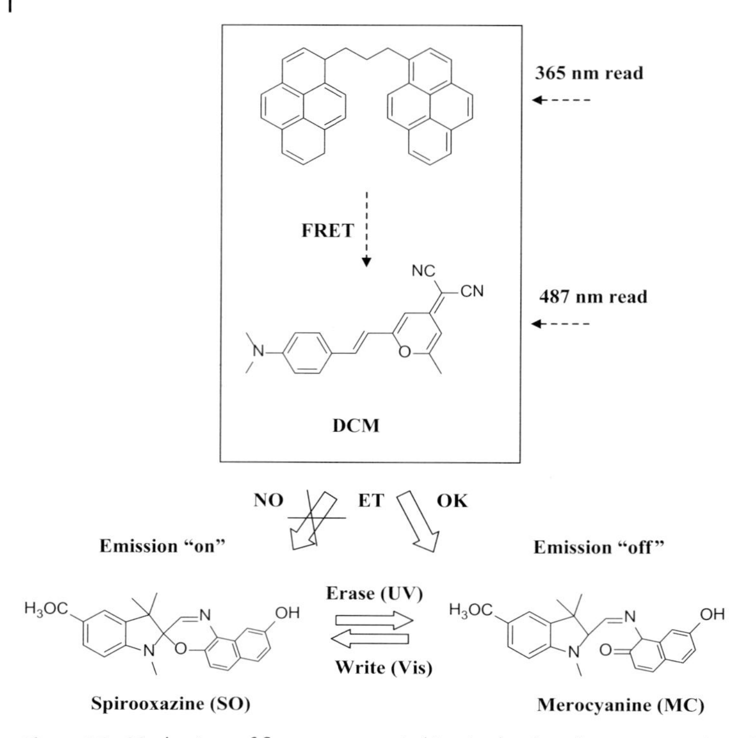

Figure 6.7 Mechanism of fluorescence switching in the doped nanocomposites. FRET: Fluorescence resonance energy transfer.

In some recent investigations, Yao and coworkers described a versatile and convenient approach to achieving fluorescence modulation by the preparation of composite nanoparticles. These were based on a derivative of photochromic spirooxazines (SO), a typical fluorescent dye DCM, and an emissive assistant molecule of 1,3-bis(pyrene) propane (BPP; see Figure 6.7 for the chemical structures), employing doping techniques [26]. It is well known that a photoisomerization occurs between the closed-ring form SO and the open-ring form merocyanine (MC), SO↔MC, as shown in Figure 6.7. When the composite nanoparticles were irradiated with ultraviolet light, some of the SO isomerized to the MC form, which quenched the emission of DCM via an intermolecular energy transfer. However, when MC reverted to SO, induced by visible light irradiation, the DCM fluorescence in the nanoparticles recovered fully. The doping of BPP not only enhanced the contrast between the fluorescence ON and OFF signals, but also provided away to tune the excitation wavelength for reading the fluorescence signals. The mechanism of the fluorescence switch is illustrated in Figure 6.7.

Photochromic dye-conducting polymer core–shell nanomaterials have been fabricated using microemulsion micelles as nanoreactors. For example, pyrene–

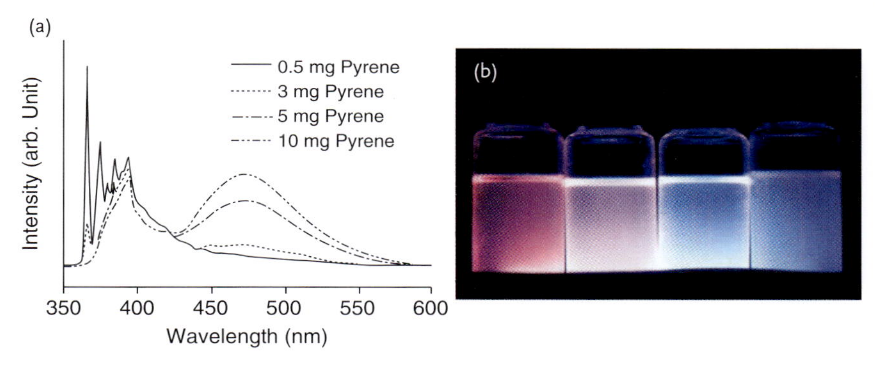

Figure 6.8 (a) The UV-visible spectra of pyrene-embedded PPy nanoparticles, for different pyrene loadings; (b) Photograph of the photoluminescence from pyrene-embedded PPy nanoparticles (the pyrene amounts are, from left to right: 0.5, 3, 5, and 10 mg).

polypyrrole (PPy) nanoparticles (of 7–13 nm diameter) that had been embedded with a pyrene core were successfully synthesized [27]. Subsequently, as the pyrrole monomers were polymerized, the pyrene molecules were phase-separated and gathered together towards the interior of the micelle. The adsorption state of pyrene was tunable over a wide range with a small amount of pyrene, because of the nanosized reaction site of the micelles and the packing constraint of pyrene crystal. The emission colors of the nanohybrids were controllable from violet to blue by altering the amount of embedded pyrene (Figure 6.8). This methodology can provide a facile and effective means of controlling the adsorption state of organic dyes, and also presents a new concept for use of the hole-transporting layer and emitting layer nanohybrids in electroluminescence devices.

6.5 Outlook

In this chapter, the details have been presented of recent progress with organic solids for photonics based on organic nanostructures, including size-dependent optical properties, unique photoluminescent properties such as fluorescent narrowing, multicolor emission, tunable emission, and switchable emission. Although, currently, research in this field is still very much in its infancy, many challenging topics remain. First, methods to create organic nanomaterials with desired morphologies and structures represent a key task, although some progress – for example, with adsorbent-assisted physical vapor deposition – has been made recently. Second, details of the optical and photonic properties of single organic nanoparticles and single nanowires remain to be determined. Third, the design and synthesis of molecules with unique optoelectronic properties, together with the controllable construction of nanostructures with deliberately designed molecules represent not only a major challenge but also a great opportunity. In addition, theoretical calculations have proved to be an effective means of designing target molecules and explaining the electronic processes in this area. Finally, with regards to applications, novel performances such as optical waveguides and optically or electrically pumped lasers are currently under investigation, as the fabrication of thin films and patterns composed of nanostructured building blocks on solid substrates will be essential if practical devices are to be realized.

References

1 Mueller, A.H., Petruska, M.A., Achermann, M., Werder, D.J., Akhadov, E.A., Koleske, D.D., Hoffbauer, M.A., and Klimov, V.I. (2005) Multicolor light-emitting diodes based on semiconductor nanocrystals encapsulated in GaN charge injection layers. *Nano Lett.*, **5** (6), 1039–1044.

2 van der Wielen, M.W.J., Cohen Stuart, M.A., and Fleer, G.J. (1999) Controlled nanometer-scale surface roughening and its effect on the ordering and stability of liquid-crystalline polymer films. *Adv. Mater.*, **11** (11), 918–923.

3 (a) Barrelet, C. J., Greytak, A. B., and Lieber, C. M. (2004) Nanowire photonic circuit elements. *Nano Lett.*, **4** (10), 1981–1985; (b) Johnson, J.C., Choi, H.-J., Knutsen, K.P., Schaller, R.D., Yang, P., and Saykally, R.J. (2002) Single gallium nitride nanowire lasers. *Nat. Mater.*, **1** (2), 106–110.

4 Campbell, I.H. and Crone, B.K. (2006) Quantum-dot/organic semiconductor composites for radiation detection. *Adv. Mater.*, **18** (1), 77–79.

5 Peng, A.D., Xiao, D.B., Ma, Y., Yang, W.S., and Yao, J.N. (2005) Tunable emission from doped 1,3,5-triphenyl-2-pyrazoline organic nanoparticles. *Adv. Mater.*, **17** (17), 2070–2073.

6 An, B.-K., Kwon, S.-K., Jung, S.-D., and Park, S.Y. (2002) Enhanced emission and its switching in fluorescent organic nanoparticles. *J. Am. Chem. Soc.*, **124** (48), 14410–14415.

7 An, B.-K., Kwon, S.-K., and Park, S.Y. (2007) Photopatterned arrays of fluorescent organic nanoparticles. *Angew. Chem. Int. Ed.*, **46** (12), 1978–1982.

8 Kasai, H., Nalwa, H.S., Oikawa, H., Okada, S., Matsuda, H., Minami, N., Kakuta, A., Ono, K., Mukoh, A., and Nakanishi, H. (1992) A novel preparation method of organic microcrystals. *Jpn. J. Appl. Phys.*, **31** (Part 2, 8A), L221–L223.

9 Chernyak, V., Meier, T., Tsiper, E., and Mukamel, S. (1999) Scaling of fluorescence stokes shift and superradiance coherence size in disordered molecular aggregates. *J. Phys. Chem. A*, **103** (49), 10294–10299.

10 Spano, F.C. and Mukamel, S. (1989) Nonlinear susceptibilities of molecular aggregates: enhancement of $\chi^{(3)}$ by size. *Phys. Rev. A*, **40** (10), 5783–5801.

11 Jiang, Q. (1999) Melting thermodynamics of organic nanocrystals. *J. Chem. Phys.*, **111** (5), 2176.

12 Kasai, H., Kamatani, H., Okada, S., Oikawa, H., Matsuda, H., and Nakanishi, H. (1996) Size- dependent colors and luminescences of organic microcrystals. *Jpn. J. Appl. Phys.*, **35** (Part 2, 2B), L221–L223.

13 Rasika Dias, H.V. and Fianchini, M. (2007) A classical silver carbonyl complex [{MeB [3-(Mes)pz]3} Ag(CO)] and the related silver ethylene adduct [{MeB[3-(Mes)pz]3} Ag (C_2H_4)]. *Angew. Chem. Int. Ed.*, **46** (13), 2188–2191.

14 Fu, H.B. and Yao, J.N. (2001) Size effects on the optical properties of organic nanoparticles. *J. Am. Chem. Soc.*, **123** (7), 1434–1439.

15 Xiao, D., Xi, L., Yang, W., Fu, H., Shuai, Z., Fang, Y., and Yao, J. (2003) Size-tunable emission from 1,3-diphenyl-5-(2-anthryl)-2-pyrazoline nanoparticles. *J. Am. Chem. Soc.*, **125** (22), 6740–6745.

16 Zhao, Y.S., Xiao, D., Yang, W., Peng, A., and Yao, J. (2006) 2,4,5-Triphenylimidazole nanowires with fluorescence narrowing spectra prepared through the adsorbent-assisted physical vapor deposition method. *Chem. Mater.*, **18** (9), 2302–2306.

17 Fu, H., Loo, B., Xiao, D., Xie, R., Ji, X., Yao, J., Zhang, B., and Zhang, L. (2002) Multiple emissions from 1,3-diphenyl-5-pyrenyl-2-pyrazoline nanoparticles: evolution from molecular to nanoscale to bulk materials. *Angew. Chem. Int. Ed.*, **41** (6), 962–965.

18 Gao, X.-C., Cao, H., Zhang, L.-Q., Zhang, B.-W., Cao, Y., and Huang, C.-H. (1999) Properties of a new pyrazoline derivative and its application in electroluminescence. *J. Mater. Chem.*, **9** (5), 1077–1080.

19 (a) Deans, R., Kim, J., Machacek, M. R., and Swager, T. M. (2000) A poly(*p*-phenyleneethynylene) with a highly emissive aggregated phase. *J. Am. Chem. Soc.*, **122** (35), 8565–8566; (b) Luo, J., Xie, Z., Lam, J.W.Y., Cheng, L., Chen, H., Qiu, C., Kwok, H.S., Zhan, X., Liu, Y., Zhu, D., and Tang, B.Z. (2001) Aggregation-induced emission of 1-methyl-1,2,3,4,5- pentaphenylsilole. *Chem. Commun.*, (18), 1740–1741.

20 (a) Murata, T., Morita, Y., Fukui, K., Sato, K., Shiomi, D., Takui, T., Maesato, M., Yamochi, H., Saito, G., and Nakasuji, K. (2004) A purely organic molecular metal based on a hydrogen-bonded charge-transfer complex: crystal structure and electronic properties of TTF-imidazole-*p*-chloranil. *Angew. Chem. Int. Ed.*, **43** (46), 6343–6346; (b) Lim, S.-J., An, B.-K., Jung, S.D., Chung, M.-A., and Park, S.Y. (2004) Photoswitchable organic nanoparticles and a polymer film employing multifunctional molecules with enhanced fluorescence emission and bistable photochromism. *Angew. Chem. Int. Ed.*, **43** (46), 6346–6350.

21 (a) Tang, C. (1989) Electroluminescence of doped organic thin films. *J. Appl. Phys.*, **65** (9), 3610; (b) Kido, J., Kimura, M., and Nagai, K. (1995) Multilayer white light-emitting organic electroluminescent device. *Science*, **267** (5202), 1332–1334.

22 Mattoussi, H. (1999) Photoluminescence quantum yield of pure and molecularly doped organic solid films. *J. Appl. Phys.*, **86** (5), 2642.

23 Sun, M., Pullerits, T., Kjellberg, P., Beenken, W.J.D., and Han, K. (2006) Control of emission by intermolecular fluorescence resonance energy transfer and intermolecular charge transfer. *J. Phys. Chem. A*, **110** (19), 6324–6328.

24 Irie, M. (2000) Diarylethenes for memories and switches. *Chem. Rev.*, **100** (5), 1685–1716.

25 Norsten, T.B. and Branda, N.R. (2001) Axially coordinated porphyrinic photochromes for non-destructive information processing. *Adv. Mater.*, **13** (5), 347–349.

26 Sheng, X.H., Peng, A.D., Fu, H.B., Liu, Y.Y., Zhao, Y.S., Ma, Y., and Yao, J.N. (2007) Modulation of a fluorescence switch based on photochromic spirooxazine in composite organic nanoparticles. *Nanotechnology*, **18** (14), 145707–145713.

27 Jang, J. and Oh, J.H. (2003) Facile fabrication of photochromic dye-conducting polymer core-shell nanomaterials and their photoluminescence. *Adv. Mater.*, **15** (12), 977–980.

7
Organic Photonic Devices

Hongbing Fu

7.1
Introduction

Organic photonics potentially enables the manipulation of photons at a subwavelength scale and applications in optical computing systems, which can overcome the limitations of frequency and power dissipation in silicon electronics. Recently, active photonic building blocks have been fabricated successfully from semiconductors and organic materials, and their waveguiding properties, optical cavity effect and lasing, have been extensively studied. In active optical waveguiding materials, the strong light–matter interactions are useful to access simultaneously the propagation and processing of optical signals, compared to the passive transmission mode through an optical fiber. Therefore, materials to meet these targets are especially important for organic photonics. Among such materials, nanomaterials based on organic small molecules have attracted increasing attention during the past 10 years, due to their higher structural tunability, reactivity [1], and processability [2]. In particular, their unique optical and electronic properties [3] have led to them becoming the "workhorses" of nanoscale science and engineering, finding applications in various electronic, photonic, and sensing devices [4]. Among these organic nanomaterials, one-dimensional (1-D) nanostructures have been shown to serve as effective building blocks for miniaturized devices, because of their two-dimensional (2-D) photon confinement. In addition, crystalline 1-D organic nanostructures with a highly ordered stacking of π-conjugated molecules offer better stabilities and charge transport properties, which in turn result in better photonic performances of the final devices. Some unique applications, such as tunable color displays, field-effect transistors, chemical sensors, optical waveguides, and lasers have been achieved from such organic 1-D materials. Consequently, 1-D organic single-crystalline nanostructures are expected to play important roles as interconnects and functional units in the next generation of photonic applications. Yet, organic nanomaterials also have other advantages, including a high photoluminescence (PL) efficiency [5], good doping properties [6], a high reaction activity [1], and a high compatibility with flexible and lightweight plastic substrates, all of which make them complementary to inorganic

Organic Optoelectronics, First Edition. Edited by Wenping Hu.

materials. In addition, crystalline 1-D organic nanostructures offer better stabilities and charge transport properties, and hence better optoelectronic performances of the final devices [4]. Taking into consideration the vast number of optically and electronically active organic molecules available, the organic version of nanowires could reasonably be expected to have a similar potential and to open broader applications for the next generation of electronic and photonic devices.

However, there remain several challenges before organic photonics is fully realized, one such problem being how to construct suitable organic structures with special optoelectronic features. In this chapter, details of recent progress on the construction of unique organic nanomaterials for novel photonic applications, such as multicolor emission, tunable emission, optical waveguide, and lasing, will be introduced, together with some details of photonic devices based on organic solids.

7.2 Crystalline One-Dimensional (1-D) Organic Nanostructures

The development of facile, mild, and universal construction strategies is prerequisite to the further investigations of organic 1-D nanostructures. Most of the present approaches to inorganic 1-D nanomaterials do not apply to their organic counterparts due to their lower melting and sublimation points. However, a number of fruitful attempts have been made in this area during the past few years, and some successful examples of the achievements of organic crystalline 1-D nanostructures, allocated to either liquid- or vapor-phase-based methods, are outlined in the following subsections.

7.2.1 Self-Assembly in Liquid Phase

Almost everything has the inclination towards self-assembly which, in the classic sense, can be defined as the ". . . spontaneous and reversible organization of molecular units into ordered structures by noncovalent interactions." The first property of a self-assembled system that this definition suggests is the spontaneity of the self-assembly process. The interactions responsible for formation of the self-assembled system act on a strictly local level; in other words, the nanostructure "builds itself." Molecular self-assembly in liquid solutions is a strategy for nanofabrication that involves designing molecules and supramolecular entities. The assembly of organic molecules to nanostructures with defined morphologies requires driving forces from the molecules themselves, including hydrogen bonds, π–π stacking, and van der Waals contacts. Occasionally, the self-assembly process also requires an induction from the surroundings, such as interactions between the organic molecules and the solvents or auxiliaries. The single crystalline nanotubes and microtubes of a small organic functional compound, 2,4,5-triphenylimidazole (TPI), were synthesized by a rapid dispersion of the TPI mol-

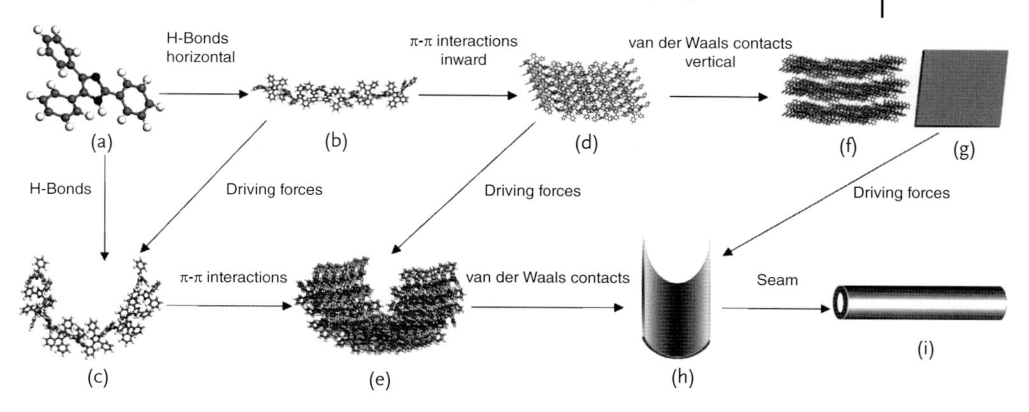

Figure 7.1 Schematic representation of the formation of TPI tubes. (a) Optimized geometry of a single TPI molecule; (b, c) Linear and curled motifs of TPI supramolecular H-bonded chains; (d, e) Flat and curled monolayer sheets from the two motifs of chains; (f) Multilayer sheet by stacking of the monolayers; (g–i) Sketch maps of the multilayer flat lamellar, curled lamellar, and tubular structures, respectively.

ecules from a good solvent into a nonsolvent, water [7]. The rapid change of the surroundings induced the self-assembly of TPI molecules, such that tubes with a high monodispersity and open-ended structures were obtained following the assembly of TPI at different temperatures. Moreover, each tube had a well-defined 1-D morphology and a single crystalline structure, while the length and diameter of the TPI tubes could be tuned simply by altering the assembly conditions. The characterization of this temporal morphological evolution indicated that the tubes were obtained by the rolling and seaming of preorganized 2-D lamellar structures. Subsequently, the results of nuclear magnetic resonance (NMR) spectroscopy, X-ray photoelectron spectroscopy (XPS) and absorption spectra revealed that the lamellar structures were constructed via the cooperation of three chemically orthogonal and spatially independent noncovalent intermolecular interactions – that is, hydrogen bonds (H-bond), π–π interactions, and van der Waals contacts, as shown in Figure 7.1.

7.2.2
Template-Induced Self-Assembly in Liquid Phase

Although a solvent change-induced assembly is a facile method for the fabrication of 1-D nanomaterials, it requires strong intermolecular interactions within the target compounds, such as H-bonds or effective π–π stacking. Yet, for those molecules without such strong noncovalent interactions, the 1-D molecular aggregations might be directed via external forces. The template method represents a straightforward route for the fabrication of nanostructures by inducing the target molecules to grow according to the patterns of the templates.

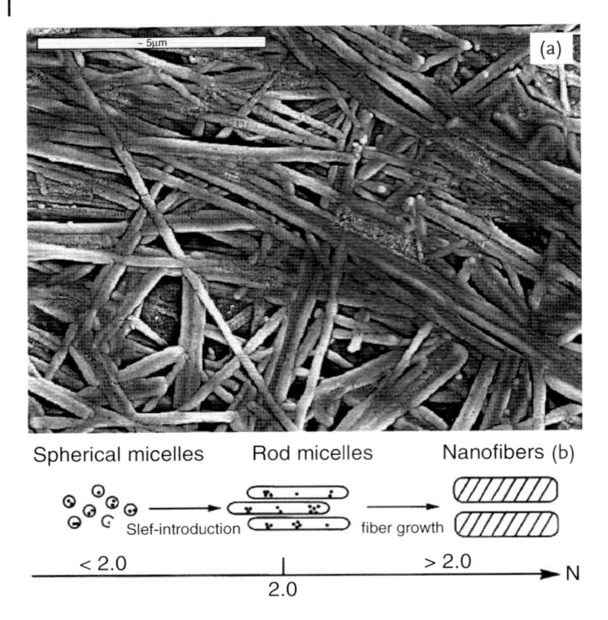

Figure 7.2 (a) SEM image of the prepared DP nanofibers (scale bar = 5 μm); (b) Representation of the formation mechanism of the DP nanofiber, in which N is the molar ratio of DP/CTAB.

It is well known that micelles with different shapes, such as spherical or rod-like, will be formed in surfactant solutions when the concentration reaches the so-called critical micelle concentration (CMC). Indeed, the rod-like micelles can be used as soft templates for the 1-D assembly of organic molecules. Nanofibers of 1,3-diphenyl-2-pyrazoline (DP) were prepared via the self-assembly of DP molecules in cetyltrimethylammonium bromide (CTAB) micelles [8]. The morphology of the nanofibers, visualized using scanning electronic microscopy (SEM), is shown in Figure 7.2a. The shape and size of the fibers could be controlled by altering the molar ratio of DP to CTAB; the mechanism of fiber formation is shown in Figure 7.2b. Notably, the CTAB tended to form spherical micelles at lower DP/CTAB molar ratios, whereas an increase in the amount of DP present induced a sphere-to-rod transition of the micelles, which could then serve as a template to direct the growth of DP in 1-D fashion. This method has proven to be applicable to a variety of functional molecules [9].

Besides micelles, porous membrane materials – including anodized aluminum oxide (AAO), silica, and nanochannel glass – have also been used as templates for inducing the 1-D growth of organic molecules via capillary forces. Thus, perylene [10] and dibenzoylmethane nanotubes were prepared by repeatedly immersing the porous alumina template into saturated organic solutions. Nanotubes of different sizes can be prepared by using alumina templates with different pore diameters, and the crystallinity of the products can be increased by subsequent thermal treatments.

7.2.3 Morphology Control with Molecular Design

It has been proven that molecular structure plays an important role in the aggregation behavior of organic materials in solutions. Therefore, it is possible to control the self-assembly manner by introducing special substitute groups in the organic molecules. The facile fabrication of nanostructures with well-defined shapes from stilbazolium-like dyes has been reported [11]. These dyes, with different substituents, were synthesized with the aim of exploring the influences of the substituent effects on the shape, and therefore the optical properties, of the resulting nanostructures. Whilst the self-assembly of the target molecules was induced with solvent exchange, the change of substituents was found to have influenced the morphology of the aggregates to a significant degree. In particular, a strong electron donor–acceptor interactions induced a 1-D aggregation in one compound, whereas spherical structures were obtained with the other compound, which had a nonplanar structure and weak electron donor–acceptor interactions.

Besides those obtained from derivatives with different substituents, nanostructures were also tuned between spheres, wires, and cubes via the self-assembly of three isomeric molecules of bis(iminopyrrole) benzene with the same substituents on different positions [12]. Despite all three isomeric precursors – *ortho-*, *meta-*, and *para-*bis(iminopyrrole) benzene – being present and providing similar strong multiple hydrogen-bonding interactions for molecular aggregation, distinctly different-shaped nanostructures were obtained. As shown in Figure 7.3a, the two iminopyrrole (IP) groups in the *o*-isomer were 60° open-armed; this allowed two monomers to embed into each other to form a dimer via quadruple hydrogen bonds; these dimers then acted as the basic units for the formation of spherical structures. In contrast, the configurations of the *m*- and *p*-isomers, with two IP groups open-armed at an angle of 120° and 180°, respectively, determined that each molecule would connect with two others via hydrogen bonds, so as to form chain-like structures. Also determined by molecular configuration, the chain of the *m*-isomer was zigzag shape (Figure 7.3b), whereas that of the *p*-isomer was almost linear (Figure 7.3c). These chains can serve as the actual building blocks for the solid states of the two isomers to form square wires and cubes, respectively. Notably, the different interactions involved in aggregate stacking at the supramolecular level, caused by the isomeric molecular structures, were found to be responsible for the evolution of the various morphological forms.

7.2.4 Physical Vapor Deposition (PVD)

As a facile and feasible method for preparing nanomaterials, PVD has achieved great success in the synthesis of inorganic 1-D nanomaterials and polymeric thin

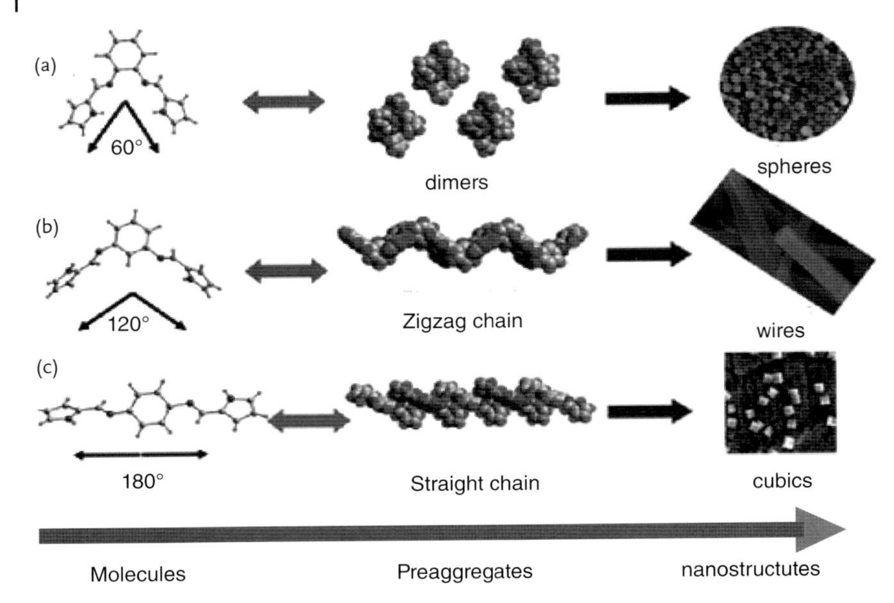

Figure 7.3 Schematic illustration of the formation processes of nanostructures from (a) *ortho*-, (b) *meta*-, and (c) *para*-bis(iminopyrrole) benzene, respectively.

films. However, the monodispersity of the products is difficult to control when small organic molecules are used as deposition sources. It is known that, with the PVD method, the degree of saturation is the predominant factor in controlling the morphology and dispersity of the products; hence, it should be possible to process most solid materials into 1-D nanostructures simply by controlling the vapor saturation at a low level.

In recent studies [13], adsorbents such as neutral aluminum oxide or silica gel that in the past have been used widely in column chromatography, were introduced into the PVD method (Figure 7.4a) to control the degree of saturation. The consideration was made that an adsorption–desorption equilibrium should exist between the adsorbents and the organic sources. In fact, adsorbent-assisted PVD has been demonstrated as a good general method for the synthesis of organic 1-D nanomaterials with high crystallinity, and that by applying this method it should be possible to prepare single crystalline nanowires from TPI. The SEM images of TPI nanowires shown in Figure 7.4b confirmed that each nanowire had a smooth surface and a uniform width through its entire length. In addition, the SEM images of some wires which grew out of the substrate surface (Figure 7.4c) showed very flat and smooth end surfaces of the nanowire quadrangular cross-sections. The transmission electronic microscopy (TEM) images in Figure 7.4d confirmed that the geometric shape of the TPI nanostructures was a solid wire, without any obvious structural defects. The electron diffraction (Figure 7.4e) pattern revealed that each wire had a single crystalline structure and grew along the b axis of the TPI crystal; this was subsequently confirmed by X-ray diffraction (XRD) measure-

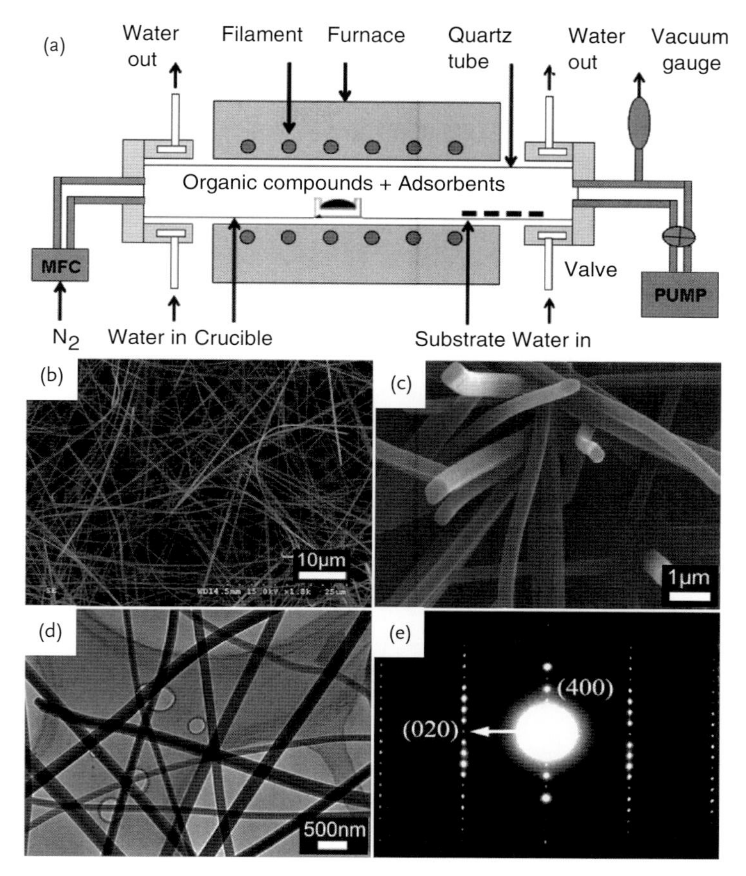

Figure 7.4 (a) Sketch of the instrument used for the adsorbent-assisted PVD; (b, c) SEM images of the TPI nanowires; (d) TEM images of some typical TPI nanowires; (e) The electron diffraction pattern of the TPI nanowire with the main diffraction spots indexed; the arrow indicates the direction of the wire.

ments. Additional measurements using thermogravimetric analysis (TGA) confirmed that the adsorbents had reduced the sublimation temperature and slowed the weight loss of TPI, such that the degree of saturation could be well controlled and the uniformity of the nanowires remarkably improved.

7.3
Organic Nanophotonics

One of the most important reasons why organic materials have attracted so much attention is their intriguing advantages compared to their inorganic counterparts; notably, they are flexible, lightweight, easily functionalized, and can be fabricated in large areas at low cost. An overview of ongoing investigations into the unique photonic and/or electronic performances of 1-D organic materials is provided in the following subsections.

7.3.1 Electroluminescence and Field Emission

In the past, tris(8-hydroxyquinoline)aluminum (Alq_3) has played an important role in the production of organic light-emitting diodes (OLEDs), due to its strong electroluminescence with a low driving voltage. In this case, the Alq_3 nanowires were prepared using an adsorbent-assisted PVD method, and then applied to LED and field-emission devices [14]. The *I–V* characteristics indicate a typical diode-like behavior of the devices. The luminance–voltage characteristics of LED devices prepared from Alq_3 nanowires of different diameters are shown in Figure 7.5a, where the performance of the Alq_3 nanowire-based LEDs demonstrated a clear size-dependent effect. The turn-on voltage of the device constructed from Alq_3 nanowires of 40 nm diameter was about 1 V lower than that of a device constructed from 60 nm diameter nanowires; moreover, the current efficiency of the former device was about 20% higher. When the driving voltage was 12.5 V, the external quantum efficiencies of the devices based on nanowires with diameters of 40, 60, and 80 nm were 0.90%, 0.58%, and 0.28%, respectively. Another important optoelectronic property of the Alq_3 nanowires is the field emission with a relatively low turn-on field, which also shows a size-dependent performance. The field emission *J–E* curves of Alq_3 nanowires with different diameters are shown in Figure 7.5b, where a decrease in diameter from 80 to 40 nm led to a decrease in the turn-on field from 10.3 to 8.5 V μm^{-1}, while the threshold field decreased from 13.5 to 11.6 V μm^{-1}. The turn-on field here was lower than that of the amorphous Alq_3 nanowires, due to the high degree of crystallinity.

7.3.2 Tunable Emission from Binary Organic Nanowires

Uniformly doped 1-D composite nanostructures were further synthesized by a simultaneous reprecipitation with an electron donor compound, 1,3,5-triphenyl-2-pyrazoline (TPP), and an electron acceptor, 1,4-dicyanonaphthalene (DCN) [15]. Exciplexes were formed between TPP and DCN at the excited state in the composite nanostructures, through which the fluorescence of TPP was quenched (Figure 7.6). The emission color of the nanocomposites can be tuned by changing the doping content of DCN, as shown in the upper inset of Figure 7.6. Theoretical calculations confirmed that photoinduced electron transfer (PET) occurs through an electron transition from the lowest unoccupied molecular orbital (LUMO) of the excited TPP to that of the ground-state DCN, as shown in the lower inset of Figure 7.6.

Binary doped organic nanowires were prepared using the above-mentioned adsorbent-assisted PVD method. Initially, a blue light emitter (TPP) and rubrene

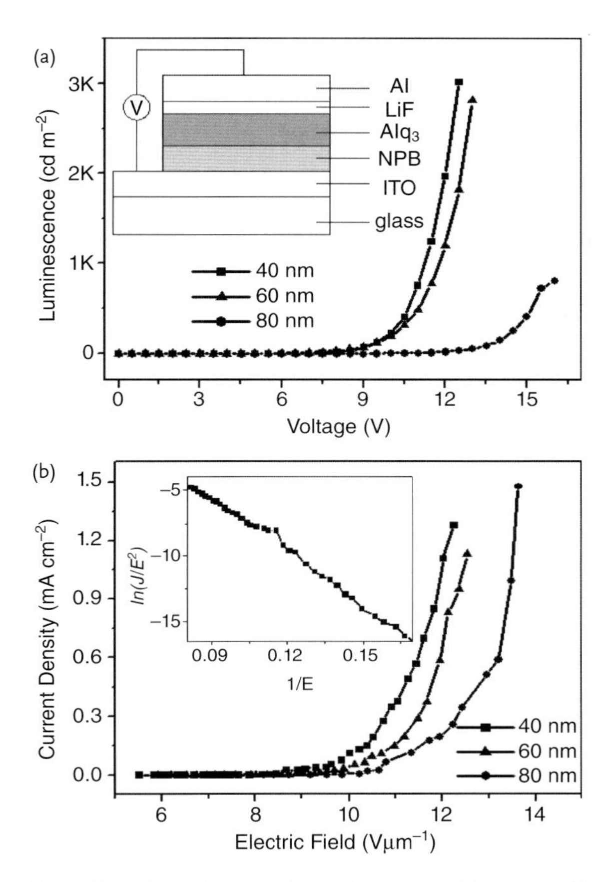

Figure 7.5 (a) Luminance–voltage characteristics of the LEDs fabricated from Alq_3 nanowires with average diameters of 40, 60, and 80 nm. The inset shows the schematic structure of the devices; (b) Field-emission *J–E* curves of the Alq_3 nanowires with different diameters. The insert shows the Fowler–Nordheim plot of nanowires with average diameter of 40 nm, which yields a straight line with a slope of about −135.

(an orange dye) were used as dopants in host/guest OLEDs with good transport properties. These were chosen as model compounds and fabricated into uniformly doped crystalline nanorods and nanowires, using an adsorbent-assisted PVD method [16]. In this case, the doping content was tuned by changing the molar ratio of the two sources. Fluorescence microscopy images of the binary nanostructures with various doping contents (see Figure 7.7a–c) confirmed a homogeneous emission of color from all of the nanorods/wires in each sample, and indicated that the rubrene crystal grains had been dispersed uniformly into the TPP matrices. The binary nanowires also showed waveguide properties, as proven by the microarea fluorescence microscopy of single nanorod (Figure 7.7c). The fluorescence microscopy images also revealed that the emission color evolved from blue

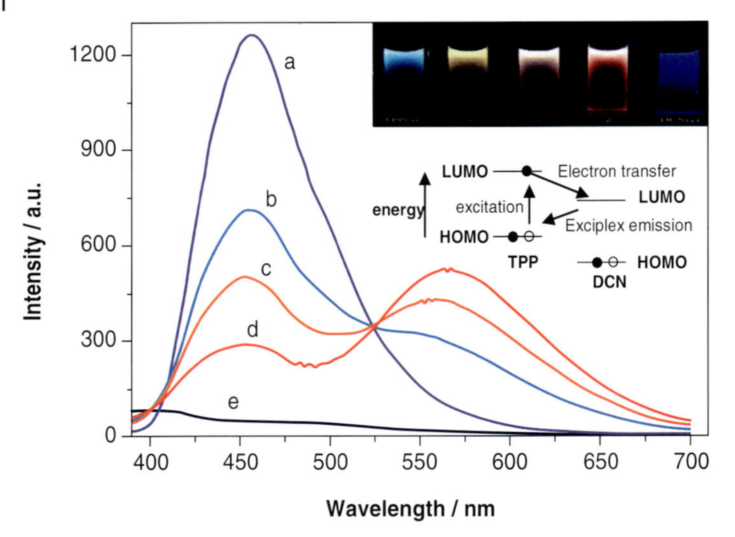

Figure 7.6 PL spectra of the 1-D nanocomposite suspensions of TPP and DCN with different molar ratios in water: (a) pure TPP; (b) TPP/DCN, 4:1; (c) TPP/DCN, 2:1; (d) TPP/DCN, 1:1; (e) pure DCN. The excitation wavelength was 365 nm. The upper inset shows images recorded under a UV lamp (355 nm) of suspensions (a) to (e); the lower inset is the energy level diagram representing the photoinduced electron transfer in the nanocomposites.

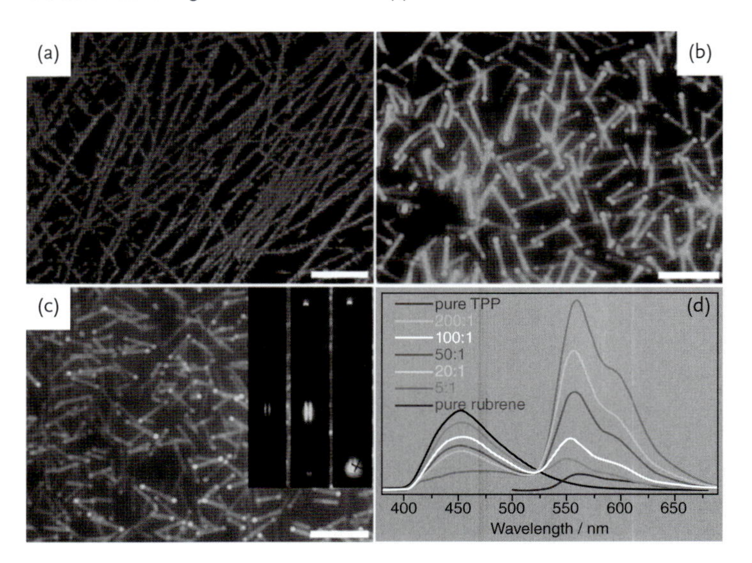

Figure 7.7 Fluorescence microscopy images of 1-D nanomaterials with different rubrene contents: (a) Pure TPP, excited with UV light (330–385 nm); (b) 100:1 TPP/rubrene, excited with UV light; (c) Pure rubrene, excited with blue light (460–490 nm). Scale bars = 5 μm. The insets show the bright-field image (left) and microarea fluorescence images of a single nanorod obtained by exciting the middle part (middle) and top area (right) using a focused laser; the crosses indicate the excited position; (d) Emission spectra of the binary nanomaterials deposited onto quartz wafers with different TPP/rubrene molar ratios.

to orange as the rubrene content was increased and, more importantly, that a white emission could be achieved when a correct TPP : rubrene molar ratio of 100 : 1 was adopted. A tunable emission was also observed from the evolution of the fluorescence spectra of nanostructures as the rubrene content was changed (Figure 7.7d). For purposes of comparison, doped amorphous thin films were also prepared using a coevaporation technique, but the film colors were unstable due to oxidation of the rubrene. In contrast, the emission colors of the binary nanowires were very stable, a property that was ascribed to the crystallinity of rubrene. In amorphous films, the emission of TPP could be quenched almost completely at a 4% rubrene doping content; this was caused by the higher energy transfer efficiency. In contrast, the emission of TPP could not be quenched completely, even when the dopant content of the nanowires was raised to 20%. Both the color stability and incomplete quenching were seen to be essential for the realization of white-light emission (WLE).

Based on the CIE chromaticity diagram, the generation of WLE can also be achieved by mixing a blue fluorescent dye with a yellow or orange dye. In this way, a supramolecular synthesis of triblock microrods of 1,3-diphenyl-2-pyroline (DP) doped with an orange dye of 4-(dicyanomethylene)- 2-methyl-6-(*p*-dimethylaminostyryl)-4*H*-pyran (DCM) selectively at both ends, was successfully carried out, induced by CTAB micelles. Although these triblock microrods demonstrated a micro area heterogeneity, they exhibited a macroscopic high-quality WLE in both a colloidal suspension and the solid state. The left inset in Figure 7.8 shows a fluorescence microscopy image of microrods placed on a quartz slide by excitation with unfocused ultraviolet (UV) light (330–380 nm). In sharp contrast to the uniform morphologies observed using SEM and TEM, triblock heterostructures with striping patterns were clearly apparent: the middle stripe exhibited blue PL,

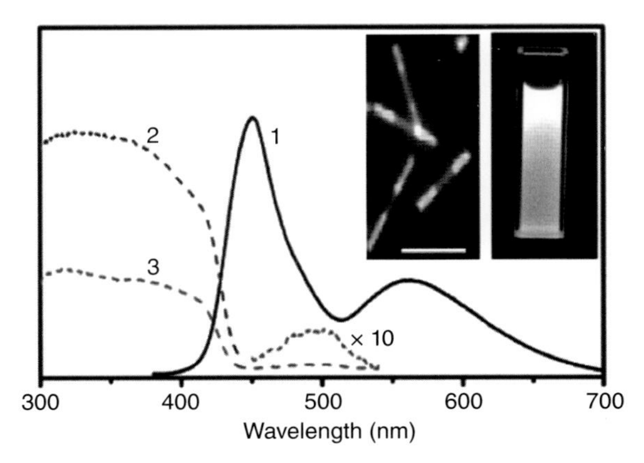

Figure 7.8 Fluorescence microscopy images (left inset) of microrods excited with unfocused UV light (330–380 nm). Steady-state emission (curve 1) and excitation (curves 2, 3) spectra of a colloidal suspension (right inset) of triblock microrods. The excitation spectra of 2 and 3 were recorded by monitoring the DP emission at 450 nm and DCM emission at 560 nm, respectively.

while the two ends presented orange PL. The length of the microrods was 21.2 ± 2.4 μm, while the lengths of the middle and end stripes were 7.6 ± 1.2 μm and 6.9 ± 1.5 μm, respectively. The absorption spectrum of the colloidal suspension of triblock microrods was similar to that of DP nanowires, including n-π* (at 320 nm), π-π* (at 375 nm), and J-aggregate (at 420 nm) bands, respectively, whereas the absorption of DCM in the region 450–550 nm was hardly observed as a result of DCM : DP = 2 : 1000. Factually, the UV light cannot excite DCM molecules because of its weak absorption between 330 and 380 nm. However, it should be noted that as there was a good spectral overlap between DP emission and DCM absorption, the orange emission from the end stripes of the triblock microrods might have been the result of a Förster energy transfer (FET) from excited DP to DCM molecules. Remarkably, although the triblock microrods showed a microarea heterogeneity of the striping pattern, the colloidal suspension exhibited a macroscopic high-quality WLE under UV illumination (365 nm), as shown in the right inset of Figure 7.8. The CIE coordinates calculated from the ensemble emission spectrum of the colloidal suspension (curve 1 of Figure 7.8) were (0.338, 0.281), and these were shifted to (0.305, 0.332) for triblock microrod solids filtered on the surface of an AAO membrane. The solid-state PL efficiency was as high as 36 ± 5%.

7.3.3
Organic 1-D Optical Waveguides

Tubular 1-D nano- and submicrostructures were synthesized from 9,10-bis(phenylethynyl)anthracene (BPEA) [17], an emitting dye with very high luminescence efficiency. The organic molecules were first exchanged into boiling water where they were dissolved, after which their self-assembly was induced by a programmed cooling of the whole system at different rates. As shown in Figure 7.9a, the PL microscopy image of the BPEA tubes on quartz substrates exhibited a characteristic green to yellow emission, with very bright luminescence spots at both tips of each tube and a relatively weaker emission from the tube bodies, which was characteristic of an optical waveguide. Almost all of the observed tubes revealed this type of waveguide behavior. The 1-D BPEA tubes could also serve as active optical waveguides to allow the locally excited PL to propagate along the length of the 1-D structure and out-couple at the tube tips (see Figure 7.9c). In addition, when the excitation light was input from one of the two crossed tubes, the emissions could be coupled to the other tubes (see Figure 7.9b,d). The refractive index of the BPEA was ~1.65, while that of the quartz substrates was ~1.5 at the measured wavelength. The small difference between the two refractive indices resulted in a considerable energy leakage in the case of the solid BPEA microrods. However, in the tubular structures the light could be propagated at the interface of BPEA, while the air inside the tube – the refractive index of which was ~1 – led to a significant elimination of the optical loss.

It is well known that the refractive indices of the external media around the waveguides are important for high-performance optical waveguiding. The indices

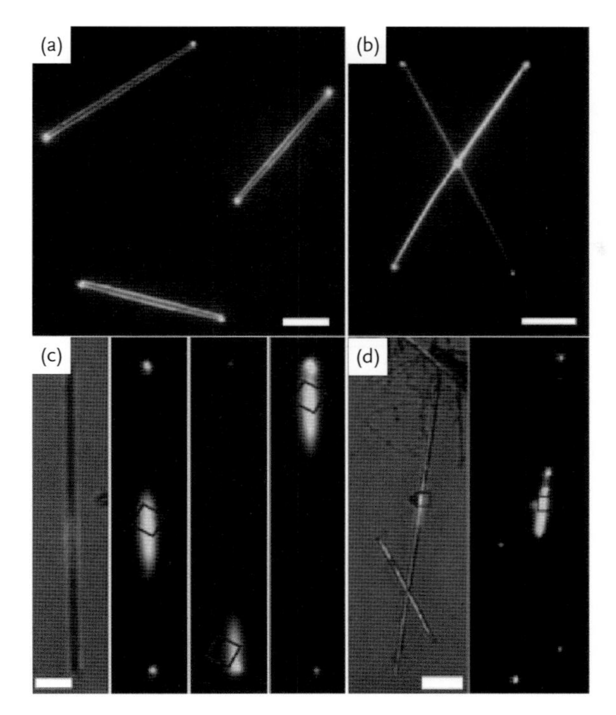

Figure 7.9 PL microscopy images of (a) some isolated BPEA tubes and (b) two crossed tubes. Both samples were excited with blue light in the 460–490 nm region. Scale bars = 20 μm; (c) Bright-field image and micro-area PL images obtained by exciting the identical tube at three different positions; scale bar = 10 μm; (d) Bright-field and micro-area PL images of two crossed tubes; scale bar = 20 μm. The squares indicate the excited positions.

of the air and glass substrate are 1 and 1.53, respectively and, as a result, there should be a considerable decrease in optical loss if the nanowires can be held vertically free-standing from the substrate (though it is difficult to fabricate such vertical organic nanowire arrays) [18]. In investigations conducted by Huang and coworkers, vertical organic nanowire arrays of 1,5-diaminoanthraquinone (DAAQ) dye molecules were prepared using a simple physical vapor transport method [1]. The DAAQ nanowires could serve as nanosized optical waveguides, and the optical loss of the vertical wires was much less compared to that of the horizontally placed wires on glass substrates [19]. Vertically aligned, single-crystalline nanowires offered the best combination of materials and geometry in supporting the low-loss waveguiding modes, which could potentially be used as optical interconnects. DAAQ is known to grow much faster on the substrates with higher surface energies, and based on this a patterned growth of the nanowire arrays was achieved by both microcontact printing and a physical scouring method. Additionally, as-prepared nanowire arrays were also integrated into sensing

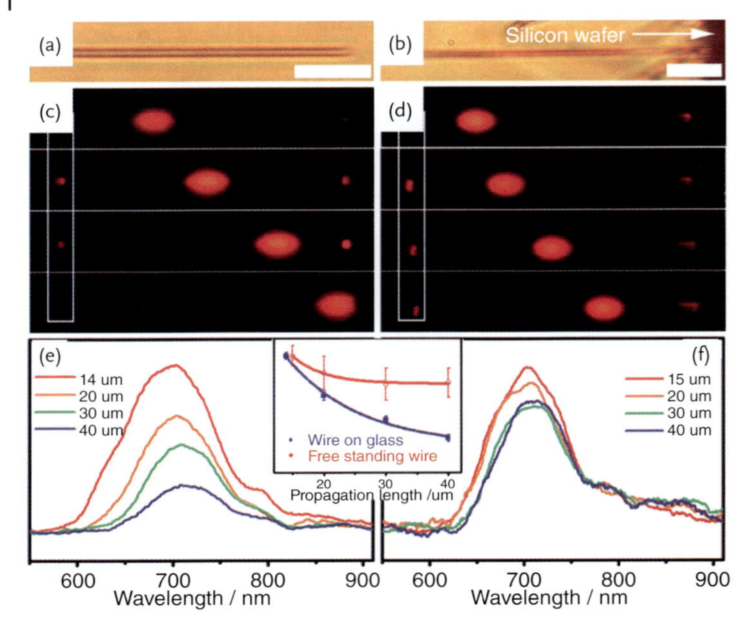

Figure 7.10 Bright-field microscopy image of (a) an isolated DAAQ nanowire on glass and (b) a single free-standing DAAQ wire vertically grown at the edge of a silicon wafer. Scale bars = 10 μm; (c, d) Micro-area excited PL images of the two wires taken by exciting each wire at four different positions; (e, f) Spatially resolved PL spectra of the two samples obtained by exciting the wires at four different positions and collecting the outcoupling light at the wire tips as marked in panels (c) and (d). The inset shows the evanescent curves of the outcoupled emissions with the propagation lengths.

devices for the detection of trace amounts of acidic vapors, with a very high sensitivity. The bright-field optical microscopy images of an isolated DAAQ nanowire on glass, and of a free-standing wire grown at the edge of a silicon wafer, are shown in Figure 7.10a,b; both wires had a diameter of about 500 nm. Fluorescence images showing the two wires excited locally with laser light focused into a 5 mm spot at four different positions, are shown in Figure 7.10c,d. In this case, the fluorescence-guided fluorescent light was scattered at the end of the nanowire. The self-guided fluorescence which propagated to the distal tips can be clearly observed in both types of nanowire, as indicated within the white boxes in Figure 7.10c,d. However, the emission intensity at the distal end of the horizontal nanowire decreased rapidly when the excitation spot was moved further away, which indicated a significant optical loss for propagating fluorescence. In contrast, the tip emission of the vertical wire exhibited a much more modest intensity variation, indicating a much lower loss of the waveguide modes. The tip emission spectra corresponding to the isolated DAAQ nanowire on glass and the free-standing DAAQ nanowire, are shown in Figure 7.10e,f. Here, the inset shows that the peak intensity of outcoupled emissions decays exponentially with the increase in propagation distance. The intensities were normalized against the emissions measured by the nearest excitation. The attenuation constant of the horizontal wire on glass (calculated from the decay curves) was $4.01 \times 10^3 cm^{-1}$, while that of the free-standing wire was $1.74 \times 10^3 cm^{-1}$. The higher optical loss of the horizontal wires on glass was due to energy leakage through the underlying substrates; this loss could be significantly minimized because of the vertical structures, which had no contact with the substrates. Hence, vertically grown nanowires were proven to serve as an alternative type of structure that can efficiently lower the optical loss by minimizing any substrate effects.

It is well known that the efficiency of electroluminescent organic light-emitting devices can be further improved by the introduction of phosphorescent dyes. The fluorescent emission can be induced only by singlet spin states, which represent only 25% of the total excited-state population (the remainder are triplet states). Phosphorescent dyes, however, offer a means of achieving improved light-emission efficiencies to 100% (in theory), as emission may result from both triplet states via phosphorescence and singlet states via intersystem crossing from singlets to triplets. Recently, the present author's group has fabricated well-defined microrods and nanowires from phosphorescent molecules fac-tris(2-phenylpyridine) iridium ($Ir(ppy)_3$) by using a liquid-phase self-assembly method [20]. In Figure 7.11a,b, it is clearly noted that as-prepared single-crystalline microrods behave as microscale active phosphorescence waveguides due to their relatively large width, whereas $Ir(ppy)_3$ nanowires show a red-shifted yellow emission and exhibit no bright spots at their tips. The differences in the emission spectra of $Ir(ppy)_3$ in PMMA film, of a single microrod, and of a single nanowire, can be observed more concretely, as shown in Figure 7.11c. Besides the difference in waveguiding and emission spectra, either the fast-decay or slow-decay component of PL lifetime in $Ir(ppy)_3$ nanowires is much faster than that in microrods, indicating a quenching of the emissive state, as shown in Figure 7.11d. Such phosphorescence waveguides may be used as building blocks for micro-scale photonic devices. However, no optically pumped lasing is observed because of intense triplet–triplet exciton annihilation in $Ir(ppy)_3$ microrod crystals. These results provide useful information for the design and fabrication of micro-scale photonic devices from organometallic molecular materials and phosphorescent materials.

Based on the above discussion, it can be seen that monocomponent waveguides based on organic micro/nanocrystals have been achieved and show a variety of optical properties. In contrast, more tunable and reliable optical properties are expected in binary systems fabricated by simple doping. In the past, the doping technique has been frequently used in OLEDs, both to tune their emission color and to improve their luminescence efficiency. Many of these doping studies have used amorphous thin films, in which charge carriers have a low mobility. Crystalline organic micro/nanomaterials may not only represent unusual optical properties induced by doping contents, but also should offer better transport properties because of their ordered crystalline natures, and hence better optoelectronic performance. Although being limited in certain matrix compounds and doping contents, binary micro/nanosized waveguide devices are believed to be more useful

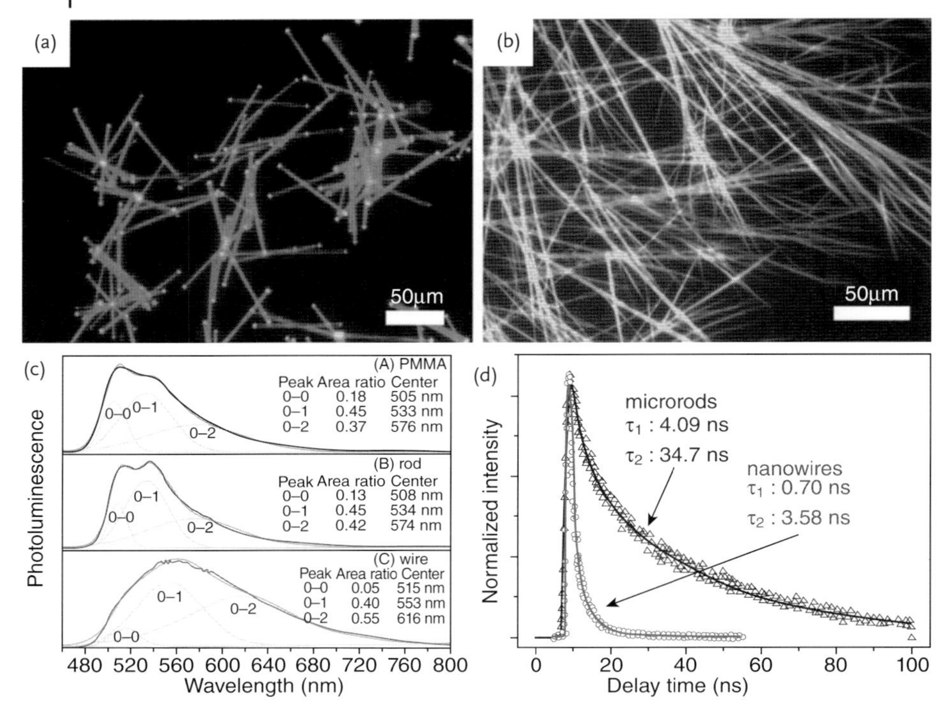

Figure 7.11 (a, b) PL microscopy images of $Ir(ppy)_3$ (a) microrods and (b) nanowires; (c) Normalized PL spectra of $Ir(ppy)_3$ in PMMA film, a single microrod and a single nanowire, excited by a continuous He–Cd laser (442 nm). Both measured and calculated (dashed line) PL spectra are shown. The sum spectrum was obtained by a summation of the calculated zero (0–0), one (0–1), and two (0–2) phonon bands; (d) The PL signal decay curves for $Ir(ppy)_3$ microrods and nanowires.

in light guiding, localization, and enhancement within wavelength volumes, which is essential for the development of micro/nanoscale photonic devices. Thus, ever-increasing attention has been paid to micro/nanostructures of multicomponent semiconductors, polymers, and functional organic molecules for their potential applications.

Optical micro/nanomaterials have been applied in several device elements in optical circuits, including LEDs, lasers, and photo-detectors. Their inherent one-dimensionality, coupled to an ability to operate above and below the diffraction limit, makes them effective building blocks to generate and propagate light in future miniaturized photonics. In order for miniaturized photonics to reach their full potential, active photonic devices such as waveguide modulators for selectively guiding photons at different wavelengths will be required as functional elements.

Following the successful fabrication of crystalline TPI nanostructures, perylene molecules as the dopant can be dispersed in the crystalline matrix of TPI. Although

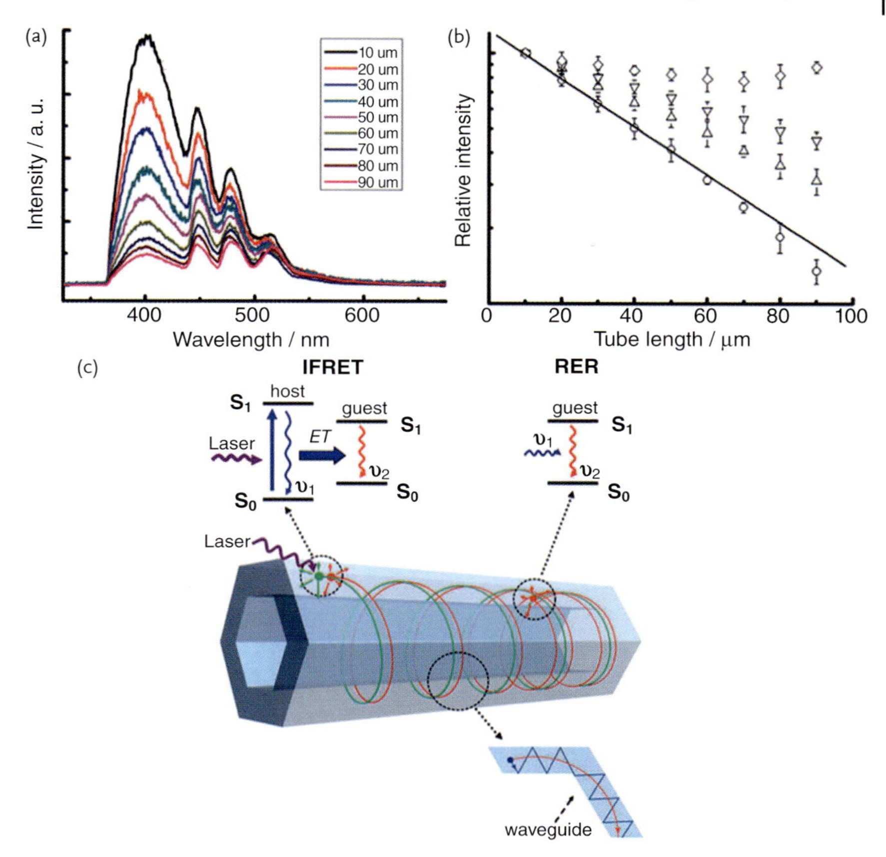

Figure 7.12 (a) Spatially resolved PL spectra of outcoupled light for a single binary microtube recorded by excitation at a distance of 10–90 mm from the tip; (b) Logarithmic plots of relative intensities of PL peaks at 402 (○), 445 (△), 478 (▽), and 515 nm (◇), respectively, for a doping microtube versus excitation position for the PL spectra. The corresponding fit curves are shown as black lines in (a). The excitation laser wavelength is 325 nm; (c) Schematic representation of the annular cavity of the binary microtube and energy diagrams for the IFRET and RER processes.

the perylene component has a doping ratio as low as 1.25%, it plays an important role as the energy acceptor in the intermolecular fluorescence resonance energy transfer (IFRET) from TPI to perylene [21]. Resulting from IFRET between TPI and perylene in a light-guiding process, both binary microtubes and rods serve as active waveguides and are capable of absorbing excitation energy and propagating the UV PL of TPI and the green PL of perylene. In the spatially resolved PL spectra shown in Figure 7.12a, the UV PL of TPI decays exponentially as a function of the propagation length. Interestingly, the PL of perylene at 445, 478, and 515 nm, when guided through the microtube, displays nonlinear losses; in particular, the overall optical loss for the PL at 515 nm was near to zero. The optical loss for the

PL at different wavelengths in microtubes are more clearly shown in Figure 7.12b, and is very different from that in the rods. As noted above, the air in tubular structures can alter the waveguide behavior and help to reduce the optical loss. Here, the annular cavity of the microtubes enables them to act as a waveguide modulator, in which wavelength redistribution occurs because of a remote energy relay (RER) process, as depicted in Figure 7.12c. Hence, an RER process has been proposed in which there is a reabsorption of the UV photon of the TPI matrix by the doping perylene component, and the subsequent re-emission of a green photon in the binary waveguide modulators. This waveguide modulator device revealed that small-content doping represents an efficient method for improving the performance of optical waveguide devices.

7.3.4
Lasing from Organic Nanowires

The nanoscale waveguides and laser sources are essential to the integrated nano-photonic systems, and are crucial for optical information processing. Inorganic semiconductor 1-D nanomaterials have been adopted as optical pumped lasers, and the microcavity effects and optically pumped lasing in nanowires based on conjugated polymers have also been studied [22]. In a recent study, it was confirmed that the nanowire waveguides and lasers can also be extended to small organic molecules [23]. The TPI nanowires described in Chapter 6 revealed obvious fluorescence narrowing spectra at room temperature. The single-crystalline wires can behave as nanoscale active optical waveguides, and optically pumped lasing was also observed from single nanowires with an increase in excitation power.

The PL microscopy image of the TPI nanowires (Figure 7.13b) shows that the wires reveal good waveguide properties. In this case, an isolated single TPI nanowire was excited with a pulsed laser using 325 nm wavelength light, and the emissions were collected from the tip area. The power-dependent PL spectra recorded from the tip of a wire with a width of 400 nm and a length of 15 μm is shown in Figure 7.13a. This indicates that PL modulations emerge on the lower energy side of the emission band (a point observed more clearly in the inset). The modes do not occur at the short wavelength side because of photon reabsorption. However, at low excitation power the structure displays a broad and featureless emission at about 375 nm. Above a certain threshold, an amplified spontaneous emission was observed, as indicated by the appearance of sharp cavity modes above an onset power. The energy dependence of the luminescent intensity and the emission spectra revealed a nonlinear gain and a threshold characteristic of the laser. Below the threshold, the PL intensity increased almost linearly with the excitation power; in fact, once the critical pumped energy had been reached, a kink emerged in the power dependence of emission intensity, which resulted in a large increased gain.

The transition from spontaneous PL to a stimulated emission is significantly influenced by the size of the nanowires. For thinner wires with a width of 200 nm, no emission was detected due to a lack of resonator property. However, if the wire length is too small – even if the width is sufficiently large – the stimulated emission

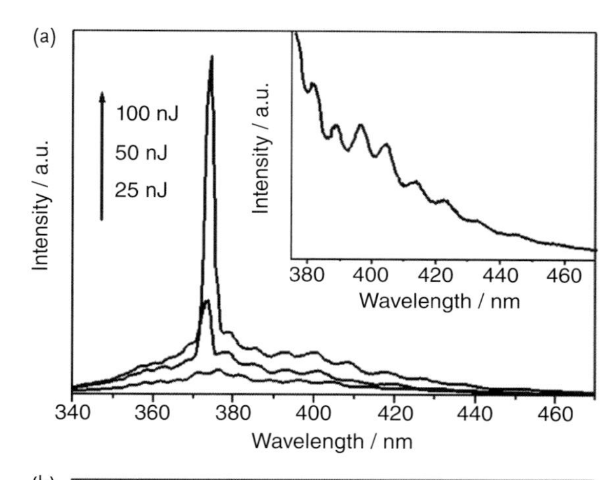

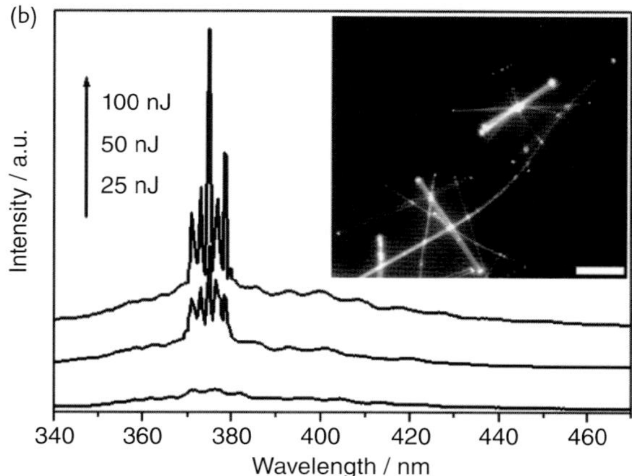

Figure 7.13 (a) Power-dependent emission spectra recorded on the tip of a TPI wire with 400 nm width and 15 μm length. The inset shows the expansion of the higher-wavelength region with resolved modes; (b) Power-dependent emission spectra recorded on the tip of a TPI wire with width of 800 nm and length of 15 μm. The inset shows the PL microscopy image of TPI nanowires deposited onto quartz wafers. Scale bar = 10 μm.

cannot be observed either, although the PL intensity is also enhanced when the pump energy is increased. This occurs because the high degree of spontaneous emission coupling in a small mode volume would result in a broad cavity mode. When both the length and the width of the wire are sufficiently large, multimode lasing can be observed, as shown in Figure 7.13b.

7.3.5 Organic Photonic Circuits

The cooperative assembly of multiple *p*-conjugated components has been recently introduced as an effective approach to the fabrication of photo- and/or electroactive nanostructures, in which adjustable electronic processes (such as electron transfer and energy transfer) between different species allow the realization of novel properties, such as p/n heterojunctions, ambipolar charge transport, and WLE. Recently, the group of Yao described a supramolecular synthesis of rectangular binary microtubes with precisely controlled lengths by the cooperative assembly of TPI and 3-(anthracen-10-yl)-1-phenylpro-2-en-1-one (APO, a green-emissive and photoactive molecule) [24]. The morphology characterization revealed that highly monodispersed 1-D structures with a rectangular cross-section were obtained. Moreover, the open-end features suggested that the structures were rectangular hollow tubes with a wall thickness of approximately 200 nm. Considering the good spectral overlap between the TPI emission and the APO absorption, the APO PL observed for binary tubes upon excitation at 320 nm might be a result of FET from excited TPI to APO molecules. Indeed, the decay of TPI PL in binary tubes (109 ps) is faster than that in pure TPI microrods (301 ps), and results in an FET efficiency of 0.64. As shown in the PL micrograph (Figure 7.14a, inset), these binary microtubes present the typical features of an active waveguide, such as bright PL spots at the tips and a weaker emission from the bodies. Propagation loss measurements were performed based on the spatially resolved PL spectra of a single tube by collecting the PL signal from one of its ends with respect to the distance traveled (Figure 7.14a,b). The chalcone derivatives such as APO are photochemically active, and might undergo either photoisomerization or photodimerization. The green fluorescence of an ensemble of APO microrods placed on a quartz plate can be quenched by exposure to light (UV400 nm) from a mercury lamp equipped with a band-pass filter (central wavelength: 400 nm, FWHM: 40 nm), most likely because of the photoisomerization reaction. The reversible recovery of green fluorescence could easily be achieved by keeping the irradiated sample in the dark. A schematic of a waveguide switch based on binary microtubes is shown in Figure 7.14c. In this case, the photoisomerization of APO components is induced locally by focusing 400 nm light from the mercury lamp down to the diffraction limit (2 mm) on a selected position on the tube, as indicated by the dashed circle in Figure 7.14c. According to the spatially resolved spectra at the irradiation point, the PL of TPI will remain the same as that before irradiation, while the PL from APO is indeed quenched. Next, the effect of UV400 nm irradiation was checked at the circled region on the waveguide behavior by excitation at the region indicated by a square by using focused 325 nm laser light (Figure 7.14c). The outcoupled spectra collected at the left tip are shown in Figure 7.14d, where the intensity of TPI PL after UV400 nm irradiation is about one-third of that before irradiation. Moreover, the PL of APO outcoupled at the left tip disappeared completely after UV400 nm irradiation at the circled region. The inset of Figure 7.14d shows the CIE coordinates calculated from the outcoupled PL spectra at the left tip before (top curve) and after UV400 nm irradiation (bottom curve). In sharp contrast, the outcoupled spectra at the right tip showed no changes in both TPI and APO features; that is, a selective switching of the waveguide to the left tip had been successfully realized by using the binary tube. Remarkably, the APO PL outcoupled at the left tip could be turned on again by keeping the irradiated sample in the dark. This switch had

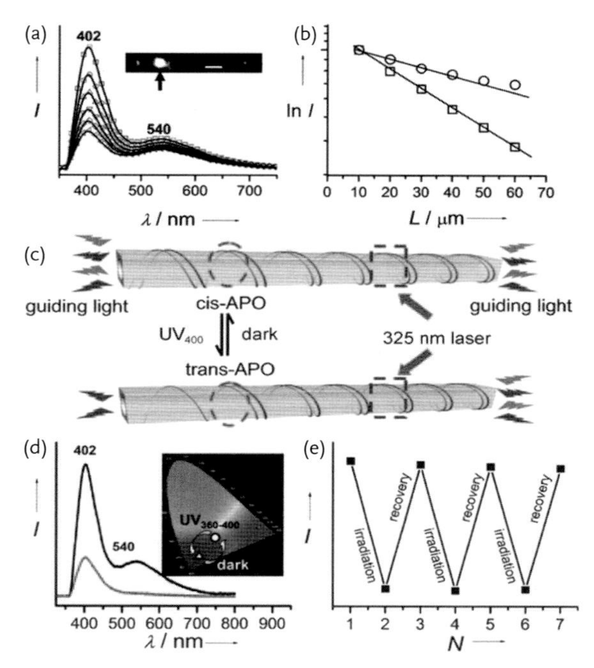

Figure 7.14 (a) Spatially resolved PL spectra of outcoupled light at a distance of 10, 20, 30, 40, 50, and 60 mm (curves from top to bottom) from the tip of a single tube, recorded by focused 325 nm laser excitation (see arrow in the inset). Scale bar = 10 μm; (b) Logarithmic plot of the relative PL intensities at 402 nm (□) and 540 nm (○) versus the propagation distance; (c) Representation of waveguide switch before (top) and after (bottom) irradiation of the region marked with a dashed circle by UV400 light (FWHM: 40 nm). The dashed squares show the excitation point of focused 325 nm laser light for waveguide measurement; (d) Outcoupled PL spectra from the left tip before (black) and after (red) UV400 nm irradiation at the region marked with a circle, upon excitation at 325 nm by focusing the laser on the region marked with a square; (e) The switch cycles (N = number of cycles) of APO PL outcoupled at the left tip upon UV400 nm irradiation at the circled region, and subsequent recovery in the dark.

an on/off ratio of approximately 20, and could be reversibly switched over at least seven cycles, without any signs of fatigue (see Figure 7.14e).

7.4 Outlook

This chapter has provided an introduction to organic photonic devices, mainly with regards to organic waveguides. It is clear that all features and applications of organic materials in photonic devices are attributable to both the molecular structure and crystal morphology. The molecular structures of organic dyes have a major impact not only on the optical properties but also on aggregation by intermolecular interactions. In typical organic dye molecules, some conjugated structures contribute to the luminescent or chromogenic phenomena, some radicals can help to adjust self-assembly behaviors, while others can modify physical properties such as melting point, crystallinity, and solubility. Based on these considerations, the design and synthesis of a new series of dye molecules is essential to the creation of organic nanocrystals for photonic use. On the basis of molecular structures, it is still possible – at least, to a certain extent – to control the shape and size of organic microcrystals for specific optical applications, mainly because numerous preparation methods have been successfully developed to break the limits of morphology in organic aggregations. Consequently, in the future more effort should be directed towards the synthesis and modification of molecular structures and crystal structures. In the next generation of photonic integrated circuits, waveguide materials will supposedly connect the light-emitting and light-detecting elements, whilst at the same time tuning the emission spectra in the guiding process. The fundamental functions of miniaturized waveguides are essentially the same as for electrical wires in electronic circuits. Moreover, the ability to tune the guided light in organic waveguides is amazing for practical applications, whether to serve as optically laser emitting/converters, as wavelength up/down-converters, and/or as homo-/polychrome emitters. As noted above, the organic molecular version of the micro/nano-optical waveguides could be expected to play important roles in miniaturized optoelectronic devices, along with their inorganic and polymer counterparts. Although currently the preparation methods and optical performances of organic molecular crystals are less developed than those of inorganic semiconductors and conjugated polymers, the vast number of optically active organic molecules and tunable optical properties from the correct doping in organic composites should be taken into consideration for future photonic applications. Investigations into these organic waveguide materials can not only provide a profound comprehension of the propagation behaviors of photons and excitons in the microcosmic view, but also offer opportunities for the design of new structures in photonic devices. Numerous breakthroughs based on the small organic molecules in the field of optical waveguides at the micro/nanoscale have been made during the past decade and, as a result, there is now a good understanding as to how these crystalline optical building blocks can be fabricated, and their optical properties modulated. This is of great interest when employing such photonic elements in applications that include biochemical sensors, light-emitting devices, photo-detectors, and nanowire lasers. Unfortunately, however, there is still a long way to go to promote these practical applications of the functional organic optical waveguides.

It is, nonetheless, possible to peer into the future with reasonable confidence, as there will continue to be increased numbers of fundamental new scientific discoveries related to this unique class of nanoscale building blocks. Clearly, new opportunities exist for a better understanding of the effects of size, shape, and dimensionality on the optical and electrical properties of micro/nanoscale waveguide materials. Moreover, binary organic systems for waveguides, such as doping or heterostructures, could lead to a combination of intermolecular interactions and energy transfers which, together with their optical properties, would render these systems more functional and available for integration. Additionally, 2-D nanomaterials for optical waveguides will continue to provide new areas of research for exploring novel optical waveguide behaviors in selected directions. Likewise, the patterns of existing 1-D nanostructures are important for specific uses, such as integrated multicolor generators. Highly sensitive biochemical sensors may be based on optical waveguiding, as the waveguide properties are heavily influenced by the dielectric constant of the environment and defects on the surface of materials. Clearly, the future of photonic technologies will be heavily dependent on a balance between issues of cost, stability, and the performance of nanowire-based devices or systems. Thus, optical waveguides fabricated by organic micro/nanocrystals, with their features of low-cost, easy preparation, and multi-functionality, might become a good choice.

This chapter has provided an overview of recent development in the synthesis, subsequent characterization, and optical properties of optical waveguides, and especially of those based on small organic molecules. Many significant attempts have been made to develop a series of multifunctional waveguide devices using either vapor-phase or liquid-phase methods, and for their application in future photonic circuits. Although integrated electronic circuits are ubiquitous in modern life, fundamental and technical constraints alike will eventually test the limits of Moore's law. However, photonic circuits, which involve the control of photons in free spaces or in light matters, may break these limits in terms of both size and performance. Micro/nanoscale photonic circuits constructed from myriad building blocks can offer increased opportunities for the development of next-generation optical information processors. Indeed, the fabrication of photonic integrations might represent one of the most disruptive and transformative technologies in the twenty-first century.

References

1 Zhao, Y.S., Wu, J., and Huang, J. (2009) Vertical organic nanowire arrays: controlled synthesis and chemical sensors. *J. Am. Chem. Soc.*, **131** (9), 3158–3159.

2 An, B.-K., Kwon, S.-K., and Park, S.Y. (2007) Photopatterned arrays of fluorescent organic nanoparticles. *Angew. Chem. Int. Ed.*, **46** (12), 1978–1982.

3 Zhao, Y.S., Fu, H., Peng, A., Ma, Y., Xiao, D., and Yao, J. (2008) Low-dimensional nanomaterials based on small organic molecules: preparation and optoelectronic properties. *Adv. Mater.*, **20** (15), 2859–2876.

4 Fardy, M. and Yang, P. (2008) Materials science: Lilliputian light sticks. *Nature*, **451** (7177), 408–409.

5 An, B.-K., Kwon, S.-K., Jung, S.-D., and Park, S.Y. (2002) Enhanced emission and its switching in fluorescent organic nanoparticles. *J. Am. Chem. Soc.*, **124** (48), 14410–14415.

6 Peng, A.D., Xiao, D.B., Ma, Y., Yang, W.S., and Yao, J.N. (2005) Tunable emission from doped 1,3,5-triphenyl-2-pyrazoline organic nanoparticles. *Adv. Mater.*, **17** (17), 2070–2073.

7 Zhao, Y.S., Yang, W., Xiao, D., Sheng, X., Yang, X., Shuai, Z., Luo, Y., and Yao, J. (2005) Single crystalline submicrotubes

from small organic molecules. *Chem. Mater.*, **17** (25), 6430–6435.

8 Fu, H., Xiao, D., Yao, J., and Yang, G. (2003) Nanofibers of 1,3-diphenyl-2-pyrazoline induced by cetyltrimethylammonium bromide micelles. *Angew. Chem. Int. Ed.*, **42** (25), 2883–2886.

9 Hu, J.-S., Guo, Y.-G., Liang, H., Wan, L.-J., and Jiang, L. (2005) Three-dimensional self- organization of supramolecular self-assembled porphyrin hollow hexagonal nanoprisms. *J. Am. Chem. Soc.*, **127** (48), 17090–17095.

10 Anderson, J.L., Pino, V., Hagberg, E.C., Sheares, V.V., and Armstrong, D.W. (2003) Surfactant solvation effects and micelle formation in ionic liquids. *Chem. Commun.*, **19**, 2444–2445.

11 Tian, Z., Chen, Y., Yang, W., Yao, J., Zhu, L., and Shuai, Z. (2004) Low-dimensional aggregates from stilbazolium-like dyes. *Angew. Chem. Int. Ed.*, **43** (31), 4060–4063.

12 Wang, Y.B., Fu, H.B., Peng, A.D., Zhao, Y.S., Ma, J.S., Ma, Y., and Yao, J.N. (2007) Distinct nanostructures from isomeric molecules of bis(iminopyrrole) benzenes: effects of molecular structures on nanostructural morphologies. *Chem. Commun.*, **16**, 1623–1625.

13 Zhao, Y.S., Xiao, D., Yang, W., Peng, A., and Yao, J. (2006) 2,4,5-triphenylimidazole nanowires with fluorescence narrowing spectra prepared through the adsorbent-assisted physical vapor deposition method. *Chem. Mater.*, **18** (9), 2302–2306.

14 Zhao, Y.S., Di, C., Yang, W., Yu, G., Liu, Y., and Yao, J. (2006) Photoluminescence and electroluminescence from tris(8-hydroxyquinoline)aluminum nanowires prepared by adsorbent-assisted physical vapor deposition. *Adv. Funct. Mater.*, **16** (15), 1985–1991.

15 Shen, F., Peng, A., Chen, Y., Dong, Y., Jiang, Z., Wang, Y., Fu, H., and Yao, J. (2008) Photoinduced electron transfer in coaggregates of dicyanonaphthalene and pyrazoline. *J. Phys. Chem. A*, **112** (11), 2206–2210.

16 Zhao, Y.S., Fu, H.B., Hu, F.Q., Peng, A.D., Yang, W.S., and Yao, J.N. (2008) Tunable emission from binary organic one-dimensional nanomaterials: an alternative approach to white-light emission. *Adv. Mater.*, **20** (1), 79–83.

17 Zhao, Y.S., Xu, J., Peng, A., Fu, H., Ma, Y., Jiang, L., and Yao, J. (2008) Optical waveguide based on crystalline organic microtubes and microrods. *Angew. Chem. Int. Ed.*, **120** (38), 7411–7415.

18 Chung, J.W., An, B.-K., Kim, J.W., Kim, J.-J., and Park, S.Y. (2008) Self-assembled perpendicular growth of organic nanoneedles via simple vapor-phase deposition: one-step fabrication of a superhydrophobic surface. *Chem. Commun.*, **26**, 2998–3000.

19 Zhao, Y.S., Zhan, P., Kim, J., Sun, C., and Huang, J. (2010) Patterned growth of vertically aligned organic nanowire waveguide arrays. *ACS Nano*, **4** (3), 1630–1636.

20 Wang, H., Liao, Q., Fu, H., Zeng, Y., Jiang, Z., Ma, J., and Yao, J. (2009) $Ir(ppy)_3$ phosphorescent microrods and nanowires: promising micro-phosphors. *J. Mater. Chem.*, **19** (1), 89–96.

21 Liao, Q., Fu, H., and Yao, J. (2009) Waveguide modulator by energy remote relay from binary organic crystalline microtubes. *Adv. Mater.*, **21** (41), 4153–4157.

22 O'Carroll, D., Lieberwirth, I., and Redmond, G. (2007) Microcavity effects and optically pumped lasing in single conjugated polymer nanowires. *Nat. Nanotechnol.*, **2** (3), 180–184.

23 Zhao, Y.S., Peng, A., Fu, H., Ma, Y., and Yao, J. (2008) Nanowire waveguides and ultraviolet lasers based on small organic molecules. *Adv. Mater.*, **20** (9), 1661–1665.

24 Liao, Q., Fu, H., Wang, C., and Yao, J. (2011) Inside cover: cooperative assembly of binary molecular components into tubular structures for multiple photonic applications. *Angew. Chem. Int. Ed.*, **50** (21), 4716–4716.

8
Organic Solar Cells Based on Small Molecules

Yuze Lin, Xiaowei Zhan

8.1
Introduction

8.1.1
Solar Energy and Solar Cells

Nowadays, the production and use of fossil fuels such as coal, oil, and gas not only gives rise to a mass of environmental problems, but the supplies of these materials are also rapidly diminishing. Clearly, the need to develop renewable energy sources is becoming increasingly urgent, with research groups now focusing their attention on the creation of renewable energies generated from natural resources such as sunlight, wind, rain, tides, and geothermal heat. Among these resources, the sun has the potential to make the largest energy contribution: a 1-h period of sunshine provides a total power of 3.8×10^{23} kW, which is far more than the highest human demands for energy in one year (ca. 1×10^{10} kW). Moreover, solar energy is a clean and renewable energy resource that offers an ideal alternative to "traditional" energy resources.

The development of the photovoltaic (PV) cell, which transforms inexhaustible solar energy into electricity, represents one of the most promising long-term solutions for providing clean, renewable energy. The inorganic semiconductor materials used to produce PV cells include: crystalline, multicrystalline, and amorphous Si; the Group III–V compounds and alloys (such as GaAs and CdTe); and chalcopyrite compounds such as copper indium gallium diselenide (CIGS). Unfortunately, with most currently available commercial solar cells being based on crystalline and multicrystalline Si, the main barrier that prevents PV cell technology from providing a large fraction of today's energy requirements is the high cost of silicon-based PV cells.

Organic photovoltaic (OPV) cells not only represent a promising and cost-effective alternative to silicon-based solar cells, but they also possess clear benefits over solar cells of low cost, light weight, and flexibility. Contemporary OPV cells are based on a heterojunction that results from the contact between electron donor (D) and electron acceptor (A) materials. Within the OPV cell, the absorption of

Organic Optoelectronics, First Edition. Edited by Wenping Hu.

solar photons creates excitons; these diffuse to the D/A interface, where they are dissociated into free holes and electrons. Opposite-polarity carriers (holes and electrons) are then transported in the D and A channels to anodes and cathodes, respectively such that, when the charges are collected at the electrodes, this results in the generation of electrical power. D/A heterojunctions can be created with two main types of architecture: (i) bilayer heterojunctions [1]; and (ii) bulk heterojunctions (BHJs) [2].

In OPV devices, the principal figures-of-merit include the power conversion efficiency (PCE), the short-circuit current density (J_{SC}), the open-circuit voltage (V_{OC}), and the fill factor (FF). These define, respectively: the ratio between the output device electrical energy versus the input solar energy; the device current density when no reverse bias is applied; the device voltage when no current flows through the cell; and the ratio between maximum power of the device and $J_{SC} \times V_{OC}$.

8.1.2
Materials Features for Solar Cells

The photoactive materials, including polymeric and small-molecule semiconductors, play a key role in influencing the physical processes involved in energy conversion, which in turn determine the electrical characteristics of the solar cell, such as J_{SC}, V_{OC}, and FF and, ultimately, the PCE. The basic requirements of specific intrinsic properties necessary for an ideal small molecular donor, acceptor or donor–acceptor dyad materials include:

- A low optical bandgap for a broad absorption range matching with the solar spectrum and a high extinction coefficient for harvesting more solar energy.
- Long exciton diffusion lengths for an effective migration of excitons to the D/A interface.
- High hole or electron mobilities for efficient charge transport, which in turn allows the thicker active layer that is required for increased light harvesting, as well as reduced charge recombination and series resistance.
- Suitable HOMO/LUMO energy levels to ensure a large V_{OC} and a downhill energy offset for exciton dissociation.
- Excellent thermal stability for vacuum deposition, or sufficient solubility to guarantee solution processability.

Recently, OPV devices based on a *p*-type conjugated polymer as a donor and fullerene derivatives as an acceptor have undergone a rapid development, with the highest reported PCEs of this type of OPV device to date being up to 8.3% [3], though this is below the value of 10% that often is considered a prerequisite for large-scale commercial applications. On the other hand, small-molecule semiconductors for OPV devices have attracted increasing attention, due to their advantages over their polymer counterparts, which include well-defined molecular structures, definite molecular weights, and high purities without batch-to-batch variations [4]. Recently, an increasing number of reports have been made relating to OPV devices based on small molecules and fabricated by either vacuum deposition or solution processing, with both approaches leading to PCEs in excess of 5% [5–9]. Indeed, when comparing small-molecule acceptors to fullerene derivatives such as [6]-phenyl-C_{61}-butyric acid methyl ester ($PC_{61}BM$), the development of nonfullerene small-molecule acceptors has resulted in only a relatively poor performance. Recently, however, investigations into nonfullerene acceptors has become more active due to the more interesting potential of these materials, notably their convenient synthesis, low cost, easy tunability of energy levels and, perhaps most importantly, a better absorption in the visible spectrum.

The synthesis and application of conjugated polymeric active materials [10–25], small-molecule donors [26–32], fullerene acceptors [33–35], and small-molecule nonfullerene acceptors [36, 37] in both OPVs and device physics [38–46] have been summarized in several reviews. In this chapter, attention is focused on the representative small-molecule donors, acceptors and donor–acceptor dyads in OPV cells.

8.1.3 Device Configurations of Solar Cells

8.1.3.1 Hamburger Structure

In 1986, when Tang fabricated a bilayer heterojunction with an efficiency that approached 1%, this was considered a milestone in the development of OPV cells [1]. Although bilayer heterojunctions have undergone intensive investigation, and remain an invaluable tool for the evaluation of new active materials, their performances are somewhat limited by the short exciton diffusion length in organic materials (typically 5–20 nm) [47]. As the exciton dissociation process is confined to the D/A interfacial zone, only those excitons produced at a distance shorter than their diffusion length have a good probability of reaching the interfacial zone and generating free charge carriers. Consequently, the exciton diffusion length will limit the maximum thickness of the active layer and, in turn, the maximum fraction of the incident light that the cell can absorb and convert into electricity. In order to improve the PCEs of bilayer heterojunction solar cells, a new structure termed the planar-mixed heterojunction (PMHJ) was created, in which a mixed layer consisting of donor and acceptor was sandwiched between homogeneous donor and acceptor layers.

In 1992, Sariciftci *et al.* [48] demonstrated that the photoexcitation of a mixture of a conjugated polymer and fullerene (C_{60}) resulted in an ultrafast, highly efficient photoinduced electron transfer. Subsequently, Yu *et al.* [2] and Halls *et al.* [49] pioneered the concept of the BHJ, which is a blend of bicontinuous and interpenetrating donor and acceptor components in a bulk volume. Such a nanoscale network exhibits a D/A phase separation on a 5–20 nm length scale, which is within a distance close to the exciton diffusion length. When compared to a bilayer heterojunction, the BHJ significantly increases the D/A interfacial area, leading to an enhanced efficiency of the OPV devices [50].

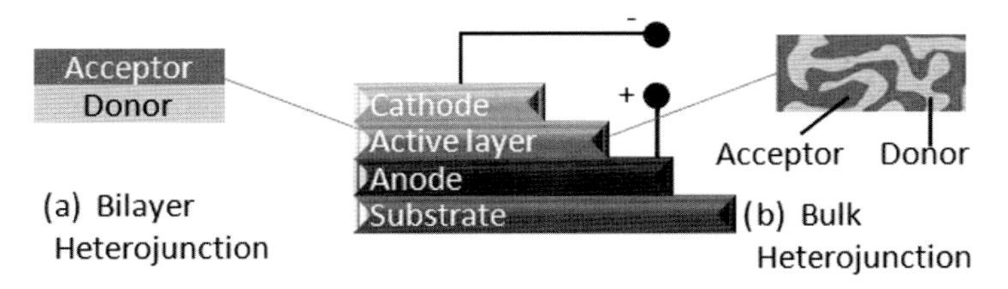

Figure 8.1 The architecture structure of (a) bilayer heterojunction and (b) bulk heterojunction (BHJ) OPV devices.

The arrangements of the bilayer heterojunction and BHJ OPV device structures are shown in Figure 8.1. In the two devices, the photoactive layers are both sandwiched between a high-work-function anode, typically a transparent indium tin oxide (ITO) layer, and a relatively low-work-function metal cathode, such as Ca or Al. In the bilayer heterojunction device, the donor materials adhere to the anode and the acceptor materials to the cathode, while in BHJ device the active layer is a blend of donor and acceptor materials. In principle, two processing techniques are available for the fabrication of OPV devices, namely vacuum deposition and solution processing. In general, the bilayer heterojunction was fabricated by vacuum deposition, because it is difficult to identify suitable solvents for the donor and acceptor layers without destroying the D/A interface. Both processing techniques are suitable for the BHJ devices, however. Some small molecules such as metal phthalocyanine and C_{60} can be deposited under high-vacuum conditions by thermal evaporation, after which BHJ layers can be obtained by coevaporation of the donor and acceptor materials. On the other hand, soluble materials can be deposited from solution, by spin-coating, inkjet printing, gravure, or flexographic printing.

8.1.3.2 **Tandem Structure**

The use of doped, wide-gap charge transport layers with a high conductivity and a low absorption in the visible range enables a high internal quantum efficiency to be achieved, and for devices to be optimized with respect to optical interference effects by adjusting the film thickness. This inspired the concept of stacking several cells on top of each other to form "tandem structures," which enables two limiting factors that exist intrinsically among organic semiconductor molecules, and have a poor charge carrier mobility and a narrow light absorption range, to be resolved. By using method, two or even more organic solar cells can be stacked together, and the components of each device then tailored to exhibit complementary absorption spectra.

8.2
Small-Molecule Donors

Among the many small-molecule *p*-type semiconductors that have been studied for decades [51], only a small fraction has been applied successfully as donors in

OPV devices due to the various optical, electrical, and stability requirements demanded of the chosen materials. The properties of these materials, such as hole mobility, exciton diffusion length, thin-film morphology, frontier energy level alignment, bandgap, and absorption coefficient, all have significant effects on the performance of OPV devices. Some representative small-molecule donors, such as dyes, oligothiophenes, and triphenylamine-based molecules used in the active layer of OPV devices, are described and discussed in the following subsections.

8.2.1 Dyes

Several well-known classes of dyes are potential materials in OPV cells, including phthalocyanine (Pc), subphthalocyanine (SubPc), merocyanine (MC), squaraine (SQ), and diketopyrrolopyrrole (DPP). The functionalization of a dye molecule has confirmed to be a successful approach to donor design. A summary of the electronic properties and OPV data for representative dye-based donors is provided in Table 8.1 (see also Figure 8.2).

Phthalocyanine, which comprises four isoindole units connected by 1,3-aza linkages, is a planar and highly aromatic 18-π-electron macrocycle. Pc derivatives typically exhibit excellent thermal and chemical stability [72], and also offer flexibility in their optical and electronic properties through synthetic modifications, such as attaching functional groups to the molecular skeleton. The optoelectronic properties and stacking in the solid state of Pc derivatives can be tuned by replacing of the two protons in the molecular cavity with a metal such as Cu or Zn. Although many metal–Pc complexes have been used in OPV cells, CuPc (**a1**) and ZnPc (**a2**) have been the most common choices to date for applications in Pc-based OPV devices, due to their longer exciton diffusion length as compared to the other Pcs [73].

In 1986, Tang used CuPc as a donor material in a bilayer heterojunction OPV, which showed a PCE value of 1% [1]. Inspired by these pioneering studies, Pc has been used widely in vacuum-deposited OPV devices in combination with fullerene acceptors, due to the high absorption coefficient and long exciton diffusion length [52, 53, 74–76]. The CuPc/C_{60} combination is among the most common for small-molecule OPV active layers, and devices deriving from this system have exhibited a high performance [5]. Bilayer heterojunction or BHJ (via vacuum codeposition) OPVs based on CuPc/C_{60} exhibited PCEs of 3.6% or 3.5% under 150 or 100 mW cm^{-2} simulated AM1.5G illumination, respectively [52, 53]. Planar-mixed heterojunction (PMHJ) devices based on CuPc and C_{60} afforded a maximum PCE of 5.0% at 120 mW cm^{-2} under simulated AM1.5G solar illumination [5]. Furthermore, stacking these two cells together in a tandem cell resulted in an approximate 15% increase in PCE (5.7% at 100 mW cm^{-2} AM1.5G simulated solar illumination) [76].

Planar Pcs have a relatively narrow absorption band and cannot effectively absorb low-energy photons. Nonplanar Pc molecules, such as chloroaluminum phthalocyanine (AlClPc, **a3**), exhibited an absorption peak around 755 nm (125 nm red shift relative to CuPc) and harvested a larger percentage of infrared photons [54]. Additionally, ultraviolet photoelectron spectroscopy (UPS) measurements

Table 8.1 Optical and electronic properties, mobilities, and OPV performance of dye-based donors.

Compound	λ_{max} [a] (nm)	E_g^{opt} (eV)	μ_h [cm^2/(V·s)]	HOMO/LUMO(eV)	Active layer [b]	J_{SC} (mA cm^{-2})	V_{OC} (V)	FF	PCE (%)	Reference
a1					a1/d1	2.3	0.45	0.65	0.95 [c]	[1]
a1					a1/C_{60}	18.8	0.58	0.52	3.6 [d]	[52]
a1					a1:C_{60} (1:1 vac)	15.4	0.50	0.46	3.5	[53]
a1					a1/a1:C_{60} (1:1 vac)/C_{60}	15.0	0.54	0.61	5.0 [e]	[5]
a3	755			−5.4/-	a3/C_{60}	–	0.68	0.50	2.1 [f]	[54]
a4	850			−5.2/-	a4/C_{60}	15.1	0.57	0.53	4.2	[55]
a5	590			−5.6/−3.6	a5/C_{60}	3.36	0.97	0.57	2.1	[56]
a6	688		2×10^{-5}	−5.4/−3.6	a6(sol)/C_{60}	5.6	0.55	0.49	1.5	[57]
a6		1.7			a6(vac)/C_{60}	6.1	0.79	0.49	2.5	[58]
a7	607		1×10^{-5}	−5.80/−3.76	a7:$PC_{61}BM$ (1:3)	5.3	0.90	0.32	1.54	[59]
a8	649			−5.59/−3.68	a8:$PC_{61}BM$ (3:7)	6.3	0.76	0.36	1.74	[59]
a9	616		5×10^{-5}	−5.75/−3.59	a9:$PC_{61}BM$ (9:11)	8.24	0.94	0.34	2.59	[60]
					a9:C_{60} (1:1 vac)	11.5	0.80	0.48	4.9 [g]	[61]
					a9:C_{60} (9:11 vac)	12.6	0.96	0.47	6.1	[7]
a10	595		0.7×10^{-5}	−5.69/−3.54	a10:$PC_{71}BM$ (9:11)	10.2	1.0	0.44	4.5	[62]
a11	760		10^{-5}–10^{-4}	−5.0/−3.3	a11:$PC_{61}BM$ (1:3)	5.70	0.62	0.35	1.24	[63]
a12	770		1.2×10^{-4}	−5.0/−3.3	a12:$PC_{71}BM$ (1:3)	9.32	0.57	0.37	1.99	[64]
a13	700			5.3/−3.4	a13(sol)/C_{60}	9.71	0.78	0.54	4.1	[65]
					a13:$PC_{71}BM$ (1:6)	12.0	0.92	0.5	5.5	[8]
a14	710			−5.3/−3.7	a14(sol)/C_{60}	10.0	0.90	0.64	5.7	[66]
a15	742	2.01	5×10^{-7}	−5.03/−3.0	a15:$PC_{61}BM$ (7:3)	8.42	0.67	0.45	2.33	[67]
a16	720		1.0×10^{-4}	−5.2/−3.7	a16:$PC_{71}BM$ (1:1)	9.2	0.75	0.44	3.0	[68]
a17	660	1.7	3×10^{-5}	−5.2/−3.4	a17:$PC_{71}BM$ (3:2)	10	0.92	0.48	4.4	[69]
a18			2.5×10^{-3}		a18:$PC_{71}BM$ (2:1)	8.3	0.76	0.58	3.7	[70]
a19	676	1.72	7.18×10^{-3}	−5.40/−3.68	a19:$PC_{61}BM$ (3:2)	11.27	0.84	0.42	4.06	[71]

a) In film.
b) Donor/acceptor: bilayer by vacuum deposition unless stated otherwise; donor:acceptor: blend by solution process unless stated otherwise; vac: vacuum deposition; sol: solution process.
c) AM2, 75 mW cm^{-2}.
d) 150 mW cm^{-2}.
e) 120 mW cm^{-2}.
f) 119 mW cm^{-2}.
g) 88 mW cm^{-2}.

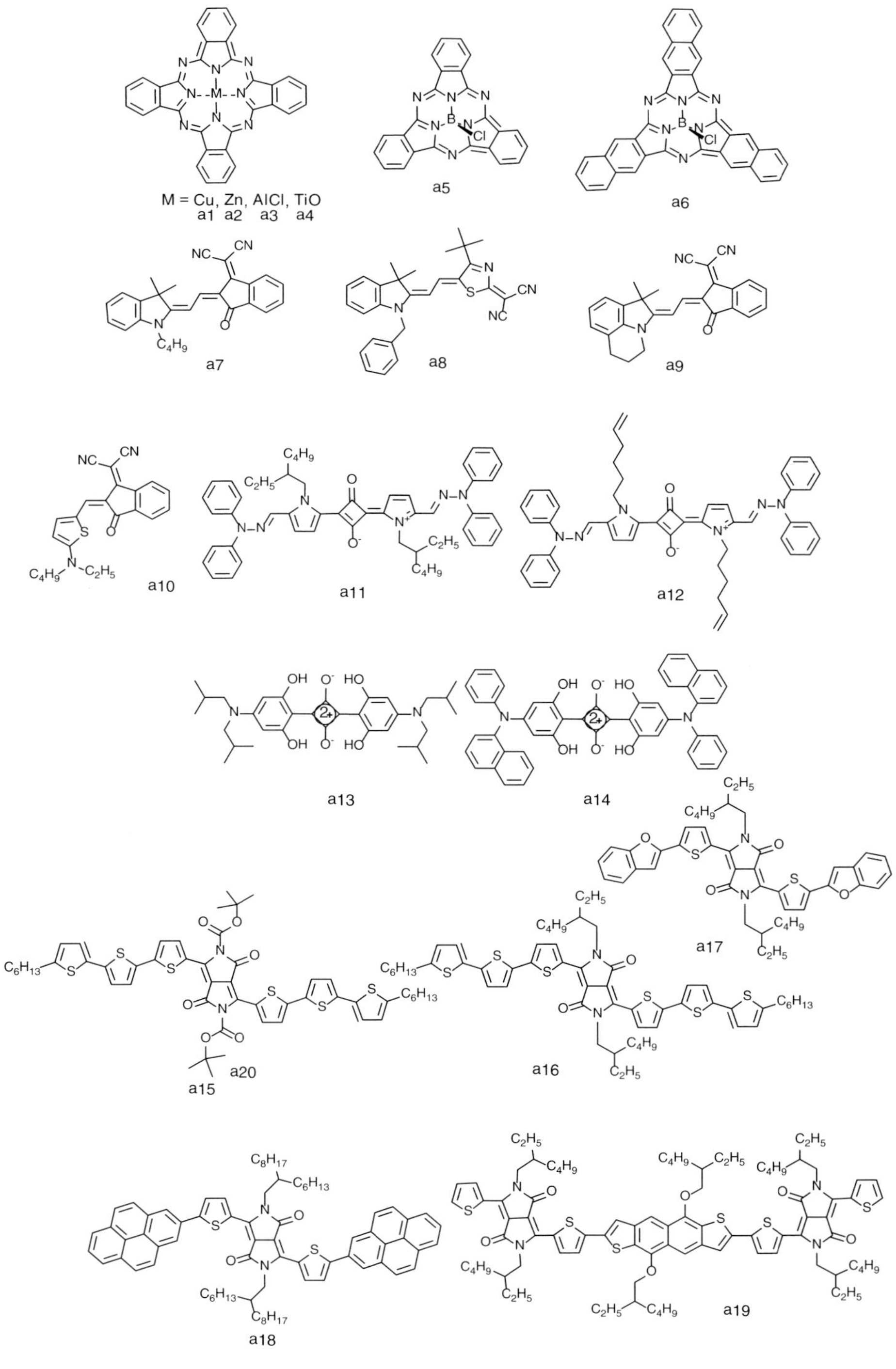

M = Cu, Zn, AlCl, TiO
a1 a2 a3 a4

a5

a6

a7

a8

a9

a10

a11

a12

a13

a14

a17

a20

a15

a16

a18

a19

Figure 8.2 Chemical structures of dye-based donors.

indicated that the highest occupied molecular orbital (HOMO) energy of **a3** was ca. 0.1 eV lower than that of **a1**, which could increase the V_{OC}. The optimized **a3**/C_{60} devices exhibited an improvement in PCE, from 1.8% for **a1**/C_{60} devices to 2.1% [54].

Oxo-titanium phthalocyanine (**a4**) has also been used in OPV devices as a donor material, leading to both a higher absorbance at long wavelengths and an increased V_{OC} in bilayer **a4**/C_{60} OPV devices, as compared to **a1** [55, 77]. In 2009, Armstrong and coworkers reported that polymorph **a4** film changed to a crystalline phase that could absorb light at the longer wavelengths by exposure of the as-deposited thin film to concentrated solvent vapors (solvent annealing). The optimized OPVs based on **a4**/C_{60} showed a relatively high PCE of 4.2% [55].

The OPVs based on Pcs showed a relatively low V_{OC} (generally < 0.6 V), which limits the PCEs of the OPV devices. The V_{OC} is generally related to the energy difference between the lowest unoccupied molecular orbital energy (LUMO) of the acceptor and the HOMO energy of the donor [78]. Thus, one way to increase the V_{OC} is to down-shift the HOMO level of donor materials. In 2006, As a result of a deeper HOMO level of subphthalocyanine (**a5**), an **a5**/C_{60}-based bilayer heterojunction OPV exhibited a much higher V_{OC} (0.97 V) as compared to CuPc/C_{60} (0.42 V), and this resulted in an enhancement of the PCE, from 0.9% to 2.1% [56]. Gommans *et al.* reported a higher J_{SC} (5.4 mA cm^{-2}) and a higher PCE (3.0%) from this device architecture [79].

Subnaphthalocyanine (SubNc, **a6**) has a good solubility, a low tendency to aggregate, and a strong light absorption in the visible region; amorphous films with good charge-transporting and light-harvesting properties can be prepared via simple solution-casting [57]. The bilayer device based on solution-processed **a6**/vacuum-deposited C_{60} demonstrated a PCE of 1.5%, with a J_{SC} of 5.6 mA cm^{-2}, V_{OC} of 0.55 V, and FF of 0.49 after thermal annealing at 120 °C for 40 min [57]. In contrast, bilayer OPV devices based on vacuum-deposited **a6**/C_{60} produced a PCE of 2.5%, with a J_{SC} of 6.1 mA cm^{-2}, V_{OC} of 0.79 V, and FF of 0.49 [58]. The reduced V_{OC}, as compared to that of **a5**-based devices, resulted from a greater conjugation imparted by the additional benzene rings in **a6**, which raised the HOMO energy level [80]. However, the benzene rings were also responsible for the shift in absorption to longer wavelengths and a resultant improvement of J_{SC}, compared to **a5**-based OPV devices.

Merocyanine dye-based molecules offer high absorption coefficients (usually ~10^5 M^{-1} cm^{-1}) and a large variability in the position of the HOMO and LUMO levels [59, 60]. In 2008, Würthner, Meerholz and coworkers used MC dyes in solution-processed BHJ OPV devices; the PCEs obtained were up to 1.54% or 1.74% for blends of MC dyes (**a7** or **a8**):$PC_{61}BM$ [59]. Later, the same group modified the push–pull dye **a7** with a flexible alkyl chain to achieve more efficient photovoltaic molecules. This was achieved by bridging the electron-donating unit with a propylene group in **a9** to diminish the flexibility of the structure, and to ensure a more planar geometry. Thus, **a9** exhibited a relatively high hole mobility of 5×10^{-5} cm^2/(V·s), which was fivefold that for **a7**. Solution-processed devices based on an **a9**:$PC_{61}BM$ blend showed a J_{SC} of 8.24 mA cm^{-2}, which was larger than that of an **a7**:$PC_{61}BM$ (5.30 mA cm^{-2}) and resulted in a remarkably improved PCE of 2.59% [60]. Vacuum-deposited BHJ OPV devices based on **a9**:C_{60} showed a higher PCE of 4.9% [61]. An even higher PCE of 6.1% was achieved by introducing MoO_3 instead of PEDOT:PSS as the hole-collecting contact; this was sandwiched between the active layer and anode after solvent annealing in CH_2Cl_2 for 10 min [7]. A new dye **(a10)** was also developed for application in solution-processed OPV devices, with the optimized unit showing a PCE of 4.5% with a V_{OC} of 1.0 V [62].

The squaraine (SQ) dye-based molecules showed broad absorption that ranged from 500 to 900 nm in the film, as well as high absorption coefficients (~10^5 M^{-1} cm^{-1}), good photochemical, and thermal stability, all of which were employed in OPV devices [8, 63–66, 81]. In 2008, Marks and coworkers reported that a series of SQ-based molecules exhibited a promising performance in solution-processed BHJ OPV devices, with the optimized **a11**-based device exhibiting a PCE of 1.24% [63]. Subsequently, a structural modification by using hexenyl groups (**a12**) instead of 2-ethylhexyl side chains improved the device PCE up to 1.99%, after thermal annealing at 50 °C for 30 min [64]. Thompson, Forrest and coworkers also reported an SQ-based molecule (**a13**) as donor in OPV devices [8, 65]. In this case, bilayer devices based on solution-processed **a13**/vacuum-evaporated C_{60} exhibited a higher PCE (4.1%) than BHJ devices (2.9%) [65]. Recently, it was found that post-annealing through an additional extended exposure of the blend to dichloromethane can be used to control the nanoscale phase separation of **a13**:$PC_{71}BM$ (1 : 6) blend films, and to realize a more optimized morphology that would reduce the series resistance [8]. By optimizing the morphology and effecting a molecular ordering of the **a13**:$PC_{71}BM$ BHJ OPV devices, a peak PCE of 5.5% was achieved, with a best cell performance obtained when the exciton diffusion length was approximately equal to the mean **a13** crystallite size. Furthermore, a series of new SQ-based dyes was developed [66, 81] whereby the optimized bilayer device with **a14**(sol)/C_{60} showed the best OPV device performance after thermal annealing at 90 °C, with a PCE of 5.7%, J_{SC} of 10.0 mA cm^{-2}, V_{OC} of 0.90 V, and FF of 0.64 [66]. In the past, crystallographic data have suggested that the intermolecular stacking of **a14** molecules is closer than that of **a13**, thereby reducing the device series resistance and increasing its fill factor.

As diketopyrrolopyrrole (DPP) dyes have a strong light absorption, good photochemical stability, and facile synthetic modification, Nguyen *et al.* used a series of DPP-based materials in solution-processed BHJ OPV devices [67–69]. In 2008, Nguyen and colleagues explored the first soluble DPP-based molecule (**a15**) with terthiophene arms as a donor blending with $PC_{61}BM$; the BHJ device exhibited a J_{SC} of 8.42 mA cm^{-2}, V_{OC} of 0.67 V, FF of 0.45, and an overall PCE of 2.3% [67]. However, various inherent aspects of this material limited the device performance, including a high HOMO energy level, a potential morphological instability due to a thermally labile alkyl group, and an imbalance of carrier mobilities (hole mobility of 5×10^{-7} $cm^2/(V \cdot s)$ and an electron mobility of 3×10^{-4} $cm^2/(V \cdot s)$, as measured using single-carrier diodes. By using a 2-ethylhexyl group instead of a *t*-Boc substituent in **a15**, the new DPP-based molecule (**a16**) showed a deeper HOMO level (−5.2 eV), an improved morphological and thermal stability, and balanced carrier mobilities when blending with $PC_{71}BM$ (hole mobility of $1.0 \times 10^{-4} cm^2/(V \cdot s)$ and an electron mobility of $4.8 \times 10^{-4} cm^2/(V \cdot s)$. The OPV devices based on **a16**:$PC_{71}BM$ (1 : 1) exhibited a PCE of 3.0%, with J_{SC} of $9.2 mA cm^{-2}$, V_{OC} of 0.75 V, and FF of 0.44 [68]. The OPV performance of DPP-based materials was further improved by replacing hexylbithiophene end groups with benzofuran (**a17**). By blending **a17** with $PC_{71}BM$, very little phase separation was apparent in the as-cast film. However, thermal annealing led to suitable phase separation so that effective BHJ morphologies were obtained. The degree of phase separation can be controlled by adjusting the annealing temperature; for example, 110 °C yielded optimum device properties of J_{SC} $10 mA cm^{-2}$, V_{OC} 0.9 V, FF 0.48, and PCE 4.4% [69]. Recently, Fréchet and coworkers reported a series of DPP-based donors with different end groups, and showed that efficient OPV materials can be constructed by attaching planar, symmetric end groups to electroactive small molecules [70]. The π–π interaction of molecule **a18** dictated a tight, aligned crystal packing, a favorable morphology, and also promoted an intermolecular connectivity. Consequently, OPV devices based on a blend of **a18**:$PC_{71}BM$ (2 : 1) exhibited a maximum PCE of 4.1% with a very high FF approaching 0.6. Marks and coworkers implemented naphtho[2,3-*b*:6,7-*b*′]dithiophene (NDT) in donor materials for BHJ OPV devices [71]. The molecule (**a19**), with NDT as core and DPP as arms, was synthesized and showed a high absorption coefficient of $1.1 \times 10^5 M^{-1} cm^{-1}$ at a maximum absorption of 624 nm, an appropriate HOMO energy level (−5.4 eV), and a relatively high hole mobility of $7.18 \times 10^{-3} cm^2/(V \cdot s)$. BHJ OPVs based on **a19**:$PC_{61}BM$ gave a PCE of 4.06% after annealing at 110 °C for 10 min.

8.2.2 Oligothiophenes

One-dimensional (1-D), two-dimensional (2-D), and three-dimensional (3-D) conjugated oligothiophenes, which are among the largest families of organic semiconductors, have been widely used in OPV devices due to their high charge-carrier mobility and facile synthesis to tune energy levels [82]. The electronic properties and OPV data for representative oligothiophenes are listed in Table 8.2 (see also Figure 8.3).

BHJ OPV devices based on a blend of α-sexithiophene (**b1**) and C_{70} were fabricated using a vacuum coevaporation method [83]. AS **b1** was easy to aggregate and crystallize, it was difficult to mix homogeneously **b1** and C_{70} at a blend ratio of 1 : 1 (w/w), and this led to an insufficient formation of charge transport network and charge separation. When the **b1**:C_{70} ratio was 1 : 5 (w/w), however, an excess of C_{70} prevented **b1** from crystallizing, and this led to the formation of amorphous structure, so that charge separation efficiency was improved and a desirable charge transport interpenetrating network was formed. After thermal annealing at 140 °C for 20 min, the OPV devices exhibited a PCE of 2.38%. However, narrow absorption and relatively high HOMO level of **b1** limited the J_{SC} and V_{OC} of the OPV devices, respectively. In order to achieve high-efficiency OPV devices, conjugated oligothiophenes with a low bandgap, a broad absorption and appropriate energy

Table 8.2 Optical and electronic properties, mobilities, and OPV performance of oligothiophene-based donors.

Compound	λ_{max}[a] (nm)	E_g^{opt} (eV)	μ_h [cm²/(V·s)]	HOMO/ LUMO (eV)	Active layer[b]	J_{SC} (mA cm⁻²)	V_{OC} (V)	FF	PCE (%)	Reference
b1				−5.3/−3.1	b1:C_{70} (1:5) (vac.)	9.2	0.58	0.45	2.38	[83]
b2	573	1.77		−5.6/−	b2/C_{60}	10.6	0.98	0.49	3.4[c]	[84]
b3	579	1.68		−5.43/−3.87	b3:C_{60} (2:1) (vac.)	11.1	0.97	0.49	5.2	[9]
b4	614	1.68	1.5×10^{-4}	−5.13/−3.42	b4:$PC_{61}BM$ (1:1.4)	12.4	0.88	0.34	3.7	[85]
b5	580	1.74	3.3×10^{-4}	−5.13/−3.29	b5:$PC_{61}BM$ (2:1)	10.74	0.86	0.55	5.08	[6]
b6	563	1.83	4.5×10^{-4}	−5.11/−3.54	b6:$PC_{61}BM$ (2:1)	9.77	0.93	0.60	5.44	[86]
b7	650	1.73		−4.95/−3.26	b7:$PC_{61}BM$ (1:0.8)	11.51	0.80	0.64	5.84	[87]
b8	503	2.00		−5.50/−3.34	b8/C_{60}	3.1	0.98	0.57	1.73	[88]
b9	548	1.87		−5.65/−3.64	b9/C_{60}	4.7	1.00	0.67	3.15	[88]
b10	415	1.9	2.5×10^{-4}	−5.28/−3.38	b10:$PC_{71}BM$ (1:4)	8.45	0.82	0.43	3.0	[89]
b11	720	1.51		−5.16/−3.60	b11:$PC_{71}BM$ (3:2)	10.9	0.7	0.42	3.2	[90]
b12	390			−5.26/−2.66	b12:$PC_{61}BM$ (1:1.2)	3.65	0.85	0.26	0.80	[91]
b13	426	2.13	1.1×10^{-4}	−5.29/−3.16	b13:$PC_{71}BM$ (1:2)	4.61	0.94	0.36	1.54	[92]

a) In film.
b) Donor/acceptor: bilayer by vacuum deposition unless stated otherwise; donor:acceptor: blend by solution process unless stated otherwise; vac: vacuum deposition; sol: solution process.
c) 118 mW cm⁻².

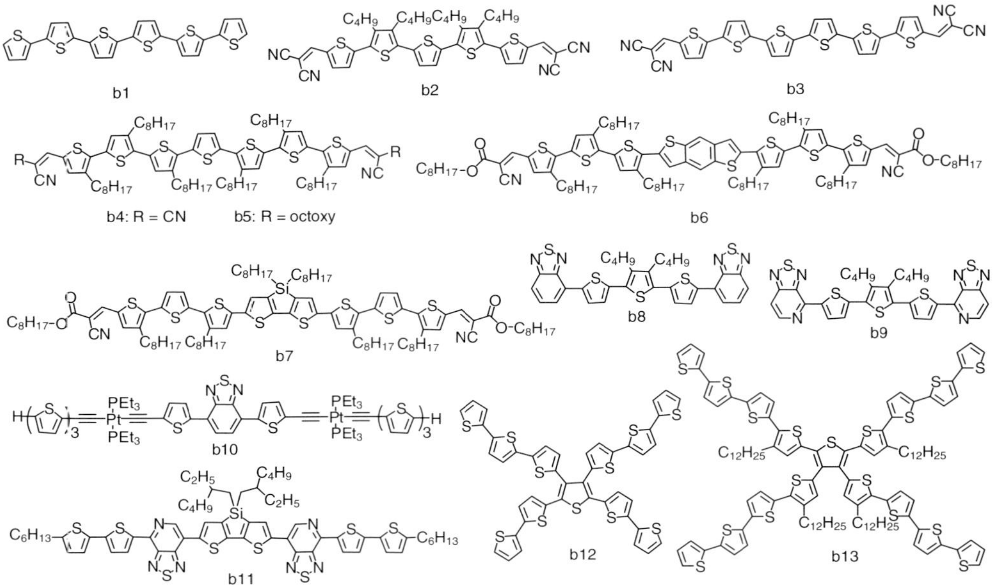

Figure 8.3 Chemical structures of oligothiophene donors.

levels are required. One successful approach would be to introduce electron-withdrawing units into the conjugated backbone so as to form D–A π-electron systems; these can extend the absorption spectrum of the donor towards longer wavelengths, and thus have a good match with the solar spectrum.

The strong electron-withdrawing group dicyanovinyl (DCV) was introduced into oligothiophenes [9, 84, 85, 93–96]. In 2006, Bäuerle, Leo and coworkers fabricated bilayer heterojunction OPV devices based on terminally DCV-substituted oligothiophenes bearing butyl side chains (**b2**) as donor and C_{60} as acceptor [84]. Due to a low HOMO level and a red-shifted absorption of **b2**, these OPVs afforded PCEs of up to 3.4%, with high V_{OC} of 0.98 V, a J_{SC} of 10.6 mA cm^{-2} at 118 mW cm^{-2} simulated sunlight. Later, the same group synthesized a series of terminally DCV-substituted oligothiophenes, without solubilizing the side chains via a convergent approach, and used these as electron donors in vacuum-processed bilayer heterojunction and BHJ OPV devices [9]. OPV devices incorporating **b3** and C_{60} showed PCEs up to 2.8% for bilayer heterojunction and 5.2% for BHJs under a simulated AM 1.5G 100 mW cm^{-2} illumination.

In 2010, Chen and coworkers reported the synthesis of a DCV-substituted oligothiophene with six solubilizing side chains (**b4**) and its application in solution-processed BHJ OPV devices [85, 96]. The absorption spectra of **b4** films showed a good solar spectral coverage, while the hole mobility of pristine **b4** was 1.5×10^{-4} cm^2/(V·s), as measured by a space charge limited current (SCLC) model [96]. Under an illumination of 100 mW cm^{-2}, the devices based on a **b4**:$PC_{61}BM$ blend (1 : 1.4) displayed a PCE of 3.7%, with J_{SC} of 12.4 mA cm^{-2}, V_{OC} of 0.88 V, and

FF of 0.34 [85]. The replacement of DCV with an alkyl cyanoacetate group led to higher PCEs [6]; OSCs based on **b5**:$PC_{61}BM$ (2 : 1) exhibited a PCE of 5.08%. The higher PCEs benefited from an improved FF (>50%), due to a better film quality and morphology. Replacing the central thiophene unit in **b5** with a more electron-rich and planar structure such as benzodithiophene (**b6**) and dithienosilole (**b7**) led to an improved mobility, absorption, and PCEs (5.44% for **b6** [86] and 5.84% for **b7** [87]).

Other electron-withdrawing units such as benzothiadiazole (BT), thiadiazolopyridine (TP), and trifluoroacetyl (TFA) were also introduced into the oligothiophene systems [88–90, 97, 98]. Bäuerle and coworkers reported two linear oligothiophenes **b8** and **b9** that had been end-capped with BT and TP acceptor units, respectively [88]. The bilayer heterojunction OPV devices based on **b9** and C_{60} showed a higher PCE (3.15%) with a very high FF (0.67) compared to that of **b8**-based devices (1.73%). The excellent FF, which was among the highest values reported for small molecule-based OPV devices, might be due to a better stacking in thin film caused by intermolecular hydrogen-bonding interactions by the nitrogen atom of the pyridine ring. In 2010, Fréchet and coworkers reported a series of platinum-acetylide linear oligothiophenes that contained a thiophene–BT–thiophene core, and various oligothiophenes were connected to control the molecular packing by changing the number of thiophene units from two to four [89]. The best device based on the oligomer with terthiophene (**b10**) blending with $PC_{71}BM$ (1 : 4) after annealing at 70 °C for 30 min exhibited a J_{SC} of 8.45 mA cm^{-2}, a V_{OC} of 0.82 V, FF of 0.43, and PCE of 3.0%. Recently, when Bazan and coworkers introduced dithienosilole (DTS) into oligothiophenes [90], these materials showed a broad absorption that extended beyond 700 nm, due to an intramolecular charge transfer. After annealing at 110 °C for 2 min, BHJ OPV devices based on **b11**:$PC_{71}BM$ (3 : 2) showed a J_{SC} of 10.9 mA cm^{-2} and PCE of 3.2%.

Aside from 1-D oligothiophenes, 2-D oligothiophene systems have also been reported. Early in 2006, Liu, Tian and coworkers described a series of X-shaped conjugated systems with four linear oligothiophene arms connected to a central thiophene core [91]. Of these, **b12**, with the longest arms, had the lowest bandgap with an absorption onset of 520 nm. BHJ OPV devices based on **b12**:$PC_{61}BM$ gave a PCE of 0.80% under simulated solar illumination. Recently, Zhan and coworkers reported an X-shaped oligothiophene (**b13**) with four longer arms than **b12** [92]. Owing to the longer conjugation, however, **b13** exhibited a red-shifted absorption with the maximum at 426 nm and the onset of 582 nm in film relative to **b12**. BHJ OPV devices based on a blend of **b13**:$PC_{61}BM$ (1 : 2) gave a V_{OC} of 0.93 V, J_{SC} of 2.71 mA cm^{-2}, FF of 0.40, and PCE of 1.02%. Replacing $PC_{61}BM$ with $PC_{71}BM$ led to an improved J_{SC} of 4.61 mA cm^{-2} and a higher PCE of 1.54%.

8.2.3
Triphenylamine Derivatives

Triphenylamine (TPA) has been regarded as a promising unit for organic semiconductor materials due to its good hole-transporting and electron-donating capabilities [51, 99]. TPA-based small molecules, including push–pull molecules with TPA as terminal group and star-shaped molecules with TPA as core, have been widely investigated for applications in OPV devices, and have exhibited good photovoltaic performances. A summary of the electronic properties and OPV data for representative linear and star-shaped TPA-based donors is provided in Table 8.3 (see also Figure 8.4).

To date, a series of linear or X-shaped push–pull chromophores with TPA as donor and BT as acceptor have been reported [100, 101, 111–114]. Zhan and coworkers synthesized a D–A–D molecule with TPA as donor, BT as acceptor, and thiophene as bridge (**c1**); as a result, BHJ OPV devices based on **c1**:$PC_{71}BM$ (1 : 3) gave a high V_{oc} (0.93 V) and PCE (2.21%) [100]. The same group then used 1% 1,8-octanedithiol as an additive to further improve the PCE to 2.86%. Such a 30% enhancement in PCE was attributed to aggregated domain formation, enhanced absorption, improved hole mobility, and a more balanced charge transport. Later, the BT in **c1** was replaced with thiazolothiazole (**c2**); the BHJ OPV devices based on **c2**:$PC_{71}BM$ (1 : 4) afforded a J_{SC} of 9.39 mA cm^{-2}, V_{OC} of 0.91 V, FF of 0.44, and a PCE of 3.73% after thermal annealing at 110 °C for 10 min [101].

Cyano-containing electron-withdrawing units, such as dicyanomethylenepyran (DCP) and DCV were also combined with TPA in order to synthesize push–pull molecules [102–104, 115–119]. In 2010, Li and coworkers designed and synthesized a D–A–D molecule with TPA as donor, DCP as acceptor, and divinylthiophene as bridge (**c3**) [102]. The V_{OC}, J_{SC}, FF, and PCE of the optimized device based on **c3**:$PC_{71}BM$ (1 : 3) reached 0.79 V, 5.94 mA cm^{-2}, 0.44, and 2.06%, respectively. Subsequently, Lin, Wong and coworkers reported a molecule (**c4**) that adopted a coplanar diphenylsubstituted dithienosilole as a central π-bridge between TPA and DCV, with a cut-off absorption wavelength of 650 nm; vacuum-deposited planar-mixed heterojunction (PMHJ) OPV devices incorporating C_{60} or C_{70} as an acceptor showed an appreciable PCE of 2.69% or 3.82%, respectively [103]. More recently, the same group reported a D–A–A molecule (**c5**) with ditolylaminothienyl as donor and BT and DCV as acceptors for vacuum-deposited OPV devices [104]; the PMHJ OPV devices using the C_{70} acceptor delivered a very high J_{sc} of 14.68 mA cm^{-2} and a PCE of 5.81%, which was attributed to the solar spectral response extending to the near-infrared region and the compact absorption dipole stacking of the thin film.

In benefitting from a special propeller starburst molecular structure, 3-D, amorphous materials with isotropic optical and charge-transporting properties could be expected when combining a TPA core with linear π-conjugated arms. Early in 2006, a series of 3-D, star-shaped molecules with TPA as core for OPV devices were first reported by Roncali and coworkers [105–107, 120]. In this case, a series of star-shaped molecules was synthesized based on a TPA core modified by various combinations of thienylenevinylene-conjugated branches and electron-withdrawing dicyanovinyl groups (**c6**–**c9**) [105]. Compared to **c6**, **c7**–**c9** with electron-withdrawing end groups showed a broadening and red shifting of the absorption and IPCE spectra, as well as a down-shifting of HOMO levels. Under the same conditions, the PCE of bilayer OPV devices was increased from 0.46% (**c6**) to 1.17% (**c8**).

Table 8.3 Optical and electronic properties, mobilities, and OPV performance of TPA-based donors.

Compound	λ_{max}[a] (nm)	E_g^{opt} (eV)	μ_h [cm^2/(V·s)]	HOMO/LUMO (eV)	Active layer[b]	J_{SC} (mA cm^{-2})	V_{OC} (V)	FF	PCE (%)	Reference
c1	534	2.03	3.5×10^{-7}	−5.16/−2.99	c1:$PC_{71}BM$ (1:3)	7.49	0.93	0.41	2.86	[100]
c2	436	2.31	1.3×10^{-6}	−5.39/−2.91	c2:$PC_{71}BM$ (1:4)	9.39	0.91	0.437	3.73	[101]
c3	536	1.79		−5.16/−3.37	c3:$PC_{71}BM$ (1:3)	5.94	0.79	0.44	2.06	[102]
c4	542	1.91		−5.4/−	c4:C_{70} (1:1) (vac)	9.53	0.83	0.48	3.82	[103]
c5	684			−5.15/−3.71	c5:C_{70} (1:1)	14.68	0.79	0.50	5.81	[104]
c6	435	2.38			c6/C_{60}	2.33	0.48	0.41	0.46	[105]
c7	544	1.91			c7/C_{60}	1.97	0.72	0.34	0.49	[105]
c8	540	1.84			c8/C_{60}	3.65	0.89	0.36	1.17	[105]
c9	538	1.78	2.9×10^{-5}	−6.02/−	c9/C_{60}	4.59	1.15	0.28	1.85[c]	[106]
c10	429		0.011	−5.50/−	c10/C_{60}	1.7	0.67	0.3	0.32	[107]
c11		1.78	3.9×10^{-5}	−5.72/−	c11:$PC_{61}BM$ (1:2)	5.30	0.87	0.39	1.80[d]	[108]
c12		1.78	5.6×10^{-5}	−5.78/−	c12:$PC_{61}BM$ (1:2)	5.83	1.07	0.31	2.02[d]	[108]
c13	538	1.9	4.9×10^{-4}	−5.28/−3.11	c13:$PC_{71}BM$ (1:3)	9.51	0.87	0.52	4.3	[109]
c14	585	1.65		−5.03/−3.42	c14:$PC_{71}BM$ (1:2)	7.66	0.88	0.439	3.0	[110]

a) In film.
b) Donor/acceptor: bilayer by vacuum deposition unless stated otherwise; donor:acceptor: blend by solution process unless stated otherwise; vac: vacuum deposition; sol: solution process.
c) 80 mW cm^{-2}.
d) 95 mW cm^{-2}.

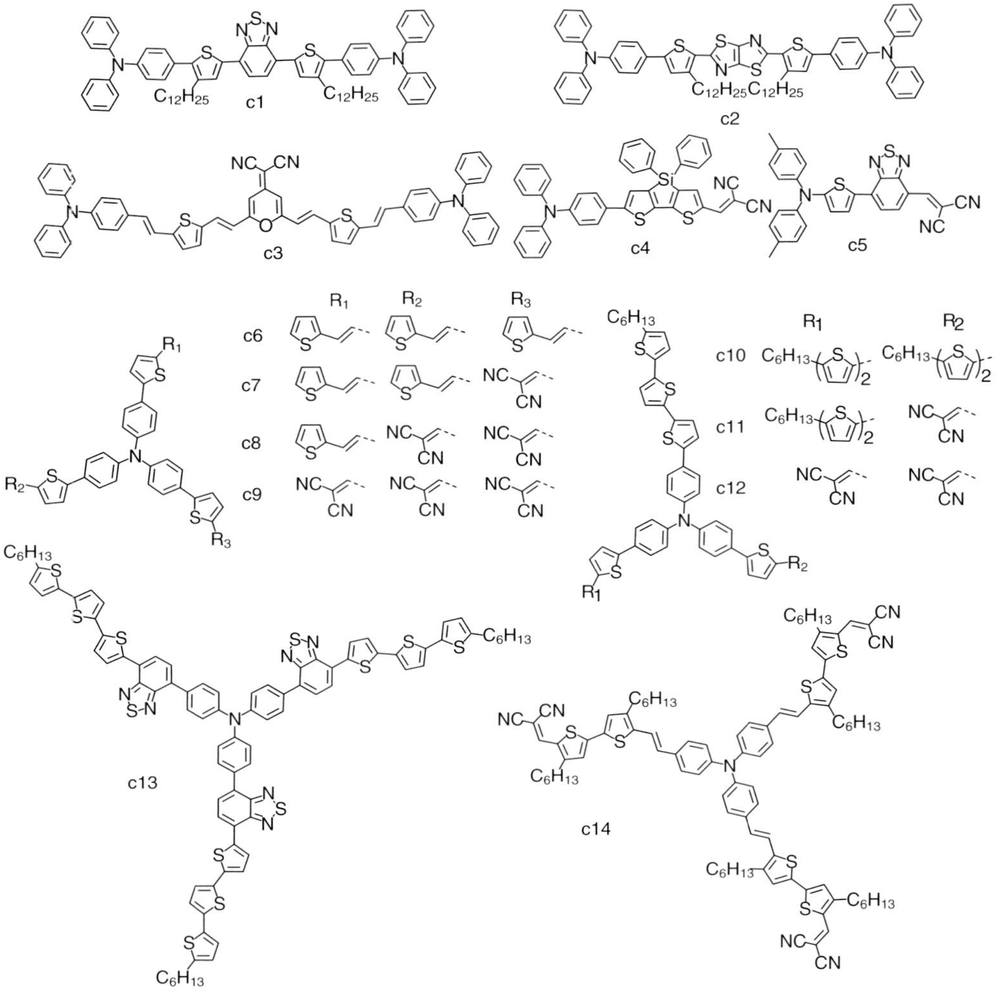

Figure 8.4 Chemical structures of triphenylamine-based donors.

Optimized devices based on **c9** exhibited a PCE of 1.85% with a very high V_{OC} of 1.15 V [106]. In 2006, a star-shaped molecule was synthesized with TPA as core and terthiophene as the arms (**c10**) [107].; this exhibited a high mobility [0.011 cm^2/(V·s)] but a narrow absorption (λ_{max} = 429 nm), leading to a low PCE (0.32%). Replacing the terminal bithiophene in **c10** with either one (**c11**) or two (**c12**) dicyanovinyl groups extended the absorption bands and lowered the HOMO levels as a result of intramolecular charge transfer [108], and this resulted in a major improvement of the V_{OC} and J_{SC}. The BHJ OPV devices based on **c11** or **c12**:$PC_{61}BM$ (1 : 2) showed PCEs of 1.80% and 2.02%, respectively.

Recently, Zhan and coworkers described a new 3-D, star-shaped, D–A–D small molecule (**c13**) with TPA as core, BT as bridge, and terthiophene as arm [109]. Relative to its counterpart **c10** without BT, **c13** exhibited a broader absorption as a result of intramolecular charge transfer. The **c13**:$PC_{71}BM$ (1 : 2) blend film exhibited nanoscale aggregated domains, which were beneficial to charge separation and enhanced the efficiency of the OPV devices. Without any post-treatment, the BHJ OPV devices exhibited a J_{SC} of 9.51 mA cm^{-2}, a V_{OC} of 0.87 V, FF of 0.52, and PCE of 4.3%. The latter PCE value was the highest reported for solution-processed BHJ OPV devices based on star-shaped small molecules, and indicated that **c13** is the most promising 3-D donor molecule for OPV devices reported to date.

Li and coworkers also synthesized a series of star-shaped, D–A molecule with TPA as core, electron-withdrawing units as arm, and vinylene as bridge [110, 121–123]. In order to improve the absorption and solution processability of **c9**, Li *et al.* replaced thiophene with 4,4′-dihexyl-2,2′-bithiophene to produce **c14** [110]; the absorption spectrum of a **c14** film covered a broad wavelength range in the visible region from 380 to 750 nm. BHJ OPV devices based on **c14**:$PC_{71}BM$ (1 : 2) showed a J_{SC} of 7.76 mA cm^{-2}, a V_{OC} of 0.88 V, FF of 0.439, and PCE of 3.0%.

8.3 Small-Molecule Acceptors

During the rapid development of donor materials – including polymers and small molecules – the PCEs of the OPV devices created have reached over 8%. Although the acceptors bear the same importance as the donors for high-performance OPV devices, research efforts devoted to acceptors have been much less than have been applied to donors. Although, to date, fullerenes and their derivatives still dominate the acceptors, nonfullerene-based acceptors have attracted increasing attention during recent years. Without question, fullerene derivatives have been the most successful acceptors in OPV devices so far. In fact, during the past decade significant advances have been made with regards to fullerene-based acceptors, and appropriate chemical modifications have led to an up-shift of their LUMO energy levels. Yet, during this time the cost of fullerene production has decreased, while the synthetic yields and material purities have increased [33]. Nonetheless, there remain many incentives to develop nonfullerene acceptors that will not only retain the favorable electron-accepting and -transporting properties of these compounds, but will also overcome their insufficiencies, such as a limited spectral breadth and bandgap variability.

8.3.1 Rylene Diimides

On the basis of recent developments in high-performance electron-transporting materials for organic field-effect transistors (OFETs), some research groups have begun to investigate nonfullerene acceptors for use in OPV devices. In this respect,

Table 8.4 Optical and electronic properties, mobilities, and OPV performance of rylene diimide-based acceptors.

Compound	λ_{max} [a)] (nm)	E_g^{opt} (eV)	μ_e [cm^2/(V·s)]^1	HOMO/LUMO (eV)	Active layer [b)]	J_{SC} (mA cm^{-2})	V_{OC} (V)	FF	PCE (%)	Reference
d1					ZnPc/d1				1.3	[130]
d2					CuPc/d2	3.66			0.93 [c)]	[131]
d3			8.8×10^{-4}	−6.0/−3.85	X:d3 (1 : 1)	6.8	0.88	0.47	2.85	[132]
d4			5.6×10^{-4}	−5.9/−3.8	Y:d4 (1 : 3.5)	6.3	0.95	0.53	3.17	[133]
d5	455	2.13	4.6×10^{-4}	−5.90/−3.95	Z:d5 (1 : 1)	8.30	0.90	0.52	3.88	[134]
d6				−5.78/−3.87	ZnPc/d6	2.11	0.50	0.51	0.54	[135]

a) In film.
b) Donor/acceptor: bilayer by vacuum deposition unless stated otherwise; donor:acceptor: blend by solution process unless stated otherwise; vac: vacuum deposition; sol: solution process.
c) 94 mW cm^{-2}.

rylene diimides have attracted much interest as alternative acceptor materials, as they exhibit excellent photostability, easy alteration of HOMO and LUMO energies, large absorption coefficients, high electron mobilities, and electron affinities similar to those of fullerenes. Moreover, each of these properties can be readily tailored through a variation of substituents either on the imide nitrogen atoms or on the rylene core [124–129]. A summary of electronic properties and OPV data for representative rylene diimides and their analogs is provided in Table 8.4 (see also Figure 8.5).

The perylene diimides (PDIs) are among the earliest and most common non-fullerene acceptors investigated in OPV devices. Many early studies of these devices incorporating PDIs consisted of layered structures that had been fabricated by vacuum deposition. The first bilayer heterojunction OPV device was reported by Tang, in which a PDI-related small molecule (**d1**) was used as the acceptor along with CuPc as the donor [1]. When ZnPc was employed as a donor in place of CuPc in a simple bilayer device, the PCE was improved from 0.95% to 1.3% [130]. Later, when the *cis*-isomer **d2** was prepared, a bilayer heterojunction device based on **d2**/CuPc showed a PCE of 0.93%, slightly lower than that (1.1%) of **d1** [131]. This lower efficiency was attributed to a less-efficient packing in acceptor **d2**, leading to shorter exciton diffusion lengths in the system.

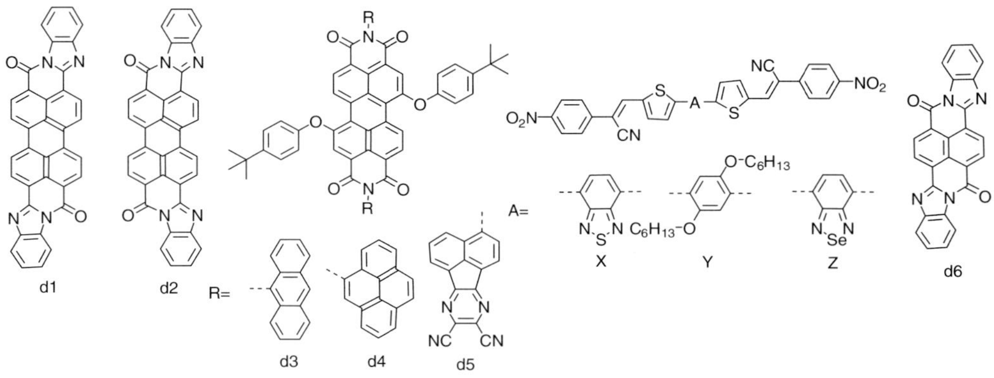

Figure 8.5 Chemical structures of rylene diimide acceptors.

PDI-based small-molecule acceptors did not show promising results in OPV devices until Sharma and colleagues developed a series of high-performance acceptors by attaching *tert*-butylphenoxy groups to the bay-region of PDIs and varying the imide substituent [132–134]. Thus, the acceptors **d3–d5**, with different fused-ring substituents on the imide, exhibited electron mobilities of $4\sim9 \times 10^{-4}\,cm^2/(V\cdot s)$ when blended with small-molecule donors **X**, **Y**, or **Z** (Figure 8.5). The devices based on **X**:**d3**, **Y**:**d4**, and **Z**:**d5** blends gave PCEs of 2.85% [132], 3.17% [133], and 3.88% [134], respectively, after annealing and/or inserting a ZnO layer between the active layer and the cathode. The PCE of 3.88% was the highest ever reported for nonfullerene-based BHJ OPV devices.

Compared to the PDI-based molecules, the smaller fused-ring unit naphthalene diimide (NDI) derivatives were less successful as acceptors in OPV devices, because they possessed a larger bandgap and thus absorbed poorly in the visible spectrum (generally onset less than 400 nm). For example, bilayer OPV devices based on ZnPc/**d6** exhibited a PCE of 0.54%, lower than that (1.3%) of its analog, **d1** [135].

8.3.2 Other Nonfullerene Acceptors

Apart from the rylene-based acceptors, some other *n*-type molecules have been reported as nonfullerene acceptors for OPV devices. A summary of electronic properties and OPV data for other representative nonfullerene small-molecule acceptors is provided in Table 8.5 (see also Figure 8.6). One method of increasing the *n*-type character of molecules is to introduce electron-withdrawing units, such as fluorine and cyano, to the periphery of the aromatic rings [141]. For example, hexadecafluorinated CuPc **e1** showed electron mobilities of up to $5 \times 10^{-3}\,cm^2/(V\cdot s)$ in OFETs [142]. When incorporating **e1** into bilayer heterojunction OPV devices as an acceptor, and sexiphenyl (p-6P) as a donor, a PCE of 0.18% was

Table 8.5 Optical and electronic properties, mobilities, and OPV performance of other nonfullerene-based acceptors.

Compound	λ_{max} [a] (nm)	E_g^{opt} (eV)	μ_e [cm^2/(V·s)]	HOMO/LUMO (eV)	Active layer [b]	J_{SC} (mA cm^{-2})	V_{OC} (V)	FF	PCE (%)	Reference
e1	780	1.5			p-6P/e1	0.96	0.42		0.18	[136]
				−6.4/−4.9	SubPc/e1	2.54	0.40	0.55	0.56	[137]
e2					SubPc/e2	2.1	0.94	0.49	0.96	[138]
e3					SubPc/e3	3.53	1.31	0.58	2.68	[139]
e4	710			−5.7/−3.95	SubPc/e4	7.8	0.95	0.54	4.0	[140]

a) In film.
b) Donor/acceptor: bilayer by vacuum deposition unless stated otherwise; donor:acceptor: blend by solution process unless stated otherwise; vac: vacuum deposition; sol: solution process.

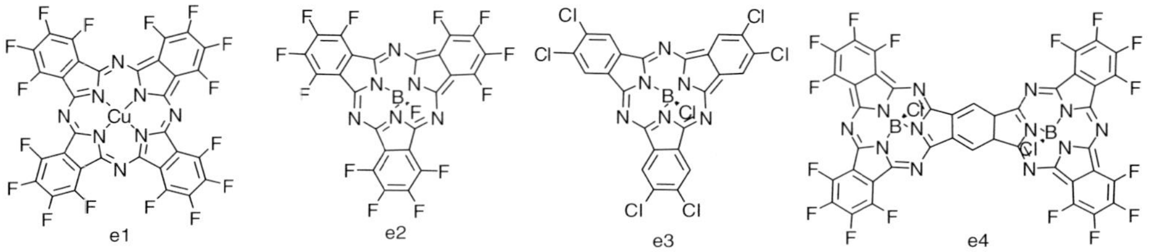

Figure 8.6 Chemical structures of other nonfullerene acceptors.

obtained [136], while the use of SubPc instead of p-6P led to an improved PCE of 0.56% [137]. Torres and coworkers synthesized a series of fluorinated SubPcs as acceptors in OPV devices [138]. Bilayer OPV devices of the acceptors with a variety of donors, such as pentacene, CuPc, AlClPc, SubPc and SubNc, were prepared by vacuum evaporation. In these devices, the SubPc/**e2**-based device showed the highest PCE of 0.96%, with a J_{SC} of 2.1 mA cm^{-2}, V_{OC} of 0.94 V, and FF of 0.49. However, the PCE value of 0.96% was clearly lower than that (3.0%) of the SubPc/C_{60}-based control device. The moderate efficiencies could be limited by the low electron mobility of the amorphous fluorinated SubPc, as well as series resistance effects in the active layer. Jones and coworkers found that the replacement of fluorinated SubPc (**e2**) with chlorinated SubPc (**e3**) led to an improved performance; the SubPc/**e3**-based device showed a promising PCE of 2.68% with a very high V_{OC} of 1.31 V [139]. Following selective halogenation to tune the energy levels of SubPc, **e3** provided sufficient interfacial HOMO and LUMO offsets for an efficient exciton dissociation, while maximizing the interface gap. Recently, Verreet *et al.* described a fluorinated fused SubPc dimer (**e4**, a mixture of two isomers) with strong and complementary absorption to the donor material SubPc [140]. The optimized SubPc/**e4**-based bilayer device exhibited a high PCE of 4%, with J_{SC} of 7.8 mA cm^{-2}, V_{OC} of 0.95 V, and FF of 0.54.

8.4
Donor–Acceptor Dyad Molecules for Single-Component OPVs

For BHJ OPV devices, one of the main problems is to fine-tune the complicated physical interactions between donor–donor, acceptor–acceptor, and donor–acceptor, in order to obtain an ideal and stable morphology with a well-defined nanostructure. To solve this problem, acceptors (e.g., fullerene, PDI) were attached to donor molecules (e.g., oligothiophene, oligophenylenevinylene, and TPA) as pendant side chains or end groups to form donor–acceptor dyad molecules (Figure 8.7), which could be regarded as a molecular heterojunction and used to fabricate single-component OPV devices. Such a structure facilitates exciton dissociation and homogeneous distribution to prevent severe phase separation. During recent years, the donor–acceptor dyad molecules have demonstrated a promising performance [143].

Hashimoto and coworkers developed a series of fullerene–dyad molecules for single-component OPV devices. For example, in 2009 they reported the synthesis of an oligophenylenevinylene that contained five phenyl rings, where the C_{60} was connected to the middle phenyl ring in the conjugated chain via a polyether linker (**f1**) [144]. After thermal annealing at 150 °C for 1 min under an inert atmosphere, the **f1**-based single-component OPV device gave a J_{SC} of 3.30 mA cm^{-2}, a V_{OC} of 0.88 V, FF of 0.44, and PCE of 1.28% under white-light irradiation at 100 mW cm^{-2}. More recently, the same group reported an oligothiophene containing eight thiophene rings and an electron-withdrawing unit DPP in the middle, with C_{60} connected to the DPP unit via a polyether linker (**f2**) [145]. In this case, **f2** showed a

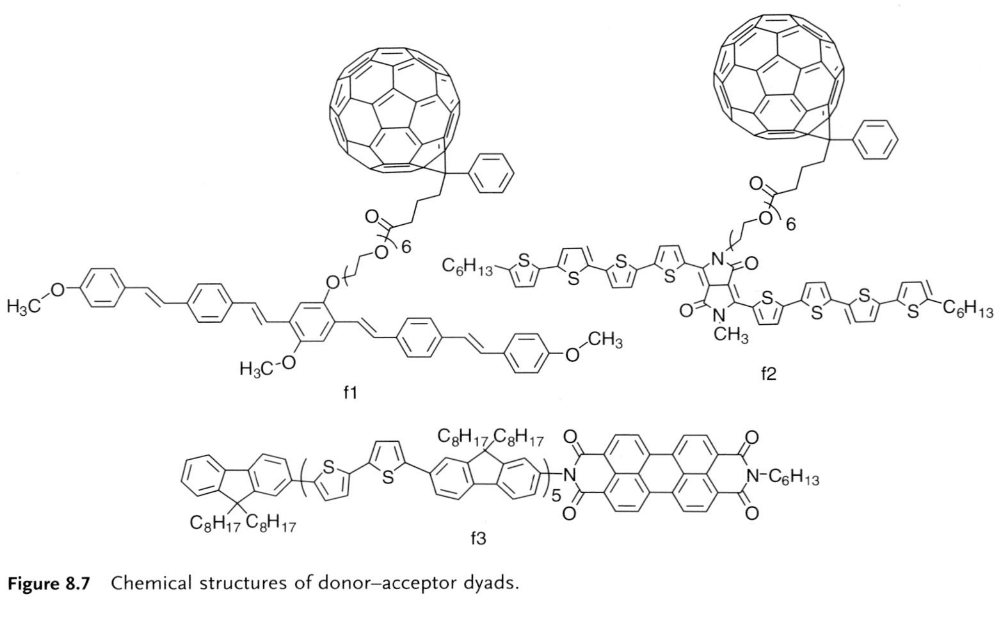

Figure 8.7 Chemical structures of donor–acceptor dyads.

low bandgap and the photocurrent response was extended to 850 nm, so that the **f2**-based single-component OPV device gave a high J_{SC} (4.79 mA cm^{-2}). This high J_{SC}, together with a V_{OC} of 0.51 V and FF of 0.46, led to a total PCE of 1.1%.

In 2009, Geng *et al.* synthesized a series of extended D–A cooligomers in which a substituted fluorene-*alt*-bithiophene oligomer was selected as donor block, and PDI as acceptor block [146]. In this case, solvent vapor annealing was found to significantly improve the order of the cooligomers, with an increase in the persistent length of the lamellae. After solvent annealing, the **f3**-based single-component OPV device showed a J_{SC} of 4.49 mA cm^{-2}, V_{OC} of 0.87 V, FF of 0.38, and PCE of 1.50%, which was the best value reported for a single-component OPV device to date. The TPA–DCV-based, star-shaped D–A molecule (**c17**, Figure 8.5) reported by Roncali *et al.* also exhibited promising results in single-component OPV devices, with a J_{SC} of 1.7 mA cm^{-2}, V_{OC} of 0.70 V, FF of 0.30, and PCE of 0.40% [106].

The rapid charge recombination and inefficient charge hopping and transport in single-component OPV devices is most likely responsible for the relatively lower PCEs of single-component devices, when compared to donor–acceptor blend or bilayer solar cells [11].

8.5 Conclusions and Outlook

In this chapter, a review has been provided of some representative and promising small-molecule donors, acceptors, and donor–acceptor dyad systems for high-performance bilayer, BHJ, and single-component OPV devices. High-efficiency small-molecule OPV devices have been achieved as a result of innovations of photovoltaic materials and device fabrication technology, and this bodes for a very bright future in this exciting field of research. In order to achieve further improvements in device performance by modifying the molecular structure of active materials, it is essential that the relationships between chemical structures and optical, electronic and device properties are clearly understood:

- Extending the absorption of active materials to match solar radiation is one way to improve J_{SC} and the efficiencies of OPV devices. An OPV molecule with a bandgap <2 eV is necessary, but not sufficient, for a high PCE. Oligomers with fused-ring units, a push–pull structure, and a long conjugation length generally possess broad and strong absorption.

- Since the V_{OC} of devices is related directly to the difference between the HOMO level of the donor and the LUMO level of the acceptor, a lower HOMO of a donor and a higher LUMO of an acceptor would help to achieve a higher V_{OC}. In general, a donor with a HOMO below −5.3 eV, or an acceptor with a LUMO above −3.7 eV, tends to give $V_{oc} > 0.8$ V. The introduction of electron-withdrawing units, such as cyano and fluorine, would down-shift the energy levels of the materials. In contrast, the main benefits of using electron-donating units such as TPA and thiophene are to raise the energy levels.

- High mobilities of donors and acceptors [$>10^{-4}\,cm^2/(V\cdot s)$] would help to achieve a high J_{SC}. Crystal structure and π–π stacking in the film are the key factors to determining the mobility of materials. In addition, mobilities matching of the donor and acceptor would also benefit the FF and PCE of devices.
- The side-chain nature affects not only the solubility and intermolecular interaction of the molecules, but also the absorption, energy levels, and charge transport properties. In particular, the side chain affects the morphology of blend films, and ultimately the photovoltaic performance of devices. Thereby, a balanced choice of a suitable solubilizing group at an appropriate location is crucially important for fine-tuning the structure–property relationship.
- The morphology and phase separation scale in BHJ films is related not only to the nature of the materials, but also to the device fabrication conditions, such as spin-casting solvent and speed, donor/acceptor radio, and concentration, thermal, or solvent annealing. With optimal morphology on the nanoscale, it becomes easy to translate the microscopic intrinsic properties of the photovoltaic materials into macroscopic OPV device performance.
- It should be emphasized that the PCE is more of a device parameter than an intrinsic photovoltaic material parameter. To achieve a high efficiency is a systematic combination of material properties with judicious and careful optimization of the various device fabrication conditions. So far, the best performances (>6%) of OPV devices have been achieved with vacuum- or solution-deposited small-molecule donors, with fullerene acceptors. An interdisciplinary approach, such as novel photovoltaic materials and new advanced device concepts, will most likely bring high-efficiency (>10%), low-cost OPV devices to future commercialization.

References

1 Tang, C.W. (1986) Two layer organic photovoltaic cell. *Appl. Phys. Lett.*, **48** (2), 183–185.

2 Yu, G., Gao, J., Hummelen, J.C., Wudl, F., and Heeger, A.J. (1995) Polymer photovoltaic cells – enhanced efficiencies via a network of internal donor-acceptor heterojunctions. *Science*, **270** (5243), 1789–1791.

3 He, Z., Zhong, C., Huang, X., Wong, W.-Y., Wu, H., Chen, L., Su, S., and Cao, Y. (2011) Simultaneous enhancement of open-circuit voltage, short-circuit current density, and fill factor in polymer solar cells. *Adv. Mater.*, **23** (40), 4636–4643.

4 Zhu, X.-H., Peng, J., Cao, Y., and Roncali, J. (2011) Solution-processable single-material molecular emitters for organic light-emitting devices. *Chem. Soc. Rev.*, **40** (7), 3509–3524.

5 Xue, J.G., Rand, B.P., Uchida, S., and Forrest, S.R. (2005) A hybrid planar-mixed molecular heterojunction photovoltaic cell. *Adv. Mater.*, **17** (1), 66–71.

6 Liu, Y., Wan, X., Wang, F., Zhou, J., Long, G., Tian, J., You, J., Yang, Y., and Chen, Y. (2011) Spin-coated small molecules for high performance solar cells. *Adv. Energy Mater.*, **1** (5), 771–775.

7 Steinmann, V., Kronenberg, N.M., Lenze, M.R., Graf, S.M., Hertel, D., Meerholz, K., Bürckstümmer, H., Tulyakova, E.V., and Würthner, F.

(2011) Simple, highly efficient vacuum-processed bulk heterojunction solar cells based on merocyanine dyes. *Adv. Energy Mater.*, **1** (5), 888–893.

8 Wei, G., Wang, S., Sun, K., Thompson, M.E., and Forrest, S.R. (2011) Solvent-annealed crystalline squaraine: PC70BM (1 : 6) solar cells. *Adv. Energy Mater.*, **1** (2), 184–187.

9 Fitzner, R., Reinold, E., Mishra, A., Mena-Osteritz, E., Ziehlke, H., Körner, C., Leo, K., Riede, M., Weil, M., Tsaryova, O., *et al.* (2011) Dicyanovinyl-substituted oligothiophenes: structure–property relationships and application in vacuum-processed small molecule organic solar cells. *Adv. Funct. Mater.*, **21** (5), 897–910.

10 Cheng, Y.-J., Yang, S.-H., and Hsu, C.-S. (2009) Synthesis of conjugated polymers for organic solar cell applications. *Chem. Rev.*, **109** (11), 5868–5923.

11 Zhan, X. and Zhu, D. (2010) Conjugated polymers for high-efficiency organic photovoltaics. *Polym. Chem.*, **1** (4), 409.

12 Zhao, X. and Zhan, X. (2011) Electron-transporting semiconducting polymers in organic electronics. *Chem. Soc. Rev.*, **40** (7), 3728–3743.

13 Chen, J. and Cao, Y. (2009) Development of novel conjugated donor polymers for high-efficiency bulk-heterojunction photovoltaic devices. *Acc. Chem. Res.*, **42** (11), 1709–1718.

14 Günes, S., Neugebauer, H., and Sariciftci, N.S. (2007) Conjugated polymer-based organic solar cells. *Chem. Rev.*, **107** (4), 1324–1338.

15 Bundgaard, E. and Krebs, F.C. (2007) Low band gap polymers for organic photovoltaics. *Sol. Energy Mater. Sol. Cells*, **91** (11), 954–985.

16 Winder, C. and Sariciftci, N.S. (2004) Low bandgap polymers for photon harvesting in bulk heterojunction solar cells. *J. Mater. Chem.*, **14** (7), 1077–1086.

17 Roncali, J. (2005) Linear pi-conjugated systems derivatized with C60-fullerene as molecular heterojunctions for organic photovoltaics. *Chem. Soc. Rev.*, **34** (6), 483–495.

18 Coakley, K.M. and McGehee, M.D. (2004) Conjugated polymer photovoltaic cells. *Chem. Mater.*, **16** (23), 4533–4542.

19 Huo, L. and Hou, J. (2011) Benzo[1,2-b:4,5-b′]dithiophene-based conjugated polymers: band gap and energy level control and their application in polymer solar cells. *Polym. Chem.*, **2** (11), 2453–2461.

20 Brabec, C.J., Gowrisanker, S., Halls, J.J.M., Laird, D., Jia, S., and Williams, S.P. (2010) Polymer-fullerene bulk-heterojunction solar cells. *Adv. Mater.*, **22** (34), 3839–3856.

21 Inganäs, O., Zhang, F., Tvingstedt, K., Andersson, L.M., Hellström, S., and Andersson, M.R. (2010) Polymer photovoltaics with alternating copolymer/fullerene blends and novel device architectures. *Adv. Mater.*, **22** (20), E100–E116.

22 Li, C., Liu, M., Pschirer, N.G., Baumgarten, M., and Müllen, K. (2010) Polyphenylene-based materials for organic photovoltaics. *Chem. Rev.*, **110** (11), 6817–6855.

23 Blouin, N. and Leclerc, M. (2008) Poly(2,7-carbazole)s: structure–property relationships. *Acc. Chem. Res.*, **41** (9), 1110–1119.

24 Inganäs, O., Zhang, F., and Andersson, M.R. (2009) Alternating polyfluorenes collect solar light in polymer photovoltaics. *Acc. Chem. Res.*, **42** (11), 1731–1739.

25 Liang, Y. and Yu, L. (2010) A new class of semiconducting polymers for bulk heterojunction solar cells with exceptionally high performance. *Acc. Chem. Res.*, **43** (9), 1227–1236.

26 Roncali, J. (2009) Molecular bulk heterojunctions: an emerging approach to organic solar cells. *Acc. Chem. Res.*, **42** (11), 1719–1730.

27 Walker, B., Kim, C., and Nguyen, T.-Q. (2011) Small molecule solution-processed bulk heterojunction solar cells. *Chem. Mater.*, **23** (3), 470–482.

28 Li, Y., Guo, Q., Li, Z., Pei, J., and Tian, W. (2010) Solution processable D-A small molecules for bulk-heterojunction solar cells. *Energy Environ. Sci.*, **3** (10), 1427–1436.

29 Hains, A.W., Liang, Z., Woodhouse, M.A., and Gregg, B.A. (2010) Molecular semiconductors in organic photovoltaic cells. *Chem. Rev.*, **110** (11), 6689–6735.

30 Lloyd, M., Anthony, J., and Malliaras, G. (2007) Photovoltaics from soluble small molecules. *Mater. Today*, **10** (11), 34–41.

31 Tang, W., Hai, J., Dai, Y., Huang, Z., Lu, B., Yuan, F., Tang, J., and Zhang, F. (2010) Recent development of conjugated oligomers for high-efficiency bulk-heterojunction solar cells. *Sol. Energy Mater. Sol. Cells*, **94** (12), 1963–1979.

32 Würthner, F. and Meerholz, K. (2010) Systems chemistry approach in organic photovoltaics. *Chem. Eur. J.*, **16** (31), 9366–9373.

33 He, Y. and Li, Y. (2011) Fullerene derivative acceptors for high performance polymer solar cells. *Phys. Chem. Chem. Phys.*, **13** (6), 1970–1983.

34 Anthony, J.E., Facchetti, A., Heeney, M., Marder, S.R., and Zhan, X. (2010) *n*-Type organic semiconductors in organic electronics. *Adv. Mater.*, **22** (34), 3876–3892.

35 Delgado, J.L., Bouit, P.-A., Filippone, S., Herranz, M., and Martín, N. (2010) Organic photovoltaics: a chemical approach. *Chem. Commun.*, **46** (27), 4853–4865.

36 Sonar, P., Fong Lim, J.P., and Chan, K.L. (2011) Organic non-fullerene acceptors for organic photovoltaics. *Energy Environ. Sci.*, **4** (5), 1558–1574.

37 Anthony, J.E. (2011) Small-molecule, nonfullerene acceptors for polymer bulk heterojunction organic photovoltaics. *Chem. Mater.*, **23** (3), 583–590.

38 Pivrikas, A., Sariciftci, N.S., Juska, G., and Osterbacka, R. (2007) A review of charge transport and recombination in polymer/fullerene organic solar cells. *Prog. Photovolt.*, **15** (8), 677–696.

39 Thompson, B.C. and Fréchet, J.M.J. (2008) Polymer–fullerene composite solar cells. *Angew. Chem. Int. Ed.*, **47** (1), 58–77.

40 Yang, X. and Loos, J. (2007) Toward high-performance polymer solar cells? The importance of morphology control. *Macromolecules*, **40** (5), 1353–1362.

41 Blom, P.W.M., Mihailetchi, V.D., Koster, L.J.A., and Markov, D.E. (2007) Device physics of polymer:fullerene bulk heterojunction solar cells. *Adv. Mater.*, **19** (12), 1551–1566.

42 Kippelen, B. and Bredas, J.-L. (2009) Organic photovoltaics. *Energy Environ. Sci.*, **2** (3), 251–261.

43 Chen, L.-M., Hong, Z., Li, G., and Yang, Y. (2009) Recent progress in polymer solar cells: manipulation of polymer:fullerene morphology and the formation of efficient inverted polymer solar cells. *Adv. Mater.*, **21** (14-15), 1434–1449.

44 Dennler, G., Scharber, M.C., and Brabec, C.J. (2009) Polymer-fullerene bulk-heterojunction solar cells. *Adv. Mater.*, **21** (13), 1323–1338.

45 Jørgensen, M., Norrman, K., and Krebs, F.C. (2008) Stability/degradation of polymer solar cells. *Sol. Energy Mater. Sol. Cells*, **92** (7), 686–714.

46 Forrest, S.R. (2005) The limits to organic photovoltaic cell efficiency. *MRS Bull.*, **30** (1), 28–32.

47 Lunt, R.R., Giebink, N.C., Belak, A.A., Benziger, J.B., and Forrest, S.R. (2009) Exciton diffusion lengths of organic semiconductor thin films measured by spectrally resolved photoluminescence quenching. *J. Appl. Phys.*, **105** (5), 053711.

48 Sariciftci, N.S., Smilowitz, L., Heeger, A.J., and Wudl, F. (1992) Photoinduced electron transfer from a conducting polymer to buckminster fullerene. *Science*, **258** (5087), 1474–1476.

49 Halls, J.J.M., Walsh, C.A., Greenham, N.C., Marseglia, E.A., and Friend, R.H. (1995) Efficient photodiodes from interpenetrating polymer networks. *Nature*, **376** (6540), 498–500.

50 Brabec, C.J., Sariciftci, N.S., and Hummelen, J.C. (2001) Plastic solar cells. *Adv. Funct. Mater.*, **11** (1), 15–26.

51 Shirota, Y. and Kageyama, H. (2007) Charge carrier transporting molecular materials and their applications in devices. *Chem. Rev.*, **107** (4), 953–1010.

52 Peumans, P. and Forrest, S.R. (2001) Very-high-efficiency double-heterostructure copper phthalocyanine/C-60 photovoltaic cells. *Appl. Phys. Lett.*, **79** (1), 126–128.

53 Uchida, S., Xue, J., Rand, B.P., and Forrest, S.R. (2004) Organic small molecule solar cells with a homogeneously mixed copper phthalocyanine: C60 active layer. *Appl. Phys. Lett.*, **84** (21), 4218.

54 Bailey-Salzman, R.F., Rand, B.P., and Forrest, S.R. (2007) Near-infrared sensitive small molecule organic photovoltaic cells based on chloroaluminum phthalocyanine. *Appl. Phys. Lett.*, **91** (1), 013508.

55 Placencia, D., Wang, W., Shallcross, R.C., Nebesny, K.W., Brumbach, M., and Armstrong, N.R. (2009) Organic photovoltaic cells based on solvent-annealed, textured titanyl phthalocyanine/C_{60} heterojunctions. *Adv. Funct. Mater.*, **19** (12), 1913–1921.

56 Mutolo, K.L., Mayo, E.I., Rand, B.P., Forrest, S.R., and Thompson, M.E. (2006) Enhanced open-circuit voltage in subphthalocyanine/C_{60} organic photovoltaic cells. *J. Am. Chem. Soc.*, **128** (25), 8108–8109.

57 Ma, B., Woo, C.H., Miyamoto, Y., and Fréchet, J.M.J. (2009) Solution processing of a small molecule, subnaphthalocyanine, for efficient organic photovoltaic cells. *Chem. Mater.*, **21** (8), 1413–1417.

58 Verreet, B., Schols, S., Cheyns, D., Rand, B.P., Gommans, H., Aernouts, T., Heremans, P., and Genoe, J. (2009) The characterization of chloroboron(III) subnaphthalocyanine thin films and their application as a donor material for organic solar cells. *J. Mater. Chem.*, **19** (30), 5295–5297.

59 Kronenberg, N.M., Deppisch, M., Wurthner, F., Lademann, H.W.A., Deing, K., and Meerholz, K. (2008) Bulk heterojunction organic solar cells based on merocyanine colorants. *Chem. Commun.*, **48**, 6489–6491.

60 Burckstummer, H., Kronenberg, N.M., Gsanger, M., Stolte, M., Meerholz, K., and Wurthner, F. (2010) Tailored merocyanine dyes for solution-processed BHJ solar cells. *J. Mater. Chem.*, **20** (2), 240–243.

61 Kronenberg, N.M., Steinmann, V., Bürckstümmer, H., Hwang, J., Hertel, D., Würthner, F., and Meerholz, K. (2010) Direct comparison of highly efficient solution- and vacuum-processed organic solar cells based on merocyanine dyes. *Adv. Mater.*, **22** (37), 4193–4197.

62 Bürckstümmer, H., Tulyakova, E.V., Deppisch, M., Lenze, M.R., Kronenberg, N.M., Gsänger, M., Stolte, M., Meerholz, K., and Würthner, F. (2011) Efficient solution-processed bulk heterojunction solar cells by antiparallel supramolecular arrangement of dipolar donor–acceptor dyes. *Angew. Chem. Int. Ed.*, **50** (49), 11628–11632.

63 Silvestri, F., Irwin, M.D., Beverina, L., Facchetti, A., Pagani, G.A., and Marks, T.J. (2008) Efficient squaraine-based solution processable bulk-heterojunction solar cells. *J. Am. Chem. Soc.*, **130** (52), 17640–17641.

64 Bagnis, D., Beverina, L., Huang, H., Silvestri, F., Yao, Y., Yan, H., Pagani, G.A., Marks, T.J., and Facchetti, A. (2010) Marked alkyl- vs alkenyl-substitutent effects on squaraine dye solid-state structure, carrier mobility, and bulk-heterojunction solar cell efficiency. *J. Am. Chem. Soc.*, **132** (12), 4074–4075.

65 Wei, G., Wang, S., Renshaw, K., Thompson, M.E., and Forrest, S.R. (2010) Solution-processed squaraine bulk heterojunction photovoltaic cells. *ACS Nano*, **4** (4), 1927–1934.

66 Wei, G., Xiao, X., Wang, S., Zimmerman, J.D., Sun, K., Diev, V.V., Thompson, M.E., and Forrest, S.R. (2011) Arylamine-based squaraine donors for use in organic solar cells. *Nano Lett.*, **11** (10), 4261–4264.

67 Tamayo, A.B., Walker, B., and Nguyen, T.Q. (2008) A low band gap, solution processable oligothiophene with a diketopyrrolopyrrole core for use in organic solar cells. *J. Phys. Chem. C*, **112** (30), 11545–11551.

68 Tamayo, A.B., Dang, X.D., Walker, B., Seo, J., Kent, T., and Nguyen, T.Q. (2009) A low band gap, solution processable oligothiophene with a dialkylated diketopyrrolopyrrole chromophore for use in bulk heterojunction solar cells. *Appl. Phys. Lett.*, **94** (10), 103301.

69 Walker, B., Tamayo, A.B., Dang, X.-D., Zalar, P., Seo, J.H., Garcia, A., Tantiwiwat, M., and Nguyen, T.-Q.

(2009) Nanoscale phase separation and high photovoltaic efficiency in solution-processed, small-molecule bulk heterojunction solar cells. *Adv. Funct. Mater.*, **19** (19), 3063–3069.

70 Lee, O.P., Yiu, A.T., Beaujuge, P.M., Woo, C.H., Holcombe, T.W., Millstone, J.E., Douglas, J.D., Chen, M.S., and Fréchet, J.M.J. (2011) Efficient small molecule bulk heterojunction solar cells with high fill factors via pyrene-directed molecular self-assembly. *Adv. Mater.*, **23** (45), 5359–5363.

71 Loser, S., Bruns, C.J., Miyauchi, H., Ortiz, R.O.P., Facchetti, A., Stupp, S.I., and Marks, T.J. (2011) A naphthodithiophene-diketopyrrolopyrrole donor molecule for efficient solution-processed solar cells. *J. Am. Chem. Soc.*, **133** (21), 8142–8145.

72 Senge, M.O., Fazekas, M., Notaras, E.G.A., Blau, W.J., Zawadzka, M., Locos, O.B., and NiMhuircheartaigh, E.M. (2007) Nonlinear optical properties of porphyrins. *Adv. Mater.*, **19** (19), 2737–2774.

73 Terao, Y., Sasabe, H., and Adachi, C. (2007) Correlation of hole mobility, exciton diffusion length, and solar cell characteristics in phthalocyanine/fullerene organic solar cells. *Appl. Phys. Lett.*, **90** (10), 103515.

74 Stubinger, T. and Brutting, W. (2001) Exciton diffusion and optical interference in organic donor-acceptor photovoltaic cells. *J. Appl. Phys.*, **90** (7), 3632–3641.

75 Xue, J.G., Uchida, S., Rand, B.P., and Forrest, S.R. (2004) 4.2% efficient organic photovoltaic cells with low series resistances. *Appl. Phys. Lett.*, **84** (16), 3013–3015.

76 Xue, J.G., Uchida, S., Rand, B.P., and Forrest, S.R. (2004) Asymmetric tandem organic photovoltaic cells with hybrid planar-mixed molecular heterojunctions. *Appl. Phys. Lett.*, **85** (23), 5757–5759.

77 Brumbach, M., Placencia, D., and Armstrong, N.R. (2008) Titanyl phthalocyanine/C_{60} heterojunctions: band-edge offsets and photovoltaic device performance. *J. Phys. Chem. C*, **112** (8), 3142–3151.

78 Brabec, C.J., Cravino, A., Meissner, D., Sariciftci, N.S., Fromherz, T., Rispens, M.T., Sanchez, L., and Hummelen, J.C. (2001) Origin of the open circuit voltage of plastic solar cells. *Adv. Funct. Mater.*, **11** (5), 374–380.

79 Gommans, H., Cheyns, D., Aernouts, T., Girotto, C., Poortmans, J., and Heremans, P. (2007) Electro-optical study of subphthalocyanine in a bilayer organic solar cell. *Adv. Funct. Mater.*, **17** (15), 2653–2658.

80 Martín, G., Rojo, G., Agulló-López, F., Ferro, V.R., García de la Vega, J.M., Martínez-Díaz, M.V., Torres, T., Ledoux, I., and Zyss, J. (2002) Subphthalocyanines and subnaphthalocyanines: nonlinear quasi-planar octupolar systems with permanent polarity. *J. Phys. Chem. B*, **106** (51), 13139–13145.

81 Wang, S., Hall, L., Diev, V.V., Haiges, R., Wei, G., Xiao, X., Djurovich, P.I., Forrest, S.R., and Thompson, M.E. (2011) *N,N*-diarylanilinosquaraines and their application to organic photovoltaics. *Chem. Mater.*, **23** (21), 4789–4798.

82 Zhang, F., Wu, D., Xu, Y., and Feng, X. (2011) Thiophene-based conjugated oligomers for organic solar cells. *J. Mater. Chem.*, **21** (44), 17590–17600.

83 Sakai, J., Taima, T., Yamanari, T., and Saito, K. (2009) Annealing effect in the sexithiophene:C_{70} small molecule bulk heterojunction organic photovoltaic cells. *Sol. Energy Mater. Sol. Cells*, **93** (6-7), 1149–1153.

84 Schulze, K., Uhrich, C., Schüppel, R., Leo, K., Pfeiffer, M., Brier, E., Reinold, E., and Bäuerle, P. (2006) Efficient vacuum-deposited organic solar cells based on a new low-bandgap oligothiophene and fullerene C_{60}. *Adv. Mater.*, **18** (21), 2872–2875.

85 Yin, B., Yang, L., Liu, Y., Chen, Y., Qi, Q., Zhang, F., and Yin, S. (2010) Solution-processed bulk heterojunction organic solar cells based on an oligothiophene derivative. *Appl. Phys. Lett.*, **97** (2), 023303.

86 Liu, Y., Wan, X., Wang, F., Zhou, J., Long, G., Tian, J., and Chen, Y. (2011) High-performance solar cells using a

solution-processed small molecule containing benzodithiophene unit. *Adv. Mater.*, **23** (45), 5387–5391.

87 Zhou, J., Wan, X., Liu, Y., Long, G., Wang, F., Li, Z., Zuo, Y., Li, C., and Chen, Y. (2011) A planar small molecule with dithienosilole core for high efficiency solution-processed organic photovoltaic cells. *Chem. Mater.*, **23** (21), 4666–4668.

88 Steinberger, S., Mishra, A., Reinold, E., Levichkov, J., Uhrich, C., Pfeiffer, M., and Bauerle, P. (2011) Vacuum-processed small molecule solar cells based on terminal acceptor-substituted low-band gap oligothiophenes. *Chem. Commun.*, **47** (7), 1982–1984.

89 Zhao, X., Piliego, C., Kim, B., Poulsen, D.A., Ma, B., Unruh, D.A., and Fréchet, J.M.J. (2010) Solution-processable crystalline platinum-acetylide oligomers with broadband absorption for photovoltaic cells. *Chem. Mater.*, **22** (7), 2325–2332.

90 Welch, G.C., Perez, L.A., Hoven, C.V., Zhang, Y., Dang, X.-D., Sharenko, A., Toney, M.F., Kramer, E.J., Nguyen, T.-Q., and Bazan, G.C. (2011) A modular molecular framework for utility in small-molecule solution-processed organic photovoltaic devices. *J. Mater. Chem.*, **21** (34), 12700–12709.

91 Sun, X., Zhou, Y., Wu, W., Liu, Y., Tian, W., Yu, G., Qiu, W., Chen, S., and Zhu, D. (2006) X-Shaped oligothiophenes as a new class of electron donors for bulk-heterojunction solar cells. *J. Phys. Chem. B*, **110** (15), 7702–7707.

92 Shang, H., Fan, H., Liu, Y., Hu, W., Li, Y., and Zhan, X. (2011) New X-shaped oligothiophenes for solution-processed solar cells. *J. Mater. Chem.*, **21** (6), 9667–9673.

93 Schulze, K., Riede, M., Brier, E., Reinold, E., Bauerle, P., and Leo, K. (2008) Dicyanovinyl-quinquethiophenes with varying alkyl chain lengths: investigation of their performance in organic devices. *J. Appl. Phys.*, **104** (7), 074511.

94 Mishra, A., Uhrich, C., Reinold, E., Pfeiffer, M., and Bäuerle, P. (2011) Synthesis and characterization of acceptor-substituted oligothiophenes for solar cell applications. *Adv. Energy Mater.*, **1** (2), 265–273.

95 Haid, S., Mishra, A., Uhrich, C., Pfeiffer, M., and Bäuerle, P. (2011) Dicyanovinylene-substituted selenophene–thiophene co-oligomers for small-molecule organic solar cells. *Chem. Mater.*, **23** (20), 4435–4444.

96 Liu, Y., Wan, X., Yin, B., Zhou, J., Long, G., Yin, S., and Chen, Y. (2010) Efficient solution processed bulk-heterojunction solar cells based a donor–acceptor oligothiophene. *J. Mater. Chem.*, **20** (12), 2464.

97 Ah Kong, J., Lim, E., Lee, K.K., Lee, S., and Hyun Kim, S. (2010) A benzothiadiazole-based oligothiophene for vacuum-deposited organic photovoltaic cells. *Sol. Energy Mater. Sol. Cells*, **94** (12), 2057–2063.

98 Steinberger, S., Mishra, A., Reinold, E., Müller, C.M., Uhrich, C., Pfeiffer, M., and Bäuerle, P. (2010) A-D-A-D-A-type oligothiophenes for vacuum-deposited organic solar cells. *Org. Lett.*, **13** (1), 90–93.

99 Ning, Z. and Tian, H. (2009) Triarylamine: a promising core unit for efficient photovoltaic materials. *Chem. Commun.*, **37**, 5483–5495.

100 Fan, H., Shang, H., Li, Y., and Zhan, X. (2010) Efficiency enhancement in small molecule bulk heterojunction organic solar cells via additive. *Appl. Phys. Lett.*, **97** (13), 133302.

101 Shi, Q., Cheng, P., Li, Y., and Zhan, X. (2012) A solution processable D-A-D molecule based on thiazolothiazole for high performance organic solar cells. *Adv. Energy Mater.*, **2** (1), 63–67.

102 Zhang, J., Wu, G., He, C., Deng, D., and Li, Y. (2011) Triphenylamine-containing D-A-D molecules with (dicyanomethylene)pyran as an acceptor unit for bulk-heterojunction organic solar cells. *J. Mater. Chem.*, **21** (11), 3768–3774.

103 Lin, H.-W., Lin, L.-Y., Chen, Y.-H., Chen, C.-W., Lin, Y.-T., Chiu, S.-W., and Wong, K.-T. (2011) A new donor-acceptor molecule with uniaxial anisotropy for efficient vacuum-deposited organic solar cells. *Chem. Commun.*, **47** (27), 7872–7874.

104 Lin, L.-Y., Chen, Y.-H., Huang, Z.-Y., Lin, H.-W., Chou, S.-H., Lin, F., Chen, C.-W., Liu, Y.-H., and Wong, K.-T. (2011) A low-energy-gap organic dye for high-performance small- molecule organic solar cells. *J. Am. Chem. Soc.*, **133** (40), 15822–15825.

105 Roquet, S., Cravino, A., Leriche, P., Aleveque, O., Frere, P., and Roncali, J. (2006) Triphenylamine-thienylenevinylene hybrid systems with internal charge transfer as donor materials for heterojunction solar cells. *J. Am. Chem. Soc.*, **128** (10), 3459–3466.

106 Cravino, A., Leriche, P., Alévêque, O., Roquet, S., and Roncali, J. (2006) Light-emitting organic solar cells based on a 3D conjugated system with internal charge transfer. *Adv. Mater.*, **18** (22), 3033–3037.

107 Cravino, A., Roquet, S., Aleveque, O., Leriche, P., Frere, P., and Roncali, J. (2006) Triphenylamine-oligothiophene conjugated systems as organic semiconductors for opto-electronics. *Chem. Mater.*, **18** (10), 2584–2590.

108 Ripaud, E., Rousseau, T., Leriche, P., and Roncali, J. (2011) Unsymmetrical triphenylamine- oligothiophene hybrid conjugated systems as donor materials for high-voltage solution-processed organic solar cells. *Adv. Energy Mater.*, **1** (4), 540–545.

109 Shang, H., Fan, H., Liu, Y., Hu, W., Li, Y., and Zhan, X. (2011) A solution-processable star-shaped molecule for high-performance organic solar cells. *Adv. Mater.*, **23** (13), 1554–1557.

110 Zhang, J., Deng, D., He, C., He, Y., Zhang, M., Zhang, Z.-G., Zhang, Z., and Li, Y. (2011) Solution-processable star-shaped molecules with triphenylamine core and dicyanovinyl endgroups for organic solar cells. *Chem. Mater.*, **23** (3), 817–822.

111 He, C., He, Q., He, Y., Li, Y., Bai, F., Yang, C., Ding, Y., Wang, L., and Ye, J. (2006) Organic solar cells based on the spin-coated blend films of TPA-th-TPA and PCBM. *Sol. Energy Mater. Sol. Cells*, **90** (12), 1815–1827.

112 Deng, D., Yang, Y., Zhang, J., He, C., Zhang, M., Zhang, Z.-G., Zhang, Z., and Li, Y. (2011) Triphenylamine-containing linear D-A-D molecules with benzothiadiazole as acceptor unit for bulk-heterojunction organic solar cells. *Org. Electron.*, **12** (4), 614–622.

113 Shang, H., Fan, H., Shi, Q., Li, S., Li, Y., and Zhan, X. (2010) Solution processable D-A-D molecules based on triphenylamine for efficient organic solar cells. *Sol. Energy Mater. Sol. Cells*, **94** (3), 457–464.

114 Li, W., Du, C., Li, F., Zhou, Y., Fahlman, M., Bo, Z., and Zhang, F. (2009) Benzothiadiazole-based linear and star molecules: design, synthesis, and their application in bulk heterojunction organic solar cells. *Chem. Mater.*, **21** (21), 5327–5334.

115 He, C., He, Q., Yang, X., Wu, G., Yang, C., Bai, F., Shuai, Z., Wang, L., and Li, Y. (2007) Synthesis and photovoltaic properties of a solution-processable organic molecule containing triphenylamine and DCM moieties. *J. Phys. Chem. C*, **111** (24), 8661–8666.

116 Xue, L., He, J., Gu, X., Yang, Z., Xu, B., and Tian, W. (2009) Efficient bulk-heterojunction solar cells based on a symmetrical D-π-A-π-D organic dye molecule. *J. Phys. Chem. C*, **113** (29), 12911–12917.

117 Leliège, A., Blanchard, P., Rousseau, T.O., and Roncali, J. (2011) Triphenylamine/tetracyanobutadiene-based D-A-D π-conjugated systems as molecular donors for organic solar cells. *Org. Lett.*, **13** (12), 3098–3101.

118 Liu, Q., Jiang, K., Guan, B., Tang, Z., Pei, J., and Song, Y. (2011) A novel bulk heterojunction solar cell based on a donor–acceptor conjugated triphenylamine dye. *Chem. Commun.*, **47** (2), 740.

119 Tang, Z.-M., Lei, T., Jiang, K.-J., Song, Y.-L., and Pei, J. (2010) Benzothiadiazole containing D-π-A conjugated compounds for dye-sensitized solar cells: synthesis, properties, and photovoltaic performances. *Chem. Asian J.*, **5** (8), 1911–1917.

120 Cravino, A., Roquet, S., Leriche, P., Aleveque, O., Frere, P., and Roncali, J. (2006) A star-shaped triphenylamine pi-conjugated system with internal charge-transfer as donor material for

hetero-junction solar cells. *Chem. Commun.*, **13**, 1416–1418.

121 Wu, G., Zhao, G., He, C., Zhang, J., He, Q., Chen, X., and Li, Y. (2009) Synthesis and photovoltaic properties of a star-shaped molecule with triphenylamine as core and benzo[1,2,5] thiadiazol vinylene as arms. *Sol. Energy Mater. Sol. Cells*, **93** (1), 108–113.

122 He, C., He, Q., Yi, Y., Wu, G., Bai, F., Shuai, Z., and Li, Y. (2008) Improving the efficiency of solution processable organic photovoltaic devices by a star-shaped molecular geometry. *J. Mater. Chem.*, **18** (34), 4085–4090.

123 Zhang, J., Yang, Y., He, C., He, Y., Zhao, G., and Li, Y. (2009) Solution-processable star-shaped photovoltaic organic molecule with triphenylamine core and benzothiadiazole–thiophene arms. *Macromolecules*, **42** (20), 7619–7622.

124 Zhan, X., Facchetti, A., Barlow, S., Marks, T.J., Ratner, M.A., Wasielewski, M.R., and Marder, S.R. (2011) Rylene and related diimides for organic electronics. *Adv. Mater.*, **23** (2), 268–284.

125 Zhan, X., Tan, Z.A., Domercq, B., An, Z., Zhang, X., Barlow, S., Li, Y., Zhu, D., Kippelen, B., and Marder, S.R. (2007) A high-mobility electron-transport polymer with broad absorption and its use in field-effect transistors and all-polymer solar cells. *J. Am. Chem. Soc.*, **129** (23), 7246–7247.

126 Zhan, X., Tan, Z.A., Zhou, E., Li, Y., Misra, R., Grant, A., Domercq, B., Zhang, X.-H., An, Z., Zhang, X., Barlow, S., Kippelen, B., and Marder, S.R. (2009) Copolymers of perylene diimide with dithienothiophene and dithienopyrrole as electron-transport materials for all-polymer solar cells and field-effect transistors. *J. Mater. Chem.*, **19** (32), 5794–5803.

127 Huang, J., Wu, Y., Fu, H., Zhan, X., Yao, J., Barlow, S., and Marder, S.R. (2009) Photoinduced intramolecular electron transfer in conjugated perylene bisimide-dithienothiophene systems: a comparative study of a small molecule and a polymer. *J. Phys. Chem. A*, **113** (17), 5039–5046.

128 Meng, L., Shang, Y., Li, Q., Li, Y., Zhan, X., Shuai, Z., Kimber, R.G.E., and Walker, A.B. (2009) Dynamic Monte Carlo simulation for highly efficient polymer blend photovoltaics. *J. Phys. Chem. B*, **114** (1), 36–41.

129 Tan, Z.A., Zhou, E.J., Zhan, X.W., Wang, X., Li, Y.F., Barlow, S., and Marder, S.R. (2008) Efficient all-polymer solar cells based on blend of tris(thienylenevinylene)-substituted polythiophene and poly[perylene diimide-alt- bis(dithienothiophene)]. *Appl. Phys. Lett.*, **93** (7), 073309.

130 Erten, S., Meghdadi, F., Gunes, S., Koeppe, R., Sariciftci, N.S., and Icli, S. (2006) Donor-acceptor heterojunction solar cells based on perylene dimide and perylene bisbenzimidazole. *Eur. Phys. J.–Appl. Phys.*, **36** (3), 225–229.

131 Rim, S.B., Fink, R.F., Schoneboom, J.C., Erk, P., and Peumans, P. (2007) Effect of molecular packing on the exciton diffusion length in organic solar cells. *Appl. Phys. Lett.*, **91** (17), 173504.

132 Sharma, G.D., Balraju, P., Mikroyannidis, J.A., and Stylianakis, M.M. (2009) Bulk heterojunction organic photovoltaic devices based on low band gap small molecule BTD-TNP and perylene-anthracene diimide. *Sol. Energy Mater. Sol. Cells*, **93** (11), 2025–2028.

133 Sharma, G.D., Suresh, P., Mikroyannidis, J.A., and Stylianakis, M.M. (2010) Efficient bulk heterojunction devices based on phenylenevinylene small molecule and perylene–pyrene bisimide. *J. Mater. Chem.*, **20** (3), 561–567.

134 Mikroyannidis, J.A., Suresh, P., and Sharma, G.D. (2010) Synthesis of a perylene bisimide with acetonaphthopyrazine dicarbonitrile terminal moieties for photovoltaic applications. *Synth. Met.*, **160** (9–10), 932–938.

135 Erten, S. and Icli, S. (2008) Bilayer heterojunction solar cell based on naphthalene bis-benzimidazole. *Inorg. Chim. Acta*, **361** (3), 595–600.

136 Jiang, X., Dai, J., Wang, H., Geng, Y., and Yan, D. (2007) Organic photovoltaic cells using

hexadecafluorophthalocyaninatocopper ($F_{16}CuPc$) as electron acceptor material. *Chem. Phys. Lett.*, **446** (4–6), 329–332.

137 Yang, J.L., Schumann, S., Hatton, R.A., and Jones, T.S. (2010) Copper hexadecafluorophthalocyanine ($F_{16}CuPc$) as an electron-accepting material in bilayer small molecule organic photovoltaic cells. *Org. Electron.*, **11** (8), 1399–1402.

138 Gommans, H., Aernouts, T., Verreet, B., Heremans, P., Medina, A., Claessens, C.G., and Torres, T. (2009) Perfluorinated subphthalocyanine as a new acceptor material in a small-molecule bilayer organic solar cell. *Adv. Funct. Mater.*, **19** (21), 3435–3439.

139 Sullivan, P., Duraud, A., Hancox, L., Beaumont, N., Mirri, G., Tucker, J.H.R., Hatton, R.A., Shipman, M., and Jones, T.S. (2011) Halogenated boron subphthalocyanines as light harvesting electron acceptors in organic photovoltaics. *Adv. Energy Mater.*, **1** (3), 352–355.

140 Verreet, B., Rand, B.P., Cheyns, D., Hadipour, A., Aernouts, T., Heremans, P., Medina, A., Claessens, C.G., and Torres, T. (2011) A 4% efficient organic solar cell using a fluorinated fused subphthalocyanine dimer as an electron acceptor. *Adv. Energy Mater.*, **1** (4), 565–568.

141 Tang, M.L., Oh, J.H., Reichardt, A.D., and Bao, Z. (2009) Chlorination: a general route toward electron transport in organic semiconductors. *J. Am. Chem. Soc.*, **131** (10), 3733–3740.

142 Bao, Z., Lovinger, A.J., and Brown, J. (1998) New air-stable *n*-channel organic thin film transistors. *J. Am. Chem. Soc.*, **120** (1), 207–208.

143 Roncali, J. (2011) Single material solar cells: the next frontier for organic photovoltaics? *Adv. Energy Mater.*, **1** (2), 147–160.

144 Nishizawa, T., Lim, H.K., Tajima, K., and Hashimoto, K. (2009) Efficient dyad-based organic solar cells with a highly crystalline donor group. *Chem. Commun.*, (18), 2469–2471.

145 Izawa, S., Hashimoto, K., and Tajima, K. (2011) Efficient charge generation and collection in organic solar cells based on low band gap dyad molecules. *Chem. Commun.*, **47** (22), 6365–6367.

146 Bu, L., Guo, X., Yu, B., Qu, Y., Xie, Z., Yan, D., Geng, Y., and Wang, F. (2009) Monodisperse co-oligomer approach toward nanostructured films with alternating donor–acceptor lamellae. *J. Am. Chem. Soc.*, **131** (37), 13242–13243.

9
Polymer Solar Cells

Huitao Bai, Qinqin Shi, Xiaowei Zhan

9.1
Introduction

The photoactive layer of polymer solar cells (PSCs) is constituted of continuous and interpenetrating polymer donor and small-molecule or polymer acceptors. This structure is able to maximally increase the area of interface between the donor and acceptor, and thus improve the charge separation efficiency. Conjugated polymers can be classified as either electron donors (*p*-type) or electron acceptors (*n*-type), according to their differences in transporting holes or electrons in the cell. Among the semiconducting polymer materials, the *p*-type materials have attracted more attention, and are generally mixed with [6,6]-phenyl-C_{61} or C_{71}-butyric acid methyl ester ($PC_{61}BM$ or $PC_{71}BM$) to construct the photoactive layer of bulk heterojunction (BHJ) solar cells. Currently, the highest reported power conversion efficiency (PCE) of polymer solar cells is 8.37% [1]. The *n*-type materials develop relatively slowly, and are frequently mixed with classic donor materials such as polyphenylenevinylene (PPV) derivatives, poly(3-hexylthiophene) (P3HT), or other *p*-type materials, to construct all-polymer solar cells. To date, the highest PCE of these material is less than 3%.

The basic requirements for an ideal donor or acceptor polymer material include: (i) good solubility and film-forming properties; (ii) strong and broad bandwidth absorption; (iii) high hole or electron mobilities; (iv) suitable HOMO/LUMO energy levels; and (v) a high purity and molecular weight [2]. The aim of this chapter is to discuss the research and development of polymer solar cells fabricated from conjugated polymer donors and fullerene acceptors or conjugated polymer acceptors.

Organic Optoelectronics, First Edition. Edited by Wenping Hu.

9.2
Polymer Donor Materials

9.2.1
Polyphenylenevinylene (PPV) Derivatives

Poly[2-methoxy-5-(2′-ethyl-hexyloxy)-1,4-phenylenevinylene] (MEH-PPV) and poly[2-methoxy -5-(3,7-dimethyloctyloxy)-1,4-phenylenevinylene] (MDMO-PPV) (Figure 9.1; Table 9.1) exhibit a strong absorption in the visible region and are classic donor materials among PPV derivatives. The MEH-PPV/$PC_{61}BM$ bilayer photovoltaic device yielded a PCE of 0.46%, while the MEH-PPV/$PC_{61}BM$ BHJ device gave an improved PCE of 2.07% [3, 9]. Shaheen *et al.* reported that PSC devices based on

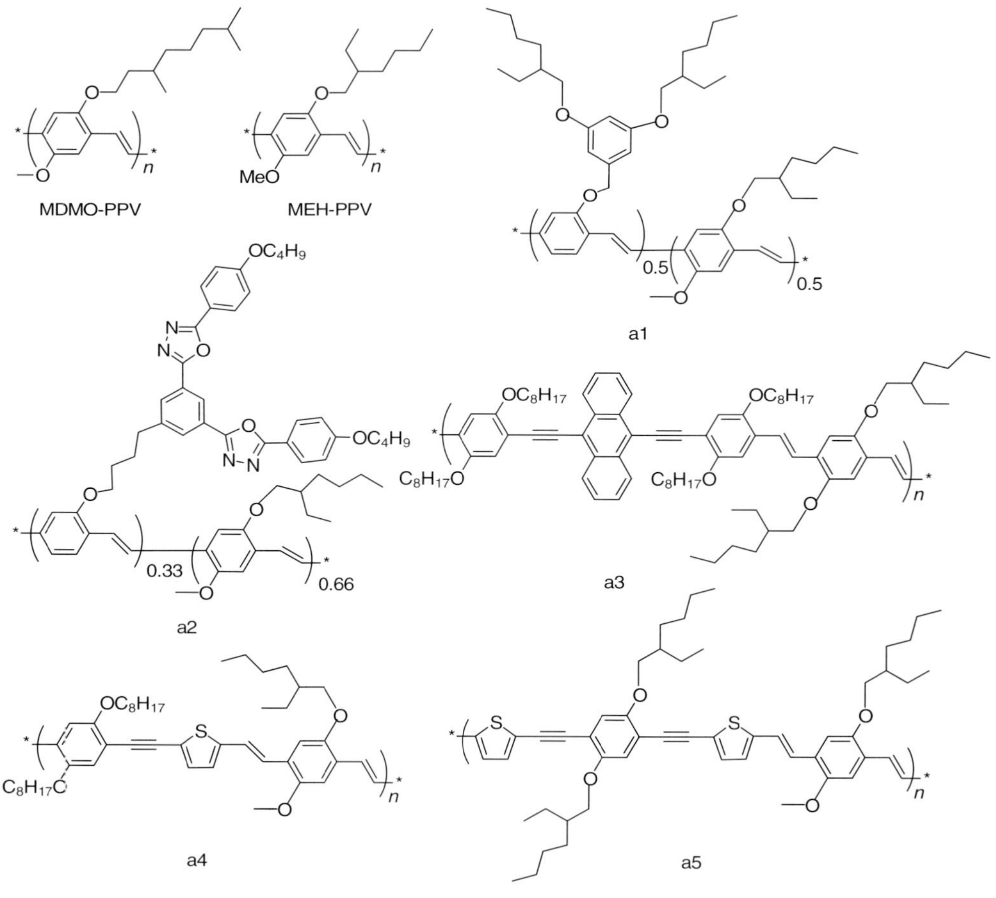

Figure 9.1 Chemical structures of PPV-based polymer donors.

Table 9.1 Electronic properties, mobilities, and PSC performance of PPV derivatives.[a)]

Polymer	λ_{max} (nm)	E_g^{opt} (eV)	μ_h [cm²/(V·s)]	HOMO/LUMO (eV)	D/PC$_{61}$BM (w/w)	J_{SC} (mA cm^{-2})	V_{OC} (V)	FF	PCE (%)	Reference
MEH-PPV	515[b)]	–	–	−5.01/−2.80[b)]	1:4	4.79	0.79	0.44	2.07[c)]	[3]
MDMO-PPV	540	–	3×10^{-10}	−5.4/–	1:2	6.2	0.71	0.7	3.1	[4]
a1	494	–	–	−5.02/−2.69	1:3	3.37	0.81	0.42	1.41	[5]
a2	513	2.10	–	−5.05/−2.86	1:4	4.93	0.81	0.4	1.6	[6]
a3	575	1.90	–	−5.36/−3.31	1:2	4.3	0.81	0.59	2.0	[7]
a4	538	2.07	–	−5.43/−3.14	1:3	3.65	0.9	0.53	1.74	[8]
a5	526	2.07	–	−5.45/−3.14	1:3	4.45	0.8	0.42	1.50	[8]

a) λ_{max} = absorption peak in film; E_g = optical band gap; μ_h = hole mobility; D = donor; J_{SC} = short-circuit current density; V_{OC} = open-circuit voltage; FF = fill factor; PCE = power conversion efficiency.
b) Ref. [5].
c) Under 80 mW cm^{-2} illumination.

MDMO-PPV/PC$_{61}$BM (1:4, w/w) exhibited a PCE of 2.5% [10]. When Tajima *et al.* synthesized a fully regioregular MDMO-PPV and utilized this for PSCs [4], a significant improvement in short-circuit current (J_{SC}) and fill factor (FF) was achieved, while the PCE was improved to 3.1%–the highest efficiency of the PPV:PC$_{61}$BM system achieved to date. The regioregular polymer chain was also beneficial for crystallization, which in turn would lead to a higher hole mobility and a better blend morphology when blending with [6,6]-phenyl-C$_{61}$-butyric acid methyl ester (PCBM). These two classic materials not only showed good performance as donor materials in PSC applications, but also played an important role in screening novel electron-acceptor materials. Subsequently, many new electron-acceptor materials, such as fullerene derivatives, polymer acceptors, inorganic electron acceptors and quantum dots, have been blended with MDMO-PPV or MEH-PPV to prepare photovoltaic devices [11].

Li *et al.* synthesized a PPV derivative with side chains of 3,5-dialkoxyl benzene (**a1**) [5], while Tian *et al.* modified the PPV side chains with electron-withdrawing oxadiazole groups (**a2**) [6]. The electronic properties of these two polymers can be tuned by adjusting ratio of the two comonomers. Those PSCs based on **a1** or **a2**:PC$_{61}$BM exhibited PCEs of 1.41% and 1.6%, respectively, which were higher than that of MEH-PPV under the same test conditions.

A number of PPV-type copolymers containing alternating ethylene and ethynylene moieties have also been synthesized. The coplanar and rigid structure may lead to a highly ordered molecular chain arrangement, which would be beneficial for an improved photovoltaic performance. Furthermore, the incorporation of electron-rich groups such as thiophene and anthracene could lead to an intramolecular charge transfer and broaden the optical absorption [7, 8]. For example, **a3** showed a relatively broad absorption, a low HOMO level (−5.36 eV), and a narrow optical band gap (1.9 eV).

A photovoltaic device based on **a3**/$PC_{61}BM$ (1 : 2, w/w) afforded a PCE of 2% [7]. Compared to all-benzene-structured polymers, the introduction of thiophene units improved the PCEs of both **a4** and **a5** [8].

9.2.2 Polythiophene Derivatives

The polythiophenes – and especially regioregular poly(3-alkylthiophene)s (P3ATs) – are a class of polymer donor materials that are widely used in PSCs, due to their excellent thermal and chemical stabilities, as well as good light-harvesting and charge-transporting properties. The length of the alkyl chain in P3ATs plays an important role in determining the solubility, crystallinity, and morphology. When Nguyen *et al.* studied the effect of alkyl groups (butyl, hexyl, octyl, decyl, and dodecyl) on the performance of P3AT/fullerene BHJ PSCs [12], longer alkyl lengths (≥C8) were found to accelerate diffusion of the soluble PCBM in the polymer matrix, leading ultimately to a larger scale of phase separation, a reduced interfacial area, and a lower PCE. In contrast, butyl group led to a poor solubility, a decreased molecular weight, poor crystallinity, decreased mobility, and a reduced PCE [13].

P3HT achieved the highest PCE among the P3AT family (Figure 9.2, Table 9.2). To date, the highest efficiency of any P3HT/PCBM system has been about 5%, with the device performance being largely related to P3HT's molecular weight, regularity, and the device fabrication methods employed. It was found that only P3HT with a molecular weight greater than 10^4 could achieve a PCE above 2.5%. In addition, regularity is an important factor, as an appropriate crystallinity will facilitate charge transfer and increase the interface area between P3HT and PCBM, which is beneficial for exciton dissociation. However, certain post-treatments – such as long-term thermal annealing – would lead to an excessive crystallization and increase the phase separation scale of P3HT/PCBM blend, which would not be favorable for an efficient exciton dissociation. The success of the P3HT/PCBM system has been largely attributed to a careful control and optimization of the active layer morphology. For example, the optimization of a different casting solvent, the film-forming rate, the thermal annealing temperature, additives, optical spacer, tandem structure, anode and cathode interfacial layer, would greatly improve the PCE value.

A relatively deep LUMO level of PCBM leads to an open-circuit voltage (V_{OC}) loss in P3HT/PCBM devices, and further limits the PCEs. Li and coworkers have developed an indene-C_{60} bisadduct ($IC_{60}BA$) [25, 26] and indene-C_{70} bisadduct ($IC_{70}BA$) [27] which exhibited LUMO up-shifts of 0.17 and 0.19 eV, respectively, relative to $PC_{61}BM$ and $PC_{71}BM$. Based on the LUMO up-shift, the V_{OC} of devices

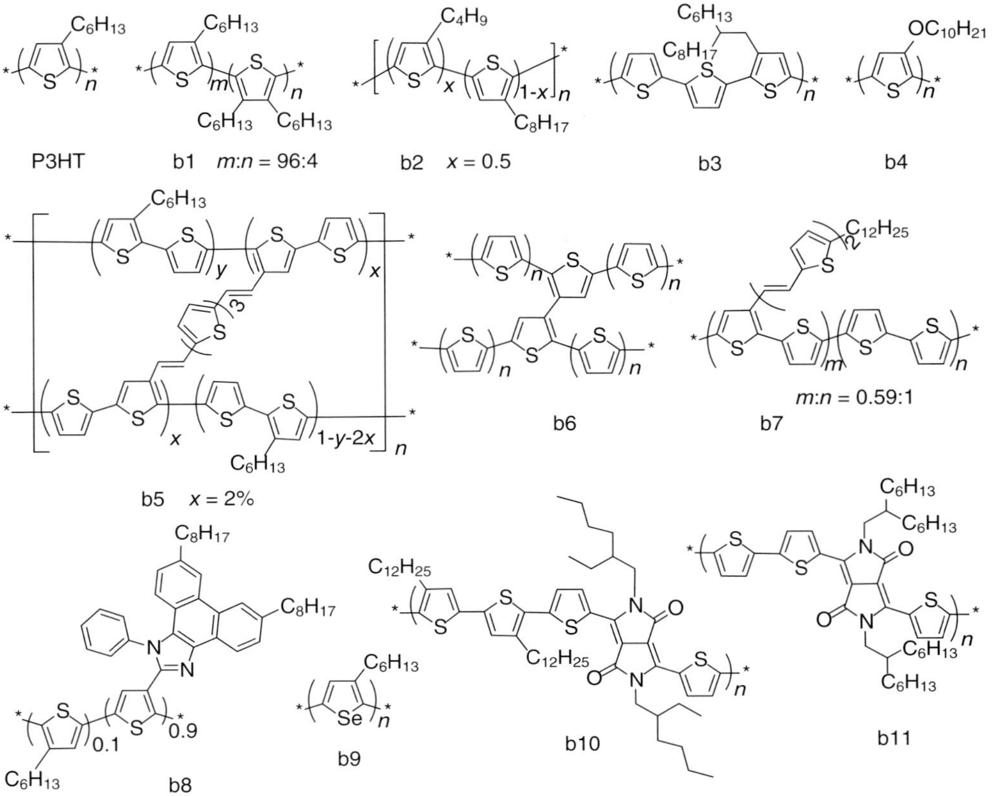

Figure 9.2 Chemical structures of polythiophene-based donors.

based on $IC_{60}BA$ or $IC_{70}BA$/P3HT reached 0.84 V, which was 0.26 V higher than that (0.58 V) of the devices with $PC_{61}BM$ as acceptor. Moreover, the PCE values of the devices were raised to 5.44% and 5.64%, respectively, which represented a >40% enhancement compared to that (3.88%) of the BHJ OPVs based on P3HT and $PC_{61}BM$. Following thermal annealing at 150 °C for 10 min, BHJ organic photovoltaic (OPV) devices based on P3HT and $IC_{60}BA$ provided a PCE of 6.48% [26]. By additive (3-hexylthiophene) processing, BHJ OPV devices based on P3HT and $IC_{70}BA$ showed PCE-values of up to 6.69% [27]–the highest values in OPV devices based on P3HT that have been reported to date.

Fréchet *et al.* introduced a small amount of 3,4-dihexylthiophene into P3HT in order to reduce the regioregularity and to weaken the crystallization formation [14]. Polymer **b1** exhibited a similar device efficiency but a better thermal stability compared to P3HT. This improvement was ascribed to a suppression of the crystallization-driven phase separation by the introduction of a controlled amount

Table 9.2 Electronic properties, mobilities, and PSC performance of polythiophene derivatives.[a)]

Polymer	λ_{max} (nm)	E_g^{opt} (eV)	μ_h [cm²/(V·s)]	HOMO/LUMO (eV)	D/PC$_{61}$BM (w/w)	J_{SC} (mA cm^{-2})	V_{OC} (V)	FF	PCE (%)	Reference
b1	–	–	–	–	1 1	9.77	0.61	0.62	3.73	[14]
b2	605	–	1.8×10^{-4}	–	4:3[b)]	7.92	0.56	0.67	3.0	[15]
b3	537	1.9	–	−5.3/–	1:1	6.33	0.82	0.66	3.40	[16]
b4	624	1.60	–	−4.47/−2.87	1:1	0.14	0.02	0.26	0.0007	[17]
b5	502	–	4.7×10^{-3}	−5.06/−2.77	1:1	6.82	0.67	0.38	1.72	[18]
b6	610	–	3.1×10^{-4}	−5.15/−3.10	1:1	–	0.55	0.47	2.0	[19]
b7	540	1.82	–	−4.93/−2.96	1:1	10.3	0.72	0.43	3.18	[20]
b8	634	1.80	–	−4.65/–	1:1	13.7	0.68	0.37	3.45	[21]
b9	690	1.6	7×10^{-5}	−4.8/−3.0	12:13	9.9	0.52	–	2.7	[22][c)]
b10	650[d)]	1.4	–	−5.10/−3.30	1:2[b)]	11.5	0.61	0.58	4.0	[23]
b11	800	1.30	0.04	−5.17/−3.61	1:2[b)]	11.8	0.65	0.60	4.7	[24]

a) λ_{max} = absorption peak in film; E_g = optical band gap; μ_h = hole mobility; D = donor; J_{SC} = short-circuit current density; V_{OC} = open-circuit voltage; FF = fill factor; PCE = power conversion efficiency.
b) Blend with PC$_{71}$BM.
c) Under 50 mW cm^{-2} illumination.
d) In chloroform.

of disorder into the polymer backbone. In order to investigate the effect of crystallinity of P3AT-based polymers, Jenekhe *et al.* synthesized copolymers of butylthiophene and octylthiophene, and showed that some polymers could form lamellar crystals [15, 28]. Thus, an improved hole mobility and a best PCE of 3% was achieved from **b2**.

One disadvantage of P3HT is its high HOMO level, leading to a low V_{OC}. Yang *et al.* downshifted the HOMO level of P3HT from −4.9 eV to −5.3 eV (**b3**) while keeping **the** band gap unchanged by reducing the number of alkyl chains. As a result, the V_{OC} was improved to 0.82 V, and a PCE of 3.4% was achieved that was higher than that of P3HT [16]. The introduction of an electron-donating alkoxyl group onto the 3-position of thiophene unit not only adjusted the solubility, but also affected the absorption, band gap, and HOMO level of polythiophenes. For example, the band gap of **b4** was decreased to 1.6 eV and the absorption maximum red-shifted to 624 nm. However, the HOMO level of **b4** was increased to −4.47 eV and the V_{OC} decreased to 0.02 V. The poor solubility of **b4** also destroyed the film morphology and finally limited the PCE [17].

The mobility of conjugated polymers plays an important role in efficient photovoltaic cells, with intermolecular charge hopping between neighboring main chains being the key factor that limits charge transport. Li *et al.* synthesized **b5** by using a vinylene–terthiophene–vinylene segment as a conjugated bridge [18]. These bridge segments provided paths for transporting charges between the main chains such that a mobility of 4.7×10^{-3} cm^2/(V·s) was achieved, which was much higher than that of polythiophenes. However, the introduction of these bridges caused a distortion of the main chain that led to the absorption spectra being blue-shifted and, finally, to the PCE being reduced. Subsequently, Scherf *et al.* reported a similar polymer (**b6**), which gave a PCE of 2% [19].

In an attempt to enhance absorbance in the short-wavelength region (<500 nm) of polythiophene, Li *et al.* attached a bi(thienylenevinylene) conjugated side chain to the polythiophene backbone (**b7**) [20, 29]. This polymer showed a broad and strong absorption in 350–650 nm region but, surprisingly, the conjugated side chains led to a 0.2 eV decrease in the HOMO level and a higher V_{OC}. A higher PCE of 3.18% was achieved compared to P3HT (2.41%) under the same test conditions, however. Wei *et al.* introduced electron-withdrawing conjugated groups onto the polythiophene side chain (**b8**), and achieved a best PCE of 3.45% [21]. The electron-withdrawing ability of the side chains induced an intramolecular charge transfer from the donor thiophene main chain to the acceptor side chain. Thus, charge separation could be facilitated through a sequential electron transfer from the main chain to the side chain, and then to $PC_{61}BM$. Such a faster electron transfer would lead to a higher PCE compared to P3HT.

Poly(3-hexylselenophene) (**b9**), an analog of P3HT in which selenium replaces the sulfur atom, exhibited a lower LUMO level but a similar HOMO level, a smaller bandgap, and a significantly red-shifted absorption compared to P3HT [22, 30]. Moreover, **b9** displayed a crystalline morphology and field-effect transistor (FET) mobility [0.02–0.04 $cm^2/(V \cdot s)$] that were similar to those of P3HT. The solar cell device based on **b9**/$PC_{61}BM$ (1 : 1, w/w) produced a PCE of 2.7% after thermal annealing, which was slightly lower than that of P3HT/$PC_{61}BM$ (3.0%).

In order to red-shift the absorption spectrum and to lower the HOMO level of polythiophene, Janssen *et al.* copolymerized electron-rich quaterthiophene or ter-thiophene with an electron-deficient diketopyrrolo pyrrole unit to afford the donor–acceptor (D–A) copolymers **b10** and **b11** [23, 24]. These two copolymers exhibited lower HOMO and LUMO levels, a smaller bandgap, and significantly red-shifted absorption compared to P3HT. The best device created by using **b10** or **b11**/$PC_{71}BM$ (1 : 2, w/w) gave PCEs of 4.0% and 4.7%, respectively.

9.2.3
Polyfluorene Derivatives

Molecules based on fused rings intend to maximize the π-orbital overlap, which is favorable for π–π stacking, and thus to facilitate charge transport by intermolecular hopping. Moreover, a fused ring system can make the polymer backbone more rigid and coplanar, which will lead in turn to a more effective conjugation, a narrower band gap, and a broadened absorption. Currently, fused-ring donor units (e.g., fluorene, carbazole, cyclopentadithiophene, benzodithiophene) and fused ring acceptor units (e.g., benzothiadiazole, quinoxaline, pyrazine) are most commonly used in the synthesis of photovoltaic polymers.

Polyfluorenes possess good thermal and photochemical stability, a low-lying HOMO level, and high mobility. However, their optical bandgap is ca. 3.3 eV, which is too large for the efficient harvesting of sunlight. However, the incorporation of

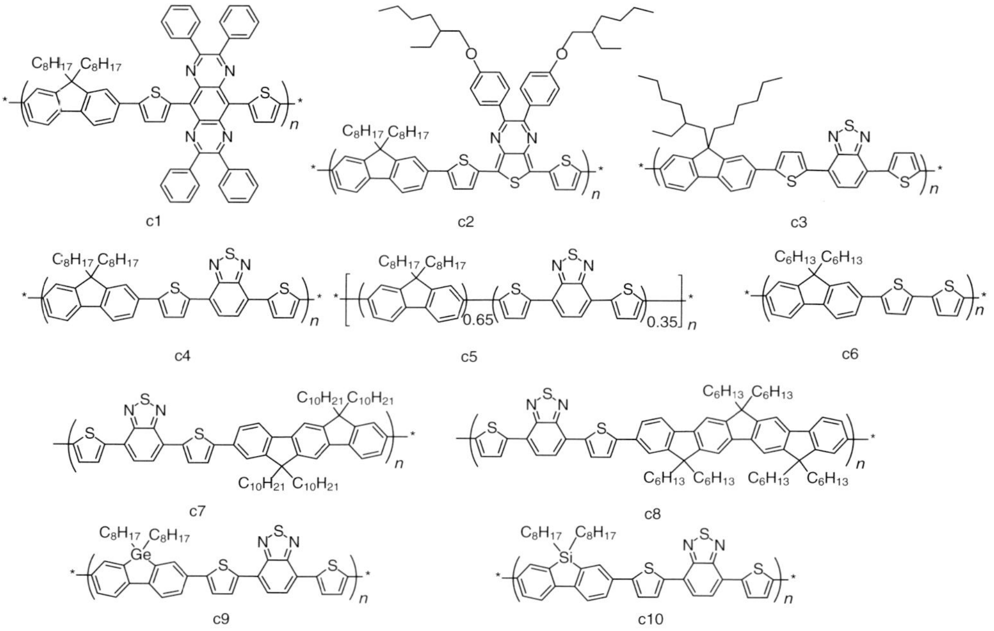

Figure 9.3 Chemical structures of polyfluorene-based donors.

electron-donating and/or -accepting units into the main chain can lower the bandgap and extend the absorption of these polymers [31–39].

Andersson *et al.* reported a series of polymers based on fluorene and quinoxaline derivatives (Figure 9.3, Table 9.3) [31, 32]. Among these, **c1** achieved a PCE of 1.7% when blended with $PC_{61}BM$, but when blended with $PC_{71}BM$, a higher PCE of 2.3% was achieved due to the improved absorption from $PC_{71}BM$. Compared to **c1**, polymer **c2** exhibited a lower optical band gap and higher HOMO and LUMO levels, which facilitated electron transfer from **c2** to PCBM. Therefore, PSCs based on **c2**/$PC_{61}BM$ exhibited an improved PCE of 2.2% [33].

Andersson *et al.* synthesized a fluorene–dithienylbenzothiadiazole copolymer (**c3**) with a bandgap of ca. 1.9 eV [34]. The PSC based on this polymer, in combination with $PC_{61}BM$, gave a PCE of 2.2% with a high V_{OC} of 1.04 V. A similar polymer with dioctyl groups at the 9-position of fluorene (**c4**) gave a higher PCE of 2.84% [35]. Replacing dioctyl with didecyl led to a further enhancement of PCE to 4.2%, which was the highest reported for fluorene-based polymers/$PC_{61}BM$ devices [36]. Thus, the substituents at 9-position of fluorene exhibited significant impacts on device performance, although the fluorene–dithienylbenzothiadiazole copolymers had similar electronic properties. A random copolymer of fluorene and dithienylbenzothiadiazole (**c5**) gave a PCE (2.24%) which was lower than that of its alternating copolymer **c4** [37, 38]. Removal of the electron-withdrawing group

Table 9.3 Electronic properties, mobilities, and PSC performance of polyfluorene derivatives.[a)]

Polymer	λ_{max} (nm)	E_g^{opt} (eV)	μ_h [cm^2/(V·s)]	HOMO/LUMO (eV)	D/PC$_{61}$BM (w/w)	J_{SC} (mA cm^{-2})	V_{OC} (V)	FF	PCE (%)	Reference
c1	710	1.75	–	−5.7/−3.9	1:3[b)]	6.5	0.81	0.44	2.3	[31]
c2	660	1.6	–	−5.0/−3.4	1:3	8.88	0.59	0.42	2.20	[33]
c3	545	–	–	–	1:4	4.66	1.04	0.46	2.2	[34]
c4	–	–	–	–	1:3	6.3	1.01	0.44	2.84	[35]
c5	551	2.01	–	−5.47/−3.46	1:2	5.18	0.95	–	2.24[c)]	[37]
c6	456	2.41	1.0×10^{-3}	−5.41/−2.52	1:4	4.96	1.03	0.53	2.7	[39]
c7	543	1.97	4.2×10^{-3}	−5.47/−3.44	1:3	6.10	1.00	0.40	2.44	[40]
c8	531	1.96	9.7×10^{-3}	−5.45/−3.45	1:4[b)]	10.3	1.04	0.42	4.50	[40]
c9	580	1.79	1.1×10^{-4}	−5.58/−3.91	–[b)]	6.9	0.79	0.51	2.8	[41]
c10	–	1.82	1.0×10^{-3}	−5.39/–	1:2	9.5	0.90	0.51	5.4	[42]

a) λ_{max} = absorption peak in film; E_g = optical band gap; μ_h = hole mobility; D = donor; J_{SC} = short-circuit current density; V_{OC} = open-circuit voltage; FF = fill factor; PCE = power conversion efficiency.
b) Blend with PC$_{71}$BM.
c) Under 78.2 mW cm^{-2} illumination.

benzothiadiazole from **c4** led to a narrow absorption, a wide band gap, and a high mobility [1.0×10^{-3} cm^2/(V·s)] (**c6**) [39]. PSCs based on **c6**:PC$_{61}$BM afforded a PCE of 2.7%.

Katz *et al.* synthesized ladder-type oligo-*p*-phenylene-containing D–A copolymers **c7** and **c8** [40]. When compared to fluorene, these ladder-type fluorene derivatives possessed a larger coplanar structure which led to a stronger absorption and electron-donating ability. In addition, the introduction of more alkyl chains onto the fused ring unit improved the solubility and processibility of these materials. Both, **c7** and **c8** exhibited almost the same band gaps, HOMO and LUMO levels, but PSCs based on **c8**:PC$_{61}$BM afforded a higher PCE of 3.67% compared to that of **c7**:PC$_{61}$BM (2.44%); this was attributed to the higher mobility of **c8**. Under the same conditions, a polyfluorene derivative with a similar main-chain structure gave a PCE of only 1.75%. On blending **c8** with PC$_{71}$BM, the PCE was improved to 4.5% [40].

Leclerc *et al.* synthesized a D–A-type copolymer based on germafluorene (**c9**) which was similar to **c4** [41]. Relative to **c4**, **c9** exhibited a red-shifted absorption, a lower band gap, and lower HOMO and LUMO levels. Devices based on **c9**:PC$_{71}$BM afforded a best PCE of 2.8%. Cao *et al.* described a PCE of 5.4% under illumination of AM1.5, 80 mW cm^{-2} from a copolymer of 2,7-silafluorene and dithienylbenzothiadiazole (**c10**):PC$_{61}$BM blend [42]. The high V_{oc} (0.9 V) was attributed to the low-lying HOMO (−5.39 eV), while the high J_{sc} was attributed to the higher mobility [1×10^{-3} cm^2/(V·s)] and broader absorption compared to its fluorene counterpart.

9.2.4 Polycarbazole Derivatives

Carbazole is an electron-rich and widely used building block for the synthesis of conjugated polymers. It has been shown that 2,7-linked carbazole polymers have a better π-conjugation and charge migration along the main chain compared to their 3,6-linked counterparts. Bo *et al.* described a copolymer that was composed of alternating 2,7-carbazole and dithienyl-benzothiadiazole with dioctoxy groups on benzothiadiazole (**d1**) which gave a PCE of 5.4% in combination with $PC_{71}BM$ (Figure 9.4, Table 9.4) [43]. The PSC based on **d1**/$PC_{71}BM$ exhibited a very high FF (0.69), indicating a balanced charge transport in the device. Indeed, FET measurements on the blend used in optimized solar cells confirmed a quite balanced mobility of $1 \times 10^{-4}\,cm^2/(V\cdot s)$ for hole and $3 \times 10^{-4}\,cm^2/(V\cdot s)$ for electron under

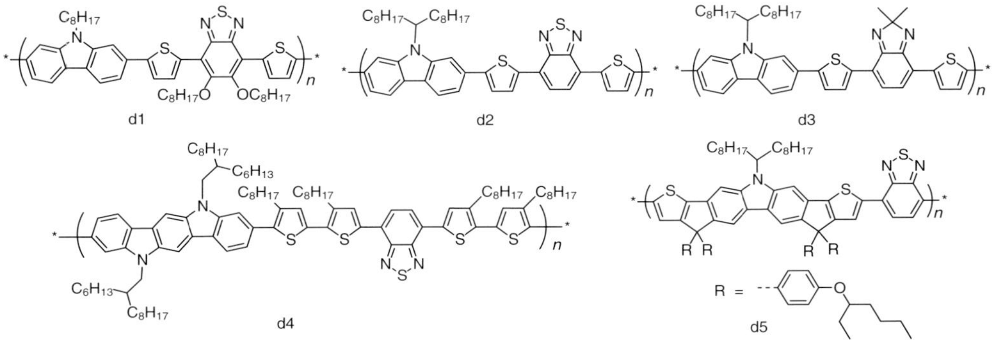

Figure 9.4 Chemical structures of carbazole-based polymer donors.

Table 9.4 Electronic properties, mobilities, and PSC performance of polycarbazole derivatives.[a)]

Polymer	λ_{max} (nm)	E_g^{opt} (eV)	μ_h [cm²/(V·s)]	HOMO/LUMO (eV)	D/$PC_{61}BM$ (w/w)	J_{SC} (mA cm^{-2})	V_{OC} (V)	FF	PCE (%)	Reference
d1	579	1.95	1.0×10^{-4}	−5.21/−3.35	1:2.5	9.6	0.81	0.69	5.4	[43]
d2	–	1.88	1.0×10^{-3}	−5.45/−3.60	1:4[b)]	10.6	0.88	0.66	6.1	[44]
d3	638	1.64	2.2×10^{-3}	−5.47/−3.82	1:4[b)]	10.0	0.65	0.48	3.12	[45]
d4	538[c)]	1.89	–	−5.17/−3.15	1:2	9.17	0.69	0.57	3.6	[46]
d5	607	1.66	1.0×10^{-4}	−5.38/−3.61	1:2[b)]	10.7	0.80	0.43	3.70	[47]

a) λ_{max} = absorption peak in film; E_g = optical band gap; μ_h = hole mobility; D = donor; J_{SC} = short-circuit current density; V_{OC} = open-circuit voltage; FF = fill factor; PCE = power conversion efficiency.
b) Blend with $PC_{71}BM$.
c) In $CHCl_3$ solution.

vacuum. When Leclerc *et al.* synthesized a copolymer composed of alternating 2,7-carbazole and dithienyl-benzothiadiazole without alkoxy groups on benzothiadiazole (**d2**) [48, 49], the PSC device based on **d2**:$PC_{61}BM$ showed a PCE of 3.6% at AM1.5, 90 mW cm^{-2}. The PCE of solar cells based on **d2** was increased from 3.6% to 6.1% by replacing $PC_{61}BM$ with $PC_{71}BM$, together with titanium oxide as the optical spacer and hole-blocking layer [44].

Suh *et al.* synthesized a copolymer composed of alternating 2,7-carbazole and dithienyl-dimethyl benzoimidazole (**d3**) [45] which exhibited a higher mobility of 2.2×10^{-3} cm^2/(V·s), a lower band gap (1.64 eV) and a lower LUMO level (−3.82) relative to its benzothiadiazole counterpart (**d2**). The blending of **d3** with $PC_{71}BM$ afforded a PCE of 3.12%, which was lower than that of **d2**.

Indolo[3,2-*b*]carbazole, which can be regarded as two carbazoles condensed together, is a large coplanar π-conjugated structure that represents a promising electron-donating building block for the synthesis of conjugated polymers. Lu *et al.* synthesized a copolymer consisting of benzothiadiazole, oligothiophene and indolocarbazole units (**d4**) [46]. The absorption edge in the solid state was significantly red-shifted (by 100 nm) compared to that in solution, which suggested that a strong intermolecular interaction had been induced by the crystallinity. The PSC based on a blend of **d4**/$PC_{61}BM$ exhibited a PCE of 3.6%. Cheng *et al.* reported another larger coplanar carbazole-related unit which contained seven fused rings and could be copolymerized with benzothiadiazole to yield **d5**, which gave a PCE of 3.7% when blended with $PC_{71}BM$ [47].

9.2.5
Polybenzodithiophene Derivatives

Benzo[1,2-*b*:4,5-*b*′]dithiophene (BDT) possesses a relatively large coplanar structure which is favorable for π–π stacking. Recently, BDT has been recognized as a very promising donor unit in the synthesis of photovoltaic polymers and, indeed, the highest PCE worldwide has been achieved with PSCs based on BDT polymers (Figure 9.5, Table 9.5). Yu *et al.* synthesized a copolymer of BDT with thienothiophene (**e1**) [50], in which the thienothiophene promoted a stabilization of the quinoid structure and planarity along the polymer backbone. The ester-substituent on thienothiophene enhanced the solubility and oxidation stability of the polymer. PSCs based on **e1**/$PC_{61}BM$ yielded a PCE of 4.76%, but this was improved to 5.3% by using $PC_{71}BM$ instead of $PC_{61}BM$. These high PCEs were related to the rigidity and planarity of the fused-ring backbone. Later, an electron-withdrawing fluorine atom was introduced into thienothiophene to produce **e2**, with a lowering of the HOMO level [51]. The V_{oc} and PCE of PSCs based on **e2**/$PC_{61}BM$ were increased to 0.74 V and 5.9%, respectively. In order to further downshift the HOMO level, Hou *et al.* replaced the ester group in **e1** and **e2** with a ketone group; this led to a reduction in the HOMO levels of both **e1** and **e2** to 0.2 eV (**e3**) [52] and 0.1 eV (**e4**) [53], respectively. The average PCE of **e3**:$PC_{71}BM$ reached 6.3%, with a best value of 6.58% [52]. In particular, PSCs based on **e4**:$PC_{71}BM$ gave a very high J_{sc} (15.2 mA cm^{-2}), a high V_{oc} (0.76 V), a high FF (0.67), and a very high PCE (7.73%)

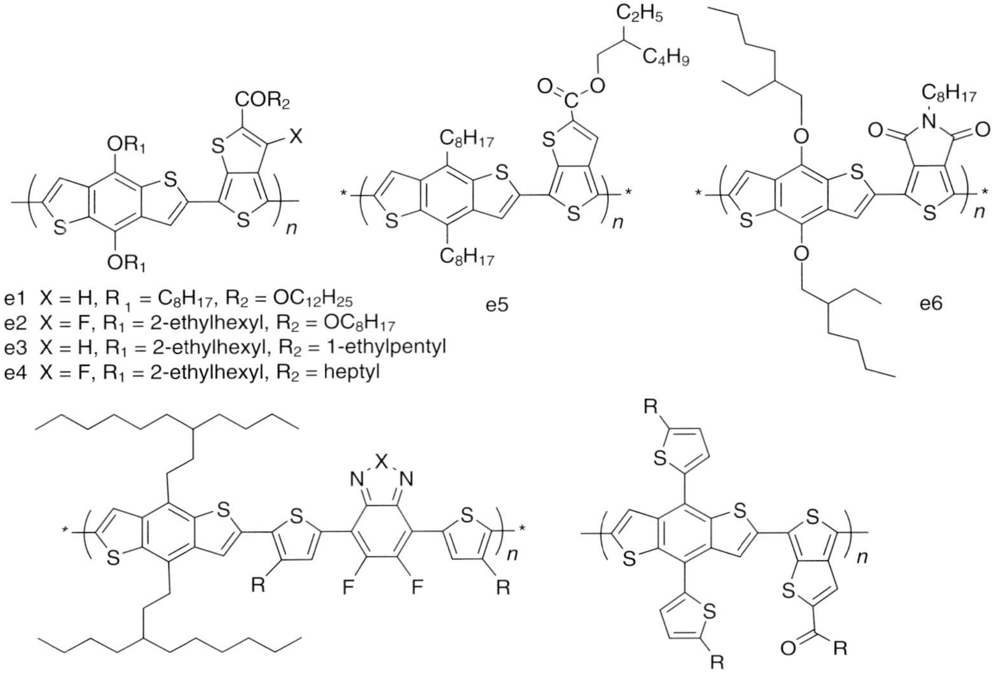

e1 X = H, R_1 = C_8H_{17}, R_2 = $OC_{12}H_{25}$
e2 X = F, R_1 = 2-ethylhexyl, R_2 = OC_8H_{17}
e3 X = H, R_1 = 2-ethylhexyl, R_2 = 1-ethylpentyl
e4 X = F, R_1 = 2-ethylhexyl, R_2 = heptyl

e5

e6

e7 X = S, R = 2-ethylhexyl
e8 X = N-2-butyloctyl, R = H

e9 R = 2-ethylhexyl

Figure 9.5 Chemical structures of benzodithiophene-based polymer donors.

Table 9.5 Electronic properties, mobilities, and PSC performance of polybenzodithiophene derivatives.[a)]

Polymer	λ_{max} (nm)	E_g^{opt} (eV)	μ_h [cm^2/(V·s)]	HOMO/LUMO (eV)	D/PC$_{61}$BM (w/w)	J_{SC} (mA cm^{-2})	V_{OC} (V)	FF	PCE (%)	Reference
e1	690	1.6	4×10^{-4}	−4.9/−3.2	1:1	12.5	0.58	0.65	4.76	[50]
e2	682	1.6	8×10^{-4}	−5.1/−3.3	1:1	13.0	0.74	0.61	5.90	[51]
e3	690	1.6	2×10^{-4}	−5.1/−3.6	1:1.5[b)]	14.7	0.70	0.64	6.58	[52]
e4	–	1.61	7×10^{-4}	−5.22/−3.45	1:1.5[b)]	15.2	0.76	0.67	7.73	[53]
e5	682	1.60	7×10^{-4}	−5.04/−3.29	1:1	13.1	0.73	0.57	5.53	[51]
e6	627	1.7	–	−5.56/−3.75	1:1.5	11.5	0.85	0.68	6.8	[54]
e7	700	1.7	–	−5.54/−3.33	1:1	12.91	0.91	0.61	7.2	[55]
e8	–	2.0	1×10^{-3}	−5.36/−3.05	1:2	11.83	0.79	0.73	7.10	[56]
e9	692	1.58	1×10^{-2}	−5.11/−3.25	1:1.5[b)]	17.48	0.74	0.59	7.59	[57]

a) λ_{max} = absorption peak in film; E_g = optical band gap; μ_h = hole mobility; D = donor; J_{sc} = short-circuit current density; V_{oc} = open-circuit voltage; FF = fill factor; PCE = power conversion efficiency.
b) Blend with PC$_{71}$BM.

[53]. Cao *et al.* incorporated an alcohol/water-soluble conjugated polymer, poly [(9,9-bis(3′-(*N,N*-dimethylamino)propyl)-2,7- fluorene)-*alt*-2,7-(9,9-dioctylfluorene)] (PFN) as a novel cathode interlayer in PSCs based on $PC_{71}BM$ and a thieno[3,4-*b*]-thiophene/benzodithiophene copolymer (using 2-ethylhexyl instead of heptyl in **e4**) [1]. Those PSCs which incorporated the PFN interlayer showed a significant and simultaneous enhancement in J_{sc}, V_{oc}, and FF, leading to a PCE of 8.37%, which was the best reported value to date for PSCs. Yu *et al.* found that replacing the octyl group on BDT in **e5** with an octyloxy group led to a 0.1 eV upshift of the HOMO level, a 0.14 V decrease in V_{oc}, and a 0.43% decrease in PCE [51].

The electron-withdrawing unit thienopyrroledione possesses a simple, compact, symmetry and coplanar structure which is favorable for electron delocalization and intramolecular charge transfer. Indeed, its electron-withdrawing ability is able to lower HOMO levels and to improve V_{OC}. Thienopyrroledione can also be alkylated with various types of alkyl group in order to tune the solubility, and this led to the synthesis of copolymers of BDT and thienopyrroledione [54, 58]. As an example, compound **e6** exhibited a low band gap (1.7 eV) and a low-lying HOMO level (−5.56 eV). PSCs based on **e6**/$PC_{61}BM$ blend yielded a high V_{OC} of 0.85 V, a high FF of 0.68, and a best PCE of 6.8% [54].

You *et al.* reported a copolymer of BDT and dithienyl-benzothiadiazole (**e7**), and attached two electron-withdrawing fluorine atoms onto benzothiadiazole to lower the HOMO level [55]. Compound **e7** exhibited a low band gap (1.7 eV) and a low-lying HOMO level (−5.54 eV). PSCs based on an **e7**/$PC_{61}BM$ blend yielded a high V_{OC} of 0.91 V and a best PCE of 7.2%. The replacement of fluorated benzothiadiazole in **e7** with fluorated benzo[*d*][1,2,9]triazole (**e8**) led to larger band gap, a higher HOMO level and a lower V_{OC}, but to a higher FF and a similar PCE [56].

Huo *et al.* reported copolymers of 4,8-bis(2-thienyl)-benzo[1,2-*b*:4,5-*b*′]dithiophene [57, 59]. Relative to the alkoxy-substituted BDT polymer **e3**, thiophene-substituted BDT polymer **e9** exhibited a similar band gap and HOMO level but a higher mobility, which led to a higher J_{sc} (17.48 mA cm^{-2}) and a higher PCE (7.59%) [57].

9.2.6 Polycyclopentadithiophene Derivatives

Dithienocyclopentadiene (DTC) derivatives have attracted considerable attention due to their fully coplanar structure, low bandgap, and strong intermolecular interactions. Moreover, as a bridging atom, the carbon atom of DTC can be replaced by silicon, germanium, nitrogen, and sulfur atoms to adjust the electronic and steric structures (Figure 9.6, Table 9.6). Mühlbacher *et al.* synthesized a copolymer of DTC and benzothiadiazole (**f1**) [69] which exhibited a narrow bandgap (1.4 eV), a low HOMO (−5.3 eV), and a high mobility [2×10^{-2} cm^2/(V·s)]. When blended with $PC_{61}BM$ and $PC_{71}BM$, these PSC devices gave PCEs of 2.67% and 3.5%, respectively; however, by adding a small amount of 1,8-octanedithiols into the **f1**:$PC_{71}BM$ solution prior to spin-coating, the device efficiency was further improved to 5.5%. This was due to the formation of a better bulk morphology

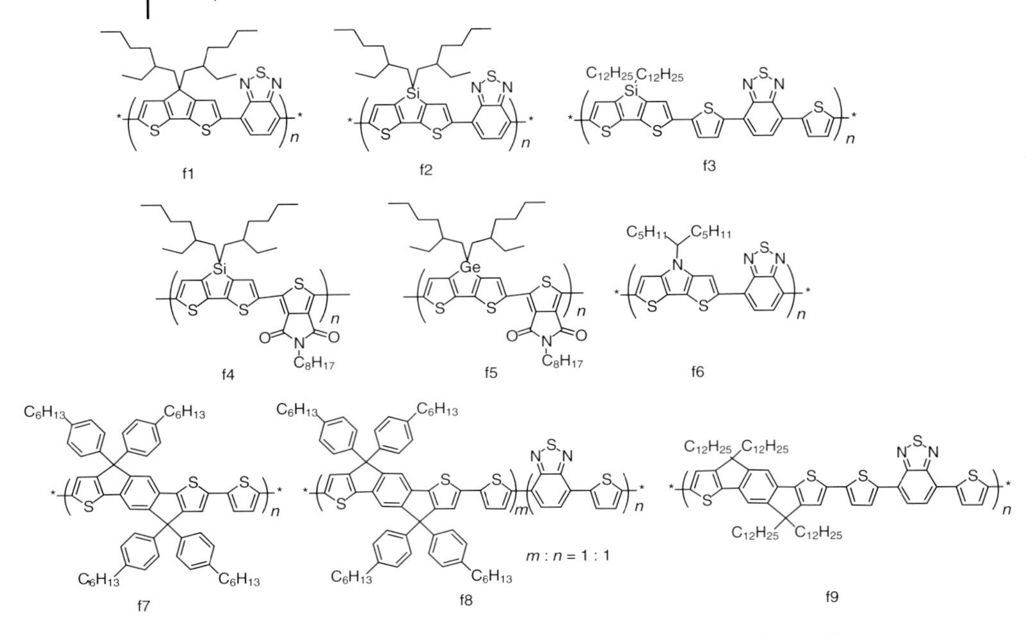

Figure 9.6 Chemical structures of cyclopentadithiophene-based polymer donors.

Table 9.6 Electronic properties, mobilities, and PSC performance of polycyclopentadithiophene derivatives.[a)]

Polymer	λ_{max} (nm)	E_g^{opt} (eV)	μ_h [cm²/(V·s)]	HOMO/LUMO (eV)	D/PC$_{61}$BM (w/w)	J_{SC} (mA cm^{-2})	V_{OC} (V)	FF	PCE (%)	Reference
f1	775	1.45	2×10^{-2}	−5.3/−3.6	1:2[b)]	16.2	0.62	0.55	5.5	[60]
f2	740	1.45	3.0×10^{-3}	−5.05/−3.27	1:1[b)]	12.7	0.68	0.55	5.1	[61]
f3	644	1.51	3.6×10^{-6}	−5.02/−3.19	1:1[b)]	10.67	0.62	0.52	3.43	[62]
f4	670	1.73	1×10^{-4}	−5.57/−3.88	1:2[b)]	12.2	0.88	0.68	7.3	[63]
f5	679	1.69	–	−5.60/−3.50	–[b)]	12.6	0.85	0.68	7.3	[64]
f6	771	1.43	–	−4.81/−3.08	1:3	11.9	0.54	0.44	2.80	[65]
f7	510	2.08	8.3×10^{-4}	−5.10/−3.02	1:3[b)]	7.6	0.80	0.54	3.3	[66]
f8	590	1.70	3.4×10^{-3}	−5.43/−3.66	1:3[b)]	10.1	0.80	0.53	4.3	[67]
f9	615	1.68	2.2×10^{-3}	−5.24/−3.10	1:3[b)]	13.27	0.82	0.57	6.17	[68]

a) λ_{max} = absorption peak in film; E_g = optical band gap; μ_h = hole mobility; D = donor; J_{SC} = short-circuit current density; V_{OC} = open-circuit voltage; FF = fill factor; PCE = power conversion efficiency.
b) Blend with PC$_{71}$BM.

which enhanced both the photoconductivity and charge carrier lifetime [60]. Heeger *et al.* reported an efficient tandem cell (PCE = 6.5%) with two active layers composed of P3HT and **f1** [70].

Replacing the carbon atom with silicon in DTC yields a dithienosilole (DTS), in which a silole ring is condensed with dithiophene [61, 71–73]. Yang *et al.* described a DTS–benzothiadiazole copolymer (**f2**) which was structurally similar to **f1** [61] but exhibited a low optical band gap (1.45 eV) and a high hole mobility [3×10^{-3} cm^2/(V·s)]. The PSC based on **f2**:$PC_{71}BM$ exhibited a PCE of 5.1%. Replacing 2-ethylhexyl group with *n*-dodecyl in **f2** also gave a high PCE (5.0% with $PC_{71}BM$) [73]. A copolymer of DTS and dithienylbenzothiadiazole (**f3**) exhibited a significantly blue-shifted (ca. 100 nm) absorption, a three orders of magnitude lower mobility, and a much lower PCE (3.43%) compared to its analog **f2** [62].

Replacing benzothiadiazole in **f2** with thienopyrroledione (**f4**) led to a larger band gap (1.73 eV), a lower HOMO level (–5.57 eV), a higher V_{OC} (0.88 V), a higher FF (0.68) and a higher PCE (7.3%) [63]. Replacing Si atom in **f4** with the larger Ge atom (**f5**) led to a red-shifted absorption, higher HOMO level, higher J_{sc}, and higher PCEs (7.3% for **f5**; 6.6% for **f4**) [64].

In addition to the tricyclic DTC and DTS blocks used as excellent donor components in low-bandgap polymers, dithieno[3,2-*b*:2′,3′-*d*]pyrrole (DTP) has attracted considerable attention as another fused bithiopene member having a pyrrole ring condensed with dithiophene due to strong electron-donating and charge-transporting properties. Geng *et al.* synthesized a copolymer which consisted of alternating DTP and benzothiadiazole units (**f6**) [65]. Relative to DTC and DTS, the strong electron-donating DTP upshifted the HOMO level and reduced the V_{OC}. Although **f6** showed a strong absorption in the wavelength range of 600–900 nm, the PSC based on **f6**:$PC_{61}BM$ gave a lower PCE of 2.80%.

Ko *et al.* synthesized a copolymer of pentacyclic fused thiophene–phenylene–thiophene (TPT) unit with thiophene (**f7**) [66]. This rigid planar structure facilitated an effective π-conjugation and strong intermolecular interactions. The FET hole mobility of **f7** was 8.3×10^{-4} cm^2/(V·s). **f7**, in combination with $PC_{61}BM$ and $PC_{71}BM$, displayed PCEs of 2.2% and 3.3%, respectively. The introduction of benzothiadiazole into the main chain of **f7** led to a lower bandgap (1.7 eV) and a higher hole mobility of 3.4×10^{-3} cm^2/(V·s) (**f8**) [67]. The blending of **f8** with $PC_{61}BM$ and $PC_{71}BM$ gave PCEs of 2.5% and 4.3%, respectively. Li *et al.* synthesized an alternating copolymer of TPT with dodecyl side chains and dithienylbenzothiadiazole (**f9**) [68]. Relative to **f8**, **f9** exhibited a red-shifted absorption, a higher HOMO level, a higher J_{sc}, and a higher PCE (6.17%).

9.2.7
Metallic Conjugated Polymers

Metallic conjugated polymers have several advantages:

- A metal atom is favorable for the assembly of building blocks.
- A metal atom is able to provide an oxidation center and a paramagnetic center which could transfer charges, and thus affect π systems and absorption properties.
- HOMO and LUMO levels can be adjusted through the *d*-orbital and coordination orbital of a metal atom.

- Due to the difference of metal valence bonds and coordination numbers, the molecular backbone configuration could be varied.
- Spin–orbital coupling facilitates the generation of triplet excitons, which have a long lifetime and a long diffusion length.

Porphyrin derivatives possess large π-conjugated system and good photochemical and thermal stabilities; consequently they have attracted much attention in optoelectronic applications. To date, few reports have been made on porphyrin in photovoltaic material applications, and indeed the photovoltaic properties obtained have not been good. The main reason for the low PCEs is the limited light absorption of porphyrin, the absorption spectra of which exhibits a narrow and strong Soret band (410–430 nm), weak Q-bands (530–540 nm), and a blank between the Soret and Q-bands. Zhan *et al.* utilized porphyrin-based conjugated polymers (**g1** and **g2**) for photovoltaic applications, and achieved a PCE of 0.3% which, at the time, was a leading efficiency of porphyrin-based polymers in photovoltaic applications (Figure 9.7, Table 9.7) [74].

The introduction of platinum into the main chain of poly(aryleneethynylene)s (PAEs) enhances π-conjugation and electron delocalization along the polymer

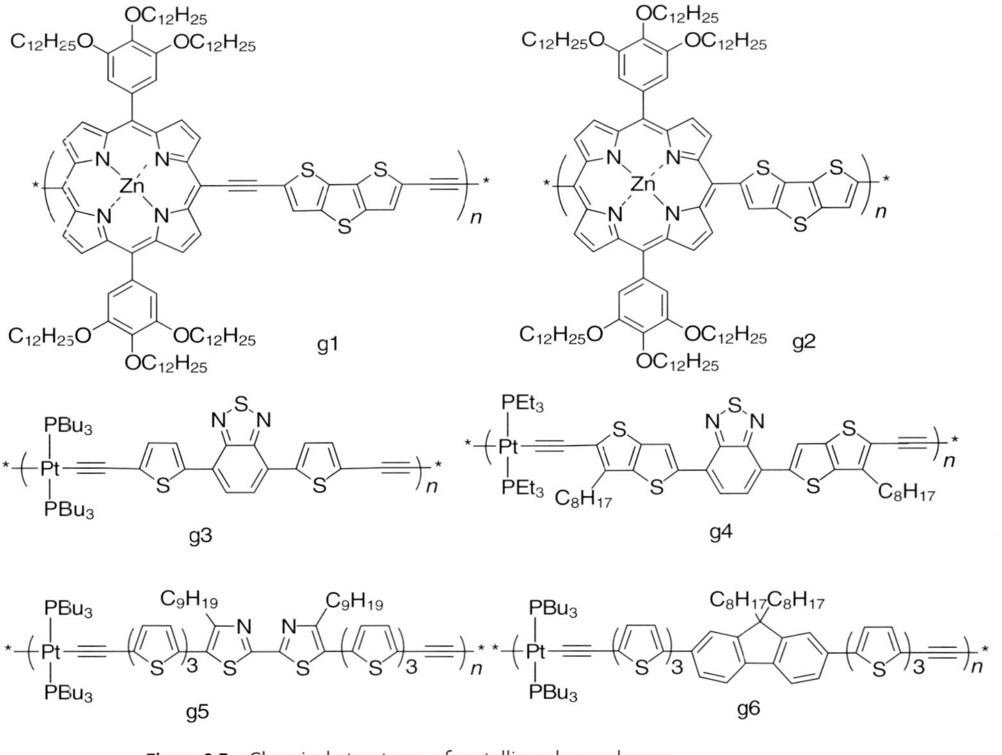

Figure 9.7 Chemical structures of metallic polymer donors.

Table 9.7 Electronic properties, mobilities, and PSC performance of metallic conjugated polymers.[a)]

Polymer	λ_{max} (nm)	E_g^{opt} (eV)	μ_h [cm^2/(V·s)]	HOMO/LUMO (eV)	D/PC$_{61}$BM (w/w)	J_{SC} (mA cm^{-2})	V_{OC} (V)	FF	PCE (%)	Reference
g1	–	–	2.1×10^{-4}	−5.5/−3.6	1:2	1.29	0.46	0.26	0.15	[74]
g2	–	–	1.1×10^{-4}	−5.2/−3.3	1:3	1.52	0.58	0.34	0.30	[74]
g3	554	1.85	–	−5.37/−3.14	1:4	15.43	0.82	0.39	4.93	[75]
g4	611	1.81	0.01	−5.14/−3.33	1:4[b)]	9.61	0.79	0.49	3.73	[76]
g5	603	2.06	–	−5.71/−3.65	1:5	6.50	0.88	0.44	2.50	[77]
g6	457[c)]	2.33	–	−5.73/−3.89	1:5	7.56	0.89	0.43	2.88	[78]

a) λ_{max} = absorption peak in film; E_g = optical band gap; μ_h = hole mobility; D = donor; J_{SC} = short-circuit current density; V_{OC} = open-circuit voltage; FF = fill factor; PCE = power conversion efficiency.
b) Blend with PC$_{71}$BM.
c) In CH_2Cl_2 solution.

chain, due to an overlap of the *d*-orbital of the Pt with the *p*-orbital of the alkyne unit. On the other hand, efficient intersystem crossing facilitates the formation of triplet excited states due to a strong spin–orbital coupling in this system. It is well known that triplet excited states have longer lifetimes, and thus allow longer exciton diffusion lengths compared to the singlet excited states [75, 79, 80]. Wong *et al.* described a platinum-containing DβA PAE polymer (**g3**) [75]; subsequently, PSCs based on **g3**/PC$_{61}$BM gave a J_{sc} of 15.43 mA cm^{-2}, a V_{oc} of 0.82 V, and a PCE of 4.93%. Jen *et al.* synthesized a similar polymer (**g4**) by using thieno[3,2-*b*]thiophene instead of thiophene in **g3** [76]. Despite this polymer being amorphous, a high FET mobility of 1×10^{-2} cm^2/(V·s) was achieved. The average PCEs of PSCs based on **g4**:PC$_{61}$BM or PC$_{71}$BM were 2.22% and 3.73% (best PCE, 4.13%), respectively, without thermal annealing.

Jenekhe *et al.* synthesized a series of Pt-bridged organometallic D–A-conjugated PAE polymers, incorporating various electron acceptors. By varying the electron-accepting strength in the D–A architecture, it was possible to demonstrate the molecular engineering of the absorption bands, as well as the HOMO/LUMO levels, charge transport, and photovoltaic properties. The highest PCE (2.41%) was observed in blends of the polymers with PC$_{71}$BM [81]. When Wong *et al.* synthesized platinum-containing D–A PAE polymers based on oligothiophenes (**g5** and **g6**) [77, 78, 82], the band gap became narrower and the PCE higher with increasing thiophene numbers. The PCEs of **g5** or **g6**:PC$_{61}$BM-based PSCs were 2.5% and 2.88%, respectively.

9.3 Polymer Acceptor Materials

With the rapid development of polymer donor materials, the PCEs of the PSCs have now reached over 8%. Although the acceptors are equally important as the donors in high-performance PSCs, much less research effort has been devoted to the acceptors than to the donors. Currently, fullerenes and their derivatives still dominate the acceptors, due to their strong tendency to accept electrons from donor semiconducting materials, their high electron mobilities in the films (even in composite form), and their favorable nanoscale morphological network with donors. Nevertheless, there remain incentives for the development of nonfullerene acceptors that not only are able to retain the favorable electron-accepting and -transporting properties of fullerenes, but also will overcome their insufficiencies, such as the limited spectral breadth and bandgap variability.

Perylene diimide (PDI) small molecules and polymers have attracted interest as alternative acceptor materials, as they exhibit large absorptivities, high electron mobilities, and electron affinities that are similar to those of the fullerenes. Zhan *et al.* reported the synthesis of the first soluble rylene-based fully conjugated polymer; this copolymer of PDI and dithienothiophene (**h1**; see Figure 9.8 and Table 9.8) exhibited abroad absorption (300–850 nm), a low band gap (1.7 eV), and a high electron mobility [$1.3 \times 10^{-2}\,cm^2/(V \cdot s)cm^2$] [83]. All-polymer solar cells based on **h1** acceptor in conjunction with a bis(thienylenevinylene)-substituted polythiophene (2TV-PT) donor gave a PCE of 1.03%. All-polymer solar cells based on a related donor tris(thienylenevinylene)-substituted polythiophene (3TV-PT) and a related acceptor PDI-bis(dithienothiophene) polymer (**h2**) exhibited a PCE as high as 1.48% by optimizing the donor/acceptor ratio [84]. Increasing the dithienothiophene number from one (**h1**) to three (**h3**) in the main chain of the PDI copolymer led to a lower PCE for PDI polymer:3TV-PT device (0.77%) [85].

Hou *et al.* reported a similar PDI copolymer (**h4**) which contained cyclopenta [2,1-*b*:3,4-*b*′] dithiophene [86] and exhibited similar HOMO/LUMO levels and optical band (E_g) to the PDI–DTT polymers. A PSC device with **h4** as electron acceptor and poly{4,4′-bis(2-ethylhexyl) dithieno [3,2-*b*:2′,3′-*d*]silole-2,6-diyl-*alt*-(2,1,3-benzothiadiazole-4,7-diyl)} (PDTSBTD) as electron donor exhibited a PCE of 0.43%. Hashimoto *et al.* used a PDI–DTP copolymer (**h5**) as an acceptor in PSC [87]. A PCE of 0.93% was achieved in a PSC device based on the blend of **h5** and 3TV-PT. Replacing DTP in **h5** with carbazole (**h6**) led to a higher PCE (1.15%), which was further improved to 2.33% after solvent optimization [88]. Copolymers of PDI with bithiophene (**h7**) or terthiophene (**h8**) were also used as acceptors in PSCs [89]. All-polymer solar cells based on the blend of **h8** and poly(3-hexylthiophene) (P3HT) showed a PCE of 0.8%. Replacing terthiophene in **h8** with bithiophene led to blue shift of the absorption maximum and a lower PCE (0.4%).

PDI polymers **h9** and **h10** are connected through the imide groups rather than its bay position. By using an alternating PDI–phenylenevinylene copolymer (**h9**) acceptor and poly(3-phenylhydrazone thiophene) (PPHT) donor in PSCs, Mikroyannidis *et al.* obtained a PCE of 2.3% under white-light illumination calibrated to an AM1.5 intensity of $30\,mW\,cm^{-2}$ after annealing at 80 °C for 10 min [90]. Jenekhe *et al.* reported a high-electron mobility [$0.1\,cm^2/(V \cdot s)$] ladder polymer (**h10**), that was closely related to PDI polymers and which was used to fabricate bilayer PSCs in conjunction with a PPV donor, with a PCE of 1.5% at AM1.5, $80\,mW\,cm^{-2}$ [91].

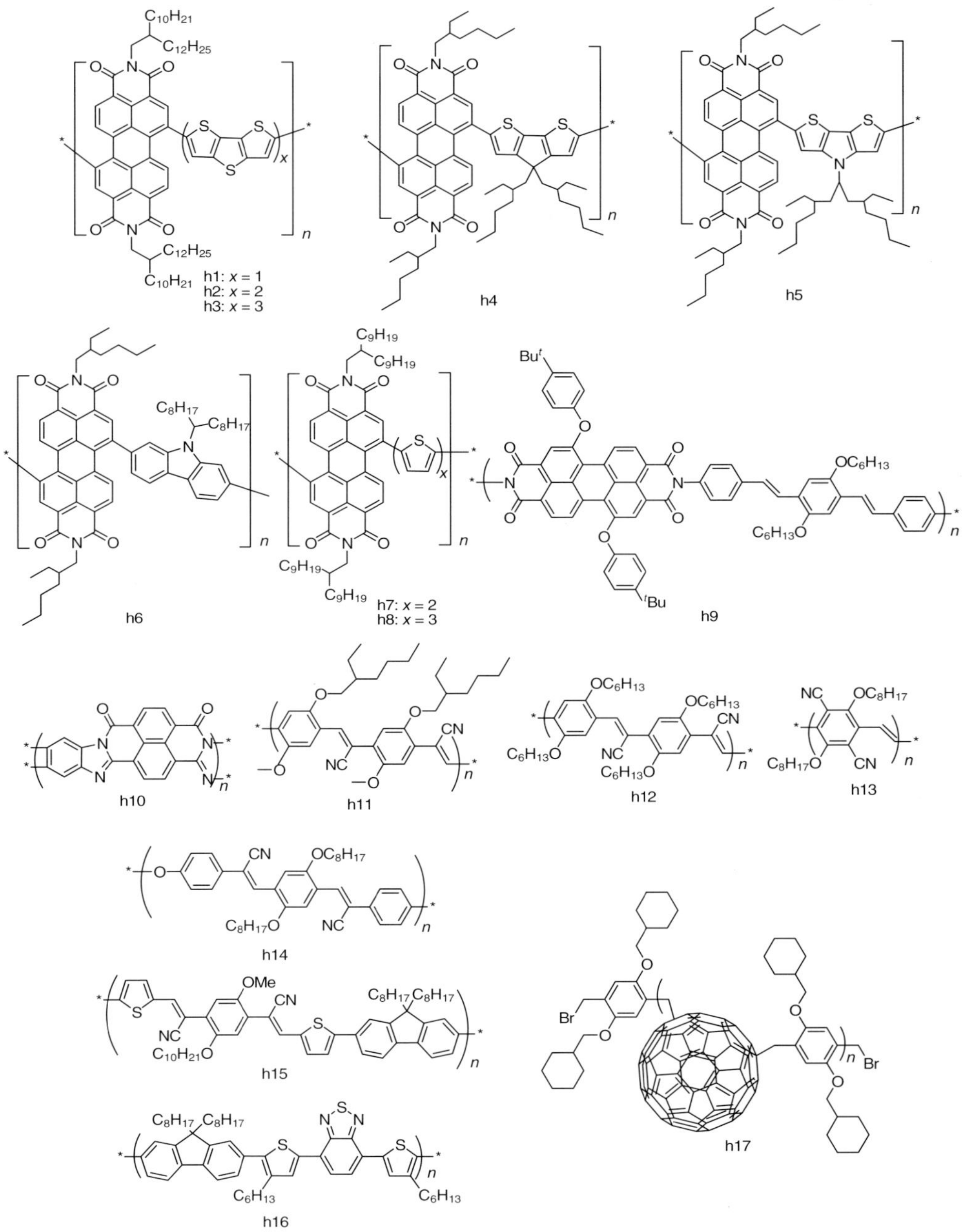

h1: $x = 1$
h2: $x = 2$
h3: $x = 3$

h4

h5

h6

h7: $x = 2$
h8: $x = 3$

h9

h10

h11

h12

h13

h14

h15

h16

h17

Figure 9.8 Chemical structures of polymer acceptors.

Table 9.8 Electronic properties, mobilities, and PSC performance of polymer acceptors.[a)]

Polymer	λ_{max} (nm)	E_g^{opt} (eV)	μ_h [cm^2/(V·s)]	HOMO/LUMO (eV)	Donor	J_{SC} (mA cm^{-2})	V_{OC} (V)	FF	PCE (%)	Reference
h1	630	1.7	1.0×10^{-2}	−5.8/−3.9	2TV-PT	4.2	0.63	0.39	1.03	[83]
h2	647	1.5	–	−5.5/−3.8	3TV-PT	5.02	0.69	0.43	1.48	[84]
h3	678	1.5	–	−5.4/−4.0	3TV-PT	2.80	0.69	0.40	0.77	[85]
h4	708	1.5	–	−5.52/−4.02	PDTSBTD	1.4	0.68	–	0.43	[86]
h5	715	1.32	–	−5.49/−3.83	3TV-PT	3.05	0.66	0.46	0.93	[87]
h6	–	–	–	−5.83/−3.66	3TV-PT	6.35	0.70	0.50	2.23	[88]
h7	600	1.65	–	−5.89/−4.23	P3HT	2.33	0.54	0.32	0.40	[89]
h8	620	1.6	–	−5.94/−4.34	P3HT	2.81	0.56	0.51	0.80	[89]
h9	504	1.7	8×10^{-3}	−5.8/−4.0	PPHT	2.98	0.60	0.39	2.32[b)]	[90]
h10	590	1.7	0.1	−5.9/−4.0	PPV	2.15	1.10	0.50	1.5[c)]	[91]
h11	–	–	–	−/3.7	POPT	–	–	–	2.0	[92]
h12	–	–	–	–	POPT	–	–	–	1.9[d)]	[93]
h13	–	2.0	–	−5.8/−3.7	PV-PT	3.14	0.85	0.29	0.8	[94]
h14	–	–	–	−6.1/−3.4	PPV2	3.57	1.36	0.35	1.70	[95]
h15	473	2.2	–	−5.4/−3.1	PPV3	3.0	1.40	0.37	1.5	[96]
h16	–	–	8.0×10^{-5}	−5.4/−3.2	P3HT	–	–	–	1.80	[97]
h17	450	–	–	−/−4.0	P3HT	–	–	–	1.6	[98]

a) λ_{max} = absorption peak in film; E_g = optical band gap; μ_h = hole mobility; J_{SC} = short-circuit current density; V_{OC} = open-circuit voltage; FF = fill factor; PCE = power conversion efficiency; 2TV-PT = bis(thienylenevinylene)-substituted polythiophene; 3TV-PT = tris(thienylenevinylene)-substituted polythiophene; PDTSBTD = poly[(4,4′-bis(2-ethylhexyl) dithieno[3,2-*b*:2′,3′-d]silole-2,6-diyl-*alt*-2,1,3-benzothiadiazole-4,7-diyl); P3HT = poly(3-hexylthiophene); PPHT = poly(3-phenylhydrazone thiophene); PPV = poly(*p*-phenylenevinylene); POPT = poly[3-(4-*n*-octyl)-phenylthiophene]; PV-PT = poly [3-(10-*n*-octyl-3-phenothiazinevinylene) thiophene-*co*-2,5-thiophene]; PPV2 = poly[2,5-dimethoxy-1,4-phenylene-1,2-ethenylene-2-methoxy-5-(2-ethylhexyloxy)-(1,4-phenylene-1,2-ethenylene)](M3EH-PPV); PPV3 = poly[2-methoxy-5-(3,7-dimethyloctyloxy)-1,4-phenylenevinylene] (MDMO-PPV).

b) Under 30 mW cm^{-2} white-light illumination.

c) Under 80 mW cm^{-2} white-light illumination.

d) Under 77 mW cm^{-2} white-light illumination.

PPV-type polymers have also played an important role in the electron-acceptor material field. The replacement of vinylene with cyanovinylene in poly[2-methoxy-5-(2′-ethylhexyloxy)-1,4-phenylenevinylene] (MEH-PPV) lowered both the HOMO and LUMO levels by 0.5 eV, which in turn increased the ease of reduction and decreased the ease of oxidation, and had only minor effects on the absorption and fluorescence spectra and band gap. Cyano-substituted PPVs (CN-PPVs) displayed relatively high electron affinities and considerable electron-transport properties as a result of the electron-withdrawing effect of the cyano group. CN-PPVs can function as a suitable electron acceptor in all-polymer solar cells in either bilayer or BHJ configuration [92, 93, 99]. For example, Fréchet *et al.* reported a bilayer device using **h11** as acceptor and regioregular poly[3-(4-*n*-octyl)-phenylthiophene] (POPT) as donor by spin-coating to achieve a PCE of 2% [92]. Friend *et al.* fabricated a bilayer device using **h12** as acceptor and POPT as donor through a lamination technique, achieving a PCE of 1.9% at AM1.5, 77 mW cm^{-2} [93]. Li *et al.* synthesized a soluble *n*-type-conjugated polymer poly(1,4-dioctyloxyl-2,5-dicyano-*p*-phenylenevinylene) (**h13**) and fabricated all-polymer solar cells using a polymer blend of poly[3- (10-*n*-octyl-3-phenothiazinevinylene)thiophene-*co*-2,5-thiophene] (PV-PT) as donor and **h13** as acceptor; a PCE of 0.8% was achieved after thermal annealing [94].

When Carter *et al.* fabricated a polymer blend device using poly[2,5-dimethoxy-1,4-phenylene-1,2-ethenylene-2-methoxy-5-(2-ethylhexyloxy)-(1,4-phenylenevinylene-1,2-ethenylene)] (M3EH-PPV) as donor and poly[oxa-1,4-phenylene-1,2-(1-cyano) ethenylene-2,5-dioctyloxy-1,4-phenylene- 1,2-(2-cyano)ethenylene-1,4-phenylene] (**h14**) as acceptor, a PCE of 1% was obtained at AM1.5, 80 mW cm^{-2} [100]. Later, Kietzke *et al.* described the creation of all-polymer solar cells based on M3EH-PPV:**h14** with a much higher PCE (1.7%) [95]. These authors proposed that, due to the much lower solubility of M3EH-PPV in chlorobenzene compared to **h14**, a vertically composition graded layer other than a homogeneous blend was formed during spin-coating which facilitated both exciton separation and an efficient transport of the separated charges to the electrodes.

Koetse *et al.* reported a polymer/polymer BHJ solar cell based on a mixture of MDMO-PPV as the donor and an alternating copolymer poly{9,9-dioctylfluorene-2,7-diyl-*alt*-1,4-bis[2-(5- thienyl)-1-cyanovinyl]-2-methoxy-5-(3,7-dimethyl-octyloxy) benzene} (**h15**) as the acceptor; a maximum external quantum efficiency of 52% at 530 nm and a PCE of 1.5% was achieved [96]. McNeill *et al.* reported that the copolymer of fluorene and 4,7-bis(3-hexylthienyl)-2,1,3-benzothiadiazole (**h16**) showed an ambipolar nature and was capable of functioning as an efficient electron acceptor in blends with donor P3HT [97]. A PCE of 1.8% was achieved under simulated sunlight for optimized **h16**/P3HT devices.

Hiorns *et al.* synthesized a main-chain C_{60}-containing polymer, poly{(1,4-fullerene)-*alt*-[1,4-dimethylene-2,5-bis(cyclohexylmethyl ether)phenylene]} (**h17**), using atom-transfer radical addition [98]. Using atomic force microscopy (AFM), a blend film of P3HT:**h17** was shown to yield nanoclusters (ca. 20 nm) that were favorable to exciton dissociation. This prototype device gave a promising PCE of 1.6%.

It is well known that mixtures of polymers tend to undergo phase segregation due to the low entropy mixing. The feature size of phase separation in polymer/polymer blend generally is hundreds of nanometers, whereas in the case of polymer/PCBM systems the phase separation is only tens of nanometers. Thus, the donor/acceptor interfacial area for charge separation in the polymer/PCBM systems is much larger than that in the polymer/polymer system. Given the fact that the typical exciton diffusion length in a disordered blend layer is about 10 nm, the large-scale phase separation in polymer/polymer system is not favorable for efficient exciton dissociation, leading to lower efficiencies. Very recently, Shuai *et al.* carried out a dynamic Monte Carlo simulation for all-polymer solar cells based on the **h1**/2TV-PT blend. The simulations indicates that a 5% PCE could be achieved with an optimum combination of charge mobility and morphology (the feature size is about 10 nm) [101].

9.4 Conclusions and Outlook

The creation of high-efficiency polymer solar cells has been achieved on the basis of innovations of photovoltaic materials and device fabrication technology, and indicates a bright future for this area of research. One key challenge that is related to organic photovoltaics is to control the electronic structures, the film morphology, and the device properties of organic semiconductors by modifying their chemical structures. Consequently, in this chapter a survey has been provided of the current knowledge of the relationships between chemical structures and optical, electronic, and device properties for most important donor and acceptor polymers for high-performance PSCs. Such an understanding is a basic requirement for the rational design of photovoltaic materials. To summarize:

1) Extending the absorption and decreasing the bandgap of a polymer donor to match solar radiation represents major approaches to improving J_{sc} and efficiencies. A polymer donor with bandgap <2 eV is necessary, but not sufficient, for a high PCE. D–A polymers based on fused dithiophene tricyclic blocks such as DTC, DTS, DTP, and BDT can exhibit relatively low E_g (<1.6 eV), broad absorption, high J_{sc} (>10 mA cm^{-2}), and high PCEs (>5% with PC$_{71}$BM).

2) Given that V_{oc} is related directly to the energy difference between the HOMO level of the donor and the LUMO level of the acceptor, a lower HOMO of a donor would help to achieve a higher V_{oc} and a higher PCE. In general, a polymer with a HOMO below −5.4 eV tends to give $V_{oc} > 0.8$ V. Polymers based on fused biphenyl tricyclic blocks such as fluorene, carbazole, and silafluorene, can exhibit relatively large E_g (~1.9 eV), low HOMOs (<−5.5 eV), high V_{oc} (>0.9 V), and high PCEs (>3.5% with PC$_{61}$BM).

3) A high mobility [$>1 \times 10^{-3}$ cm^2/(V·s)] of a donor would help to achieve a high J_{sc} and a high PCE. Polymers based on fused ring blocks such as DTC, DTS,

DTP, BDT, fluorene, carbazole, and silafluorene, tend to exhibit relatively high mobilities and high PCEs.

4) Among the acceptor polymers, those based on PDI and CN–PPV are promising for all-polymer solar cells. However, the large-scale phase separation and smaller donor/acceptor interfacial area in polymer/polymer system may lead to an inefficient exciton dissociation and lower efficiencies compared to polymer/fullerene systems.

5) The PCE is more of a device parameter than an intrinsic photovoltaic material parameter. A high efficiency achievement is a systematic engineering of excellent properties of materials with careful optimization of the various device fabrication conditions. To date, the best performances (>7%) of PSCs have been achieved from blends of fused-ring-based polymer donors with $PC_{71}BM$. An interdisciplinary approach such as novel photovoltaic materials and new advanced device concepts will most likely bring high-efficiency (>10%) and low-cost plastic solar cells to final commercialization.

References

1 He, Z., Zhong, C., Huang, X., Wong, W.Y., Wu, H., Chen, L., Su, S., and Cao, Y. (2011) Simultaneous enhancement of open-circuit voltage, short-circuit current density, and fill factor in polymer solar cells. *Adv. Mater.*, **23** (40), 4636–4643.

2 Zhan, X. and Zhu, D. (2010) Conjugated polymers for high-efficiency organic photovoltaics. *Polym. Chem.*, **1** (4), 409–419.

3 Chang, E.C., Chao, C.I., and Lee, R.H. (2006) Enhancing the efficiency of MEH-PPV and PCBM based polymer solar cells via optimization of device configuration and processing conditions. *J. Appl. Polym. Sci.*, **101** (3), 1919–1924.

4 Tajima, K., Suzuki, Y., and Hashimoto, K. (2008) Polymer photovoltaic devices using fully regioregular poly[(2-methoxy-5-(3′,7′-dimethyl octyloxy))-1,4-phenylenevinylene]. *J. Phys. Chem. C*, **112** (23), 8507–8510.

5 Tan, Z., Tang, R., Zhou, E., He, Y., Yang, C., Xi, F., and Li, Y. (2008) Electroluminescence and photovoltaic properties of poly(*p*-phenylene vinylene) derivatives with dendritic pendants. *J. Appl. Polym. Sci.*, **107** (1), 514–521.

6 Wen, S., Pei, J., Zhou, Y., Xue, L., Xu, B., Li, Y., and Tian, W. (2009) Synthesis and photovoltaic properties of poly(*p*-phenylenevinylene) derivatives containing oxadiazole. *J. Polym. Sci. Part A: Polym. Chem.*, **47** (4), 1003–1012.

7 Hoppe, H., Egbe, D.A.M., Mühlbacher, D., and Sariciftci, N.S. (2004) Photovoltaic action of conjugated polymer/fullerene bulk heterojunction solar cells using novel PPE-PPV copolymers. *J. Mater. Chem.*, **14** (23), 3462–3467.

8 Egbe, D.A.M., Nguyen, L.H., Schmidtke, K., Wild, A., Sieber, C., Guenes, S., and Sariciftci, N.S. (2007) Combined effects of conjugation pattern and alkoxy side chains on the photovoltaic properties of thiophene-containing PPE-PPVs. *J. Polym. Sci. Part A: Polym. Chem.*, **45** (9), 1619–1631.

9 Zhang, F., Johansson, M., Andersson, M.R., Hummelen, J.C., and Inganäs, O. (2002) Polymer photovoltaic cells with conducting polymer anodes. *Adv. Mater.*, **14** (9), 662–665.

10 Shaheen, S.E., Brabec, C.J., Sariciftci, N.S., Padinger, F., Fromherz, T., and Hunnelen, J.C. (2001) 2.5% efficient

organic plastic solar cells. *Appl. Phys. Lett.*, **78** (6), 841–843.

11 Li, J., Sun, N., Guo, Z.X., Li, C., Li, Y., Dai, L., Zhu, D., Sun, D., Cao, Y., and Fan, L. (2002) Photovoltaic devices with methanofullerenes as electron acceptors. *J. Phys. Chem. B*, **106** (44), 11509–11514.

12 Nguyen, L.H., Hoppe, H., Erb, T., Gunes, S., Gobsch, G., and Sariciftci, N.S. (2007) Effects of annealing on the nanomorphology and performance of poly(alkylthiophene):fullerene bulk-heterojunction solar cells. *Adv. Funct. Mater.*, **17** (7), 1071–1078.

13 Xin, H., Kim, F.S., and Jenekhe, S.A. (2008) Highly efficient solar cells based on poly(3-butylthiophene) nanowires. *J. Am. Chem. Soc.*, **130** (16), 5424–5425.

14 Sivula, K., Luscombe, C.K., Thompson, B.C., and Fréchet, J.M.J. (2006) Enhancing the thermal stability of polythiophene: fullerene solar cells by decreasing effective polymer regioregularity. *J. Am. Chem. Soc.*, **128** (43), 13988–13989.

15 Wu, P.T., Ren, G.Q., and Jenekhe, S.A. (2010) Crystalline random conjugated copolymers with multiple side chains: tunable intermolecular interactions and enhanced charge transport and photovoltaic properties. *Macromolecules*, **43** (7), 3306–3313.

16 Hou, J.H., Chen, T.L., Zhang, S.Q., Huo, L.J., Sista, S., and Yang, Y. (2009) An easy and effective method to modulate molecular energy level of poly(3-alkylthiophene) for high-V_{oc} polymer solar cells. *Macromolecules*, **42** (23), 9217–9219.

17 Shi, C.J., Yao, Y., Yang, Y., and Pei, Q.B. (2006) Regioregular copolymers of 3-alkoxythiophene and their photovoltaic application. *J. Am. Chem. Soc.*, **128** (27), 8980–8986.

18 Zhou, E., Tan, Z., Yang, Y., Huo, L., Zou, Y., Yang, C., and Li, Y. (2007) Synthesis, hole mobility, and photovoltaic properties of cross-linked polythiophenes with vinylene-terthiophene-vinylene as conjugated bridge. *Macromolecules*, **40** (6), 1831–1837.

19 Tu, G., Bilge, A., Adamczyk, S., Forster, M., Heiderhoff, R., Balk, L.J., Mühlbacher, D., Morana, M., Koppe, M., Scharber, M.C., Choulis, S.A., Brabec, C.J., and Scherf, U. (2007) The influence of interchain branches on solid state packing, hole mobility and photovoltaic properties of poly(3-hexylthiophene) (P3HT). *Macromol. Rapid Commun.*, **28** (17), 1781–1785.

20 Hou, J., Tan, Z., Yan, Y., He, Y., Yang, C., and Li, Y. (2006) Synthesis and photovoltaic properties of two-dimensional conjugated polythiophenes with bi(thienylenevinylene) side chains. *J. Am. Chem. Soc.*, **128** (14), 4911–4916.

21 Chang, Y.T., Hsu, S.L., Chen, G.Y., Su, M.H., Singh, T.A., Diau, E.W.G., and Wei, K.H. (2008) Intramolecular donor-acceptor regioregular poly(3-hexylthiophene)s presenting octylphenanthrenyl-imidazole moieties exhibit enhanced charge transfer for heterojunction solar cell applications. *Adv. Funct. Mater.*, **18** (16), 2356–2365.

22 Ballantyne, A.M., Chen, L., Nelson, J., Bradley, D.D.C., Astuti, Y., Maurano, A., Shuttle, C.G., Durrant, J.R., Heeney, M., Duffy, W., and McCulloch, I. (2007) Studies of highly regioregular poly(3-hexylselenophene) for photovoltaic applications. *Adv. Mater.*, **19** (24), 4544–4547.

23 Wienk, M.M., Turbiez, M., Gilot, J., and Janssen, R.A.J. (2008) Narrow-bandgap diketo-pyrrolo-pyrrole polymer solar cells: the effect of processing on the performance. *Adv. Mater.*, **20** (13), 2556–2560.

24 Bijleveld, J.C., Zoombelt, A.P., Mathijssen, S.G.J., Wienk, M.M., Turbiez, M., de Leeuw, D.M., and Janssen, R.A.J. (2009) Poly(diketopyrrolopyrrole-terthiophene) for ambipolar logic and photovoltaics. *J. Am. Chem. Soc.*, **131** (46), 16616–16617.

25 He, Y., Chen, H.Y., Hou, J., and Li, Y. (2010) Indene-C(60) Bisadduct: a new acceptor for high-performance polymer solar cells. *J. Am. Chem. Soc.*, **132** (4), 1377–1382.

26 Zhao, G., He, Y., and Li, Y. (2010) 6.5% efficiency of polymer solar cells based on poly(3-hexylthiophene) and indene-C(60) bisadduct by device

optimization. *Adv. Mater.*, **22** (39), 4355–4358.

27 Sun, Y., Cui, C., Wang, H., and Li, Y. (2011) Efficiency enhancement of polymer solar cells based on poly(3-hexylthiophene)/indene-C70 bisadduct via methylthiophene additive. *Adv. Energy. Mater.*, **1** (6), 1058–1061.

28 Ren, G.Q., Wu, P.T., and Jenekhe, S.A. (2010) Enhanced performance of bulk heterojunction solar cells using block copoly(3-alkylthiophene)s. *Chem. Mater.*, **22** (6), 2020–2026.

29 Li, Y. and Zou, Y. (2008) Conjugated polymer photovoltaic materials with broad absorption band and high charge carrier mobility. *Adv. Mater.*, **20** (15), 2952–2958.

30 Heeney, M., Zhang, W., Crouch, D.J., Chabinyc, M.L., Gordeyev, S., Hamilton, R., Higgins, S.J., McCulloch, I., Skabara, P.J., Sparrowe, D., and Tierney, S. (2007) Regioregular poly(3-hexyl) selenophene: a low band gap organic hole transporting polymer. *Chem. Commun.*, **47** (17), 5061–5063.

31 Zhang, F., Bijleveld, J., Perzon, E., Tvingstedt, K., Barrau, S., Inganäs, O., and Andersson, M.R. (2008) High photovoltage achieved in low band gap polymer solar cells by adjusting energy levels of a polymer with the LUMOs of fullerene derivatives. *J. Mater. Chem.*, **18** (45), 5468–5474.

32 Lindgren, L.J., Zhang, F.L., Andersson, M., Barrau, S., Hellstrom, S., Mammo, W., Perzon, E., Inganas, O., and Andersson, M.R. (2009) Synthesis, characterization, and devices of a series of alternating copolymers for solar cells. *Chem. Mater.*, **21** (15), 3491–3502.

33 Zhang, F., Mammo, W., Andersson, L.M., Admassie, S., Andersson, M.R., and Inganäs, O. (2006) Low-bandgap alternating fluorene copolymer/methanofullerene heterojunctions in efficient near-infrared polymer solar cells. *Adv. Mater.*, **18** (16), 2169–2173.

34 Svensson, M., Zhang, F., Veenstra, S.C., Verhees, W.J.H., Hummelen, J.C., Kroon, J.M., Inganäs, O., and Andersson, M.R. (2003) High performance polymer solar cells of an alternating polyfluorene copolymer and a fullerene derivative. *Adv. Mater.*, **15** (12), 988–991.

35 Zhang, F., Jespersen, K.G., Björström, C., Svensson, M., Andersson, M.R., Sundström, V., Magnusson, K., Moons, E., Yartsev, A., and Inganäs, O. (2006) Influence of solvent mixing on the morphology and performance of solar cells based on polyfluorene copolymer/fullerene blends. *Adv. Funct. Mater.*, **16** (5), 667–674.

36 Slooff, L.H., Veenstra, S.C., Kroon, J.M., Moet, D.J.D., Sweelssen, J., and Koetse, M.M. (2007) Determining the internal quantum efficiency of highly efficient polymer solar cells through optical modeling. *Appl. Phys. Lett.*, **90** (14), 143506- 143506.

37 Zhou, Q., Hou, Q., Zheng, L., Deng, X., Yu, G., and Cao, Y. (2004) Fluorene-based low band-gap copolymers for high performance photovoltaic devices. *Appl. Phys. Lett.*, **84** (10), 1653–1655.

38 Hou, Q., Xu, Y., Yang, W., Yuan, M., Peng, J., and Cao, Y. (2002) Novel red-emitting fluorene-based copolymers. *J. Mater. Chem.*, **12** (10), 2887–2892.

39 Schulz, G.L., Chen, X., and Holdcroft, S. (2009) High band gap poly(9,9-dihexylfluorene-*alt*-bithiophene) blended with [6,6]-phenyl C_{61} butyric acid methyl ester for use in efficient photovoltaic devices. *Appl. Phys. Lett.*, **94** (2), 023302–023303.

40 Zheng, Q.D., Jung, B.J., Sun, J., and Katz, H.E. (2010) Ladder-type oligo-*p*-phenylene-containing copolymers with high open-circuit voltages and ambient photovoltaic activity. *J. Am. Chem. Soc.*, **132** (15), 5394–5404.

41 Allard, N., Aich, R.B., Gendron, D., Boudreault, P.L.T., Tessier, C., Alem, S., Tse, S.C., Tao, Y., and Leclerc, M. (2010) Germafluorenes: new heterocycles for plastic electronics. *Macromolecules*, **43** (5), 2328–2333.

42 Wang, E.G., Wang, L., Lan, L.F., Luo, C., Zhuang, W.L., Peng, J.B., and Cao, Y. (2008) High-performance polymer heterojunction solar cells of a polysilafluorene derivative. *Appl. Phys. Lett.*, **92** (3), 033307.

43 Qin, R.P., Li, W.W., Li, C.H., Du, C., Veit, C., Schleiermacher, H.F.,

Andersson, M., Bo, Z.S., Liu, Z., Inganäs, O., Wuerfel, U., and Zhang, F.L. (2009) A planar copolymer for high efficiency polymer solar cells. *J. Am. Chem. Soc.*, **131** (41), 14612–14613.

44 Park, S.H., Roy, A., Beaupre, S., Cho, S., Coates, N., Moon, J.S., Moses, D., Leclerc, M., Lee, K., and Heeger, A.J. (2009) Bulk heterojunction solar cells with internal quantum efficiency approaching 100%. *Nat. Photonics*, **3** (5), 297–302.

45 Song, S., Jin, Y., Park, S.H., Cho, S., Kim, I., Lee, K., Heeger, A.J., and Suh, H. (2010) A low-bandgap alternating copolymer containing the dimethylbenzimidazole moiety. *J. Mater. Chem.*, **20** (31), 6517–6523.

46 Lu, J.P., Liang, F.S., Drolet, N., Ding, J.F., Tao, Y., and Movileanu, R. (2008) Crystalline low band-gap alternating indolocarbazole and benzothiadiazole-cored oligothiophene copolymer for organic solar cell applications. *Chem. Commun.*, **42**, 5315–5317.

47 Wu, J.S., Cheng, Y.J., Dubosc, M., Hsieh, C.H., Chang, C.Y., and Hsu, C.S. (2010) Donor-acceptor polymers based on multi-fused heptacyclic structures: synthesis, characterization and photovoltaic applications. *Chem. Commun.*, **46** (42), 3259–3261.

48 Blouin, N., Michaud, A., Gendron, D., Wakim, S., Blair, E., Neagu-Plesu, R., Belletete, M., Durocher, G., Tao, Y., and Leclerc, M. (2008) Toward a rational design of poly(2,7-carbazole) derivatives for solar cells. *J. Am. Chem. Soc.*, **130** (2), 732–742.

49 Blouin, N. and Leclerc, M. (2008) Poly(2,7-carbazole)s: structure–property relationships. *Acc. Chem. Res.*, **41** (9), 1110–1119.

50 Liang, Y.Y., Wu, Y., Feng, D.Q., Tsai, S.T., Son, H.J., Li, G., and Yu, L.P. (2009) Development of new semiconducting polymers for high performance solar cells. *J. Am. Chem. Soc.*, **131** (1), 56–57.

51 Liang, Y.Y., Feng, D.Q., Wu, Y., Tsai, S.T., Li, G., Ray, C., and Yu, L.P. (2009) Highly efficient solar cell polymers developed via fine-tuning of structural and electronic properties. *J. Am. Chem. Soc.*, **131** (22), 7792–7799.

52 Hou, J.H., Chen, H.Y., Zhang, S.Q., Chen, R.I., Yang, Y., Wu, Y., and Li, G. (2009) Synthesis of a low band gap polymer and its application in highly efficient polymer solar cells. *J. Am. Chem. Soc.*, **131** (43), 15586–15587.

53 Chen, H.Y., Hou, J.H., Zhang, S.Q., Liang, Y.Y., Yang, G.W., Yang, Y., Yu, L.P., Wu, Y., and Li, G. (2009) Polymer solar cells with enhanced open-circuit voltage and efficiency. *Nat. Photonics*, **3** (11), 649–653.

54 Piliego, C., Holcombe, T.W., Douglas, J.D., Woo, C.H., Beaujuge, P.M., and Fréchet, J.M.J. (2010) Synthetic control of structural order in *N*-alkylthieno[3,4-*c*]pyrrole-4,6-dione-based polymers for efficient solar cells. *J. Am. Chem. Soc.*, **132** (22), 7595–7597.

55 Zhou, H.X., Yang, L.Q., Stuart, A.C., Price, S.C., Liu, S.B., and You, W. (2011) Development of fluorinated benzothiadiazole as a structural unit for a polymer solar cell of 7% efficiency. *Angew. Chem. Int. Edit.*, **50** (13), 2995–2998.

56 Price, S.C., Stuart, A.C., Yang, L., Zhou, H., and You, W. (2011) Fluorine substituted conjugated polymer of medium band gap yields 7% efficiency in polymer-fullerene solar cells. *J. Am. Chem. Soc.*, **133** (12), 4625–4631.

57 Huo, L.J., Zhang, S.Q., Guo, X., Xu, F., Li, Y.F., and Hou, J.H. (2011) Replacing alkoxy groups with alkylthienyl groups: a feasible approach to improve the properties of photovoltaic polymers. *Angew. Chem. Int. Ed.*, **50** (41), 9697–9702.

58 Zou, Y.P., Najari, A., Berrouard, P., Beaupre, S., Aich, B.R., Tao, Y., and Leclerc, M. (2010) A thieno[3,4-*c*]pyrrole-4,6-dione-based copolymer for efficient solar cells. *J. Am. Chem. Soc.*, **132** (15), 5330–5331.

59 Huo, L.J., Hou, J.H., Zhang, S.Q., Chen, H.Y., and Yang, Y. (2010) A polybenzo[1,2-*b*:4,5-*b*′] dithiophene derivative with deep HOMO level and its application in high-performance polymer solar cells. *Angew. Chem. Int. Edit.*, **49** (8), 1500–1503.

60 Peet, J., Kim, J.Y., Coates, N.E., Ma, W.L., Moses, D., Heeger, A.J., and Bazan, G.C. (2007) Efficiency enhancement in low-bandgap polymer solar cells by processing with alkane dithiols. *Nat. Mater.*, **6** (7), 497–500.

61 Hou, J.H., Chen, H.Y., Zhang, S.Q., Li, G., and Yang, Y. (2008) Synthesis, characterization, and photovoltaic properties of a low band gap polymer based on silole-containing polythiophenes and 2,1,3-benzothiadiazole. *J. Am. Chem. Soc.*, **130** (48), 16144–16145.

62 Huo, L.J., Chen, H.Y., Hou, J.H., Chen, T.L., and Yang, Y. (2009) Low band gap dithieno[3,2-*b*:2′,3′-*d*]silole-containing polymers, synthesis, characterization and photovoltaic application. *Chem. Commun.*, **37**, 5570–5572.

63 Chu, T.-Y., Lu, J., Beaupré, S., Zhang, Y., Pouliot, J.-R.M., Wakim, S., Zhou, J., Leclerc, M., Li, Z., Ding, J., and Tao, Y. (2011) Bulk heterojunction solar cells using thieno[3,4-*c*]pyrrole-4,6-dione and dithieno[3,2-*b*:2′,3′-*d*]silole copolymer with a power conversion efficiency of 7.3%. *J. Am. Chem. Soc.*, **133** (12), 4250–4253.

64 Amb, C.M., Chen, S., Graham, K.R., Subbiah, J., Small, C.E., So, F., and Reynolds, J.R. (2011) Dithienogermole as a fused electron donor in bulk heterojunction solar cells. *J. Am. Chem. Soc.*, **133** (26), 10062–10065.

65 Yue, W., Zhao, Y., Shao, S.Y., Tian, H.K., Xie, Z.Y., Geng, Y.H., and Wang, F.S. (2009) Novel NIR-absorbing conjugated polymers for efficient polymer solar cells: effect of alkyl chain length on device performance. *J. Mater. Chem.*, **19** (15), 2199–2206.

66 Chan, S.H., Chen, C.P., Chao, T.C., Ting, C., Lin, C.S., and Ko, B.T. (2008) Synthesis, characterization, and photovoltaic properties of novel semiconducting polymers with thiophene-phenylene-thiophene (TPT) as coplanar units. *Macromolecules*, **41** (15), 5519–5526.

67 Chen, C.P., Chan, S.H., Chao, T.C., Ting, C., and Ko, B.T. (2008) Low-bandgap poly(thiophene-phenylene-thiophene)derivatives with broaden absorption spectra for use in high-performance bulk-heterojunction polymer solar cells. *J. Am. Chem. Soc.*, **130** (38), 12828–12833.

68 Zhang, M., Guo, X., Wang, X., Wang, H., and Li, Y. (2011) Synthesis and photovoltaic properties of D-A copolymers based on alkyl-substituted indacenodithiophene donor unit. *Chem. Mater.*, **23** (18), 4264–4270.

69 Mühlbacher, D., Scharber, M., Zhengguo, M.M., Zhu, M.M.Z., Waller, D., Gaudiana, R., and Brabec, C. (2006) High photovoltaic performance of a low-bandgap polymer. *Adv. Mater.*, **18** (21), 2884–2889.

70 Kim, J.Y., Lee, K., Coates, N.E., Moses, D., Nguyen, T.-Q., Dante, M., and Heeger, A.J. (2007) Efficient tandem polymer solar cells fabricated by all-solution processing. *Science*, **317** (5835), 222–225.

71 Chen, H.Y., Hou, J.H., Hayden, A.E., Yang, H., Houk, K.N., and Yang, Y. (2010) Silicon atom substitution enhances interchain packing in a thiophene-based polymer system. *Adv. Mater.*, **22** (3), 371–375.

72 Morana, M., Azimi, H., Dennler, G., Egelhaaf, H.J., Scharber, M., Forberich, K., Hauch, J., Gaudiana, R., Waller, D., Zhu, Z.H., Hingerl, K., van Bavel, S.S., Loos, J., and Brabec, C.J. (2010) Nanomorphology and charge generation in bulk heterojunctions based on low-bandgap dithiophene polymers with different bridging atoms. *Adv. Funct. Mater.*, **20** (7), 1180–1188.

73 Coffin, R.C., Peet, J., Rogers, J., and Bazan, G.C. (2009) Streamlined microwave-assisted preparation of narrow-bandgap conjugated polymers for high-performance bulk heterojunction solar cells. *Nat. Chem.*, **1** (8), 657–661.

74 Huang, X.B., Zhu, C.L., Zhang, S.M., Li, W.W., Guo, Y.L., Zhan, X.W., Liu, Y.Q., and Bo, Z.S. (2008) Porphyrin-dithienothiophene π-conjugated copolymers: synthesis and their applications in field-effect transistors and solar cells. *Macromolecules*, **41** (19), 6895–6902.

75 Wong, W.Y., Wang, X.Z., He, Z., Djurišić, A.B., Yip, C.T., Cheung, K.Y., Wang, H., Mak, C.S.K., and Chan, W.K. (2007) Metallated conjugated polymers as a new avenue towards high-efficiency polymer solar cells. *Nat. Mater.*, **6** (7), 521–527.

76 Baek, N.S., Hau, S.K., Yip, H.L., Acton, O., Chen, K.S., and Jen, A.K.Y. (2008) High performance amorphous metallated pi-conjugated polymers for field-effect transistors and polymer solar cells. *Chem. Mater.*, **20** (18), 5734–5736.

77 Wong, W.Y., Wang, X.Z., He, Z., Chan, K.K., Djurišić, A.B., Cheung, K.Y., Yip, C.T., Ng, A.M.C., Xi, Y.Y., Mak, C.S.K., and Chan, W.K. (2007) Tuning the absorption, charge transport properties, and solar cell efficiency with the number of thienyl rings in platinum-containing poly(aryleneethynylene)s. *J. Am. Chem. Soc.*, **129** (46), 14372–14380.

78 Liu, L., Ho, C.L., Wong, W.Y., Cheung, K.Y., Fung, M.K., Lam, W.T., Djurišić, A.B., and Chan, W.K. (2008) Effect of oligothienyl chain length on tuning the solar cell performance in fluorene-based polyplatinynes. *Adv. Funct. Mater.*, **18** (18), 2824–2833.

79 Wang, X.Z., Wong, W.Y., Cheung, K.Y., Fung, M.K., Djurišić, A.B., and Chan, W.K. (2008) Polymer solar cells based on very narrow-bandgap polyplatinynes with photocurrents extended into the near-infrared region. *Dalton Trans.*, **40**, 5484–5494.

80 Wong, W.Y., Wang, X.Z., Zhang, H.L., Cheung, K.Y., Fung, M.K., Djurišić, A.B., and Chan, W.K. (2008) Synthesis, characterization and photovoltaic properties of a low-bandgap platinum(II) polyyne functionalized with a 3,4-ethylenedioxythiophene-benzothiadiazole hybrid spacer. *J. Organomet. Chem.*, **693** (24), 3603–3612.

81 Wu, P.T., Bull, T., Kim, F.S., Luscombe, C.K., and Jenekhe, S.A. (2009) Organometallic donor-acceptor conjugated polymer semiconductors: tunable optical, electrochemical, charge transport, and photovoltaic properties. *Macromolecules.*, **42** (3), 671–681.

82 Wong, W.Y., Chow, W.C., Fung, M.K., Djurišić, A.B., and Chan, W.K. (2009) Harvesting solar energy using conjugated metallopolyyne donors containing electron-rich phenothiazine–oligothiophene moieties. *J. Organomet. Chem.*, **694** (17), 2717–2726.

83 Zhan, X., Tan, Z., Domercq, B., An, Z., Zhang, X., Barlow, S., Li, Y., Zhu, D., Kippelen, B., and Marder, S.R. (2007) A high-mobility electron-transport polymer with broad absorption and its use in field-effect transistors and all-polymer solar cells. *J. Am. Chem. Soc.*, **129** (23), 7246–7247.

84 Tan, Z.A., Zhou, E.J., Zhan, X.W., Wang, X., Li, Y.F., Barlow, S., and Marder, S.R. (2008) Efficient all-polymer solar cells based on blend of tris(thienylenevinylene)-substituted polythiophene and poly[perylene diimide-*alt*-bis (dithienothiophene)]. *Appl. Phys. Lett.*, **93** (7), 073309.

85 Zhan, X., Tan, Z., Zhou, E., Li, Y., Misra, R., Grant, A., Domercq, B., Zhang, X., An, Z., Zhang, X., Barlow, S., Kippelen, B., and Marder, S.R. (2009) Copolymers of perylene diimide with dithienothiophene and dithienopyrrole as electron-transport materials for all-polymer solar cells and field-effect transistors. *J. Mater. Chem.*, **19** (32), 5794–5803.

86 Hou, J.H., Zhang, S.Q., Chen, T.L., and Yang, Y. (2008) A new *n*-type low bandgap conjugated polymer P-*co*-CDT: synthesis and excellent reversible electrochemical and electrochromic properties. *Chem. Commun.*, **45**, 6034–6036.

87 Zhou, E., Tajima, K., Yang, C.H., and Hashimoto, K. (2010) Band gap and molecular energy level control of perylene diimide-based donor–acceptor copolymers for all-polymer solar cells. *J. Mater. Chem.*, **20** (12), 2362–2368.

88 Zhou, E., Cong, J., Wei, Q., Tajima, K., Yang, C.H., and Hashimoto, K. (2011) All-polymer solar cells from perylene diimide based copolymers: material design and phase separation control. *Angew. Chem. Int. Ed.*, **50** (12), 2799–2803.

89 Kozma, E., Kotowski, D., Bertini, F., Luzzati, S., and Catellani, M. (2010) Synthesis of donor–acceptor poly(perylene diimide-*alt*-oligothiophene) copolymers as *n*-type materials for polymeric solar cells. *Polymer*, **51** (11), 2264–2270.

90 Mikroyannidis, J.A., Stylianakis, M.M., Sharma, G.D., Balraju, P., and Roy, M.S. (2009) A novel alternating phenylenevinylene copolymer with perylene bisimide units: synthesis, photophysical, electrochemical, and photovoltaic properties. *J. Phys. Chem. C*, **113** (18), 7904–7912.

91 Alam, M.M. and Jenekhe, S.A. (2004) Efficient solar cells from layered nanostructures of donor and acceptor conjugated polymers. *Chem. Mater.*, **16** (23), 4647–4656.

92 Holcombe, T.W., Woo, C.H., Kavulak, D.F.J., Thompson, B.C., and Fréchet, J.M.J. (2009) All-polymer photovoltaic devices of poly(3-(4-*n*-octyl)-phenylthiophene) from Grignard metathesis (GRIM) polymerization. *J. Am. Chem. Soc.*, **131** (40), 14160–14161.

93 Granström, M., Petritsch, K., Arias, A.C., Lux, A., Andersson, M.R., and Friend, R.H. (1998) Laminated fabrication of polymeric photovoltaic diodes. *Nature*, **395** (6699), 257–260.

94 Sang, G., Zou, Y., Huang, Y., Zhao, G., Yang, Y., and Li, Y. (2009) All-polymer solar cells based on a blend of poly[3-(10-*n*-octyl-3-phenothiazine-vinylene) thiophene-*co*-2,5-thiophene] and poly[1,4-dioctyloxyl-*p*-2,5-dicyanophenylenevinylene]. *Appl. Phys. Lett.*, **94** (19), 193302.

95 Kietzke, T., Horhold, H.H., and Neher, D. (2005) Efficient polymer solar cells based on M3EH-PPV. *Chem. Mater.*, **17** (26), 6532–6537.

96 Koetse, M.M., Sweelssen, J., Hoekerd, K.T., Schoo, H.F.M., Veenstra, S.C., Kroon, J.M., Yang, X., and Loos, J. (2006) Efficient polymer:polymer bulk heterojunction solar cells. *Appl. Phys. Lett.*, **88** (8), 083504.

97 McNeill, C.R., Abrusci, A., Zaumseil, J., Wilson, R., McKiernan, M.J., Halls, J.J.M., Greenham, N.C., and Friend, R.H. (2007) Dual electron donor/ electron acceptor character of a conjugated polymer in efficient photovoltaic diodes. *Appl. Phys. Lett.*, **90** (19), 193506.

98 Hiorns, R.C., Cloutet, E., Ibarboure, E., Vignau, L., Lemaitre, N., Guillerez, S., Absalon, C., and Cramail, H. (2009) Main-chain fullerene polymers for photovoltaic devices. *Macromolecules*, **42** (10), 3549–3558.

99 Gupta, D., Kabra, D., Kolishetti, N., Ramakrishnan, S., and Narayan, K.S. (2007) An efficient bulk-heterojunction photovoltaic cell based on energy transfer in graded-bandgap polymers. *Adv. Funct. Mater.*, **17** (2), 226–232.

100 Breeze, A.J., Schlesinger, Z., Carter, S.A., Tillmann, H., and Horhold, H.H. (2004) Improving power efficiencies in polymer–polymer blend photovoltaics. *Sol. Energy Mater. Sol. Cells.*, **83** (2-3), 263–271.

101 Meng, L., Shang, Y., Li, Q., Li, Y., Zhan, X., Shuai, Z., Kimber, R.G.E., and Walker, A.B. (2010) Dynamic Monte Carlo simulation for highly efficient polymer blend photovoltaics. *J. Phys. Chem. B*, **114** (1), 36–41.

10
Dye-Sensitized Solar Cells (DSSCs)

Lanchao Ma, Xiaowei Zhan

10.1
Introduction

Among the many devices which utilize solar energy, dye-sensitized solar cells (DSSCs) have a perceptibly high performance-to-price ratio [1]. Following the initial breakthrough with DSSCs made by Grätzel and O'Regan, who applied nanocrystalline TiO_2 in DSSCs to achieve a power-to-electricity conversion efficiency of 7.9% [2], investigations into DSSCs have since attracted significant interest worldwide. Yet today, despite materials such as electrodes, dyes, electrolytes and redox couples that constitute DSSCs continuing to undergo rapid development [3], the concept of the most efficient device has remained essentially unchanged (Figure 10.1) [4].

A typical *n*-type DSSC will contain five components: (i) a fluorine-doped SnO_2 (FTO) glass substrate; (ii) a nanocrystalline TiO_2 thin film as the semiconductor; (iii) dye sensitizers; (iv) an electrolyte (redox mediator); and (v) a platinum-coated glass substrate. In *n*-type DSSCs, the incoming light is absorbed by the sensitizer, which is anchored to the surface of the semiconducting TiO_2 nanocrystals. A charge separation then takes place at the interface through a photoinduced electron injection from the excited dye into the conduction band of the TiO_2. As a consequence, holes are created at the dye ground state, which is further regenerated through reduction by the redox couple (reduction state) or hole-transport material (HTM), while the oxidation state itself is regenerated at the counterelectrode by electrons, through an external circuit [5].

Unlike *n*-type DSSCs, *p*-type DSSCs comprise a photocathode and a passive anode [6]. The photo-to-electricity conversion begins with the injection of the hole to a *p*-type semiconductor from the highest occupied molecular orbital (HOMO) of the sensitizer where the exciton is formed upon photoexcitation; the electron in the lowest unoccupied molecular orbital (LUMO) level is then transferred to the oxidized species in the electrolyte. Subsequently, the injected hole diffuses to the back contact and the reduction-state species in the electrolyte is recovered on the counterelectrode.

Organic Optoelectronics, First Edition. Edited by Wenping Hu.

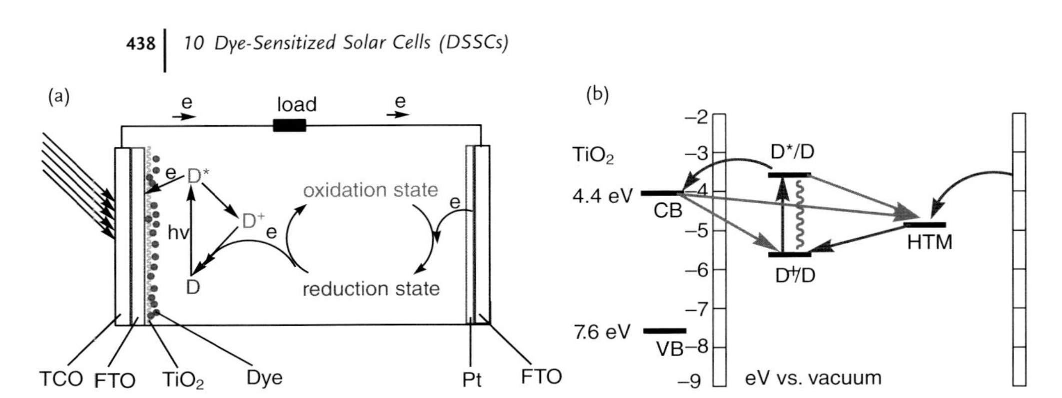

Figure 10.1 (a) Schematic representation of a dye-sensitized TiO_2 solar cell; (b) Energy level diagram and the electron-transfer processes involved in energy conversion. D represents the dye sensitizer; the charge mediator in common use is I^-/I_3^-.

The main difference in the operating mechanism between these two types of DSSC, based on the electrolyte-containing redox couple and the HTM, is the dye-regeneration process. In the former case, in a liquid-state DSSC the oxidized dye molecule is regenerated by accepting an electron from a reduction state species (iodide) in the electrolyte. The oxidation state species (triiodide) that is formed in the electrolyte then diffuses to the counterelectrode (which usually is platinized fluorine-doped tin oxide conductive glass, FTO), where it is reduced. In the latter case, the oxidized dye molecule is regenerated by injecting a hole into the HTM. The hole then moves through the HTM by hopping to a metal electrode (usually Ag or Au).

A driving force is essential to keep the DSSC operating smoothly. The reduction potential of the dye should be more negative than the conduction band edge of TiO_2, providing a suitable thermodynamic driving force for electron injection. The oxidation potential of the dye should be sufficiently more positive than the iodine/iodide redox potential value, ensuring that the oxidized dyes formed after electron injection into the conduction band of TiO_2 could accept electrons from I^- ions thermodynamically [7]. Side reactions may also occur, with the excited dye most likely decaying (either radiatively or nonradiatively) or deoxidizing the oxidation state to a reduction state of the redox couple before it injects an electron. The injected electron in the conduction band of TiO_2 can then recombine with the oxidized dye before the dye is regenerated; alternatively, the redox shuttle can intercept an electron from the photoanode before it is collected [4].

The pursuit of a high efficiency for DSSCs has entailed a great deal of effort that is, unfortunately, limited by the constituent elements. In the case of the dye, it is almost impossible to absorb all of the incident light because absorbance is limited by the band gap between the LUMO and the HOMO of the sensitizer. This occurs because the LUMO must be higher than the conduction band of an *n*-type semiconductor [usually nanocrystalline TiO_2, −0.5 V versus normal hydrogen electrode (NHE) [7]] for the effective injection. Likewise, the HOMO must be sufficiently low to ensure that the oxidation potential of D^0/D^+ is more positive than that of the redox shuffle (usually I^-/I_3^- redox couple ca. 0.4 V versus NHE). As a result, it is impossible to improve the open-circuit voltage (V_{oc}) and short-circuit current density (J_{sc}) at the same time. For the redox shuffle, the concentration is considered to be appropriate. In the case of nanocrystalline TiO_2, there is a trade-off between light harvesting and series resistance, with DSSCs based on thinner nanocrystalline TiO_2 having a lower series resistance but showing a limited light-absorption capability due to the reduced amount of absorbed dye [8]. The fill factor (FF), which is associated with the fabrication of a solar cell, is attenuated by the total series resistance of the cell, which includes the sheet resistances of the substrate and counterelectrode, the electron transport resistance through the photoanode, ion transport resistance, and the charge-transfer resistance at the counterelectrode. Consequently, careful engineering is very important in new device designs [4].

As a key part of DSSCs, the dyes perform the function of light absorption and injection of the photoexcited electrons to the conduction band of the semiconductor in the anode. Two types of dye have been identified, namely metal–organic complexes and metal-free organic dyes. Those DSSCs based on ruthenium dyes achieve a power conversion efficiency (PCE) of 11.1%, measured using an air mass of 1.5 (AM 1.5) [9–13]; indeed, a new record PCE of 11.9% was achieved by YD2-*o*-C8, a porphyrin–zinc complex based on a cobalt(II/III) redox electrolyte under an illumination of 99.5 $mW\,cm^{-2}$ [14]; by comparison, DSSCs based on organic dyes achieved PCEs of 10.1% [15].

The characteristics pertaining to ruthenium dyes include:

- A broad absorption that extends to >700 nm and allows efficient light-harvesting in the red to near-infrared region due to the metal-to-ligand charge transfer (MLCT) [16].
- The back electron-transfer process in ruthenium dyes is slower [17]; in fact, several ruthenium polypyridyl complexes have demonstrated an ability to withstand thermal or light-soaking stress tests for at least 1000 h, while retaining an efficiency above 7% [18, 19]. However, the molar extinction coefficient of the best Ru-based sensitizers is less than $2 \times 10^4\,l\,mol^{-1}\,cm^{-1}$, which limits their light-harvesting role.

The synthesis of ruthenium dyes must be conducted under harsh conditions, and the effort required for their purification is substantial. Clearly, the limited availability of Ru complexes will become problematic if, eventually, DSSCs are applied to an extensive degree.

Metal-free organic dyes have greatly widened the choice of sensitizers in DSSCs:

- The molar extinction coefficient of organic dyes is relatively high due to intramolecular $\pi \rightarrow \pi^*$ transitions [20]. This reduces the amounts of dye used and, as a result, the application of a thinner nanocrystalline TiO_2 is permitted, so as to reduce the film's roughness factors [21] and improve its physical strength. Thin photoactive layers are required in DSSCs composed of ionic liquid electrolyte, as their fluidity is weak and the charge collection yield in DSSCs will be decreased due to a shortened electron diffusion length [22]. Thin TiO_2 films are also required in solid-state devices where the diffusion length is shorter and an insufficient pore filling limits the photovoltaic performance [8]. DSSCs employing thin films usually produce a higher V_{oc} due to the smaller dark current [23].

- Compared to ruthenium dyes, the molecular engineering of organic sensitizers is much more versatile, providing a greater structural diversity to tune the absorption wavelength, the molar extinction coefficient, and the HOMO/LUMO energy levels [5].
- The raw materials used to synthesize organic dyes are easily accessible and inexpensive. The synthetic procedures are also relatively simple and convenient.
- Metal-free dyes are environment-friendly [5].

The basic requirements for an ideal dye are that:

- The absorption spectrum should have a great overlap with solar irradiation.
- The energy level of an excited photosensitizer should be higher than the conduction band edge of an *n*-type semiconductor (*n*-type DSSCs), in order to ensure an efficient electron-transfer process from the excited dye to the conduction band (CB) of the semiconductor [24]. In contrast, for *p*-type DSSCs, the HOMO level of the photosensitizer should be at more positive potential than the valance band (VB) level of a *p*-type semiconductor. The oxidized state level of the photosensitizer must be more positive than the redox potential of the electrolyte in order to guarantee dye regeneration [3].
- The sensitizers also require photostability, as well as electrochemical and thermal stabilities.

Additives such as 4-*tert*-butylpyridine (TBP), deoxycholic acid (DCA), and chenodeoxycholic acid (CDCA) play important roles in DSSCs (Figure 10.2). A relatively large energy gap between the LUMO of the dye and the CB of TiO_2 provides a

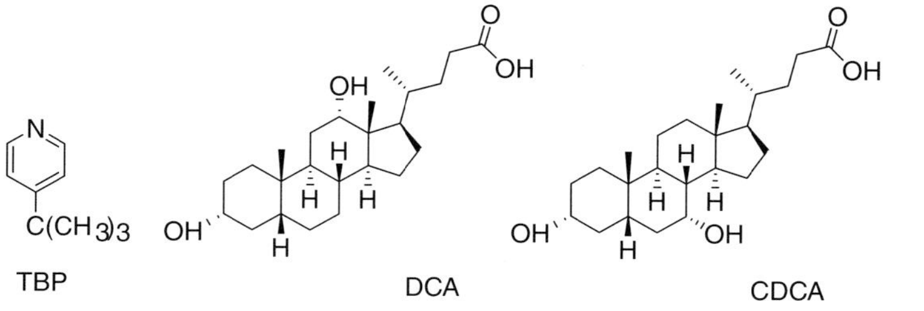

Figure 10.2 Chemical structures of some additives. TBP, 4-tert-butylpyridine; DCA, deoxycholic acid; CDCA, chenode-oxycholic acid.

possibility for the addition of TBP to the electrolyte, which in turn improves the photovoltage [25] because the basicity generated by TBP negatively shifts the CB level of TiO_2 [26], while TBP forms several complexes with I_3, and this leads to a decrease in the concentration of I_3^- at the TiO_2 surface [27]. In addition, a blocking effect of TBP on the TiO_2 surface prevents recombination between the injected electrons and I_3^- ions, which results in an improved photovoltage [28]. Conversely, employing TBP in a system where the energy gap between the LUMO of a dye and the CB of TiO_2 is small leads to a substantial decrease in the photocurrent due to a decreased electron injection yield. Both DCA and CDCA have been used to prevent aggregation of the dye on the TiO_2 surface, and this has resulted in a high electron-injection yield from the dye into the TiO_2 [29]. However, DCA and CDCA compete with the dye for surface adsorption on the TiO_2, thus reducing the dye adsorption.

In terms of the nature of their electrolytes, DSSCs can be generally classified as liquid, ion liquid, and quasi-solid state. The demands on the liquid redox electrolyte are that they should be chemically stable, have a low viscosity in order to minimize transport problems, and serve as a good solvent for the redox couple components and various additives, while at the same time not causing any significant dissolution of the adsorbed dye or of the semiconducting material of the electrodes [3]. The two major types of solvent that simultaneously fulfill most of these criteria are based on relatively polar organic solvents, such as acetonitrile and ionic liquids.

Whereas, organic solvents have been shown as impractical, with solvent leakage and electrode erosion blocking their commercial application, ionic liquid electrolytes have proved to be of value in terms of their usage. Consequently, liquid DSSCs with a high efficiency have also been tested under ionic liquid conditions. However, when using a low-fluidity ionic liquid electrolyte the charge collection yield in DSSCs becomes very low due to the shortened electron diffusion length [22]. Owing to the encapsulation and stability issues that arise at higher temperatures, nonvolatile or ionic-liquid electrolytes are generally preferred over their volatile analogs. The use of ionic liquids, in particular, is attractive because of their negligible vapor pressure under photovoltaic operating conditions, as well as their high conductivity and thermal stability. However, their conversion efficiency still lags behind those of organic solvent-containing DSSCs. The main reason for a worse performance is the high viscosity of ionic liquids, which causes mass-transfer limitations on the photocurrent under full sunlight. The thickness of the nanocrystalline TiO_2 film also greatly affected the photovoltaic characteristics, particularly for the ionic liquid, owing to a limitation of the photocurrent by the diffusion of I_3^- ions.

Of prime concern in liquid devices are electrode corrosion and electrolyte leakage, as these may seriously limit the device's stability as well as its long-term operation. The replacement of the liquid electrolyte by a HTM or quasi-solid-state electrolyte allows for an easier sealing and series connection of DSSC modules, and in this respect 2,2′,7,7′-tetrakis-(*N*,*N*-di-*p*-methoxyphenylamine)-9,9′-spirobifluorene (spiro-MeOTAD), CuSCN, or even conducting polymers such as

HTM have attracted considerable attention. A quasi-solid-state electrolyte usually refers to a mixing redox couple in the quasi-polymer. Although solid-state DSSCs have a lower efficiency compared to their liquid-type counterparts, investigations on solid-state DSSCs have attracted considerable attention because they represent an attractive means of realizing flexible photovoltaic cells in a roll-to-roll production. Considering the charge diffusion length L_D, where $L_D = (D\tau)^{1/2}$ (D is the charge diffusion coefficient and τ the charge lifetime), the electron lifetime is much shorter in solid-state devices than in liquid devices. In liquid cells, the L_D is much greater than the film thickness, whereas for the solid-state device it is comparable to the film thickness. Thus, a trade-off between film thickness and light harvesting must be made in the solid-state devices, and typically a film of only ~2 μm is used, compared to the 8 to 15 μm thick films used in liquid solar cells. A thicker film will also lead to the pore filling of solid-state DSSCs. Ru-based sensitizers have demonstrated a solar energy PCE of over 5% with spiro-OMeTAD as the hole conductor in solid-state DSSCs. Organic dyes are attractive materials in this respect, because of their high optical absorption extinction coefficient, adjustable spectral response, and environmentally benign nature. Indeed, it is these properties that have led to organic dyes becoming promising materials as the light-absorbing layer in solid-state DSSCs.

10.2 Small-Molecule Dyes in DSSCs

10.2.1 Coumarin Dyes

Hara *et al.* reported the details of a series of coumarin dyes (**a1–a3**) with different vinylene unit numbers (Figure 10.3; Table 10.1) [30, 31]. The absorption spectra

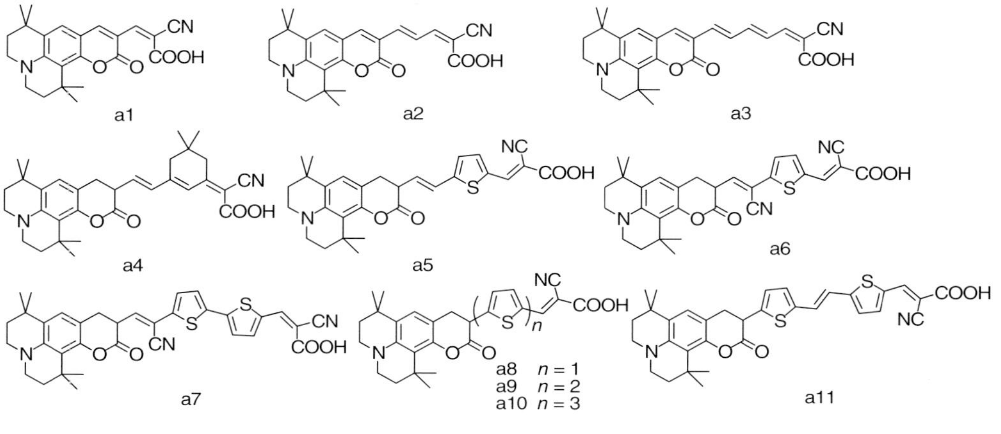

a1 a2 a3 a4 a5 a6 a7 a8 $n = 1$ a9 $n = 2$ a10 $n = 3$ a11

Figure 10.3 Chemical structures of coumarin dyes.

Table 10.1 Electronic properties and liquid device performance of coumarin dyes.

Dye	λ_{abs}[a] (nm)	ε (10^4 l mol^{-1} cm^{-1})	$E_{0\text{-}0}$ (eV)	E_{ox}[b] (V)	E_{red}[b] (V)	J_{sc} (mA cm^{-2})	V_{oc} (V)	FF	PCE (%)	Reference
a1						12.9	0.50	0.64	4.1	[31]
a2						14.0	0.60	0.71	6.0	[31]
a3						15.1	0.47	0.50	3.5	[31]
a4	492	5.01	1.77	0.9	−0.87	16.1	0.60	0.69	6.7	[32]
a5	507	5.57	1.84	0.92	−0.91	14.7	0.67	0.73	7.2	[33]
a6	566	5.76	1.68	1.08	−0.60	14.32	0.51	0.73	5.3	[34]
a7	552	9.74	1.69	0.97	−0.69	16.50	0.61	0.76	7.6	[34]
a8	507	5.43	1.86	1.01	−0.85	12.1	0.66	0.73	5.8	[35]
a9	511	6.43	1.82	0.93	−0.89	14.8	0.71	0.77	8.1	[35]
a10	501	7.33	1.68	0.91	−0.77	14.3	0.70	0.64	6.4	[35]
a11	525	7.0	1.77	0.82	−0.95	15.9	0.69	0.75	8.2	[36]

a) In solution.
b) versus NHE.

of coumarin dyes adsorbed onto the TiO_2 surface were remarkably broadened relative to those in solution, but this may be due to dye–dye and/or dye–TiO_2 interactions. The absorption threshold of **a2** was shifted to 630 nm and the absorption spectrum broadened owing to the introduction of two conjoining vinylene units and the electron-withdrawing moieties of carboxyl and cyano groups connected directly to the former, which also played an important role in the effective electron injection into the conduction band of TiO_2. As a result, DSSC based on **a2** exhibited a PCE of 6% and an incident photon-to-current conversion efficiency (IPCE) of >70% in the range from 460 to 600 nm, with a maximum value of 80% at 470 nm. The reduction of one vinylene unit in **a2** gave **a1**, the PCE of which was decreased (4.1%) when compared to **a2**, due to the blue-shifted absorption. However, expansion of the vinylene unit in **a2** gave **a3**, the efficiency of which fell due to a decreased electron-injection yield caused by intermolecular energy transfer because the dyes were prone to aggregate on TiO_2. An appropriate length of conjugation (–CH=CH–) was important in dye molecule design. Although such an extension would contribute to a red shift in the absorption spectrum, it would simultaneously cause two problems: (i) it would complicate the synthetic procedure; and (ii) it would increase the instability of the dye molecule, owing to the possibility of isomer formation [33]. A coumarin dye **a4** with a side ring to prevent dye aggregation was synthesized, ensuring a high surface concentration of dye because there was no use for DCA as coadsorbent [32]. The IPCE of **a4** reached a maximum of 84% at 540 nm. A PCE of 6.7% was obtained in the presence of guanidinium thiocyanate (GT); otherwise, the PCE was 5.7%.

The introduction of thiophene between two vinylene units in **a2**, rather than by extending the ethylene unit itself, broadened IPCE spectrum, thus leading to an improved photocurrent and PCE (7.2%) for **a5** [33]. Compared to the dye containing one CN group (**a5**), the addition of one more CN group to the π-conjugation bridge (**a6**) positively shifted the LUMO and thus red-shifted the absorption band, harvesting more photons in the long-wavelength region for photoelectric conversion [34]. However, a downshift of the LUMO led to a decrease in the driving force for electron injection (i.e., the difference between LUMO and the conduction band edge), decrease in IPCE and decrease in PCE (5.3%). Compared to the dye containing one thiophene unit (**a6**), the addition of one more thiophene unit to the π-conjugation bridge (**a7**) upshifted the LUMO and enhanced the absorption coefficient, leading to improved J_{sc} and PCE (7.6%) [34].

The removal of one vinylene unit from **a5** gave **a8** which, relative to **a5**, exhibited a similar absorption but downshifted HOMO and LUMO levels, leading to decreased J_{sc} and PCE (5.8%), most likely due to a decrease in the driving force for electron injection [35] Across the series of **a8–a10** with one to three thiophene units, an increase in the thiophene number led to a red-shifted absorption, an enhanced molar extinction coefficient, and an upshifted HOMO level [35]. **a9** afforded the highest PCE of 8.1% with 1.0 M TBP as an additive in the electrolyte. The insertion of one vinylene unit between two thiophenes in **a9** gave **a11** which, relative to **a9**, exhibited a red-shifted absorption and upshifted HOMO and LUMO levels, leading to a slightly improved PCE (8.2%) [36]. **a11** has one more thiophene than **a5**, and one more thiophene vinylene unit than **a8**. Relative to **a5** and **a8**, **a11** exhibited a red-shifted absorption and upshifted HOMO and LUMO levels, leading to an improved PCE (8.2%) [36].

10.2.2 Triphenylamine Dyes

As the donor unit, triphenylamine (TPA) can improve the hole-transporting ability of the materials, and its nonplanar structure can also prevent the formation of dye aggregates [37]. In the meantime, the TPA moiety is expected to greatly locate the cationic charge from the TiO_2 surface and to efficiently restrict recombination between the photoelectron and the oxidized sensitizer [38], thus achieving a higher rates of charge separation and collection compared to interfacial charge- recombination processes [39].

Across the series of **b1–b3** with one to three thiophene units, an increase in the thiophene number led to a red-shifted absorption and an enhanced molar extinction coefficient (Figure 10.4; Table 10.2) [40, 41]. **b2** afforded the highest PCE, of 6.15% [41]. The extra hydrophobic hexyloxy side chains on TPA in **b4** retarded diffusion of the electrolyte to the surface of the TiO_2, and this resulted in a suppression of the dark current and blocking of the cations, Li^+ and H^+, from approaching the TiO_2 surface. Thus, the V_{oc}, FF and PCE of **b4** were improved relative to **b2** [42]. Solid-state DSSCs based on **b4**, using 2,2′,7,7′-tetrakis(*N,N*-dimethoxyphenylamine)-9,9′-spirobifluorene (spiro-OMeTAD) hole conductor as a redox couple, exhibited a J_{sc} of 9.64 mA cm^{-2}, a V_{oc} of 0.798 V, FF of 0.57, and PCE of 4.44% [42]. Replacement of the phenyl group in **b1** with a fluorenyl group

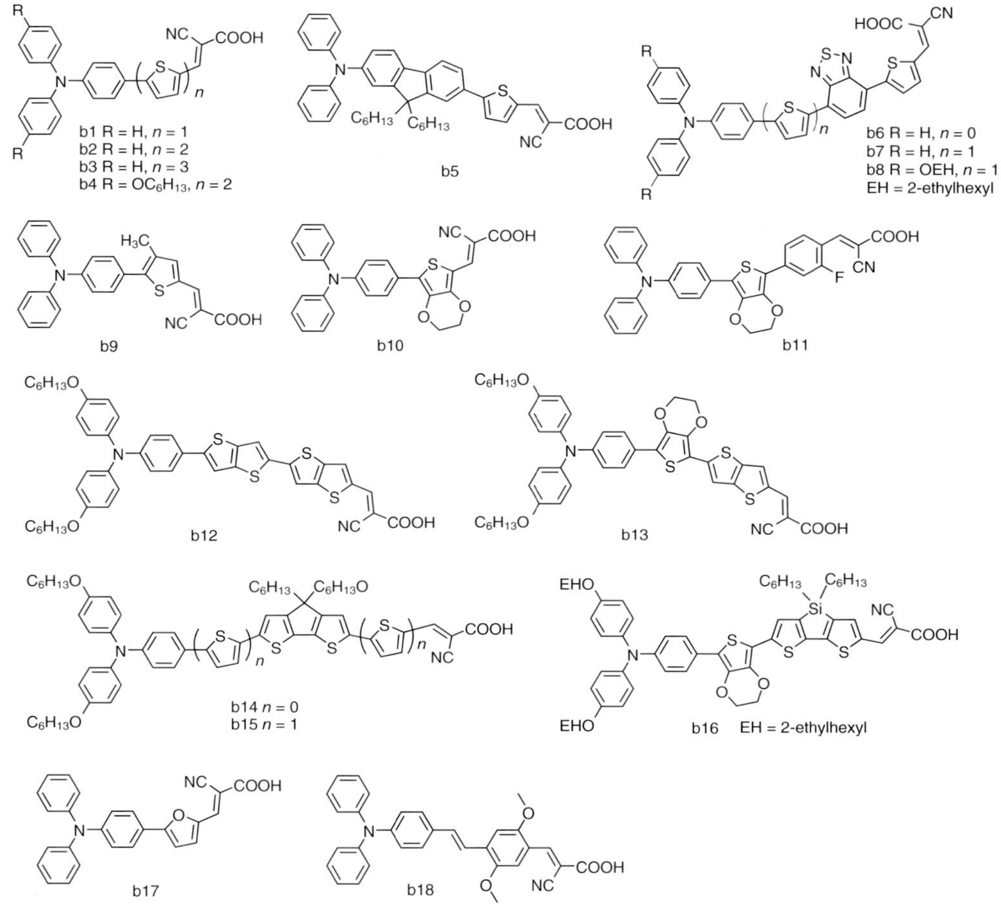

Figure 10.4 Chemical structures of triphenylamine dyes.

produced **b5**, which exhibited **a** red-shifted absorption, an enhanced molar extinction coefficient, and an improved PCE (from 5.2% to 6.78%) [16].

An early application of the TPA dye **b6** with push–pull–pull architecture, comprising the low-band-gap building block benzothiadiazole which was easily polarizable and electron-deficient, gave a PCE of 3.77% [43]. The insertion of thiophene between benzene and benzothiadiazole in **b6** gave **b7**, which exhibited a red-shifted absorption, an enhanced molar extinction coefficient, and an improved PCE (5.72%) at a coadsorbent DCA concentration of 80 mM [44]. **b8** was obtained by attaching bulky alkoxy groups to **b7**, a strong electron-donating group that was responsible for the red-shifted absorption and a reduction of dye aggregation on TiO_2 film, thus giving an improved PCE of 7.30% [44]. Replacement of the

Table 10.2 Electronic properties and liquid device performance of TPA dyes.

Dye	λ_{abs}[a] (nm)	ε ($10^4\,l\,mol^{-1}\,cm^{-1}$)	E_{0-0} (eV)	E_{ox}[b] (V)	E_{red}[b] (V)	J_{sc} ($mA\,cm^{-2}$)	V_{oc} (V)	FF	PCE (%)	Reference
b1	410	2.58	2.40	1.06	−1.34	12.8	0.620	0.66	5.20	[40]
b2	473	4.35	2.20	–	–	16.0	0.63	0.61	6.15	[41]
b3	480	4.62	2.17	–	–	15.2	0.61	0.58	5.41	[41]
b4	458	3.70	–	0.98	−1.35	14.10	0.728	0.71	7.25	[42]
b5	427	4.26	2.38	–	−1.26	15.82	0.68	0.63	6.78	[16]
b6	491	2.75	2.12	–	–	10.44	0.546	0.66	3.77	[43]
b7	533	3.08	1.97	1.05	−0.92	15.0	0.58	0.65	5.72	[44]
b8	542	3.10	1.93	0.98	−0.95	17.9	0.62	0.66	7.30	[44]
b9	466	2.20	2.24	1.19	−1.05	15.76	0.72	0.73	8.27	[45]
b10	426	2.45	2.53	0.98	−1.55	15.5	0.690	0.683	7.3	[40]
b11	420	4.22	2.53	0.94	−1.59	15.58	0.787	0.67	8.22	[46]
b12	524	4.70	2.046	–	–	15.2	0.72	0.733	8.02	[47]
b13	552	–	–	–	–	16.1	0.803	0.759	9.8	[48]
b14	555	6.27	–	–	–	15.84	0.769	0.735	8.95	[23]
b15	–	–	–	–	–	15.31	0.85	0.73	9.4	[49]
b16	493	5.75	2.19	–	–	17.94	0.77	0.730	10.1	[15]
b17	456	3.35	–	–	–	14.16	0.68	0.66	6.30	[50]
b18	433	4.61	2.49	–	–	20.85	0.697	0.666	9.67	[51]

a) In solution.
b) versus NHE.

thiophene bridge in **b1** with methylthiophene gave **b9**, and raised the PCE as high as 8.27% due to the red-shifted absorption [45]. Replacement of the thiophene bridge in **b1** with the electron-donating 3,4-ethylenedioxythiophene (EDOT) gave **b10**, which exhibited a higher PCE (7.3%), as the introduction of EDOT increased the spectral response and perhaps also rendered a better degree of charge separation [40]. Relative to **b10**, the dye **b11** achieved a higher extinction coefficient by incorporating an extended π-conjugation into the framework of the parent structure [46]. **B11** had a larger uptake ($3.88 \times 10^{-7}\,mol\,cm^{-2}$) than its non-fluorine counterpart, due to the resonance conjugation between fluorine and carbonyl groups, which increased the ligand basicity and hence the TiO_2 affinity. DSSCs based on **b11** exhibited a higher PCE (8.22%). Solid-state DSSCs based on **b11**, using spiro-OMeTAD as a redox couple, exhibited a J_{sc} of $7.75\,mA\,cm^{-2}$, V_{oc} of 0.95 V, FF of 0.62, and PCE of 4.53% [46].

Based on the thienothiophene unit, Wang *et al.* developed **b12** [47] and **b13** [48] successively. The charge-transfer transition absorption of **b12** was not only bathochromic but also enhanced compared to that of its counterpart with only one thienothiophene unit. Due to the incorporation of the bisthienothiophene linker, the PCE was improved from 7.05% to 8.02% [47]. A fresh cell employing the **b12** dye in combination with a solvent-free ionic liquid electrolyte exhibited a J_{sc} of

13 mA cm^{-2}, V_{oc} of 0.664 V, FF of 0.75, and PCE of 6.5%; moreover, after one week of aging the device PCE was even higher, up to 6.8%. This cell showed good stability, retaining 92% of its initial efficiency after 1000 h of full sunlight-soaking at 60 °C [47].

The replacement of one thienothiophene in **b12** with EDOT gave **b13**, which exhibited a 28 nm red shift of absorption in contrast to that of **b12** [48]. In addition, the spatial orientation of HOMO and LUMO in **b13** not only facilitated an ultrafast interfacial electron injection from the excited dye molecules to the TiO_2 conduction band, but also slowed down the recombination of injected electrons in TiO_2 with the oxidized dye molecules, due to their remoteness. It also facilitated a rapid dye regeneration, because the hole localized on the TPA unit was spatially convenient for the electron donor to approach, giving an IPCE >90% from 440 to 590 nm. Thus, **b13** gave higher PCEs than **b12** in both liquid DSSC (9.8%) and ionic liquid DSSC (8.1%) [48].

Dye **b14** was obtained by replacing 2,2′-bithiophene in **b4** with a stronger electron-donating unit, 4,4-dihexyl-4*H*-cyclopenta[2,1-*b*:3,4-*b*′]dithiophene [23]. Compared to **b4** [42], the absorption peak of **b14** was red-shifted by ca. 100 nm, and the maximum molar absorption coefficient in chloroform was doubled. DSSCs based on **b14** showed a higher PCE (8.95%) than **b4** (7.25%). The ionic liquid cell with **b14** featured a red-shifted photocurrent response with respect to the acetonitrile cell, rationalizing its higher photocurrent and high PCE (8.17%) [23]. The solvent-free cell retained about 98% of its initial efficiency after a 1000 h period of accelerated aging under full sunlight soaking at 60 °C, which highlighted the robustness of **b14** itself and of the titania/dye/electrolyte interface. The introduction of thiophene on each side of 4,4-dihexyl-4*H*-cyclopenta[2,1-*b*:3,4-*b*′] dithiophene in **b14** gave **b15**, which proved to be a suitable dye for the tris(1,10-phenanthroline)cobalt(II/III) redox couple. DSSC based on **b15** afforded a high PCE of 9.4% [49].

DSSCs based on the dye **b16** with high molar absorption coefficient, gave a strikingly high PCE of 10.0–10.3% due to meticulous molecule engineering from both energy level and structure [15]. An EDOT unit was connected to the donor TPA to uplift the HOMO, while dihexyl-substituted dithienosilole conjugated with an acceptor maintained a suitable LUMO without affecting the planarity of the π-conjugated spacer, although this may significantly reduce the π–π stacking of dye molecules on nanocrystals. DSSCs based on dye **b16**, in combination with a solvent-free ionic liquid electrolyte, exhibited a J_{sc} of 14.96 mA cm^{-2}, V_{oc} of 0.693 V, FF of 0.736, and PCE of 7.6%.

The replacement of thiophene in **b1** with the furan ring with a smaller resonance energy gave **b17** [50]. Compared to **b1** [40], the absorption peak of **b17** was red-shifted by ca. 46 nm. DSSCs based on **b17** showed a higher PCE (6.3%) than that of **b1** (5.2%). The insertion of one vinylene unit between TPA and furan in **b17** further red-shifted the absorption and enhanced the PCE (7.36%). The replacement of thiophene in **b1** with dimethoxyphenylenevinylene gave **b18** [51]; compared to **b1** [40], the absorption peak of **b18** was red-shifted by ca. 23 nm. DSSCs based on **b18** showed a higher PCE (9.67%) than **b1**.

10.2.3
Bisfluorenylaniline Dyes

The bis(dimethylfluorenyl)anilines suppress aggregate formation because their nonplanar structure disfavors molecular stacking and ensures a great resistance to degradation when exposed to light and high temperature. The latter effects are due to the possession of a bipolar character that allows the formation of both stable cation and anion radicals [52]. However, the adsorption density of difluoreneaniline is smaller due to the great steric congestion (Figure 10.5; Table 10.3).

The incorporation of one more thiophene unit into **c1** extended the conjugation length, the absorption spectrum was broadened and red-shifted by 16 nm, and the molar extinction coefficient was also improved from **c1** to **c2** [52]. Solid-state DSSCs based on **c2**, using spiro-OMeTAD as a redox couple, exhibited a J_{sc} of 3.85 mA cm^{-2}, V_{oc} of 1.0875 V, FF of 0.677, and PCE of 3.17% [61]. The very high photovoltage of 1.0875 V was due to the negative conduction band shift caused by the favorable dipolar field exerted by the **c2** sensitizer to the TiO_2 surface [61]. The PCE value (8.39%) of the **c3**-based cells was higher than that (7.63%) of **c2**-based

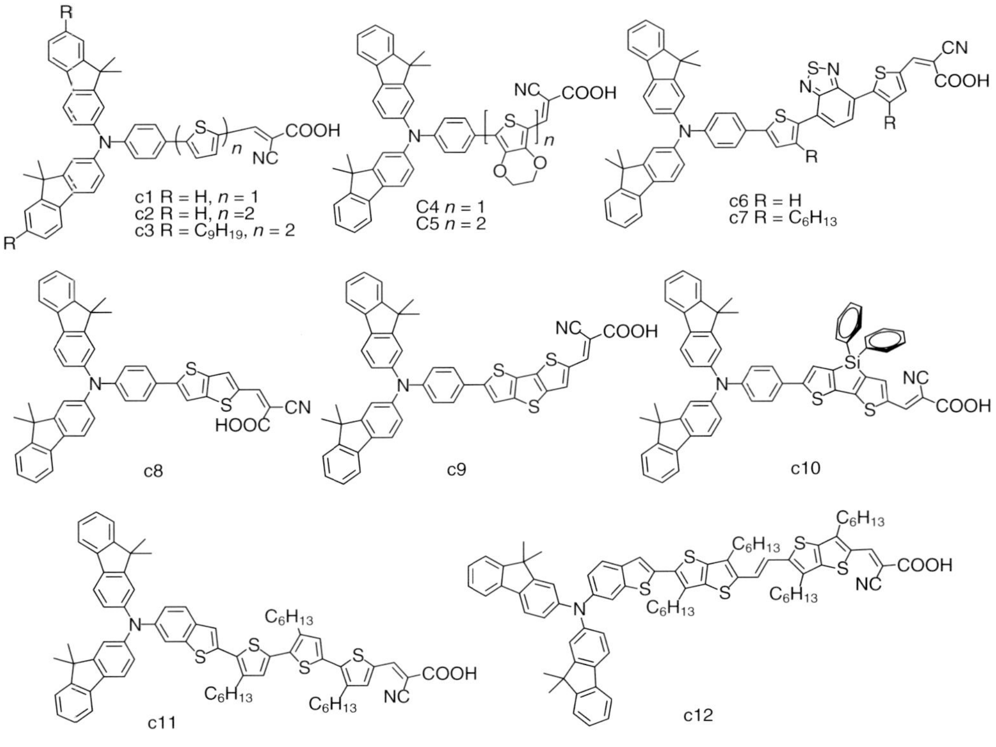

Figure 10.5 Chemical structures of bisfluorenylaniline dyes.

Table 10.3 Electronic properties and liquid device performance of bisfluorenylaniline dyes.

Dye	λ_{abs} [a)] (nm)	ε (10^4 l mol^{-1} cm^{-1})	E_{0-0} (eV)	E_{ox} [b)] (V)	E_{red} [b)] (V)	J_{sc} (mA cm^{-2})	V_{oc} (V)	FF	PCE (%)	Reference
c1	436	3.00	–	1.06	−1.34	12.20	0.764	0.77	7.20	[52]
c2	452	3.90	–	1.01	−1.30	14.0	0.753	0.77	8.01	[52]
c3	447	3.66	2.37	1.06	−1.31	15.43	0.74	0.74	8.39	[53]
c4	525	3.35	2.08	1.00	−0.87	13.82	0.762	0.751	7.92	[54]
c5	544	3.85	2.00	0.89	−0.87	15.68	0.746	0.711	8.32	[54]
c6	534	1.97	1.93	0.98	−0.95	17.10	0.61	0.72	7.51	[55]
c7	468	1.93	2.07	1.08	−0.99	14.98	0.77	0.77	8.19	[55]
c8	514	4.12	2.13	1.0	−0.78	13.35	0.777	0.749	7.8	[56]
c9	525	4.48	2.04	0.987	−0.762	14.33	0.734	0.76	8.0	[57]
c10	459	2.41	2.27	1.07	−1.20	13.9	0.739	0.73	7.5	[58]
c11	430	2.92	2.35	–	−1.36	17.45	0.664	0.742	8.6	[59]
c12	490	8.50	2.24	–	−1.18	14.9	0.810	0.79	9.5	[60]

a) In solution.
b) versus NHE.

cells; this important contribution derived from the 40 mV increase in V_{oc}, as the introduction of long alkyl chains to the dye decreased the electronic coupling between the TiO_2 surface and the oxidizing species in electrolyte, thus increasing the electron lifetime [53]. A quasi-solid-state DSSC comprising 5 wt% poly (vinylidenefluoride-*co*-hexafluoropropylene) (PVDF-HFP), 0.6 M 1,2-dimethyl-3-propyl-imidazolium iodide (DMPII), 0.1 M I_2, and 0.5 M *N*-methylbenzimidazole (NMBI) in 3-methoxypropionitrile(MPN), which was based on **c3**, yielded a remarkably high conversion efficiency of 7.31% which, after 1000 h of light soaking at 60 °C, was decreased to 6.21%. In contrast, the PCE of a quasi-solid-state DSSC based on **c2** was decreased from 6.31% to 4.99%, under the same conditions. This enhanced long-term stability of **c3** may be attributed to the introduction of long alkyl chains to the bis-dimethylfluorenylamino unit, which led to a retardation of charge recombination. Relative to their thiophene and bithiophene counterparts **c1** and **c2** [52], the sensitizers **c4** and **c5** with EDOT and biEDOT exhibited remarkably red-shifted spectral and photocurrent responses and higher PCEs (7.92% and 8.32%, respectively) [54]. The ionic liquid cells based on **c4** and **c5** afforded PCEs of 7.31 and 7.61%, respectively [54].

The incorporation of the acceptor unit benzothiadizole between two thiophene units in **c2** yielded **c6**, which exhibited a red-shifted absorption, an upshifted HOMO level and a downshifted LUMO level [55]. A DSSC based on **c6**, using an acetonitrile electrolyte comprising 0.6 M DMPImI, 0.05 M I_2, 0.1 M LiI, and 0.5 M *tert*-butylpyridine, afforded **a** higher J_{sc} but a lower V_{oc} and PCE (7.51%) than that based on **c2**. The introduction of hexyl groups substituted at the thiophene units in **c7** led to a significant increase in V_{oc}, from 0.61 to 0.77 V, and an increase in

PCE, from 7.51% to 8.19% [55]. The alkyl chains were quite effective in increasing the electron lifetime by preventing the dark current, and improving the V_{oc}. Quasi-solid DSSCs based on **c6** and **c7** using a polymer gel electrolyte gave J_{sc}-values of 9.58 and 12.03 mA cm^{-2}, V_{oc}s of 0.643 and 0.720 V, and FFs of 0.755 and 0.762, respectively, corresponding to overall conversion efficiencies of 4.66% and 6.61%. The device employing **c7** showed an excellent long-term stability, with an initial PCE of 6.61% being only slightly reduced to 6.0% during a 1000 h light-soaking test. The enhanced stability of **c7** compared to **c6** could be attributed to the substituted hexyl chains, which prevented the approach of acceptors to the TiO_2 surface, so as to prevent the dark current [55].

The replacement of thiophene bridge in **c1** with a fused thienothiophene gave dye **c8** [56] which, relative to **c1**, exhibited a red-shifted absorption and a higher PCE (7.8%). Ionic liquid DSSCs based on **c8** exhibited a J_{sc} of 12.40 mA cm^{-2}, V_{oc} of 0.723 V, and FF of 0.779, corresponding to an overall PCE of 7.0%. The device which was based on **c8** and fabricated with an ionic liquid electrolyte showed excellent light-soaking stability, with the PCE of the device falling by only 7.5% after >1000 h aging at 60 °C. An all-solid-state device based on **c8**, with spiro-OMeTAD as the hole-transporting material, showed **a** J_{sc} of 9.06 mA cm^{-2}, V_{oc} of 0.86 V, and FF of 0.61, resulting in an impressive PCE of 4.8% [56]. **c9**, with fused dithienothiophene that had a low free energy of solvation in the high-polarity electrolytes normally used by DSSCs, showed a PCE of 8.0% [57]. A comparable PCE (7.5%) was also achieved by **c10**, where dithienothiophene was replaced by 4,4-diphenyldithienosilole (DTS) as the bridge [58].

c11 adopted aliphatic chains that were attached to the thiophene in order to enhance its tolerance towards water in the electrolytes [59]. Liquid DSSCs based on **c11** exhibited a PCE of 8.6%. Long-term accelerated aging experiments indicated a remarkable stability of the ionic-liquid DSSC with **c11**, with the initial PCE of 6.82% being slightly increased (to 7.03%) during a 1000 h light-soaking test. A liquid DSSC based on **c12** with a fused thienothiophene bridge displayed an even higher efficiency of 9.5%, using a TiO_2 electrode stained with 0.15 mm dye in the presence of 1 mM tetrabutylammonium deoxycholic acid salt (TBA-DCA) in tetrahydrofuran (THF) [60]. The ionic-liquid DSSC with **c12** exhibited an initial PCE of 7.9% that this was only slightly decreased (to 7.4%) during a 1000 h light-soaking test [60].

10.2.4
Other Dyes

In **d1** and **d2**, carbazole was connected to the oligothiophene bridge via the 3-position (Figure 10.6; Table 10.4) [62]. Under the same conditions, the V_{oc} (0.71 V) for DSSCs based on **d1** *with* long alkyl chains, was higher than that of its counterpart *without* alkyl chains (0.63 V), since the electron lifetime (τ) of the former DSSC was remarkably longer than that of the latter. The alkyl chains on the thiophenes in **d1** may help to form a blocking layer that would keep I_3^- ions away from the TiO_2 electrode surface. The addition of one more thiophene unit

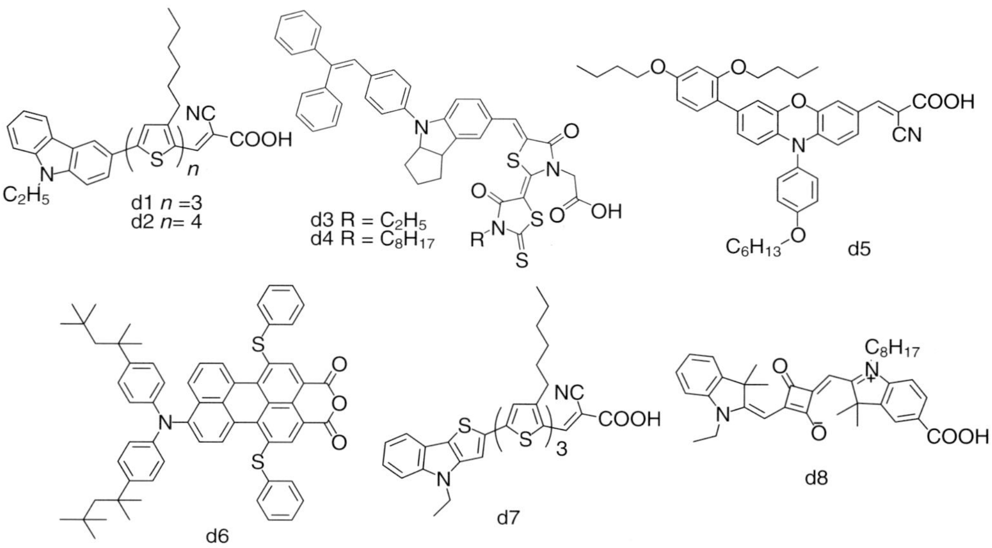

Figure 10.6 Chemical structures of other dyes.

Table 10.4 Electronic properties and liquid device performance of other dyes.

Dye	λ_{abs}[a] (nm)	ε (10^4 l mol^{-1} cm^{-1})	$E_{0\text{-}0}$ (eV)	E_{ox}[b] (V)	E_{red}[b] (V)	J_{sc} (mA cm^{-2})	V_{oc} (V)	FF	PCE (%)	Reference
d1	463	4.14	–	–	–	11.1	0.71	0.71	5.6	[62]
d2	473	3.58	–	–	–	12.3	0.72	0.71	6.3	[62]
d3	526	6.87	–	–	–	18.5	0.693	0.624	8.0	[63]
d4	554	7.47	–	–	–	18.68	0.710	0.707	9.40	[64]
d5	500	2.03	2.19	0.967	–	13.09	0.80	0.70	7.40	[65]
d6	620	2.27	–	–	–	12.60	0.728	0.74	6.8	[66]
d7	496	4.50	1.89	0.89	–	14.6	0.70	0.76	7.8	[67]
d8	636	15.85	1.92	0.98	−0.78	10.50	0.603	0.71	4.5	[68]

a) In solution.
b) versus NHE.

in **d1** gave **d2** which, relative to **d1**, exhibited a red-shifted absorption and an improved PCE. Following optimization of the electrolyte, **d2** produced a PCE of 8.3% [69]. An excellent combination of **d2** dye and ionic liquid electrolytes was realized by the high coverage of TiO_2 surface by **d2** and the moderate viscosity of ionic liquid electrolytes, leading to a high PCE of 7.6%. In addition, stability testing indicated that the DSSC based on **d2** and an ionic liquid electrolyte could survive for 2000 h with an almost constant PCE >7%, under continuous visible light-soaking conditions (100 mW cm^{-2} AM 1.5 G simulated solar light) at temperatures of 50 °C [70].

The indoline moiety was shown to possess a stronger electron-donating ability than the TPA unit in a series of isophorone-bridged organic dyes; moreover, indoline dyes were also found to be highly stable to photoredox processes [71]. Indolene **d3** achieved a PCE of 8.00% under optimized evaluation conditions [63]. With liquid-based electrolytes, DSSCs based on **d3** achieved PCEs as high as 9.03% when the thickness of the nanocrystalline TiO_2 film was optimized as 12.6 μm [72], whereas with ionic-liquid-based electrolytes, DSSCs based on **d3** achieved PCEs of only up to 6.67% [72]. **d3** also performed remarkably well in thin, solid-state DSSCs, delivering over 6% PCE under simulated solar conditions at low to moderate intensity (1 and 10 mW cm^{-2}) [73]. Polymer-hole-transporting material (HTM)-based DSSCs, prepared by the *in-situ* polymerization of bis-EDOT in a thin-layer electrolytic cell using **d3** dye as the sensitizer, showed an average PCE of 6.1% [74], which represented a remarkable improvement for polymer-HTM-based DSSCs.

Replacement of the ethyl group on the terminal rhodamine unit of **d3** by an octyl chain yielded **d4**. Liquid DSSCs based on **d4** achieved a mean PCE of 9.40 ± 0.12% on the addition of CDCA, which was higher than that obtained with **d3** (8.85 ± 0.18%) [64]. The combination of CDCA and the *n*-octyl chain improved the V_{oc} to 0.710 V. CDCA also enhanced the FF for **d4**, while following the use of an ionic-liquid-based electrolyte, DSSCs based on **d4** achieved a PCE of up to 7.18%, which was higher than that of **d3**-based DSSCs (6.38%) [75]. Endowing the sensitizer with an octyl chain was shown to be highly effective in suppressing electron recombination, as verified by the enhanced electron lifetimes of 10.9 and 23.0 ms for **d3** and **d4**, respectively [75].

The phenoxazine (POZ) unit was shown to be a strong electron-donating group, because of its electron-rich oxygen and nitrogen heteroatoms. The POZ unit displayed a stronger electron-donating ability than the TPA unit (0.88 and 1.04 V, respectively versus the normal hydrogen electrode; NHE), which implied that POZ could also be incorporated as a dye sensitizer. Furthermore, the POZ system was nonplanar, with a rare butterfly conformation in the ground state which reduced the aggregation of the dye molecule. With hexyloxy-substituted benzene ring at N-position in POZ, the structure of **d5** resembled that of TPA; subsequently, the introduction of an additional electron donor in the 7-position on the POZ lifted both the HOMO and LUMO levels, which allowed the addition of 4-*tert*-butylpyridine, giving a V_{oc} of 0.80 V and PCE of 7.4% [65].

Diphenylamino perylene anhydride **d6**, which bore two thiophenol groups at the 1- and 6-positions, had a broad IPCE spectrum and a high plateau value of 87% from 520 to 600 nm, and showed a PCE of up to 6.8% [66]. The electron-donating ability of thieno[3,2-*b*]indole in **d7** was stronger than that of carbazole; consequently, thieno[3,2-*b*]indole as the donor part could hold the dye molecule in a more planar conformation compared to the carbazole-based dyes, thus providing a higher molar extinction coefficient [67]. DSSCs based on thieno[3,2-*b*]indole dyes showed a shorter electron lifetime than a DSSC based on a conventional carbazole dye, which resulted in a lower V_{oc}. Relative to **d1**, **d7** exhibited a red-shifted absorption and an improved PCE (7.8%). In squaraine dye **d8**, the carboxylic acid group formed part of the conjugated π-system of the dye, and provided a strong electronic coupling to the conduction band of TiO_2; moreover, the asymmetry created by the octyl chain prevented surface aggregation and limited the self-quenching of the excited state [68]. A DSSC based on **d8** showed a PCE of up to 4.5%.

10.3
Polymer Dyes in DSSCs

While polymer have a broadened and red-shifted light response, the absorption band of a small molecule is narrow. Compared to poly(3-hexylthiophene) **e1**, polythiophene carboxylic acid **e2** showed a much higher efficiency because the carboxylic group in the branch of the latter dye provided an efficient adsorption onto the surface of the TiO_2 and also red-shifted the absorption spectrum by interacting with each other upon adsorption [76]. Preparation of the photoanode by dipping the TiO_2 thin film into polymer solutions rather than spin-coating resulted in a higher efficiency (**e1**, 0.18–0.42%; **e2**, 1.4%), due to there being a larger area of contact between the polymer and TiO_2. In contrast, spin-coating caused the liquid electrolytes to be blocked from entering the nanoporous TiO_2, and this resulted in a relatively low efficiency (**e2**, 0.27%). In the presence of the electron-donating ionic liquid 1-methyl-3-*n*-hexylimidazolium iodide, in an electrolyte composed of the redox couple I_3^-/I^- , a DSSC based on **e2** generated a J_{sc} of ~9.74 mA cm^{-2}, a V_{oc} of ~405 mV, and a PCE of 2.4% [77]. In the case of **e3**, an increase in either the benzothiadiazole ratio or in the molecular weight led to a decrease in PCE [78]. In the former situation the reduced number of carboxylic acid groups affected the dye–TiO_2 interaction, whereas in the latter situation it was difficult for a polymer with a large volume to penetrate the TiO_2 porous structure. Thus, a DSSC based on **e3** generated a J_{sc} of 4.03 mA cm^{-2}, a V_{oc} of 523 mV, a FF of 0.66, and a PCE of 1.39%. A quasi-solid-state DSSC based on **e4** showed a PCE of 3.78% following the addition of TBP to the quasi-solid-state polymer gel electrolyte [79] (Figure 10.7; Table 10.5).

A judicious design was included in molecular engineering of **e5**–**e8**, whereby the alternating D–A structure led to a charge separation in the excited state and a red-shift in the absorption spectrum. In this case, the carboxylic acid groups were in close proximity to the 2,1,3-benzothiadiazole acceptor unit, which may have facilitated electronic coupling between the polymer and the TiO_2 surface, so as to facilitate charge injection [80]. The DSSCs based on **e6** exhibited the highest efficiencies among **e6–e8**; this was due to the largest extents of adsorption and the highest charge injection yields, both of which could be attributed to the smallest molecular weights. However, the device efficiency was shown to decrease in line with the increasing molecular weight of the polymer.

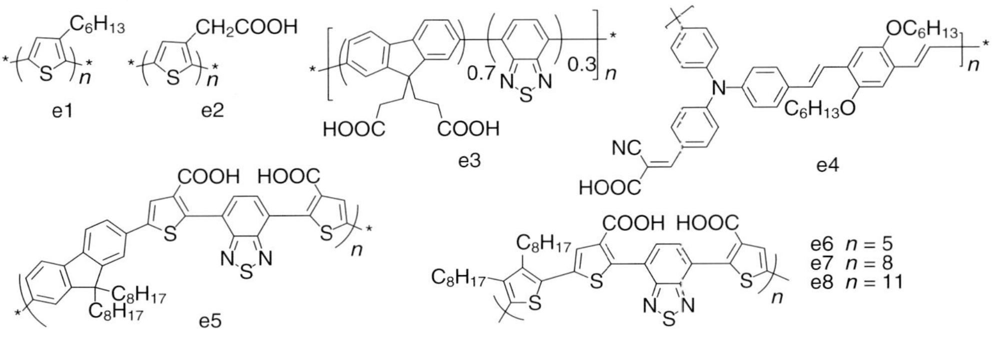

Figure 10.7 Chemical structures of polymer dyes.

Table 10.5 Electronic properties and liquid device performance of polymer dyes.

Dye	λ_{abs} a) (nm)	ε (10^4 l mol^{-1} cm^{-1})	E_{0-0} (eV)	E_{ox} b) (V)	E_{red} b) (V)	J_{sc} (mA cm^{-2})	V_{oc} (V)	FF	PCE (%)	Reference
e1	–	–	–	–	–	2	–	–	0.42	[76]
e2	–	–	–	–	–	9.74	0.405	0.60	2.4	[77]
e3	444	–	2.44	1.25 b)	−1.19 b)	4.03	0.523	0.66	1.39	[78]
e4	390	–	1.94	–	–	9.45	0.74	0.54	3.78	[79]
e5	475	3.10	2.07	0.95 c)	−1.12 c)	7.02	0.55	0.438	1.68	[80]
e6	466	1.30	1.90	0.82 c)	−1.08 c)	12.58	0.54	0.449	2.99	[80]
e7	472	1.40	1.89	0.76 c)	−1.13 c)	9.82	0.52	0.475	2.45	[80]
e8	476	1.20	1.88	0.70 c)	−1.18 c)	6.80	0.52	0.475	1.70	[80]

a) In solution.
b) versus Ag/Ag^+.
c) versus Fc^+/Fc.

10.4 Dyes in p-Type DSSCs

Although dyes in *p*-type DSSCs share certain structural similarities with dyes in *n*-type DSSCs, they use different materials and have different charge-transfer kinetics. In general, the HOMO level of the sensitizer is below the energy level of the top of the valence band, whereas the LUMO level is above the potential of the redox system (e.g., I^-/I_3^-) and far below the bottom of the conduction band of NiO. In order for a dye to be a good sensitizer, its energy levels should match well with the energy of the valence band of the semiconductor and the redox potential of the electrolyte, so as to provide a sufficient driving force for hole injection and dye regeneration. A long-lived, charge-separated state and an avoidance of aggregation are also preferable to achieve higher efficiencies. The pioneering investigations in this area were reported by Lindquist and coworkers (Figure 10.8;

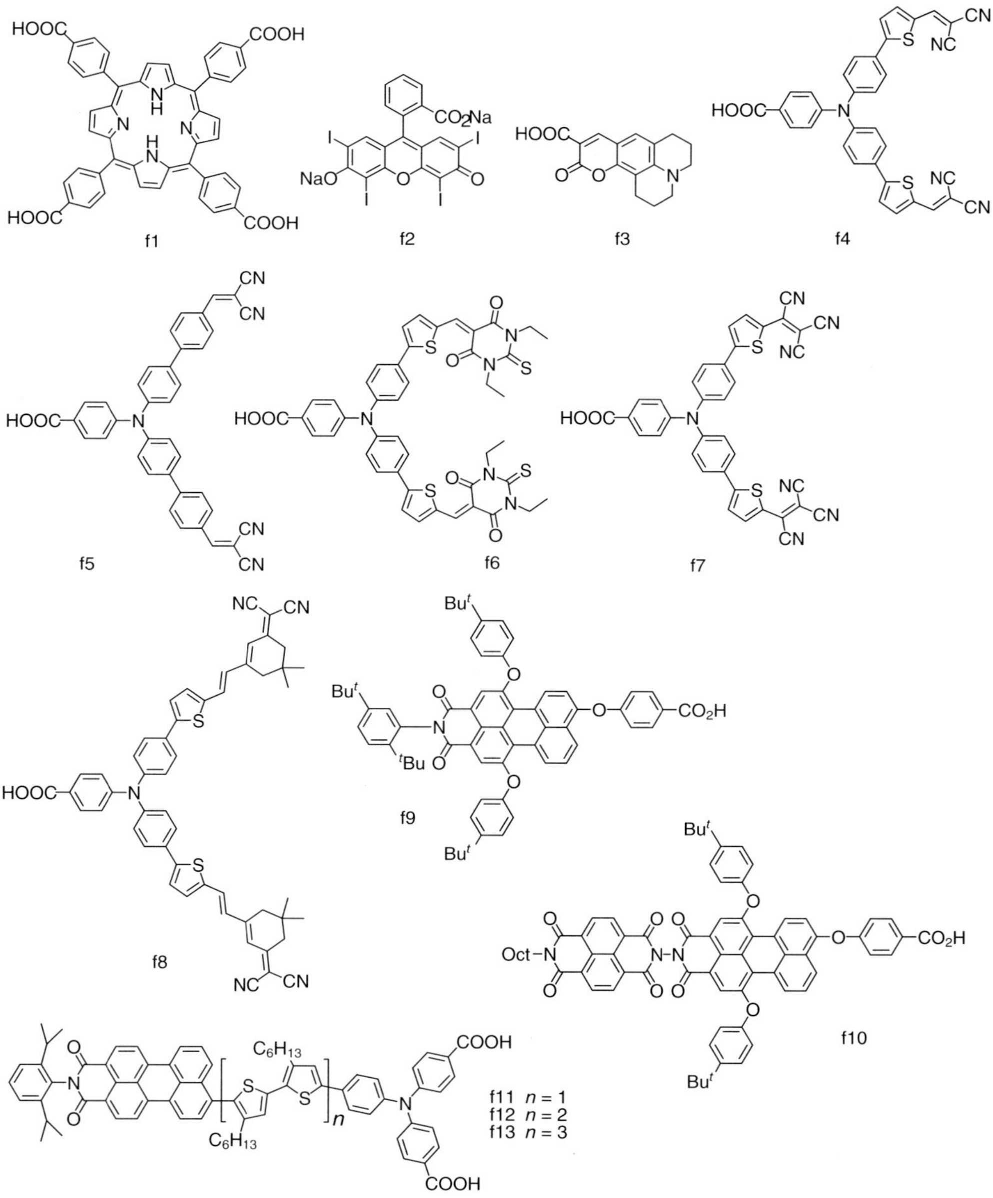

f1 f2 f3 f4

f5 f6 f7

f8 f9 f10

f11 $n = 1$
f12 $n = 2$
f13 $n = 3$

Figure 10.8 Chemical structures of dyes in *p*-DSSCs.

Table 10.6 Electronic properties and liquid device performance of dyes in *p*-type DSSCs.

Dye	λ_{abs}[a] (nm)	ε ($10^4\,l\,mol^{-1}\,cm^{-1}$)	E_{0-0} (eV)	E_{ox}[b] (V)	E_{red}[b] (V)	J_{sc} ($mA\,cm^{-2}$)	V_{oc} (V)	FF	PCE (%)	Reference
f1	–	–	–	–	–	0.079	0.0985	0.285	0.0033	[81]
f2	–	–	–	–	–	0.232	0.0828	0.27	0.0076	[81]
f3	444	–	–	1.15	−1.54	0.86	0.101	0.362	0.031	[82]
f4	481	5.79	2.25	1.32	−0.83	5.48	0.084	0.33	0.15	[85]
f5	430	4.5		1.32	−1.06	2.48	0.10	0.36	0.09	[84]
f6	531	6.40	2.04	1.30	−0.70	3.37	0.063	0.31	0.07	[85]
f7	574	4.10	1.86	1.40	−0.36	1.36	0.055	0.34	0.03	[85]
f8	490	7.50	2.07	1.28	−0.66	3.37	0.080	0.35	0.09	[85]
f9	515	3.20	2.24	1.12	−1.05	–	–	–	–	[86]
f10	525	3.20	2.20	1.18	−0.9	1.3	0.370	0.34	0.16	[87]
f13	–	–	–	–	–	5.35	0.218	–	–	[88]

a) In solution.
b) versus NHE.

Table 10.6) [81]. The highest IPCEs of tetrakis(4-carboxyphenyl)porphyrin (**f1**) and erythrosin B (**f2**)-coated NiO films were 0.24% and 3.44%, respectively, under a light intensity of 68 $mW\,cm^{-2}$. The **f1**-coated NiO cell exhibited a J_{SC} of 0.079 $mA\,cm^{-2}$, V_{OC} of 98.5 mV, FF of 28.5%, and overall PCE of 0.0033%. The **f2**-NiO cell exhibited a J_{SC} of 0.232 $mA\,cm^{-2}$, V_{OC} of 82.8 mV, FF of 27.0%, and PCE of 0.0076%. It should be noted that the overall conversion efficiency was very low because of the small photocurrent and photovoltage [81].

The PCE of DSSCs based on coumarin **f3** reached 0.031% with NiO [82]. In **f4**, the TPA moiety was adopted as the electron donor, the malononitrile moiety as the electron acceptor, and a thiophene unit as the bridge [83]. The main difference between **f4** and the normal organic dyes used in *n*-type DSSCs was that the attaching group was connected via the electron donor region. Thus, by introducing two electron-acceptor groups and having the anchoring group on the donor moiety, light excitation was associated with electron flow from the donor part of the dye to the acceptor part, thus providing an efficient electron-transfer pathway from the dye to the electrolyte. The highest IPCE achieved for **f4** was 18%, and the overall PCE was 0.05% [83]. The IPCE and PCE for **f4** were further improved to 35% and 0.08%, respectively, when the thickness of the NiO film was increased from typically 600 nm to 1–1.4 μm. This produced a higher internal surface area and thus a higher dye loading; as a result, the electrolyte propylene carbonate was displaced by acetonitrile, the viscosity of which was lower than propylene carbonate, and this led to an improved ion transportation in the electrolyte [84]. The efficiency of DSSCs based on **f4** was further improved to 0.15% in a series of studies [85]. The substitution of thiophene in the linker part of **f4** with a phenyl ring gave **f5** [84]; subsequently, DSSCs based on **f5** yielded the highest IPCE and PCE of 44% and 0.09%, respectively, both of which were superior to values obtained with the **f4**-

based counterparts. However, the efficiencies of DSSCs based on **f6–f8**, which were obtained by replacing malononitrile in **f4** with other electron acceptors, were remarkably decreased due to an insufficient driving force and aggregation of the dye on the NiO film [85].

Three *O*-aryl substituents were introduced on the perylene ring of **f9**, in which the reduction potential could be tuned and the aggregation of dyes limited [86]. Compared to **f9**, the introduction of a naphthalene diimide unit in a DSSC device based on **f10** caused a threefold increase in PCE. On its addition, the naphthalene diimide served as an electron acceptor, which meant that the charge-separated state could become sufficiently long-lived for the electron transfer to the electrolyte and hole collection at the photocathode back contact to occur at a higher efficiency. As a result, the PCE for DSSCs based on **f10** rose to 0.16% [87]. **f11–f13** incorporated a perylene monoimide as the acceptor and an oligothiophene of variable length coupled to triphenylamine as the donor. Among these dyes, **f13** showed the highest IPCE of 62% over a spectral range of 400–500 nm; this effect was due to a retardation of recombination caused by the longer distance between the acceptor functionality and the attachment group of the sensitizer [88].

The recombination of the hole in the semiconductor with the reduced dye has been identified as a major problem in *p*-type DSSCs [89]. Electron injection from the dye to the valence band of NiO, which occurs just after the hole injection, represents a major factor in the relatively low efficiency reported, even in the case of large ΔE (the energy difference between the valence band edge of NiO and the ground state of the dyes) [82]. Consequently, a judicious molecular engineering is urgently required in any future investigations.

The following sequence represents the an ideal situation. Light excitation is first associated with an electron flow from the donor part to the acceptor part of the dye; this causes the electron density to be moved far away from the semiconductor surface, and results in a longer-lived charge-separated state. As a consequence, electron transfer to the electrolyte and hole collection at the photocathode back contact can occur with a high efficiency. In order to alleviate a rapid charge recombination between the hole in the valence band of NiO and the reduced sensitizer, one possible strategy might be to increase the distance of the hole/electron pair when the sensitizer has been excited.

The low photovoltage of these *p*-type systems was due to the small difference between the quasi-Fermi level of the NiO electrode and the electrochemical potential of the I^-/I_3^- redox system, which was only 0.1 V. However, when the I^-/I_3^- redox system was replaced with Co(II)/Co(III) redox couple, the **f10**-sensitized *p*-DSSC exhibited a V_{oc} of 0.35 V. This increase was attributed to the more negative redox potential of the cobalt redox couple compared to I^-/I_3^- [90].

10.5
Summary and Outlook

The donor–(π-spacer)–acceptor (D–π–A) system is commonly adopted when constructing metal-free dyes, due to the effective photoinduced intramolecular charge-transfer properties [91]. HOMO is mainly populated over the donor moiety, while LUMO is mainly populated on the acceptor (anchoring) moiety. This orientationally spatial separation of HOMO and LUMO is an ideal condition for dye-sensitized solar cells, which not only facilitates the ultrafast interfacial electron injection from excited dyes to the TiO_2 conduction band, but also slows down the recombination of injected electrons in TiO_2 with oxidized sensitizers, due to their remoteness. In addition, the hole localized on the donor unit will be spatially convenient for the electron donor to approach, thus facilitating a rapid dye regeneration [48]. Yet, D–π–A means much more in molecule engineering. A combination of donor and acceptor with different strengths can provide various energy levels, such that photophysical properties can be easily tuned by varying the donor, spacer, and acceptor moieties in these dyes [92]. In addition, the separation of acceptor and linker provides much more versatility to molecule engineering, though the injection efficiency may become problematic [93].

D–π–A can be developed into a D–π–π–A structure by introducing additional π-conjugation segments between the donor and acceptor [94, 95]. However, the prolonged rod-like molecules may facilitate the recombination of electrons to the triiodide, and also the formation of aggregation between molecules. A close π–π aggregation may lead not only to a self-quenching and reduction of electron injection into TiO_2, but also to an instability of the organic dyes due to the formation of excited triplet states and unstable radicals under light irradiation [96].

A starburst configuration [97] has emerged which inhibits dye aggregation as well as increasing the electron lifetime (τ) of the dyes in the conduction band of TiO_2, and this has resulted in an increase of V_{oc}. Aggregation – usually J- and H-aggregation – upon adsorption is also an issue, and has changed the absorption spectrum. Whilst J-aggregates demonstrate a shift of the absorption to a longer wavelength region, the H-aggregates show a shift to a shorter wavelength region. As more of a common-sense issue, aggregation is one of the main causes of instability. Although J-aggregation has been reported as detrimental to device efficiency (based on coumarin dyes [98]), it is sometimes beneficial to the efficiency of DSSCs. The latter suggestion has been based on merocyanine, as red-shifted J-aggregates facilitated by long alkyl chain substitution in their merocyanine dyes lead to an enhancement in photosensitization efficiency [99, 100]. The relationship between dye aggregation on the oxide semiconductor electrode and the photoelectrochemical properties is very interesting; thus, the judicious design of a dye should take advantage of both J- and H-aggregation. Furthermore, the ideal situation is that dyestuffs are grafted controllably on to a nanostructured semiconducting film with favorable stacking modes and energy alignments, which would in turn have significant influences on multichannel charge-transfer kinetics, determining the ultimate cell efficiency [15]. A careful interfacial engineering is also required since, in order to develop high-performance DSSCs, it is first necessary to develop a novel molecular design of organic dye sensitizers that are capable of controlling not only the photophysical and electrochemical properties of the dyes themselves but also the molecular orientation and arrangement of the dyes on a TiO_2 surface. For example, dyes with a rigid structure are prone to acquire a higher efficiency than flexible dyes with a stronger light-absorption ability, due to the favorable arrangement on TiO_2 surface [95].

Cosensitization is effective for broadening spectrum responses [101, 102]. Although I^-/I_3^- is the most widely used redox couple in electrolytes, the substitution of I^-/I_3^- with Br^-/Br_3^- [103] or Co(II)/Co(III) [90] may occasionally be more effective, as this mainly affects V_{oc}. Discovering suitable redox couples with a good match to dyes represents an accessible means of improving the efficiency of these systems.

The theoretical upper limit for a cell with only one photoactive dye-sensitized electrode is about 30%, but the corresponding limit for a tandem device with two photoactive semiconductors is about 43%. An apparently simple means of improving the efficiency of a DSSC would, therefore, be to substitute a passive cathode with a photoactive counterpart, thereby extending it to a tandem cell [104]. Unfortunately, however, the main drawbacks to this proposal are a rather low efficiency of the photoactive cathode and the match between two photoactive electrodes.

References

1 Lewis, N.S. (2007) Toward cost-effective solar energy use. *Science*, **315** (5813), 798–801.
2 Oregan, B. and Grätzel, M. (1991) A low-cost, high-efficiency solar-cell based on dye-sensitized colloidal TiO_2 Films. *Nature*, **353** (6346), 737–740.
3 Hagfeldt, A., Boschloo, G., Sun, L., Kloo, L., and Pettersson, H. (2010) Dye-sensitized solar cells. *Chem. Rev.*, **110** (11), 6595–6663.
4 Hamann, T.W., Jensen, R.A., Martinson, A.B.F., Van Ryswyk, H., and Hupp, J.T. (2008) Advancing beyond current generation dye-sensitized solar cells. *Energy Environ. Sci.*, **1** (1), 66–78.
5 Mishra, A., Fischer, M.K.R., and Bäuerle, P. (2009) Metal-free organic dyes for dye-sensitized solar cells: from structure: property relationships to design rules. *Angew. Chem. Int. Ed.*, **48** (14), 2474–2499.
6 Morandeira, A., Boschloo, G., Hagfeldt, A., and Hammarström, L. (2005) Photoinduced ultrafast dynamics of coumarin 343 sensitized *p*-type-nanostructured NiO films. *J. Phys. Chem. B*, **109** (41), 19403–19410.
7 Hagfeldt, A. and Grätzel, M. (1995) Light- induced redox reactions in nanocrystalline systems. *Chem. Rev*, **95** (1), 49–68.
8 Schmidt-Mende, L., Bach, U., Humphry-Baker, R., Horiuchi, T., Miura, H., Ito, S., Uchida, S., and Grätzel, M. (2005) Organic dye for highly efficient solid-state dye-sensitized solar cells. *Adv. Mater.*, **17** (7), 813–815.
9 Chiba, Y., Islam, A., Watanabe, Y., Komiya, R., Koide, N., and Han, L.Y. (2006) Dye-sensitized solar cells with conversion efficiency of 11.1%. *Jpn. J. Appl. Phys. Part 2–Lett. Express Lett.*, **45** (24–28), L638–L640.
10 Nazeeruddin, M.K., Kay, A., Rodicio, I., Humphry-Baker, R., Mueller, E., Liska, P., Vlachopoulos, N., and Grätzel, M. (1993) Conversion of light to electricity by *cis*-X2bis(2,2′-bipyridyl-4,4′-dicarboxylate)ruthenium(II) charge-transfer sensitizers (X = Cl-, Br-, I-, CN-, and SCN-) on nanocrystalline titanium dioxide electrodes. *J. Am. Chem. Soc.*, **115** (14), 6382–6390.
11 Péchy, P., Renouard, T., Zakeeruddin, S.M., Humphry-Baker, R., Comte, P., Liska, P., Cevey, L., Costa, E., Shklover, V., Spiccia, L., Deacon, G.B., Bignozzi, C.A., and Grätzel, M. (2001) Engineering of efficient panchromatic sensitizers for nanocrystalline

TiO_2-based solar cells. *J. Am. Chem. Soc.*, **123** (8), 1613–1624.

12 Nazeeruddin, M.K., De Angelis, F., Fantacci, S., Selloni, A., Viscardi, G., Liska, P., Ito, S., Takeru, B., and Grätzel, M. (2005) Combined experimental and DFT-TDDFT computational study of photoelectrochemical cell ruthenium sensitizers. *J. Am. Chem. Soc.*, **127** (48), 16835–16847.

13 Grätzel, M. (2004) Conversion of sunlight to electric power by nanocrystalline dye-sensitized solar cells. *J. Photochem. Photobiol. A: Chem.*, **164** (1–3), 3–14.

14 Yella, A., Lee, H.-W., Tsao, H.N., Yi, C., Chandiran, A.K., Nazeeruddin, M.K., Diau, E.W.-G., Yeh, C.-Y., Zakeeruddin, S.M., and Grätzel, M. (2011) Porphyrin-sensitized solar cells with cobalt (II/III)-based redox electrolyte exceed 12 percent efficiency. *Science*, **334** (6056), 629–634.

15 Zeng, W., Cao, Y., Bai, Y., Wang, Y., Shi, Y., Zhang, M., Wang, F., Pan, C., and Wang, P. (2010) Efficient dye-sensitized solar cells with an organic photosensitizer featuring orderly conjugated ethylenedioxythiophene and dithienosilole blocks. *Chem. Mater.*, **22** (5), 1915–1925.

16 Chen, C.-H., Hsu, Y.-C., Chou, H.-H., Thomas, K.R.J., Lin, J.T., and Hsu, C. (2010) Dipolar compounds containing fluorene and a heteroaromatic ring as the conjugating bridge for high-performance dye-sensitized solar cells. *Chem. Eur. J.*, **16** (10), 3184–3193.

17 Oapos;Regan, B., Moser, J., Anderson, M., and Grätzel, M. (1990) Vectorial electron injection into transparent semiconductor membranes and electric field effects on the dynamics of light-induced charge separation. *J. Phys. Chem.*, **94** (24), 8720–8726.

18 Wang, P., Zakeeruddin, S.M., Humphry-Baker, R., Moser, J.E., and Grätzel, M. (2003) Molecular-scale interface engineering of TiO_2 nanocrystals: improve the efficiency and stability of dye-sensitized solar cells. *Adv. Mater.*, **15** (24), 2101–2104.

19 Wang, P., Klein, C., Humphry-Baker, R., Zakeeruddin, S.M., and Grätzel, M. (2004) A High molar extinction coefficient sensitizer for stable dye-sensitized solar cells. *J. Am. Chem. Soc.*, **127** (3), 808–809.

20 Wang, Z.S., Cui, Y., Hara, K., Dan-oh, Y., Kasada, C., and Shinpo, A. (2007) A high-light-harvesting-efficiency coumarin dye for stable dye-sensitized solar cells. *Adv. Mater.*, **19** (8), 1138–1141.

21 Konno, A., Kumara, G.R.A., Kaneko, S., Onwona-Agyeman, B., and Tennakone, K. (2007) Solid-state solar cells sensitized with indoline dye. *Chem. Lett.*, **36** (6), 716–717.

22 Gao, F., Wang, Y., Shi, D., Zhang, J., Wang, M., Jing, X., Humphry-Baker, R., Wang, P., Zakeeruddin, S.M., and Grätzel, M. (2008) Enhance the optical absorptivity of nanocrystalline TiO_2 film with high molar extinction coefficient ruthenium sensitizers for high performance dye-sensitized solar cells. *J. Am. Chem. Soc.*, **130** (32), 10720–10728.

23 Li, R., Liu, J., Cai, N., Zhang, M., and Wang, P. (2010) Synchronously reduced surface states, charge recombination, and light absorption length for high-performance organic dye-sensitized solar cells. *J. Phys. Chem. B*, **114** (13), 4461–4464.

24 Nakajima, K., Ohta, K., Katayanagi, H., and Mitsuke, K. (2011) Photoexcitation and electron injection processes in azo dyes adsorbed on nanocrystalline TiO(2) films. *Chem. Phys. Lett.*, **510** (4–6), 228–233.

25 Hara, K., Kurashige, M., Ito, S., Shinpo, A., Suga, S., Sayama, K., and Arakawa, H. (2003) Novel polyene dyes for highly efficient dye-sensitized solar cells. *Chem. Commun.*, **39** (2), 252–253.

26 Boschloo, G., Haggman, L., and Hagfeldt, A. (2006) Quantification of the effect of 4-*tert*-butylpyridine addition to I-/I-3(-) redox electrolytes in dye-sensitized nanostructured TiO_2 solar cells. *J. Phys. Chem. B*, **110** (26), 13144–13150.

27 Greijer, H., Lindgren, J., and Hagfeldt, A. (2001) Resonance Raman scattering of a dye-sensitized solar cell: mechanism of thiocyanato ligand exchange. *J. Phys. Chem. B*, **105** (27), 6314–6320.

28 Hagfeldt, A. and Grätzel, M. (2000) Molecular photovoltaics. *Acc. Chem. Res.*, **33** (5), 269–277.

29 Kay, A. and Grätzel, M. (1993) Artificial photosynthesis. 1. Photosensitization of titania solar cells with chlorophyll derivatives and related natural porphyrins. *J. Phys. Chem.*, **97** (23), 6272–6277.

30 Hara, K., Sayama, K., Ohga, Y., Shinpo, A., Suga, S., and Arakawa, H. (2001) A coumarin-derivative dye sensitized nanocrystalline TiO_2 solar cell having a high solar-energy conversion efficiency up to 5.6%. *Chem. Commun.*, **37** (6), 569–570.

31 Hara, K., Tachibana, Y., Ohga, Y., Shinpo, A., Suga, S., Sayama, K., Sugihara, H., and Arakawa, H. (2003) Dye-sensitized nanocrystalline TiO_2 solar cells based on novel coumarin dyes. *Solar Energy Sol. Energy Mater. Sol. Cells*, **77** (1), 89–103.

32 Wang, Z.S., Hara, K., Dan-oh, Y., Kasada, C., Shinpo, A., Suga, S., Arakawa, H., and Sugihara, H. (2005) Photophysical and (photo) electrochemical properties of a coumarin dye. *J. Phys. Chem. B*, **109** (9), 3907–3914.

33 Hara, K., Kurashige, M., Dan-oh, Y., Kasada, C., Shinpo, A., Suga, S., Sayama, K., and Arakawa, H. (2003) Design of new coumarin dyes having thiophene moieties for highly efficient organic-dye-sensitized solar cells. *New J. Chem.*, **27** (5), 783–785.

34 Wang, Z.-S., Cui, Y., Dan-oh, Y., Kasada, C., Shinpo, A., and Hara, K. (2008) Molecular design of coumarin dyes for stable and efficient organic dye-sensitized solar cells. *J. Phys. Chem. C*, **112** (43), 17011–17017.

35 Hara, K., Wang, Z.-S., Sato, T., Furube, A., Katoh, R., Sugihara, H., Dan-oh, Y., Kasada, C., Shinpo, A., and Suga, S. (2005) Oligothiophene-containing coumarin dyes for efficient dye-sensitized solar cells. *J. Phys. Chem. B*, **109** (32), 15476–15482.

36 Wang, Z.-S., Cui, Y., Dan-oh, Y., Kasada, C., Shinpo, A., and Hara, K. (2007) Thiophene-functionalized coumarin dye for efficient dye-sensitized solar cells: electron lifetime improved by coadsorption of deoxycholic acid. *J. Phys. Chem. C*, **111** (19), 7224–7230.

37 Ning, Z. and Tian, H. (2009) Triarylamine: a promising core unit for efficient photovoltaic materials. *Chem. Commun.*, **45** (37), 5483–5495.

38 Hagberg, D.P., Edvinsson, T., Marinado, T., Boschloo, G., Hagfeldt, A., and Sun, L. (2006) A novel organic chromophore for dye-sensitized nanostructured solar cells. *Chem. Commun.*, **42** (21), 2245–2247.

39 Karthikeyan, C.S., Wietasch, H., and Thelakkat, M. (2007) Highly efficient solid-state dye-sensitized TiO_2 solar cells using donor-antenna dyes capable of multistep charge-transfer cascades. *Adv. Mater.*, **19** (8), 1091–1095.

40 Liu, W.-H., Wu, I.C., Lai, C.-H., Lai, C.-H., Chou, P.-T., Li, Y.-T., Chen, C.-L., Hsu, Y.-Y., and Chi, Y. (2008) Simple organic molecules bearing a 3,4-ethylenedioxythiophene linker for efficient dye-sensitized solar cells. *Chem. Commun.*, **44** (41), 5152–5154.

41 Thomas, K.R.J., Hsu, Y.-C., Lin, J.T., Lee, K.-M., Ho, K.-C., Lai, C.-H., Cheng, Y.-M., and Chou, P.-T. (2008) 2,3-disubstituted thiophene-based organic dyes for solar cells. *Chem. Mater.*, **20** (5), 1830–1840.

42 Yum, J.-H., Hagberg, D.P., Moon, S.-J., Karlsson, K.M., Marinado, T., Sun, L., Hagfeldt, A., Nazeeruddin, M.K., and Grätzel, M. (2009) A light-resistant organic sensitizer for solar-cell applications. *Angew. Chem. Int. Ed.*, **48** (9), 1576–1580.

43 Velusamy, M., Justin Thomas, K.R., Lin, J.T., Hsu, Y.-C., and Ho, K.-C. (2005) Organic dyes incorporating low-band-gap chromophores for dye-sensitized solar cells. *Org. Lett.*, **7** (10), 1899–1902.

44 Lee, D.H., Lee, M.J., Song, H.M., Song, B.J., Seo, K.D., Pastore, M., Anselmi, C., Fantacci, S., De Angelis, F., Nazeeruddin, M.K., Grätzel, M., and Kim, H.K. (2011) Organic dyes incorporating low-band-gap chromophores based on Π-extended benzothiadiazole for dye-sensitized solar cells. *Dyes Pigments*, **91** (2), 192–198.

45 Tian, Z., Huang, M., Zhao, B., Huang, H., Feng, X., Nie, Y., Shen, P., and Tan, S. (2010) Low-cost dyes based on methylthiophene for high-performance dye-sensitized solar cells. *Dyes Pigments*, **87** (3), 181–187.

46 Chen, B.-S., Chen, D.-Y., Chen, C.-L., Hsu, C.-W., Hsu, H.-C., Wu, K.-L., Liu, S.-H., Chou, P.-T., and Chi, Y. (2011) Donor-acceptor dyes with fluorine substituted phenylene spacer for dye-sensitized solar cells. *J. Mater. Chem.*, **21** (6), 1937–1945.

47 Zhang, G., Bai, Y., Li, R., Shi, D., Wenger, S., Zakeeruddin, S.M., Grätzel, M., and Wang, P. (2009) Employ a bisthienothiophene linker to construct an organic chromophore for efficient and stable dye-sensitized solar cells. *Energy Environ. Sci.*, **2** (1), 92–95.

48 Zhang, G., Bala, H., Cheng, Y., Shi, D., Lv, X., Yu, Q., and Wang, P. (2009) High efficiency and stable dye-sensitized solar cells with an organic chromophore featuring a binary π-conjugated spacer. *Chem. Commun.*, **45** (16), 2198–2200.

49 Bai, Y., Zhang, J., Zhou, D., Wang, Y., Zhang, M., and Wang, P. (2011) Engineering organic sensitizers for iodine-free dye-sensitized solar cells: red-shifted current response concomitant with attenuated charge recombination. *J. Am. Chem. Soc.*, **133** (30), 11442–11445.

50 Lin, J.T., Chen, P.-C., Yen, Y.-S., Hsu, Y.-C., Chou, H.-H., and Yeh, M.-C.P. (2009) Organic dyes containing furan moiety for high-performance dye-sensitized solar cells. *Org. Lett.*, **11** (1), 97–100.

51 Im, H., Kim, S., Park, C., Jang, S.-H., Kim, C.-J., Kim, K., Park, N.-G., and Kim, C. (2010) High performance organic photosensitizers for dye-sensitized solar cells. *Chem. Commun.*, **46** (8), 1335–1337.

52 Kim, S., Lee, J.K., Kang, S.O., Ko, J., Yum, J.H., Fantacci, S., De Angelis, F., Di Censo, D., Nazeeruddin, M.K., and Grätzel, M. (2006) Molecular engineering of organic sensitizers for solar cell applications. *J. Am. Chem. Soc.*, **128** (51), 16701–16707.

53 Kim, S., Kim, D., Choi, H., Kang, M.-S., Song, K., Kang, S.O., and Ko, J. (2008) Enhanced photovoltaic performance and long-term stability of quasi-solid-state dye-sensitized solar cells via molecular engineering. *Chem. Commun.*, **44** (40), 4951–4953.

54 Xu, M., Wenger, S., Bala, H., Shi, D., Li, R., Zhou, Y., Zakeeruddin, S.M., Grätzel, M., and Wang, P. (2009) Tuning the energy level of organic sensitizers for high-performance dye-sensitized solar cells. *J. Phys. Chem. C*, **113** (7), 2966–2973.

55 Kim, J.-J., Choi, H., Lee, J.-W., Kang, M.-S., Song, K., Kang, S.O., and Ko, J. (2008) A polymer gel electrolyte to achieve ≥ 6% power conversion efficiency with a novel organic dye incorporating a low-band-gap chromophore. *J. Mater. Chem.*, **18** (43), 5223–5229.

56 Wang, M., Xu, M., Shi, D., Li, R., Gao, F., Zhang, G., Yi, Z., Humphry-Baker, R., Wang, P., Zakeeruddin, S.M., and Grätzel, M. (2008) High-performance liquid and solid dye-sensitized solar cells based on a novel metal-free organic sensitizer. *Adv. Mater.*, **20** (23), 4460–4463.

57 Qin, H., Wenger, S., Xu, M., Gao, F., Jing, X., Wang, P., Zakeeruddin, S.M., and Grätzel, M. (2008) An organic sensitizer with a fused dithienothiophene unit for efficient and stable dye-sensitized solar cells. *J. Am. Chem. Soc.*, **130** (29), 9202–9203.

58 Ko, S., Choi, H., Kang, M.-S., Hwang, H., Ji, H., Kim, J., Ko, J., and Kang, Y. (2010) Silole-spaced triarylamine derivatives as highly efficient organic sensitizers in dye-sensitized solar cells (DSSCs). *J. Mater. Chem.*, **20** (12), 2391–2399.

59 Choi, H., Baik, C., Kang, S.O., Ko, J., Kang, M.-S., Nazeeruddin, M.K., and Grätzel, M. (2008) Highly efficient and thermally stable organic sensitizers for solvent-fee dye-sensitized solar cells. *Angew. Chem. Int. Ed.*, **47** (2), 327–330.

60 Choi, H., Raabe, I., Kim, D., Teocoli, F., Kim, C., Song, K., Yum, J.-H., Ko, J., Nazeeruddin, M.K., and Grätzel, M. (2010) High molar extinction coefficient

organic sensitizers for efficient dye-sensitized solar cells. *Chem. Eur. J.*, **16** (4), 1193–1201.

61 Chen, P., Yum, J.H., Angelis, F.D., Mosconi, E., Fantacci, S., Moon, S.-J., Baker, R.H., Ko, J., Nazeeruddin, M.K., and Grätzel, M. (2009) High open-circuit voltage solid-state dye-sensitized solar cells with organic dye. *Nano Lett.*, **9** (6), 2487–2492.

62 Koumura, N., Wang, Z.-S., Mori, S., Miyashita, M., Suzuki, E., and Hara, K. (2006) Alkyl-functionalized organic dyes for efficient molecular photovoltaics. *J. Am. Chem. Soc.*, **128** (44), 14256–14257.

63 Horiuchi, T., Miura, H., Sumioka, K., and Uchida, S. (2004) High efficiency of dye-sensitized solar cells based on metal-free indoline dyes. *J. Am. Chem. Soc.*, **126** (39), 12218–12219.

64 Ito, S., Miura, H., Uchida, S., Takata, M., Sumioka, K., Liska, P., Comte, P., Pechy, P., and Grätzel, M. (2008) High-conversion-efficiency organic dye-sensitized solar cells with a novel indoline dye. *Chem. Commun.*, **44** (41), 5194–5196.

65 Karlsson, K.M., Jiang, X., Eriksson, S.K., Gabrielsson, E., Rensmo, H., Hagfeldt, A., and Sun, L. (2011) Phenoxazine dyes for dye-sensitized solar cells: relationship between molecular structure and electron lifetime. *Chem. Eur. J*, **17** (23), 6415–6424.

66 Li, C., Yum, J.-H., Moon, S.-J., Herrmann, A., Eickemeyer, F., Pschirer, N.G., Erk, P., Schoeneboom, J., Muellen, K., Grätzel, M., and Nazeeruddin, M. (2008) An improved perylene sensitizer for solar cell applications. *ChemSusChem*, **1** (7), 615–618.

67 Zhang, X.-H., Cui, Y., Katoh, R., Koumura, N., and Hara, K. (2010) Organic dyes containing thieno[3,2-*b*] indole donor for efficient dye-sensitized solar cells. *J. Phys. Chem. C*, **114** (42), 18283–18290.

68 Yum, J.-H., Walter, P., Huber, S., Rentsch, D., Geiger, T., Nueesch, F., De Angelis, F., Grätzel, M., and Nazeeruddin, M.K. (2007) Efficient far red sensitization of nanocrystalline TiO_2 films by an unsymmetrical squaraine dye. *J. Am. Chem. Soc.*, **129** (34), 10320.

69 Wang, Z.-S., Koumura, N., Cui, Y., Takahashi, M., Sekiguchi, H., Mori, A., Kubo, T., Furube, A., and Hara, K. (2008) Hexylthiophene-functionalized carbazole dyes for efficient molecular photovoltaics: tuning of solar-cell performance by structural modification. *Chem. Mater.*, **20** (12), 3993–4003.

70 Wang, Z.-S., Koumura, N., Cui, Y., Miyashita, M., Mori, S., and Hara, K. (2009) Exploitation of ionic liquid electrolyte for dye-sensitized solar cells by molecular modification of organic-dye sensitizers. *Chem. Mater.*, **21** (13), 2810–2816.

71 Horiuchi, T., Miura, H., and Uchida, S. (2003) Highly-efficient metal-free organic dyes for dye-sensitized solar cells. *Chem. Commun.*, **39** (24), 3036–3037.

72 Ito, S., Zakeeruddin, S.M., Humphry-Baker, R., Liska, P., Charvet, R., Comte, P., Nazeeruddin, M.K., Péchy, P., Takata, M., Miura, H., Uchida, S., and Grätzel, M. (2006) High-efficiency organic-dye-sensitized solar cells controlled by nanocrystalline-TiO_2 electrode thickness. *Adv. Mater.*, **18** (9), 1202–1205.

73 Snaith, H.J., Petrozza, A., Ito, S., Miura, H., and Grätzel, M. (2009) Charge generation and photovoltaic operation of solid-state dye-sensitized solar cells incorporating a high extinction coefficient indolene-based sensitizer. *Adv. Funct. Mater.*, **19** (11), 1810–1818.

74 Liu, X.Z., Zhang, W., Uchida, S., Cai, L.P., Liu, B., and Ramakrishna, S. (2010) An efficient organic-dye-sensitized solar cell with in situ polymerized poly(3,4-ethylenedioxythiophene) as a hole-transporting material. *Adv. Mater.*, **22** (20), E150–E155.

75 Kuang, D., Uchida, S., Humphry-Baker, R., Zakeeruddin, S.M., and Grätzel, M. (2008) Organic dye-sensitized ionic liquid based solar cells: remarkable enhancement in performance through molecular design of indoline sensitizers. *Angew. Chem.*, **120** (10), 1949–1953.

76 Kim, Y.-G., Walker, J., Samuelson, L.A., and Kumar, J. (2003) Efficient light harvesting polymers for nanocrystalline

TiO_2 photovoltaic cells. *Nano Lett.*, **3** (4), 523–525.

77 Yanagida, S., Senadeera, G.K.R., Nakamura, K., Kitamura, T., and Wada, Y. (2004) Polythiophene-sensitized TiO_2 solar cells. *J. Photochem. Photobiol., A: Chem.*, **166** (1–3), 75–80.

78 Liu, X., Zhu, R., Zhang, Y., Liu, B., and Ramakrishna, S. (2008) Anionic benzothiadiazole containing polyfluorene and oligofluorene as organic sensitizers for dye-sensitized solar cells. *Chem. Commun.*, **44** (32), 3789–3791.

79 Mikroyannidis, J.A., Tsagkournos, D.V., Balraju, P., and Sharma, G.D. (2011) Synthesis and photovoltaic properties of an alternating phenylenevinylene copolymer with substituted-triphenylamine units along the backbone for bulk heterojunction and dye-sensitized solar cells. *J. Power Sources*, **196** (4), 2364–2372.

80 Fang, Z., Eshbaugh, A.A., and Schanze, K.S. (2011) Low-bandgap donor–acceptor conjugated polymer sensitizers for dye-sensitized solar cells. *J. Am. Chem. Soc.*, **133** (9), 3063–3069.

81 He, J., Lindström, H., Hagfeldt, A., and Lindquist, S.-E. (1999) Dye-sensitized nanostructured *p*-type nickel oxide film as a photocathode for a solar cell. *J. Phys. Chem. B*, **103** (42), 8940–8943.

82 Mori, S., Fukuda, S., Sumikura, S., Takeda, Y., Tamaki, Y., Suzuki, E., and Abe, T. (2008) Charge-transfer processes in dye-sensitized NiO solar cells. *J. Phys. Chem. C*, **112** (41), 16134–16139.

83 Qin, P., Zhu, H., Edvinsson, T., Boschloo, G., Hagfeldt, A., and Sun, L. (2008) Design of an organic chromophore for *p*-type dye-sensitized solar cells. *J. Am. Chem. Soc.*, **130** (27), 8570–8571.

84 Qin, P., Linder, M., Brinck, T., Boschloo, G., Hagfeldt, A., and Sun, L. (2009) High incident photon-to-current conversion efficiency of *p*-type dye-sensitized solar cells based on NiO and organic chromophores. *Adv. Mater.*, **21** (29), 2993–2996.

85 Qin, P., Wiberg, J., Gibson, E.A., Linder, M., Li, L., Brinck, T., Hagfeldt, A., Albinsson, B., and Sun, L. (2010) Synthesis and mechanistic studies of organic chromophores with different energy levels for *p*-type dye-sensitized solar cells. *J. Phys. Chem. C*, **114** (10), 4738–4748.

86 Morandeira, A., Fortage, J., Edvinsson, T., Le Pleux, L., Blart, E., Boschloo, G., Hagfeldt, A., Hammarstrom, L., and Odobel, F. (2008) Improved photon-to-current conversion efficiency with a nanoporous *p*-type NiO electrode by the use of a sensitizer-acceptor dyad. *J. Phys. Chem. C*, **112** (5), 1721–1728.

87 Le Pleux, L., Chavillon, B., Pellegrin, Y., Blart, E., Cario, L., Jobic, S., and Odobel, F. (2009) Simple and reproducible procedure to prepare self-nanostructured NiO films for the fabrication of *p*-type dye-sensitized solar cells. *Inorg. Chem.*, **48** (17), 8245–8250.

88 Nattestad, A., Mozer, A.J., Fischer, M.K.R., Cheng, Y.B., Mishra, A., Bauerle, P., and Bach, U. (2010) Highly efficient photocathodes for dye-sensitized tandem solar cells. *Nat. Mater.*, **9** (1), 31–35.

89 Zhu, H., Hagfeldt, A., and Boschloo, G. (2007) Photoelectrochemistry of mesoporous NiO electrodes in iodide/triiodide electrolytes. *J. Phys. Chem. C*, **111** (47), 17455–17458.

90 Gibson, E.A., Smeigh, A.L., Le Pleux, L., Fortage, J., Boschloo, G., Blart, E., Pellegrin, Y., Odobel, F., Hagfeldt, A., and Hammarstrom, L. (2009) A *p*-type NiO-based dye-sensitized solar cell with an open-circuit voltage of 0.35 V. *Angew. Chem. Int. Ed.*, **48** (24), 4402–4405.

91 Tian, H., Yang, X., Chen, R., Zhang, R., Hagfeldt, A., and Sun, L. (2008) Effect of different dye baths and dye-structures on the performance of dye-sensitized solar cells based on triphenylamine dyes. *J. Phys. Chem. C*, **112** (29), 11023–11033.

92 Fischer, M.K.R., Wenger, S., Wang, M., Mishra, A., Zakeeruddin, S.M., Gra?tzel, M., and Ba?uerle, P. (2010) D-π-A sensitizers for dye-sensitized solar cells: linear vs branched oligothiophenes. *Chem. Mater.*, **22** (5), 1836–1845.

93 Hao, Y., Yang, X., Cong, J., Tian, H., Hagfeldt, A., and Sun, L. (2009) Efficient near infrared D-π-A sensitizers with

lateral anchoring group for dye-sensitized solar cells. *Chem. Commun.*, **45** (27), 4031–4033.

94 Li, S.-L., Jiang, K.-J., Shao, K.-F., and Yang, L.-M. (2006) Novel organic dyes for efficient dye-sensitized solar cells. *Chem. Commun.*, **42** (26), 2792–2794.

95 Chen, R., Yang, X., Tian, H., Wang, X., Hagfeldt, A., and Sun, L. (2007) Effect of tetrahydroquinoline dyes structure on the performance of organic dye-sensitized solar cells. *Chem. Mater.*, **19** (16), 4007–4015.

96 Liu, D., Fessenden, R.W., Hug, G.L., and Kamat, P.V. (1997) Dye capped semiconductor nanoclusters. Role of back electron transfer in the photosensitization of SnO_2 nanocrystallites with cresyl violet aggregates. *J. Phys. Chem. B*, **101** (14), 2583–2590.

97 Ning, Z., Zhang, Q., Pei, H., Luan, J., Lu, C., Cui, Y., and Tian, H. (2009) Photovoltage improvement for dye-sensitized solar cells via cone-shaped structural design. *J. Phys. Chem. C*, **113** (23), 10307–10313.

98 Hara, K., Sato, T., Katoh, R., Furube, A., Ohga, Y., Shinpo, A., Suga, S., Sayama, K., Sugihara, H., and Arakawa, H. (2003) Molecular design of coumarin dyes for efficient dye-sensitized solar cells. *J. Phys. Chem. B*, **107** (2), 597–606.

99 Sayama, K., Hara, K., Mori, N., Satsuki, M., Suga, S., Tsukagoshi, S., Abe, Y., Sugihara, H., and Arakawa, H. (2000) Photosensitization of a porous TiO_2 electrode with merocyanine dyes containing a carboxyl group and a long alkyl chain. *Chem. Commun.*, **36** (13), 1173–1174.

100 Sayama, K., Tsukagoshi, S., Hara, K., Ohga, Y., Shinpou, A., Abe, Y., Suga, S., and Arakawa, H. (2002) Photoelectrochemical properties of J-aggregates of benzothiazole merocyanine dyes on a nanostructured TiO_2 film. *J. Phys. Chem. B*, **106** (6), 1363–1371.

101 Cid, J.-J., Yum, J.-H., Jang, S.-R., Nazeeruddin, M.K., Martínez-Ferrero, E., Palomares, E., Ko, J., Grätzel, M., and Torres, T. (2007) Molecular cosensitization for efficient panchromatic dye-sensitized solar cells. *Angew. Chem. Int. Ed.*, **46** (44), 8358–8362.

102 Robertson, N. (2008) Catching the rainbow: light harvesting in dye-sensitized solar cells. *Angew. Chem. Int. Ed*, **47** (6), 1012–1014.

103 Teng, C., Yang, X., Li, S., Cheng, M., Hagfeldt, A., Wu, L.-Z., and Sun, L. (2010) Tuning the HOMO energy levels of organic dyes for dye-sensitized solar cells based on Br^-/Br_3^- electrolytes. *Chem. Eur. J.*, **16** (44), 13127–13138.

104 He, J., Lindström, H., Hagfeldt, A., and Lindquist, S.-E. (2000) Dye-sensitized nanostructured tandem cell-first demonstrated cell with a dye-sensitized photocathode. *Sol. Energy Mater. Sol. Cells*, **62** (3), 265–273.

11
Organic Thermoelectric Power Devices

Martin Leijnse, Karsten Flensberg, Thomas Bjørnholm

11.1
Introduction

One of the main scientific and technological challenges during the twenty-first century will surely be to ensure a sustainable supply of energy. The primary product of most power-sources is heat, and it is therefore of major importance to convert this heat efficiently into electric energy. Today, this is usually achieved by first converting the heat into mechanical work, typically by using the heat to boil water, thereby generating steam which drives a turbine that in turn produces electric energy. Although this conversion is highly efficient and works well in large-scale power plants, it requires rather bulky machinery and large operating temperature differences, and is therefore unsuitable when weight is an issue (e.g., in cars or other vehicles) or for low-grade heat "waste" (e.g., from home appliances). In such cases, *thermoelectric* energy converters could instead be used. The thermoelectric effect involves the direct conversion of a heat difference into electric energy. In thermoelectric materials, an electric current is generated as a result of a temperature gradient, allowing–at least in principle–the use of arbitrarily small energy-converting devices that are highly reliable, have no moving parts, and which can operate at arbitrarily small temperature differences. The thermoelectric effect can also be used for refrigeration applications, whereby the driving of a current through a thermoelectric material will generate a temperature difference.

The thermoelectric effect is currently exploited in a number of niche applications. For example, thermoelectric refrigerators are used to spot-cool electronic components, while power converters are used in space probes and satellites. However, in order to attract a broader range of applications, it is important that more efficient low-cost thermoelectric materials are made available. The search for such materials initially experienced rapid progress during the 1950s, when the thermoelectric narrow-gap semiconducting material $Bi_2Te_3$was identified and developed commercially. However, since then only minor improvements have been made, primarily using the $(Bi_{1-x}Sb_x)_2(Se_{1-y}Te_y)_3$ alloy family. However, recent developments in nanotechnology have given rise to a variety of new and

Organic Optoelectronics, First Edition. Edited by Wenping Hu.

revolutionary materials which seem especially well suited for thermoelectric applications.

In this chapter, attention will be focused on thermoelectric devices based on the types of organic material discussed in previous chapters. The stage is set in Section 11.2, where a generic introduction to the thermoelectric effect is provided, and the details of the material properties required for efficient energy conversion are discussed. Possible realizations in terms of materials and devices are then outlined in Section 11.3, where developments in new types of inorganic nanostructured materials are discussed, and the properties required for the identification of organic compounds are pinpointed. Also discussed are single-molecule devices which, though not useful for practical devices (due to their low output power), may serve as an important proof-of-principle study. Devices based on organic solids are then described, initially by discussing polymer materials, followed by small-molecule materials and, finally, hybrid materials, including additional organic or inorganic compounds. The chapter concludes with a discussion, in Section 11.4, of the possible directions for future research in this area.

11.2 Basic Thermoelectric Principles

Here, the basic principles of the thermoelectric effect and of thermoelectric devices will be discussed, important terminology will be introduced, and details provided as to which material properties are desirable for thermoelectric devices.

11.2.1 The Thermoelectric Effect

In thermoelectric materials an electric current is generated as a result of an applied temperature gradient. The underlying physical mechanism is perhaps easiest to understand in a so-called ballistic device, where electrons can be transported between two metallic contacts through a central region, without losing energy (see Figure 11.1). The transport mechanism can, for example, be tunneling through the central region, which may be anything from a simple insulating layer to a single molecule.

If one electrode is heated compared to the other, the larger "smearing" of the Fermi surface on the hot side results in a net current of high-energy electrons tunneling from the hot electrode to the cold electrode (see Figure 11.1a). Analogously, however, there will be a net current of low-energy electrons tunneling from the cold to the hot side. Thus, the total electric current will be zero, and only a heat current can flow from the hot to the cold electrode. However, if it is imagined that either high- or low-energy electrons are more easily transported through the central region, then the situation will change (see Figure 11.1b). Now, the currents involving electrons above and below the Fermi level no longer cancel each other

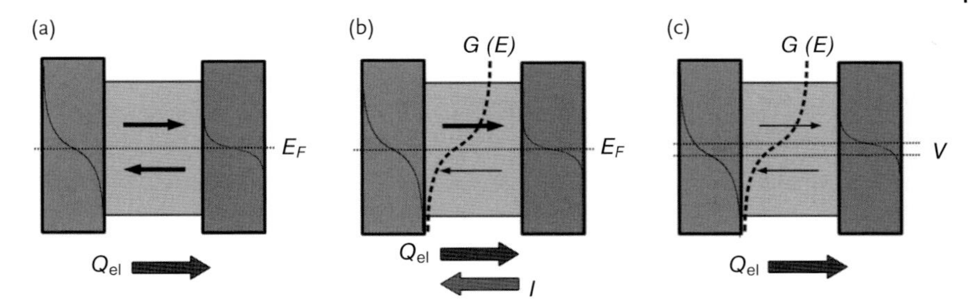

Figure 11.1 Explanation of the thermoelectric effect. Left and right rectangles indicate the conduction bands of the metallic hot and cold electrodes, respectively, and the curved line represents the thermal smearing of the Fermi-surface around the Fermi energy, E_F. (a) The transport properties of the central region are the same for electrons above and below E_F, and only an (electronic) heat current Q_{el} flows, but no net electric current; (b) The conductance $G(E)$ of the central region is larger for $E > E_F$, which results in a net (thermo)electric current, I; (c) If this current is allowed to flow for some time in an open circuit, a (thermoelectric) voltage V builds up, which eventually cancels the electric current.

out and, in addition to the heat current, an electric current will flow as a result of the temperature difference: this is the *thermoelectric effect*! Clearly, a thermoelectric material must have different transport properties for the electrons above and below the Fermi energy. By imagining that the central region consists of a single molecule, the type of energy asymmetry seen in Figure 11.1b could originate as a result of the lowest unoccupied molecular orbital (LUMO) being much closer to the metal Fermi levels than the highest occupied molecular orbital (HOMO). An analogous situation occurs in (diffusive) bulk transport, where a finite thermoelectric effect requires an energy asymmetry in the conductivity of electrons above and below the Fermi energy; in other words, it requires different electron and hole conductivities.

In the following sections, it will be necessary to consider both the ballistic (tunneling) transport regime, which for example is relevant in single-molecule devices, and the diffusive transport regime, which is relevant for bulk organic and inorganic devices. Despite such details of the transport mechanisms being relevant when attempting to determine (theoretically) the thermoelectric properties of a device, they are not important for the basic thermoelectric phenomenology. As far as possible, both cases will be considered together in the following subsections.

Imagine that the electrodes in Figure 11.1b are finite pieces of metal, isolated from the outside world. Then, the thermoelectric current will lead to a charge accumulation in the electrodes, as the charge carriers have nowhere to go, thus producing a thermoelectric voltage, V, which eventually will be large enough to stop the net electric current (see Figure 11.1c). This is termed the *Seebeck effect*, and in the limit of an infinitely small temperature difference, ΔT, the *Seebeck coefficient*, S (also called thermopower or thermoelectric power), is defined as:

$$S(T) = -\lim_{\Delta T \to 0} \frac{V}{\Delta T}, \tag{11.1}$$

where T is the average electrode temperature. In other words, $V = -S\Delta T$ is the voltage which must be applied to compensate for the small temperature difference ΔT and completely cancel the electric current. The sign of the voltage which builds up depends on the direction of the thermally generated current, and therefore on whether the electron or the hole conductance is largest. For electron-dominated transport, $V > 0$ and therefore $S < 0$, whereas for hole-dominated transport, $V < 0$ and $S > 0$. Note that $S(T)$ is here defined only for infinitesimal temperature differences. In the example in Figure 11.1, the temperatures in the two electrodes can differ significantly and the charge carriers in the central region have no well-defined temperature, since ballistic transport means that they do not equilibrate. Therefore, $S(T)$ as defined in Eq. (11.1) does not necessarily predict the voltage resulting from a large temperature difference. In contrast, in bulk materials there will always be a smooth temperature gradient, and $S(T)$ is therefore sufficient to predict the voltage resulting even from a large temperature difference. For example, if a homogeneous bulk rod is considered (as shown in Figure 11.2a), with one cold end (temperature T_c) and one hot end (temperature T_h), the total voltage which builds up is obtained from

$$V = -\int_{T_c}^{T_h} dT\, S(T). \tag{11.2}$$

Note also that the wire connecting the two ends, which is used to measure the voltage, will experience a temperature gradient and, if it has a non-zero Seebeck coefficient, will also contributes to the voltage. Usually, however, devices based on two different materials are used, for example, two rods A and B (as shown in

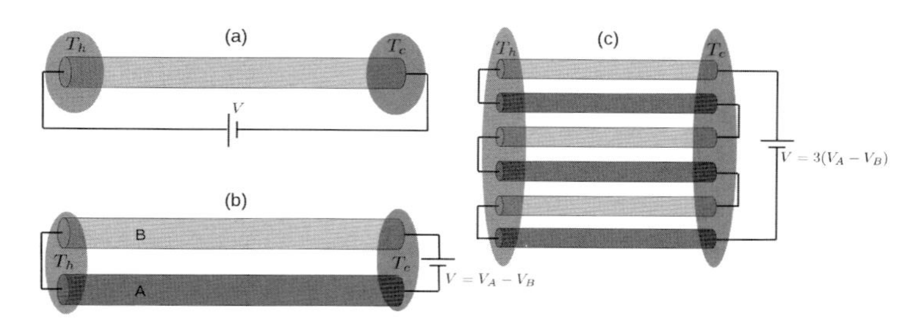

Figure 11.2 (a) Thermoelectric effect in a rod. One end is kept hot $T_h = T + \Delta T/2$ (usually by being connected to a heat source) and the other end is kept cold $T_c = T - \Delta T/2$ (usually, by being connected to a heat sink, T_c is most commonly the ambient temperature); (b) Sketch of a thermocouple. The voltage between the two rods is given by the difference between the thermovoltages, and rods A and B should ideally have opposite signs of the Seebeck coefficient to maximize the voltage; (c) Sketch of a thermopile (i.e., a number of thermocouples connected electrically in series but thermally in parallel). In such a set-up, the voltages of the individual thermocouples are summed.

Figure 11.2b). In such a *thermocouple*, the voltage generated between the two materials is given by the difference of the thermovoltages:

$$V = -\int_{T_c}^{T_h} dT\,(S_A(T) - S_B(T)). \tag{11.3}$$

To maximize V, the materials A and B should be chosen to have Seebeck coefficients with opposite signs; that is, in one material transport should be electron-dominated and the other hole-dominated. Typical values for the Seebeck coefficient are in the range of $\mu V\,K^{-1}$ in metals and up to $mV\,K^{-1}$ in semiconductors. The larger value for semiconductors occurs because they can be almost completely electron or hole conductors, whereas metals always have both a significant electron and hole conductivity. However, since even the best semiconductors are capable of producing at best a few hundred mV at typical operating temperature differences, real devices almost always use *thermopiles* to obtain a significant output voltage. A thermopile consists simply of several thermocouples connected electrically in series, but thermally in parallel (see Figure 11.2c). In such a set-up the thermoelectric voltages associated with each thermocouple add up, and in principle an arbitrarily large output voltage can be achieved. Thermopiles can be used as thermometers: if, for example, T_c and the Seebeck coefficients $S_A(T)$ and $S_B(T)$ are known, then T_h can be measured by measuring the thermovoltage, V. Thermopiles can also be used to convert heat into electric energy; in this case, the cold end is maintained at ambient temperature, while the heat source is used to heat the hot end. However, in order to actually generate useful electric power, the voltmeter in the set-up in Figure 11.2c must be replaced with the electric device which is to be powered, and a finite current must be allowed to flow.

Cooling can also be achieved by using a thermopile set-up, though the voltmeter must then be replaced with a current source. For this, a single rod (as shown in Figure 11.2a) is first considered, but with the two ends maintained at the same temperature T, while driving a current I through the rod. The *Peltier heat* Q_{el}^P is then the electronic heat current passed through the rod, which is given by

$$Q_{el}^P = \Pi(T)I, \tag{11.4}$$

where Π is the Peltier coefficient. Note that such a linear relationship only strictly holds for an infinitely small current, I. Analogously, in a thermocouple

$$Q_{el}^P = (\Pi_A(T) - \Pi_B(T))I, \tag{11.5}$$

and to maximize the effect, one of the materials A and B should again be chosen as an electron-dominated conductor, and the other to be hole-dominated, so that the signs of Π_A and Π_B are different. In a thermopile, the total Peltier heat scales with the number of thermocouples.

A refrigeration device, or heat pump, is obtained only when the temperature of the hot end is fixed by coupling it to a heat sink, while the Peltier heat is allowed to cool the other end. However, in this case, with both a finite current and a finite temperature gradient, the so-called *Thomson effect* will also be relevant. The total

Thompson heat, $Q_{\rm el}^{T}$, generated (or absorbed) in for example, a homogeneous rod (as in Figure 11.2a) is given by

$$Q_{\rm el}^{T} = I\int_{T_c}^{T_h} dT\ \mu(T), \tag{11.6}$$

where μ is the Thompson coefficient. The three thermoelectric effects (Seebeck, Peltier, and Thompson) are all consequences of the same physics, but are realized under different conditions. Therefore, the corresponding coefficients, S, Π, and μ, are not independent, but rather are related through the Kelvin (or Thompson) relations:

$$\Pi(T) = S(T)T, \tag{11.7}$$

$$\mu(T) = T\frac{dS(T)}{dT}. \tag{11.8}$$

In many materials, the Seebeck coefficient has a rather weak temperature-dependence, and the Thomson effect can be neglected.

It should be noted that, in addition to the three thermoelectric effects, there are other forms of interplay between heat and electric current, the most prominent probably being Joule heating, or resistive heating. In contrast to the three thermoelectric effects, Joule heating cannot even in the ideal case be thermodynamically reversible, and is basically an additional loss mechanism in nonideal thermoelectric devices.

11.2.2
Thermoelectric Efficiency and Figure of Merit

At this point, the concept of efficiency in thermoelectric devices will be investigated. For this, attention will be focused on the case of thermoelectric energy converters, although the desirable material properties for efficient thermoelectric heat pumps are the same. As noted above, in order to produce useful electric work, the Seebeck set-up of Figure 11.1c must be modified to include an external electric circuit making use of the extracted power, which can simply be considered as a resistor (see Figure 11.3). (For the general discussion, a single thermoelectric element can be considered, bearing in mind that it can finally be used in a thermopile.)

The efficiency of a heat to electric power converter is given by the generated output electric power, P, divided by the heat power $Q_{\rm in}$, which must be supplied to the hot electrode in order to keep it hot:

$$\eta = \frac{P}{Q_{\rm in}} = \frac{IV}{Q_{\rm in}}. \tag{11.9}$$

The goal is therefore to achieve values of η as close as possible to 1. However, fundamental thermodynamics sets a limit on how close to this limit one can come,

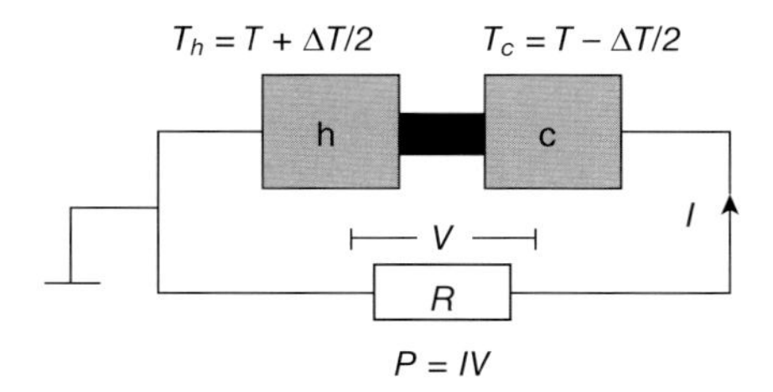

Figure 11.3 Sketch of thermoelectric circuit converting heat power, Q_{in}, supplied to the hot electrode, into electric power, P, in an electrically driven device (represented by a simple resistor, R).

as no heat engine operating between a hot and a cold heat bath can ever be more efficient than the ideal Carnot process, which has the efficiency

$$\eta_C = 1 - \frac{T_c}{T_h}. \tag{11.10}$$

Thus, the theoretical upper limit of the efficiency is set by the ratio of the temperatures of the hot and cold electrodes, and only a device operating at large ΔT can achieve a high efficiency.

Commonly, only the linear regime is considered, which means that the applied temperature and voltage differences are assumed small enough that the resulting current is directly proportional to them:

$$I = GV + G_T \Delta T, \tag{11.11}$$

where G is the electric conductance and G_T is the thermal conductance of the thermoelectric device. In the Seebeck set-up discussed above $I = 0$, and therefore the Seebeck coefficient is given by:

$$S = -\frac{V}{\Delta T} = \frac{G_T}{G}. \tag{11.12}$$

In addition to the electric current, there will be a heat current, which at the condition that V is adjusted such that $I = 0$ is given by:

$$Q = (\kappa_{el} + \kappa_{ph})\Delta T, \tag{11.13}$$

where κ_{el} and κ_{ph} are the electron and phonon contributions to the thermal conductance, respectively. If other heat losses are neglected (e.g., of radiative origin), then the required heat input power Q_{in} in Eq. (11.9) is equal to the heat current Q [actually, additional heat losses could simply be added as a third term in Eq. (11.13), without affecting the following discussion, as long as these losses are assumed proportional to ΔT]. It is then possible to show that maximizing the thermoelectric power-conversion efficiency in Eq. (11.9) is equivalent to maximizing the so-called dimensionless *thermoelectric figure of merit*, ZT, given by

$$ZT = \frac{GS^2T}{\kappa_{\text{el}} + \kappa_{\text{ph}}}. \tag{11.14}$$

A derivation of the exact relation between the efficiency and ZT is beyond the scope of this chapter (see for example, Ref [1]). The result for small temperature differences is

$$\eta = \frac{\Delta T}{T_h} \frac{\sqrt{1+ZT} - 1}{\sqrt{1+ZT} + T_c / T_h}. \tag{11.15}$$

The prefactor $\Delta T/T_h$ is recognized as the Carnot efficiency in the limit of small ΔT, and it can be seen that when $ZT \to 0$, the efficiency vanishes, whereas when $ZT \to \infty$ at a fixed T, the efficiency approaches the Carnot value. It is easy to understand why a high figure of merit translates into a high efficiency: A large S means that a large voltage builds up as a response to the temperature gradient, a large G means that a large thermoelectric current can flow, while $\kappa = \kappa_{\text{el}} + \kappa_{\text{ph}}$ being small is essential to minimize the heat losses. In a thermocouple (or thermopile) it is essential that both components have similar ZT (but opposite signs of S) to achieve a high efficiency. It is therefore important to identify good thermoelectric materials with both electron- and hole-dominated transport.

11.2.3
Optimizing the Figure of Merit

Attention will now be focused on the material properties required to achieve a high thermoelectric figure of merit, and thereby a high efficiency. The materials currently in use, where the highest ZT have been achieved, are narrow bandgap semiconductors which have been heavily doped [2] to allow very high carrier concentrations (typically in the range of 10^{19}–10^{21} carriers per cm^3). Most thermoelectric devices in use today operate at relatively small temperature differences, $T_c \approx$ room temperature and $T_h \approx 200\,°C$. In this regime, Bi_2Te_3—which was discovered back in the 1950s—is a highly efficient thermoelectric material (see Refs [2, 3], and references therein). Alloying with Sb_2Te_3 and Bi_2Se_3 allows the carrier concentration to be fine-tuned, such that both electron- and hole-dominated transport can be achieved (remember that both are needed to make a thermocouple), in both cases reaching $ZT \approx 1$. At higher operating temperature differences, elements such as PbTe, SnTe, or GeTe are typically used, or alloys of Si and Ge at very large temperature differences ($T_h \approx 900\,K$).

Inserting $ZT = 1$ in Eq. (11.15) and assuming that $T_h \approx T_c$, η is found to be approximately 17% of the Carnot efficiency. It should be pointed out, however, that this is a maximum efficiency set by the material constraints, and that an actual device would show a lower value. Achieving $ZT \approx 3$–4 would allow thermoelectric devices to compete with today's mechanically based power converters and refrigerators, although any increase in ZT would increase the number of possible applications for thermoelectric devices. However, finding bulk materials with high ZTs has proven to be more difficult than was initially anticipated, one reason being that the ratio of the electronic thermal conductance and the electric conductance in bulk materials follows the Wiedemann–Franz law [4]

$$\frac{\kappa_{\rm el}}{GT} = \frac{\pi^2 k_{\rm B}^2}{3e^2}, \tag{11.16}$$

and increasing G is therefore accompanied by an increase in $\kappa_{\rm el}$. Efforts to maximize the efficiency have instead focused on reducing the phonon contribution to the heat loss – that is, on reducing $\kappa_{\rm ph}$. The challenge is to achieve this without affecting G too much, but this can be done by introducing effective phonon-scattering centers in the form of heavy-ion species with large vibrational amplitudes (so-called "rattlers"), one example being partially filled skutterudites [5]. It has been said that the ideal thermoelectric material should be an electron crystal, but a phonon glass [2].

Strictly speaking, Eq. (11.16) holds only for free electrons, though it is approximately valid in most bulk systems. Fortunately, the results of several recent studies (e.g., Refs [6–9]) have shown that the Wiedemann–Franz law completely breaks down in nanoscale systems, making it much easier to achieve large values of ZT in such systems. The reason for such a breakdown is either electron–electron interaction, which is increasingly important when electrons are confined in a small device, or quantum confinement, which leads effectively to one-dimensional (1-D) or two-dimensional (2-D) structures. The subject of recent advances in such nano-sized or nanostructured devices will be returned to shortly.

First, however, it is interesting to ask the question of which material properties would make the perfect thermoelectric. That is, if a free wish could be granted, how could ZT be maximized as defined in Eq. (14), and how large would this ZT be? From a purely theoretical point of view, the phonon heat current is usually considered independent of the electronic properties, and for the sake of discussion it can be assumed that the phonon thermal conductance is zero ($\kappa_{\rm ph} \rightarrow 0$), since it is clear that this is the best choice for maximizing ZT. The electronic heat conductance, $\kappa_{\rm el}$, is however intimately connected to the electrical conductance and cannot simply be taken as arbitrarily small. Therefore, the identification of the optimal electronic structure which maximizes ZT proved to be a nontrivial problem that was investigated and resolved some 15 years ago by Mahan and Soho [10]. These authors studied diffusive transport, and showed that in this case G, G_T, and $\kappa_{\rm el}$ could all be formulated as integrals involving the so-called "transport distribution function," $\Sigma(E)$, which essentially describes how easily charge carriers with a certain energy can be transported through the system. Mahan and Soho showed that the ideal material could be characterized by having a transport distribution function which was infinitely high and sharp; that is, it was proportional to Dirac's δ-function, where the peak should be localized exactly $2.4k_{\rm B}T$ above or below the Fermi level (giving an electron-type or a hole-type thermoelectric device). In this case, ZT goes to infinity and Carnot efficiency can be achieved (of course, in practice the phonon thermal conductance does not vanish and ZT remains finite).

A δ-like transport distribution function could be achieved if the density of states were to be nonzero only for one specific energy, or if, for some reason, only electrons with one specific energy could be transported through the material. Clearly, this is a very idealized situation which is not found in any real material. Mahan and Soho suggested using rare-earth compounds with very sharp electronic *f*-levels, but a correct alignment of these levels with respect to the Fermi energy would be problematic, as would be the phonon contributions to the heat current. Molecular devices as an even more promising route to a high ZT, will be discussed later, where the δ-function behavior of the transport distribution function arises naturally from the molecular HOMO or LUMO levels, the position of which can be chemically controlled, and where the phonon heat current may be minimized by a clever choice of molecules and anchoring groups.

The achievement of a high ZT is not all that matters for thermoelectric materials. For example, if there is an ample supply of heat to use as input power, achieving a high output power can be more important than a high efficiency. That is, it would be preferable to maximize the numerator in Eq. (11.9), without caring so much about the denominator. Basically, this corresponds to maximizing the numerator of the figure of merit, the so-called *power factor*, GS^2. On the other hand, it is worth mentioning that even with an infinite figure of merit, Carnot efficiency can only be reached in the limit of infinitely slow, reversible operation. In this limit, the device also produces zero output power. It is also interesting, therefore, to investigate quantities such as maximum output power and efficiency at maximum output power, both of which are intrinsically nonequilibrium properties which cannot easily be related to the figure of merit. Nonetheless, the results of various studies have indicated [11] that also in such cases a δ-like transport distribution function would be desirable. Finally, in cooling applications it is not only the cooling efficiency and cooling power that are of interest; an ability to achieve a high temperature difference at a given cooling current is also desirable. A large maximum temperature difference may, however, also be related to a large figure of merit [2].

11.3
Thermoelectric Materials and Devices

11.3.1
Inorganic Nanostructured Materials

Before discussing organic thermoelectric materials, a very brief review will be provided of present-day, state-of-the-art inorganic nanostructured materials that have demonstrated promising properties not available in bulk compounds. The aim is to pinpoint targets in organic compounds, as the development of high-ZT thin-film devices is based on various key ideas and approaches.

Perhaps the simplest way to take advantage of nanostructuring is to use quantum confinement to enhance the Seebeck coefficient, S. By confining charge carriers to move in a reduced number of dimensions, the density of states is made less smooth (see Figure 11.4), which leads to an increase in S.

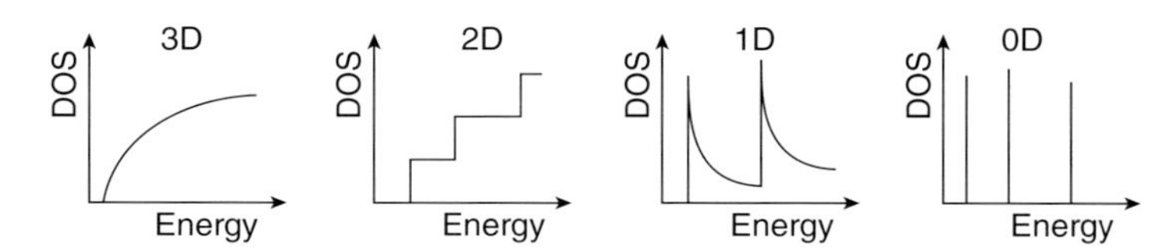

Figure 11.4 Density of states in systems with different dimensionalities.

The initial experiments in this area were focused on 2-D quantum wells (e.g., Ref. [12]), where the density of states as a function of energy is constant, with steps whenever a new 2-D band becomes accessible. In one dimension, the density of states has sharp peaks corresponding to the bottom of the 1-D bands. The thermoelectric properties of 1-D nanowires [13, 14] and carbon nanotubes (CNTs) [15, 16] have been investigated. In a zero-dimensional system (so-called quantum dots; QDs), the charge carriers are confined in all spatial directions and can exist only at specific energies. The density of states is, therefore, highly peaked and resembles the δ-function shape of Mahan and Soho [10]. However, QDs prepared from inorganic materials have level spacings on the order of one or a few meV at the most, and the density of states therefore appears rather smooth at energy scales relevant at room temperature. As will be discussed later, single molecules that are weakly coupled to electrodes behave as QDs, but with a level spacing (HOMO–LUMO gap) that is much larger than k_BT also at room temperature.

An alternative approach would be to use superlattice structures – that is, periodic structures of two (or more) layers. The key idea here is to utilize an acoustic mismatch between the different layers, which results in interfaces that scatter phonons more efficiently than electrons, thus reducing κ_{ph} more than G. This technique has been used successfully in several studies (e.g., Ref. [17]), and has provided impressive results for the figure of merit ($ZT \approx 2.4$ in Ref. [17]).

One problem with superlattice structures is that they require precise nanoscale engineering, and are thus not suitable for large-scale production. Fortunately, however, it appears that the same positive effects can be achieved in non-periodic structures, such as nanocomposite materials consisting of a host material filled with nanoparticles [3, 18]. It has also been observed [14] that the introduction of roughness into Si nanowires leads to a dramatic reduction in phonon thermal conductance, without substantially modifying the electrical conductance.

11.3.2
Single-Molecule Devices

The discussion of organic thermoelectric materials commences with the consideration of single-molecule devices. Clearly, the output power of a single-molecule device is far too low for real applications; however, to discuss such proof-of-principle studies may serve as a useful starting point, as well as demonstrating the clear promise of organic thermoelectrics.

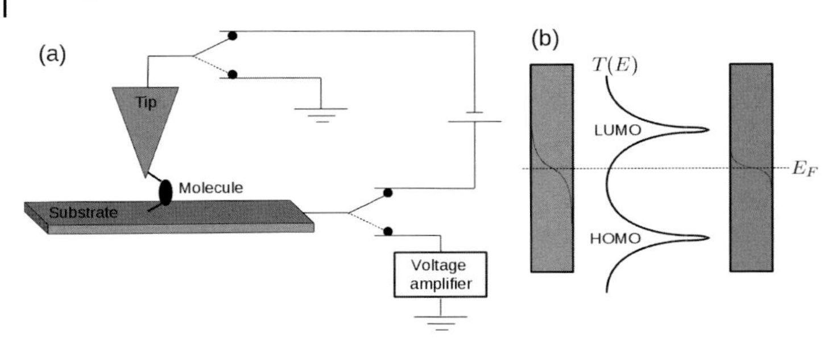

Figure 11.5 (a) Sketch of scanning transmission microscopy set-up used to create single-molecule thermoelectric junctions. That there is a molecule bridging the tip–substrate gap is verified by a conductance measurement, whereafter the tip–substrate electric contact is broken and the thermoelectric voltage measured (see text for details); (b) Energy diagram of hot (left) and cold (right) electrode (cf. Figure 11.1). Between the electrodes the transmission function of the molecule, $T(E)$, is sketched, which shows peaks at the positions of the HOMO and LUMO. The Seebeck coefficient is sensitive to both the value and the slope of $T(E)$ at the Fermi energy.

To date, the Seebeck effect has been measured in single-molecule junctions, using scanning transmission microscopy (STM), as shown schematically in Figure 11.5a (see also Refs [19–21]). In this procedure, a gold substrate is covered with the molecules to be measured, which included benzenedithiol, dibenzenedithiol, and tribenzenedithiol [21], benzenedithiol modified by various substituent groups [19], or alkenedithiols of varying length [20]. A current is passed through the substrate to heat it (Joule heating), while the microscope tip is coupled to a heat bath so that it is maintained at ambient temperature. While continuously measuring the tip–substrate conductance, the microscope tip is made to approach the substrate. When the conductance exceeds a threshold value (ca. 0.1 G_0), this is interpreted as one or a few molecules bridging the tip–substrate gap (the exact number of molecules contacted in parallel is of less interest, as this does not influence the Seebeck voltage). When a molecular junction has been established, the tip–substrate voltage bias and current amplifier are disconnected and replaced by a voltage amplifier, which measures the induced thermoelectric voltage as the tip is slowly retracted from the substrate.

When coupled to the tip and substrate contacts, the HOMO and LUMO of the molecules studied in Refs [19–21] are far from the electrode Fermi levels. Consequently, transport through the molecular wires can basically be seen as an elastic (energy-conserving) tunneling through the HOMO or LUMO (i.e., ballistic transport), and is well described by the so-called Landauer–Büttiker formalism [6, 22]. The conductance through a single molecule is then proportional to the transmission function at the Fermi energy, $G \propto T(E_F)$, where $T(E)$ describes the probability of an incident electron with energy E tunneling through the molecule to the opposite contact. Often, $T(E)$ is roughly proportional to the molecular density of states, and shows peaks at the positions of the HOMO and LUMO (see Figure 11.5b). As discussed above, the thermoelectric effect relies on electrons above and below the Fermi level having different conductances; indeed, it can be shown that:

$$S \propto \frac{1}{T(E_F)} \frac{dT(E_F)}{dE}. \tag{11.17}$$

Therefore, as shown in Figure 11.5b, the sign of the Seebeck coefficient gives an indication of whether the HOMO or LUMO lies closest to the Fermi level. In this way, it was concluded in Ref [21] from the positive Seebeck coefficient that the HOMO was closest to the Fermi level, and thus that transport was hole-dominated. It was also shown that the modification of benzenedithiol by adding electron-donating and -withdrawing substituent groups resulted in a shift of both HOMO and LUMO to higher and lower energies, respectively – an effect that was reflected by the resultant change in the Seebeck coefficient [19].

The Seebeck coefficients determined in Refs [19–21] were rather modest ($|S| < 10\,\mu V\,K^{-1}$), and much higher values could be expected if either the HOMO or the LUMO were to lie closer to the Fermi level. This can be achieved by a chemical doping of the used molecules, for example, by adding electron-accepting/-donating side groups to move the LUMO/HOMO closer to the Fermi level. Furthermore, a more narrow width of the transmission peaks would increase the derivative of the transmission function in Eq. (11.17), and thereby the Seebeck coefficient, provided that the HOMO or LUMO position was adjusted appropriately. The width of the transmission peaks is set by the strength of the tunnel couplings to the electrodes, and therefore a weak coupling is desirable. In fact, in the limit of vanishing tunnel coupling and large HOMO–LUMO gap (where both should be compared to the thermal energy scale k_BT), a single-molecule junction is a realization of Mahan and Soho's [10] δ-function form of the transport distribution function. Thus, a single molecule, when weakly coupled to two metallic electrodes can, by disregarding the phonon degrees of freedom, achieve Carnot efficiency and is thus, in principle, the *perfect thermoelectric device*. The ideal thermocouple would then consist of two different molecular species, where one species has the HOMO at about $2.4\,k_BT$ below the Fermi level, and the other the LUMO at around $2.4\,k_BT$ above the Fermi level. That a single-molecule junction can indeed (in theory) operate at an ideal efficiency in the equilibrium limit was the conclusion of Refs [23] and [11] confirmed that such a device would have a high efficiency even away from equilibrium, when operated under conditions to maximize output power rather than efficiency. If it proves too difficult to move the HOMO or LUMO close enough to the Fermi level, or if sufficiently small tunnel couplings cannot be achieved, an alternative would be to use molecules with sharp features in the transmission function, such as Fano resonances [24], "transmission supernodes" [7], or interference-related dips [25], which also result in a strong energy-dependence of the conductance.

Although the electronic properties of molecular junctions seem to promise an efficient thermoelectric energy conversion, their vibrational degrees of freedom might be cause for concern. At this point, two different effects can be identified:

- **Electron–phonon coupling:** This is a coupling between the charge on the molecule and its vibrational motion. Due to electron–phonon coupling, electrons tunneling through a molecule can excite that molecule vibrationally [26, 27]. Essentially, such inelastic contributions to the current destroy the δ-like character of the transmission function, as electrons can tunnel through the molecule at energies other than the conducting HOMO or LUMO, either by giving off excess energy into the molecular vibrations, or by absorbing vibrational energy from them. In a thermoelectric device, the relevant modes are those with a vibrational energy ω around the energy scale set by the operating temperatures and voltages (which are both roughly the same). Vibrational modes with much larger energies cannot be excited, while those with much smaller energies contribute less to the heat losses. Thus, in a good thermoelectric molecule, all vibrational modes with $\omega \approx k_B T$ should have a small electron–phonon coupling [28].

- **Direct coupling:** This may occur between substrate phonon modes in the electrodes and molecular vibrations. Essentially, the chemical bond between the molecular anchoring groups and the electrodes acts as a spring, which can transfer vibrational energy from the phonon modes in the hot electrode, into the molecular vibrations, and finally out again into the cold electrode phonons. This effect is what gives rise to a finite phonon heat conductance, κ_{ph}, in molecular junctions. To minimize this phonon heat current, molecules should be chosen which form strong chemical bonds to the electrodes (meaning a stiff spring), such that all molecular vibrations with a significant amplitude at this bond have frequencies above the highest acoustical phonons in the electrodes. A strong chemical bond could be combined with a weak tunnel coupling (as would be needed to obtain sharp electronic resonances), for example by connecting the bonding atoms to the rest of the molecule through saturated carbon atoms, as in a methylene spacer [29]. The combined effects of electron–phonon and phonon–phonon coupling on power conversion efficiency were recently investigated [28]. This highlighted the importance of minimizing both of these couplings, in particular for vibrational modes with energies around the operating temperature.

Clearly, single-molecule devices would produce far too little electric (or cooling) power to be practically useful. However, it might be possible to fabricate devices based on molecular monolayers sandwiched between metallic electrodes. Such a device could show a behavior similar to many single-molecule junctions in parallel if the molecules within the layer were made to interact only weakly, for example, by using a submonolayer coverage, or by separating the thermoelectrically active molecules by electrically and thermally inactive "spacer" molecules. Contacting both sides of a molecular monolayer with metallic electrodes, without creating defects leading to a short-circuit, is certainly a challenging task but would be greatly rewarding if successful.

11.3.3 Devices Based on Polymers

Although inorganic semiconducting materials can be used for efficient thermoelectric devices, especially when improved by engineering their structure on the nanoscale, they often involve materials that are toxic and also often rare (e.g., Te), which leads to high production costs. Conducting polymer materials, on the other hand, are cheap and based on carbon and other elements that exist in abundance in Nature. Furthermore, polymeric materials are both lightweight and cheap, which makes them more attractive for many thermoelectric applications.

However, the most exciting aspect of polymers as thermoelectric materials is probably their very low thermal conductivity, down to $\kappa \approx 0.1\,\mathrm{W\,mK^{-1}}$ [30, 31] compared to $\kappa \approx 1.2\,\mathrm{W\,mK^{-1}}$ in Bi_2Te_3. By appropriate doping, either electron- or hole-type conduction can be favored, such that Seebeck coefficients of $|S| \approx 10$–$200\,\mu\mathrm{V\,K^{-1}}$ have been reported [30, 31], which is not too far from that of the best inorganic semiconductors ($|S| \approx 100$–$250\,\mu\mathrm{V\,K^{-1}}$ in Bi_2Te_3). The main problem with polymer materials is their low conductivity and, as a result, the highest values reported for the figure of merit in air-stable polymers are $ZT \approx 0.1$. Thus, it seems that the "only" factor required to make polymer materials competitive with inorganic semiconductors in thermoelectric applications is to increase their conductivity by a factor of about 10 (as $ZT \propto G$). However, it is known from Eq. (11.12) that $S = G_T/G$, and hence G must be increased in such a way that G_T is also increased; otherwise the Seebeck coefficient will be decreased, which is highly undesirable as $ZT \propto S^2$. For example, attempting to improve the conductivity by very heavy doping will cause a dramatic decrease in S, as the Fermi energy will then be pushed inside the conduction band, destroying the energy-asymmetry of carriers around E_F. Another problem is that many polymers become unstable at temperatures only slightly above room temperature, and so would not be practical at greater operating temperature differences, where the efficiency would be higher.

Although none of the polymer materials known today appear to be a realistic candidates to compete with thermoelectric devices based on inorganic semiconductors, their great promise – as noted above – confirms that the quest for new candidates is clearly worthwhile. Some of the materials investigated in this respect are discussed briefly below.

The earliest investigations included studies of the thermoelectric properties of *polyacetylene* [32, 33], doped for example, with iodine vapor. Although, with $ZT \approx 0.4$, polyacetylene has achieved the highest thermoelectric efficiency among polymers [31], it is both insoluble and unstable in air, and is therefore not an option in real devices. *Polyaniline* is a more stable polymer and has a rather high conductivity [31, 34], but a much lower S and ZT than polyacetylene. The situation is even worse for *polypyrrole*, which also has a rather small Seebeck coefficient, but in addition a much lower conductivity. Indeed, a study of the correlation between conductance and Seebeck coefficient at different doping levels in polyaniline and polypyrrole [35] appeared to indicate substantial transport by both electrons and holes in both of these materials. In polyacetylene, on the other hand, it seems to be possible – by appropriate doping – to achieve transport with one carrier type only, which explains the higher Seebeck coefficients in this material. The thermoelectric efficiency of *polythiophene* appears equally poor ($ZT < 10^{-2}$; [31]), although more recent studies of poly(3-methylthiophene) [36] and poly(3-hexylthiophene) [31] have shown slightly more promising values. Even more promising has been a recent study in which layers with different blends of polythiophene derivatives were used [37], and where varying the different constituents was shown to lead to a simultaneous increase in both S and G. Although the actual ZT deduced from the measurements was still low, this demonstrated an important proof of principle. Devices based on pellets of poly(3,4-ethylenedioxythiophene):poly(styrenesulf onate) (PEDOT:PSS) were also assessed [38] and showed very low thermal conductivities; however, the ZT remained small due to the small S and G.

A more detailed account of the many more polymer materials that have been investigated is available in Ref. [30].

11.3.4 Devices Based on Small Molecules

Very few studies have been conducted on thermoelectric applications using small-molecule materials. In one case, organic charge transfer salts and graphite powder were embedded in a PVC blend, thick-film devices were fabricated, and their Seebeck coefficients measured [39]. Although some of these compounds were found to exhibit very large Seebeck coefficients, their conductivity was much too low for an efficient thermoelectric operation (ZT could not be determined as the thermal conductivity was not measured).

As pentacene is known to have good charge-transport properties [40], two studies were carried out to investigate the thermoelectric properties of pentacene thin-film devices [41, 42]. In these studies, pentacene was found to have a thermal conductivity of $\kappa \approx 0.5\,\mathrm{W\,mK^{-1}}$, which was significantly lower than for inorganic semiconductors, but higher than in the best polymer materials. The Seebeck coefficient found in one study [41] was rather high, at $S \approx 400\,\mu\mathrm{V\,K^{-1}}$ in almost undoped films. Although higher doping concentrations led to a significant reduction in S, too-high doping concentrations were shown to reduce the conductivity. The latter effect was ascribed to a decrease in the mobility of charge carriers due to scattering off the dopant atoms. To overcome this problem, further devices were fabricated in which an undoped pentacene layer was capped with a thin acceptor layer that provided the charge carriers. Although these devices demonstrated $S \approx 200\,\mu\mathrm{V}$, the conductivity was still much too low for an acceptable figure of merit.

11.3.5 Hybrid and Composite Materials

One means of moving forward and hopefully overcoming the low efficiency of standard organic compounds would be to mix them with other materials. Thus, fully organic composite materials have been obtained by incorporating CNTs (i.e., a "rolled-up" sheet of graphene [43]) into conjugated polymer materials. Studies of poly(vinyl acetate)–CNT composites [44] showed that the inclusion of CNTs caused a significant increase in conductivity, but left S and κ unchanged. This was encouraging, even though the room temperature $ZT \approx 0.006$ of the devices was still very low [44]. To date, $ZT \approx 0.02$ has been the highest value reported for CNT–polymer composites [5] in which PEDOT:PSS was used.

Alternatively, composite materials with both polymers and inorganic constituents can be used. The role of the inorganic component is, as with CNTs, to increase the intrinsically low conductivity of the polymer, hopefully without destroying its better properties such as low κ and large S. As an example, promising results have been obtained by including a powder of Bi_2Te_3 into PEDOT:PSS [45], although the contact resistance between Bi_2Te_3 and PEDOT:PSS was found to be a limiting factor for improving the thermoelectric efficiency. In another study, composite nanocrystals composed of a Te core functionalized with PEDOT:PSS were synthesized [46]. This material combination was found to enhance the electric conductivity compared to pure PEDOT:PSS, while retaining a low thermal conductivity, giving $ZT \approx 0.1$.

11.4 Conclusions and Outlook

The aim of this chapter has been to provide an introduction to the basics of thermoelectric energy conversion and cooling. Although thermoelectric devices have many advantages compared to standard mechanically based devices (such as turbines and compressors), they have so far been limited by their relatively low efficiency which, despite considerable effort, has not been significantly improved during the past 50 years. Recent progress in nanostructured materials has, however, allowed some of the fundamental obstacles of large-scale bulk thermoelectric materials to be overcome, and has given rise to new hope. Clearly, the race is now on to find ever-more efficient thermoelectric materials. On the other hand, nanostructured materials suffer problems with scalability and prize. Moreover, the types of inorganic semiconductors used both in old-type bulk thermoelectrics and in new nanostructured materials continue to experience problems in terms of their toxicity and price.

Organic materials, on the other hand, can be produced cheaply and in large quantities, and also have advantages such as flexibility and light weight.

Devices based on single molecules, or monolayers of weakly interacting molecules contacted in parallel, appear to serve as ideal thermoelectric devices. Their highly peaked density of states, when weakly coupled to the electrodes, provide the perfect energy filtering of charge carriers required for efficient thermoelectric energy conversion or cooling. Phononic heat losses could perhaps be minimized by using strong chemical bonds to the electrodes and rigid molecules, such that there would be an energy mismatch between the molecular vibrations and the acoustic phonon modes in the electrodes. Clearly, such devices present a technological problem. Although single-molecule devices have been realized experimentally, their output power is far too low to be of practical use, while monolayer devices are very difficult to produce.

Although devices based on organic bulk or thin-film materials have very low thermal conductivities, which appears promising for thermoelectric applications, they suffer from much too low conductivities compared to their inorganic counterparts. Whilst some progress has been made in improving the conductivity, the situation is significantly complicated by a central problem of thermoelectric materials: an improvement in conductivity will result in a lowering of the Seebeck coefficient, and thereby to a minimal increase – or even a decrease – in the overall thermoelectric efficiency. To date, efforts at combining polymer materials with other organic compounds (e.g., CNTs) or with inorganic materials have been only partly successful.

Different thermoelectric applications require very different material properties, depending on the thermoelectric functions (e.g., energy conversion, cooling, thermometry), on the operating temperatures (which may in turn depend on the heat source), on size and weight limits (e.g., when converting waste heat from vehicles), on the most relevant quantity to optimize (e.g., efficiency or output – or cooling – power), and finally of course on the acceptable cost of the devices. At present, there is no such thing as *the* best thermoelectric material; rather, different materials must be used for different applications. Therefore, it is plausible that future advances in organic materials will prove useful for at least some of these applications.

References

1 Muller, K.-H. (2008) Thermoelectrics in an array of molecular junctions. *J. Chem. Phys.*, **129** (4), 044708.

2 Snyder, G.J. and Toberer, E.S. (2008) Complex thermoelectric materials. *Nat. Mater.*, **7** (2), 105–114.

3 Dresselhaus, M.S., Chen, G., Tang, M.Y., Yang, R.G., Lee, H., Wang, D.Z., Ren, Z.F., Fleurial, J.P., and Gogna, P. (2007) New directions for low-dimensional thermoelectric materials. *Adv. Mater.*, **19** (8), 1043–1053.

4 Ashcroft, N.W. and Mermin, N.D. (1976) *Solid State Physics*, Holt, Rinehart and Winston, New York City.

5 Kim, H., Kaviany, M., Thomas, J.C., Van der Ven, A., Uher, C., and Huang, B. (2010) Structural order-disorder transitions and phonon conductivity of partially filled skutterudites. *Phys. Rev. Lett.*, **105** (26), 265901.

6 Appleyard, N.J., Nicholls, J.T., Pepper, M., Tribe, W.R., Simmons, M.Y., and Ritchie, D.A. (2000) Direction-resolved transport and possible many-body effects in one-dimensional thermopower. *Phys. Rev. B*, **62** (24), R16275–R16278.

7 Bergfield, J.P., Solis, M.A., and Stafford, C.A. (2010) Giant thermoelectric effect from transmission supernodes. *ACS Nano*, **4** (9), 5314–5320.

8 Boese, D. and Fazio, R. (2001) Thermoelectric effects in Kondo-correlated quantum dots. *Eur. Phys. Lett.*, **56** (4), 576–582.

9 Kubala, B., König, J., and Pekola, J. (2008) Violation of the Wiedemann–Franz law in a single-electron transistor. *Phys. Rev. Lett.*, **100** (6), 066801.

10 Mahan, G.D. and Sofo, J.O. (1996) The best thermoelectric. *Proc. Natl Acad. Sci. USA*, **93** (15), 7436–7439.

11 Esposito, M., Lindenberg, K., and Van den Broeck, C. (2009) Thermoelectric efficiency at maximum power in a quantum dot. *Eur. Phys. Lett.*, **85** (6), 60010.

12 Hicks, L.D., Harman, T.C., Sun, X., and Dresselhaus, M.S. (1996) Experimental study of the effect of quantum-well structures on the thermoelectric figure of merit. *Phys. Rev. B*, **53** (16), R10493–R10496.

13 Boukai, A.I., Bunimovich, Y., Tahir-Kheli, J., Yu, J.-K., Goddard, W.A. III, and Heath, J.R. (2008) Silicon nanowires as efficient thermoelectric materials. *Nature*, **451** (7175), 168–171.

14 Hochbaum, A.I., Chen, R., Delgado, R.D., Liang, W., Garnett, E.C., Najarian, M., Majumdar, A., and Yang, P. (2008) Enhanced thermoelectric performance of rough silicon nanowires. *Nature*, **451** (7175), 163–167.

15 Llaguno, M.C., Fischer, J.E., Johnson, A.T., and Hone, J. (2004) Observation of thermopower oscillations in the Coulomb blockade regime in a semiconducting carbon nanotube. *Nano Lett.*, **4** (1), 45–49.

16 Small, J.P., Perez, K.M., and Kim, P. (2003) Modulation of thermoelectric power of individual carbon nanotubes. *Phys. Rev. Lett.*, **91** (25), 256801.

17 Venkatasubramanian, R., Siivola, E., Colpitts, T., and O'Quinn, B. (2001) Thin-film thermoelectric devices with high room-temperature figures of merit. *Nature*, **413** (6856), 597–602.

18 Yang, R. and Chen, G. (2004) Thermal conductivity modeling of periodic two-dimensional nanocomposites. *Phys. Rev. B*, **69** (19), 195316.

19 Baheti, K., Malen, J.A., Doak, P., Reddy, P., Jang, S.Y., Tilley, T.D., Majumdar, A., and Segalman, R.A. (2008) Probing the chemistry of molecular heterojunctions using thermoelectricity. *Nano Lett.*, **8** (2), 715–719.

20 Malen, J.A., Doak, P., Baheti, K., Tilley, T.D., Segalman, R.A., and Majumdar, A. (2009) Identifying the length dependence of orbital alignment and contact coupling in molecular heterojunctions. *Nano Lett.*, **9** (3), 1164–1169.

21 Reddy, P., Jang, S.-Y., Segalman, R.A., and Majumdar, A. (2007) Thermoelectricity in molecular junctions. *Science*, **315** (5818), 1568–1571.

22 Landauer, R. (1957) Spatial variation of currents and fields due to localized scatterers in metallic conduction. *IBM J. Res. Dev.*, **1** (3), 223–231.

23 Murphy, P., Mukerjee, S., and Moore, J. (2008) Optimal thermoelectric figure of merit of a molecular junction. *Phys. Rev. B*, **78** (16), 161406.

24 Finch, C.M., Garcia-Suarez, V.M., and Lambert, C.J. (2009) Giant thermopower and figure of merit in single-molecule devices. *Phys. Rev. B*, **79** (3), 033405.

25 Solomon, G.C., Andrews, D.Q., Hansen, T., Goldsmith, R.H., Wasielewski, M.R., Duyne, R.P.V., and Ratner, M.A. (2008) Understanding quantum interference in coherent molecular conduction. *J. Chem. Phys.*, **129** (5), 054701.

26 Braig, S. and Flensberg, K. (2003) Vibrational sidebands and dissipative tunneling in molecular transistors. *Phys. Rev. B*, **68** (20), 205324.

27 Park, H., Park, J., Lim, A.K.L., Anderson, E.H., Alivisatos, A.P., and McEuen, P.L. (2000) Nanomechanical oscillations in a single-C60 transistor. *Nature*, **407** (6800), 57–60.

28 Leijnse, M., Wegewijs, M.R., and Flensberg, K. (2010) Nonlinear thermoelectric properties of molecular junctions with vibrational coupling. *Phys. Rev. B*, **82** (4), 045412.

29 Moth-Poulsen, K. and Bjørnholm, T. (2009) Molecular electronics with single molecules in solid-state devices. *Nat. Nanotechnol.*, **4** (9), 551–556.

30 Dubey, N. and Leclerc, M. (2011) Conducting polymers: efficient thermoelectric materials. *J. Polym. Sci. Part B: Polym. Phys.*, **49** (7), 467–475.

31 Xuan, Y., Liu, X., Desbief, S., Leclère, P., Fahlman, M., Lazzaroni, R., Berggren, M., Cornil, J., Emin, D., and Crispin, X. (2010) Thermoelectric properties of conducting polymers: the case of poly(3-hexylthiophene). *Phys. Rev. B*, **82** (11), 115454.

32 Kaiser, A.B. (1989) Thermoelectric power and conductivity of heterogeneous conducting polymers. *Phys. Rev. B*, **40** (5), 2806–2813.

33 Zuzok, R., Kaiser, A.B., Pukacki, W., and Roth, S. (1991) Thermoelectric power and conductivity of iodine-doped "new" polyacetylene. *J. Chem. Phys.*, **95** (2), 1270–1275.

34 Holland, E.R., Pomfret, S.J., Adams, P.N., Abell, L., and Monkman, A.P. (1997) Doping-dependent transport properties of polyaniline-CSA films. *Synth. Met.*, **84** (1–3), 777–778.

35 Mateeva, N., Niculescu, H., Schlenoff, J., and Testardi, L.R. (1998) Correlation of Seebeck coefficient and electric conductivity in polyaniline and polypyrrole. *J. Appl. Phys.*, **83** (6), 3111–3117.

36 Lu, B.-Y., Liu, C.-C., Lu, S., Xu, J.-K., Jiang, F.-X., Li, Y.-Z., and Zhang, Z. (2010) Thermoelectric performances of free-standing polythiophene and poly(3-methylthiophene) nanofilms. *Chin. Phys. Lett.*, **27** (5), 057201.

37 Sun, J., Yeh, M.L., Jung, B.J., Zhang, B., Feser, J., Majumdar, A., and Katz, H.E. (2010) Simultaneous increase in Seebeck coefficient and conductivity in a doped poly(alkylthiophene) blend with defined density of states. *Macromolecules*, **43** (6), 2897–2903.

38 Jiang, F.-X., Xu, J.-K., Lu, B.-Y., Xie, Y., Huang, R.-J., and Li, L.-F. (2008) Thermoelectric performance of poly(3,4-ethylenedioxythiophene): poly(styrenesulfonate). *Chin. Phys. Lett.*, **25** (6), 2202–2205.

39 Wüsten, J. and Potje-Kamloth, K. (2008) Organic thermogenerators for energy autarkic systems on flexible substrates. *J. Phys. D: Appl. Phys.*, **41** (13), 135113.

40 Jurchescu, O.D., Baas, J., and Palstra, T.T.M. (2004) Effect of impurities on the mobility of single crystal pentacene. *Appl. Phys. Lett.*, **84** (16), 3061–3063.

41 Harada, K., Sumino, M., Adachi, C., Tanaka, S., and Miyazaki, K. (2010) Improved thermoelectric performance of organic thin-film elements utilizing a bilayer structure of pentacene and 2,3,5,6-tetrafluoro-7,7,8,8-tetracyanoquinodimethane (F(4)-TCNQ). *Appl. Phys. Lett.*, **96** (25), 253304.

42 Kim, N., Domercq, B., Yoo, S., Christensen, A., Kippelen, B., and Graham, S. (2005) Thermal transport properties of thin films of small molecule organic semiconductors. *Appl. Phys. Lett.*, **87** (24), 241908.

43 Iijima, S. (1991) Helical microtubules of graphitic carbon. *Nature*, **354** (6348), 56–58.

44 Yu, C., Kim, Y.S., Kim, D., and Grunlan, J.C. (2008) Thermoelectric behavior of segregated-network polymer nanocomposites. *Nano Lett.*, **8** (12), 4428–4432.

45 Zhang, B., Sun, J., Katz, H.E., Fang, F., and Opila, R.L. (2010) Promising thermoelectric properties of commercial PEDOT:PSS materials and their Bi2Te3 powder composites. *ACS Appl. Mater. Interfaces*, **2** (11), 3170–3178.

46 See, K.C., Feser, J.P., Chen, C.E., Majumdar, A., Urban, J.J., and Segalman, R.A. (2010) Water-processable polymer–nanocrystal hybrids for thermoelectrics. *Nano Lett.*, **10** (11), 4664–4667.

Glossary of the book

[6,6]-phenyl-C_{61}-butyric acid methyl ester (PC_{61}BM)

[6,6]-phenyl-C_{71}-butyric acid methyl ester (PC_{71}BM)

1,2-bisthienylethene (BTE)

1,2-dimethyl-3-propyl-imidazolium iodide (DMPII)

1,3,5-triphenyl-2-pyrazoline (TPP)

1,3-diphenyl-2-pyrazoline (DP)

1,3-diphenyl-5-(2-anthryl)-2-pyrazoline (DAP)

1,3-diphenyl-5-pyrenyl-2-pyrazoline (DPP)

1,4-dicyanonaphthalene (DCN)

1,5-diaminoanthraquinone (DAAQ)

1-cyano-*trans*-1,2-bis-(40-methylbiphenyl)ethylene (CN-MBE)

1-phenyl-3-((dimethylamino)-styryl)-5-((dimethylamino) phenyl)-2-pyrazoline (PDDP)

2,2′,7,7′-tetrakis-(*N,N*-di-*p*-methoxyphenylamine)-9,9′-spirobifluorene (spiro-MeOTAD)

2,2′,7,7′-tetrakis(*N,N*-dimethoxyphenylamine)-9,9′-spirobifluorene (spiro-OMeTAD)

2,4,5-triphenylimidazole (TPI)

2,4,5-triphenylimidazole (TPI)

3-(anthracen-10-yl)-1-phenylpro-2-en-1-one (APO)

3,4-ethylenedioxythiophene (EDOT)

3-methoxypropionitrile (MPN)

Organic Optoelectronics, First Edition. Edited by Wenping Hu.

4-(5-(4-*tert*-butylphenyl)-1,3,4-oxadiazole-2-yl)-biphenyl-4′-yl sulfonic sodium (*t*-Bu-PBD-SO_3Na)

4-(dicyanomethylene)-2-methyl-6-(*p*-dimethyl-aminostyryl)-4H-pyran (DCM)

4-*tert*-butylpyridine (TBP)

9,10-bis(phenylethynyl)anthracene (BPEA)

Acceptor (A)

Aggregation-induced enhanced emission (AIEE)

Al-doped zinc oxide (AZO)

Aluminum gallium arsenide (AlGaAs)

Anodized aluminum oxide (AAO)

Anthracene-tetracyanobenzene (TCNB)

Arithmetic and logical unit (ALU)

Atomic force microscopy (AFM)

Benzenedithiol (BDT)

Benzothiadiazole (BT)

Bipyridyl oligophenylene-ethynylene dithiol (BP-DT)

Bipyridyl-dinitrooligophenylene-ethynylene dithiol (BPDN-DT)

Bis(thienylenevinylene)-substituted polythiophene (2TV-PT)

Bottom-gate–bottom-contact (BGBC)

Bottom-gate–top-contact (BGTC)

Bravais–Friedel–Donnay–Harker (BFDH)

Bulk heterojunction (BHJ)

Butyl lithium (*n*-BuLi)

Carbon nanotube (CNT)

Cetyltrimethylammonium bromide (CTAB)

Charge modulation spectroscopy (CMS)

Charge transfer (CT)

Charge-coupled device (CCD)

Charge-transfer state (CT state)

Chemical vapor deposition (CVD)

Chenodeoxycholic acid (CDCA)

Chloroaluminum phthalocyanine (AlClPc)

Chlorobenzene (CB)

Color rendering index (CRI)

Color temperature (CT)

Commission Internationale d'Eclairage (CIE)

Compact fluorescent lamp (CFL)

Conducting atomic force microscopy (C-AFM)

Conduction band (CB)

Copper indium gallium diselenide (CIGS)

Copper phthalocyanine (CuPc)

Correlated Color Temperature (CCT)

Critical micelle concentration (CMC)

Cyanoethylpullulan (CYEPL)

Cyano-substituted PPVs (CN-PPVs)

Density functional theory (DFT)

Deoxycholic acid (DCA)

Dicyanomethylene-2-methyl-6-(*p*-dimethyl-aminostyryl)-4*H*-pyran (DCM)

Dicyanomethylenepyran (DCP)

Dicyanovinyl (DCV)

Diketopyrrolopyrrole (DPP)

Dithieno[3,2-*b*:2′,3′-*d*]pyrrole (DTP)

Dithienocyclopentadiene (DTC)

Dithienosilole (DTS)

Dithiol-functionalized polyethylene glycol (SH-PEG-SH)

Divinyltetramethylsiloxane-bis(benzocyclobutene) (BCB)

Donor (D)

Donor–acceptor (D–A)

Double-ended aryl dithiols [α,α′-xylyldithiol] (XYL)

Dye-sensitized solar cell (DSSC)

Electroluminescent (EL)

Electron beam-induced decomposition (EBID)

Electronic excited state of A (A*)

Electronic excited state of D (D*)

Electron-transport layer (ETL)

Energy gap (E_g)

Ethylhexyl (PTVPhI-Eh)

Field-emission scanning electron microscopy (FESEM)

Fill factor (FF)

Fluorescence quantum yield (Φ)

Fluorescence resonance energy transfer (FRET)

Fluorine-doped tin oxide (FTO)

Focused ion beam (FIB)

Förster energy transfer (FET)

Fullerene (C_{60})

Gate voltage (V_G)

Grazing incidence X-ray diffraction (GIXD)

Guanidinium thiocynate (GT)

Head-to-head (HH)

Head-to-tail (HT)

Hexamethyldisilazane (HMDS)

Hexathiapentacene (HTP)

Highest occupied molecular orbital (HOMO)

Hole-transport material (HTM)

Hole-transporting layer (HTL)

Incident photon-to-current conversion efficiency (IPCE)

Indene-C_{60} bisadduct ($IC_{60}BA$)

Indene-C_{70} bisadduct ($IC_{70}BA$)

Indium tin oxide (ITO)

Inelastic electron tunneling spectroscopy (IETS)

Injection limited current (ILC)

Intermolecular fluorescence resonance energy transfer (IFRET)

Intramolecular charge-transfer state (ICT state)

Langmuir–Blodgett (LB)

Light-induced electron spin resonance (LESR)

Lithium diisopropylamide (LDA)

Lowest unoccupied molecular orbital (LUMO)

Mechanical controllable break (MCB)

Merocyanine (MC)

Metal-insulator-metal (MIM)

Metal-ligand charge transfer (MLCT)

Mobility (μ)

Molecular orbital (MO)

Multiple trapping and release (MTR)

Multiwalled carbon nanotube (MWNT)

N,*N'*-di[2,4-difluorophenyl]-3,4,9,10-perylenetetracarboxylic diimide (PTCDI)

Naphthalene diimide (NDI)

Naphtho[2,3-*b*:6,7-*b'*]dithiophene (NDT)

Near-edge X-ray absorption fine structure (NEXAFS)

N-methyl-benzimidazole (NMBI)

Non-equilibrium Green's function (NEGF)

Normal hydrogen electrode (NHE)

Octadecylphosphonic acid (ODPA)

Octadecyltrichlorosilane (ODTS)

Octadecyltrimethoxysilane (OTMS)

Octanedithiol (ODT)

Oligophenyleneimine (OPI)

On/off current ratio (I_{on}/I_{off})

One-dimensional (1-D)

Open-circuit voltage (V_{OC})

Organic field-effect transistor (OFET)

Organic light-emitting diode (OLED)

Organic photovoltaic (OPV)

Organic thin-film transistor (OTFT)

Perylene diimide (PDI)

Perylene-3,4,9,10-tetracarboxylicdiimide (PTCDI)

Phenoxazine (POZ)

Phenoxyoctadecyltrichlorosilane (PhO-OTS)

Photoinduced electron transfer (PET)

Photoinduced electron transfer (PIET)

Photoluminescence (PL)

Photovoltaic (PV)

Phthalocyanine (Pc)

Physical vapor deposition (PVD)

Physical vapor transport (PVT)

Planar-mixed heterojunction (PMHJ)

Poly(3,4-ethylenedioxythiophene): poly(styrenesulfonate) (PEDOT:PSS)

Poly(3-alkylthiophene) (P3AT)

Poly(3-hexylthiophene) (P3HT)

Poly(3-phenylhydrazone thiophene) (PPHT)

Poly(9,9-dihexylfluorene)-*co*-2,5-dicyanophenylene (PF_3CNP_1)

Poly(9,9-dioctylfluorenyl-2,7-diyl) (PFO)

Poly(*p*-phenylene ethynylene) (PPE)

Poly(*p*-phenylene vinylene) (PPV)

Poly(*p*-phenylene) (PPP)

Poly(vinylcarbazole) (PVK)

Poly(vinylcarbazole) sulfonic lithium ($PVK\text{-}SO_3Li$)

Poly(vinylidenefluoride-*co*-hexafluoropropylene) (PVDF-HFP)

Poly[(9,9-bis(3′-(*N,N*-dimethylamino)propyl)-2,7-fluorene)-*alt*-2,7-(9,9–dioctylfluorene)] (PFN)

Poly[2,5-dimethoxy-1,4-phenylene-1,2-ethenylene-2-methoxy-5-(2-ethylhexyloxy)-(1,4-phenylenevinylene-1,2-ethenylene)] (M3EH-PPV)

Poly[2-methoxy-5-(2′-ethyl-hexyloxy)-1,4-phenylenevinylene] (MEH-PPV)

Poly[2-methoxy-5-(3,7-dimethyloctyloxy)-1,4-phenylenevinylene] (MDMO-PPV)

Poly[3-(10-*n*-octyl-3-phenothiazinevinylene)thiophene-*co*-2,5-thiophene] (PV-PT)

Poly[3-(4-*n*-octyl)-phenylthiophene] (POPT)

Poly{4,4′-bis(2-ethylhexyl) dithieno[3,2-*b*:2′,3′-*d*]silole-2,6-diyl-*alt*-(2,1,3-benzothiadiazole-4,7-diyl)} (PDTSBTD)

Polyacetylene (PA)

Polyaniline (PANi)

Polycyclic aromatic hydrocarbon/polyaromatic hydrocarbon (PAH)

Polydimethylsiloxane (PDMS)

Polyfluorene (PF)

Polyimide (PI)

Polymer solar cell (PSC)

Polymeric light-emitting diode (PLED)

Polymethylmethacrylate (PMMA)

Polypyrrole (PPy)

Polystyrene (PS)

Polythiophene (PT)

Polyvinylphenol (PVP)

Power conversion efficiency (PCE)

Pulse-radiolysis time-resolved microwave conductivity (PR-TRMC)

Radiofrequency identification (RFID)

Remote energy relay (RER)

Roll-to-roll (R2R)

Scanning electron microscopy (SEM)

Scanning probe lithography (SPL)

Scanning tunneling microscopy (STM)

Selected area electron diffraction (SAED)

Self-assembled monolayer (SAM)

Sexiphenyl (p-6P)

Short-circuit current density (J_{SC})

Single-crystal field-effect transistor (SCFET)

Singlet electronic excited state (S_1)

Single-walled carbon nanotube (SWNT)

Solid-state lighting (SSL)

Source-drain current (I_{DS})

Source-drain electrode (V_{DS})

Space charge limited current (SCLC)

Spirooxazine (SO)

Squaraine (SQ)

Subnaphthalocyanine (SubNc)

Subphthalocyanine (SubPc)

Surface-enhanced Raman scattering (SERS)

Tail-to-tail (TT)

Tert-butyl lithium (*t*-BuLi)

tert-butylphenyl-1,3,4-oxadiazole-2-yl-biphenyl-4′-yl sulfonic sodium (*t*-Bu-PBD-SO_3Na)

Tetrabutylammonium deoxycholic acid salt (DCA·TBA)

Tetramethylpentacene (TMPC)

Tetrathiafulvalene (TTF)

Tetrathiafulvalene-tetracyanoquinodimethane (TTF:TCNQ)

Thermogravimetric analysis (TGA)

Thiadiazolopyridine (TP)

Thiophene–phenylene–thiophene (TPT)

Threshold voltage (V_{Th})

Time-of-flight (TOF)

Top-gate–bottom-contact (TGBC)

Top-gate–top-contact (TGTC)

Transition voltage spectroscopy (TVS)

Transmission electron microscopy (TEM)

Transparent conductive oxide (TCO)

Trans-polyacetylene (t-PA)

Trifluoroacetyl (TFA)

Triisopropylsilylethynyl (TIPS)

Triphenylamine (TPA)

Tris(2-phenylpyridine) iridium (III)(Ir(ppy)$_3$)

Tris(8-hydroxyquinolinato) aluminum (Alq$_3$)

Tris(thienylenevinylene)-substituted polythiophene (3TV-PT)

Tris[2,5-bis-2′-(9,9′-dihexylfluorene)iridium] (Ir(HFP)$_3$)

Tris[2,5-bis-2′-(9′,9′-dihexylfluorene)pyridine-κ2NC3′]Iridium(III) (Ir(HFP)$_3$)

Tris-[9,9-dihexyl-2-(phenyl-4′-(-pyridin-2″-yl)) fluorene] iridium(III) (Ir(DPPF)$_3$)

Tris-[9,9-dihexyl-2-(pyridinyl-2′) fluorene] iridium (III)(Ir(DPF)$_3$)

Twisted intramolecular charge transfer (TICT)

Two-dimensional (2-D)

Ultraviolet photoelectron spectroscopy (UPS)

Valence band (VB)

White organic light-emitting diode (WOLED)

White-light-emitting diode (WLED)

X-ray diffraction (XRD)

Index

Page numbers in italics refer to illustrations and tables.

Organic Optoelectronics, First Edition. Edited by Wenping Hu.

t

u

v

w

x

z